Probability
and Statistics

Second Edition

Morris H. DeGroot
Carnegie-Mellon University

ADDISON-WESLEY PUBLISHING COMPANY

Reading, Massachusetts/Menlo Park, California
Don Mills, Ontario/Wokingham, England/Amsterdam/Sydney
Singapore/Tokyo/Mexico City/Bogotá/Santiago/San Juan

This book is in the Addison-Wesley Series in Statistics.
Frederick Mosteller   Consulting Editor

Library of Congress Cataloging in Publication Data

DeGroot, Morris H.
  Probability and statistics.

  Bibliography: p.
  Includes index.
  1. Probabilities.   2. Mathematical statistics.
I. Title.
QA273.D35 1984   519.2   84-6269
ISBN 0-201-11366-X

Reprinted with corrections, August 1987

FGHIJ-MA-898

To Jenny and Jeremy

Preface

This book contains enough material for a one-year course in probability and statistics. The mathematical requirements for the course are a knowledge of the elements of calculus and a familiarity with the concepts and elementary properties of vectors and matrices. No previous knowledge of probability or statistics is assumed.

The book has been written with both the student and the teacher in mind. Special care has been taken to make sure that the text can be read and understood with few obscure passages or other stumbling blocks. Theorems and proofs are presented where appropriate, and illustrative examples are given at almost every step of the way. More than 1100 exercises are included in the book. Some of these exercises provide numerical applications of results presented in the text, and others are intended to stimulate further thought about these results. A new feature of this second edition is the inclusion of approximately 20 or 25 exercises at the end of each chapter that supplement the exercises given at the end of most of the individual sections of the book.

The first five chapters are devoted to probability and can serve as the text for a one-semester course on that topic. The elementary concepts of probability are illustrated by such famous examples as the birthday problem, the tennis tournament problem, the matching problem, the collector's problem, and the game of craps. Standard material on random variables and probability distributions is highlighted by discussions of statistical swindles, the use of a table of random digits, the elementary notions of life testing, a comparison of the relative advantages of the mean and the median as predictors, the importance of the central limit theorem, and the correction for continuity. Also included as special features of these chapters are sections on Markov chains, the Gambler's Ruin problem, choosing the best, utility and preferences among gambles, and the Borel-Kolmogorov paradox. These topics are treated in a completely elementary fashion, but they can be omitted without loss of continuity if time is limited. Sections of the book that can be so omitted are indicated, in the traditional way, by asterisks in the Contents and in the text.

The last five chapters of the book are devoted to statistical inference. The coverage here is modern in outlook. Both classical and Bayesian statistical methods are developed in an integrated presentation. No single school of thought is treated in a dogmatic fashion. My goal is to equip the student with the theory and methodology that have proved to be useful in the past and promise to be useful in the future.

These chapters contain a comprehensive but elementary survey of estimation, testing hypotheses, nonparametric methods, multiple regression, and the analysis of variance. The strengths and weaknesses and the advantages and disadvantages of such basic concepts as maximum likelihood estimation, Bayesian decision procedures, unbiased estimation, confidence intervals, and levels of significance are discussed from a contemporary viewpoint. Special features of these chapters include discussions of prior and posterior distributions, sufficient statistics, Fisher information, the delta method, the Bayesian analysis of samples from a normal distribution, unbiased tests, multidecision problems, tests of goodness-of-fit, contingency tables, Simpson's paradox, inferences about the median and other quantiles, robust estimation and trimmed means, confidence bands for a regression line, and the regression fallacy. If time does not permit complete coverage of the contents of these chapters, any of the following sections can be omitted without loss of continuity: 7.6, 7.8, 8.3, 9.6, 9.7, 9.8, 9.9, and 9.10.

In summary, the main changes in this second edition are new sections or subsections on statistical swindles, choosing the best, the Borel-Kolmogorov paradox, the correction for continuity, the delta method, unbiased tests, Simpson's paradox, confidence bands for a regression line, and the regression fallacy, as well as a new section of supplementary exercises at the end of each chapter. The material introducing random variables and their distributions has been thoroughly revised, and minor changes, additions, and deletions have been made throughout the text.

Although a computer can be a valuable adjunct in a course in probability and statistics such as this one, none of the exercises in this book requires access to a computer or a knowledge of programming. For this reason, the use of this book is not tied to a computer in any way. Instructors are urged, however, to use computers in the course as much as is feasible. A small calculator is a helpful aid for solving some of the numerical exercises in the second half of the book.

One further point about the style in which the book is written should be emphasized. The pronoun "he" is used throughout the book in reference to a person who is confronted with a statistical problem. This usage certainly does not mean that only males calculate probabilities and make decisions, or that only males can be statisticians. The word "he" is used quite literally as defined in Webster's Third New International Dictionary to mean "that one whose sex is unknown or immaterial." The field of statistics should certainly be as accessible to women as it is to men. It should certainly be as accessible to members of minority groups as it is to the majority. It is my sincere hope that this book will help create among all groups an awareness and appreciation of probability and statistics as an interesting, lively, and important branch of science.

I am indebted to the readers, instructors, and colleagues whose comments have strengthened this edition. Marion Reynolds, Jr., of Virginia Polytechnic Institute and James Stapleton of Michigan State University reviewed the manuscript for the publisher and made many valuable suggestions. I am grateful to the Literary Executor of the late Sir Ronald A. Fisher, F.R.S., to Dr. Frank Yates, F.R.S., and the Longman Group Ltd., London, for permission to adapt Table III of their book *Statistical Tables for Biological, Agricultural and Medical Research* (6th Edition, 1974).

The field of statistics has grown and changed since I wrote a Preface for the first edition of this book in November, 1974, and so have I. The influence on my life and work of those who made that first edition possible remains vivid and undiminished; but with growth and change have come new influences as well, both personal and professional. The love, warmth, and support of my family and friends, old and new, have sustained and stimulated me, and enabled me to write a book that I believe reflects contemporary probability and statistics.

Pittsburgh, Pennsylvania   M. H. D.
October 1985

Contents

6

Estimation

7

Sampling Distributions of Estimators

8

Testing Hypotheses

9   Categorical Data and Nonparametric Methods

10   Linear Statistical Models

Tables

Introduction to Probability

1

1.1. THE HISTORY OF PROBABILITY

The concepts of chance and uncertainty are as old as civilization itself. People have always had to cope with uncertainty about the weather, their food supply, and other aspects of their environment, and have strived to reduce this uncertainty and its effects. Even the idea of gambling has a long history. By about the year 3500 B.C., games of chance played with bone objects that could be considered precursors of dice were apparently highly developed in Egypt and elsewhere. Cubical dice with markings virtually identical to those on modern dice have been found in Egyptian tombs dating from 2000 B.C. We know that gambling with dice has been popular ever since that time and played an important part in the early development of probability theory.

It is generally believed that the mathematical theory of probability was started by the French mathematicians Blaise Pascal (1623–1662) and Pierre Fermat (1601–1665) when they succeeded in deriving exact probabilities for certain gambling problems involving dice. Some of the problems that they solved had been outstanding for about 300 years. However, numerical probabilities of various dice combinations had been calculated previously by Girolamo Cardano (1501–1576) and by Galileo Galilei (1564–1642).

The theory of probability has been developed steadily since the seventeenth century and has been widely applied in diverse fields of study. Today, probability theory is an important tool in most areas of engineering, science, and management. Many research workers are actively engaged in the discovery and establishment of new applications of probability in fields such as medicine, meteorology, photography from spaceships, marketing, earthquake prediction, human behavior, the design of computer systems, and law. In most legal proceedings involving antitrust violations or employment discrimination, both sides often present probability and statistical calculations to help support their cases.

References

The ancient history of gambling and the origins of the mathematical theory of probability are discussed by David (1962), Ore (1960), and Todhunter (1865).

Some introductory books on probability theory, which discuss many of the same topics that will be studied in this book, are Feller (1968); Hoel, Port, and Stone (1971); Meyer (1970); and Olkin, Gleser, and Derman (1980). Other introductory books, which discuss both probability theory and statistics at about the same level as they will be discussed in this book, are Brunk (1975); Devore (1982); Fraser (1976); Freund and Walpole (1980); Hogg and Craig (1978); Kempthorne and Folks (1971); Larson (1974); Lindgren (1976); Mendenhall, Scheaffer, and Wackerly (1981); and Mood, Graybill, and Boes (1974).

1.2. INTERPRETATIONS OF PROBABILITY

In addition to the many formal applications of probability theory, the concept of probability enters our everyday life and conversation. We often hear and use such expressions as: "It probably will rain tomorrow afternoon"; "It is very likely that the plane will arrive late"; or "The chances are good that he will be able to join us for dinner this evening." Each of these expressions is based on the concept of the probability, or the likelihood, that some specific event will occur.

Despite the fact that the concept of probability is such a common and natural part of our experience, no single scientific interpretation of the term probability is accepted by all statisticians, philosophers, and other authorities. Through the years, each interpretation of probability that has been proposed by some authorities has been criticized by others. Indeed, the true meaning of probability is still a highly controversial subject and is involved in many current philosophical discussions pertaining to the foundations of statistics. Three different interpretations of probability will be described here. Each of these interpretations can be very useful in applying probability theory to practical problems.

The Frequency Interpretation of Probability

In many problems, the probability that some specific outcome of a process will be obtained can be interpreted to mean the *relative frequency* with which that outcome would be obtained if the process were repeated a large number of times under similar conditions. For example, the probability of obtaining a head when a coin is tossed is considered to be $1/2$ because the relative frequency of heads should be approximately $1/2$ when the coin is tossed a large number of times under similar conditions. In other words, it is assumed that the proportion of tosses on which a head is obtained would be approximately $1/2$.

Of course, the conditions mentioned in this example are too vague to serve as the basis for a scientific definition of probability. First, a "large number" of tosses of the coin is specified, but there is no definite indication of an actual number that would be considered large enough. Second, it is stated that the coin should be tossed each time "under similar conditions," but these conditions are not described precisely. The conditions under which the coin is tossed must not be completely identical for each toss because the outcomes would then be the same, and there would be either all heads or all tails. In fact, a skilled person can toss a coin into the air repeatedly and catch it in such a way that a head is obtained on almost every toss. Hence, the tosses must not be completely controlled but must have some "random" features.

Furthermore, it is stated that the relative frequency of heads should be "approximately $1/2$," but no limit is specified for the permissible variation from $1/2$. If a coin were tossed 1,000,000 times, we would not expect to obtain exactly 500,000 heads. Indeed, we would be extremely surprised if we obtained exactly 500,000 heads. On the other hand, neither would we expect the number of heads to be very far from 500,000. It would be desirable to be able to make a precise statement of the likelihoods of the different possible numbers of heads, but these likelihoods would of necessity depend on the very concept of probability that we are trying to define.

Another shortcoming of the frequency interpretation of probability is that it applies only to a problem in which there can be, at least in principle, a large number of similar repetitions of a certain process. Many important problems are not of this type. For example, the frequency interpretation of probability cannot be applied directly to the probability that a specific acquaintance will get married within the next two years or to the probability that a particular medical research project will lead to the development of a new treatment for a certain disease within a specified period of time.

The Classical Interpretation of Probability

The classical interpretation of probability is based on the concept of *equally likely outcomes*. For example, when a coin is tossed, there are two possible outcomes: a head or a tail. If it may be assumed that these outcomes are equally likely to occur, then they must have the same probability. Since the sum of the probabilities must be 1, both the probability of a head and the probability of a tail must be $1/2$. More generally, if the outcome of some process must be one of n different outcomes, and if these n outcomes are equally likely to occur, then the probability of each outcome is $1/n$.

Two basic difficulties arise when an attempt is made to develop a formal definition of probability from the classical interpretation. First, the concept of equally likely outcomes is essentially based on the concept of probability that we are trying to define. The statement that two possible outcomes are equally likely to occur is the same as the statement that two outcomes have the same probability. Second, no systematic method is given for assigning probabilities to outcomes that are not assumed to be equally likely. When a coin is tossed, or a well-balanced die is rolled, or a card is chosen from a well-shuffled deck of cards, the different possible outcomes can usually be regarded as equally likely because of the nature of the process. However, when the problem is to guess whether an acquaintance will get married or whether a research project will be successful, the possible outcomes would not typically be considered to be equally likely, and a different method is needed for assigning probabilities to these outcomes.

The Subjective Interpretation of Probability

According to the subjective, or personal, interpretation of probability, the probability that a person assigns to a possible outcome of some process represents his own judgment of the likelihood that the outcome will be obtained. This judgment will be based on that person's beliefs and information about the process. Another person, who may have different beliefs or different information, may assign a different probability to the same outcome. For this reason, it is appropriate to speak of a certain person's *subjective probability* of an outcome, rather than to speak of the *true probability* of that outcome.

As an illustration of this interpretation, suppose that a coin is to be tossed once. A person with no special information about the coin or the way in which it is tossed might regard a head and a tail to be equally likely outcomes. That person would then assign a subjective probability of $1/2$ to the possibility of obtaining a head. The person who is actually tossing the coin, however, might feel that a head is much more likely to be obtained than a tail. In order that this person may be able to assign subjective probabilities to the outcomes, he must express the strength of his belief in numerical terms. Suppose, for example, that he regards the likelihood of obtaining a head to be the same as the likelihood of obtaining a red card when one card is chosen from a well-shuffled deck containing four red cards and one black card. Since the person would assign a probability of $4/5$ to the possibility of obtaining a red card, he should also assign a probability of $4/5$ to the possibility of obtaining a head when the coin is tossed.

This subjective interpretation of probability can be formalized. In general, if a person's judgments of the relative likelihoods of various combinations of outcomes satisfy certain conditions of consistency, then it can be shown that his subjective probabilities of the different possible events can be uniquely determined. However, there are two difficulties with the subjective interpretation. First, the requirement that a person's judgments of the relative likelihoods of an infinite number of events be completely consistent and free from contradictions does not seem to be humanly attainable. Second, the subjective interpretation provides no "objective" basis for two or more scientists working together to reach a common evaluation of the state of knowledge in some scientific area of common interest.

On the other hand, recognition of the subjective interpretation of probability has the salutary effect of emphasizing some of the subjective aspects of science. A particular scientist's evaluation of the probability of some uncertain outcome must ultimately be his own evaluation based on all the evidence available to him. This evaluation may well be based in part on the frequency interpretation of probability, since the scientist may take into account the relative frequency of occurrence of this outcome or similar outcomes in the past. It may also be based in part on the classical interpretation of probability, since the scientist may take into account the total number of possible outcomes that he considers equally likely to occur. Nevertheless, the final assignment of numerical probabilities is the responsibility of the scientist himself.

The subjective nature of science is also revealed in the actual problem that a particular scientist chooses to study from the class of problems that might have been chosen, in the experiments that he decides to perform in carrying out this study, and in the conclusions that he draws from his experimental data. The mathematical theory of probability and statistics can play an important part in these choices, decisions, and conclusions. Moreover, this theory of probability and statistics can be developed, and will be presented in this book, without regard to the controversy surrounding the different interpretations of the term probability. This theory is correct and can be usefully applied, regardless of which interpretation of probability is used in a particular problem. The theories and techniques that will be presented in this book have served as valuable guides and tools in almost all aspects of the design and analysis of effective experimentation.

1.3. EXPERIMENTS AND EVENTS

Types of Experiments

The theory of probability pertains to the various possible outcomes that might be obtained and the possible events that might occur when an experiment is performed. The term "experiment" is used in probability theory to describe virtually any process whose outcome is not known in advance with certainty. Some examples of experiments will now be given.

1. In an experiment in which a coin is to be tossed 10 times, the experimenter might want to determine the probability that at least 4 heads will be obtained.

2. In an experiment in which a sample of 1000 transistors is to be selected from a large shipment of similar items and each selected item is to be inspected, a person might want to determine the probability that not more than one of the selected transistors will be defective.

3. In an experiment in which the air temperature at a certain location is to be observed every day at noon for 90 successive days, a person might want to determine the probability that the average temperature during this period will be less than some specified value.

4. From information relating to the life of Thomas Jefferson, a certain person might want to determine the probability that Jefferson was born in the year 1741.

5. In evaluating an industrial research and development project at a certain time, a person might want to determine the probability that the project will result in the successful development of a new product within a specified number of months.

It can be seen from these examples that the possible outcomes of an experiment may be either random or nonrandom, in accordance with the usual meanings of those terms. The interesting feature of an experiment is that each of its possible outcomes can be specified before the experiment is performed, and probabilities can be assigned to various combinations of outcomes that are of interest.

The Mathematical Theory of Probability

As was explained in Section 1.2, there is controversy in regard to the proper meaning and interpretation of some of the probabilities that are assigned to the outcomes of many experiments. However, once probabilities have been assigned to some simple outcomes in an experiment, there is complete agreement among all authorities that the mathematical theory of probability provides the appropriate methodology for the further study of these probabilities. Almost all work in the mathematical theory of probability, from the most elementary textbooks to the most advanced research, has been related to the following two problems: (i) methods for determining the probabilities of certain events from the specified probabilities of each possible outcome of an experiment and (ii) methods for revising the probabilities of events when additional relevant information is obtained.

These methods are based on standard mathematical techniques. The purpose of the first five chapters of this book is to present these techniques which, together, form the mathematical theory of probability.

1.4. SET THEORY

The Sample Space

The collection of all possible outcomes of an experiment is called the *sample space* of the experiment. In other words, the sample space of an experiment can be thought of as a *set*, or collection, of different possible outcomes; and each outcome can be thought of as a *point*, or an *element*, in the sample space. Because of this interpretation, the language and concepts of set theory provide a natural context for the development of probability theory. The basic ideas and notation of set theory will now be reviewed.

Relations of Set Theory

Let S denote the sample space of some experiment. Then any possible outcome s of the experiment is said to be a member of the space S, or to belong to the space S. The statement that s is a member of S is denoted symbolically by the relation $s \in S$.

When an experiment has been performed and we say that some *event* has occurred, we mean that the outcome of the experiment satisfied certain conditions which specified that event. In other words, some outcomes in the space S signify that the event occurred, and all other outcomes in S signify that the event did not occur. In accordance with this interpretation, any event can be regarded as a certain subset of possible outcomes in the space S.

For example, when a six-sided die is rolled, the sample space can be regarded as containing the six numbers $1, 2, 3, 4, 5, 6$. Symbolically, we write

$$S = \{1, 2, 3, 4, 5, 6\}.$$

The event A that an even number is obtained is defined by the subset $A = \{2, 4, 6\}$. The event B that a number greater than 2 is obtained is defined by the subset $B = \{3, 4, 5, 6\}$.

It is said that an event A *is contained in* another event B if every outcome that belongs to the subset defining the event A also belongs to the subset defining the event B. This relation between two events is expressed symbolically by the relation $A \subset B$. The relation $A \subset B$ is also expressed by saying that A is a subset of B. Equivalently, if $A \subset B$, we may say that B contains A and may write $B \supset A$.

In the example pertaining to the die, suppose that A is the event that an even number is obtained and C is the event that a number greater than 1 is obtained. Since $A = \{2, 4, 6\}$ and $C = \{2, 3, 4, 5, 6\}$, it follows that $A \subset C$. It should be noted that $A \subset S$ for any event A.

If two events A and B are so related that $A \subset B$ and $B \subset A$, it follows that A and B must contain exactly the same points. In other words, $A = B$.

If A, B, and C are three events such that $A \subset B$ and $B \subset C$, then it follows that $A \subset C$. The proof of this fact is left as an exercise.

The Empty Set

Some events are impossible. For example, when a die is rolled, it is impossible to obtain a negative number. Hence, the event that a negative number will be obtained is defined by the subset of S that contains no outcomes. This subset of S is called the *empty set*, or *null set*, and it is denoted by the symbol $\emptyset$.

Now consider any arbitrary event A. Since the empty set $\emptyset$ contains no points, it is logically correct to say that any point belonging to $\emptyset$ also belongs to A, or $\emptyset \subset A$. In other words, for any event A, it is true that $\emptyset \subset A \subset S$.

Operations of Set Theory

Unions.   If A and B are any two events, the *union* of A and B is defined to be the event containing all outcomes that belong to A alone, to B alone, or to both A and B. The notation for the union of A and B is $A \cup B$. The event $A \cup B$ is sketched in Fig. 1.1. A sketch of this type is called a *Venn diagram*.

For any events A and B, the union has the following properties:

$$A \cup B = B \cup A, \qquad A \cup A = A,$$
$$A \cup \emptyset = A, \qquad A \cup S = S.$$

Furthermore, if $A \subset B$, then $A \cup B = B$.

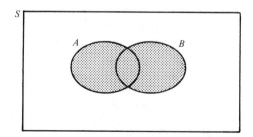

Figure 1.1   The event $A \cup B$.

The union of n events $A_1, \ldots, A_n$ is defined to be the event that contains all outcomes which belong to at least one of these n events. The notation for this union is either $A_1 \cup A_2 \cup \cdots \cup A_n$ or $\bigcup_{i=1}^{n} A_i$. Similarly, the notation for the union of an infinite sequence of events $A_1, A_2, \ldots$ is $\bigcup_{i=1}^{\infty} A_i$. The notation for the union of an arbitrary collection of events A_i, where the values of the subscript i belong to some index set I, is $\bigcup_{i \in I} A_i$.

The union of three events A, B, and C can be calculated either directly from the definition of $A \cup B \cup C$ or by first evaluating the union of any two of the events and then forming the union of this combination of events and the third event. In other words, the following *associative* relations are satisfied:

$$A \cup B \cup C = (A \cup B) \cup C = A \cup (B \cup C).$$

Intersections. If A and B are any two events, the *intersection* of A and B is defined to be the event that contains all outcomes which belong *both to A and to B*. The notation for the intersection of A and B is $A \cap B$. The event $A \cap B$ is sketched in a Venn diagram in Fig. 1.2. It is often convenient to denote the intersection of A and B by the symbol AB instead of $A \cap B$, and we shall use these two types of notation interchangeably.

For any events A and B, the intersection has the following properties:

$$A \cap B = B \cap A, \quad A \cap A = A,$$
$$A \cap \emptyset = \emptyset, \quad A \cap S = A.$$

Furthermore, if $A \subset B$, then $A \cap B = A$.

The intersection of n events $A_1, \ldots, A_n$ is defined to be the event that contains the outcomes which are common to all these n events. The notation for this intersection is $A_1 \cap A_2 \cap \cdots \cap A_n$, or $\bigcap_{i=1}^{n} A_i$, or $A_1 A_2 \cdots A_n$. Similar notations are used for the intersection of an infinite sequence of events or for the intersection of an arbitrary collection of events.

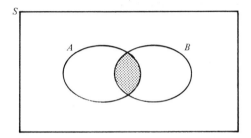

Figure 1.2   The event $A \cap B$.

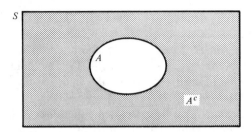

Figure 1.3 The event A^c.

For any three events A, B, and C, the following associative relations are satisfied:

$$A \cap B \cap C = (A \cap B) \cap C = A \cap (B \cap C).$$

Complements. The *complement* of an event A is defined to be the event that contains all outcomes in the sample space S which *do not* belong to A. The notation for the complement of A is A^c. The event A^c is sketched in Fig. 1.3.

For any event A, the complement has the following properties:

$$(A^c)^c = A, \qquad \emptyset^c = S, \qquad S^c = \emptyset,$$
$$A \cup A^c = S, \qquad A \cap A^c = \emptyset.$$

Disjoint Events. It is said that two events A and B are *disjoint*, or *mutually exclusive*, if A and B have no outcomes in common. It follows that A and B are disjoint if and only if $A \cap B = \emptyset$. It is said that the events in an arbitrary collection of events are disjoint if no two events in the collection have any outcomes in common.

As an illustration of these concepts, a Venn diagram for three events A_1, A_2, and A_3 is presented in Fig. 1.4. This diagram indicates that the various intersections of A_1, A_2, and A_3 and their complements will partition the sample space S into eight disjoint subsets.

Example 1: Tossing a Coin. Suppose that a coin is tossed three times. Then the sample space S contains the following eight possible outcomes $s_1, \ldots, s_8$:

s_1: HHH,
s_2: THH,
s_3: HTH,
s_4: HHT,
s_5: HTT,
s_6: THT,
s_7: TTH,
s_8: TTT.

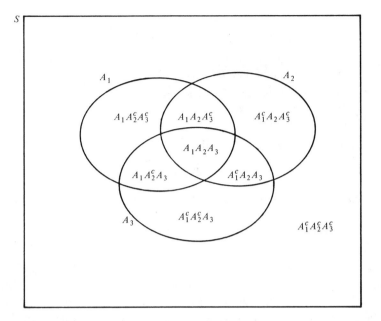

Figure 1.4 Partition of S determined by three events A_1, A_2, A_3.

In this notation, H indicates a head and T indicates a tail. The outcome s_3, for example, is the outcome in which a head is obtained on the first toss, a tail is obtained on the second toss, and a head is obtained on the third toss.

To apply the concepts introduced in this section, we shall define four events as follows: Let A be the event that at least one head is obtained in the three tosses; let B be the event that a head is obtained on the second toss; let C be the event that a tail is obtained on the third toss; and let D be the event that *no* heads are obtained. Accordingly,

$$A = \{s_1, s_2, s_3, s_4, s_5, s_6, s_7\},$$

$$B = \{s_1, s_2, s_4, s_6\},$$

$$C = \{s_4, s_5, s_6, s_8\},$$

$$D = \{s_8\}.$$

Various relations among these events can be derived. Some of these relations are $B \subset A$, $A^c = D$, $BD = \emptyset$, $A \cup C = S$, $BC = \{s_4, s_6\}$, $(B \cup C)^c = \{s_3, s_7\}$, and $A(B \cup C) = \{s_1, s_2, s_4, s_5, s_6\}$. □

EXERCISES

1. Suppose that $A \subset B$. Show that $B^c \subset A^c$.

2. For any three events A, B, and C, show that

   $$A(B \cup C) = (AB) \cup (AC).$$

3. For any two events A and B, show that

   $$(A \cup B)^c = A^c \cap B^c \text{ and } (A \cap B)^c = A^c \cup B^c.$$

4. For any collection of events A_i $(i \in I)$, show that

   $$\left(\bigcup_{i \in I} A_i \right)^c = \bigcap_{i \in I} A_i^c \text{ and } \left(\bigcap_{i \in I} A_i \right)^c = \bigcup_{i \in I} A_i^c.$$

5. Suppose that one card is to be selected from a deck of twenty cards that contains ten red cards numbered from 1 to 10 and ten blue cards numbered from 1 to 10. Let A be the event that a card with an even number is selected; let B be the event that a blue card is selected; and let C be the event that a card with a number less than 5 is selected. Describe the sample space S and describe each of the following events both in words and as subsets of S:
   (a) ABC,   (b) BC^c,   (c) $A \cup B \cup C$,
   (d) $A(B \cup C)$,   (e) $A^c B^c C^c$.

6. Suppose that a number x is to be selected from the real line S, and let A, B, and C be the events represented by the following subsets of S, where the notation $\{x: \text{----}\}$ denotes the set containing every point x for which the property presented following the colon is satisfied:

   $$A = \{x: 1 \leqslant x \leqslant 5\},$$

   $$B = \{x: 3 < x \leqslant 7\},$$

   $$C = \{x: x \leqslant 0\}.$$

   Describe each of the following events as a set of real numbers:
   (a) A^c,   (b) $A \cup B$,   (c) BC^c,
   (d) $A^c B^c C^c$,   (e) $(A \cup B)C$.

7. Let S be a given sample space and let $A_1, A_2, \ldots$ be an infinite sequence of events. For $n = 1, 2, \ldots$, let $B_n = \bigcup_{i=n}^{\infty} A_i$ and let $C_n = \bigcap_{i=n}^{\infty} A_i$.
   (a) Show that $B_1 \supset B_2 \supset \cdots$ and that $C_1 \subset C_2 \subset \cdots$.
   (b) Show that an outcome in S belongs to the event $\bigcap_{n=1}^{\infty} B_n$ if and only if it belongs to an infinite number of the events $A_1, A_2, \ldots$.

(c) Show that an outcome in S belongs to the event $\bigcup_{n=1}^{\infty} C_n$ if and only if it belongs to all the events $A_1, A_2, \ldots$ except possibly a finite number of those events.

1.5. THE DEFINITION OF PROBABILITY

Axioms and Basic Theorems

In this section we shall present the mathematical, or axiomatic, definition of probability. In a given experiment, it is necessary to assign to each event A in the sample space S a number $\Pr(A)$ which indicates the probability that A will occur. In order to satisfy the mathematical definition of probability, the number $\Pr(A)$ that is assigned must satisfy three specific axioms. These axioms ensure that the number $\Pr(A)$ will have certain properties which we intuitively expect a probability to have under any of the various interpretations described in Section 1.2.

The first axiom states that the probability of every event must be nonnegative.

Axiom 1.   *For any event A, $\Pr(A) \geq 0$.*

The second axiom states that if an event is certain to occur, then the probability of that event is 1.

Axiom 2. $\Pr(S) = 1$.

Before stating Axiom 3, we shall discuss the probabilities of disjoint events. If two events are disjoint, it is natural to assume that the probability that one or the other will occur is the sum of their individual probabilities. In fact, it will be assumed that this *additive property* of probability is also true for any finite number of disjoint events and even for any infinite sequence of disjoint events. If we assume that this additive property is true only for a finite number of disjoint events, we cannot then be certain that the property will be true for an infinite sequence of disjoint events as well. However, if we assume that the additive property is true for every infinite sequence of disjoint events, then (as we shall prove) the property must also be true for any finite number of disjoint events. These considerations lead to the third axiom.

Axiom 3.   *For any infinite sequence of disjoint events $A_1, A_2, \ldots$,*

$$\Pr\left(\bigcup_{i=1}^{\infty} A_i \right) = \sum_{i=1}^{\infty} \Pr(A_i).$$

The mathematical definition of probability can now be given as follows: A *probability distribution*, or simply a *probability*, on a sample space S is a specification of numbers $\Pr(A)$ which satisfy Axioms 1, 2, and 3.

We shall now derive two important consequences of Axiom 3. First, we shall show that if an event is impossible, its probability must be 0.

Theorem 1. $\Pr(\emptyset) = 0$.

Proof. Consider the infinite sequence of events $A_1, A_2, \ldots$ such that $A_i = \emptyset$ for $i = 1, 2, \ldots$. In other words, each of the events in the sequence is just the empty set $\emptyset$. Then this sequence is a sequence of disjoint events, since $\emptyset \cap \emptyset = \emptyset$. Furthermore, $\bigcup_{i=1}^{\infty} A_i = \emptyset$. Therefore, it follows from Axiom 3 that

$$\Pr(\emptyset) = \Pr\left(\bigcup_{i=1}^{\infty} A_i\right) = \sum_{i=1}^{\infty} \Pr(A_i) = \sum_{i=1}^{\infty} \Pr(\emptyset).$$

This equation states that when the number $\Pr(\emptyset)$ is added repeatedly in an infinite series, the sum of that series is simply the number $\Pr(\emptyset)$. The only real number with this property is $\Pr(\emptyset) = 0$. $\square$

We can now show that the additive property assumed in Axiom 3 for an infinite sequence of disjoint events is also true for any finite number of disjoint events.

Theorem 2.   *For any finite sequence of n disjoint events $A_1, \ldots, A_n$,*

$$\Pr\left(\bigcup_{i=1}^{n} A_i\right) = \sum_{i=1}^{n} \Pr(A_i).$$

Proof. Consider the infinite sequence of events $A_1, A_2, \ldots$, in which $A_1, \ldots, A_n$ are the n given disjoint events and $A_i = \emptyset$ for $i > n$. Then the events in this infinite sequence are disjoint and $\bigcup_{i=1}^{\infty} A_i = \bigcup_{i=1}^{n} A_i$. Therefore, by Axiom 3,

$$\Pr\left(\bigcup_{i=1}^{n} A_i\right) = \Pr\left(\bigcup_{i=1}^{\infty} A_i\right) = \sum_{i=1}^{\infty} \Pr(A_i)$$

$$= \sum_{i=1}^{n} \Pr(A_i) + \sum_{i=n+1}^{\infty} \Pr(A_i)$$

$$= \sum_{i=1}^{n} \Pr(A_i) + 0$$

$$= \sum_{i=1}^{n} \Pr(A_i). \quad \square$$

Further Properties of Probability

From the axioms and theorems just given, we shall now derive four other general properties of probability distributions. Because of the fundamental nature of these four properties, they will be presented in the form of four theorems, each one of which is easily proved.

Theorem 3.   *For any event A, $\Pr(A^c) = 1 - \Pr(A)$.*

Proof. Since A and A^c are disjoint events and $A \cup A^c = S$, it follows from Theorem 2 that $\Pr(S) = \Pr(A) + \Pr(A^c)$. Since $\Pr(S) = 1$ by Axiom 2, then $\Pr(A^c) = 1 - \Pr(A)$. □

Theorem 4.   *For any event A, $0 \leqslant \Pr(A) \leqslant 1$.*

Proof. It is known from Axiom 1 that $\Pr(A) \geqslant 0$. If $\Pr(A) > 1$, then it follows from Theorem 3 that $\Pr(A^c) < 0$. Since this result contradicts Axiom 1, which states that the probability of every event must be nonnegative, it must also be true that $\Pr(A) \leqslant 1$. □

Theorem 5.   *If $A \subset B$, then $\Pr(A) \leqslant \Pr(B)$.*

Proof. As illustrated in Fig. 1.5, the event B may be treated as the union of the two disjoint events A and BA^c. Therefore, $\Pr(B) = \Pr(A) + \Pr(BA^c)$. Since $\Pr(BA^c) \geqslant 0$, then $\Pr(B) \geqslant \Pr(A)$. □

Theorem 6.   *For any two events A and B,*

$$\Pr(A \cup B) = \Pr(A) + \Pr(B) - \Pr(AB).$$

Proof. As illustrated in Fig. 1.6,

$$A \cup B = (AB^c) \cup (AB) \cup (A^cB).$$

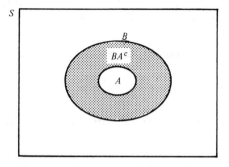

Figure 1.5 $B = A \cup (BA^c)$.

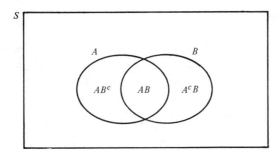

Figure 1.6   Partition of $A \cup B$.

Since the three events on the right side of this equation are disjoint, it follows from Theorem 2 that

$$\Pr(A \cup B) = \Pr(AB^c) + \Pr(AB) + \Pr(A^cB).$$

Furthermore, it is seen from Fig. 1.6 that

$$\Pr(A) = \Pr(AB^c) + \Pr(AB)$$

and

$$\Pr(B) = \Pr(A^cB) + \Pr(AB).$$

The theorem now follows from these relations.   □

EXERCISES

1. A student selected from a class will be either a boy or a girl. If the probability that a boy will be selected is 0.3, what is the probability that a girl will be selected?

2. One ball is to be selected from a box containing red, white, blue, yellow, and green balls. If the probability that the selected ball will be red is 1/5 and the probability that it will be white is 2/5, what is the probability that it will be blue, yellow, or green?

3. If the probability that student A will fail a certain statistics examination is 0.5, the probability that student B will fail the examination is 0.2, and the probability that both student A and student B will fail the examination is 0.1, what is the probability that at least one of these two students will fail the examination?

4. For the conditions of Exercise 3, what is the probability that neither student A nor student B will fail the examination?

5. For the conditions of Exercise 3, what is the probability that exactly one of the two students will fail the examination?

6. Consider two events A and B such that $\Pr(A) = 1/3$ and $\Pr(B) = 1/2$. Determine the value of $\Pr(BA^c)$ for each of the following conditions: (a) A and B are disjoint; (b) $A \subset B$; (c) $\Pr(AB) = 1/8$.

7. If 50 percent of the families in a certain city subscribe to the morning newspaper, 65 percent of the families subscribe to the afternoon newspaper, and 85 percent of the families subscribe to at least one of the two newspapers, what proportion of the families subscribe to both newspapers?

8. Consider two events A and B with $\Pr(A) = 0.4$ and $\Pr(B) = 0.7$. Determine the maximum and minimum possible values of $\Pr(AB)$ and the conditions under which each of these values is attained.

9. Prove that for any two events A and B, the probability that exactly one of the two events will occur is given by the expression

$$\Pr(A) + \Pr(B) - 2\Pr(AB).$$

10. A point (x, y) is to be selected from the square S containing all points (x, y) such that $0 \leqslant x \leqslant 1$ and $0 \leqslant y \leqslant 1$. Suppose that the probability that the selected point will belong to any specified subset of S is equal to the area of that subset. Find the probability of each of the following subsets: (a) the subset of points such that $\left(x - \frac{1}{2}\right)^2 + \left(y - \frac{1}{2}\right)^2 \geqslant \frac{1}{4}$; (b) the subset of points such that $\frac{1}{2} < x + y < \frac{3}{2}$; (c) the subset of points such that $y \leqslant 1 - x^2$; (d) the subset of points such that $x = y$.

11. Let $A_1, A_2, \ldots$ be any infinite sequence of events, and let $B_1, B_2, \ldots$ be another infinite sequence of events defined as follows: $B_1 = A_1$, $B_2 = A_1^c A_2$, $B_3 = A_1^c A_2^c A_3$, $B_4 = A_1^c A_2^c A_3^c A_4, \ldots$. Prove that

$$\Pr\left(\bigcup_{i=1}^{n} A_i\right) = \sum_{i=1}^{n} \Pr(B_i) \text{ for } n = 1, 2, \ldots,$$

and that

$$\Pr\left(\bigcup_{i=1}^{\infty} A_i\right) = \sum_{i=1}^{\infty} \Pr(B_i).$$

12. For any events $A_1, \ldots, A_n$, prove that

$$\Pr\left(\bigcup_{i=1}^{n} A_i\right) \leqslant \sum_{i=1}^{n} \Pr(A_i).$$

1.6. FINITE SAMPLE SPACES

Requirements of Probabilities

In this section we shall consider experiments for which there are only a finite number of possible outcomes. In other words, we shall consider experiments for which the sample space S contains only a finite number of points $s_1, \ldots, s_n$. In an experiment of this type, a probability distribution on S is specified by assigning a probability p_i to each point $s_i \in S$. The number p_i is the probability that the outcome of the experiment will be s_i ($i = 1, \ldots, n$). In order to satisfy the axioms for a probability distribution, the numbers $p_1, \ldots, p_n$ must satisfy the following two conditions:

$$p_i \geqslant 0 \text{ for } i = 1, \ldots, n$$

and

$$\sum_{i=1}^{n} p_i = 1.$$

The probability of any event A can then be found by adding the probabilities p_i of all outcomes s_i that belong to A.

Example 1: Fiber Breaks. Consider an experiment in which five fibers having different lengths are subjected to a testing process to learn which fiber will break first. Suppose that the lengths of the five fibers are 1 inch, 2 inches, 3 inches, 4 inches, and 5 inches, respectively. Suppose also that the probability that any given fiber will be the first to break is proportional to the length of that fiber. We shall determine the probability that the length of the fiber that breaks first is not more than 3 inches.

In this example, we shall let s_i be the outcome in which the fiber whose length is i inches breaks first ($i = 1, \ldots, 5$). Then $S = \{s_1, \ldots, s_5\}$ and $p_i = \alpha i$ for $i = 1, \ldots, 5$, where α is a proportionality factor. Since it must be true that $p_1 + \cdots + p_5 = 1$, the value of α must be $1/15$. If A is the event that the length of the fiber that breaks first is not more than 3 inches, then $A = \{s_1, s_2, s_3\}$. Therefore,

$$\Pr(A) = p_1 + p_2 + p_3 = \frac{1}{15} + \frac{2}{15} + \frac{3}{15} = \frac{2}{5}. \quad \square$$

Simple Sample Spaces

A sample space S containing n outcomes $s_1, \ldots, s_n$ is called a simple sample space if the probability assigned to each of the outcomes $s_1, \ldots, s_n$ is $1/n$. If an event A in this simple sample space contains exactly m outcomes, then

$$\Pr(A) = \frac{m}{n}.$$

Example 2: Tossing Coins. Suppose that three fair coins are tossed simultaneously. We shall determine the probability of obtaining exactly two heads.

Regardless of whether or not the three coins can be distinguished from each other by the experimenter, it is convenient for the purpose of describing the sample space to assume that the coins can be distinguished. We can then speak of the result for the first coin, the result for the second coin, and the result for the third coin; and the sample space will comprise the eight possible outcomes listed in Example 1 of Section 1.4.

Furthermore, because of the assumption that the coins are fair, it is reasonable to assume that this sample space is simple and that the probability assigned to each of the eight outcomes is $1/8$. As can be seen from the listing in Section 1.4, exactly two heads will be obtained in three of these outcomes. Therefore, the probability of obtaining exactly two heads is $3/8$. □

It should be noted that if we had considered the only possible outcomes to be no heads, one head, two heads, and three heads, it would have been reasonable to assume that the sample space contains just these four outcomes. This sample space would not be simple because the outcomes *would not be equally probable*.

Example 3: Rolling Two Dice. We shall now consider an experiment in which two balanced dice are rolled, and we shall calculate the probability of each of the possible values of the sum of the two numbers that may appear.

Although the experimenter need not be able to distinguish the two dice from one another in order to observe the value of their sum, the specification of a simple sample space in this example will be facilitated if we assume that the two dice are distinguishable. If this assumption is made, each outcome in the sample space S can be represented as a pair of numbers (x, y), where x is the number that appears on the first die and y is the number that appears on the second die. Therefore, S comprises the following 36 outcomes:

$$
\begin{array}{cccccc}
(1,1) & (1,2) & (1,3) & (1,4) & (1,5) & (1,6) \\
(2,1) & (2,2) & (2,3) & (2,4) & (2,5) & (2,6) \\
(3,1) & (3,2) & (3,3) & (3,4) & (3,5) & (3,6) \\
(4,1) & (4,2) & (4,3) & (4,4) & (4,5) & (4,6) \\
(5,1) & (5,2) & (5,3) & (5,4) & (5,5) & (5,6) \\
(6,1) & (6,2) & (6,3) & (6,4) & (6,5) & (6,6)
\end{array}
$$

It is natural to assume that S is a simple sample space and that the probability of each of these outcomes is $1/36$.

Let P_i denote the probability that the sum of the two numbers is i for $i = 2, 3, \ldots, 12$. The only outcome in S for which the sum is 2 is the outcome $(1, 1)$. Therefore, $P_2 = 1/36$. The sum will be 3 for either of the two outcomes $(1, 2)$ and $(2, 1)$. Therefore, $P_3 = 2/36 = 1/18$. By continuing in this manner we obtain the following probability for each of the possible values of the sum:

$$P_2 = P_{12} = \frac{1}{36}, \qquad P_5 = P_9 = \frac{4}{36},$$

$$P_3 = P_{11} = \frac{2}{36}, \qquad P_6 = P_8 = \frac{5}{36},$$

$$P_4 = P_{10} = \frac{3}{36}, \qquad P_7 = \frac{6}{36}. \quad \square$$

EXERCISES

1. A school contains students in grades 1, 2, 3, 4, 5, and 6. Grades 2, 3, 4, 5, and 6 all contain the same number of students, but there are twice this number in grade 1. If a student is selected at random from a list of all the students in the school, what is the probability that he will be in grade 3?

2. For the conditions of Exercise 1, what is the probability that the selected student will be in an odd-numbered grade?

3. If three fair coins are tossed, what is the probability that all three faces will be the same?

4. If two balanced dice are rolled, what is the probability that the sum of the two numbers that appear will be odd?

5. If two balanced dice are rolled, what is the probability that the sum of the two numbers that appear will be even?

6. If two balanced dice are rolled, what is the probability that the difference between the two numbers that appear will be less than 3?

7. Consider an experiment in which a fair coin is tossed once and a balanced die is rolled once. (a) Describe the sample space for this experiment. (b) What is the probability that a head will be obtained on the coin and an odd number will be obtained on the die?

1.7. COUNTING METHODS

We have seen that in a simple sample space S, the probability of an event A is the ratio of the number of outcomes in A to the total number of outcomes in S. In many experiments, the number of outcomes in S is so large that a complete listing of these outcomes is too expensive, too slow, or too likely to be incorrect to be useful. In such an experiment, it is convenient to have a method of determining the total number of outcomes in the space S and in various events in S without compiling a list of all these outcomes. In this section, some of these methods will be presented.

Multiplication Rule

Consider an experiment that has the following two characteristics:

(i) The experiment is performed in two parts.
(ii) The first part of the experiment has m possible outcomes $x_1, \ldots, x_m$ and, regardless of which one of these outcomes x_i occurs, the second part of the experiment has n possible outcomes $y_1, \ldots, y_n$.

Each outcome in the sample space S of the experiment will therefore be a pair having the form (x_i, y_j), and S will be composed of the following pairs:

$$(x_1, y_1)(x_1, y_2) \cdots (x_1, y_n)$$
$$(x_2, y_1)(x_2, y_2) \cdots (x_2, y_n)$$
$$\cdots \cdots \cdots \cdots \cdots \cdots \cdots$$
$$(x_m, y_1)(x_m, y_2) \cdots (x_m, y_n).$$

Since each of the m rows in this array contains n pairs, it follows that the sample space S contains exactly mn outcomes.

For example, suppose that there are three different routes from city A to city B and five different routes from city B to city C. Then the number of different routes from A to C that pass through B is $3 \times 5 = 15$. As another example, suppose that two dice are rolled. Since there are six possible outcomes for each die, the number of possible outcomes for the experiment is $6 \times 6 = 36$.

This multiplication rule can be extended to experiments with more than two parts. Suppose that an experiment has k parts ($k \geq 2$), that the ith part of the experiment can have n_i possible outcomes ($i = 1, \ldots, k$), and that each of the outcomes in any part can occur regardless of which specific outcomes have occurred in the other parts. Then the sample space S of the experiment will contain all vectors of the form $(u_1, \ldots, u_k)$, where u_i is one of the n_i possible outcomes of part i ($i = 1, \ldots, k$). The total number of these vectors in S will be equal to the product $n_1 n_2 \cdots n_k$.

For example, if six coins are tossed, each outcome in S will consist of a sequence of six heads and tails, such as HTTHHH. Since there are two possible outcomes for each of the six coins, the total number of outcomes in S will be $2^6 = 64$. If head and tail are considered equally likely for each coin, then S will be a simple sample space. Since there is only one outcome in S with six heads and no tails, the probability of obtaining heads on all six coins is $1/64$. Since there are six outcomes in S with one head and five tails, the probability of obtaining exactly one head is $6/64 = 3/32$.

Permutations

Sampling without Replacement.   Consider an experiment in which a card is selected and removed from a deck of n different cards, a second card is then selected and removed from the remaining $n - 1$ cards, and finally a third card is selected from the remaining $n - 2$ cards. A process of this kind is called *sampling without replacement*, since a card that is drawn is not replaced in the deck before the next card is selected. In this experiment, any one of the n cards could be selected first. Once this card has been removed, any one of the other $n - 1$ cards could be selected second. Therefore, there are $n(n - 1)$ possible outcomes for the first two selections. Finally, for any given outcome of the first two selections, there are $n - 2$ other cards that could possibly be selected third. Therefore, the total number of possible outcomes for all three selections is $n(n - 1)(n - 2)$. Thus, each outcome in the sample space S of this experiment will be some arrangement of three cards from the deck. Each different arrangement is called a *permutation*. The total number of possible permutations for the described experiment will be $n(n - 1)(n - 2)$.

This reasoning can be generalized to any number of selections without replacement. Suppose that k cards are to be selected one at a time and removed from a deck of n cards ($k = 1, 2, \ldots, n$). Then each possible outcome of this experiment will be a permutation of k cards from the deck, and the total number of these permutations will be $P_{n,k} = n(n - 1) \cdots (n - k + 1)$. This number $P_{n,k}$ is called the *number of permutations of n elements taken k at a time.*

When $k = n$, the number of possible outcomes of the experiment will be the number $P_{n,n}$ of different permutations of all n cards. It is seen from the equation just derived that

$$P_{n,n} = n(n - 1) \cdots 1 = n!.$$

The symbol $n!$ is read *n factorial*. In general, the number of permutations of n different items is $n!$.

The expression for $P_{n,k}$ can be rewritten in the following alternate form for $k = 1, \ldots, n - 1$:

$$P_{n,k} = n(n - 1) \cdots (n - k + 1) \frac{(n - k)(n - k - 1) \cdots 1}{(n - k)(n - k - 1) \cdots 1} = \frac{n!}{(n - k)!}.$$

Here and elsewhere in the theory of probability, it is convenient to define 0! by the relation

$$0! = 1.$$

With this definition, it follows that the relation $P_{n,k} = n!/(n-k)!$ will be correct for the value $k = n$ as well as for the values $k = 1, \ldots, n-1$.

Example 1: Choosing Officers. Suppose that a club consists of 25 members, and that a president and a secretary are to be chosen from the membership. We shall determine the total possible number of ways in which these two positions can be filled.

Since the positions can be filled by first choosing one of the 25 members to be president and then choosing one of the remaining 24 members to be secretary, the possible number of choices is $P_{25,2} = (25)(24) = 600$. □

Example 2: Arranging Books. Suppose that six different books are to be arranged on a shelf. The number of possible permutations of the books is $6! = 720$. □

Sampling with Replacement. We shall now consider the following experiment: A box contains n balls numbered $1, \ldots, n$. First, one ball is selected at random from the box and its number is noted. This ball is then put back in the box and another ball is selected (it is possible that the same ball will be selected again). As many balls as desired can be selected in this way. This process is called *sampling with replacement*. It is assumed that each of the n balls is equally likely to be selected at each stage and that all selections are made independently of each other.

Suppose that a total of k selections are to be made, where k is a given positive integer. Then the sample space S of this experiment will contain all vectors of the form $(x_1, \ldots, x_k)$, where x_i is the outcome of the ith selection $(i = 1, \ldots, k)$. Since there are n possible outcomes for each of the k selections, the total number of vectors in S is n^k. Furthermore, from our assumptions it follows that S is a simple sample space. Hence, the probability assigned to each vector in S is $1/n^k$.

Example 3: Obtaining Different Numbers. For the experiment just described, we shall determine the probability that each of the k balls that are selected will have a different number.

If $k > n$, it is impossible for all the selected balls to have different numbers because there are only n different numbers. Suppose, therefore, that $k \leqslant n$. The number of vectors in S for which all k components are different is $P_{n,k}$, since the first component x_1 of each vector can have n possible values, the second component x_2 can then have any one of the other $n-1$ values, and so on. Since S is a simple sample space containing n^k vectors, the probability p that k

different numbers will be selected is

$$p = \frac{P_{n,k}}{n^k} = \frac{n!}{(n-k)! \, n^k}. \quad \square$$

The Birthday Problem

In the following problem, which is often called the birthday problem, it is required to determine the probability p that at least two people in a group of k people ($2 \leqslant k \leqslant 365$) will have the same birthday, that is, will have been born on the same day of the same month but not necessarily in the same year. In order to solve this problem, we must assume that the birthdays of the k people are unrelated (in particular, we must assume that twins are not present) and that each of the 365 days of the year is equally likely to be the birthday of any person in the group. Thus, we must ignore the fact that the birth rate actually varies during the year and we must assume that anyone actually born on February 29 will consider his birthday to be another day, such as March 1.

When these assumptions are made, this problem becomes similar to the one in Example 3. Since there are 365 possible birthdays for each of k people, the sample space S will contain 365^k outcomes, all of which will be equally probable. Furthermore, the number of outcomes in S for which all k birthdays will be different is $P_{365,k}$, since the first person's birthday could be any one of the 365 days, the second person's birthday could then be any of the other 364 days, and so on. Hence, the probability that all k persons will have different birthdays is

$$\frac{P_{365,k}}{365^k}.$$

The probability p that at least two of the people will have the same birthday is, therefore,

$$p = 1 - \frac{P_{365,k}}{365^k} = 1 - \frac{(365)!}{(365-k)! \, 365^k}.$$

Numerical values of this probability p for various values of k are given in Table 1.1. These probabilities may seem surprisingly large to anyone who has not thought about them before. Many persons would guess that in order to obtain a value of p greater than $1/2$, the number of people in the group would have to be about 100. However, according to Table 1.1, there would have to be only 23 people in the group. As a matter of fact, for $k = 100$ the value of p is 0.9999997.

Table 1.1
The probability p that at least two people in a
group of k people will have the same birthday

k	p	k	p
5	0.027	25	0.569
10	0.117	30	0.706
15	0.253	40	0.891
20	0.411	50	0.970
22	0.476	60	0.994
23	0.507		

EXERCISES

1. Three different classes contain 20, 18, and 25 students, respectively, and no student is a member of more than one class. If a team is to be composed of one student from each of these three classes, in how many different ways can the members of the team be chosen?

2. In how many different ways can the five letters a, b, c, d, and e be arranged?

3. If a man has six different sportshirts and four different pairs of slacks, how many different combinations can he wear?

4. If four dice are rolled, what is the probability that each of the four numbers that appear will be different?

5. If six dice are rolled, what is the probability that each of the six different numbers will appear exactly once?

6. If 12 balls are thrown at random into 20 boxes, what is the probability that no box will receive more than one ball?

7. An elevator in a building starts with five passengers and stops at seven floors. If each passenger is equally likely to get off at any floor and all the passengers leave independently of each other, what is the probability that no two passengers will get off at the same floor?

8. Suppose that three runners from team A and three runners from team B participate in a race. If all six runners have equal ability and there are no ties, what is the probability that the three runners from team A will finish first, second, and third, and the three runners from team B will finish fourth, fifth, and sixth?

9. A box contains 100 balls, of which r are red. Suppose that the balls are drawn from the box one at a time, at random, without replacement. De-

termine (a) the probability that the first ball drawn will be red; (b) the probability that the fiftieth ball drawn will be red; and (c) the probability that the last ball drawn will be red.

1.8. COMBINATORIAL METHODS

Combinations

Suppose that there is a set of n distinct elements from which it is desired to choose a subset containing k elements ($1 \leqslant k \leqslant n$). We shall determine the number of different subsets that can be chosen. In this problem, the arrangement of the elements in a subset is irrelevant and each subset is treated as a unit. Such a subset is called a *combination*. No two combinations will consist of exactly the same elements. We shall let $C_{n,k}$ denote the number of combinations of n elements taken k at a time. The problem, then, is to determine the value of $C_{n,k}$.

For example, if the set contains four elements a, b, c, and d and if each subset is to consist of two of these elements, then the following six different combinations can be obtained:

$$\{a,b\}, \quad \{a,c\}, \quad \{a,d\}, \quad \{b,c\}, \quad \{b,d\}, \quad \text{and} \quad \{c,d\}.$$

Hence, $C_{4,2} = 6$. When combinations are considered, the subsets $\{a,b\}$ and $\{b,a\}$ are identical and only one of these subsets is counted.

The numerical value of $C_{n,k}$ for given integers n and k ($1 \leqslant k \leqslant n$) will now be derived. It is known that the number of *permutations* of n elements taken k at a time is $P_{n,k}$. A list of these $P_{n,k}$ permutations could be constructed as follows: First, a particular combination of k elements is selected. Each different permutation of these k elements will yield a permutation on the list. Since there are $k!$ permutations of these k elements, this particular combination will produce $k!$ permutations on the list. When a different combination of k elements is selected, $k!$ other permutations on the list will be obtained. Since each combination of k elements will yield $k!$ permutations on the list, the total number of permutations on the list must be $k!C_{n,k}$. Hence, it follows that $P_{n,k} = k!C_{n,k}$, from which

$$C_{n,k} = \frac{P_{n,k}}{k!} = \frac{n!}{k!(n-k)!}.$$

Example 1: Selecting a Committee. Suppose that a committee composed of 8 people is to be selected from a group of 20 people. The number of different groups of people that might be on the committee is

$$C_{20,8} = \frac{20!}{8!\,12!} = 125{,}970. \quad \square$$

Binomial Coefficients

Notation. The number $C_{n,k}$ is also denoted by the symbol $\binom{n}{k}$. When this notation is used, this number is called a *binomial coefficient* because it appears in the *binomial theorem*, which may be stated as follows: *For any numbers x and y and any positive integer n,*

$$(x + y)^n = \sum_{k=0}^{n} \binom{n}{k} x^k y^{n-k}.$$

Thus, for $k = 0, 1, \ldots, n$,

$$\binom{n}{k} = \frac{n!}{k!\,(n-k)!}.$$

Since $0! = 1$, the value of the binomial coefficient $\binom{n}{k}$ for $k = 0$ or $k = n$ is 1. Thus,

$$\binom{n}{0} = \binom{n}{n} = 1.$$

It can be seen from these relations that for $k = 0, 1, \ldots, n$,

$$\binom{n}{k} = \binom{n}{n-k}.$$

ALSO, $\binom{n}{k} = \binom{n-1}{k-1} + \binom{n-1}{k}$ $1 \leq k \leq n$

This equation can also be derived from the fact that selecting k elements to form a subset is equivalent to selecting the remaining $n - k$ elements to form the complement of the subset. Hence, the number of combinations containing k elements is equal to the number of combinations containing $n - k$ elements.

It is sometimes convenient to use the expression "n choose k" for the value of $C_{n,k}$. Thus, the same quantity is represented by the two different notations $C_{n,k}$ and $\binom{n}{k}$; and we may refer to this quantity in three different ways: as the number of combinations of n elements taken k at a time, as the binomial coefficient of n and k, or simply as "n choose k."

Arrangements of Elements of Two Distinct Types. When a set contains only elements of two distinct types, a binomial coefficient can be used to represent the number of different arrangements of all the elements in the set. Suppose, for example, that k similar red balls and $n - k$ similar green balls are to be arranged in a row. Since the red balls will occupy k positions in the row, each different arrangement of the n balls corresponds to a different choice of the k positions occupied by the red balls. Hence, the number of different arrangements of the n balls will be equal to the number of different ways in which k positions can be selected for the red balls from the n available positions. Since this number of ways is specified by the binomial coefficient $\binom{n}{k}$, the number of different arrangements of the n balls is also $\binom{n}{k}$. In other words, the number of different arrangements of n objects consisting of k similar objects of one type and $n - k$ similar objects of a second type is $\binom{n}{k}$.

Example 2: Tossing a Coin. Suppose that a fair coin is to be tossed ten times, and it is desired to determine (a) the probability p of obtaining exactly three heads and (b) the probability p' of obtaining three or fewer heads.

(a) The total possible number of different sequences of ten heads and tails is 2^{10}, and it may be assumed that each of these sequences is equally probable. The number of these sequences that contain exactly three heads will be equal to the number of different arrangements that can be formed with three heads and seven tails. Since this number is $\binom{10}{3}$, the probability of obtaining exactly three heads is

$$p = \frac{\binom{10}{3}}{2^{10}} = 0.1172.$$

(b) Since, in general, the number of sequences in the sample space that contain exactly k heads ($k = 0, 1, 2, 3$) is $\binom{10}{k}$, the probability of obtaining three or fewer heads is

$$p' = \frac{\binom{10}{0} + \binom{10}{1} + \binom{10}{2} + \binom{10}{3}}{2^{10}}$$

$$= \frac{1 + 10 + 45 + 120}{2^{10}} = \frac{176}{2^{10}} = 0.1719. \quad \square$$

Example 3: Sampling without Replacement. Suppose that a class contains 15 boys and 30 girls, and that 10 students are to be selected at random for a special assignment. We shall determine the probability p that exactly 3 boys will be selected.

The number of different combinations of the 45 students that might be obtained in the sample of 10 students is $\binom{45}{10}$, and the statement that the 10 students are selected at random means that each of these $\binom{45}{10}$ possible combinations is equally probable. Therefore, we must find the number of these combinations that contain exactly 3 boys and 7 girls.

When a combination of 3 boys and 7 girls is formed, the number of different combinations in which 3 boys can be selected from the 15 available boys is $\binom{15}{3}$, and the number of different combinations in which 7 girls can be selected from the 30 available girls is $\binom{30}{7}$. Since each of these combinations of 3 boys can be paired with each of the combinations of 7 girls to form a distinct sample, the number of combinations containing exactly 3 boys is $\binom{15}{3}\binom{30}{7}$. Therefore, the desired probability is

$$p = \frac{\binom{15}{3}\binom{30}{7}}{\binom{45}{10}} = 0.2904. \quad \square$$

Example 4: Playing Cards. Suppose that a deck of 52 cards containing four aces is shuffled thoroughly and the cards are then distributed among four players so that each player receives 13 cards. We shall determine the probability that each player will receive one ace.

The number of possible different combinations of the four positions in the deck occupied by the four aces is $\binom{52}{4}$, and it may be assumed that each of these $\binom{52}{4}$ combinations is equally probable. If each player is to receive one ace, then there must be exactly one ace among the 13 cards that the first player will receive and one ace among each of the remaining three groups of 13 cards that the other three players will receive. In other words, there are 13 possible positions for the ace that the first player is to receive, 13 other possible positions for the ace that the second player is to receive, and so on. Therefore, among the $\binom{52}{4}$ possible combinations of the positions for the four aces, exactly 13^4 of these combinations will lead to the desired result. Hence, the probability p that each player will receive one ace is

$$p = \frac{13^4}{\binom{52}{4}} = 0.1055. \quad \square$$

The Tennis Tournament

We shall now present a difficult problem that has a simple and elegant solution. Suppose that n tennis players are entered in a tournament. In the first round the players are paired one against another at random. The loser in each pair is eliminated from the tournament, and the winner in each pair continues into the second round. If the number of players n is odd, then one player is chosen at random before the pairings are made for the first round, and he automatically continues into the second round. All the players in the second round are then paired at random. Again, the loser in each pair is eliminated, and the winner in each pair continues into the third round. If the number of players in the second round is odd, then one of these players is chosen at random before the others are paired, and he automatically continues into the third round. The tournament continues in this way until only two players remain in the final round. They then play against each other, and the winner of this match is the winner of the tournament. We shall assume that all n players have equal ability, and we shall determine the probability p that two specific players A and B will play against each other at any time during the tournament.

We shall first determine the total number of matches that will be played during the tournament. After each match has been played, one player—the loser of that match—is eliminated from the tournament. The tournament ends when everyone has been eliminated from the tournament except the winner of the final match. Since exactly $n - 1$ players must be eliminated, it follows that exactly $n - 1$ matches must be played during the tournament.

The number of possible pairs of players is $\binom{n}{2}$. Each of the two players in any match is equally likely to win that match and all initial pairings are made in a random manner. Therefore, before the tournament begins, each possible pair of players is equally likely to appear in any particular one of the $n - 1$ matches to be played during the tournament. Accordingly, the probability that players A and B will meet in some particular match which is specified in advance is $1/\binom{n}{2}$. If A and B do meet in that particular match, one of them will lose and be eliminated. Therefore, these same two players cannot meet in more than one match.

It follows from the preceding explanation that the probability p that players A and B will meet at some time during the tournament is equal to the product of the probability $1/\binom{n}{2}$ that they will meet in any particular specified match and the total number $n - 1$ of different matches in which they might possibly meet. Hence,

$$p = \frac{n - 1}{\binom{n}{2}} = \frac{2}{n}.$$

EXERCISES

1. Which of the following two numbers is larger: $\binom{93}{30}$ or $\binom{93}{31}$?

2. Which of the following two numbers is larger: $\binom{93}{30}$ or $\binom{93}{63}$?

3. A box contains 24 light bulbs, of which 4 are defective. If a person selects 4 bulbs from the box at random, without replacement, what is the probability that all 4 bulbs will be defective?

4. Prove that the following number is an integer:

$$\frac{4155 \times 4156 \times \cdots \times 4250 \times 4251}{2 \times 3 \times \cdots \times 96 \times 97}.$$

5. Suppose that n people are seated in a random manner in a row of n theater seats. What is the probability that two particular people A and B will be seated next to each other?

6. If k people are seated in a random manner in a row containing n seats $(n > k)$, what is the probability that the people will occupy k adjacent seats in the row?

7. If k people are seated in a random manner in a circle containing n chairs $(n > k)$, what is the probability that the people will occupy k adjacent chairs in the circle?

8. If n people are seated in a random manner in a row containing $2n$ seats, what is the probability that no two people will occupy adjacent seats?

9. A box contains 24 light bulbs, of which 2 are defective. If a person selects 10 bulbs at random, without replacement, what is the probability that both defective bulbs will be selected?

10. Suppose that a committee of 12 people is selected in a random manner from a group of 100 people. Determine the probability that two particular people A and B will both be selected.

11. Suppose that 35 people are divided in a random manner into two teams in such a way that one team contains 10 people and the other team contains 25 people. What is the probability that two particular people A and B will be on the same team?

12. A box contains 24 light bulbs of which 4 are defective. If one person selects 10 bulbs from the box in a random manner, and a second person then takes the remaining 14 bulbs, what is the probability that all 4 defective bulbs will be obtained by the same person?

13. Prove that, for any positive integers n and k $(n \geqslant k)$,

$$\binom{n}{k} + \binom{n}{k-1} = \binom{n+1}{k}.$$

14. (a) Prove that

$$\binom{n}{0} + \binom{n}{1} + \binom{n}{2} + \cdots + \binom{n}{n} = 2^n.$$

(b) Prove that

$$\binom{n}{0} - \binom{n}{1} + \binom{n}{2} - \binom{n}{3} + \cdots + (-1)^n \binom{n}{n} = 0.$$

Hint: Use the binomial theorem.

15. The United States Senate contains two senators from each of the 50 states. (a) If a committee of 8 senators is selected at random, what is the probability that it will contain at least one of the two senators from a certain specified state? (b) What is the probability that a group of 50 senators selected at random will contain one senator from each state?

16. A deck of 52 cards contains 4 aces. If the cards are shuffled and distributed in a random manner to four players so that each player receives 13 cards, what is the probability that all 4 aces will be received by the same player?

17. Suppose that 100 mathematics students are divided into five classes, each containing 20 students, and that awards are to be given to 10 of these students. If each student is equally likely to receive an award, what is the probability that exactly 2 students in each class will receive awards?

1.9.   MULTINOMIAL COEFFICIENTS

Suppose that n distinct elements are to be divided into k different groups ($k \geqslant 2$) in such a way that, for $j = 1, \ldots, k$, the jth group contains exactly n_j elements, where $n_1 + n_2 + \cdots + n_k = n$. It is desired to determine the number of different ways in which the n elements can be divided into the k groups. The n_1 elements in the first group can be selected from the n available elements in $\binom{n}{n_1}$ different ways. After the n_1 elements in the first group have been selected, the n_2 elements in the second group can be selected from the remaining $n - n_1$ elements in $\binom{n - n_1}{n_2}$ different ways. Hence, the total number of different ways of selecting the elements for both the first group and the second group is $\binom{n}{n_1}\binom{n - n_1}{n_2}$. After the $n_1 + n_2$ elements in the first two groups have been selected, the number of different ways in which the n_3 elements in the third group can be selected is $\binom{n - n_1 - n_2}{n_3}$. Hence, the total number of different ways of selecting the elements for the first three groups is

$$\binom{n}{n_1}\binom{n-n_1}{n_2}\binom{n-n_1-n_2}{n_3}.$$

It follows from the preceding explanation that after the first $k-2$ groups have been formed, the number of different ways in which the n_{k-1} elements in the next group can be selected from the remaining $n_{k-1}+n_k$ elements is $\binom{n_{k-1}+n_k}{n_{k-1}}$, and the remaining n_k elements must then form the last group. Hence, the total number of different ways of dividing the n elements into the k groups is

$$\binom{n}{n_1}\binom{n-n_1}{n_2}\binom{n-n_1-n_2}{n_3}\cdots\binom{n_{k-1}+n_k}{n_{k-1}}.$$

When these binomial coefficients are expressed in terms of factorials, this product can be written in the simple form

$$\frac{n!}{n_1!n_2!\cdots n_k!}.$$

The number that has just been obtained is called a *multinomial coefficient* because it appears in the *multinomial theorem*, which can be stated as follows: *For any numbers $x_1,\ldots,x_k$ and any positive integer n,*

$$(x_1+\cdots+x_k)^n = \sum \frac{n!}{n_1!n_2!\cdots n_k!}x_1^{n_1}x_2^{n_2}\cdots x_k^{n_k}.$$

The summation in this equation is extended over all possible combinations of nonnegative integers $n_1,\ldots,n_k$ such that $n_1+n_2+\cdots+n_k=n$.

A multinomial coefficient is a generalization of the binomial coefficient discussed in Sec. 1.8. For $k=2$, the multinomial theorem is the same as the binomial theorem and the multinomial coefficient becomes a binomial coefficient.

Example 1: Choosing Committees. Suppose that 20 members of an organization are to be divided into three committees A, B, and C in such a way that each of the committees A and B is to have 8 members and committee C is to have 4 members. We shall determine the number of different ways in which members can be assigned to these committees.

Here, the required number is the multinomial coefficient for which $n=20$, $k=3$, $n_1=n_2=8$, and $n_3=4$. Hence the answer is

$$\frac{20!}{(8!)^2 4!} = 62,355,150. \quad \square$$

Arrangements of Elements of More Than Two Distinct Types.   Just as binomial coefficients can be used to represent the number of different arrangements of the elements of a set containing elements of only two distinct types, multinomial coefficients can be used to represent the number of different arrangements of the elements of a set containing elements of k different types ($k \geqslant 2$). Suppose, for example, that n balls of k different colors are to be arranged in a row and that there are n_j balls of color j ($j = 1, \ldots, k$), where $n_1 + n_2 + \cdots + n_k = n$. Then each different arrangement of the n balls corresponds to a different way of dividing the n available positions in the row into a group of n_1 positions to be occupied by the balls of color 1, a second group of n_2 positions to be occupied by the balls of color 2, and so on. Hence, the total number of different possible arrangements of the n balls must be

$$\frac{n!}{n_1! n_2! \cdots n_k!}.$$

Example 2: Rolling Dice.   Suppose that 12 dice are to be rolled. We shall determine the probability p that each of the six different numbers will appear twice.

Each outcome in the sample space S can be regarded as an ordered sequence of 12 numbers, where the ith number in the sequence is the outcome of the ith roll. Hence, there will be 6^{12} possible outcomes in S and each of these outcomes can be regarded as equally probable. The number of these outcomes that would contain each of the six numbers $1, 2, \ldots, 6$ exactly twice will be equal to the number of different possible arrangements of these 12 elements. This number can be determined by evaluating the multinomial coefficient for which $n = 12$, $k = 6$, and $n_1 = n_2 = \cdots = n_6 = 2$. Hence, the number of such outcomes is

$$\frac{12!}{(2!)^6}$$

and the required probability p is

$$p = \frac{12!}{2^6 6^{12}} = 0.0034. \quad \square$$

Example 3: Playing Cards.   A deck of 52 cards contains 13 hearts. Suppose that the cards are shuffled and distributed among four players A, B, C, and D so that each player receives 13 cards. We shall determine the probability p that player A will receive 6 hearts, player B will receive 4 hearts, player C will receive 2 hearts, and player D will receive 1 heart.

The total number N of different ways in which the 52 cards can be distributed among the four players so that each player receives 13 cards is

$$N = \frac{52!}{(13!)^4}.$$

It may be assumed that each of these ways is equally probable. We must now calculate the number M of ways of distributing the cards so that each player receives the required number of hearts. The number of different ways in which the hearts can be distributed to players A, B, C, and D so that the numbers of hearts they receive are 6, 4, 2, and 1, respectively, is

$$\frac{13!}{6!\,4!\,2!\,1!}.$$

Also, the number of different ways in which the other 39 cards can then be distributed to the four players so that each will have a total of 13 cards is

$$\frac{39!}{7!\,9!\,11!\,12!}.$$

Therefore,

$$M = \frac{13!}{6!\,4!\,2!\,1!} \cdot \frac{39!}{7!\,9!\,11!\,12!}$$

and the required probability p is

$$p = \frac{M}{N} = \frac{13!\,39!\,(13!)^4}{6!\,4!\,2!\,1!\,7!\,9!\,11!\,12!\,52!} = 0.00196.$$

There is another approach to this problem along the lines indicated in Example 4 of Sec. 1.8. The number of possible different combinations of the 13 positions in the deck occupied by the hearts is $\binom{52}{13}$. If player A is to receive 6 hearts, there are $\binom{13}{6}$ possible combinations of the six positions these hearts occupy among the 13 cards that A will receive. Similarly, if player B is to receive 4 hearts, there are $\binom{13}{4}$ possible combinations of their positions among the 13 cards that B will receive. There are $\binom{13}{2}$ possible combinations for player C, and there are $\binom{13}{1}$ possible combinations for player D. Hence,

$$p = \frac{\binom{13}{6}\binom{13}{4}\binom{13}{2}\binom{13}{1}}{\binom{52}{13}},$$

which produces the same value as the one obtained by the first method of solution. $\square$

EXERCISES

1. Suppose that 18 red beads, 12 yellow beads, 8 blue beads, and 12 black beads are to be strung in a row. How many different arrangements of the beads can be formed?

2. Suppose that two committees are to be formed in an organization that has 300 members. If one committee is to have 5 members and the other committee is to have 8 members, in how many different ways can these committees be selected?

3. If the letters $s, s, s, t, t, t, i, i, a, c$ are arranged in a random order, what is the probability that they will spell the word "statistics"?

4. Suppose that n balanced dice are rolled. Determine the probability that the number j will appear exactly n_j times ($j = 1, \ldots, 6$), where $n_1 + n_2 + \cdots + n_6 = n$.

5. If seven balanced dice are rolled, what is the probability that each of the six different numbers will appear at least once?

6. Suppose that a deck of 25 cards contains 12 red cards. Suppose also that the 25 cards are distributed in a random manner to three players A, B, and C in such a way that player A receives 10 cards, player B receives 8 cards, and player C receives 7 cards. Determine the probability that player A will receive 6 red cards, player B will receive 2 red cards, and player C will receive 4 red cards.

7. A deck of 52 cards contains 12 picture cards. If the 52 cards are distributed in a random manner among four players in such a way that each player receives 13 cards, what is the probability that each player will receive 3 picture cards?

8. Suppose that a deck of 52 cards contains 13 red cards, 13 yellow cards, 13 blue cards, and 13 green cards. If the 52 cards are distributed in a random manner among four players in such a way that each player receives 13 cards, what is the probability that each player will receive 13 cards of the same color?

9. Suppose that 2 boys named Davis, 3 boys named Jones, and 4 boys named Smith are seated at random in a row containing 9 seats. What is the probability that the Davis boys will occupy the first two seats in the row, the Jones boys will occupy the next three seats, and the Smith boys will occupy the last four seats?

1.10.   THE PROBABILITY OF A UNION OF EVENTS

The Union of Three Events

We shall now consider again an arbitrary sample space S that may contain either a finite number of outcomes or an infinite number, and we shall develop some further general properties of the various probabilities that might be specified for the events in S. In this section, we shall study in particular the probability of the union $\bigcup_{i=1}^{n} A_i$ of n events $A_1, \ldots, A_n$.

If the events $A_1, \ldots, A_n$ are disjoint, we know that

$$\Pr\left(\bigcup_{i=1}^{n} A_i \right) = \sum_{i=1}^{n} \Pr(A_i).$$

Furthermore, for any two events A_1 and A_2, regardless of whether or not they are disjoint, we know from Theorem 6 of Sec. 1.5 that

$$\Pr(A_1 \cup A_2) = \Pr(A_1) + \Pr(A_2) - \Pr(A_1 A_2).$$

We shall now extend this result, first to three events and then to an arbitrary finite number of events.

Theorem 1.   *For any three events A_1, A_2, and A_3,*

$$\Pr(A_1 \cup A_2 \cup A_3) = \Pr(A_1) + \Pr(A_2) + \Pr(A_3)$$

$$- \left[\Pr(A_1 A_2) + \Pr(A_2 A_3) + \Pr(A_1 A_3) \right] \qquad (1)$$

$$+ \Pr(A_1 A_2 A_3).$$

This equation indicates that the value of $\Pr(A_1 \cup A_2 \cup A_3)$ can be found by taking the sum of the probabilities of each of the three individual events, subtracting the sum of the probabilities of the intersections of the three possible pairs of events, and then adding the probability of the intersection of all three events.

Proof. In Fig. 1.4 it was shown that the union $A_1 \cup A_2 \cup A_3$ can be represented as the union of seven disjoint events. We shall denote the probabilities of these seven disjoint events by the values $p_1, \ldots, p_7$, as indicated in Fig. 1.7. Then $\Pr(A_1 \cup A_2 \cup A_3) = \sum_{i=1}^{7} p_i$, and we must show that the right side of Eq. (1) is also equal to $\sum_{i=1}^{7} p_i$.

It can be seen from Fig. 1.7 that the following three relations are correct:

$$\Pr(A_1) + \Pr(A_2) + \Pr(A_3) = (p_1 + p_4 + p_6 + p_7) + (p_2 + p_4 + p_5 + p_7)$$

$$+ (p_3 + p_5 + p_6 + p_7),$$

$$\Pr(A_1 A_2) + \Pr(A_2 A_3) + \Pr(A_1 A_3) = (p_4 + p_7) + (p_5 + p_7) + (p_6 + p_7),$$

$$\Pr(A_1 A_2 A_3) = p_7.$$

It follows from these relations that the right side of Eq. (1) is equal to $\sum_{i=1}^{7} p_i$. □

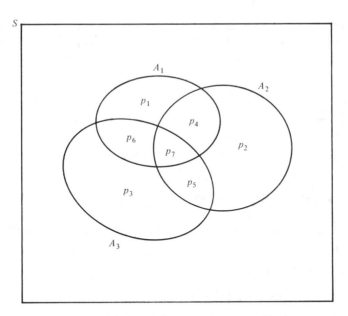

Figure 1.7   Probabilities of the seven events partitioning $A_1 \cup A_2 \cup A_3$.

Example 1: Student Enrollment. Among a group of 200 students, 137 students are enrolled in a mathematics class, 50 students are enrolled in a history class, and 124 students are enrolled in a music class. Furthermore, the number of students enrolled in both the mathematics and history classes is 33; the number enrolled in both the history and music classes is 29; and the number enrolled in both the mathematics and music classes is 92. Finally, the number of students enrolled in all three classes is 18. We shall determine the probability that a student selected at random from the group of 200 students will be enrolled in at least one of the three classes.

Let A_1 denote the event that the selected student is enrolled in the mathematics class; let A_2 denote the event that he is enrolled in the history class; and let A_3 denote the event that he is enrolled in the music class. To solve the problem, we must determine the value of $\Pr(A_1 \cup A_2 \cup A_3)$. From the given numbers,

$$\Pr(A_1) = \frac{137}{200}, \qquad \Pr(A_2) = \frac{50}{200}, \qquad \Pr(A_3) = \frac{124}{200},$$

$$\Pr(A_1 A_2) = \frac{33}{200}, \qquad \Pr(A_2 A_3) = \frac{29}{200}, \qquad \Pr(A_1 A_3) = \frac{92}{200},$$

$$\Pr(A_1 A_2 A_3) = \frac{18}{200}.$$

It follows from Eq. (1) that $\Pr(A_1 \cup A_2 \cup A_3) = 175/200 = 7/8$. $\square$

The Union of a Finite Number of Events

An equation similar to Eq. (1) holds for any arbitrary finite number of events, as shown by the following theorem:

Theorem 2.   *For any n events* $A_1, \ldots, A_n$,

$$
\Pr\left(\bigcup_{i=1}^{n} A_i \right) = \sum_{i=1}^{n} \Pr(A_i) - \sum_{i<j} \Pr(A_i A_j) + \sum_{i<j<k} \Pr(A_i A_j A_k)
$$

$$
- \sum_{i<j<k<l} \Pr(A_i A_j A_k A_l) + \cdots \tag{2}
$$

$$
+ (-1)^{n+1} \Pr(A_1 A_2 \cdots A_n).
$$

The steps in evaluating $\Pr(\bigcup_{i=1}^{n} A_i)$ can be outlined as follows: First, take the sum of the probabilities of the n individual events. Second, subtract the sum of the probabilities of the intersections of all possible pairs of events; in this step, there will be $\binom{n}{2}$ different pairs for which the probabilities are included. Third, add the probabilities of the intersections of all possible groups of three of the events; there will be $\binom{n}{3}$ intersections of this type. Fourth, subtract the sum of the probabilities of the intersections of all possible groups of four of the events; there will be $\binom{n}{4}$ intersections of this type. Continue in this way until, finally, the probability of the intersection of all n events is either added or subtracted, depending on whether n is an odd number or an even number.

Proof.   The proof of Eq. (2) is similar to that given for Eq. (1), but we cannot rely on a Venn diagram in this case. Suppose that $\bigcup_{i=1}^{n} A_i$ is partitioned into disjoint subsets, such that each subset contains the outcomes which are common to certain specified events among $A_1, \ldots, A_n$ and which do not belong to any of the other events. Then $\Pr(\bigcup_{i=1}^{n} A_i)$ will be the sum of the probabilities of these disjoint subsets. To prove Theorem 2, we must show that the probability of each subset is counted exactly once on the right side of Eq. (2).

For a given value of k $(1 \leqslant k \leqslant n)$, we shall consider first the subset B of outcomes which belong to each of the events $A_1, \ldots, A_k$ but which do not belong to any of the events $A_{k+1}, \ldots, A_n$. That is, we shall consider the event

$$
B = A_1 \cap \cdots \cap A_k \cap A_{k+1}^c \cap \cdots \cap A_n^c.
$$

Since B is a subset of exactly k of the n events $A_1, \ldots, A_n$, then $\Pr(B)$ will contribute to k terms in the first summation on the right side of Eq. (2).

Furthermore, since B is a subset of the intersection of every pair of the events $A_1, \ldots, A_k$, then $\Pr(B)$ will contribute to $\binom{k}{2}$ terms in the second summation on the right side of Eq. (2). Therefore, $\Pr(B)$ will be subtracted $\binom{k}{2}$ times. Similarly, in the third summation, $\Pr(B)$ will be added $\binom{k}{3}$ times. By continuing in this way, we find that $\Pr(B)$ will be counted on the right side of Eq. (2) the following number of times:

$$k - \binom{k}{2} + \binom{k}{3} - \binom{k}{4} + \cdots + (-1)^{k+1}\binom{k}{k}.$$

By Exercise 14(b) of Sec. 1.8, this number is 1.

The same result can be obtained by using the same reasoning for the subset of outcomes which belong to any specified group of exactly k events among $A_1, \ldots, A_n$. Since k is arbitrary ($1 \leqslant k \leqslant n$), this argument establishes that the probability of each disjoint subset of $\bigcup_{i=1}^{n} A_i$ is counted exactly once on the right side of Eq. (2).   □

The Matching Problem

Suppose that all the cards in a deck of n different cards are placed in a row, and that the cards in another similar deck are then shuffled and placed in a row on top of the cards in the original deck. It is desired to determine the probability p_n that there will be at least one match between the corresponding cards from the two decks. The same problem can be expressed in various entertaining contexts. For example, we could suppose that a person types n letters, types the corresponding addresses on n envelopes, and then places the n letters in the n envelopes in a random manner. It could be desired to determine the probability p_n that at least one letter will be placed in the correct envelope. As another example, we could suppose that the photographs of n famous film actors are paired in a random manner with n photographs of the same actors taken when they were babies. It could then be desired to determine the probability p_n that the photograph of at least one actor will be paired correctly with this actor's own baby photograph.

Here we shall discuss this matching problem in the context of letters being placed in envelopes. Thus, we shall let A_i be the event that letter i is placed in the correct envelope ($i = 1, \ldots, n$) and we shall determine the value of $p_n = \Pr(\bigcup_{i=1}^{n} A_i)$ by using Eq. (2). Since the letters are placed in the envelopes at random, the probability $\Pr(A_i)$ that any particular letter will be placed in the correct envelope is $1/n$. Therefore, the value of the first summation on the right side of Eq. (2) is

$$\sum_{i=1}^{n} \Pr(A_i) = n \cdot \frac{1}{n} = 1.$$

Furthermore, since letter 1 could be placed in any one of n envelopes and letter 2 could then be placed in any one of the other $n - 1$ envelopes, the probability $\Pr(A_1 A_2)$ that both letter 1 and letter 2 will be placed in the correct envelopes is $1/[n(n - 1)]$. Similarly, the probability $\Pr(A_i A_j)$ that any two specific letters i and j ($i \neq j$) will both be placed in the correct envelopes is $1/[n(n - 1)]$. Therefore, the value of the second summation on the right side of Eq. (2) is

$$\sum_{i<j} \Pr(A_i A_j) = \binom{n}{2} \frac{1}{n(n-1)} = \frac{1}{2!}.$$

By similar reasoning it can be determined that the probability $\Pr(A_i A_j A_k)$ that any three specific letters i, j, and k ($i < j < k$) will be placed in the correct envelopes is $1/[n(n-1)(n-2)]$. Therefore, the value of the third summation is

$$\sum_{i<j<k} \Pr(A_i A_j A_k) = \binom{n}{3} \frac{1}{n(n-1)(n-2)} = \frac{1}{3!}.$$

This procedure can be continued until it is found that the probability $\Pr(A_1 A_2 \cdots A_n)$ that all n letters will be placed in the correct envelopes is $1/(n!)$. It now follows from Eq. (2) that the probability p_n that at least one letter will be placed in the correct envelope is

$$p_n = 1 - \frac{1}{2!} + \frac{1}{3!} - \frac{1}{4!} + \cdots + (-1)^{n+1} \frac{1}{n!}. \tag{3}$$

This probability has the following interesting features. As $n \to \infty$, the value of p_n approaches the following limit:

$$\lim_{n \to \infty} p_n = 1 - \frac{1}{2!} + \frac{1}{3!} - \frac{1}{4!} + \cdots.$$

It is shown in books on elementary calculus that the sum of the infinite series on the right side of this equation is $1 - (1/e)$, where $e = 2.71828\ldots$. Hence, $1 - (1/e) = 0.63212\ldots$. It follows that for a large value of n, the probability p_n that at least one letter will be placed in the correct envelope is approximately 0.63212.

The exact values of p_n, as given in Eq. (3), will form an oscillating sequence as n increases. As n increases through the even integers $2, 4, 6, \ldots$, the values of p_n will increase toward the limiting value 0.63212; and as n increases through the odd integers $3, 5, 7, \ldots$, the values of p_n will decrease toward this same limiting value.

The values of p_n converge to the limit very rapidly. In fact, for $n = 7$ the exact value p_7 and the limiting value of p_n agree to four decimal places. Hence, regardless of whether seven letters are placed at random in seven envelopes or seven million letters are placed at random in seven million envelopes, the probability that at least one letter will be placed in the correct envelope is 0.6321.

EXERCISES

1. In a certain city, three newspapers A, B, and C are published. Suppose that 60 percent of the families in the city subscribe to newspaper A, that 40 percent of the families subscribe to newspaper B, and that 30 percent subscribe to newspaper C. Suppose also that 20 percent of the families subscribe to both A and B, that 10 percent subscribe to both A and C, that 20 percent subscribe to both B and C, and that 5 percent subscribe to all three newspapers A, B, and C. What percentage of the families in the city subscribe to at least one of the three newspapers?

2. For the conditions of Exercise 1, what percentage of the families in the city subscribe to exactly one of the three newspapers?

3. Suppose that three phonograph records are removed from their jackets, and that after they have been played, they are put back into the three empty jackets in a random manner. Determine the probability that at least one of the records will be put back into the proper jacket.

4. Suppose that four guests check their hats when they arrive at a restaurant, and that these hats are returned to them in a random order when they leave. Determine the probability that no guest will receive the proper hat.

5. A box contains 30 red balls, 30 white balls, and 30 blue balls. If 10 balls are selected at random, without replacement, what is the probability that at least one color will be missing from the selection?

6. Suppose that a school band contains 10 students from the freshman class, 20 students from the sophomore class, 30 students from the junior class, and 40 students from the senior class. If 15 students are selected at random from the band, what is the probability that at least one student will be selected from each of the four classes? *Hint*: First determine the probability that at least one of the four classes will not be represented in the selection.

7. If n letters are placed at random in n envelopes, what is the probability that exactly $n - 1$ letters will be placed in the correct envelopes?

8. Suppose that n letters are placed at random in n envelopes, and let q_n denote the probability that no letter is placed in the correct envelope. For which of the following four values of n is q_n largest: $n = 10$, $n = 21$, $n = 53$, or $n = 300$?

9. If three letters are placed at random in three envelopes, what is the probability that exactly one letter will be placed in the correct envelope?

10. Suppose that 10 cards, of which 5 are red and 5 are green, are placed at random in 10 envelopes, of which 5 are red and 5 are green. Determine the probability that exactly x envelopes will contain a card with a matching color ($x = 0, 1, \ldots, 10$).

11. Let $A_1, A_2, \ldots$ be an infinite sequence of events such that $A_1 \subset A_2 \subset \cdots$. Prove that

$$\Pr\left(\bigcup_{i=1}^{\infty} A_i \right) = \lim_{n \to \infty} \Pr(A_n).$$

Hint: Let the sequence $B_1, B_2, \ldots$ be defined as in Exercise 11 of Sec. 1.5, and show that

$$\Pr\left(\bigcup_{i=1}^{\infty} A_i \right) = \lim_{n \to \infty} \Pr\left(\bigcup_{i=1}^{n} B_i \right) = \lim_{n \to \infty} \Pr(A_n).$$

12. Let $A_1, A_2, \ldots$ be an infinite sequence of events such that $A \supset A_2 \supset \cdots$. Prove that

$$\Pr\left(\bigcap_{i=1}^{\infty} A_i \right) = \lim_{n \to \infty} \Pr(A_n).$$

Hint: Consider the sequence $A_1^c, A_2^c, \ldots$, and apply Exercise 11.

1.11. INDEPENDENT EVENTS

Suppose that two events A and B occur independently of one another in the sense that the occurrence or nonoccurrence of either of them has no relation to, and no influence on, the occurrence or nonoccurrence of the other. We shall show that under these conditions it is natural to assume that $\Pr(AB) = \Pr(A)\Pr(B)$. In other words, it is natural to assume that the probability that both A and B will occur is equal to the product of their individual probabilities.

This result can easily be justified in terms of the frequency interpretation of probability. For example, suppose that A is the event that a head is obtained when a fair coin is tossed, and B is the event that either the number 1 or the number 2 is obtained when a balanced die is rolled. Then the event A will occur with a relative frequency of $1/2$ when the coin is tossed repeatedly, and the event B will occur with a relative frequency of $1/3$ when the die is rolled repeatedly. Therefore, $\Pr(A) = 1/2$ and $\Pr(B) = 1/3$.

Consider now a composite experiment in which the coin is tossed and the die is rolled simultaneously. If this experiment is performed repeatedly, then the relative frequency of the event A, in which a head is obtained, will be $1/2$. Since the outcomes for the coin and the outcomes for the die are unrelated, it is reasonable to assume that among those experiments in which the event A occurs, the relative frequency of the event B, in which the number 1 or the number 2 is obtained, will be $1/3$. Hence, in a sequence of composite experiments, the relative frequency with which both A and B occur simultaneously will be $(1/2)(1/3) = 1/6$. Thus, in this experiment,

$$\Pr(AB) = \frac{1}{6} = \frac{1}{2} \cdot \frac{1}{3} = \Pr(A)\Pr(B).$$

This relationship can also be justified in terms of the classical interpretation of probability. There are two equally probable outcomes of the toss of the coin and six equally probable outcomes of the roll of the die. Since the outcome for the coin and the outcome for the die are unrelated, it is natural to assume also that the $6 \times 2 = 12$ possible outcomes of the composite experiment in which both the coin is tossed and the die is rolled are equally probable. In this composite experiment, it will again be true that $\Pr(AB) = 2/12 = 1/6 = \Pr(A)\Pr(B)$.

Independence of Two Events

As a result of the foregoing discussion, the mathematical definition of the independence of two events is stated as follows: Two events A and B are independent if $\Pr(AB) = \Pr(A)\Pr(B)$.

If two events A and B are considered to be independent because the events are physically unrelated, and if the probabilities $\Pr(A)$ and $\Pr(B)$ are known, then this definition can be used to assign a value to $\Pr(AB)$.

Example 1: Machine Operation. Suppose that two machines 1 and 2 in a factory are operated independently of each other. Let A be the event that machine 1 will become inoperative during a given 8-hour period; let B be the event that machine 2 will become inoperative during the same period; and suppose that $\Pr(A) = 1/3$ and $\Pr(B) = 1/4$. We shall determine the probability that at least one of the machines will become inoperative during the given period.

The probability $\Pr(AB)$ that both machines will become inoperative during the period is

$$\Pr(AB) = \Pr(A)\Pr(B) = \left(\frac{1}{3}\right)\left(\frac{1}{4}\right) = \frac{1}{12}.$$

Therefore, the probability $\Pr(A \cup B)$ that at least one of the machines will become inoperative during the period is

$$Pr(A \cup B) = Pr(A) + Pr(B) - Pr(AB)$$

$$= \frac{1}{3} + \frac{1}{4} - \frac{1}{12} = \frac{1}{2}. \quad \square$$

The next example shows that two events A and B which are physically related can, nevertheless, satisfy the definition of independence.

Example 2: Rolling a Die. Suppose that a balanced die is rolled. Let A be the event that an even number is obtained, and let B be the event that one of the numbers 1, 2, 3, or 4 is obtained. We shall show that the events A and B are independent.

In this example, $Pr(A) = 1/2$ and $Pr(B) = 2/3$. Furthermore, since AB is the event that either the number 2 or the number 4 is obtained, $Pr(AB) = 1/3$. Hence, $Pr(AB) = Pr(A)Pr(B)$. It follows that the events A and B are independent events, even though the occurrence of each event depends on the same roll of a die. $\square$

The independence of the events A and B in Example 2 can also be interpreted as follows: Suppose that a person must bet on whether the number obtained on the die will be even or odd, i.e., on whether or not the event A will occur. Since three of the possible outcomes of the roll are even and the other three are odd, the person will typically have no preference between betting on an even number and betting on an odd number.

Suppose also that after the die has been rolled, but before the person has learned the outcome and before he has decided whether to bet on an even outcome or on an odd outcome, he is informed that the actual outcome was one of the numbers 1, 2, 3, or 4, i.e., that the event B has occurred. The person now knows that the outcome was 1, 2, 3, or 4. However, since two of these numbers are even and two are odd, the person will typically still have no preference between betting on an even number and betting on an odd number. In other words, the information that the event B has occurred is of no help to the person who is trying to decide whether or not the event A has occurred. This matter will be discussed in a more general way in Chapter 2.

In the foregoing discussion of independent events, we stated that if A and B are independent, then the occurrence or nonoccurrence of A should not be related to the occurrence or nonoccurrence of B. Hence, if A and B satisfy the mathematical definition of independent events, then it should also be true that A and B^c are independent events, that A^c and B are independent events, and that A^c and B^c are independent events. One of these results is established in the next theorem.

Theorem 1.   *If two events A and B are independent, then the events A and B^c are also independent.*

Proof. It is always true that

$$\Pr(AB^c) = \Pr(A) - \Pr(AB).$$

Furthermore, since A and B are independent events,

$$\Pr(AB) = \Pr(A)\Pr(B).$$

It now follows that

$$\Pr(AB^c) = \Pr(A) - \Pr(A)\Pr(B) = \Pr(A)[1 - \Pr(B)]$$
$$= \Pr(A)\Pr(B^c).$$

Therefore, the events A and B^c are independent. □

The proof of the analogous result for the events A^c and B is similar, and the proof for the events A^c and B^c is required in Exercise 1 at the end of this section.

Independence of Several Events

The discussion that has just been given for two events can be extended to any number of events. If k events $A_1, \ldots, A_k$ are independent in the sense that they are physically unrelated to each other, then it is natural to assume that the probability $\Pr(A_1 \cap \cdots \cap A_k)$ that all k events will occur is the product $\Pr(A_1) \cdots \Pr(A_k)$. Furthermore, since the events $A_1, \ldots, A_k$ are unrelated, this product rule should hold not only for the intersection of all k events, but also for the intersection of any two of them, any three of them, or any other number of them. These considerations lead to the following definition: The k events $A_1, \ldots, A_k$ are *independent* if, for every subset $A_{i_1}, \ldots, A_{i_j}$ of j of these events $(j = 2, 3, \ldots, k)$,

$$\Pr\left(A_{i_1} \cap \cdots \cap A_{i_j}\right) = \Pr\left(A_{i_1}\right) \cdots \Pr\left(A_{i_j}\right).$$

In particular, in order for three events A, B, and C to be independent, the following four relations must be satisfied:

$$\Pr(AB) = \Pr(A)\Pr(B),$$

$$\Pr(AC) = \Pr(A)\Pr(C), \tag{1}$$

$$\Pr(BC) = \Pr(B)\Pr(C),$$

and

$$\Pr(ABC) = \Pr(A)\Pr(B)\Pr(C). \tag{2}$$

It is possible that Eq. (2) will be satisfied but one or more of the three relations (1) will not be satisfied. On the other hand, as is shown in the next example, it is also possible that each of the three relations (1) will be satisfied but Eq. (2) will not be satisfied.

Example 3: Pairwise Independence. Consider an experiment in which the sample space S contains four outcomes $\{s_1, s_2, s_3, s_4\}$, and suppose that the probability of each outcome is $1/4$. Let the three events A, B, and C be defined as follows:

$$A = \{s_1, s_2\}, \quad B = \{s_1, s_3\}, \quad \text{and} \quad C = \{s_1, s_4\}.$$

Then $AB = AC = BC = ABC = \{s_1\}$. Hence,

$$\Pr(A) = \Pr(B) = \Pr(C) = 1/2$$

and

$$\Pr(AB) = \Pr(AC) = \Pr(BC) = \Pr(ABC) = 1/4.$$

It follows that each of the three relations (1) is satisfied, but Eq. (2) is not satisfied. These results can be summarized by saying that the events A, B, and C are *pairwise independent* but all three events are not independent. □

We shall now present some examples that will illustrate the power and scope of the concept of independence in the solution of probability problems.

Example 4: Inspecting Items. Suppose that a machine produces a defective item with probability p $(0 < p < 1)$ and produces a nondefective item with probability $q = 1 - p$. Suppose further that six items produced by the machine are selected at random and inspected, and that the outcomes for these six items are independent. We shall determine the probability that exactly two of the six items are defective.

It can be assumed that the sample space S contains all possible arrangements of six items, each one of which might be either defective or nondefective. For $j = 1, \ldots, 6$, we shall let D_j denote the event that the jth item in the sample is defective and shall let N_j denote the event that this item is nondefective. Since the outcomes for the six different items are independent, the probability of obtaining any particular sequence of defective and nondefective items will simply be the product of the individual probabilities for the items. For example,

$$\Pr(N_1 D_2 N_3 N_4 D_5 N_6) = \Pr(N_1)\Pr(D_2)\Pr(N_3)\Pr(N_4)\Pr(D_5)\Pr(N_6)$$

$$= qpqqpq = p^2 q^4.$$

It can be seen that the probability of any other particular sequence in S containing two defective items and four nondefective items will also be p^2q^4. Hence, the probability that there will be exactly two nondefectives in the sample of six items can be found by multiplying the probability p^2q^4 of any particular sequence containing two defectives by the possible number of such sequences. Since there are $\binom{6}{2}$ distinct arrangements of two defective items and four nondefective items, the probability of obtaining exactly two defectives is $\binom{6}{2}p^2q^4$.   □

Example 5: Obtaining a Defective Item. For the conditions of Example 4, we shall now determine the probability that at least one of the six items in the sample will be defective.

Since the outcomes for the different items are independent, the probability that all six items will be nondefective is q^6. Therefore, the probability that at least one item will be defective is $1 - q^6$.   □

Example 6: Tossing a Coin Until a Head Appears. Suppose that a fair coin is tossed until a head appears for the first time, and assume that the outcomes of the tosses are independent. We shall determine the probability p_n that exactly n tosses will be required.

The desired probability is equal to the probability of obtaining $n - 1$ tails in succession and then obtaining a head on the next toss. Since the outcomes of the tosses are independent, the probability of this particular sequence of n outcomes is $p_n = (1/2)^n$.

The probability that a head will be obtained sooner or later (or, equivalently, that tails will not be obtained forever) is

$$\sum_{n=1}^{\infty} p_n = \frac{1}{2} + \frac{1}{4} + \frac{1}{8} + \cdots = 1.$$

Since the sum of the probabilities p_n is 1, it follows that the probability of obtaining an infinite sequence of tails without ever obtaining a head must be 0.   □

Example 7: Inspecting Items One at a Time. Consider again a machine that produces a defective item with probability p and produces a nondefective item with probability $q = 1 - p$. Suppose that items produced by the machine are selected at random and inspected one at a time until exactly five defective items have been obtained. We shall determine the probability p_n that exactly n items ($n \geqslant 5$) must be selected to obtain the five defectives.

The fifth defective item will be the nth item that is inspected if and only if there are exactly four defectives among the first $n - 1$ items and then the nth item is defective. By reasoning similar to that given in Example 4, it can be shown that the probability of obtaining exactly 4 defectives and $n - 5$ nondefectives among the first $n - 1$ items is $\binom{n-1}{4} p^4 q^{n-5}$. The probability that the nth item will be defective is p. Since the first event refers to outcomes for only the first $n - 1$ items and the second event refers to the outcome for only the nth item, these two events are independent. Therefore, the probability that both events will occur is equal to the product of their probabilities. It follows that

$$p_n = \binom{n-1}{4} p^5 q^{n-5}. \quad \square$$

The Collector's Problem

Suppose that n balls are thrown in a random manner into r boxes ($r \leqslant n$). We shall assume that the n throws are independent and that each of the r boxes is equally likely to receive any given ball. The problem is to determine the probability p that every box will receive at least one ball. This problem can be reformulated in terms of a collector's problem as follows: Suppose that each package of bubble gum contains the picture of a baseball player; that the pictures of r different players are used; that the picture of each player is equally likely to be placed in any given package of gum; and that pictures are placed in different packages independently of each other. The problem now is to determine the probability p that a person who buys n packages of gum ($n \geqslant r$) will obtain a complete set of r different pictures.

For $i = 1, \ldots, r$, let A_i denote the event that the picture of player i is missing from all n packages. Then $\bigcup_{i=1}^{r} A_i$ is the event that the picture of at least one player is missing. We shall find $\Pr(\bigcup_{i=1}^{r} A_i)$ by applying Eq. (2) of Sec. 1.10.

Since the picture of each of the r players is equally likely to be placed in any particular package, the probability that the picture of player i will not be obtained in any particular package is $(r - 1)/r$. Since the packages are filled independently, the probability that the picture of player i will not be obtained in any of the n packages is $[(r - 1)/r]^n$. Hence,

$$\Pr(A_i) = \left(\frac{r-1}{r}\right)^n \qquad \text{for } i = 1, \ldots, r.$$

Now consider any two players i and j. The probability that neither the picture of player i nor the picture of player j will be obtained in any particular package is $(r - 2)/r$. Therefore, the probability that neither picture will be obtained in any of the n packages is $[(r - 2)/r]^n$. Thus,

$$\Pr(A_i A_j) = \left(\frac{r-2}{r}\right)^n.$$

If we next consider any three players i, j, and k, we find that

$$\Pr(A_i A_j A_k) = \left(\frac{r-3}{r}\right)^n.$$

By continuing in this way, we finally arrive at the probability $\Pr(A_1 A_2 \cdots A_r)$ that the pictures of all r players are missing from the n packages. Of course, this probability is 0. Therefore, by Eq. (2) of Section 1.10,

$$\Pr\left(\bigcup_{i=1}^{r} A_i\right) = r\left(\frac{r-1}{r}\right)^n - \binom{r}{2}\left(\frac{r-2}{r}\right)^n + \cdots + (-1)^r \binom{r}{r-1}\left(\frac{1}{r}\right)^n$$

$$= \sum_{j=1}^{r-1} (-1)^{j+1}\binom{r}{j}\left(1 - \frac{j}{r}\right)^n.$$

Since the probability p of obtaining a complete set of r different pictures is equal to $1 - \Pr(\bigcup_{i=1}^{r} A_i)$, it follows from the foregoing derivation that p can be written in the form

$$p = \sum_{j=0}^{r-1} (-1)^j \binom{r}{j}\left(1 - \frac{j}{r}\right)^n.$$

EXERCISES

1. Assuming that A and B are independent events, prove that the events A^c and B^c are also independent.

2. Suppose that A is an event such that $\Pr(A) = 0$ and that B is any other event. Prove that A and B are independent events.

3. Suppose that a person rolls two balanced dice three times in succession. Determine the probability that on each of the three rolls, the sum of the two numbers which appear will be 7.

4. Suppose that the probability that the control system used in a spaceship will malfunction on a given flight is 0.001. Suppose further that a duplicate, but completely independent, control system is also installed in the spaceship to take control in case the first system malfunctions. Determine the probability that the spaceship will be under the control of either the original system or the duplicate system on a given flight.

5. Suppose that 10,000 tickets are sold in one lottery and 5000 tickets are sold in another lottery. If a person owns 100 tickets in each lottery, what is the probability that he will win at least one first prize?

6. Two students A and B are both registered for a certain course. If student A attends class 80 percent of the time and student B attends class 60 percent of the time, and if the absences of the two students are independent, what is the probability that at least one of the two students will be in class on a given day?

7. If three balanced dice are rolled, what is the probability that all three numbers will be the same?

8. Consider an experiment in which a fair coin is tossed until a head is obtained for the first time. If this experiment is performed three times, what is the probability that exactly the same number of tosses will be required for each of the three performances?

9. Suppose that A, B, and C are three independent events such that $\Pr(A) = 1/4$, $\Pr(B) = 1/3$, and $\Pr(C) = 1/2$. (a) Determine the probability that none of these three events will occur. (b) Determine the probability that exactly one of these three events will occur.

10. Suppose that the probability that any particle emitted by a radioactive material will penetrate a certain shield is 0.01. If ten particles are emitted, what is the probability that exactly one of the particles will penetrate the shield?

11. Consider again the conditions of Exercise 10. If ten particles are emitted, what is the probability that at least one of the particles will penetrate the shield?

12. Consider again the conditions of Exercise 10. How many particles must be emitted in order for the probability to be at least 0.8 that at least one particle will penetrate the shield?

13. In the World Series of baseball, two teams A and B play a sequence of games against each other and the first team that wins a total of four games becomes the winner of the World Series. If the probability that team A will win any particular game against team B is $1/3$, what is the probability that team A will win the World Series?

14. Two boys A and B throw a ball at a target. Suppose that the probability that boy A will hit the target on any throw is $1/3$ and the probability that boy B will hit the target on any throw is $1/4$. Suppose also that boy A throws first and the two boys take turns throwing. Determine the probability that the target will be hit for the first time on the third throw of boy A.

15. For the conditions of Exercise 14, determine the probability that boy A will hit the target before boy B does.

16. A box contains 20 red balls, 30 white balls, and 50 blue balls. Suppose that 10 balls are selected at random one at a time, with replacement; i.e., each selected ball is replaced in the box before the next selection is made.

Determine the probability that at least one color will be missing from the 10 selected balls.

17. Suppose that $A_1, \ldots, A_k$ are a sequence of k independent events. Let $B_1, \ldots, B_k$ be another sequence of k events such that for each value of j ($j = 1, \ldots, k$), either $B_j = A_j$ or $B_j = A_j^c$. Prove that $B_1, \ldots, B_k$ are also independent events. *Hint:* Use an induction argument based on the number of events B_j for which $B_j = A_j^c$.

1.12. STATISTICAL SWINDLES

Misleading Use of Statistics

The field of statistics has a poor image in the minds of many people because there is a widespread belief that statistical data and statistical analyses can easily be manipulated in an unscientific and unethical fashion in an effort to show that a particular conclusion or point of view is correct. We all have heard the sayings that "there are lies, damned lies, and statistics" (often attributed to Mark Twain) and that "you can prove anything with statistics."

One benefit of studying probability and statistics is that the knowledge we gain enables us to analyze statistical arguments that we read in newspapers, magazines, or elsewhere. We can then evaluate these arguments on their merits, rather than accepting them blindly. In this section, we shall describe two schemes that have been used to induce consumers to send money to the operators of the schemes in exchange for certain types of information. Although neither scheme is strictly statistical in nature, both schemes are strongly based on undertones of probability.

Perfect Forecasts

Suppose that one Monday morning you receive in the mail a letter from a firm with which you are not familiar, stating that the firm sells forecasts about the stock market for very high fees. To indicate the firm's ability in forecasting, it predicts that a particular stock, or a particular portfolio of stocks, will rise in value during the coming week. You do not respond to this letter, but you do watch the stock market during the week and notice that the prediction was correct. On the following Monday morning you receive another letter from the same firm containing another prediction, this one specifying that a particular stock will drop in value during the coming week. Again the prediction proves to be correct.

This routine continues for seven weeks. Every Monday morning you receive a prediction in the mail from the firm, and each of these seven predictions proves to be correct. On the eighth Monday morning you receive another letter from the firm. This letter states that for a large fee the firm will provide another prediction, on the basis of which you can presumably make a large amount of money on the stock market. How should you respond to this letter?

Since the firm has made seven successive correct predictions, it would seem that it must have some special information about the stock market and is not simply guessing. After all, the probability of correctly guessing the outcomes of seven successive tosses of a fair coin is only $(1/2)^7 = 0.008$. Hence, if the firm had only probability $1/2$ of making a correct prediction each week, and if the outcomes of successive predictions were independent, then the firm had a probability less than 0.01 of being correct seven weeks in a row.

The fallacy here is that you may have seen only a relatively small number of the forecasts that the firm made during the seven-week period. Suppose, for example that the firm started the entire process with a list of $2^7 = 128$ potential clients. On the first Monday the firm could send the forecast that a particular stock will rise in value to half of these clients and send the forecast that the same stock will drop in value to the other half. On the second Monday the firm could continue writing to those 64 clients for whom the first forecast proved to be correct. It could again send a new forecast to half of those 64 clients and the opposite forecast to the other half. At the end of seven weeks, the firm (which usually consists of only one person and a typewriter) must necessarily have one client (and only one client) for whom all seven forecasts were correct.

By following this procedure with several different groups of 128 clients, and starting new groups each week, the firm may be able to generate enough positive responses from clients for it to realize significant profits.

Guaranteed Winners

There is another scheme that is somewhat related to the one just described but that is even more elegant because of its simplicity. In this scheme, a firm advertises that for a fixed fee, usually 10 or 20 dollars, it will send the client its forecast of the winner of any upcoming baseball game, football game, boxing match, or other sports event that the client might specify. Furthermore, the firm offers a money-back guarantee that this forecast will be correct; that is, if the team or person designated as the winner in the forecast does not actually turn out to be the winner, the firm will return the full fee to the client.

How should you react to such an advertisement? At first glance, it would appear that the firm must have some special knowledge about these sports events, because otherwise it could not afford to guarantee its forecasts. Further reflection reveals, however, that the firm simply cannot lose, because its only expenses are those for advertising and postage. In effect, when this scheme is used, the firm holds the client's fee until the winner has been decided. If the forecast was correct, the firm keeps the fee; otherwise, it simply returns the fee to the client.

On the other hand, the client can very well lose. He presumably purchases the firm's forecast because he desires to bet on the sports event. If the forecast proves to be wrong, the client will not have to pay any fee to the firm, but he will have lost any money that he bet on the predicted winner.

Thus, when there are "guaranteed winners," only the firm is guaranteed to win. In fact, the firm knows that it will be able to keep the fees from all the clients for whom the forecasts were correct.

If the firm restricts its forecasts to just one specified football game, it can actually offer to give back more than the full fee. Even if the firm gives the client one-and-one-half times his fee when the forecasted winner is not correct, the firm will still be guaranteed a profit. In this case, the firm simply sends half of its clients the forecast that one particular team will be the winner and sends the other half the forecast that the other team will be the winner. Regardless of which team wins, the firm will return to each loser his fee plus half of the fee collected from a winner, but the firm will retain the other half of each winner's fee.

1.13. SUPPLEMENTARY EXERCISES

1. Suppose that a fair coin is tossed repeatedly until both a head and a tail have appeared at least once. (a) Describe the sample space of this experiment. (b) What is the probability that exactly three tosses will be required?

2. Suppose that a coin is tossed seven times. Let A denote the event that a head is obtained on the first toss, and let B denote the event that a head is obtained on the fifth toss. Are A and B disjoint?

3. Suppose that the events A and B are disjoint and that each has positive probability. Are A and B independent?

4. (a) Suppose that the events A and B are disjoint. Under what conditions are A^c and B^c disjoint? (b) Suppose that the events A and B are independent. Under what conditions are A^c and B^c independent?

5. If A, B, and D are three events such that $\Pr(A \cup B \cup D) = 0.7$, what is the value of $\Pr(A^c \cap B^c \cap D^c)$?

6. Suppose that A, B, and C are three events such that A and B are disjoint, A and C are independent, and B and C are independent. Suppose also that $4\Pr(A) = 2\Pr(B) = \Pr(C) > 0$ and $\Pr(A \cup B \cup C) = 5\Pr(A)$. Determine the value of $\Pr(A)$.

7. Suppose that each of two dice is loaded so that when either die is rolled, the probability that the number k will appear is 0.1 for $k = 1, 2, 5,$ or 6 and is 0.3 for $k = 3$ or 4. If the two loaded dice are rolled, what is the probability that the sum of the two numbers that appear will be 7?

8. Suppose that there is a probability of $1/50$ that you will win a certain game. If you play the game 50 times, independently, what is the probability that you will win at least once?

9. Suppose that a certain precinct contains 350 voters, of which 250 are Democrats and 100 are Republicans. If 30 voters are chosen at random from the precinct, what is the probability that exactly 18 Democrats will be selected?

10. Three students A, B, and C are enrolled in the same class. Suppose that A attends class 30 percent of the time, B attends class 50 percent of the time, and C attends class 80 percent of the time. If these students attend class independently of each other, what is (a) the probability that at least one of them will be in class on a particular day and (b) the probability that exactly one of them will be in class on a particular day?

11. Suppose that a balanced die is rolled three times, and let X_i denote the number that appears on the ith roll ($i = 1, 2, 3$). Evaluate $\Pr(X_1 > X_2 > X_3)$.

12. Consider the World Series of baseball, as described in Exercise 13 of Sec. 1.11. If there is probability p that team A will win any particular game, what is the probability that it will be necessary to play seven games in order to determine the winner of the Series?

13. Suppose that in a deck of 20 cards, each card has one of the numbers 1, 2, 3, 4, or 5 and there are 4 cards with each number. If 10 cards are chosen from the deck at random, without replacement, what is the probability that each of the numbers 1, 2, 3, 4, and 5 will appear exactly twice?

14. Suppose that three red balls and three white balls are thrown at random into three boxes, and all throws are independent. What is the probability that each box contains one red ball and one white ball?

15. If five balls are thrown at random into n boxes, and all throws are independent, what is the probability that no box contains more than two balls?

16. Bus tickets in a certain city contain four numbers, U, V, W, and X. Each of these numbers is equally likely to be any of the ten digits $0, 1, \ldots, 9$, and the four numbers are chosen independently. A bus rider is said to be lucky if $U + V = W + X$. What proportion of the riders are lucky?

17. Suppose that a box contains r red balls and w white balls. Suppose also that balls are drawn from the box one at a time, at random, without replacement. (a) What is the probability that all r red balls will be obtained before any white balls are obtained? (b) What is the probability that all r red balls will be obtained before two white balls are obtained?

18. Suppose that a box contains r red balls, w white balls, and b blue balls. Suppose also that balls are drawn from the box one at a time, at random, without replacement. What is the probability that all r red balls will be obtained before any white balls are obtained?

19. Suppose that 10 cards, of which 7 are red and 3 are green, are put at random into 10 envelopes, of which 7 are red and 3 are green, so that each envelope contains one card. Determine the probability that exactly k envelopes will contain a card with a matching color ($k = 0, 1, \ldots, 10$).

20. Suppose that 10 cards, of which 5 are red and 5 are green, are put at random into 10 envelopes, of which 7 are red and 3 are green, so that each envelope contains one card. Determine the probability that exactly k envelopes will contain a card with a matching color ($k = 0, 1, \ldots, 10$).

21. A certain group has eight members. In January, three members are selected at random to serve on a committee. In February, four members are selected at random and independently of the first selection to serve on another committee. In March, five members are selected at random and independently of the previous two selections to serve on a third committee. Determine the probability that each of the eight members serves on at least one of the three committees.

22. For the conditions of Exercise 21, determine the probability that two particular members A and B will serve together on at least one of the three committees.

23. Suppose that two players A and B take turns rolling a pair of balanced dice, and that the winner is the first player who obtains the sum of 7 on a given roll of the two dice. If A rolls first, what is the probability that B will win?

24. Three players A, B, and C take turns tossing a fair coin. Suppose that A tosses the coin first, B tosses second, and C tosses third; and this cycle is repeated indefinitely until someone wins by being the first player to obtain a head. Determine the probability that each of three players will win.

25. Let A_1, A_2, and A_3 be three arbitrary events. Show that the probability that exactly one of these three events will occur is

$$\Pr(A_1) + \Pr(A_2) + \Pr(A_3)$$

$$- 2 \Pr(A_1 A_2) - 2 \Pr(A_1 A_3) - 2 \Pr(A_2 A_3)$$

$$+ 3 \Pr(A_1 A_2 A_3).$$

26. Let $A_1, \ldots, A_n$ be n arbitrary events. Show that the probability that exactly one of these n events will occur is

$$\sum_{i=1}^{n} \Pr(A_i) - 2 \sum_{i<j} \Pr(A_i A_j) + 3 \sum_{i<j<k} \Pr(A_i A_j A_k)$$

$$- \cdots + (-1)^{n+1} n \Pr(A_1 A_2 \cdots A_n).$$

Conditional Probability

2

2.1. THE DEFINITION OF CONDITIONAL PROBABILITY

Suppose that an experiment is to be performed for which the sample space of all possible outcomes is S and also that probabilities have been specified for all the events in S. We shall now study the way in which the probability of an event A changes after it has been learned that some other event B has occurred. This new probability of A is called the *conditional probability of the event A given that the event B has occurred*. The notation for this conditional probability is $\Pr(A|B)$. For convenience, this notation is read simply as the conditional probability of A given B.

If we know that the event B has occurred, then we know that the outcome of the experiment is one of those included in B. Hence, to evaluate the probability that A will occur, we must consider the set of those outcomes in B that also result in the occurrence of A. As sketched in Fig. 2.1, this set is precisely the set AB. It is therefore natural to define the conditional probability $\Pr(A|B)$ as the proportion of the total probability $\Pr(B)$ that is represented by the probability $\Pr(AB)$. These considerations lead to the following definition: If A and B are any two events such that $\Pr(B) > 0$, then

$$\Pr(A|B) = \frac{\Pr(AB)}{\Pr(B)}.$$

The conditional probability $\Pr(A|B)$ is not defined if $\Pr(B) = 0$.

The conditional probability $\Pr(A|B)$ has a simple meaning in terms of the frequency interpretation of probability presented in Sec. 1.2. According to that

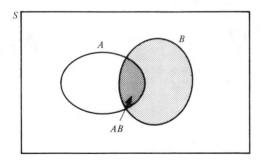

Figure 2.1 The outcomes in the event B that also belong to the event A.

interpretation, if an experimental process is repeated a large number of times, then the proportion of repetitions in which the event B will occur is approximately $\Pr(B)$ and the proportion of repetitions in which both the event A and the event B will occur is approximately $\Pr(AB)$. Therefore, among those repetitions in which the event B occurs, the proportion of repetitions in which the event A will also occur is approximately equal to

$$\Pr(A|B) = \frac{\Pr(AB)}{\Pr(B)}.$$

Example 1: Rolling Dice. Suppose that two dice were rolled and it was observed that the sum T of the two numbers was odd. We shall determine the probability that T was less than 8.

   If we let A be the event that $T < 8$ and let B be the event that T is odd, then AB is the event that T is 3, 5, or 7. From the probabilities for two dice given at the end of Sec. 1.6, we can evaluate $\Pr(AB)$ and $\Pr(B)$ as follows:

$$\Pr(AB) = \frac{2}{36} + \frac{4}{36} + \frac{6}{36} = \frac{12}{36} = \frac{1}{3},$$

$$\Pr(B) = \frac{2}{36} + \frac{4}{36} + \frac{6}{36} + \frac{4}{36} + \frac{2}{36} = \frac{18}{36} = \frac{1}{2}.$$

Hence,

$$\Pr(A|B) = \frac{\Pr(AB)}{\Pr(B)} = \frac{2}{3}. \quad \square$$

Example 2: Rolling Dice Repeatedly. Suppose that two dice are to be rolled repeatedly and the sum T of the two numbers is to be observed for each roll. We shall determine the probability p that the value $T = 7$ will be observed before the value $T = 8$ is observed.

The desired probability p could be calculated directly as follows: We could assume that the sample space S contains all sequences of outcomes which terminate as soon as either the sum $T = 7$ or the sum $T = 8$ is obtained. Then we could find the sum of the probabilities of all the sequences which terminate when the value $T = 7$ is obtained.

However, there is a simpler approach in this example. We can consider the simple experiment in which two dice are rolled. If we repeat the experiment until either the sum $T = 7$ or the sum $T = 8$ is obtained, the effect is to restrict the outcome of the experiment to one of these two values. Hence, the problem can be restated as follows: Given that the outcome of the experiment is either $T = 7$ or $T = 8$, determine the probability p that the outcome is actually $T = 7$.

If we let A be the event that $T = 7$ and let B be the event that the value of T is either 7 or 8, then $AB = A$ and

$$p = \Pr(A|B) = \frac{\Pr(AB)}{\Pr(B)} = \frac{\Pr(A)}{\Pr(B)}.$$

From the probabilities for two dice given in Example 3 of Sec. 1.6, $\Pr(A) = 6/36$ and $\Pr(B) = (6/36) + (5/36) = 11/36$. Hence, $p = 6/11$. $\square$

Conditional Probability for Independent Events

If two events A and B are independent, then $\Pr(AB) = \Pr(A)\Pr(B)$. Hence, if $\Pr(B) > 0$, it follows from the definition of conditional probability that

$$\Pr(A|B) = \frac{\Pr(A)\Pr(B)}{\Pr(B)} = \Pr(A).$$

In other words, if two events A and B are independent, then the conditional probability of A when it is known that B has occurred is the same as the unconditional probability of A when no information about B is available. The converse of this statement is also true. If $\Pr(A|B) = \Pr(A)$, then the events A and B must be independent.

Similarly, if A and B are two independent events and if $\Pr(A) > 0$, then $\Pr(B|A) = \Pr(B)$. Conversely, if $\Pr(B|A) = \Pr(B)$, then the events A and B are independent. These properties of conditional probabilities for independent events reinforce the interpretations of the concept of independence that were given in Chapter 1.

The Multiplication Rule for Conditional Probabilities

In an experiment involving two events A and B that are not independent, it is often convenient to compute the probability $\Pr(AB)$ that both events will occur by applying one of the following two equations:

$$\Pr(AB) = \Pr(B)\Pr(A|B)$$

or

$$\Pr(AB) = \Pr(A)\Pr(B|A).$$

Example 3: Selecting Two Balls. Suppose that two balls are to be selected at random, without replacement, from a box containing r red balls and b blue balls. We shall determine the probability p that the first ball will be red and the second ball will be blue.

Let A be the event that the first ball is red, and let B be the event that the second ball is blue. Obviously, $\Pr(A) = r/(r + b)$. Furthermore, if the event A has occurred, then one red ball has been removed from the box on the first draw. Therefore, the probability of obtaining a blue ball on the second draw will be

$$\Pr(B|A) = \frac{b}{r + b - 1}.$$

It follows that

$$\Pr(AB) = \frac{r}{r + b} \cdot \frac{b}{r + b - 1}. \quad \square$$

The principle that has just been applied can be extended to any finite number of events, as stated in the following theorem:

Theorem 1. *Suppose that $A_1, A_2, \ldots, A_n$ are any events such that $\Pr(A_1 A_2 \cdots A_{n-1}) > 0$. Then*

$$\Pr(A_1 A_2 \cdots A_n) \tag{1}$$
$$= \Pr(A_1)\Pr(A_2|A_1)\Pr(A_3|A_1 A_2) \cdots \Pr(A_n|A_1 A_2 \cdots A_{n-1}).$$

Proof. The product of probabilities on the right side of Eq. (1) is equal to

$$\Pr(A_1) \cdot \frac{\Pr(A_1 A_2)}{\Pr(A_1)} \cdot \frac{\Pr(A_1 A_2 A_3)}{\Pr(A_1 A_2)} \cdots \frac{\Pr(A_1 A_2 \cdots A_n)}{\Pr(A_1 A_2 \cdots A_{n-1})}.$$

Since $\Pr(A_1 A_2 \cdots A_{n-1}) > 0$, each of the denominators in this product must be positive. All of the terms in the product cancel each other except the final numerator $\Pr(A_1 A_2 \cdots A_n)$, which is the left side of Eq. (1).   □

Example 4: Selecting Four Balls. Suppose that four balls are selected one at a time, without replacement, from a box containing r red balls and b blue balls ($r \geqslant 2$, $b \geqslant 2$). We shall determine the probability of obtaining the sequence of outcomes red, blue, red, blue.

If we let R_j denote the event that a red ball is obtained on the jth draw and let B_j denote the event that a blue ball is obtained on the jth draw ($j = 1, \ldots, 4$), then

$$\Pr(R_1 B_2 R_3 B_4) = \Pr(R_1)\Pr(B_2|R_1)\Pr(R_3|R_1 B_2)\Pr(B_4|R_1 B_2 R_3)$$

$$= \frac{r}{r+b} \cdot \frac{b}{r+b-1} \cdot \frac{r-1}{r+b-2} \cdot \frac{b-1}{r+b-3}. \quad □$$

The Game of Craps

We shall conclude this section by discussing a popular gambling game called craps. One version of this game is played as follows: A player rolls two dice, and the sum of the two numbers that appear is observed. If the sum on the first roll is 7 or 11, the player wins the game immediately. If the sum on the first roll is 2, 3, or 12, the player loses the game immediately. If the sum on the first roll is 4, 5, 6, 8, 9, or 10, then the two dice are rolled again and again until the sum is either 7 or the original value. If the original value is obtained a second time before 7 is obtained, then the player wins. If the sum 7 is obtained before the original value is obtained a second time, then the player loses.

We shall now compute the probability P that the player will win. The probability π_0 that the sum on the first roll will be either 7 or 11 is

$$\pi_0 = \frac{6}{36} + \frac{2}{36} = \frac{8}{36} = \frac{2}{9}.$$

If the sum obtained on the first roll is 4, the probability q_4 that the player will win is equal to the conditional probability that the sum 4 will be obtained again before the sum 7 is obtained. As described in Example 2, this probability is the same as the probability of obtaining the sum 4 when the outcome must be either 4 or 7. Hence,

$$q_4 = \frac{3/36}{(3/36) + (6/36)} = \frac{1}{3}.$$

Since the probability p_4 of obtaining the sum 4 on the first roll is $p_4 = 3/36 = 1/12$, it follows that the probability π_4 of obtaining the sum 4 on the first roll and then winning the game is

$$\pi_4 = p_4 q_4 = \frac{1}{12} \cdot \frac{1}{3} = \frac{1}{36}.$$

Similarly, the probability p_{10} of obtaining the sum 10 on the first roll is $p_{10} = 1/12$, and the probability q_{10} of winning the game when the sum 10 has been obtained on the first roll is $q_{10} = 1/3$. Hence, the probability π_{10} of obtaining the sum 10 on the first roll and then winning the game is

$$\pi_{10} = p_{10} q_{10} = \frac{1}{12} \cdot \frac{1}{3} = \frac{1}{36}.$$

The values of p_i, q_i, and π_i can be defined similarly for $i = 5, 6, 8,$ and 9. It will be found that

$$\pi_5 = p_5 q_5 = \frac{4}{36} \cdot \frac{4/36}{(4/36) + (6/36)} = \frac{2}{45}$$

and

$$\pi_9 = p_9 q_9 = \frac{2}{45}.$$

Also,

$$\pi_6 = p_6 q_6 = \frac{5}{36} \cdot \frac{5/36}{(5/36) + (6/36)} = \frac{25}{396}$$

and

$$\pi_8 = p_8 q_8 = \frac{25}{396}.$$

Since the probability P that the player will win in some way is the sum of all the probabilities just found, we obtain

$$P = \pi_0 + (\pi_4 + \pi_{10}) + (\pi_5 + \pi_9) + (\pi_6 + \pi_8)$$

$$= \frac{2}{9} + 2 \cdot \frac{1}{36} + 2 \cdot \frac{2}{45} + 2 \cdot \frac{25}{396} = \frac{244}{495} = 0.493.$$

Thus, the probability of winning in the game of craps is slightly less than $1/2$.

EXERCISES

1. If A and B are disjoint events and $\Pr(B) > 0$, what is the value of $\Pr(A|B)$?

2. If S is the sample space of an experiment and A is any event in that space, what is the value of $\Pr(A|S)$?

3. If A and B are independent events and $\Pr(B) < 1$, what is the value of $\Pr(A^c|B^c)$?

4. A box contains r red balls and b blue balls. One ball is selected at random and its color is observed. The ball is then returned to the box and k additional balls of the same color are also put into the box. A second ball is then selected at random, its color is observed, and it is returned to the box together with k additional balls of the same color. Each time another ball is selected, the process is repeated. If four balls are selected, what is the probability that the first three balls will be red and the fourth ball will be blue?

5. Each time a shopper purchases a tube of toothpaste, he chooses either brand A or brand B. Suppose that for each purchase after the first, the probability is $1/3$ that he will choose the same brand that he chose on his preceding purchase and the probability is $2/3$ that he will switch brands. If he is equally likely to choose either brand A or brand B on his first purchase, what is the probability that both his first and second purchases will be brand A and both his third and fourth purchases will be brand B?

6. A box contains three cards. One card is red on both sides, one card is green on both sides, and one card is red on one side and green on the other. One card is selected from the box at random, and the color on one side is observed. If this side is green, what is the probability that the other side of the card is also green?

7. Consider again the conditions of Exercise 6 of Sec. 1.11. Two students A and B are both registered for a certain course. Student A attends class 80 percent of the time and student B attends class 60 percent of the time, and the absences of the two students are independent. If at least one of the two students is in class on a given day, what is the probability that A is in class that day?

8. Consider again the conditions of Exercise 1 of Sec. 1.10. If a family selected at random from the city subscribes to newspaper A, what is the probability that the family also subscribes to newspaper B?

9. Consider again the conditions of Exercise 1 of Sec. 1.10. If a family selected at random from the city subscribes to at least one of the three newspapers A, B, and C, what is the probability that the family subscribes to newspaper A?

10. Suppose that a box contains one blue card and four red cards, which are labeled A, B, C, and D. Suppose also that two of these five cards are selected at random, without replacement.

    (a) If it is known that card A has been selected, what is the probability that both cards are red?

    (b) If it is known that at least one red card has been selected, what is the probability that both cards are red?

11. The probability that any child in a certain family will have blue eyes is $1/4$, and this feature is inherited independently by different children in the family. If there are five children in the family and it is known that at least one of these children has blue eyes, what is the probability that at least three of the children have blue eyes?

12. Consider the family with five children described in Exercise 11.

    (a) If it is known that the youngest child in the family has blue eyes, what is the probability that at least three of the children have blue eyes?

    (b) Explain why the answer in part (a) is different from the answer in Exercise 11.

13. Consider the following version of the game of craps: The player rolls two dice. If the sum on the first roll is 7 or 11, the player wins the game immediately. If the sum on the first roll is 2, 3, or 12, the player loses the game immediately. However, if the sum on the first roll is 4, 5, 6, 8, 9, or 10, then the two dice are rolled again and again until the sum is either 7 or 11 or the original value. If the original value is obtained a second time before either 7 or 11 is obtained, then the player wins. If either 7 or 11 is obtained before the original value is obtained a second time, then the player loses. Determine the probability that the player will win this game.

2.2. BAYES' THEOREM

Probability and Partitions

Let S denote the sample space of some experiment, and consider k events $A_1, \ldots, A_k$ in S such that $A_1, \ldots, A_k$ are disjoint and $\bigcup_{i=1}^{k} A_i = S$. It is said that these events form a *partition* of S.

If the k events $A_1, \ldots, A_k$ form a partition of S and if B is any other event in S, then the events $A_1 B, A_2 B, \ldots, A_k B$ will form a partition of B, as illustrated in Fig. 2.2. Hence, we can write

$$B = (A_1 B) \cup (A_2 B) \cup \cdots \cup (A_k B).$$

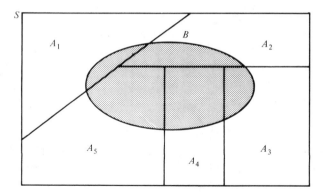

Figure 2.2 The intersections of B with events $A_1, \ldots, A_5$ of a partition.

Furthermore, since the k events on the right side of this equation are disjoint,

$$\Pr(B) = \sum_{j=1}^{k} \Pr(A_j B).$$

Finally, if $\Pr(A_j) > 0$ for $j = 1, \ldots, k$, then $\Pr(A_j B) = \Pr(A_j)\Pr(B|A_j)$ and it follows that

$$\Pr(B) = \sum_{j=1}^{k} \Pr(A_j)\Pr(B|A_j).$$

In summary, we have derived the following result: Suppose that the events $A_1, \ldots, A_k$ form a partition of the space S and that $\Pr(A_j) > 0$ for $j = 1, \ldots, k$. Then, for any event B in S,

$$\Pr(B) = \sum_{j=1}^{k} \Pr(A_j)\Pr(B|A_j).$$

Example 1: Selecting Bolts. Two boxes contain long bolts and short bolts. Suppose that one box contains 60 long bolts and 40 short bolts, and that the other box contains 10 long bolts and 20 short bolts. Suppose also that one box is selected at random and a bolt is then selected at random from that box. We shall determine the probability that this bolt is long.

Let A_1 be the event that the first box is selected; let A_2 be the event that the second box is selected; and let B be the event that a long bolt is selected. Then

$$\Pr(B) = \Pr(A_1)\Pr(B|A_1) + \Pr(A_2)\Pr(B|A_2).$$

Since a box is selected at random, we know that $\Pr(A_1) = \Pr(A_2) = 1/2$. Furthermore, the probability of selecting a long bolt from the first box is $\Pr(B|A_1) = 60/100 = 3/5$ and the probability of selecting a long bolt from the second box is $\Pr(B|A_2) = 10/30 = 1/3$. Hence,

$$\Pr(B) = \frac{1}{2} \cdot \frac{3}{5} + \frac{1}{2} \cdot \frac{1}{3} = \frac{7}{15}. \quad \square$$

Example 2: Achieving a High Score. Suppose that a person plays a game in which his score must be one of the 50 numbers $1, 2, \ldots, 50$ and that each of these 50 numbers is equally likely to be his score. The first time he plays the game, his score is X. He then continues to play the game until he obtains another score Y such that $Y \geqslant X$. It may be assumed that all plays of the game are independent. We shall determine the probability that Y will be 50.

For any given value x of X, the value of Y is equally likely to be any one of the numbers $x, x + 1, \ldots, 50$. Since each of these $(51 - x)$ possible values for Y is equally likely, it follows that

$$\Pr(Y = 50 | X = x) = \frac{1}{51 - x}.$$

Furthermore, since the probability of each of the 50 values of X is $1/50$, it follows that

$$\Pr(Y = 50) = \sum_{x=1}^{50} \frac{1}{50} \cdot \frac{1}{51 - x} = \frac{1}{50}\left(1 + \frac{1}{2} + \frac{1}{3} + \cdots + \frac{1}{50}\right) = 0.0900. \quad \square$$

Statement and Proof of Bayes' Theorem

We can now state the following result, which is known as *Bayes' theorem*:

Bayes' Theorem.   *Let the events $A_1, \ldots, A_k$ form a partition of the space S such that $\Pr(A_j) > 0$ for $j = 1, \ldots, k$, and let B be any event such that $\Pr(B) > 0$. Then, for $i = 1, \ldots, k$,*

$$\Pr(A_i|B) = \frac{\Pr(A_i)\Pr(B|A_i)}{\sum\limits_{j=1}^{k} \Pr(A_j)\Pr(B|A_j)}. \tag{1}$$

Proof. By the definition of conditional probability,

$$\Pr(A_i|B) = \frac{\Pr(A_i B)}{\Pr(B)}.$$

The numerator on the right side of Eq. (1) is equal to $\Pr(A_i B)$ and the denominator is equal to $\Pr(B)$. □

Bayes' theorem provides a simple rule for computing the conditional probability of each event A_i given B from the conditional probability of B given each event A_i and the unconditional probability of each A_i.

Example 3: Identifying the Source of a Defective Item. Three different machines M_1, M_2, and M_3 were used for producing a large batch of similar manufactured items. Suppose that 20 percent of the items were produced by machine M_1, 30 percent by machine M_2, and 50 percent by machine M_3. Suppose further that 1 percent of the items produced by machine M_1 are defective, that 2 percent of the items produced by machine M_2 are defective, and that 3 percent of the items produced by machine M_3 are defective. Finally, suppose that one item is selected at random from the entire batch and it is found to be defective. We shall determine the probability that this item was produced by machine M_2.

Let A_i be the event that the selected item was produced by machine M_i ($i = 1, 2, 3$), and let B be the event that the selected item is defective. We must evaluate the conditional probability $\Pr(A_2|B)$.

The probability $\Pr(A_i)$ that an item selected at random from the entire batch was produced by machine M_i is as follows, for $i = 1, 2, 3$:

$$\Pr(A_1) = 0.2, \qquad \Pr(A_2) = 0.3, \qquad \Pr(A_3) = 0.5.$$

Furthermore, the probability $\Pr(B|A_i)$ that an item produced by machine M_i will be defective is:

$$\Pr(B|A_1) = 0.01, \qquad \Pr(B|A_2) = 0.02, \qquad \Pr(B|A_3) = 0.03.$$

It now follows from Bayes' theorem that

$$\Pr(A_2|B) = \frac{\Pr(A_2)\Pr(B|A_2)}{\sum_{j=1}^{3} \Pr(A_j)\Pr(B|A_j)}$$

$$= \frac{(0.3)(0.02)}{(0.2)(0.01) + (0.3)(0.02) + (0.5)(0.03)} = 0.26. \quad □$$

Prior and Posterior Probabilities

In Example 3, a probability like $\Pr(A_2)$ is often called the *prior probability* that the selected item will have been produced by machine M_2, because $\Pr(A_2)$ is the probability of this event before the item is selected and before it is known whether the selected item is defective or nondefective. A probability like $\Pr(A_2|B)$ is then called the *posterior probability* that the selected item was produced by machine M_2, because it is the probability of this event after it is known that the selected item is defective.

Thus, in Example 3, the prior probability that the selected item will have been produced by machine M_2 is 0.3. After an item has been selected and has been found to be defective, the posterior probability that the item was produced by machine M_2 is 0.26. Since this posterior probability is smaller than the prior probability that the item was produced by machine M_2, the posterior probability that the item was produced by one of the other machines must be larger than the prior probability that it was produced by one of those machines (see Exercises 4 and 5 at the end of this section).

Computation of Posterior Probabilities in More Than One Stage

Suppose that a box contains one fair coin and one coin with a head on each side. Suppose also that one coin is selected at random and that when it is tossed, a head is obtained. We shall determine the probability that the coin is the fair coin.

Let A_1 be the event that the coin is fair; let A_2 be the event that the coin has two heads; and let H_1 be the event that a head is obtained when the coin is tossed. Then, by Bayes' theorem,

$$\Pr(A_1|H_1) = \frac{\Pr(A_1)\Pr(H_1|A_1)}{\Pr(A_1)\Pr(H_1|A_1) + \Pr(A_2)\Pr(H_1|A_2)}$$

$$= \frac{(1/2)(1/2)}{(1/2)(1/2) + (1/2)(1)} = \frac{1}{3}.$$

Thus, after the first toss, the posterior probability that the coin is fair is $1/3$.

Now suppose that the same coin is tossed again and that another head is obtained. There are two ways of determining the new value of the posterior probability that the coin is fair.

The first way is to return to the beginning of the experiment and to assume again that the prior probabilities are $\Pr(A_1) = \Pr(A_2) = 1/2$. We shall let $H_1 H_2$ denote the event in which heads are obtained on two tosses of the coin, and we shall calculate the posterior probability $\Pr(A_1|H_1H_2)$ that the coin is fair after we have observed the event H_1H_2. By Bayes' theorem,

$$\Pr(A_1|H_1H_2) = \frac{\Pr(A_1)\Pr(H_1H_2|A_1)}{\Pr(A_1)\Pr(H_1H_2|A_1) + \Pr(A_2)\Pr(H_1H_2|A_2)}$$

$$= \frac{(1/2)(1/4)}{(1/2)(1/4) + (1/2)(1)} = \frac{1}{5}.$$

The second way of determining this same posterior probability is to use the information that after the first head has been obtained, the posterior probability of A_1 is $1/3$ and the posterior probability of A_2 is therefore $2/3$. These posterior probabilities can now serve as the prior probabilities for the next stage of the experiment, in which the coin is tossed a second time. Thus, we can now assume that the probabilities of A_1 and A_2 are $\Pr(A_1) = 1/3$ and $\Pr(A_2) = 2/3$, and we can compute the posterior probability $\Pr(A_1|H_2)$ that the coin is fair after we have observed a head on the second toss. In this way we obtain

$$\Pr(A_1|H_2) = \frac{(1/3)(1/2)}{(1/3)(1/2) + (2/3)(1)} = \frac{1}{5}.$$

The posterior probability of the event A_1 obtained in the second way is the same as that obtained in the first way. We can make the following general statement: If an experiment is carried out in more than one stage, then the posterior probability of any event can also be calculated in more than one stage. After each stage has been carried out, the posterior probability of the event found from that stage serves as the prior probability for the next stage.

Now suppose that in the experiment we have been considering here, the selected coin is tossed a third time and another head is obtained. By our previous calculations, the prior probabilities for this stage of the experiment are $\Pr(A_1) = 1/5$ and $\Pr(A_2) = 4/5$. Therefore, the posterior probability of A_1 after the third head has been observed will be

$$\Pr(A_1|H_3) = \frac{(1/5)(1/2)}{(1/5)(1/2) + (4/5)(1)} = \frac{1}{9}.$$

Finally, suppose that the coin is tossed a fourth time and that a tail is observed. In this case the posterior probability that the coin is fair becomes 1, and no further tosses of the coin can affect this probability.

Combining a Prior Probability with an Observation

Suppose that you are walking down the street and notice that the Department of Public Health is giving a free medical test for a certain disease. The test is 90 percent reliable in the following sense: If a person has the disease, there is a probability of 0.9 that the test will give a positive response; whereas, if a person does not have the disease, there is a probability of only 0.1 that the test will give a positive response.

Data indicate that your chances of having the disease are only 1 in 10,000. However since the test costs you nothing, and is fast and harmless, you decide to stop and take the test. A few days later you learn that you had a positive response to the test. What is now the probability that you have the disease?

Some readers may feel that this probability should be about 0.9. However, this feeling completely ignores the small prior probability of 0.0001 that you had the disease. We shall let A denote the event that you have the disease and let B denote the event that the response to the test is positive. Then, by Bayes' theorem,

$$\Pr(A|B) = \frac{\Pr(B|A)\Pr(A)}{\Pr(B|A)\Pr(A) + \Pr(B|A^c)\Pr(A^c)}$$

$$= \frac{(0.9)(0.0001)}{(0.9)(0.0001) + (0.1)(0.9999)} = 0.00090.$$

Thus, the posterior probability that you have the disease is approximately only 1 in 1000. Of course, this posterior probability is approximately 10 times as great as the prior probability before you were tested, but even the posterior probability is quite small.

Another way to explain this result is as follows: Only one person in every 10,000 actually has the disease, but the test gives a positive response for approximately one person in every 10. Hence, the number of positive responses is approximately 1000 times the number of persons who actually have the disease. In other words, out of every 1000 persons for whom the test gives a positive response, only one person actually has the disease.

EXERCISES

1. A box contains three coins with a head on each side, four coins with a tail on each side, and two fair coins. If one of these nine coins is selected at random and tossed once, what is the probability that a head will be obtained?

2. The percentages of voters classed as Liberals in three different election districts are divided as follows: In the first district, 21 percent; in the second

district, 45 percent; and in the third district, 75 percent. If a district is selected at random and a voter is selected at random from that district, what is the probability that he will be a Liberal?

3. Consider again the shopper described in Exercise 5 of Sec. 2.1. On each purchase, the probability that he will choose the same brand of toothpaste that he chose on his preceding purchase is $1/3$, and the probability that he will switch brands is $2/3$. Suppose that on his first purchase the probability that he will choose brand A is $1/4$ and the probability that he will choose brand B is $3/4$. What is the probability that his second purchase will be brand B?

4. Suppose that k events $A_1, \ldots, A_k$ form a partition of the sample space S. For $i = 1, \ldots, k$, let $\Pr(A_i)$ denote the prior probability of A_i. Also, for any event B such that $\Pr(B) > 0$, let $\Pr(A_i|B)$ denote the posterior probability of A_i given that the event B has occurred. Prove that if $\Pr(A_1|B) < \Pr(A_1)$, then $\Pr(A_i|B) > \Pr(A_i)$ for at least one value of i ($i = 2, \ldots, k$).

5. Consider again the conditions of Example 3 in this section, in which an item was selected at random from a batch of manufactured items and was found to be defective. For which values of i ($i = 1, 2, 3$) is the posterior probability that the item was produced by machine M_i larger than the prior probability that the item was produced by machine M_i?

6. Suppose that in Example 3 in this section, the item selected at random from the entire lot is found to be nondefective. Determine the posterior probability that it was produced by machine M_2.

7. A new test has been devised for detecting a particular type of cancer. If the test is applied to a person who has this type of cancer, the probability that the person will have a positive reaction is 0.95 and the probability that the person will have a negative reaction is 0.05. If the test is applied to a person who does not have this type of cancer, the probability that the person will have a positive reaction is 0.05 and the probability that the person will have a negative reaction is 0.95. Suppose that in the general population, one person out of every 100,000 people has this type of cancer. If a person selected at random has a positive reaction to the test, what is the probability that he has this type of cancer?

8. In a certain city, 30 percent of the people are Conservatives, 50 percent are Liberals, and 20 percent are Independents. Records show that in a particular election, 65 percent of the Conservatives voted, 82 percent of the Liberals voted, and 50 percent of the Independents voted. If a person in the city is selected at random and it is learned that he did not vote in the last election, what is the probability that he is a Liberal?

9. Suppose that when a machine is adjusted properly, 50 percent of the items produced by it are of high quality and the other 50 percent are of medium

quality. Suppose, however, that the machine is improperly adjusted during 10 percent of the time and that, under these conditions, 25 percent of the items produced by it are of high quality and 75 percent are of medium quality.

(a) Suppose that five items produced by the machine at a certain time are selected at random and inspected. If four of these items are of high quality and one item is of medium quality, what is the probability that the machine was adjusted properly at that time?

(b) Suppose that one additional item, which was produced by the machine at the same time as the other five items, is selected and found to be of medium quality. What is the new posterior probability that the machine was adjusted properly?

10. Suppose that a box contains five coins, and that for each coin there is a different probability that a head will be obtained when the coin is tossed. Let p_i denote the probability of a head when the ith coin is tossed ($i = 1, \ldots, 5$), and suppose that $p_1 = 0$, $p_2 = 1/4$, $p_3 = 1/2$, $p_4 = 3/4$, and $p_5 = 1$.

(a) Suppose that one coin is selected at random from the box and that when it is tossed once, a head is obtained. What is the posterior probability that the ith coin was selected ($i = 1, \ldots, 5$)?

(b) If the same coin were tossed again, what would be the probability of obtaining another head?

(c) If a tail had been obtained on the first toss of the selected coin and the same coin were tossed again, what would be the probability of obtaining a head on the second toss?

11. Consider again the box containing the five different coins described in Exercise 10. Suppose that one coin is selected at random from the box and is tossed repeatedly until a head is obtained.

(a) If the first head is obtained on the fourth toss, what is the posterior probability that the ith coin was selected ($i = 1, \ldots, 5$)?

(b) If we continue to toss the same coin until another head is obtained, what is the probability that exactly three additional tosses will be required?

*2.3. MARKOV CHAINS

Stochastic Processes

Suppose that a certain business office has five telephone lines and that any number of these lines may be in use at any given time. During a certain period of time, the telephone lines are observed at regular intervals of 2 minutes and the number of lines that are being used at each time is noted. Let X_1 denote the number of lines that are being used when the lines are first observed at the beginning of the period; let X_2 denote the number of lines that are being used when they are observed the second time, 2 minutes later; and in general, for $n = 1, 2, \ldots$, let X_n denote the number of lines that are being used when they are observed for the nth time.

The sequence of observations $X_1, X_2, \ldots$ is called a *stochastic process*, or *random process*, because the values of these observations cannot be predicted precisely beforehand but probabilities can be specified for each of the different possible values at any particular time. A stochastic process like the one just described is called a process with a *discrete time parameter* because the lines are observed only at discrete or separated points in time, rather than continuously in time.

In a stochastic process the first observation X_1 is called the *initial state* of the process; and for $n = 2, 3, \ldots$, the observation X_n is called the *state of the process at time n*. In the preceding example, the state of the process at any time is the number of lines being used at that time. Therefore, each state must be an integer between 0 and 5. In the remainder of this chapter we shall consider only stochastic processes for which there are just a finite number of possible states at any given time.

In a stochastic process with a discrete time parameter, the state of the process varies in a random manner from time to time. To describe a complete probability model for a particular process, it is necessary to specify a probability for each of the possible values of the initial state X_1 and also to specify for each subsequent state X_{n+1} ($n = 1, 2, \ldots$) every conditional probability of the following form:

$$\Pr(X_{n+1} = x_{n+1} | X_1 = x_1, X_2 = x_2, \ldots, X_n = x_n).$$

In other words, for every time n, the probability model must specify the conditional probability that the process will be in state x_{n+1} at time $n + 1$, given that at times $1, \ldots, n$ the process was in states $x_1, \ldots, x_n$.

Markov Chains

Definition. A Markov chain is a special type of stochastic process, which may be described as follows: At any given time n, when the current state X_n and all previous states $X_1, \ldots, X_{n-1}$ of the process are known, the probabilities of all future states X_j ($j > n$) depend only on the current state X_n and do not depend on the earlier states $X_1, \ldots, X_{n-1}$. Formally, a *Markov chain* is a stochastic process such that for $n = 1, 2, \ldots$ and for any possible sequence of states $x_1, x_2, \ldots, x_{n+1}$,

$$\Pr(X_{n+1} = x_{n+1} | X_1 = x_1, X_2 = x_2, \ldots, X_n = x_n)$$
$$= \Pr(X_{n+1} = x_{n+1} | X_n = x_n).$$

It follows from the multiplication rule for conditional probabilities given in Sec. 2.1 that the probabilities in a Markov chain must satisfy the relation

$$\Pr(X_1 = x_1, X_2 = x_2, \ldots, X_n = x_n)$$

$$= \Pr(X_1 = x_1)\Pr(X_2 = x_2 | X_1 = x_1)\Pr(X_3 = x_3 | X_2 = x_2) \cdots$$

$$\Pr(X_n = x_n | X_{n-1} = x_{n-1}).$$

Finite Markov Chains with Stationary Transition Probabilities. We shall now consider a Markov chain for which there are only a finite number k of possible states $s_1, \ldots, s_k$, and at any time the chain must be in one of these k states. A Markov chain of this type is called a *finite Markov chain*.

The conditional probability $\Pr(X_{n+1} = s_j | X_n = s_i)$ that the Markov chain will be in state s_j at time $n + 1$ if it is in state s_i at time n is called a *transition probability*. If for a certain Markov chain this transition probability has the same value for every time n ($n = 1, 2, \ldots$), then it is said that the Markov chain has stationary transition probabilities. In other words, a Markov chain has *stationary transition probabilities* if, for any states s_i and s_j, there is a transition probability p_{ij} such that

$$\Pr(X_{n+1} = s_j | X_n = s_i) = p_{ij} \qquad \text{for } n = 1, 2, \ldots.$$

To illustrate the application of these definitions we shall consider again the example involving the office with five telephone lines. In order for this stochastic process to be a Markov chain, the specified probability for each possible number of lines that may be in use at any time must depend only on the number of lines that were in use when the process was observed most recently 2 minutes earlier and must not depend on any other observed values previously obtained. For example, if three lines were in use at time n, then the probability specified for time $n + 1$ must be the same regardless of whether 0, 1, 2, 3, 4, or 5 lines were in use at time $n - 1$. In reality, however, the observation at time $n - 1$ might provide some information in regard to the length of time for which each of the three lines in use at time n had been occupied, and this information might be helpful in determining the probability for time $n + 1$. Nevertheless, we shall suppose now that this process is a Markov chain. If this Markov chain is to have stationary transition probabilities, it must be true that the rates at which incoming and outgoing telephone calls are made and the average duration of these telephone calls do not change during the entire period covered by the process. This requirement means that the overall period cannot include busy times when more calls are expected or quiet times when fewer calls are expected. For example, if only one line is in use at a particular observation time, regardless of when this time occurs during the entire period covered by the process, then there must be a specific probability p_{1j} that exactly j lines will be in use 2 minutes later.

The Transition Matrix

The Transition Matrix for a Single Step. Consider a finite Markov chain with k possible states $s_1, \ldots, s_k$ and stationary transition probabilities. For $i = 1, \ldots, k$ and $j = 1, \ldots, k$, we shall again let p_{ij} denote the conditional probability that the process will be in state s_j at a given observation time if it is in state s_i at the preceding observation time. The *transition matrix* of the Markov chain is defined to be the $k \times k$ matrix P with elements p_{ij}. Thus,

$$P = \begin{bmatrix} p_{11} \cdots p_{1k} \\ p_{21} \cdots p_{2k} \\ \cdots \cdots \cdots \\ p_{k1} \cdots p_{kk} \end{bmatrix}. \tag{1}$$

Since each number p_{ij} is a probability, then $p_{ij} \geq 0$. Furthermore, $\sum_{j=1}^{k} p_{ij} = 1$ for $i = 1, \ldots, k$, because if the chain is in state s_i at a given observation time, then the sum of the probabilities that it will be in each of the states $s_1, \ldots, s_k$ at the next observation time must be 1.

A square matrix for which all elements are nonnegative and the sum of the elements in each row is 1 is called a *stochastic matrix*. It is seen that the transition matrix P for any finite Markov chain with stationary transition probabilities must be a stochastic matrix. Conversely, any $k \times k$ stochastic matrix can serve as the transition matrix of a finite Markov chain with k possible states and stationary transition probabilities.

Example 1: A Transition Matrix for the Number of Occupied Telephone Lines. Suppose that in the example involving the office with five telephone lines, the numbers of lines being used at times $1, 2, \ldots$ form a Markov chain with stationary transition probabilities. This chain has six possible states $b_0, b_1, \ldots, b_5$, where b_i is the state in which exactly i lines are being used at a given time ($i = 0, 1, \ldots, 5$). Suppose that the transition matrix P is as follows:

$$P = \begin{array}{c} \\ b_0 \\ b_1 \\ b_2 \\ b_3 \\ b_4 \\ b_5 \end{array} \begin{array}{cccccc} b_0 & b_1 & b_2 & b_3 & b_4 & b_5 \\ \begin{bmatrix} 0.1 & 0.4 & 0.2 & 0.1 & 0.1 & 0.1 \\ 0.2 & 0.3 & 0.2 & 0.1 & 0.1 & 0.1 \\ 0.1 & 0.2 & 0.3 & 0.2 & 0.1 & 0.1 \\ 0.1 & 0.1 & 0.2 & 0.3 & 0.2 & 0.1 \\ 0.1 & 0.1 & 0.1 & 0.2 & 0.3 & 0.2 \\ 0.1 & 0.1 & 0.1 & 0.1 & 0.4 & 0.2 \end{bmatrix} \end{array}. \tag{2}$$

(a) Assuming that all five lines are in use at a certain observation time, we shall determine the probability that exactly four lines will be in use at the next observation time. (b) Assuming that no lines are in use at a certain time, we shall determine the probability that at least one line will be in use at the next observation time.

(a) This probability is the element in the matrix P in the row corresponding to the state b_5 and the column corresponding to the state b_4. Its value is seen to be 0.4.

(b) If no lines are in use at a certain time, then the element in the upper left corner of the matrix P gives the probability that no lines will be in use at the next observation time. Its value is seen to be 0.1. Therefore, the probability that at least one line will be in use at the next observation time is $1 - 0.1 = 0.9$.   □

The Transition Matrix for Several Steps. Consider again an arbitrary Markov chain with k possible states $s_1, \ldots, s_k$ and the transition matrix P given by Eq. (1), and assume that the chain is in state s_i at a given time n. We shall now determine the probability that the chain will be in state s_j at time $n + 2$. In other words, we shall determine the probability of moving from state s_i to state s_j in two steps. The notation for this probability is $p_{ij}^{(2)}$.

For $n = 1, 2, \ldots$, let X_n denote the state of the chain at time n. Then, if s_r denotes the state to which the chain has moved at time $n + 1$,

$$p_{ij}^{(2)} = \Pr\left(X_{n+2} = s_j | X_n = s_i \right)$$

$$= \sum_{r=1}^{k} \Pr\left(X_{n+1} = s_r \text{ and } X_{n+2} = s_j | X_n = s_i \right)$$

$$= \sum_{r=1}^{k} \Pr\left(X_{n+1} = s_r | X_n = s_i \right) \Pr\left(X_{n+2} = s_j | X_{n+1} = s_r \right)$$

$$= \sum_{r=1}^{k} p_{ir} p_{rj}.$$

The value of $p_{ij}^{(2)}$ can be determined in the following manner: If the transition matrix P is squared, i.e., if the matrix $P^2 = PP$ is constructed, then the element in the ith row and the jth column of the matrix P^2 will be $\sum_{r=1}^{k} p_{ir} p_{rj}$. Therefore, $p_{ij}^{(2)}$ will be the element in the ith row and the jth column of P^2.

By a similar argument the probability that the chain will move from the state s_i to the state s_j in three steps, or $p_{ij}^{(3)} = \Pr\left(X_{n+3} = s_j | X_n = s_i \right)$, can be found by constructing the matrix $P^3 = P^2 P$. Then the probability $p_{ij}^{(3)}$ will be the element in the ith row and the jth column of the matrix P^3.

In general, for any value of m ($m = 2, 3, \ldots$), the mth power $\boldsymbol{P}^m$ of the matrix $\boldsymbol{P}$ will specify the probability $p_{ij}^{(m)}$ that the chain will move from any state s_i to any state s_j in m steps. For this reason the matrix $\boldsymbol{P}^m$ is called the *m-step transition matrix* of the Markov chain.

Example 2: The Two-Step and Three-Step Transition Matrices for the Number of Occupied Telephone Lines. Consider again the transition matrix $\boldsymbol{P}$ given by Eq. (2) for the Markov chain based on five telephone lines. We shall assume first that i lines are in use at a certain time, and we shall determine the probability that exactly j lines will be in use two time periods later.

If we multiply the matrix $\boldsymbol{P}$ by itself, we obtain the following two-step transition matrix:

$$
\boldsymbol{P}^2 = \begin{array}{c} \\ b_0 \\ b_1 \\ b_2 \\ b_3 \\ b_4 \\ b_5 \end{array}
\begin{array}{cccccc}
b_0 & b_1 & b_2 & b_3 & b_4 & b_5 \\
\left[\begin{array}{cccccc}
0.14 & 0.23 & 0.20 & 0.15 & 0.16 & 0.12 \\
0.13 & 0.24 & 0.20 & 0.15 & 0.16 & 0.12 \\
0.12 & 0.20 & 0.21 & 0.18 & 0.17 & 0.12 \\
0.11 & 0.17 & 0.19 & 0.20 & 0.20 & 0.13 \\
0.11 & 0.16 & 0.16 & 0.18 & 0.24 & 0.15 \\
0.11 & 0.16 & 0.15 & 0.17 & 0.25 & 0.16
\end{array}\right]
\end{array} \quad (3)
$$

From this matrix we can find any two-step transition probability for the chain, such as the following:

(i) If two lines are in use at a certain time, then the probability that four lines will be in use two time periods later is 0.17.

(ii) If three lines are in use at a certain time, then the probability that three lines will again be in use two time periods later is 0.20.

We shall now assume that i lines are in use at a certain time, and we shall determine the probability that exactly j lines will be in use three time periods later.

If we construct the matrix $\boldsymbol{P}^3 = \boldsymbol{P}^2\boldsymbol{P}$, we obtain the following three-step transition matrix:

$$
\boldsymbol{P}^3 = \begin{array}{c} \\ b_0 \\ b_1 \\ b_2 \\ b_3 \\ b_4 \\ b_5 \end{array}
\begin{array}{cccccc}
b_0 & b_1 & b_2 & b_3 & b_4 & b_5 \\
\left[\begin{array}{cccccc}
0.123 & 0.208 & 0.192 & 0.166 & 0.183 & 0.128 \\
0.124 & 0.207 & 0.192 & 0.166 & 0.183 & 0.128 \\
0.120 & 0.197 & 0.192 & 0.174 & 0.188 & 0.129 \\
0.117 & 0.186 & 0.186 & 0.179 & 0.199 & 0.133 \\
0.116 & 0.181 & 0.177 & 0.176 & 0.211 & 0.139 \\
0.116 & 0.180 & 0.174 & 0.174 & 0.215 & 0.141
\end{array}\right]
\end{array} \quad (4)
$$

From this matrix we can find any three-step transition probability for the chain, such as the following:

(i) If all five lines are in use at a certain time, then the probability that no lines will be in use three time periods later is 0.116.

(ii) If one line is in use at a certain time, then the probability that exactly one line will again be in use three time periods later is 0.207. □

The Initial Probability Vector

Suppose that a finite Markov chain with stationary transition probabilities has k possible states $s_1, \ldots, s_k$ and that the chain might be in any one of these k states at the initial observation time $n = 1$. Suppose also that, for $i = 1, \ldots, k$, the probability that the chain will be in state s_i at the beginning of the process is v_i, where $v_i \geqslant 0$ and $v_1 + \cdots + v_k = 1$.

Any vector $w = (w_1, \ldots, w_k)$ such that $w_i \geqslant 0$ for $i = 1, \ldots, k$ and also $\sum_{i=1}^{k} w_i = 1$ is called a *probability vector*. The probability vector $v = (v_1, \ldots, v_k)$, which specifies the probabilities of the various states of a chain at the initial observation time, is called the *initial probability vector* for the chain.

The initial probability vector and the transition matrix together determine the probability that the chain will be in any particular state at any particular time. If v is the initial probability vector for a chain, then $\Pr(X_1 = s_i) = v_i$ for $i = 1, \ldots, k$. If the transition matrix of the chain is the $k \times k$ matrix P having the elements p_{ij} indicated by Eq. (1), then for $j = 1, \ldots, k$,

$$\Pr(X_2 = s_j) = \sum_{i=1}^{k} \Pr(X_1 = s_i \text{ and } X_2 = s_j)$$

$$= \sum_{i=1}^{k} \Pr(X_1 = s_i)\Pr(X_2 = s_j | X_1 = s_i)$$

$$= \sum_{i=1}^{k} v_i p_{ij}.$$

Since $\sum_{i=1}^{k} v_i p_{ij}$ is the jth component of the vector vP, this derivation shows that the probabilities for the state of the chain at the observation time 2 are specified by the probability vector vP.

More generally, suppose that at some given time n the probability that the chain will be in state s_i is $\Pr(X_n = s_i) = w_i$, for $i = 1, \ldots, k$. Then

$$\Pr(X_{n+1} = s_j) = \sum_{i=1}^{k} w_i p_{ij} \qquad \text{for } j = 1, \ldots, k.$$

In other words, if the probabilities of the various states at time n are specified by the probability vector w, then the probabilities at time $n + 1$ are specified by the probability vector wP. It follows that if the initial probability vector for a chain with stationary transition probabilities is v, then the probabilities of the various states at time $n + 1$ are specified by the probability vector vP^n.

Example 3: Probabilities for the Number of Occupied Telephone Lines. Consider again the office with five telephone lines and the Markov chain for which the transition matrix P is given by Eq. (2). Suppose that at the beginning of the observation process at time $n = 1$, the probability that no lines will be in use is 0.5, the probability that one line will be in use is 0.3, and the probability that two lines will be in use is 0.2. Then the initial probability vector is $v = (0.5, 0.3, 0.2, 0, 0, 0)$. We shall first determine the probability that exactly j lines will be in use at time 2, one period later.

By an elementary computation it will be found that

$$vP = (0.13, 0.33, 0.22, 0.12, 0.10, 0.10).$$

Since the first component of this probability vector is 0.13, the probability that no lines will be in use at time 2 is 0.13; since the second component is 0.33, the probability that exactly one line will be in use at time 2 is 0.33; and so on.

Next, we shall determine the probability that exactly j lines will be in use at time 3.

By use of Eq. (3), it will be found that

$$vP^2 = (0.133, 0.227, 0.202, 0.156, 0.162, 0.120).$$

Since the first component of this probability vector is 0.133, the probability that no lines will be in use at time 3 is 0.133; since the second component is 0.227, the probability that exactly one line will be in use at time 3 is 0.227; and so on. □

EXERCISES

1. Suppose that the weather can be only sunny or cloudy, and that the weather conditions on successive mornings form a Markov chain with stationary transition probabilities. Suppose also that the transition matrix is as follows:

	Sunny	Cloudy
Sunny	0.7	0.3
Cloudy	0.6	0.4

   (a) If it is cloudy on a given day, what is the probability that it will also be cloudy the next day?

   (b) If it is sunny on a given day, what is the probability that it will be sunny on the next two days?

   (c) If it is cloudy on a given day, what is the probability that it will be sunny on at least one of the next three days?

2. Consider again the Markov chain described in Exercise 1.

   (a) If it is sunny on a certain Wednesday, what is the probability that it will be sunny on the following Saturday?

   (b) If it is cloudy on a certain Wednesday, what is the probability that it will be sunny on the following Saturday?

3. Consider again the conditions of Exercises 1 and 2.

   (a) If it is sunny on a certain Wednesday, what is the probability that it will be sunny on both the following Saturday and Sunday?

   (b) If it is cloudy on a certain Wednesday, what is the probability that it will be sunny on both the following Saturday and Sunday?

4. Consider again the Markov chain described in Exercise 1. Suppose that the probability that it will be sunny on a certain Wednesday is 0.2 and the probability that it will be cloudy is 0.8.

   (a) Determine the probability that it will be cloudy on the next day, Thursday.

   (b) Determine the probability that it will be cloudy on Friday.

   (c) Determine the probability that it will be cloudy on Saturday.

5. Suppose that a student will be either on time or late for a particular class, and that the events that he is on time or late for the class on successive days form a Markov chain with stationary transition probabilities. Suppose also that if he is late on a given day, then the probability that he will be on time the next day is 0.8. Furthermore, if he is on time on a given day, then the probability that he will be late the next day is 0.5.

   (a) If the student is late on a certain day, what is the probability that he will be on time on each of the next three days?

   (b) If the student is on time on a given day, what is the probability that he will be late on each of the next three days?

6. Consider again the Markov chain described in Exercise 5.

   (a) If the student is late on the first day of class, what is the probability that he will be on time on the fourth day of class?

   (b) If the student is on time on the first day of class, what is the probability that he will be on time on the fourth day of class?

7. Consider again the conditions of Exercises 5 and 6. Suppose that the probability that the student will be late on the first day of class is 0.7 and that the probability that he will be on time is 0.3.

   (a) Determine the probability that he will be late on the second day of class.

   (b) Determine the probability that he will be on time on the fourth day of class.

8. Suppose that a Markov chain has four states s_1, s_2, s_3, s_4 and stationary transition probabilities as specified by the following transition matrix:

$$
\begin{array}{c c c c c}
 & s_1 & s_2 & s_3 & s_4 \\
\begin{array}{c} s_1 \\ s_2 \\ s_3 \\ s_4 \end{array} &
\left[\begin{array}{cccc}
1/4 & 1/4 & 0 & 1/2 \\
0 & 1 & 0 & 0 \\
1/2 & 0 & 1/2 & 0 \\
1/4 & 1/4 & 1/4 & 1/4
\end{array} \right]
\end{array} .
$$

   (a) If the chain is in state s_3 at a given time n, what is the probability that it will be in state s_2 at time $n + 2$?

   (b) If the chain is in state s_1 at a given time n, what is the probability that it will be in state s_3 at time $n + 3$?

9. Let X_1 denote the initial state at time 1 of the Markov chain for which the transition matrix is as specified in Exercise 8, and suppose that the initial probabilities are as follows:

$$\Pr(X_1 = s_1) = 1/8, \Pr(X_1 = s_2) = 1/4,$$

$$\Pr(X_1 = s_3) = 3/8, \Pr(X_1 = s_4) = 1/4.$$

Determine the probabilities that the chain will be in states s_1, s_2, s_3, and s_4 at time n for each of the following values of n: (a) $n = 2$; (b) $n = 3$; (c) $n = 4$.

10. Each time that a shopper purchases a tube of toothpaste, he chooses either brand A or brand B. Suppose that the probability is $1/3$ that he will choose the same brand chosen on his previous purchase, and the probability is $2/3$ that he will switch brands.

   (a) If his first purchase is brand A, what is the probability that his fifth purchase will be brand B?

   (b) If his first purchase is brand B, what is the probability that his fifth purchase will be brand B?

11. Suppose that three boys A, B, and C are throwing a ball from one to another. Whenever A has the ball he throws it to B with a probability of 0.2 and to C with a probability of 0.8. Whenever B has the ball he throws it to A

with a probability of 0.6 and to C with a probability of 0.4. Whenever C has the ball he is equally likely to throw it to either A or B.

(a) Consider this process to be a Markov chain and construct the transition matrix.

(b) If each of the three boys is equally likely to have the ball at a certain time n, which boy is most likely to have the ball at time $n + 2$?

12. Suppose that a coin is tossed repeatedly in such a way that heads and tails are equally likely to appear on any given toss and that all tosses are independent, with the following exception: Whenever either three heads or three tails have been obtained on three successive tosses, then the outcome of the next toss is always of the opposite type. At time n ($n \geq 3$) let the state of this process be specified by the outcomes on tosses $n - 2$, $n - 1$, and n. Show that this process is a Markov chain with stationary transition probabilities and construct the transition matrix.

13. There are two boxes A and B, each containing red and green balls. Suppose that box A contains one red ball and two green balls and that box B contains eight red balls and two green balls. Consider the following process: One ball is selected at random from box A, and one ball is selected at random from box B. The ball selected from box A is then placed in box B and the ball selected from box B is placed in box A. These operations are then repeated indefinitely. Show that the numbers of red balls in box A form a Markov chain with stationary transition probabilities, and construct the transition matrix of the Markov chain.

*2.4.   THE GAMBLER'S RUIN PROBLEM

Statement of the Problem

Suppose that two gamblers A and B are playing a game against each other. Let p be a given number ($0 < p < 1$), and suppose that on each play of the game, the probability that gambler A will win one dollar from gambler B is p and the probability that gambler B will win one dollar from gambler A is $q = 1 - p$. Suppose also that the initial fortune of gambler A is i dollars and the initial fortune of gambler B is $k - i$ dollars, where i and $k - i$ are given positive integers. Thus, the total fortune of the two gamblers is k dollars. Finally, suppose that the gamblers continue playing the game until the fortune of one of them has been reduced to 0 dollars.

We shall now consider this game from the point of view of gambler A. His initial fortune is i dollars and on each play of the game his fortune will either increase by one dollar with a probability of p or decrease by one dollar with a probability of q. If $p > 1/2$, the game is favorable to him; if $p < 1/2$, the game is unfavorable to him; and if $p = 1/2$, the game is equally favorable to both gamblers. The game ends either when the fortune of gambler A reaches k dollars, in which case gambler B will have no money left, or when the fortune of gambler A reaches 0 dollars. The problem is to determine the probability that the fortune of gambler A will reach k dollars before it reaches 0 dollars. Because one of the gamblers will have no money left at the end of the game, this problem is called the *Gambler's Ruin* problem.

Solution of the Problem

We shall continue to assume that the total fortune of the gamblers A and B is k dollars, and we shall let a_i denote the probability that the fortune of gambler A will reach k dollars before it reaches 0 dollars, given that his initial fortune is i dollars. If $i = 0$, then gambler A is ruined; and if $i = k$, then gambler A has won the game. Therefore, we shall assume that $a_0 = 0$ and $a_k = 1$. We shall now determine the value of a_i for $i = 1, \ldots, k - 1$.

Let A_1 denote the event that gambler A wins one dollar on the first play of the game; let B_1 denote the event that gambler A loses one dollar on the first play of the game; and let W denote the event that the fortune of gambler A ultimately reaches k dollars before it reaches 0 dollars. Then

$$\Pr(W) = \Pr(A_1)\Pr(W|A_1) + \Pr(B_1)\Pr(W|B_1)$$
$$= p \Pr(W|A_1) + q \Pr(W|B_1). \tag{1}$$

Since the initial fortune of gambler A is i dollars ($i = 1, \ldots, k - 1$), then $\Pr(W) = a_i$. Furthermore, if gambler A wins one dollar on the first play of the game, then his fortune becomes $i + 1$ dollars and the probability $\Pr(W|A_1)$ that his fortune will ultimately reach k dollars is therefore a_{i+1}. If A loses one dollar on the first play of the game, then his fortune becomes $i - 1$ dollars and the probability $\Pr(W|B_1)$ that his fortune will ultimately reach k dollars is therefore a_{i-1}. Hence, by Eq. (1),

$$a_i = pa_{i+1} + qa_{i-1}. \tag{2}$$

We shall let $i = 1, \ldots, k - 1$ in Eq. (2). Then, since $a_0 = 0$ and $a_k = 1$, we obtain the following $k - 1$ equations:

$$
\begin{aligned}
a_1 &= pa_2, \\
a_2 &= pa_3 + qa_1, \\
a_3 &= pa_4 + qa_2, \\
&\;\cdots\cdots\cdots\cdots\cdots \\
a_{k-2} &= pa_{k-1} + qa_{k-3}, \\
a_{k-1} &= p + qa_{k-2}.
\end{aligned}
\tag{3}
$$

If the value of a_i on the left side of the ith equation is rewritten in the form $pa_i + qa_i$ and some elementary algebra is performed, then these $k - 1$ equations can be rewritten as follows:

$$a_2 - a_1 = \frac{q}{p} a_1,$$

$$a_3 - a_2 = \frac{q}{p}(a_2 - a_1) \qquad = \left(\frac{q}{p}\right)^2 a_1,$$

$$a_4 - a_3 = \frac{q}{p}(a_3 - a_2) \qquad = \left(\frac{q}{p}\right)^3 a_1,$$

$$\vdots \qquad\qquad \vdots \qquad\qquad \vdots \qquad\qquad\qquad (4)$$

$$a_{k-1} - a_{k-2} = \frac{q}{p}(a_{k-2} - a_{k-3}) = \left(\frac{q}{p}\right)^{k-2} a_1,$$

$$1 - a_{k-1} = \frac{q}{p}(a_{k-1} - a_{k-2}) = \left(\frac{q}{p}\right)^{k-1} a_1.$$

By equating the sum of the left sides of these $k - 1$ equations with the sum of the right sides, we obtain the relation

$$1 - a_1 = a_1 \sum_{i=1}^{k-1} \left(\frac{q}{p}\right)^i. \qquad (5)$$

Solution for a Fair Game. Suppose first that $p = q = 1/2$. Then $q/p = 1$, and it follows from Eq. (5) that $1 - a_1 = (k - 1)a_1$, from which $a_1 = 1/k$. In turn, it follows from the first equation in (4) that $a_2 = 2/k$; it follows from the second equation in (4) that $a_3 = 3/k$; and so on. In this way, we obtain the following complete solution when $p = q = 1/2$:

$$a_i = \frac{i}{k} \qquad \text{for } i = 1, \ldots, k - 1. \qquad (6)$$

Example 1: The Probability of Winning in a Fair Game. Suppose that $p = q = 1/2$, in which case the game is equally favorable to both gamblers; and suppose that the initial fortune of gambler A is 98 dollars and the initial fortune of gambler B is just 2 dollars. In this example, $i = 98$ and $k = 100$. Therefore, it follows from Eq. (6) that there is a probability of 0.98 that gambler A will win two dollars from gambler B before gambler B wins 98 dollars from gambler A. $\square$

Solution for an Unfair Game. Suppose now that $p \neq q$. Then Eq. (5) can be rewritten in the form

$$1 - a_1 = a_1 \frac{\left(\frac{q}{p}\right)^k - \left(\frac{q}{p}\right)}{\left(\frac{q}{p}\right) - 1}. \tag{7}$$

Hence,

$$a_1 = \frac{\left(\frac{q}{p}\right) - 1}{\left(\frac{q}{p}\right)^k - 1}. \tag{8}$$

Each of the other values of a_i for $i = 2, \ldots, k-1$ can now be determined in turn from the equations in (4). In this way, we obtain the following complete solution:

$$a_i = \frac{\left(\frac{q}{p}\right)^i - 1}{\left(\frac{q}{p}\right)^k - 1} \qquad \text{for } i = 1, \ldots, k-1. \tag{9}$$

Example 2: The Probability of Winning in an Unfavorable Game. Suppose that $p = 0.4$ and $q = 0.6$, in which case the probability that gambler A will win one dollar on any given play is smaller than the probability that he will lose one dollar. Suppose also that the initial fortune of gambler A is 99 dollars and the initial fortune of gambler B is just one dollar. We shall determine the probability that gambler A will win one dollar from gambler B before gambler B wins 99 dollars from gambler A.

In this example, the required probability a_i is given by Eq. (9), in which $q/p = 3/2$, $i = 99$, and $k = 100$. Therefore,

$$a_i = \frac{\left(\frac{3}{2}\right)^{99} - 1}{\left(\frac{3}{2}\right)^{100} - 1} \approx \frac{1}{\frac{3}{2}} = \frac{2}{3}.$$

Hence, although the probability that gambler A will win one dollar on any given play is only 0.4, the probability that he will win one dollar before he loses 99 dollars is approximately 2/3. □

EXERCISES

1. Consider the following three different possible conditions in the gambler's ruin problem:
   (a) The initial fortune of gambler A is 2 dollars and the initial fortune of gambler B is 1 dollar.
   (b) The initial fortune of gambler A is 20 dollars and the initial fortune of gambler B is 10 dollars.
   (c) The initial fortune of gambler A is 200 dollars and the initial fortune of gambler B is 100 dollars.
   Suppose that $p = q = 1/2$. For which of these three conditions is there the greatest probability that gambler A will win the initial fortune of gambler B before he loses his own initial fortune?

2. Consider again the three different conditions (a), (b), and (c) given in Exercise 1, but suppose now that $p < q$. For which of these three conditions is there the greatest probability that gambler A will win the initial fortune of gambler B before he loses his own initial fortune?

3. Consider again the three different conditions (a), (b), and (c) given in Exercise 1, but suppose now that $p > q$. For which of these three conditions is there the greatest probability that gambler A will win the initial fortune of gambler B before he loses his own initial fortune?

4. Suppose that on each play of a certain game, a person is equally likely to win one dollar or lose one dollar. Suppose also that the person's goal is to win two dollars by playing this game. How large an initial fortune must the person have in order for the probability to be at least 0.99 that he will achieve his goal before he loses his initial fortune?

5. Suppose that one each play of a certain game, a person will either win one dollar with probability $2/3$ or lose one dollar with probability $1/3$. Suppose also that the person's goal is to win two dollars by playing this game. How large an initial fortune must the person have in order for the probability to be at least 0.99 that he will achieve his goal before he loses his initial fortune?

6. Suppose that on each play of a certain game, a person will either win one dollar with probability $1/3$ or lose one dollar with probability $2/3$. Suppose also that the person's goal is to win two dollars by playing this game. Show that no matter how large the person's initial fortune might be, the probability that he will achieve his goal before he loses his initial fortune is less than $1/4$.

7. Suppose that the probability of a head on any toss of a certain coin is p ($0 < p < 1$), and suppose that the coin is tossed repeatedly. Let X_n denote the total number of heads that have been obtained on the first n tosses, and

let $Y_n = n - X_n$ denote the total number of tails on the first n tosses. Suppose that the tosses are stopped as soon as a number n is reached such that either $X_n = Y_n + 3$ or $Y_n = X_n + 3$. Determine the probability that $X_n = Y_n + 3$ when the tosses are stopped.

8. Suppose that a certain box A contains 5 balls and another box B contains 10 balls. One of these two boxes is selected at random, and one ball from the selected box is transferred to the other box. If this process of selecting a box at random and transferring one ball from that box to the other box is repeated indefinitely, what is the probability that box A will become empty before box B becomes empty?

*2.5. CHOOSING THE BEST

Optimal Selection

In this section we shall describe a special problem of decision-making that illustrates in a striking fashion how the basic concepts of probability that we have developed can be applied to achieve some surprisingly strong results. Suppose that you are an employer who must hire a new person to fill an available position from n candidates for the position. We shall make the following assumptions about the hiring process.

The candidates will appear in a random order to be interviewed by you, and you will interview them sequentially, i.e., one at a time, in the order in which they appear. After interviewing each candidate, you must decide immediately whether or not you want to hire that candidate. If you decide to hire that candidate, the process terminates. If you decide not to hire that candidate, then you proceed to interview the next one. The candidate you have just interviewed leaves and, we assume, accepts some other position. Thus, once you have decided not to hire a candidate, you cannot later change your mind.

At the beginning of the process you have no information about the qualifications or abilities of any of the n candidates. However, after interviewing a candidate you can rank that candidate in relation to each of the other candidates whom you have previously interviewed. Since the candidates appear in random order, the only information that you can obtain about the first candidate (even after you have interviewed him) is that he is equally likely to have any one of the ranks $1, \ldots, n$ among all n candidates. In other words, you gain no immediate information about the relative rank of the first candidate. All you know is that he is equally likely to be the best candidate, the worst candidate, or anywhere in-between.

After interviewing the second candidate, you can determine whether he is better or worse than the first candidate. However, you do not know with certainty where either of these candidates would rank among all n candidates. After interviewing the third candidate, you can determine who is the best among the first three candidates, but again you do not know where the third candidate would rank among all n candidates. Only if you did not terminate the interviewing process until you had interviewed all n candidates, would you be able to determine all the rankings.

We shall now inject a note of excitement and risk into the process by assuming that your task is to hire that candidate who is actually the best among all n candidates. If you do not succeed in hiring the best of all n candidates, you will have failed in your task and it then does not matter which of the other $n - 1$ candidates you hire. Thus, when you interview the candidates one at a time, you will fail if you unknowingly stop too soon and hire a candidate while the best one is still waiting to be interviewed; and you will fail also if you unknowingly continue past the best candidate in the hope that there is still a better candidate waiting to be interviewed.

The Form of the Best Procedure

From the statement of the problem just presented, it follows that you should never stop the interviewing process and hire any candidate who is not better than all the previous candidates because such a candidate certainly cannot be the best among all n candidates. The only times at which you have to make a serious decision are whenever you have just interviewed a candidate who is better than all previous candidates. You must then decide whether to stop because of the likelihood that this candidate is actually the best among all n candidates or to continue because of the likelihood that the best is yet to come.

We shall refer to a candidate who is better than all previous candidates as a contender. Thus, after you have interviewed a contender, you must immediately decide whether to stop or to continue. If the contender appears among the first few candidates to be interviewed, you will not want to stop because there is a high probability that the best candidate will still be among the remaining candidates. On the other hand, if a contender appears when only a few candidates remain to be interviewed, you will want to stop because there is only a low probability that a better candidate is yet to be interviewed. It appears from these comments that you will not want to stop and hire any contender who appears before some critical stage in the interviewing process has been reached, and you will want to stop and hire a contender who appears after that critical stage.

In the light of this discussion, we shall consider a procedure of the following type: Never stop before r candidates have been interviewed. If the rth candidate is a contender, then stop the interviewing process and hire him. If the rth candidate is not a contender, continue the interviewing process until a contender is found. Then stop and hire him. Of course, if you continue the interviewing process beyond the rth candidate and never obtain another contender, then you have failed in your task.

It can be shown by advanced methods that the optimal procedure, that is, the procedure that maximizes your probability of hiring the best candidate, is actually of the form just described. In order to implement this rule, a numerical value must be chosen for r. We shall now determine the value of r for which the probability p_r of hiring the best candidate is maximized, and we shall see that this probability is surprisingly large.

The Best Procedure

Let A denote the event that you actually hire the best candidate; and for $i = 1, \ldots, n$, let B_i denote the event that the best candidate will be the ith person to be interviewed. Then

$$p_r = \Pr(A) = \sum_{i=1}^{n} \Pr(A|B_i)\Pr(B_i). \tag{1}$$

Since you will never stop before the rth candidate, there is no chance of hiring the best candidate if he is one of the first $r - 1$ candidates to be interviewed. Hence,

$$\Pr(A|B_i) = 0 \qquad \text{for } i = 1, \ldots, r - 1. \tag{2}$$

Furthermore, if the best candidate is the rth person to be interviewed, then he will surely be a contender at the time at which he is interviewed. In this case, you will correctly stop and hire him. Thus,

$$\Pr(A|B_r) = 1. \tag{3}$$

Since the candidates appear in random order, the best candidate is equally likely to appear anywhere in the interviewing process. Hence,

$$\Pr(B_i) = \frac{1}{n} \qquad \text{for } i = 1, \ldots, n. \tag{4}$$

It now follows from Eq. (1) that

$$p_r = \frac{1}{n}\left[1 + \sum_{i=r+1}^{n} \Pr(A|B_i)\right]. \tag{5}$$

Next, suppose that the event B_i occurs; that is, the best candidate is the ith person who is interviewed ($i = r + 1, \ldots, n$). In this case, if you have not stopped before interviewing the ith candidate, then you will surely stop and correctly hire the best candidate immediately after the ith interview. Hence, the event that you hire the best candidate is the same as the event that you do not stop after any of the first $i - 1$ interviews. This event is the same as the event that there are no contenders among the candidates interviewed at stages $r, r + 1, \ldots, i - 1$. In turn, this event is the same as the event that the candidate who is relatively the best among the first $i - 1$ candidates is among the first $r - 1$ candidates who are interviewed.

To analyze this last statement further, let γ denote the candidate who is actually the best among the first $i - 1$ candidates. First, suppose that γ appears at any one of the stages $r, r + 1, \ldots, i - 1$. In this case γ is a contender when he is interviewed and you will stop the interviews and hire him. If you do stop, however, you will have failed in your task because the best candidate among all n candidates will not appear until stage i. Suppose, instead, that γ appears among the first $r - 1$ candidates who are interviewed. Then you would not have hired him, since you should never stop before stage r; and you will not have stopped immediately after any of the stages $r, r + 1, \ldots, i - 1$ since γ is better than any of the candidates interviewed at those stages. In this case you will succeed in selecting the best candidate at stage i.

Since the candidate who is relatively the best among the first $i - 1$ candidates is equally likely to appear at any of the first $i - 1$ stages, the probability that he will actually appear among the first $r - 1$ stages is $(r - 1)/(i - 1)$. Thus, we have established that

$$\Pr(A|B_i) = \frac{r-1}{i-1} \qquad \text{for } i = r + 1, \ldots, n. \tag{6}$$

From Eq. (5),

$$p_r = \frac{1}{n}\left(1 + \frac{r-1}{r} + \frac{r-1}{r+1} + \cdots + \frac{r-1}{n-1}\right)$$

$$= \frac{1}{n} + \frac{r-1}{n}\left(\frac{1}{r} + \frac{1}{r+1} + \cdots + \frac{1}{n-1}\right). \tag{7}$$

We must find the value of r for which p_r is a maximum. From Eq. (7), it is found after some algebra that

$$p_{r+1} - p_r = \frac{1}{n}\left(\frac{1}{r} + \frac{1}{r+1} + \cdots + \frac{1}{n-1} - 1\right). \tag{8}$$

It can be seen from Eq. (8) that the difference ($p_{r+1} - p_r$) is a decreasing function of r. As long as this difference is positive, p_{r+1} will be larger than p_r. However, as soon as this difference becomes negative for some value of r, then p_{r+1} will be smaller than p_r and it will continue to decrease for all larger values of r. Hence, the value of r for which p_r is a maximum will be the smallest value of r for which ($p_{r+1} - p_r$) $\leqslant 0$.

It now follows from Eq. (8) that we must find the smallest value of r such that

$$\frac{1}{r} + \frac{1}{r+1} + \cdots + \frac{1}{n-1} \leqslant 1. \tag{9}$$

We shall call this value r^*. Table 2.1 gives the values of r^* for various values of n and also the probability p_{r*} of actually hiring the best candidate when the procedure based on r^* is followed.

The values of p_{r*} found in Table 2.1 are surprisingly large. When there are 10 candidates, the chances are approximately 40 percent that the best one will be hired. Out of 100 candidates, the chances of hiring the best one are 37 percent; and even out of 1000 candidates the chances of hiring the best one are almost the same 37 percent.

Note that the values of p_{r*} in Table 2.1 not only are surprisingly large but also decrease surprisingly slowly as n becomes large. It is natural to investigate the limiting behavior of p_{r*} as $n \to \infty$. We shall now show that p_{r*} decreases so

Table 2.1

n	r^*	p_{r*}
2	1 or 2	0.5
3	2	0.5
4	2	0.4583
5	3	0.4333
6	3	0.4278
7	3	0.4143
8	4	0.4098
9	4	0.4060
10	4	0.3987
15	6	0.3894
20	8	0.3842
30	12	0.3787
50	19	0.3742
100	38	0.3710
1000	369	0.3682

slowly that its lower limit remains bounded away from zero no matter how large n becomes. In fact, p_{r*} converges to a very simple but interesting limit as $n \to \infty$.

The Simple but Interesting Limiting Value

Since $r*$ is the smallest value of r for which the relation (9) is satisfied, it will be approximately true for large values of n that

$$\frac{1}{r*} + \frac{1}{r* + 1} + \cdots + \frac{1}{n - 1} \approx 1. \tag{10}$$

The sum on the left side of (10) is equal to the sum of the areas of the shaded rectangles in Fig. 2.3 and can be approximated by the following integral:

$$\int_{r*}^{n} \frac{1}{x} \, dx = \log n - \log r* = \log \frac{n}{r*}. \tag{11}$$

Here, and everywhere else in this book, the abbreviation log is used to denote a natural logarithm, that is, a logarithm with base e.

Thus, it follows from (10) that

$$\log \frac{n}{r*} \approx 1. \tag{12}$$

Hence,

$$r* \approx \frac{1}{e}(n) = 0.3679\, n. \tag{13}$$

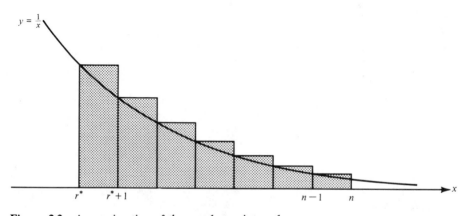

Figure 2.3   Approximation of the sum by an integral.

The interpretation of (13) is that the best procedure has approximately the following simple form for large values of n: Never stop the interviewing process until the proportion 0.3679 of all the candidates have been interviewed. Stop the process after interviewing the first contender who appears thereafter, and hire him.

Furthermore, it follows from (10), (7), and (13) that

$$p_{r*} \approx \frac{r^*}{n} \approx \frac{1}{e} = 0.3679. \tag{14}$$

Thus, the probability that the best candidate will be hired is approximately $1/e = 0.3679$, regardless of how large n is. In fact, it can be shown that $p_{r*} > 1/e$ for every finite value of n and that

$$\lim_{n \to \infty} p_{r*} = \frac{1}{e}. \tag{15}$$

Parlor Games

We shall now describe briefly two games in which the results that have just been described can be verified. One game you play by yourself; the other game you play with another person. To play the first game, write the numbers from 1 to 50 on separate slips of paper and place the 50 slips in a box. As shown in Table 2.1, the value of r^* for $n = 50$ is 19. Therefore, draw 18 slips from the box at random, without replacement, and observe and record their numbers. Then continue to draw slips from the box one at a time, without replacement, until you obtain one with a number X that is larger than any previously drawn. You win the game if $X = 50$. You lose the game either if the slip with the number 50 was included among the first 18 slips drawn from the box or if $X < 50$. As indicated in Table 2.1, the probability that you will win is approximately 0.37.

To play the second game, have another person write 50 different numbers (which are not necessarily integers) on slips of paper without revealing the numbers to you. Draw the slips from the box one at a time, without replacement, and observe and record the number on each slip as it is drawn. As before, never stop during the first 18 drawings, and then, beginning with the nineteenth slip, stop as soon as you obtain a slip with a number X that is larger than any previously drawn. You win if X is the largest of the 50 numbers on the slips. Again, the probability that you will win is approximately 0.37.

Even if you play either of these games with 500 slips or 50,000 slips, the probability that you will win is still larger than 0.36. Thus, if you play the second game repeatedly with another person, you should win approximately one out of every three games. If the other person is not aware of the effectiveness of your procedure, you may be able to make some favorable bets with him.

2.6. SUPPLEMENTARY EXERCISES

1. Suppose that A, B, and D are any three events such that $\Pr(A|D) \geqslant \Pr(B|D)$ and $\Pr(A|D^c) \geqslant \Pr(B|D^c)$. Prove that $\Pr(A) \geqslant \Pr(B)$.

2. For any three events A, B, and D, such that $\Pr(D) > 0$, prove that $\Pr(A \cup B|D) = \Pr(A|D) + \Pr(B|D) - \Pr(AB|D)$.

3. Suppose that A and B are independent events such that $\Pr(A) = 1/3$ and $\Pr(B) > 0$. What is the value of $\Pr(A \cup B^c|B)$?

4. Suppose that A and B are events such that $\Pr(A) = 1/3$, $\Pr(B) = 1/5$, and $\Pr(A|B) + \Pr(B|A) = 2/3$. Evaluate $\Pr(A^c \cup B^c)$.

5. Suppose that A, B, and D are events such that A and B are independent, $\Pr(ABD) = 0.04$, $\Pr(D|AB) = 0.25$, and $\Pr(B) = 4\Pr(A)$. Evaluate $\Pr(A \cup B)$.

6. Suppose that in ten rolls of a balanced die, the number 6 appeared exactly three times. What is the probability that the first three rolls each yielded the number 6?

7. Suppose that a balanced die is rolled repeatedly until the same number appears on two successive rolls, and let X denote the number of rolls that are required. Determine the value of $\Pr(X = x)$, for $x = 2, 3, \dots$.

8. Suppose that 80 percent of all statisticians are shy, whereas only 15 percent of all economists are shy. Suppose also that 90 percent of the people at a large gathering are economists and the other 10 percent are statisticians. If you meet a shy person at random at the gathering, what is the probability that the person is a statistician?

9. Dreamboat cars are produced at three different factories A, B, and C. Factory A produces 20 percent of the total output of Dreamboats; B produces 50 percent; and C produces 30 percent. However, 5 percent of the cars produced at A are lemons; 2 percent of those produced at B are lemons; and 10 percent of those produced at C are lemons. If you buy a Dreamboat and it turns out to be a lemon, what is the probability that it was produced at factory A?

10. Suppose that 30 percent of the bottles produced in a certain plant are defective. If a bottle is defective, the probability is 0.9 that an inspector will notice it and remove it from the filling line. If a bottle is not defective, the probability is 0.2 that the inspector will think that it is defective and remove it from the filling line. (a) If a bottle is removed from the filling line, what is

the probability that it is defective? (b) If a customer buys a bottle that has not been removed from the filling line, what is the probability that it is defective?

11. Suppose that a fair coin is tossed until a head is obtained and that this entire experiment is then performed a second time. What is the probability that the second experiment requires more tosses than the first experiment?

12. Suppose that a family has exactly n children ($n \geqslant 2$). Assume that the probability that any child will be a girl is $1/2$ and that all births are independent. Given that the family has at least one girl, determine the probability that the family has at least one boy.

13. Suppose that a fair coin is tossed independently n times. Determine the probability of obtaining exactly $n - 1$ heads, given (a) that at least $n - 2$ heads are obtained and (b) that heads are obtained on the first $n - 2$ tosses.

14. Suppose that 13 cards are selected at random from a regular deck of 52 playing cards. (a) If it is known that at least one ace has been selected, what is the probability that at least two aces have been selected? (b) If it is known that the ace of hearts has been selected, what is the probability that at least two aces have been selected?

15. Suppose that n letters are placed at random in n envelopes, as in the matching problem of Sec. 1.10, and let q_n denote the probability that no letter is placed in the correct envelope. Show that the probability that exactly one letter is placed in the correct envelope is q_{n-1}.

16. Consider again the conditions of Exercise 15. Show that the probability that exactly two letters are placed in the correct envelopes is $(1/2)q_{n-2}$.

17. Consider again the conditions of Exercise 6 of Sec. 1.11. If exactly one of the two students A and B is in class on a given day, what is the probability that it is A?

18. Consider again the conditions of Exercise 1 of Sec. 1.10. If a family selected at random from the city subscribes to exactly one of the three newspapers A, B, and C, what is the probability that it is A?

19. Three prisoners A, B, and C in a foreign land know that exactly two of them are going to be executed, but they do not know which two. Prisoner A knows that the jailer will not tell him whether or not he is going to be executed. He therefore asks the jailer to tell him the name of one prisoner other than A himself who will be executed. The jailer responds that B will be executed. Upon receiving this response, Prisoner A reasons as follows: Before he spoke to the jailer, the probability was $2/3$ that he would be one of the two prisoners executed. After speaking to the jailer, he knows that either he or prisoner C will be the other one to be executed. Hence, the probability that he will be executed is now only $1/2$. Thus, merely by asking the jailer his question, the prisoner reduced the probability that he would be executed

from 2/3 to 1/2, because he could go through exactly this same reasoning regardless of which answer the jailer gave. Discuss what is wrong with prisoner A's reasoning.

20. Three boys A, B, and C are playing table tennis. In each game, two of the boys play against each other and the third boy does not play. The winner of any given game n plays again in game $n + 1$ against the boy who did not play in game n, and the loser of game n does not play in game $n + 1$. The probability that A will beat B in any game that they play against each other is 0.3; the probability that A will beat C is 0.6; and the probability that B will beat C is 0.8. Represent this process as a Markov chain with stationary transition probabilities by defining the possible states and constructing the transition matrix.

21. Consider again the Markov chain described in Exercise 20. (a) Determine the probability that the two boys who play against each other in the first game will play against each other again in the fourth game. (b) Show that this probability does not depend on which two boys play in the first game.

22. Suppose that each of two gamblers A and B has an initial fortune of 50 dollars, and that there is probability p that gambler A will win on any single play of a game against gambler B. Also, suppose either that one gambler can win one dollar from the other on each play of the game or that they can double the stakes and one can win two dollars from the other on each play of the game. Under which of these two conditions does A have the greater probability of winning the initial fortune of B before losing his own for each of the following conditions: (a) $p < 1/2$; (b) $p > 1/2$; (c) $p = 1/2$?

Random Variables and Distributions **3**

3.1. RANDOM VARIABLES AND DISCRETE DISTRIBUTIONS

Definition of a Random Variable

Consider an experiment for which the sample space is denoted by S. A real-valued function that is defined on the space S is called a *random variable*. In other words, in a particular experiment a random variable X would be some function that assigns a real number $X(s)$ to each possible outcome $s \in S$.

Example 1: Tossing a Coin. Consider an experiment in which a coin is tossed ten times. In this experiment the sample space can be regarded as the set of outcomes consisting of the 2^{10} different sequences of ten heads and tails that are possible, and the random variable X could be the number of heads obtained on the ten tosses. For each possible sequence s consisting of ten heads and tails, this random variable would then assign a number $X(s)$ equal to the number of heads in the sequence. Thus, if s is the sequence HHTTTHTTTH, then $X(s) = 4$. □

Example 2: Choosing a Point in the Plane. Suppose that a point in the xy-plane is chosen in accordance with some specified probability distribution. Thus each outcome in the sample space is a point of the form $s = (x, y)$. If the random variable X is taken as the x-coordinate of the chosen point, then $X(s) = x$ for each outcome s. Another possible random variable Y for this experiment is the y-coordinate of the chosen point. A third possible random variable Z is the distance from the origin to the chosen point. These random variables are defined by the functions

$$Y(s) = y \quad \text{and} \quad Z(s) = \left(x^2 + y^2\right)^{1/2}. \quad □$$

Example 3: Measuring a Person's Height. Consider an experiment in which a person is selected at random from some population and his height in inches is measured. This height is a random variable. □

The Distribution of a Random Variable

When a probability distribution has been specified on the sample space of an experiment, we can determine a probability distribution for the possible values of any random variable X. Let A be any subset of the real line, and let $\Pr(X \in A)$ denote the probability that the value of X will belong to the subset A. Then $\Pr(X \in A)$ is equal to the probability that the outcome s of the experiment will be such that $X(s) \in A$. In symbols,

$$\Pr(X \in A) = \Pr\{s: X(s) \in A\}.$$

Example 4: Tossing a Coin. Consider again an experiment in which a coin is tossed ten times, and let X be the number of heads that are obtained. In this experiment the possible values of X are $0, 1, 2, \ldots, 10$; and, as explained in Example 2 of Sec. 1.8,

$$\Pr(X = x) = \binom{10}{x}\frac{1}{2^{10}} \qquad \text{for } x = 0, 1, 2, \ldots, 10. \quad \square$$

Example 5: Choosing a Point from a Rectangle. Consider an experiment in which a point $s = (x, y)$ is chosen at random from the rectangle $S = \{(x, y): 0 \leqslant x \leqslant 2$ and $0 \leqslant y \leqslant 1/2\}$ sketched in Fig. 3.1. The area of this rectangle S is 1, and we shall assume that the probability that the chosen point will lie in any given subset of S is equal to the area of that subset. If the random variable X is the x-coordinate of the chosen point, then for any numbers x_1 and x_2 such that $0 \leqslant x_1 \leqslant x_2 \leqslant 2$, it follows that $\Pr(x_1 \leqslant X \leqslant x_2)$ will be given by the area of the

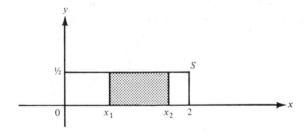

Figure 3.1   The points where $x_1 \leqslant X(s) \leqslant x_2$.

shaded portion in Fig. 3.1. Hence,

$$\Pr(x_1 \leqslant X \leqslant x_2) = \frac{1}{2}(x_2 - x_1). \quad \square$$

Discrete Distributions

It is said that a random variable X has a *discrete distribution* if X can take only a finite number k of different values $x_1, \ldots, x_k$ or, at most, an infinite sequence of different values $x_1, x_2, \ldots$. If a random variable X has a discrete distribution, the *probability function* (abbreviated p.f.) of X is defined as the function f such that for any real number x,

$$f(x) = \Pr(X = x).$$

For any point x which is not one of the possible values of X, it is evident that $f(x) = 0$. Also, if the sequence $x_1, x_2, \ldots$ includes all the possible values of X, then $\sum_{i=1}^{\infty} f(x_i) = 1$. A typical p.f. is sketched in Fig. 3.2, in which each vertical segment represents the value of $f(x)$ corresponding to a possible value x. The sum of the heights of the vertical segments in that figure must be 1.

If X has a discrete distribution, the probability of any subset A of the real line can be determined from the relation

$$\Pr(X \in A) = \sum_{x_i \in A} f(x_i).$$

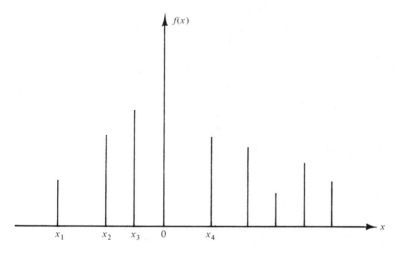

Figure 3.2 An example of a p.f.

We shall illustrate these concepts by considering two specific types of discrete distributions.

The Uniform Distribution on Integers

Suppose that the value of a random variable X is equally likely to be any one of the k integers $1, 2, \ldots, k$. Then the p.f. of X is as follows:

$$f(x) = \begin{cases} \dfrac{1}{k} & \text{for } x = 1, 2, \ldots, k, \\ 0 & \text{otherwise.} \end{cases}$$

This discrete distribution is called the *uniform distribution on the integers* $1, 2, \ldots, k$. It represents the outcome of an experiment that is often described by saying that one of the integers $1, 2, \ldots, k$ is *chosen at random*. In this context, the phrase "at random" means that each of the k integers is equally likely to be chosen. In this same sense, it is not possible to choose an integer at random from the set of *all* positive integers because it is not possible to assign the same probability to every one of the positive integers and still make the sum of these probabilities equal to 1. In other words, a uniform distribution cannot be assigned to an infinite sequence of possible values, but such a distribution can be assigned to any finite sequence.

The Binomial Distribution

Suppose that a certain machine produces a defective item with a probability of $p (0 < p < 1)$ and produces a nondefective item with a probability of $q = 1 - p$. Suppose further that n independent items produced by the machine are examined, and let X denote the number of these items that are defective. Then the random variable X will have a discrete distribution, and the possible values of X will be $0, 1, 2, \ldots, n$.

This example is similar to Example 4 in Sec. 1.11. For $x = 0, 1, \ldots, n$, the probability of obtaining any particular ordered sequence of n items containing exactly x defectives and $n - x$ nondefectives is $p^x q^{n-x}$. Since there are $\binom{n}{x}$ different ordered sequences of this type, it follows that

$$\Pr(X = x) = \binom{n}{x} p^x q^{n-x}.$$

Therefore, the p.f. of X will be as follows:

$$f(x) = \begin{cases} \binom{n}{x} p^x q^{n-x} & \text{for } x = 0, 1, \ldots, n, \\ 0 & \text{otherwise.} \end{cases} \tag{1}$$

The discrete distribution represented by this p.f. is called the *binomial distribution with parameters n and p*. This distribution is very important in probability and statistics and will be discussed further in later chapters of this book.

A short table of values of the binomial distribution is given at the end of this book. It can be found from this table, for example, that if X has a binomial distribution with parameters $n = 10$ and $p = 0.2$, then $\Pr(X = 5) = 0.0264$ and $\Pr(X \geqslant 5) = 0.0328$.

EXERCISES

1. Suppose that a random variable X has a discrete distribution with the following p.f.:

$$f(x) = \begin{cases} cx & \text{for } x = 1, 2, 3, 4, 5, \\ 0 & \text{otherwise.} \end{cases}$$

   Determine the value of the constant c.

2. Suppose that two balanced dice are rolled, and let X denote the absolute value of the difference between the two numbers that appear. Determine and sketch the p.f. of X.

3. Suppose that a fair coin is tossed ten times independently. Determine the p.f. of the number of heads that will be obtained.

4. Suppose that a box contains 7 red balls and 3 blue balls. If 5 balls are selected at random, without replacement, determine the p.f. of the number of red balls that will be obtained.

5. Suppose that a random variable X has a binomial distribution for which the parameters are $n = 15$ and $p = 0.5$. Find $\Pr(X < 6)$.

6. Suppose that a random variable X has a binomial distribution for which the parameters are $n = 8$ and $p = 0.7$. Find $\Pr(X \geqslant 5)$ by using the table given at the end of this book. *Hint:* Use the fact that $\Pr(X \geqslant 5) = \Pr(Y \leqslant 3)$, where Y has a binomial distribution with parameters $n = 8$ and $p = 0.3$.

7. If 10 percent of the balls in a certain box are red and if 20 balls are selected from the box at random, with replacement, what is the probability that more than 3 red balls will be obtained?

8. Suppose that a random variable X has a discrete distribution with the following p.f.:

$$f(x) = \begin{cases} \dfrac{c}{x^2} & \text{for } x = 1, 2, \ldots, \\ 0 & \text{otherwise.} \end{cases}$$

   Find the value of the constant c.

9. Show that there does not exist any number c such that the following function would be a p.f.:

$$f(x) = \begin{cases} \dfrac{c}{x} & \text{for } x = 1, 2, \ldots, \\ 0 & \text{otherwise.} \end{cases}$$

3.2. CONTINUOUS DISTRIBUTIONS

The Probability Density Function

It is said that a random variable X has a *continuous distribution* if there exists a nonnegative function f, defined on the real line, such that for any interval A,

$$\Pr(X \in A) = \int_A f(x)\, dx. \tag{1}$$

The function f is called the *probability density function* (abbreviated p.d.f.) of X. Thus, if a random variable X has a continuous distribution, the probability that X will belong to any subset of the real line can be found by integrating the p.d.f. of X over that subset. Every p.d.f. must satisfy the following two requirements:

$$f(x) \geqslant 0 \tag{2}$$

and

$$\int_{-\infty}^{\infty} f(x)\, dx = 1. \tag{3}$$

A typical p.d.f. is sketched in Fig. 3.3. In that figure, the total area under the curve must be 1, and the value of $\Pr(a \leqslant X \leqslant b)$ is equal to the area of the shaded region.

Nonuniqueness of the p.d.f.

If a random variable X has a continuous distribution, then $\Pr(X = x) = 0$ for every individual value x. Because of this property, the values of any p.d.f. can be changed at a finite number of points, or even at an infinite sequence of points, without changing the value of the integral of the p.d.f. over any subset A. In other words, the values of the p.d.f. of a random variable X can be changed arbitrarily

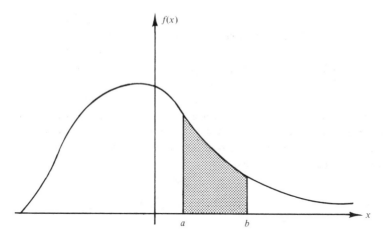

Figure 3.3   An example of a p.d.f.

at an infinite sequence of points without affecting any probabilities involving X, that is, without affecting the probability distribution of X.

To the extent just described, the p.d.f. of a random variable is not unique. In many problems, however, there will be one version of the p.d.f. that is more natural than any other because for this version the p.d.f. will, wherever possible, be continuous on the real line. For example, the p.d.f. sketched in Fig. 3.3 is a continuous function over the entire real line. This p.d.f. could be changed arbitrarily at a few points without affecting the probability distribution that it represents, but these changes would introduce discontinuities into the p.d.f. without introducing any apparent advantages.

Throughout most of this book we shall adopt the following practice: If a random variable X has a continuous distribution, we shall give only one version of the p.d.f. of X and we shall refer to that version as *the* p.d.f. of X, just as though it had been uniquely determined. It should be remembered, however, that there is some freedom in the selection of the particular version of the p.d.f. that is used to represent any continuous distribution.

We shall now illustrate these concepts with a few examples.

The Uniform Distribution on an Interval

Let a and b be two given real numbers such that $a < b$, and consider an experiment in which a point X is selected from the interval $S = \{x: a \leqslant x \leqslant b\}$ in such a way that the probability that X will belong to any subinterval of S is proportional to the length of that subinterval. This distribution of the random variable X is called the *uniform distribution on the interval* (a, b). Here X represents the outcome of an experiment that is often described by saying that a point is chosen *at random* from the interval (a, b). In this context, the phrase "at random" means that the point is just as likely to be chosen from any particular part of the interval as from any other part.

Since X must belong to the interval S, the p.d.f. $f(x)$ of X must be 0 outside of S. Furthermore, since any particular subinterval of S having a given length is as likely to contain X as is any other subinterval having the same length, regardless of the location of the particular subinterval in S, it follows that $f(x)$ must be constant throughout S. Also,

$$\int_S f(x)\, dx = \int_a^b f(x)\, dx = 1. \tag{4}$$

Therefore, the constant value of $f(x)$ throughout S must be $1/(b - a)$, and the p.d.f. of X must be as follows:

$$f(x) = \begin{cases} \dfrac{1}{b - a} & \text{for } a \leqslant x \leqslant b, \\ 0 & \text{otherwise.} \end{cases} \tag{5}$$

This p.d.f. is sketched in Fig. 3.4.

It is seen from Eq. (5) that the p.d.f. representing a uniform distribution on a given interval is constant over that interval, and the constant value of the p.d.f. is the reciprocal of the length of the interval. It is not possible to define a uniform distribution over the interval $x \geqslant a$ because the length of this interval is infinite.

Consider again the uniform distribution on the interval (a, b). Since the probability is 0 that one of the endpoints a or b will be chosen, it is irrelevant whether the distribution is regarded as a uniform distribution on the *closed*

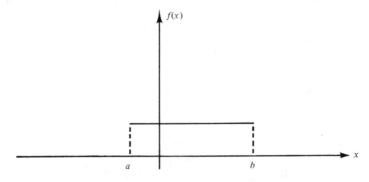

Figure 3.4 The p.d.f. for a uniform distribution.

interval $a \leqslant x \leqslant b$, or as a uniform distribution on the *open* interval $a < x < b$, or as a uniform distribution on the half-open and half-closed interval (a, b) in which one endpoint is included and the other endpoint is excluded.

For example, if a random variable X has a uniform distribution on the interval $(-1, 4)$, then the p.d.f. of X is

$$f(x) = \begin{cases} 1/5 & \text{for } -1 \leqslant x \leqslant 4, \\ 0 & \text{otherwise.} \end{cases}$$

Furthermore,

$$\Pr(0 \leqslant X < 2) = \int_0^2 f(x)\, dx = \frac{2}{5}.$$

Example 1: Calculating Probabilities from a p.d.f. Suppose that the p.d.f. of a certain random variable X has the following form:

$$f(x) = \begin{cases} cx & \text{for } 0 < x < 4, \\ 0 & \text{otherwise,} \end{cases}$$

where c is a given constant. We shall determine the value of c and also the values of $\Pr(1 \leqslant X \leqslant 2)$ and $\Pr(X > 2)$.

For every p.d.f. it must be true that $\int_{-\infty}^{\infty} f(x) = 1$. Therefore, in this example,

$$\int_0^4 cx\, dx = 8c = 1.$$

Hence, $c = 1/8$. It follows that

$$\Pr(1 \leqslant X \leqslant 2) = \int_1^2 \frac{1}{8} x\, dx = \frac{3}{16}$$

and

$$\Pr(X > 2) = \int_2^4 \frac{1}{8} x\, dx = \frac{3}{4}. \quad \square$$

Example 2: Unbounded Random Variables. It is often convenient and useful to represent a continuous distribution by a p.d.f. that is positive over an unbounded interval of the real line. For example, in a practical problem, the voltage X in a certain electrical system might be a random variable with a continuous distribution that can be approximately represented by the p.d.f.

$$f(x) = \begin{cases} 0 & \text{for } x \leqslant 0, \\ \dfrac{1}{(1 + x)^2} & \text{for } x > 0. \end{cases} \qquad (6)$$

It can be verified that the properties (2) and (3) required of all p.d.f.'s are satisfied by $f(x)$.

Even though the voltage X may actually be bounded in the real situation, the p.d.f. (6) may provide a good approximation for the distribution of X over its full range of values. For example, suppose that it is known that the maximum possible value of X is 1000, in which case $\Pr(X > 1000) = 0$. When the p.d.f. (6) is used, it is found that $\Pr(X > 1000) = 0.001$. If (6) adequately represents the variability of X over the interval $(0, 1000)$, then it may be more convenient to use the p.d.f. (6) than a p.d.f. that is similar to (6) for $x \leqslant 1000$, except for a new normalizing constant, and is 0 for $x > 1000$.   □

Example 3: Unbounded p.d.f.'s. Since a value of a p.d.f. is a probability density, rather than a probability, such a value can be larger than 1. In fact, the values of the following p.d.f. are unbounded in the neighborhood of $x = 0$:

$$f(x) = \begin{cases} \dfrac{2}{3} x^{-1/3} & \text{for } 0 < x < 1, \\ 0 & \text{otherwise.} \end{cases} \qquad (7)$$

It can be verified that even though the p.d.f. (7) is unbounded, it satisfies the properties (2) and (3) required of a p.d.f.   □

Mixed Distributions

Most distributions that are encountered in practical problems are either discrete or continuous. We shall show, however, that it may sometimes be necessary to consider a distribution that is a mixture of a discrete distribution and a continuous distribution.

Suppose that in the electrical system considered in Example 2, the voltage X is to be measured by a voltmeter which will record the actual value of X if $X \leqslant 3$ but will simply record the value 3 if $X > 3$. If we let Y denote the value recorded by the voltmeter, then the distribution of Y can be derived as follows:

First, $\Pr(Y = 3) = \Pr(X \geqslant 3) = 1/4$. Since the single value $Y = 3$ has probability $1/4$, it follows that $\Pr(0 < Y < 3) = 3/4$. Furthermore, since $Y = X$ for $0 < X < 3$, this probability $3/4$ for Y is distributed over the interval $0 < Y < 3$ according to the same p.d.f. (6) as that of X over the same interval. Thus, the distribution of Y is specified by the combination of a p.d.f. over the interval $0 < Y < 3$ and a positive probability at the point $Y = 3$.

EXERCISES

1. Suppose that the p.d.f. of a random variable X is as follows:

$$f(x) = \begin{cases} \frac{4}{3}(1 - x^3) & \text{for } 0 < x < 1, \\ 0 & \text{otherwise.} \end{cases}$$

Sketch this p.d.f. and determine the values of the following probabilities:

(a) $\Pr\left(X < \frac{1}{2}\right)$ (b) $\Pr\left(\frac{1}{4} < X < \frac{3}{4}\right)$ (c) $\Pr\left(X > \frac{1}{3}\right)$.

2. Suppose that the p.d.f. of a random variable X is as follows:

$$f(x) = \begin{cases} \frac{1}{36}(9 - x^2) & \text{for } -3 \leqslant x \leqslant 3, \\ 0 & \text{otherwise.} \end{cases}$$

Sketch this p.d.f. and determine the values of the following probabilities:

(a) $\Pr(X < 0)$ (b) $\Pr(-1 \leqslant X \leqslant 1)$ (c) $\Pr(X > 2)$.

3. Suppose that the p.d.f. of a random variable X is as follows:

$$f(x) = \begin{cases} cx^2 & \text{for } 1 \leqslant x \leqslant 2, \\ 0 & \text{otherwise.} \end{cases}$$

(a) Find the value of the constant c and sketch the p.d.f.

(b) Find the value of $\Pr(X > 3/2)$.

4. Suppose that the p.d.f. of a random variable X is as follows:

$$f(x) = \begin{cases} \frac{1}{8}x & \text{for } 0 \leqslant x \leqslant 4, \\ 0 & \text{otherwise.} \end{cases}$$

(a) Find the value of t such that $\Pr(X \leqslant t) = 1/4$.

(b) Find the value of t such that $\Pr(X \geqslant t) = 1/2$.

5. Let X be a random variable for which the p.d.f. is as given in Exercise 4. After the value of X has been observed, let Y be the integer closest to X. Find the p.f. of the random variable Y.

6. Suppose that a random variable X has a uniform distribution on the interval $(-2, 8)$. Find the p.d.f. of X and the value of $\Pr(0 < X < 7)$.

7. Suppose that the p.d.f. of a random variable X is as follows:

$$f(x) = \begin{cases} ce^{-2x} & \text{for } x > 0, \\ 0 & \text{otherwise.} \end{cases}$$

(a) Find the value of the constant c and sketch the p.d.f.

(b) Find the value of $\Pr(1 < X < 2)$.

8. Show that there does not exist any number c such that the following function $f(x)$ would be a p.d.f.:

$$f(x) = \begin{cases} \dfrac{c}{1+x} & \text{for } x > 0, \\ 0 & \text{otherwise.} \end{cases}$$

9. Suppose that the p.d.f. of a random variable X is as follows:

$$f(x) = \begin{cases} \dfrac{c}{(1-x)^{1/2}} & \text{for } 0 < x < 1, \\ 0 & \text{otherwise.} \end{cases}$$

(a) Find the value of the constant c and sketch the p.d.f.

(b) Find the value of $\Pr(X \leqslant 1/2)$.

10. Show that there does not exist any number c such that the following function $f(x)$ would be a p.d.f.:

$$f(x) = \begin{cases} \dfrac{c}{x} & \text{for } 0 < x < 1, \\ 0 & \text{otherwise.} \end{cases}$$

3.3. THE DISTRIBUTION FUNCTION

Definition and Basic Properties

The *distribution function* F of a random variable X is a function defined for each real number x as follows:

$$F(x) = \Pr(X \leqslant x) \qquad \text{for } -\infty < x < \infty. \tag{1}$$

It should be emphasized that the distribution function is defined in this way for every random variable X, regardless of whether the distribution of X is discrete, continuous, or mixed. The abbreviation for distribution function is d.f. Some authors use the term *cumulative distribution function*, instead of distribution function, and use the abbreviation c.d.f.

It follows from Eq. (1) that the d.f. of any random variable X is a function F defined on the real line. The value of $F(x)$ at any point x must be a number in the interval $0 \leqslant F(x) \leqslant 1$ because $F(x)$ is the probability of the event $\{X \leqslant x\}$.

Furthermore, it follows from Eq. (1) that the d.f. of any random variable X must have the following three properties:

Property 1.   *The function $F(x)$ is nondecreasing as x increases; that is, if $x_1 < x_2$, then $F(x_1) \leqslant F(x_2)$.*

Proof. If $x_1 < x_2$, then the occurrence of the event $\{X \leqslant x_1\}$ also implies that the event $\{X \leqslant x_2\}$ has occurred. Hence, $\Pr\{X \leqslant x_1\} \leqslant \Pr\{X \leqslant x_2\}$. □

An example of a d.f. is sketched in Fig. 3.5. It is shown in that figure that $0 \leqslant F(x) \leqslant 1$ over the entire real line. Also, $F(x)$ is always nondecreasing as x increases, although $F(x)$ is constant over the interval $x_1 \leqslant x \leqslant x_2$ and for $x \geqslant x_4$.

Property 2. $\displaystyle\lim_{x \to -\infty} F(x) = 0 \;\; and \;\; \lim_{x \to \infty} F(x) = 1.$

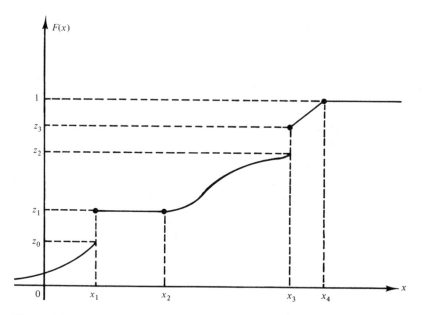

Figure 3.5   An example of a d.f.

Proof. These limiting values follow directly from the fact that $\Pr(X \leq x)$ must approach 0 as $x \to -\infty$ and $\Pr(X \leq x)$ must approach 1 as $x \to \infty$. These relations can in turn be rigorously established by using Exercises 11 and 12 of Sec. 1.10. $\square$

The limiting values specified in Property 2 are indicated in Fig. 3.5. In this figure, the value of $F(x)$ actually becomes 1 at $x = x_4$ and then remains 1 for $x > x_4$. Hence, it may be concluded that $\Pr(X \leq x_4) = 1$ and $\Pr(X > x_4) = 0$. On the other hand, according to the sketch in Fig. 3.5, the value of $F(x)$ approaches 0 as $x \to -\infty$, but does not actually become 0 at any finite point x. Therefore, for any finite value of x, no matter how small, $\Pr(X \leq x) > 0$.

A d.f. need not be continuous. In fact, the value of $F(x)$ may jump at any number of points. In Fig. 3.5, for instance, such jumps or points of discontinuity occur where $x = x_1$ and $x = x_3$. For any fixed value x, we shall let $F(x^-)$ denote the limit of the values of $F(y)$ as y approaches x from the left, that is, as y approaches x through values smaller than x. In symbols,

$$F(x^-) = \lim_{\substack{y \to x \\ y < x}} F(y).$$

Similarly, we shall define $F(x^+)$ as the limit of the values of $F(y)$ as y approaches x from the right. Thus,

$$F(x^+) = \lim_{\substack{y \to x \\ y > x}} F(y).$$

If the d.f. is continuous at a given point x, then $F(x^-) = F(x^+) = F(x)$ at that point.

Property 3. *A d.f. is always continuous from the right; that is, $F(x) = F(x^+)$ at every point x.*

Proof. Let $y_1 > y_2 > \cdots$ be a sequence of numbers that are decreasing such that $\lim_{n \to \infty} y_n = x$. Then the event $\{X \leq x\}$ is the intersection of all the events $\{X \leq y_n\}$ for $n = 1, 2, \ldots$. Hence, by Exercise 12 of Sec. 1.10,

$$F(x) = \Pr(X \leq x) = \lim_{n \to \infty} \Pr(X \leq y_n) = F(x^+). \quad \square$$

It follows from Property 3 that at any point x at which a jump occurs,

$$F(x^+) = F(x) \text{ and } F(x^-) < F(x).$$

In Fig. 3.5 this property is illustrated by the fact that at the points of discontinuity $x = x_1$ and $x = x_3$, the value of $F(x_1)$ is taken as z_1 and the value of $F(x_3)$ is taken as z_3.

Determining Probabilities from the Distribution Function

If the d.f. of a random variable X is known, then the probability that X will lie in any specified interval of the real line can be determined from the d.f. We shall derive this probability for four different types of intervals.

Theorem 1.   *For any given value x,*

$$\Pr(X > x) = 1 - F(x). \tag{2}$$

Proof. Since $\Pr(X > x) = 1 - \Pr(X \leqslant x)$, Eq. (2) follows from Eq. (1).   □

Theorem 2.   *For any given values x_1 and x_2 such that $x_1 < x_2$,*

$$\Pr(x_1 < X \leqslant x_2) = F(x_2) - F(x_1). \tag{3}$$

Proof. $\Pr(x_1 < X \leqslant x_2) = \Pr(X \leqslant x_2) - \Pr(X \leqslant x_1)$. Hence, Eq. (3) follows directly from Eq. (1).   □

For example, if the d.f. of X is as sketched in Fig. 3.5, it follows from Theorems 1 and 2 that $\Pr(X > x_2) = 1 - z_1$ and $\Pr(x_2 < X \leqslant x_3) = z_3 - z_1$. Also, since $F(x)$ is constant over the interval $x_1 \leqslant x \leqslant x_2$, then $\Pr(x_1 < X \leqslant x_2) = 0$.

It is important to distinguish carefully between the strict inequalities and the weak inequalities that appear in all the preceding relations and also in the next theorem. If there is a jump in $F(x)$ at a given value x, then the values of $\Pr(X \leqslant x)$ and $\Pr(X < x)$ will be different.

Theorem 3.   *For any given value x,*

$$\Pr(X < x) = F(x^-). \tag{4}$$

Proof. Let $y_1 < y_2 < \cdots$ be an increasing sequence of numbers such that $\lim_{n \to \infty} y_n = x$. Then it can be shown that

$$\{X < x\} = \bigcup_{n=1}^{\infty} \{X \leqslant y_n\}.$$

Therefore, it follows from Exercise 11 of Sec. 1.10 that

$$\Pr(X < x) = \lim_{n \to \infty} \Pr(X \leqslant y_n)$$

$$= \lim_{n \to \infty} F(y_n) = F(x^-). \quad \square$$

For example, for the d.f. sketched in Fig. 3.5, $\Pr(X < x_3) = z_2$ and $\Pr(X < x_4) = 1$.

Finally, we shall show that for any given value x, $\Pr(X = x)$ is equal to the amount of the jump that occurs in F at the point x. If F is continuous at the point x, that is, if there is no jump in F at x, then $\Pr(X = x) = 0$.

Theorem 4.   *For any given value x,*

$$\Pr(X = x) = F(x^+) - F(x^-). \tag{5}$$

Proof. It is always true that $\Pr(X = x) = \Pr(X \leqslant x) - \Pr(X < x)$. The relation (5) follows from the fact that $\Pr(X \leqslant x) = F(x) = F(x^+)$ at every point and from Theorem 3. $\square$

In Fig. 3.5, for example, $\Pr(X = x_1) = z_1 - z_0$, $\Pr(X = x_3) = z_3 - z_2$, and the probability of any other individual value of X is 0.

The d.f. of a Discrete Distribution

From the definition and properties of a d.f. $F(x)$ it follows that if $\Pr(a < X < b) = 0$ for two numbers a and b $(a < b)$, then $F(x)$ will be constant and horizontal over the interval $a < x < b$. Furthermore, as we have just seen, at any point x such that $\Pr(X = x) > 0$, the d.f. will jump by the amount $\Pr(X = x)$.

Suppose that X has a discrete distribution with the p.f. $f(x)$. Together, the properties of a d.f. imply that $F(x)$ must have the following form: $F(x)$ will have a jump of magnitude $f(x_i)$ at each possible value x_i of X; and $F(x)$ will be constant between any two successive jumps. The distribution of a discrete random variable X can be represented equally well by either the p.f. or the d.f. of X.

The d.f. of a Continuous Distribution

Consider now a random variable X with a continuous distribution, and let $f(x)$ and $F(x)$ denote the p.d.f. and the d.f., respectively, of X. Since the probability of any individual point x is 0, the d.f. $F(x)$ will have no jumps. Hence, $F(x)$ will be a continuous function over the entire real line. Furthermore, since

$$F(x) = \Pr(X \leq x) = \int_{-\infty}^{x} f(t)\, dt,$$ (6)

BASIC RELATIONSHIPS BETWEEN p.d.f. AND d.f.

it follows that, at any point x at which $f(x)$ is continuous,

$$F'(x) = \frac{dF(x)}{dx} = f(x).$$ (7)

Thus, the distribution of a continuous random variable X can be represented equally well by either the p.d.f. or the d.f. of X.

Example 1: Calculating a p.d.f. from a d.f. Suppose that in a certain electrical system the voltage X is a random variable for which the d.f. is as follows:

$$F(x) = \begin{cases} 0 & \text{for } x < 0, \\ \dfrac{x}{1+x} & \text{for } x \geq 0. \end{cases}$$

This function satisfies the three properties required of every d.f., as given in Sec. 3.1. Furthermore, since this d.f. is continuous over the entire real line and is differentiable at every point except $x = 0$, the distribution of X is continuous. Therefore, the p.d.f. of X can be found at any point other than $x = 0$ by the relation (7). The value of $f(x)$ at the single point $x = 0$ can be assigned arbitrarily. When the derivative $F'(x)$ is calculated, it is found that $f(x)$ is as given by Eq. (6) of Sec. 3.2. Conversely, if the p.d.f. of X is given by Eq. (6) of Sec. 3.2, then by using Eq. (6) of this section it is found that $F(x)$ is as given in this example. □

EXERCISES

1. Suppose that a random variable X can take only the values -2, 0, 1, and 4, and that the probabilities of these values are as follows: $\Pr(X = -2) = 0.4$, $\Pr(X = 0) = 0.1$, $\Pr(X = 1) = 0.3$, and $\Pr(X = 4) = 0.2$. Sketch the d.f. of X.

2. Suppose that a coin is tossed repeatedly until a head is obtained for the first time, and let X denote the number of tosses that are required. Sketch the d.f. of X.

3. Suppose that the d.f. F of a random variable X is as sketched in Fig. 3.6. Find each of the following probabilities:

(a) $\Pr(X = -1)$ (b) $\Pr(X < 0)$ (c) $\Pr(X \leqslant 0)$

(d) $\Pr(X = 1)$ (e) $\Pr(0 < X \leqslant 3)$ (f) $\Pr(0 < X < 3)$

(g) $\Pr(0 \leqslant X \leqslant 3)$ (h) $\Pr(1 < X \leqslant 2)$ (i) $\Pr(1 \leqslant X \leqslant 2)$

(j) $\Pr(X > 5)$ (k) $\Pr(X \geqslant 5)$ (l) $\Pr(3 \leqslant X \leqslant 4)$.

4. Suppose that the d.f. of a random variable X is as follows:

$$F(x) = \begin{cases} 0 & \text{for } x \leqslant 0, \\ \dfrac{1}{9}x^2 & \text{for } 0 < x \leqslant 3, \\ 1 & \text{for } x > 3. \end{cases}$$

Find and sketch the p.d.f. of X.

5. Suppose that the d.f. of a random variable X is as follows:

$$F(x) = \begin{cases} e^{x-3} & \text{for } x \leqslant 3, \\ 1 & \text{for } x > 3. \end{cases}$$

Find and sketch the p.d.f. of X.

6. Suppose, as in Exercise 6 of Sec. 3.2, that a random variable X has a uniform distribution on the interval $(-2, 8)$. Find and sketch the d.f. of X.

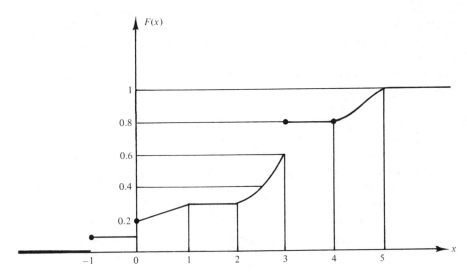

Figure 3.6 The d.f. for Exercise 3.

7. Suppose that a point in the xy-plane is chosen at random from the interior of a circle for which the equation is $x^2 + y^2 = 1$; and suppose that the probability that the point will belong to any region inside the circle is proportional to the area of that region. Let Z denote a random variable representing the distance from the center of the circle to the point. Find and sketch the d.f. of Z.

8. Suppose that X has a uniform distribution on the interval $(0, 5)$ and that the distribution of a random variable Y is such that $Y = 0$ if $X \leqslant 1$, $Y = 5$ if $X \geqslant 3$, and $Y = X$ otherwise. Sketch the d.f. of Y.

3.4. BIVARIATE DISTRIBUTIONS

In many experiments it is necessary to consider the properties of two or more random variables simultaneously. The joint probability distribution of two random variables is called a *bivariate distribution*. In this section and the next two sections we shall consider bivariate distributions. In Sec. 3.7 these considerations will be extended to the joint distribution of an arbitrary finite number of random variables.

Discrete Joint Distributions

Suppose that a given experiment involves two random variables X and Y, each of which has a discrete distribution. For example, if a sample of theater patrons is selected, one random variable might be the number X of people in the sample who are over 60 years of age and another random variable might be the number Y of people who live more than 25 miles from the theater. If both X and Y have discrete distributions with a finite number of possible values, then there will be only a finite number of different possible values (x, y) for the pair (X, Y). On the other hand, if either X or Y can take an infinite number of possible values, then there will also be an infinite number of possible values for the pair (X, Y). In either case, it is said that X and Y have a *discrete joint distribution*.

The *joint probability function*, or the *joint p.f.*, of X and Y is defined as the function f such that for any point (x, y) in the xy-plane,

$$f(x, y) = \Pr(X = x \text{ and } Y = y).$$

If (x, y) is not one of the possible values of the pair of random variables (X, Y), then it is clear that $f(x, y) = 0$. Also, if the sequence $(x_1, y_1), (x_2, y_2), \ldots$

includes all the possible values of the pair (X, Y), then

$$\sum_{i=1}^{\infty} f(x_i, y_i) = 1.$$

For any subset A of the xy-plane,

$$\Pr[(X, Y) \in A] = \sum_{(x_i, y_i) \in A} f(x_i, y_i).$$

Example 1: Specifying a Discrete Bivariate Distribution by a Table of Probabilities.
Suppose that the random variable X can take only the values 1, 2, and 3; that the
random variable Y can take only the values 1, 2, 3, and 4; and that the joint p.f.
of X and Y is as specified in the following table:

X \ Y	1	2	3	4
1	0.1	0	0.1	0
2	0.3	0	0.1	0.2
3	0	0.2	0	0

This joint p.f. is sketched in Fig. 3.7. We shall determine the values of
$\Pr(X \geq 2 \text{ and } Y \geq 2)$ and $\Pr(X = 1)$.

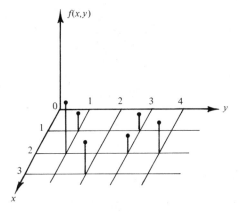

Figure 3.7 The joint p.f. of X and Y in Example 1.

By summing $f(x, y)$ over all values of $x \geqslant 2$ and $y \geqslant 2$, we obtain the value

$$\Pr(X \geqslant 2 \text{ and } Y \geqslant 2) = f(2,2) + f(2,3) + f(2,4) + f(3,2)$$

$$+ f(3,3) + f(3,4)$$

$$= 0.5.$$

By summing the probabilities in the first row of the table, we obtain the value

$$\Pr(X = 1) = \sum_{y=1}^{4} f(1, y) = 0.2. \quad \square$$

Continuous Joint Distributions

It is said that two random variables X and Y have a *continuous joint distribution* if there exists a nonnegative function f defined over the entire xy-plane such that for any subset A of the plane,

$$\Pr[(X, Y) \in A] = \int_A \int f(x, y) \, dx \, dy.$$

The function f is called the *joint probability density function*, or *joint p.d.f.*, of X and Y. Such a joint p.d.f. must satisfy the following two conditions:

$$f(x, y) \geqslant 0 \qquad \text{for } -\infty < x < \infty \text{ and } -\infty < y < \infty,$$

and

$$\int_{-\infty}^{\infty} \int_{-\infty}^{\infty} f(x, y) \, dx \, dy = 1.$$

The probability that the pair (X, Y) will belong to any specified region of the xy-plane can be found by integrating the joint p.d.f. $f(x, y)$ over that region.

If X and Y have a continuous joint distribution, then the following two statements must be true: (i) Any individual point, or any infinite sequence of points, in the xy-plane has probability 0. (ii) Any one-dimensional curve in the xy-plane has probability 0. Thus, the probability that (X, Y) will lie on any specified straight line in the plane is 0, and the probability that (X, Y) will lie on any specified circle in the plane is 0.

An example of a joint p.d.f. is presented in Fig. 3.8. The total volume beneath the surface $z = f(x, y)$ and above the xy-plane must be 1. The probability that the pair (X, Y) will belong to the rectangle A is equal to the volume of the solid

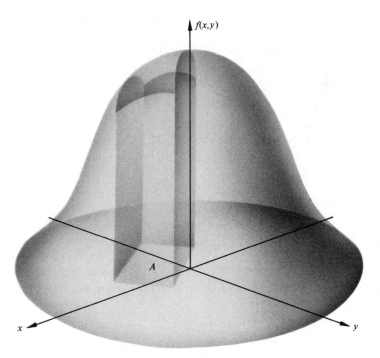

Figure 3.8   An example of a joint p.d.f.

figure with base A shown in Fig. 3.8. The top of this solid figure is formed by the surface $z = f(x, y)$.

Example 2: Calculating Probabilities from a Joint p.d.f. Suppose that the joint p.d.f. of X and Y is specified as follows:

$$f(x, y) = \begin{cases} cx^2y & \text{for } x^2 \leqslant y \leqslant 1, \\ 0 & \text{otherwise.} \end{cases}$$

We shall first determine the value of the constant c and then determine the value of $\Pr(X \geqslant Y)$.

The set S of points (x, y) for which $f(x, y) > 0$ is sketched in Fig. 3.9. Since $f(x, y) = 0$ outside S, it follows that

$$\int_{-\infty}^{\infty} \int_{-\infty}^{\infty} f(x, y) \, dx \, dy = \int_{S} \int f(x, y) \, dx \, dy$$

$$= \int_{-1}^{1} \int_{x^2}^{1} cx^2 y \, dy \, dx = \frac{4}{21} c.$$

Since the value of this integral must be 1, the value of c must be $21/4$.

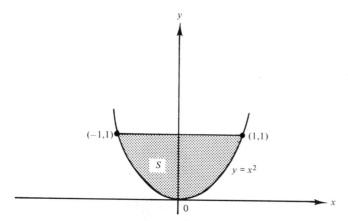

Figure 3.9   The set S where $f(x, y) > 0$ in Example 2.

The subset S_0 of S where $x \geqslant y$ is sketched in Fig. 3.10. Hence,

$$\Pr(X \geqslant Y) = \int_{S_0} \int f(x, y) \, dx \, dy = \int_0^1 \int_{x^2}^x \frac{21}{4} x^2 y \, dy \, dx = \frac{3}{20}. \quad \square$$

Example 3: Determining a Joint p.d.f. by Geometric Methods. Suppose that a point (X, Y) is selected at random from inside the circle $x^2 + y^2 \leqslant 9$. We shall determine the joint p.d.f. of X and Y.

Let S denote the set of points in the circle $x^2 + y^2 \leqslant 9$. The statement that

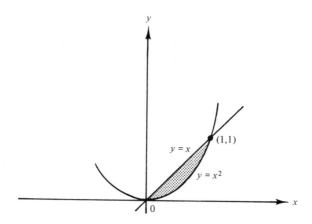

Figure 3.10   The subset S_0 where $x \geqslant y$ and $f(x, y) > 0$ in Example 2.

the point (X, Y) is selected at random from inside the circle is interpreted to mean that the joint p.d.f. of X and Y is constant over S and is 0 outside S. Thus,

$$f(x, y) = \begin{cases} c & \text{for } (x, y) \in S, \\ 0 & \text{otherwise.} \end{cases}$$

We must have

$$\int_S \int f(x, y) \, dx \, dy = c \times (\text{area of } S) = 1.$$

Since the area of the circle S is 9π, the value of the constant c must be $1/(9\pi)$. $\square$

Mixed Bivariate Distributions

So far in this section we have discussed bivariate distributions that were either discrete or continuous. Occasionally, a statistician must consider a mixed bivariate distribution in which the distribution of one of the random variables is discrete and the distribution of the other random variable is continuous. The probability that the pair (X, Y) will belong to a certain region of the xy-plane is then found by summing the values of $f(x, y)$ for one variable and integrating $f(x, y)$ for the other variable.

A more complicated type of mixed distribution can also arise in a practical problem. For example, suppose that X and Y are the times at which two specific components in an electronic system fail. There might be a certain probability p $(0 < p < 1)$ that the two components will fail at the same time and a certain probability $1 - p$ that they will fail at different times. Furthermore, if they fail at the same time, then their common failure time x might be distributed according to a certain p.d.f. $f(x)$; if they fail at different times x and y, then these times might be distributed according to a certain joint p.d.f. $g(x, y)$.

The joint distribution of X and Y in this example is not continuous for the following reason: For any continuous distribution the probability that (X, Y) will lie on the line $x = y$ must be 0, whereas in this example the value of this probability is p.

Bivariate Distribution Functions

The *joint distribution function*, or *joint d.f.*, of two random variables X and Y is defined as the function F such that for all values of x and y $(-\infty < x < \infty$ and $-\infty < y < \infty)$,

$$F(x, y) = \Pr(X \leqslant x \text{ and } Y \leqslant y).$$

If the joint d.f. of two arbitrary random variables X and Y is F, then the probability that the pair (X, Y) will lie in a specified rectangle in the xy-plane can be found from F as follows: For any given numbers $a < b$ and $c < d$,

$$
\begin{aligned}
\Pr(a < X \leqslant b \text{ and } c &< Y \leqslant d) \\
&= \Pr(a < X \leqslant b \text{ and } Y \leqslant d) - \Pr(a < X \leqslant b \text{ and } Y \leqslant c) \\
&= [\Pr(X \leqslant b \text{ and } Y \leqslant d) - \Pr(X \leqslant a \text{ and } Y \leqslant d)] \\
&\quad - [\Pr(X \leqslant b \text{ and } Y \leqslant c) - \Pr(X \leqslant a \text{ and } Y \leqslant c)] \\
&= F(b, d) - F(a, d) - F(b, c) + F(a, c).
\end{aligned} \tag{1}
$$

Hence, the probability of the rectangle A sketched in Fig. 3.11 is given by the combination of values of F just derived. It should be noted that two sides of the rectangle are included in the set A and the other two sides are excluded. Thus, if there are points on the boundary of A that have positive probability, it is important to distinguish between the weak inequalities and the strict inequalities in Eq. (1).

The d.f. F_1 of just the single random variable X can be derived from the joint d.f. F as follows, for $-\infty < x < \infty$:

$$
F_1(x) = \Pr(X \leqslant x) = \lim_{y \to \infty} \Pr(X \leqslant x \text{ and } Y \leqslant y)
$$

$$
= \lim_{y \to \infty} F(x, y).
$$

Similarly, if F_2 denotes the d.f. of Y, then for $-\infty < y < \infty$,

$$
F_2(y) = \lim_{x \to \infty} F(x, y).
$$

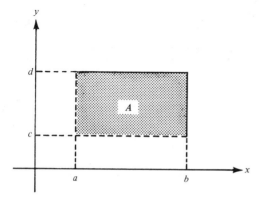

Figure 3.11 The probability of a rectangle.

Other relationships involving the univariate distribution of X, the univariate distribution of Y, and their joint bivariate distribution will be presented in the next section.

Finally, if X and Y have a continuous joint distribution with joint p.d.f. f, then the joint d.f. for any values of x and y is

$$F(x, y) = \int_{-\infty}^{y} \int_{-\infty}^{x} f(r, s) \, dr \, ds.$$

Here, the symbols r and s are used simply as dummy variables of integration. The joint p.d.f. can be derived from the joint d.f. by using the relation

$$f(x, y) = \frac{\partial^2 F(x, y)}{\partial x \, \partial y}$$

at every point (x, y) at which this second-order derivative exists.

Example 4: Determining a Joint p.d.f. from a Joint d.f. Suppose that X and Y are random variables that can only take values in the intervals $0 \leqslant X \leqslant 2$ and $0 \leqslant Y \leqslant 2$. Suppose also that the joint d.f. of X and Y, for $0 \leqslant x \leqslant 2$ and $0 \leqslant y \leqslant 2$, is as follows:

$$F(x, y) = \frac{1}{16} xy(x + y). \tag{2}$$

We shall first determine the d.f. F_1 of just the random variable X and then determine the joint p.d.f. f of X and Y.

The value of $F(x, y)$ at any point (x, y) in the xy-plane that does not represent a pair of possible values of X and Y can be calculated from (2) and the fact that $F(x, y) = \Pr(X \leqslant x \text{ and } Y \leqslant y)$. Thus, if either $x < 0$ or $y < 0$, then $F(x, y) = 0$. If both $x > 2$ and $y > 2$, then $F(x, y) = 1$. If $0 \leqslant x \leqslant 2$ and $y > 2$, then $F(x, y) = F(x, 2)$ and it follows from Eq. (2) that

$$F(x, y) = \frac{1}{8} x(x + 2).$$

Similarly, if $0 \leqslant y \leqslant 2$ and $x > 2$, then

$$F(x, y) = \frac{1}{8} y(y + 2).$$

The function $F(x, y)$ has now been specified for every point in the xy-plane.

By letting $y \to \infty$, we find that the d.f. of just the random variable X is

$$F_1(x) = \begin{cases} 0 & \text{for } x < 0, \\ \dfrac{1}{8}x(x+2) & \text{for } 0 \leqslant x \leqslant 2, \\ 1 & \text{for } x > 2. \end{cases}$$

Furthermore, for $0 < x < 2$ and $0 < y < 2$,

$$\frac{\partial^2 F(x, y)}{\partial x \, \partial y} = \frac{1}{8}(x + y).$$

Also, if $x < 0$, $y < 0$, $x > 2$, or $y > 2$, then

$$\frac{\partial^2 F(x, y)}{\partial x \, \partial y} = 0.$$

Hence, the joint p.d.f. of X and Y is

$$f(x, y) = \begin{cases} \dfrac{1}{8}(x + y) & \text{for } 0 < x < 2 \text{ and } 0 < y < 2, \\ 0 & \text{otherwise.} \quad \square \end{cases}$$

EXERCISES

1. Suppose that in an electric display sign there are three light bulbs in the first row and four light bulbs in the second row. Let X denote the number of bulbs in the first row that will be burned out at a specified time t, and let Y denote the number of bulbs in the second row that will be burned out at the same time t. Suppose that the joint p.f. of X and Y is as specified in the following table:

X \ Y	0	1	2	3	4
0	0.08	0.07	0.06	0.01	0.01
1	0.06	0.10	0.12	0.05	0.02
2	0.05	0.06	0.09	0.04	0.03
3	0.02	0.03	0.03	0.03	0.04

Determine each of the following probabilities:

(a) $\Pr(X = 2)$ (b) $\Pr(Y \geqslant 2)$ (c) $\Pr(X \leqslant 2 \text{ and } Y \leqslant 2)$

(d) $\Pr(X = Y)$ (e) $\Pr(X > Y)$.

2. Suppose that X and Y have a discrete joint distribution for which the joint p.f. is defined as follows:

$$f(x, y) = \begin{cases} c|x + y| & \text{for } x = -2, -1, 0, 1, 2 \text{ and} \\ & \quad y = -2, -1, 0, 1, 2, \\ 0 & \text{otherwise.} \end{cases}$$

Determine (a) the value of the constant c; (b) $\Pr(X = 0 \text{ and } Y = -2)$; (c) $\Pr(X = 1)$; (d) $\Pr(|X - Y| \leqslant 1)$.

3. Suppose that X and Y have a continuous joint distribution for which the joint p.d.f. is defined as follows:

$$f(x, y) = \begin{cases} cy^2 & \text{for } 0 \leqslant x \leqslant 2 \text{ and } 0 \leqslant y \leqslant 1, \\ 0 & \text{otherwise.} \end{cases}$$

Determine (a) the value of the constant c; (b) $\Pr(X + Y > 2)$; (c) $\Pr(Y < 1/2)$; (d) $\Pr(X \leqslant 1)$; (e) $\Pr(X = 3Y)$.

4. Suppose that the joint p.d.f. of two random variables X and Y is as follows:

$$f(x, y) = \begin{cases} c(x^2 + y) & \text{for } 0 \leqslant y \leqslant 1 - x^2, \\ 0 & \text{otherwise.} \end{cases}$$

Determine (a) the value of the constant c; (b) $\Pr(0 \leqslant X \leqslant 1/2)$; (c) $\Pr(Y \leqslant X + 1)$; (d) $\Pr(Y = X^2)$.

5. Suppose that a point (X, Y) is chosen at random from the region S in the xy-plane containing all points (x, y) such that $x \geqslant 0$, $y \geqslant 0$, and $4y + x \leqslant 4$.

   (a) Determine the joint p.d.f. of X and Y.

   (b) Suppose that S_0 is a subset of the region S having area α, and determine $\Pr[(X, Y) \in S_0]$.

6. Suppose that a point (X, Y) is to be chosen from the square S in the xy-plane containing all points (x, y) such that $0 \leqslant x \leqslant 1$ and $0 \leqslant y \leqslant 1$. Suppose that the probability that the chosen point will be the corner $(0, 0)$ is 0.1; the probability that it will be the corner $(1, 0)$ is 0.2; the probability that it will be the corner $(0, 1)$ is 0.4; and the probability that it will be the corner $(1, 1)$ is 0.1. Suppose also that if the chosen point is not one of the four corners of the square, then it will be an interior point of the square and will be chosen

according to a constant p.d.f. over the interior of the square. Determine (a) $\Pr(X \leqslant 1/4)$ and (b) $\Pr(X + Y \leqslant 1)$.

7. Suppose that X and Y are random variables such that (X, Y) must belong to the rectangle in the xy-plane containing all points (x, y) for which $0 \leqslant x \leqslant 3$ and $0 \leqslant y \leqslant 4$. Suppose also that the joint d.f. of X and Y at any point (x, y) in this rectangle is specified as follows:

$$F(x, y) = \frac{1}{156} xy(x^2 + y).$$

Determine (a) $\Pr(1 \leqslant X \leqslant 2$ and $1 \leqslant Y \leqslant 2)$; (b) $\Pr(2 \leqslant X \leqslant 4$ and $2 \leqslant Y \leqslant 4)$; (c) the d.f. of Y; (d) the joint p.d.f. of X and Y; (e) $\Pr(Y \leqslant X)$.

3.5. MARGINAL DISTRIBUTIONS

Deriving a Marginal p.f. or a Marginal p.d.f.

We have seen in Sec. 3.4 that if the joint d.f. F of two random variables X and Y is known, then the d.f. F_1 of the random variable X can be derived from F. In this context, where the distribution of X is derived from the joint distribution of X and Y, F_1 is called the *marginal d.f.* of X. Similarly, if the joint p.f. or joint p.d.f. f of X and Y is known, then the *marginal p.f.* or *marginal p.d.f.* of each random variable can be derived from f.

For example, if X and Y have a discrete joint distribution for which the joint p.f. is f, then the marginal p.f. f_1 of X can be found as follows:

$$f_1(x) = \Pr(X = x) = \sum_y \Pr(X = x \text{ and } Y = y) = \sum_y f(x, y). \tag{1}$$

In other words, for any given value x of X, the value of $f_1(x)$ is found by summing $f(x, y)$ over all possible values y of Y.

Similarly, the marginal p.f. f_2 of Y can be found from the relation

$$f_2(y) = \sum_x f(x, y). \tag{2}$$

Example 1: Deriving a Marginal p.f. from a Table of Probabilities. Suppose that X and Y have the joint p.f. specified by the table in Example 1 of Sec 3.4. The marginal p.f. f_1 of X can be determined by summing the values in each row of this table. In this way it is found that $f_1(1) = 0.2$, $f_1(2) = 0.6$, $f_1(3) = 0.2$, and $f_1(x) = 0$ for all other values of x. □

If X and Y have a continuous joint distribution for which the joint p.d.f. is f, then the marginal p.d.f. f_1 of X is again determined in the manner shown in Eq. (1), but the sum over all possible values of Y is now replaced by the integral over

MARGINAL OF X
INTEGRATES OVER Y.

all possible values of Y. Thus,

$$f_1(x) = \int_{-\infty}^{\infty} f(x, y)\, dy \qquad \text{for } -\infty < x < \infty.$$

Similarly, the marginal p.d.f. f_2 of Y is determined as in Eq. (2), but the sum is again replaced by an integral. Thus,

$$f_2(y) = \int_{-\infty}^{\infty} f(x, y)\, dx \qquad \text{for } -\infty < y < \infty.$$

Example 2: Deriving a Marginal p.d.f. Suppose that the joint p.d.f. of X and Y is as specified in Example 2 of Sec. 3.4. We shall determine first the marginal p.d.f. f_1 of X and then the marginal p.d.f. f_2 of Y.

It can be seen from Fig. 3.9 that X cannot take any value outside the interval $-1 \leqslant X \leqslant 1$. Therefore, $f_1(x) = 0$ for $x < -1$ or $x > 1$. Furthermore, for $-1 \leqslant x \leqslant 1$, it is seen from Fig. 3.9 that $f(x, y) = 0$ unless $x^2 \leqslant y \leqslant 1$. Therefore, for $-1 \leqslant x \leqslant 1$,

$$f_1(x) = \int_{-\infty}^{\infty} f(x, y)\, dy = \int_{x^2}^{1} \left(\frac{21}{4}\right) x^2 y\, dy = \left(\frac{21}{8}\right) x^2 (1 - x^4).$$

This marginal p.d.f. of X is sketched in Fig. 3.12.

Next, it can be seen from Fig. 3.9 that Y cannot take any value outside the interval $0 \leqslant Y \leqslant 1$. Therefore, $f_2(y) = 0$ for $y < 0$ or $y > 1$. Furthermore, for $0 \leqslant y \leqslant 1$, it is seen from Fig. 3.9 that $f(x, y) = 0$ unless $-\sqrt{y} \leqslant x \leqslant \sqrt{y}$. Therefore, for $0 \leqslant y \leqslant 1$,

$$f_2(y) = \int_{-\infty}^{\infty} f(x, y)\, dx = \int_{-\sqrt{y}}^{\sqrt{y}} \left(\frac{21}{4}\right) x^2 y\, dx = \left(\frac{7}{2}\right) y^{5/2}.$$

This marginal p.d.f. of Y is sketched in Fig. 3.13. $\square$

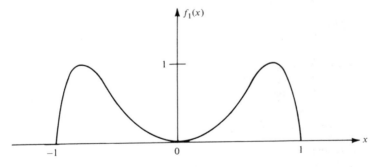

Figure 3.12   The marginal p.d.f. of X in Example 2.

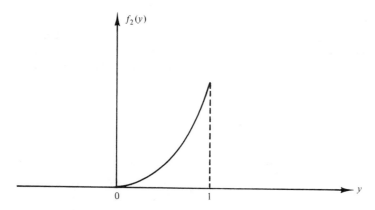

Figure 3.13 The marginal p.d.f. of Y in Example 2.

Although the marginal distributions of X and Y can be derived from their joint distribution, it is not possible to reconstruct the joint distribution of X and Y from their marginal distributions without additional information. For instance, the marginal p.d.f.'s sketched in Figs. 3.12 and 3.13 reveal no information about the relationship between X and Y. In fact, by definition, the marginal distribution of X specifies probabilities for X without regard for the values of any other random variables. This property of a marginal p.d.f. can be further illustrated by another example.

Suppose that a penny and a nickel are each tossed n times, and consider the following two definitions of X and Y: (i) X is the number of heads obtained with the penny and Y is the number of heads obtained with the nickel. (ii) Both X and Y are the number of heads obtained with the penny, so the random variables X and Y are actually identical.

In case (i), the marginal distribution of X and the marginal distribution of Y will be identical binomial distributions. The same pair of marginal distributions of X and Y will also be obtained in case (ii). However, the joint distribution of X and Y will not be the same in the two cases. In case (i) the values of X and Y will be completely unrelated, whereas in case (ii) the values of X and Y must be identical.

Independent Random Variables

It is said that two random variables X and Y are *independent* if, for any two sets A and B of real numbers,

$$\Pr(X \in A \text{ and } Y \in B) = \Pr(X \in A)\Pr(Y \in B). \tag{3}$$

In other words, let A be any event the occurrence or nonoccurrence of which depends only on the value of X, and let B be any event the occurrence or nonoccurrence of which depends only on the value of Y. Then X and Y are independent random variables if and only if A and B are independent events for all such events A and B.

If X and Y are independent, then for any real numbers x and y it must be true that

$$\Pr(X \leqslant x \text{ and } Y \leqslant y) = \Pr(X \leqslant x)\Pr(Y \leqslant y). \tag{4}$$

Moreover, since any probabilities for X and Y of the type appearing in Eq. (3) can be derived from probabilities of the type appearing in Eq. (4), it can be shown that if Eq. (4) is satisfied for all values of x and y, then X and Y must be independent. The proof of this statement is beyond the scope of this book and is omitted. In other words, two random variables X and Y are independent if and only if Eq. (4) is satisfied for all values of x and y.

If the joint d.f. of X and Y is denoted by F, the marginal d.f. of X by F_1, and the marginal d.f. of Y by F_2, then the preceding result can be restated as follows: *Two random variables X and Y are independent if and only if, for all real numbers x and y,*

$$F(x, y) = F_1(x)F_2(y).$$

Suppose now that X and Y have either a discrete joint distribution or a continuous joint distribution, for which the joint p.f. or the joint p.d.f. is f. As before, we shall let the marginal p.f.'s or the marginal p.d.f.'s of X and Y be denoted by f_1 and f_2. It then follows from the relation just given that X and Y are independent if and only if the following factorization is satisfied for all real numbers x and y:

$$f(x, y) = f_1(x)f_2(y). \tag{5}$$

As stated in Sec. 3.2, in a continuous distribution the values of a p.d.f. can be changed arbitrarily at any set of points that has probability 0. Therefore, for such a distribution it would be more precise to state that the random variables X and Y are independent if and only if it is possible to choose versions of the p.d.f.'s f, f_1, and f_2 such that Eq. (5) is satisfied for $-\infty < x < \infty$ and $-\infty < y < \infty$.

Example 3: Calculating a Probability Involving Independent Random Variables. Suppose that two independent measurements X and Y are made of the rainfall during a given period of time at a certain location and that the p.d.f. g of each measurement is as follows:

$$g(x) = \begin{cases} 2x & \text{for } 0 \leqslant x \leqslant 1, \\ 0 & \text{otherwise.} \end{cases}$$

We shall determine the value of $\Pr(X + Y \leqslant 1)$.

Since X and Y are independent and each has the p.d.f. g, it follows from Eq. (5) that for any values of x and y the joint p.d.f. $f(x, y)$ of X and Y will be specified by the relation $f(x, y) = g(x)g(y)$. Hence,

$$f(x, y) = \begin{cases} 4xy & \text{for } 0 \leqslant x \leqslant 1 \text{ and } 0 \leqslant y \leqslant 1, \\ 0 & \text{otherwise.} \end{cases}$$

The set S in the xy-plane where $f(x, y) > 0$ and the subset S_0 where $x + y \leqslant 1$ are sketched in Fig. 3.14. Thus,

$$\Pr(X + Y \leqslant 1) = \int_{S_0}\int f(x, y)\, dx\, dy = \int_0^1 \int_0^{1-x} 4xy\, dy\, dx = \frac{1}{6}. \quad \square$$

Suppose next that X and Y have discrete distributions; that X can take the values $1, 2, \ldots, r$; that Y can take the values $1, 2, \ldots, s$; and that

$$\Pr(X = i \text{ and } Y = j) = p_{ij} \qquad \text{for } i = 1, \ldots, r \text{ and } j = 1, \ldots, s.$$

Then, for $i = 1, \ldots, r$, let

$$\Pr(X = i) = \sum_{j=1}^{s} p_{ij} = p_{i+}.$$

Also, for $j = 1, \ldots, s$, let

$$\Pr(Y = j) = \sum_{i=1}^{r} p_{ij} = p_{+j}.$$

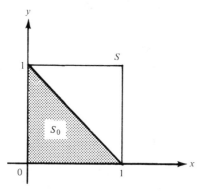

Figure 3.14 The subset S_0 where $x + y \leqslant 1$ in Example 3.

Therefore, X and Y will be independent if and only if the following relationship is satisfied for all values of i and j:

$$p_{ij} = p_{i+}p_{+j}. \tag{6}$$

Example 4: Determining Whether Random Variables Are Independent from a Table of Probabilities. Suppose that the joint p.f. of X and Y is specified by the table given in Example 1 of Sec. 3.4. We shall determine whether or not X and Y are independent.

In Eq. (6), p_{ij} is the probability in the ith row and the jth column of the table, p_{i+} is the sum of the probabilities in the ith row, and p_{+j} is the sum of the probabilities in the jth column. It is found from the table that $p_{11} = 0.1$, $p_{1+} = 0.2$, and $p_{+1} = 0.4$. Hence, $p_{11} \neq p_{1+}p_{+1}$. It follows immediately that X and Y cannot be independent.

Suppose next that the joint p.f. of X and Y is specified by the following table.

X \ Y	1	2	3	4	Total
1	0.06	0.02	0.04	0.08	0.20
2	0.15	0.05	0.10	0.20	0.50
3	0.09	0.03	0.06	0.12	0.30
Total	0.30	0.10	0.20	0.40	1.00

Since it can be found from this table that Eq. (6) is satisfied for all values of i and j, it follows that X and Y are independent. □

It should be noted from Example 4 that X and Y will be independent if and only if the rows of the table specifying their joint p.f. are proportional to one another, or equivalently, if and only if the columns of the table are proportional to one another.

Now suppose that X and Y are random variables that have a continuous joint distribution for which the joint p.d.f. is f. Then X and Y will be independent if and only if f can be represented in the following form for $-\infty < x < \infty$ and $-\infty < y < \infty$:

$$f(x, y) = g_1(x)g_2(y), \tag{7}$$

where g_1 is a nonnegative function of x alone and g_2 is a nonnegative function of y alone. In other words, it is necessary and sufficient that, for all values of x and y, f can be factored into the product of an arbitrary nonnegative function of x and an arbitrary nonnegative function of y. However, it should be emphasized that, just as in Eq. (5), the factorization in Eq. (7) must be satisfied for all values of x and $y (-\infty < x < \infty$ and $-\infty < y < \infty)$.

In one important special case, in which $f(x, y) = 0$ for all values of x and y outside a rectangle having sides parallel to the x-axis and the y-axis, it is not actually necessary to check Eq. (7) for all values of x and y. In order to verify that X and Y are independent in this case, it is sufficient to verify that Eq. (7) is satisfied for all values of x and y inside the rectangle. In particular, let a, b, c, and d be given values such that $-\infty \leqslant a < b \leqslant \infty$ and $-\infty \leqslant c < d \leqslant \infty$, and let S be the following rectangle in the xy-plane:

$$S = \{(x, y): a \leqslant x \leqslant b \text{ and } c \leqslant y \leqslant d\}. \tag{8}$$

It should be noted that any of the endpoints a, b, c, and d can be infinite. Suppose that $f(x, y) = 0$ for every point (x, y) outside S. Then X and Y will be independent if and only if f can be factored as in Eq. (7) at all points in S.

Example 5: Verifying the Factorization of a Joint p.d.f. Suppose that the joint p.d.f. f of X and Y is as follows:

$$f(x, y) = \begin{cases} ke^{-(x+2y)} & \text{for } x \geqslant 0 \text{ and } y \geqslant 0, \\ 0 & \text{otherwise.} \end{cases}$$

We shall first determine whether X and Y are independent and then determine their marginal p.d.f.'s.

In this example $f(x, y) = 0$ outside a rectangle S which has the form specified in Eq. (8) and for which $a = 0$, $b = \infty$, $c = 0$, and $d = \infty$. Furthermore, at any point inside S, $f(x, y)$ can be factored as in Eq. (7) by letting $g_1(x) = ke^{-x}$ and $g_2(y) = e^{-2y}$. Therefore, X and Y are independent.

It follows that in this case, except for constant factors, $g_1(x)$ and $g_2(y)$ must be the marginal p.d.f.'s of X and Y for $X \geqslant 0$ and $Y \geqslant 0$. By choosing constants which make $g_1(x)$ and $g_2(y)$ integrate to unity, we can conclude that the marginal p.d.f.'s f_1 and f_2 of X and Y must be as follows:

$$f_1(x) = \begin{cases} e^{-x} & \text{for } x \geqslant 0, \\ 0 & \text{otherwise,} \end{cases}$$

and

$$f_2(y) = \begin{cases} 2e^{-2y} & \text{for } y \geqslant 0, \\ 0 & \text{otherwise.} \quad \square \end{cases}$$

Example 6: Dependent Random Variables. Suppose that the joint p.d.f. of X and Y has the following form:

$$f(x, y) = \begin{cases} kx^2y^2 & \text{for } x^2 + y^2 \leqslant 1, \\ 0 & \text{otherwise.} \end{cases}$$

We shall show that X and Y are not independent.

It is evident that at any point inside the circle $x^2 + y^2 \leqslant 1$, $f(x, y)$ can be factored as in Eq. (7). However, this same factorization cannot also be satisfied at every point outside this circle. The important feature of this example is that the values of X and Y are constrained to lie inside a circle. The joint p.d.f. of X and Y is positive inside the circle and is zero outside the circle. Under these conditions X and Y cannot be independent, because for any given value y of Y, the possible values of X will depend on y. For example, if $Y = 0$, then X can have any value such that $X^2 \leqslant 1$; if $Y = 1/2$, then X must have a value such that $X^2 \leqslant 3/4$. □

EXERCISES

1. Suppose that X and Y have a discrete joint distribution for which the joint p.f. is defined as follows:

$$f(x, y) = \begin{cases} \dfrac{1}{30}(x + y) & \text{for } x = 0, 1, 2 \text{ and } y = 0, 1, 2, 3, \\ 0 & \text{otherwise.} \end{cases}$$

   (a) Determine the marginal p.f.'s of X and Y.

   (b) Are X and Y independent?

2. Suppose that X and Y have a continuous joint distribution for which the joint p.d.f. is defined as follows:

$$f(x, y) = \begin{cases} \left(\dfrac{3}{2}\right)y^2 & \text{for } 0 \leqslant x \leqslant 2 \text{ and } 0 \leqslant y \leqslant 1, \\ 0 & \text{otherwise.} \end{cases}$$

   (a) Determine the marginal p.d.f.'s of X and Y.

   (b) Are X and Y independent?

   (c) Are the event $\{X < 1\}$ and the event $\{Y \geqslant 1/2\}$ independent?

3. Suppose that the joint p.d.f. of X and Y is as follows:

$$f(x, y) = \begin{cases} \left(\dfrac{15}{4}\right)x^2 & \text{for } 0 \leqslant y \leqslant 1 - x^2, \\ 0 & \text{otherwise.} \end{cases}$$

   (a) Determine the marginal p.d.f.'s of X and Y.

   (b) Are X and Y independent?

4. A certain drugstore has three public telephone booths. For $i = 0, 1, 2, 3$, let p_i denote the probability that exactly i telephone booths will be occupied on any Monday evening at 8:00 P.M.; and suppose that $p_0 = 0.1$, $p_1 = 0.2$, $p_2 = 0.4$, and $p_3 = 0.3$. Let X and Y denote the number of booths that will be occupied at 8:00 P.M. on two independent Monday evenings. Determine: (a) the joint p.f. of X and Y; (b) $\Pr(X = Y)$; (c) $\Pr(X > Y)$.

5. Suppose that in a certain drug the concentration of a particular chemical is a random variable with a continuous distribution for which the p.d.f. g is as follows:

$$g(x) = \begin{cases} (3/8)x^2 & \text{for } 0 \leqslant x \leqslant 2, \\ 0 & \text{otherwise.} \end{cases}$$

Suppose that the concentrations X and Y of the chemical in two separate batches of the drug are independent random variables for each of which the p.d.f. is g. Determine (a) the joint p.d.f. of X and Y; (b) $\Pr(X = Y)$; (c) $\Pr(X > Y)$; (d) $\Pr(X + Y \leqslant 1)$.

6. Suppose that the joint p.d.f. of X and Y is as follows:

$$f(x, y) = \begin{cases} 2xe^{-y} & \text{for } 0 \leqslant x \leqslant 1 \text{ and } 0 < y < \infty, \\ 0 & \text{otherwise.} \end{cases}$$

Are X and Y independent?

7. Suppose that the joint p.d.f. of X and Y is as follows:

$$f(x, y) = \begin{cases} 24xy & \text{for } x \geqslant 0, \ y \geqslant 0, \text{ and } x + y \leqslant 1, \\ 0 & \text{otherwise.} \end{cases}$$

Are X and Y independent?

8. Suppose that a point (X, Y) is chosen at random from the rectangle S defined as follows:

$$S = \{(x, y): 0 \leqslant x \leqslant 2 \text{ and } 1 \leqslant y \leqslant 4\}.$$

(a) Determine the joint p.d.f. of X and Y, the marginal p.d.f. of X, and the marginal p.d.f. of Y.

(b) Are X and Y independent?

9. Suppose that a point (X, Y) is chosen at random from the circle S defined as follows:

$$S = \{(x, y): x^2 + y^2 \leqslant 1\}.$$

(a) Determine the joint p.d.f. of X and Y, the marginal p.d.f. of X, and the marginal p.d.f. of Y.

(b) Are X and Y independent?

10. Suppose that two persons make an appointment to meet between 5 P.M. and 6 P.M. at a certain location, and they agree that neither person will wait more than 10 minutes for the other person. If they arrive independently at random times between 5 P.M. and 6 P.M., what is the probability that they will meet?

3.6. CONDITIONAL DISTRIBUTIONS

Discrete Conditional Distributions

Suppose that X and Y are two random variables having a discrete joint distribution for which the joint p.f. is f. As before, we shall let f_1 and f_2 denote the marginal p.f.'s of X and Y, respectively. After the value y of the random variable Y has been observed, the probability that the random variable X will take any particular value x is specified by the following conditional probability:

$$\Pr(X = x \mid Y = y) = \frac{\Pr(X = x \text{ and } Y = y)}{\Pr(Y = y)}$$

$$= \frac{f(x, y)}{f_2(y)}.$$

(1)

In other words, if it is known that $Y = y$, then the distribution of X will be a discrete distribution for which the probabilities are specified in Eq. (1). This distribution is called the *conditional distribution of X given that* $Y = y$. It follows from Eq. (1) that for any given value y such that $f_2(y) > 0$, this conditional distribution of X can be represented by a p.f. $g_1(x|y)$ which is defined as follows:

$$g_1(x|y) = \frac{f(x, y)}{f_2(y)}.$$

(2)

The function g_1 is called the *conditional p.f. of X given that* $Y = y$. For each fixed value of y, the function $g_1(x|y)$ will be a p.f. over all possible values of X because $g_1(x|y) \geqslant 0$ and

$$\sum_x g_1(x|y) = \frac{1}{f_2(y)} \sum_x f(x, y) = \frac{1}{f_2(y)} f_2(y) = 1.$$

Similarly, if x is any given value of X such that $f_1(x) = \Pr(X = x) > 0$ and if $g_2(y|x)$ is the *conditional p.f. of Y given that X = x*, then

$$g_2(y|x) = \frac{f(x, y)}{f_1(x)}. \tag{3}$$

For any fixed value of x, the function $g_2(y|x)$ will be a p.f. over all possible values of Y.

Example 1: Calculating a Conditional p.f. from a Joint p.f. Suppose that the joint p.f. of X and Y is as specified in the table given in Example 1 of Sec. 3.4. We shall determine the conditional p.f. of Y given that $X = 2$.

From the table in Sec. 3.4, $f_1(2) = \Pr(X = 2) = 0.6$. Therefore, the conditional probability $g_2(y|2)$ that Y will take any particular value y is

$$g_2(y|2) = \frac{f(2, y)}{0.6}.$$

It should be noted that for all possible values of y, the conditional probabilities $g_2(y|2)$ must be proportional to the joint probabilities $f(2, y)$. In this example, each value of $f(2, y)$ is simply divided by the constant $f_1(2) = 0.6$ in order that the sum of the results will be equal to 1. Thus,

$$g_2(1|2) = 1/2, \quad g_2(2|2) = 0, \quad g_2(3|2) = 1/6, \quad g_2(4|2) = 1/3. \qquad \square$$

Continuous Conditional Distributions

Suppose that X and Y have a continuous joint distribution for which the joint p.d.f. is f and the marginal p.d.f.'s are f_1 and f_2. Suppose also that the value $Y = y$ has been observed and it is desired to specify probabilities for various possible sets of values of X. It should be noted that in this case $\Pr(Y = y) = 0$ for each value of y, and conditional probabilities of the form $\Pr(A|B)$ have not been defined when $\Pr(B) = 0$. Therefore, in order to make it possible to derive conditional probabilities when X and Y have a continuous joint distribution, the concept of conditional probability will be extended by considering the definition of the conditional p.f. of X given in Eq. (2) and the analogy between a p.f. and a p.d.f.

Let y be any given value such that $f_2(y) > 0$. Then the *conditional p.d.f.* g_1 *of X given that Y = y* may be defined as follows:

$$g_1(x|y) = \frac{f(x, y)}{f_2(y)} \qquad \text{for } -\infty < x < \infty. \tag{4}$$

For each fixed value of y, the function g_1 will be a p.d.f. for X over the real line, since $g_1(x|y) \geq 0$ and

$$\int_{-\infty}^{\infty} g_1(x|y) \, dx = 1.$$

It should be noted that Eq. (2) and Eq. (4) are identical. However, Eq. (2) was *derived* as the conditional probability that $X = x$ given that $Y = y$, whereas Eq. (4) was *defined* to be the value of the conditional p.d.f. of X given that $Y = y$.

The definition given in Eq. (4) has an interpretation that can be understood by considering Fig. 3.15. The joint p.d.f. f defines a surface over the xy-plane for which the height $f(x, y)$ at any point (x, y) represents the relative likelihood of that point. For instance, if it is known that $Y = y_0$, then the point (x, y) must lie on the line $y = y_0$ in the xy-plane, and the relative likelihood of any point (x, y_0) on this line is $f(x, y_0)$. Hence, the conditional p.d.f. $g_1(x|y_0)$ of X should be proportional to $f(x, y_0)$. In other words, $g_1(x|y_0)$ is essentially the same as $f(x, y_0)$, but it includes a constant factor $1/[f_2(y_0)]$ which is required to make the conditional p.d.f. integrate to unity over all values of X.

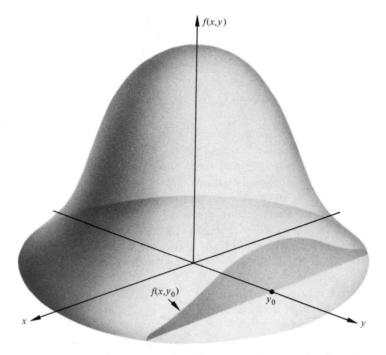

Figure 3.15   The conditional p.d.f. $g_1(x|y_0)$ is proportional to $f(x, y_0)$.

Similarly, for any value of x such that $f_1(x) > 0$, the *conditional p.d.f. of Y given that X = x* is defined as follows:

$$g_2(y|x) = \frac{f(x, y)}{f_1(x)} \qquad \text{for} -\infty < y < \infty. \tag{5}$$

This equation is identical to Eq. (3) which was derived for discrete distributions.

Example 2: Calculating a Conditional p.d.f. from a Joint p.d.f. Suppose that the joint p.d.f. of X and Y is as specified in Example 2 of Sec. 3.4. We shall first determine the conditional p.d.f. of Y given that $X = x$ and then determine some probabilities for Y given the specific value $X = 1/2$.

The set S for which $f(x, y) > 0$ was sketched in Fig. 3.9. Furthermore, the marginal p.d.f. f_1 was derived in Example 2 of Sec. 3.5 and sketched in Fig. 3.12. It can be seen from Fig. 3.12 that $f_1(x) > 0$ for $-1 < x < 1$ but not for $x = 0$. Therefore, for any given value of x such that $-1 < x < 0$ or $0 < x < 1$, the conditional p.d.f. $g_2(y|x)$ of Y will be as follows:

$$g_2(y|x) = \begin{cases} \dfrac{2y}{1 - x^4} & \text{for } x^2 \leqslant y \leqslant 1, \\ 0 & \text{otherwise.} \end{cases}$$

In particular, if it is known that $X = 1/2$, then $\Pr\left(Y \geqslant \dfrac{1}{4} \Big| X = \dfrac{1}{2}\right) = 1$ and

$$\Pr\left(Y \geqslant \frac{3}{4} \Big| X = \frac{1}{2}\right) = \int_{3/4}^{1} g_2\left(y \Big| \frac{1}{2}\right) dy = \frac{7}{15}. \quad \square$$

Construction of the Joint Distribution

Basic Relations. It follows from Eq. (4) that for any value of y such that $f_2(y) > 0$ and for any value of x,

$$f(x, y) = g_1(x|y)f_2(y). \tag{6}$$

Furthermore, if $f_2(y_0) = 0$ for some value y_0, then it can be assumed without loss of generality that $f(x, y_0) = 0$ for all values of x. In this case, both sides of Eq. (6) will be 0 and the fact that $g_1(x|y_0)$ is not defined becomes irrelevant. Hence, Eq. (6) will be satisfied for *all* values of x and y.

Similarly, it follows from Eq. (5) that the joint p.d.f. $f(x, y)$ can also be represented as follows for all values of x and y:

$$f(x, y) = f_1(x)g_2(y|x). \tag{7}$$

Example 3: Choosing Points from Uniform Distributions. Suppose that a point X is chosen from a uniform distribution on the interval $(0, 1)$; and that after the value $X = x$ has been observed $(0 < x < 1)$, a point Y is then chosen from a uniform distribution on the interval $(x, 1)$. We shall derive the marginal p.d.f. of Y.

Since X has a uniform distribution, the marginal p.d.f. of X is as follows:

$$f_1(x) = \begin{cases} 1 & \text{for } 0 < x < 1, \\ 0 & \text{otherwise.} \end{cases}$$

Similarly, for any given value $X = x$ $(0 < x < 1)$, the conditional distribution of Y is a uniform distribution on the interval $(x, 1)$. Since the length of this interval is $1 - x$, the conditional p.d.f. of Y given that $X = x$ will be

$$g_2(y|x) = \begin{cases} \dfrac{1}{1 - x} & \text{for } x < y < 1, \\ 0 & \text{otherwise.} \end{cases}$$

It follows from Eq. (7) that the joint p.d.f. of X and Y will be

$$f(x, y) = \begin{cases} \dfrac{1}{1 - x} & \text{for } 0 < x < y < 1, \\ 0 & \text{otherwise.} \end{cases} \tag{8}$$

Thus, for $0 < y < 1$, the value of the marginal p.d.f. $f_2(y)$ of Y will be

$$f_2(y) = \int_{-\infty}^{\infty} f(x, y) \, dx = \int_{0}^{y} \frac{1}{1 - x} \, dx = -\log(1 - y).$$

Furthermore, since Y cannot be outside the interval $0 < y < 1$, then $f_2(y) = 0$ for $y \leqslant 0$ or $y \geqslant 1$. This marginal p.d.f. f_2 is sketched in Fig. 3.16. It is interesting to note that in this example the function f_2 is unbounded. $\square$

Independent Random Variables. Suppose that X and Y are two random variables having a continuous joint distribution. It is known from Sec. 3.5 that X and Y are independent if and only if their joint p.d.f. $f(x, y)$ can be factored in the following form for $-\infty < x < \infty$ and $-\infty < y < \infty$:

$$f(x, y) = f_1(x)f_2(y).$$

It follows from Eq. (6) that X and Y are independent if and only if for every value of y such that $f_2(y) > 0$ and every value of x,

$$g_1(x|y) = f_1(x). \tag{9}$$

In other words, X and Y are independent if and only if the conditional p.d.f. of X for each given value of Y is the same as the marginal p.d.f. of X.

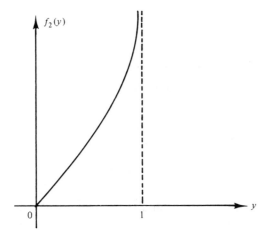

Figure 3.16 The marginal p.d.f. of Y in Example 3.

Similarly, it follows from Eq. (7) that X and Y are independent if and only if for every value of x such that $f_1(x) > 0$ and every value of y,

$$g_2(y \mid x) = f_2(y). \tag{10}$$

If the joint distribution of two random variables X and Y is discrete, then Eq. (9) and Eq. (10) are still necessary and sufficient conditions for X and Y to be independent. In this case, however, the functions f_1, f_2, g_1, and g_2 should be interpreted as marginal and conditional p.f.'s rather than p.d.f.'s.

EXERCISES

1. Each student in a certain high school was classified according to his year in school (freshman, sophomore, junior, or senior) and according to the number of times that he had visited a certain museum (never, once, or more than once). The proportions of students in the various classifications are given in the following table:

	Never	Once	More than once
Freshmen	0.08	0.10	0.04
Sophomores	0.04	0.10	0.04
Juniors	0.04	0.20	0.09
Seniors	0.02	0.15	0.10

(a) If a student selected at random from the high school is a junior, what is the probability that he has never visited the museum?

(b) If a student at random from the high school has visited the museum three times, what is the probability that he is a senior?

2. Suppose that a point (X, Y) is chosen at random from the circle S defined as follows:

$$S = \left\{ (x, y): (x - 1)^2 + (y + 2)^2 \leqslant 9 \right\}.$$

Determine (a) the conditional p.d.f. of Y for any given value of X, and (b) $\Pr(Y > 0 | X = 2)$.

3. Suppose that the joint p.d.f. of two random variables X and Y is as follows:

$$f(x, y) = \begin{cases} c(x + y^2) & \text{for } 0 \leqslant x \leqslant 1 \text{ and } 0 \leqslant y \leqslant 1, \\ 0 & \text{otherwise.} \end{cases}$$

Determine (a) the conditional p.d.f. of X for any given value of Y, and (b) $\Pr\left(X < \dfrac{1}{2} \Big| Y = \dfrac{1}{2} \right)$.

4. Suppose that the joint p.d.f. of two points X and Y chosen by the process described in Example 3 is as given by Eq. (8). Determine (a) the conditional p.d.f. of X for any given value of Y, and (b) $\Pr\left(X > \dfrac{1}{2} \Big| Y = \dfrac{3}{4} \right)$.

5. Suppose that the joint p.d.f. of two random variables X and Y is as follows:

$$f(x, y) = \begin{cases} c \sin x & \text{for } 0 \leqslant x \leqslant \pi/2 \text{ and } 0 \leqslant y \leqslant 3, \\ 0 & \text{otherwise.} \end{cases}$$

Determine (a) the conditional p.d.f. of Y for any given value of X, and (b) $\Pr(1 < Y < 2 | X = 0.73)$.

6. Suppose that the joint p.d.f. of two random variables X and Y is as follows:

$$f(x, y) = \begin{cases} \dfrac{3}{16}(4 - 2x - y) & \text{for } x > 0, \ y > 0, \text{ and } 2x + y < 4, \\ 0 & \text{otherwise.} \end{cases}$$

Determine (a) the conditional p.d.f. of Y for any given value of X, and (b) $\Pr(Y \geqslant 2 | X = 0.5)$.

7. Suppose that a person's score X on a mathematics aptitude test is a number between 0 and 1, and that his score Y on a music aptitude test is also a number between 0 and 1. Suppose further that in the population of all college

students in the United States, the scores X and Y are distributed according to the following joint p.d.f.:

$$f(x, y) = \begin{cases} \dfrac{2}{5}(2x + 3y) & \text{for } 0 \leqslant x \leqslant 1 \text{ and } 0 \leqslant y \leqslant 1, \\ 0 & \text{otherwise.} \end{cases}$$

(a) What proportion of college students obtain a score greater than 0.8 on the mathematics test?

(b) If a student's score on the music test is 0.3, what is the probability that his score on the mathematics test will be greater than 0.8?

(c) If a student's score on the mathematics test is 0.3, what is the probability that his score on the music test will be greater than 0.8?

8. Suppose that either of two instruments might be used for making a certain measurement. Instrument 1 yields a measurement whose p.d.f. h_1 is

$$h_1(x) = \begin{cases} 2x & \text{for } 0 < x < 1, \\ 0 & \text{otherwise.} \end{cases}$$

Instrument 2 yields a measurement whose p.d.f. h_2 is

$$h_2(x) = \begin{cases} 3x^2 & \text{for } 0 < x < 1, \\ 0 & \text{otherwise.} \end{cases}$$

Suppose that one of the two instruments is chosen at random and a measurement X is made with it.

(a) Determine the marginal p.d.f. of X.

(b) If the value of the measurement is $X = 1/4$, what is the probability that instrument 1 was used?

9. In a large collection of coins, the probability X that a head will be obtained when a coin is tossed varies from one coin to another and the distribution of X in the collection is specified by the following p.d.f.:

$$f_1(x) = \begin{cases} 6x(1 - x) & \text{for } 0 < x < 1, \\ 0 & \text{otherwise.} \end{cases}$$

Suppose that a coin is selected at random from the collection and tossed once, and that a head is obtained. Determine the conditional p.d.f. of X for this coin.

3.7. MULTIVARIATE DISTRIBUTIONS

We shall now extend the results that were developed in Secs. 3.4, 3.5, and 3.6 for two random variables X and Y to an arbitrary finite number n of random variables $X_1, \ldots, X_n$. In general, the joint distribution of more than two random variables is called a *multivariate distribution*.

Joint Distributions

The Joint Distribution Function.   The *joint d.f.* of n random variables $X_1, \ldots, X_n$ is defined as the function F whose value at any point $(x_1, \ldots, x_n)$ in n-dimensional space R^n is specified by the relation

$$F(x_1, \ldots, x_n) = \Pr(X_1 \leqslant x_1, \, X_2 \leqslant x_2, \ldots, X_n \leqslant x_n). \tag{1}$$

Every multivariate d.f. satisfies properties similar to those given earlier for univariate and bivariate d.f.'s.

Vector Notation.   In the study of the joint distribution of n random variables $X_1, \ldots, X_n$, it is often convenient to use the vector notation $X = (X_1, \ldots, X_n)$ and to refer to X as a *random vector*. Instead of speaking of the joint distribution of the random variables $X_1, \ldots, X_n$ with a joint d.f. $F(x_1, \ldots, x_n)$, we can simply speak of the distribution of the random vector X with d.f. $F(x)$. When this vector notation is used, it must be kept in mind that if X is an n-dimensional random vector, then its d.f. is defined as a function on n-dimensional space R^n. At any point $x = (x_1, \ldots, x_n) \in R^n$, the value of $F(x)$ is specified by Eq. (1).

Discrete Distributions.   It is said that n random variables $X_1, \ldots, X_n$ have a *discrete joint distribution* if the random vector $(X_1, \ldots, X_n)$ can have only a finite number or an infinite sequence of different possible values $(x_1, \ldots, x_n)$ in R^n. The joint p.f. of $X_1, \ldots, X_n$ is then defined as the function f such that for any point $(x_1, \ldots, x_n) \in R^n$,

$$f(x_1, \ldots, x_n) = \Pr(X_1 = x_1, \ldots, X_n = x_n).$$

In vector notation, it is said that the random vector X has a discrete distribution and that its p.f. is specified at any point $x \in R^n$ by the relation

$$f(x) = \Pr(X = x).$$

For any subset $A \subset R^n$,

$$\Pr(X \in A) = \sum_{x \in A} f(x).$$

Continuous Distributions. It is said that n random variables $X_1, \ldots, X_n$ have a continuous joint distribution if there is a nonnegative function f defined on R^n such that for any subset $A \subset R^n$,

$$\Pr[(X_1, \ldots, X_n) \in A] = \int \cdots \int_A f(x_1, \ldots, x_n) \, dx_1 \cdots dx_n. \tag{2}$$

The function f is called the *joint p.d.f.* of $X_1, \ldots, X_n$.

If the joint distribution of $X_1, \ldots, X_n$ is continuous, then the joint p.d.f. f can be derived from the joint d.f. F by using the relation

$$f(x_1, \ldots, x_n) = \frac{\partial^n F(x_1, \ldots, x_n)}{\partial x_1 \cdots \partial x_n}$$

at all points $(x_1, \ldots, x_n)$ at which the derivative in this relation exists.

In vector notation, $f(x)$ denotes the p.d.f. of the random vector X and Eq. (2) could be rewritten more simply in the form

$$\Pr(X \in A) = \int \cdots \int_A f(x) \, dx_1 \cdots dx_n.$$

Mixed Distributions. In a particular problem it is possible that some of the random variables $X_1, \ldots, X_n$ have discrete distributions while the other variables have continuous distributions. The joint distribution of $X_1, \ldots, X_n$ would then be represented by a function which might be called the *joint p.f.-p.d.f.* The probability that $(X_1, \ldots, X_n)$ lies in any subset $A \subset R^n$ would be calculated by summing over the values of some components and integrating over the values of the others.

As stated in Sec. 3.4, more complicated types of joint distributions, in which various combinations of the variables $X_1, \ldots, X_n$ have a mixed distribution, can also arise in practical problems.

Marginal Distributions

Deriving a Marginal p.d.f. If the joint distribution of n random variables $X_1, \ldots, X_n$ is known, then the marginal distribution of any single random variable X_i can be derived from this joint distribution. For example, if the joint p.d.f. of $X_1, \ldots, X_n$ is f, then the marginal p.d.f. f_1 of X_1 is specified at any value x_1 by the relation

$$f_1(x_1) = \underbrace{\int_{-\infty}^{\infty} \cdots \int_{-\infty}^{\infty}}_{n-1} f(x_1, \ldots, x_n) \, dx_2 \cdots dx_n.$$

More generally, the marginal joint p.d.f. of any k of the n random variables $X_1, \ldots, X_n$ can be found by integrating the joint p.d.f. over all possible values of the other $n - k$ variables. For example, if f is the joint p.d.f. of four random variables X_1, X_2, X_3, and X_4, then the marginal bivariate p.d.f. f_{24} of X_2 and X_4 is specified at any point (x_2, x_4) by the relation

$$f_{24}(x_2, x_4) = \int_{-\infty}^{\infty} \int_{-\infty}^{\infty} f(x_1, x_2, x_3, x_4)\, dx_1\, dx_3.$$

If n random variables $X_1, \ldots, X_n$ have a discrete joint distribution, then the marginal joint p.f. of any subset of the n variables can be obtained from relations similar to those just given. In the new relations, the integrals are replaced by sums.

Deriving a Marginal d.f.   Consider now either a discrete or a continuous joint distribution for which the joint d.f. of $X_1, \ldots, X_n$ is F. The marginal d.f. F_1 of X_1 can be obtained from the following relation:

$$F_1(x_1) = \Pr(X_1 \leqslant x_1) = \Pr(X_1 \leqslant x_1, X_2 < \infty, \ldots, X_n < \infty)$$

$$= \lim_{\substack{x_j \to \infty \\ j=2,\ldots,n}} F(x_1, x_2, \ldots, x_n).$$

More generally, the marginal joint d.f. of any k of the n random variables $X_1, \ldots, X_n$ can be found by computing the limiting value of the n-dimensional d.f. F as $x_j \to \infty$ for each of the other $n - k$ variables x_j. For example, if F is the joint d.f. of four random variables X_1, X_2, X_3, and X_4, then the marginal bivariate d.f. F_{24} of X_2 and X_4 is specified at any point (x_2, x_4) by the relation

$$F_{24}(x_2, x_4) = \lim_{\substack{x_1 \to \infty \\ x_3 \to \infty}} F(x_1, x_2, x_3, x_4).$$

Independent Random Variables.   It is said that n random variables $X_1, \ldots, X_n$ are *independent* if, for any n sets $A_1, A_2, \ldots, A_n$ of real numbers,

$$\Pr(X_1 \in A_1, X_2 \in A_2, \ldots, X_n \in A_n)$$

$$= \Pr(X_1 \in A_1)\Pr(X_2 \in A_2) \cdots \Pr(X_n \in A_n).$$

If we let F denote the joint d.f. of $X_1, \ldots, X_n$ and let F_i denote the marginal univariate d.f. of X_i for $i = 1, \ldots, n$, then it follows from the definition of independence that the variables $X_1, \ldots, X_n$ will be independent if and only if, for all points $(x_1, x_2, \ldots, x_n) \in R^n$,

$$F(x_1, x_2, \ldots, x_n) = F_1(x_1) F_2(x_2) \cdots F_n(x_n).$$

In other words, the random variables $X_1, \ldots, X_n$ are independent if and only if their joint d.f. is the product of their n individual marginal d.f.'s.

Furthermore, if the variables $X_1, \ldots, X_n$ have a continuous joint distribution for which the joint p.d.f. is f, and if f_i is the marginal univariate p.d.f. of X_i $(i = 1, \ldots, n)$, then $X_1, \ldots, X_n$ will be independent if and only if the following relation is satisfied at all points $(x_1, x_2, \ldots, x_n) \in R^n$:

$$f(x_1, x_2, \ldots, x_n) = f_1(x_1) f_2(x_2) \cdots f_n(x_n). \tag{3}$$

Similarly, if $X_1, \ldots, X_n$ have a discrete joint distribution for which the joint p.f. is f and if the marginal p.f. of X_i is f_i, for $i = 1, \ldots, n$, then these variables will be independent if and only if Eq. (3) is satisfied.

Random Samples.   Consider a given probability distribution on the real line that can be represented by either a p.f. or a p.d.f. f. It is said that n random variables $X_1, \ldots, X_n$ form a *random sample* from this distribution if these variables are independent and the marginal p.f. or p.d.f. of each of them is f. In other words, the variables $X_1, \ldots, X_n$ form a random sample from the distribution represented by f if their joint p.f. or p.d.f. g is specified as follows at all points $(x_1, x_2, \ldots, x_n) \in R^n$:

$$g(x_1, \ldots, x_n) = f(x_1) f(x_2) \cdots f(x_n).$$

Thus, the variables in a random sample are *independent and identically distributed*. This expression is often abbreviated i.i.d.

In these terms, the statement that $X_1, \ldots, X_n$ form a random sample from a distribution with p.d.f. $f(x)$ is equivalent to the statement that $X_1, \ldots, X_n$ are i.i.d. with common p.d.f. $f(x)$. The number n is called the *sample size*.

Example 1: Lifetimes of Light Bulbs.   Suppose that the lifetimes of light bulbs produced in a certain factory are distributed according to the following p.d.f.:

$$f(x) = \begin{cases} xe^{-x} & \text{for } x > 0, \\ 0 & \text{otherwise.} \end{cases}$$

We shall determine the joint p.d.f. of the lifetimes of a random sample of n light bulbs drawn from the factory's production.

The lifetimes $X_1, \ldots, X_n$ of the selected bulbs will form a random sample from the p.d.f. f. For typographical simplicity we shall use the notation $\exp(v)$ to denote the exponential e^v when the expression for v is complicated. Then the joint p.d.f. g of $X_1, \ldots, X_n$ will be as follows: If $x_i > 0$ for $i = 1, \ldots, n$,

$$g(x_1, \ldots, x_n) = \prod_{i=1}^{n} f(x_i)$$

$$= \left(\prod_{i=1}^{n} x_i \right) \exp\left(-\sum_{i=1}^{n} x_i \right).$$

Otherwise, $g(x_1, \ldots, x_n) = 0$.

Any probability involving the n lifetimes $X_1, \ldots, X_n$ can in principle be determined by integrating this joint p.d.f. over the appropriate subset of R^n. For example, if A is the subset of points $(x_1, \ldots, x_n)$ such that $x_i > 0$ for $i = 1, \ldots, n$ and $\sum_{i=1}^{n} x_i < a$, where a is a given positive number, then

$$\Pr\left(\sum_{i=1}^{n} X_i < a \right) = \int \cdots \int_A \left(\prod_{i=1}^{n} x_i \right) \exp\left(-\sum_{i=1}^{n} x_i \right) dx_1 \cdots dx_n. \quad \square$$

The evaluation of the integral given at the end of Example 1 may require a considerable amount of time without the aid of tables or a computer. Certain other probabilities, however, can easily be evaluated from the basic properties of continuous distributions and random samples. For example, suppose that for the conditions of Example 1 it is desired to find $\Pr(X_1 < X_2 < \cdots < X_n)$. Since the variables $X_1, \ldots, X_n$ have a continuous joint distribution, the probability that at least two of these random variables will have the same value is 0. In fact, the probability is 0 that the vector $(X_1, \ldots, X_n)$ will belong to any specific subset of R^n for which the n-dimensional volume is 0. Furthermore, since $X_1, \ldots, X_n$ are independent and identically distributed, each of these variables is equally likely to be the smallest of the n lifetimes, and each is equally likely to be the largest. More generally, if the lifetimes $X_1, \ldots, X_n$ are arranged in order from the smallest to the largest, any particular ordering of $X_1, \ldots, X_n$ is as likely to be obtained as any other ordering. Since there are $n!$ different possible orderings, the probability that the particular ordering $X_1 < X_2 < \cdots < X_n$ will be obtained is $1/n!$. Hence,

$$\Pr(X_1 < X_2 < \cdots < X_n) = \frac{1}{n!}.$$

Conditional Distributions

Suppose that n random variables $X_1, \ldots, X_n$ have a continuous joint distribution for which the joint p.d.f. is f, and that f_0 denotes the marginal joint p.d.f. of the $n - 1$ variables $X_2, \ldots, X_n$. Then for any values of $x_2, \ldots, x_n$ such that

$f_0(x_2, \ldots, x_n) > 0$, the conditional p.d.f. of X_1 given that $X_2 = x_2, \ldots, X_n = x_n$ is defined as follows:

$$g_1(x_1 | x_2, \ldots, x_n) = \frac{f(x_1, x_2, \ldots, x_n)}{f_0(x_2, \ldots, x_n)}.$$

More generally, suppose that the random vector $X = (X_1, \ldots, X_n)$ is divided into two subvectors Y and Z, where Y is a k-dimensional random vector comprising k of the n random variables in X, and Z is an $(n - k)$-dimensional random vector comprising the other $n - k$ random variables in X. Suppose also that the n-dimensional p.d.f. of (Y, Z) is f and that the marginal $(n - k)$-dimensional p.d.f. of Z is f_2. Then for any given point $z \in R^{n-k}$ such that $f_2(z) > 0$, the conditional k-dimensional p.d.f. g_1 of Y when $Z = z$ is defined as follows:

$$g_1(y | z) = \frac{f(y, z)}{f_2(z)} \qquad \text{for } y \in R^k. \tag{4}$$

For example, suppose that the joint p.d.f. of five random variables $X_1, \ldots, X_5$ is f, and that the marginal joint p.d.f. of X_2 and X_4 is f_{24}. If we assume that $f_{24}(x_2, x_4) > 0$, then the conditional joint p.d.f. of X_1, X_3, and X_5, given that $X_2 = x_2$ and $X_4 = x_4$, will be

$$g(x_1, x_3, x_5 | x_2, x_4) = \frac{f(x_1, x_2, x_3, x_4, x_5)}{f_{24}(x_2, x_4)} \qquad \text{for } (x_1, x_3, x_5) \in R^3.$$

If the random vectors Y and Z have a discrete joint distribution for which the joint p.f. is f, and if the marginal p.f. of Z is f_2, then the conditional p.f. $g_1(y | z)$ of Y for any given value $Z = z$ also will be specified by Eq. (4).

Example 2: Determining a Marginal Joint p.d.f. Suppose that X_1 is a random variable for which the p.d.f. f_1 is as follows:

$$f_1(x) = \begin{cases} e^{-x} & \text{for } x > 0, \\ 0 & \text{otherwise.} \end{cases}$$

Suppose, furthermore, that for any given value $X_1 = x_1$ ($x_1 > 0$), two other random variables X_2 and X_3 are independent and identically distributed and the conditional p.d.f. of each of these variables is as follows:

$$g(t | x_1) = \begin{cases} x_1 e^{-x_1 t} & \text{for } t > 0, \\ 0 & \text{otherwise.} \end{cases}$$

We shall determine the marginal joint p.d.f. of X_2 and X_3.

Since X_2 and X_3 are i.i.d. for any given value of X_1, their conditional joint p.d.f. when $X_1 = x_1$ ($x_1 > 0$) is

$$g_{23}(x_2, x_3 | x_1) = \begin{cases} x_1^2 e^{-x_1(x_2 + x_3)} & \text{for } x_2 > 0 \text{ and } x_3 > 0, \\ 0 & \text{otherwise.} \end{cases}$$

The joint p.d.f. f of X_1, X_2, and X_3 will be positive only at those points (x_1, x_2, x_3) such that $x_1 > 0$, $x_2 > 0$, and $x_3 > 0$. It now follows that, at any such point,

$$f(x_1, x_2, x_3) = f_1(x_1) g_{23}(x_2, x_3 | x_1) = x_1^2 e^{-x_1(1 + x_2 + x_3)}.$$

For $x_2 > 0$ and $x_3 > 0$, the marginal joint p.d.f. $f_{23}(x_2, x_3)$ of X_2 and X_3 can be determined as follows:

$$f_{23}(x_2, x_3) = \int_0^\infty f(x_1, x_2, x_3) \, dx_1 = \frac{2}{(1 + x_2 + x_3)^3}.$$

From this marginal joint p.d.f. we can evaluate probabilities involving X_2 and X_3, such as $\Pr(X_2 + X_3 < 4)$. We have

$$\Pr(X_2 + X_3 < 4) = \int_0^4 \int_0^{4 - x_3} f_{23}(x_2, x_3) \, dx_2 \, dx_3 = \frac{16}{25}.$$

Next, we shall determine the conditional p.d.f. of X_1 given that $X_2 = x_2$ and $X_3 = x_3$ ($x_2 > 0$ and $x_3 > 0$). For any value of x_1,

$$g_1(x_1 | x_2, x_3) = \frac{f(x_1, x_2, x_3)}{f_{23}(x_2, x_3)}.$$

For $x_1 > 0$, it follows that

$$g_1(x_1 | x_2, x_3) = \frac{1}{2}(1 + x_2 + x_3)^3 x_1^2 e^{-x_1(1 + x_2 + x_3)}.$$

For $x_1 \leqslant 0$, $g_1(x_1 | x_2, x_3) = 0$.

Finally, we shall evaluate $\Pr(X_1 \leqslant 1 | X_2 = 1, X_3 = 4)$. We have

$$\Pr(X_1 \leqslant 1 | X_2 = 1, X_3 = 4) = \int_0^1 g_1(x_1 | 1, 4) \, dx_1$$

$$= \int_0^1 108 x_1^2 e^{-6x_1} \, dx_1 = 0.938. \quad \square$$

EXERCISES

1. Suppose that three random variables X_1, X_2, and X_3 have a continuous joint distribution with the following joint p.d.f.:

$$f(x_1, x_2, x_3) = \begin{cases} c(x_1 + 2x_2 + 3x_3) & \text{for } 0 \leqslant x_i \leqslant 1 \ (i = 1, 2, 3), \\ 0 & \text{otherwise.} \end{cases}$$

Determine (a) the value of the constant c; (b) the marginal joint p.d.f. of X_1 and X_2; and (c) $\Pr\left(X_3 < \dfrac{1}{2} \middle| X_1 = \dfrac{1}{4}, X_2 = \dfrac{3}{4} \right)$.

2. Suppose that three random variables X_1, X_2, and X_3 have a continuous joint distribution with the following joint p.d.f.:

$$f(x_1, x_2, x_3) = \begin{cases} ce^{-(x_1 + 2x_2 + 3x_3)} & \text{for } x_i > 0 \ (i = 1, 2, 3), \\ 0 & \text{otherwise.} \end{cases}$$

Determine (a) the value of the constant c; (b) the marginal joint p.d.f. of X_1 and X_3; and (c) $\Pr(X_1 < 1 | X_2 = 2, X_3 = 1)$.

3. Suppose that a point (X_1, X_2, X_3) is chosen at random, that is, in accordance with a uniform p.d.f., from the following set S:

$$S = \{(x_1, x_2, x_3): 0 \leqslant x_i \leqslant 1 \quad \text{for } i = 1, 2, 3\}.$$

Determine:

(a) $\Pr\left[\left(X_1 - \dfrac{1}{2} \right)^2 + \left(X_2 - \dfrac{1}{2} \right)^2 + \left(X_3 - \dfrac{1}{2} \right)^2 \leqslant \dfrac{1}{4} \right]$

(b) $\Pr(X_1^2 + X_2^2 + X_3^2 \leqslant 1)$.

4. Suppose that an electronic system contains n components which function independently of each other; and that the probability that component i will function properly is p_i $(i = 1, \ldots, n)$. It is said that the components are connected *in series* if a necessary and sufficient condition for the system to function properly is that all n components function properly. It is said that the components are connected *in parallel* if a necessary and sufficient condition for the system to function properly is that at least one of the n components functions properly. The probability that the system will function properly is called the *reliability* of the system. Determine the reliability of the system, (a) assuming that the components are connected in series, and (b) assuming that the components are connected in parallel.

5. Suppose that the n random variables $X_1, \ldots, X_n$ form a random sample from a discrete distribution for which the p.f. is f. Determine the value of $\Pr(X_1 = X_2 = \cdots = X_n)$.

6. Suppose that the n random variables $X_1, \ldots, X_n$ form a random sample from a continuous distribution for which the p.d.f. is f. Determine the probability that at least k of these n random variables will lie in a specified interval $a \leqslant x \leqslant b$.

7. Suppose that the p.d.f. of a random variable X is as follows:

$$f(x) = \begin{cases} \dfrac{1}{n!} x^n e^{-x} & \text{for } x > 0, \\ 0 & \text{otherwise.} \end{cases}$$

Suppose also that for any given value $X = x$ $(x > 0)$, the n random variables $Y_1, \ldots, Y_n$ are i.i.d. and the conditional p.d.f. g of each of them is as follows:

$$g(y|x) = \begin{cases} \dfrac{1}{x} & \text{for } 0 < y < x, \\ 0 & \text{otherwise.} \end{cases}$$

Determine (a) the marginal joint p.d.f. of $Y_1, \ldots, Y_n$ and (b) the conditional p.d.f. of X for any given values of $Y_1, \ldots, Y_n$.

3.8. FUNCTIONS OF A RANDOM VARIABLE

Variable with a Discrete Distribution

Suppose that a random variable X has a discrete distribution for which the p.f. is f, and that another random variable $Y = r(X)$ is defined as a certain function of X. Then the p.f. g of Y can be derived from f in a straightforward manner as follows: For any possible value y of Y,

$$g(y) = \Pr(Y = y) = \Pr[r(X) = y]$$

$$= \sum_{x:r(x)=y} f(x).$$

Variable with a Continuous Distribution

If a random variable X has a continuous distribution, then the procedure for deriving the probability distribution of any function of X differs from that just given. Suppose that the p.d.f. of X is f and that another random variable is defined as $Y = r(X)$. For any real number y, the d.f. $G(y)$ of Y can be derived as follows:

$$G(y) = \Pr(Y \leqslant y) = \Pr[r(X) \leqslant y]$$

$$= \int_{\{x:r(x)\leqslant y\}} f(x)\, dx.$$

If the random variable Y also has a continuous distribution, its p.d.f. g can be obtained from the relation

$$g(y) = \frac{dG(y)}{dy}.$$

This relation is satisfied at any point y at which G is differentiable.

Example 1: Deriving the p.d.f. of X^2 When X Has a Uniform Distribution.
Suppose that X has a uniform distribution on the interval $(-1, 1)$, so

$$f(x) = \begin{cases} 1/2 & \text{for } -1 < x < 1, \\ 0 & \text{otherwise.} \end{cases}$$

We shall determine the p.d.f. of the random variable $Y = X^2$.

Since $Y = X^2$, then Y must belong to the interval $0 \leqslant Y < 1$. Thus, for any value of Y such that $0 < y < 1$, the d.f. $G(y)$ of Y is

$$G(y) = \Pr(Y \leqslant y) = \Pr(X^2 \leqslant y)$$

$$= \Pr\left(-y^{1/2} \leqslant X \leqslant y^{1/2}\right)$$

$$= \int_{-y^{1/2}}^{y^{1/2}} f(x)\, dx = y^{1/2}.$$

For $0 < y < 1$, it follows that the p.d.f. $g(y)$ of Y is

$$g(y) = \frac{dG(y)}{dy} = \frac{1}{2y^{1/2}}.$$

This p.d.f. of Y is sketched in Fig. 3.17. It should be noted that although Y is simply the square of a random variable with a uniform distribution, the p.d.f. of Y is unbounded in the neighborhood of $y = 0$. $\square$

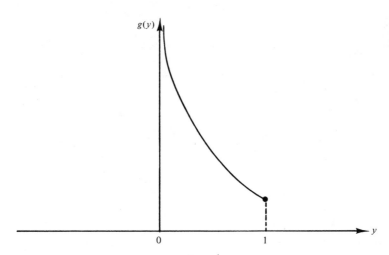

Figure 3.17   The p.d.f. of $Y = X^2$ in Example 1.

Direct Derivation of the Probability Density Function

If a random variable X has a continuous distribution and if $Y = r(X)$, then it is not necessarily true that Y will also have a continuous distribution. For example, suppose that $r(x) = c$, where c is a constant, for all values of x in some interval $a \leqslant x \leqslant b$, and that $\Pr(a \leqslant X \leqslant b) > 0$. Then $\Pr(Y = c) > 0$. Since the distribution of Y assigns positive probability to the value c, this distribution cannot be continuous. In order to derive the distribution of Y in a case like this, the d.f. of Y must be derived by applying the method just described. For certain functions r, however, the distribution of Y will be continuous; and it will then be possible to derive the p.d.f. of Y directly without first deriving its d.f.

Suppose that $Y = r(X)$ and that the random variable X must lie in a certain interval (a, b) over which the function $r(x)$ is strictly increasing. This interval, for which $\Pr(a < X < b) = 1$, might be a bounded interval, an unbounded interval, or the entire real line. As x varies over the interval $a < x < b$, the values of $y = r(x)$ will vary over an interval $\alpha < y < \beta$. Furthermore, corresponding to each value of y in the interval $\alpha < y < \beta$, there will be a unique value x in the interval $a < x < b$ such that $r(x) = y$. If we denote this value of x as $x = s(y)$, then the function s will be the inverse of the function r. In other words, for any values of x and y in the intervals $a < x < b$ and $\alpha < y < \beta$, it will be true that $y = r(x)$ if and only if $x = s(y)$. Since it is assumed that the function r is continuous and strictly increasing over the interval (a, b), the inverse function s will also be continuous and strictly increasing over the interval (α, β). Hence, for any value of y such that $\alpha < y < \beta$,

$$G(y) = \Pr(Y \leqslant y) = \Pr[r(X) \leqslant y] = \Pr[X \leqslant s(y)] = F[s(y)].$$

If it is also assumed that s is a differentiable function over the interval (α, β), then the distribution of Y will be continuous and the value of its p.d.f. $g(y)$ for $\alpha < y < \beta$ will be

$$g(y) = \frac{dG(y)}{dy} = \frac{dF[s(y)]}{dy} = f[s(y)]\frac{ds(y)}{dy}. \tag{1}$$

In other words, the p.d.f. $g(y)$ of Y can be obtained directly from the p.d.f. $f(x)$ by replacing x by its expression in terms of y and multiplying the result by dx/dy.

Similarly, if $Y = r(X)$ and it is assumed that the function r is continuous and *strictly decreasing* over the interval (a, b), then Y will vary over some interval (α, β) as X varies over the interval (a, b); and the inverse function s will be continuous and strictly decreasing over the interval (α, β). Hence, for $\alpha < y < \beta$,

$$G(y) = \Pr[r(X) \leqslant y] = \Pr[X \geqslant s(y)] = 1 - F[s(y)].$$

If it is again assumed that s is differentiable over the interval (α, β), then it follows that

$$g(y) = \frac{dG(y)}{dy} = -f[s(y)]\frac{ds(y)}{dy}.$$

Since s is strictly decreasing, $ds(y)/dy < 0$ and $g(y)$ can be expressed in the form

$$g(y) = f[s(y)]\left|\frac{ds(y)}{dy}\right|. \tag{2}$$

Since Eq. (1) is satisfied when r and s are strictly increasing functions and Eq. (2) is satisfied when r and s are strictly decreasing functions, the results that have been obtained can be summarized as follows:

Let X be a random variable for which the p.d.f. is f and for which $\Pr(a < X < b) = 1$. Let $Y = r(X)$, and suppose that $r(x)$ is continuous and either strictly increasing or strictly decreasing for $a < x < b$. Suppose also that $a < X < b$ if and only if $\alpha < Y < \beta$, and let $X = s(Y)$ be the inverse function for $\alpha < Y < \beta$. Then the p.d.f. g of Y is specified by the relation

$$g(y) = \begin{cases} f[s(y)]\left|\dfrac{ds(y)}{dy}\right| & \text{for } \alpha < y < \beta, \\ 0 & \text{otherwise.} \end{cases} \tag{3}$$

Example 2: Deriving a p.d.f. by the Direct Method. Suppose that X is a random variable for which the p.d.f. is

$$f(x) = \begin{cases} 3x^2 & \text{for } 0 < x < 1, \\ 0 & \text{otherwise.} \end{cases}$$

We shall derive the p.d.f. of $Y = 1 - X^2$.

In this example $\Pr(0 < X < 1) = 1$ and Y is a continuous and strictly decreasing function of X for $0 < X < 1$. As X varies over the interval $(0, 1)$, it is found that Y also varies over the interval $(0, 1)$. Furthermore, for $0 < Y < 1$, the inverse function is $X = (1 - Y)^{1/2}$. Hence, for $0 < y < 1$,

$$\frac{ds(y)}{dy} = -\frac{1}{2(1 - y)^{1/2}}.$$

It follows from Eq. (3) that for $0 < y < 1$, the value of $g(y)$ will be

$$g(y) = 3(1 - y) \cdot \frac{1}{2(1 - y)^{1/2}} = \frac{3}{2}(1 - y)^{1/2}.$$

Finally, $g(y) = 0$ for any value of y outside the interval $0 < y < 1$.   □

The Probability Integral Transformation

Suppose that a random variable X has a continuous d.f. F, and let $Y = F(X)$. This transformation from X to Y is called the *probability integral transformation*. We shall show that the distribution of Y must be a uniform distribution on the interval $(0, 1)$.

First, since F is the d.f. of a random variable, then $0 \leqslant F(x) \leqslant 1$ for $-\infty < x < \infty$. Therefore, $\Pr(Y < 0) = \Pr(Y > 1) = 0$. Next, for any given value of y in the interval $0 < y < 1$, let x_0 be a number such that $F(x_0) = y$. If F is strictly increasing, there will be a unique number x_0 such that $F(x_0) = y$. However, if $F(x) = y$ over an entire interval of values of x, then x_0 can be chosen arbitrarily from this interval. If G denotes the d.f. of Y, then

$$G(y) = \Pr(Y \leqslant y) = \Pr(X \leqslant x_0) = F(x_0) = y.$$

Hence, $G(y) = y$ for $0 < y < 1$. Since this function is the d.f. of a uniform distribution on the interval $(0, 1)$, this uniform distribution is the distribution of Y.

Now suppose that X is a random variable with a continuous d.f. F; that G is some other continuous d.f. on the real line; and that it is required to construct a random variable $Z = r(X)$ for which the d.f. will be G.

For any value of y in the interval $0 < y < 1$, we shall let $z = G^{-1}(y)$ be any number such that $G(z) = y$. We can then define the random variable Z as follows:

$$Z = G^{-1}[F(X)].$$

To verify that the d.f. of Z is actually G, we note that for any number z such that $0 < G(z) < 1$,

$$\Pr(Z \leqslant z) = \Pr\{G^{-1}[F(X)] \leqslant z\}$$

$$= \Pr[F(X) \leqslant G(z)].$$

It follows from the probability integral transformation that $F(X)$ has a uniform distribution and, therefore, that

$$\Pr[F(X) \leqslant G(z)] = G(z).$$

Hence, $\Pr(Z \leqslant z) = G(z)$, which means that the d.f. of Z is G, as required.

Tables of Random Digits

Uses of a Table.   Consider an experiment in which the outcome must be one of the ten digits $0, 1, 2, \ldots, 9$ and in which each of these ten digits occurs with probability $1/10$. If this experiment is repeated independently a large number of times and the sequence of outcomes is presented in the form of a table, this table is called a *table of random digits*, or a *table of random numbers*. A short table of random digits, which contains 4000 digits, is given at the end of this book.

A table of random digits can be very useful in an experiment that involves sampling. For example, any pair of random digits selected from the table has a probability of 0.01 of being equal to any one of the 100 values $00, 01, 02, \ldots, 99$. Similarly, any triple of random digits selected from the table has a probability of 0.001 of being equal to each of the 1000 values $000, 001, 002, \ldots, 999$. Thus, if an experimenter wishes to select a person at random from a list of 645 people, each of whom has been assigned a number from 1 to 645, he can simply start at any arbitrary place in a table of random digits, examine triples of digits until he finds a triple between 001 and 645, and then select the person in the list to whom that number was assigned.

Generating Values of a Random Variable Having a Uniform Distribution.   A table of random digits can be used in the following way to generate values of a random variable X with a uniform distribution on the interval $(0, 1)$. Any triple of random digits can be regarded as specifying a number between 0 and 1 to three decimal places. For instance, the triple 308 can be associated with the decimal 0.308. Since all triples of random digits are equally probable, the values of X between 0 and 1 that are generated in this way from the table of random digits will have a uniform distribution on the interval $(0, 1)$. Of course, X in this case will be specified to only three decimal places.

If additional decimal places are desired in specifying the values of X, then the experimenter could select additional digits from the table. In fact, if the entire list of 4000 random digits given in the table in this book is regarded as a single long sequence, then the entire table could be used to specify a single value between 0 and 1 to 4000 decimal places. In most problems, however, an acceptable degree of accuracy can be attained by using a sequence of only three or four random digits for each value. The table can then be used to generate several independent values from the uniform distribution.

Generating Values of a Random Variable Having a Specified Distribution.   A table of random digits can be used to generate values of a random variable Y having any specified continuous d.f. G. If a random variable X has a uniform distribution on the interval $(0, 1)$ and if the function G^{-1} is defined as before, then it follows from the probability integral transformation that the d.f. of the random variable $Y = G^{-1}(X)$ will be G. Hence, if a value of X is determined from the table of random digits by the method just described, then the corresponding value of Y will have the desired property. If n independent values $X_1, \ldots, X_n$ are determined from the table, then the corresponding values $Y_1, \ldots, Y_n$ will form a random sample of size n from the distribution with the d.f. G.

Example 3: Generating Independent Values from a Specified p.d.f. Suppose that a table of random digits is to be used to generate three independent values from the distribution for which the p.d.f. g is as follows:

$$g(y) = \begin{cases} \dfrac{1}{2}(2 - y) & \text{for } 0 < y < 2, \\ 0 & \text{otherwise.} \end{cases}$$

For $0 < y < 2$, the d.f. G of the given distribution is

$$G(y) = y - \frac{y^2}{4}.$$

Also, for $0 < x < 1$, the inverse function $y = G^{-1}(x)$ can be found by solving the equation $x = G(y)$ for y. The result is

$$y = 2\left[1 - (1 - x)^{1/2}\right]. \tag{4}$$

The next step is to use the table of random digits to specify three independent values x_1, x_2, and x_3 from the uniform distribution. We shall specify each of these values to four decimal places, as follows: Starting from an arbitrary point in the table of random digits, we select three successive groups of four digits each. Such a grouping is

4125   0894   8302.

These groups of digits can be regarded as specifying the values

$$x_1 = 0.4125, \qquad x_2 = 0.0894, \qquad x_3 = 0.8302.$$

When these values of x_1, x_2, and x_3 are substituted successively into Eq. (4), the values of y that are obtained are $y_1 = 0.47$, $y_2 = 0.09$, and $y_3 = 1.18$. These will be three independent values from the distribution for which the d.f. is G. □

Construction of a Table of Random Digits. A table of random digits is usually constructed with the aid of a large computer by employing certain numerical techniques. It is an interesting fact that the digits in such a table are often computed by methods that do not involve any random operations at all. Nevertheless, many of the properties of these digits are the same as the properties of digits that are generated by repeatedly carrying out independent experiments in which each digit occurs with probability $1/10$.

There are other computer methods for generating values from certain specified distributions that are faster and more accurate than the preceding method which is based on the probability integral transformation. These topics are discussed in the books by Kennedy and Gentle (1980) and Rubinstein (1981).

EXERCISES

1. Suppose that a random variable X can have each of the seven values -3, -2, -1, 0, 1, 2, 3 with equal probability. Determine the p.f. of $Y = X^2 - X$.

2. Suppose that the p.d.f. of a random variable X is as follows:

$$f(x) = \begin{cases} \dfrac{1}{2}x & \text{for } 0 < x < 2, \\ 0 & \text{otherwise.} \end{cases}$$

   Also, suppose that $Y = X(2 - X)$. Determine the d.f. and the p.d.f. of Y.

3. Suppose that the p.d.f. of X is as given in Exercise 2. Determine the p.d.f. of $Y = 4 - X^3$.

4. Suppose that X is a random variable for which the p.d.f. is f and that $Y = aX + b$ $(a \neq 0)$. Show that the p.d.f. of Y is as follows:

$$g(y) = \frac{1}{|a|} f\left(\frac{y - b}{a}\right) \qquad \text{for } -\infty < y < \infty.$$

5. Suppose that the p.d.f. of X is as given in Exercise 2. Determine the p.d.f. of $Y = 3X + 2$.

6. Suppose that a random variable X has a uniform distribution on the interval $(0, 1)$. Determine the p.d.f. of (a) X^2, (b) $-X^3$, and (c) $X^{1/2}$.

7. Suppose that the p.d.f. of X is as follows:

$$f(x) = \begin{cases} e^{-x} & \text{for } x > 0, \\ 0 & \text{for } x \leqslant 0. \end{cases}$$

Determine the p.d.f. of $Y = X^{1/2}$.

8. Suppose that X has a uniform distribution on the interval $(0, 1)$. Construct a random variable $Y = r(X)$ for which the p.d.f. will be

$$g(y) = \begin{cases} \dfrac{3}{8} y^2 & \text{for } 0 < y < 2, \\ 0 & \text{otherwise.} \end{cases}$$

9. Let X be a random variable for which the p.d.f. f is as given in Exercise 2. Construct a random variable $Y = r(X)$ for which the p.d.f. g is as given in Exercise 8.

10. Use the table of random digits at the end of this book to generate four independent values from a distribution for which the p.d.f. is

$$g(y) = \begin{cases} \dfrac{1}{2}(2y + 1) & \text{for } 0 < y < 1, \\ 0 & \text{otherwise.} \end{cases}$$

3.9. FUNCTIONS OF TWO OR MORE RANDOM VARIABLES

Variables with a Discrete Joint Distribution

Suppose that n random variables $X_1, \ldots, X_n$ have a discrete joint distribution for which the joint p.f. is f, and that m functions $Y_1, \ldots, Y_m$ of these n random variables are defined as follows:

$$Y_1 = r_1(X_1, \ldots, X_n),$$
$$Y_2 = r_2(X_1, \ldots, X_n),$$
$$\vdots \qquad \vdots$$
$$Y_m = r_m(X_1, \ldots, X_n).$$

For any given values $y_1, \ldots, y_m$ of the m random variables $Y_1, \ldots, Y_m$, let A denote the set of all points $(x_1, \ldots, x_n)$ such that

$$r_1(x_1, \ldots, x_n) = y_1,$$
$$r_2(x_1, \ldots, x_n) = y_2,$$
$$\vdots \qquad \vdots$$
$$r_m(x_1, \ldots, x_n) = y_m.$$

Then the value of the joint p.f. g of $Y_1, \ldots, Y_m$ is specified at the point $(y_1, \ldots, y_m)$ by the relation

$$g(y_1, \ldots, y_m) = \sum_{(x_1, \ldots, x_n) \in A} f(x_1, \ldots, x_n).$$

Variables with a Continuous Joint Distribution

If the joint distribution of $X_1, \ldots, X_n$ is continuous, then some method other than that just given must be used to derive the distribution of any function of $X_1, \ldots, X_n$ or the joint distribution of two or more such functions. If the joint p.d.f. of $X_1, \ldots, X_n$ is $f(x_1, \ldots, x_n)$ and if $Y = r(X_1, \ldots, X_n)$, then the d.f. $G(y)$ of Y can be determined from basic principles. For any given value y $(-\infty < y < \infty)$, let A_y be the subset of R^n containing all points $(x_1, \ldots, x_n)$ such that $r(x_1, \ldots, x_n) \leqslant y$. Then

$$G(y) = \Pr(Y \leqslant y) = \Pr[r(X_1, \ldots, X_n) \leqslant y]$$

$$= \int \cdots \int_{A_y} f(x_1, \ldots, x_n) \, dx_1 \, \cdots \, dx_n.$$

If the distribution of Y also is continuous, then the p.d.f. of Y can be found by differentiating the d.f. $G(y)$.

The Distribution of the Maximum and Minimum Values in a Random Sample.   As an illustration of the method just outlined, suppose that the variables $X_1, \ldots, X_n$

form a random sample of size n from a distribution for which the p.d.f. is f and the d.f. is F; and consider the random variable Y_n which is defined as follows:

$$Y_n = \max\{X_1, \ldots, X_n\}.$$

In other words, Y_n is the largest value in the random sample. We shall determine both the d.f. G_n and the p.d.f. g_n of Y_n.

For any given value of y $(-\infty < y < \infty)$,

$$G_n(y) = \Pr(Y_n \leqslant y) = \Pr(X_1 \leqslant y, X_2 \leqslant y, \ldots, X_n \leqslant y)$$

$$= \Pr(X_1 \leqslant y)\Pr(X_2 \leqslant y) \cdots \Pr(X_n \leqslant y)$$

$$= F(y)F(y) \cdots F(y) = [F(y)]^n.$$

Thus, $G_n(y) = [F(y)]^n$.

The p.d.f. g_n of Y_n can be determined by differentiating this d.f. G_n. The result is

$$g_n(y) = n[F(y)]^{n-1}f(y) \qquad \text{for } -\infty < y < \infty.$$

Consider now the random variable Y_1, which is defined as follows:

$$Y_1 = \min\{X_1, \ldots, X_n\}.$$

In other words, Y_1 is the smallest value in the random sample.

For any given value of y $(-\infty < y < \infty)$, the d.f. G_1 of Y_1 can be derived as follows:

$$G_1(y) = \Pr(Y_1 \leqslant y) = 1 - \Pr(Y_1 > y)$$

$$= 1 - \Pr(X_1 > y, X_2 > y, \ldots, X_n > y)$$

$$= 1 - \Pr(X_1 > y)\Pr(X_2 > y) \cdots \Pr(X_n > y)$$

$$= 1 - [1 - F(y)][1 - F(y)] \cdots [1 - F(y)]$$

$$= 1 - [1 - F(y)]^n.$$

Thus, $G_1(y) = 1 - [1 - F(y)]^n$.

The p.d.f. g_1 of Y_1 can be determined by differentiating this d.f. G_1. The result is

$$g_1(y) = n[1 - F(y)]^{n-1}f(y) \qquad \text{for } -\infty < y < \infty.$$

Finally, we shall demonstrate how this method can be used to determine the joint distribution of two or more functions of $X_1, \ldots, X_n$ by deriving the joint distribution of Y_1 and Y_n. If G denotes the bivariate joint d.f. of Y_1 and Y_n, then for any given values of y_1 and y_n $(-\infty < y_1 < y_n < \infty)$,

$$G(y_1, y_n) = \Pr(Y_1 \leqslant y_1 \text{ and } Y_n \leqslant y_n)$$

$$= \Pr(Y_n \leqslant y_n) - \Pr(Y_n \leqslant y_n \text{ and } Y_1 > y_1)$$

$$= \Pr(Y_n \leqslant y_n)$$

$$- \Pr(y_1 < X_1 \leqslant y_n, y_1 < X_2 \leqslant y_n, \ldots, y_1 < X_n \leqslant y_n)$$

$$= G_n(y_n) - \prod_{i=1}^{n} \Pr(y_1 < X_i \leqslant y_n)$$

$$= [F(y_n)]^n - [F(y_n) - F(y_1)]^n.$$

The bivariate joint p.d.f. g of Y_1 and Y_n can be found from the relation

$$g(y_1, y_n) = \frac{\partial^2 G(y_1, y_n)}{\partial y_1 \, \partial y_n}.$$

Thus, for $-\infty < y_1 < y_n < \infty$,

$$g(y_1, y_n) = n(n-1)[F(y_n) - F(y_1)]^{n-2} f(y_1) f(y_n). \tag{1}$$

Also, for any other values of y_1 and y_n, $g(y_1, y_n) = 0$. We shall return to this joint p.d.f. at the end of this section.

Transformation of a Multivariate Probability Density Function

We shall continue to assume that the random variables $X_1, \ldots, X_n$ have a continuous joint distribution for which the joint p.d.f. is f; and we shall now consider the joint p.d.f. g of n new random variables $Y_1, \ldots, Y_n$ which are defined as follows:

$$\begin{aligned} Y_1 &= r_1(X_1, \ldots, X_n), \\ Y_2 &= r_2(X_1, \ldots, X_n), \\ &\;\;\vdots \qquad\quad \vdots \\ Y_n &= r_n(X_1, \ldots, X_n). \end{aligned} \tag{2}$$

We shall make certain assumptions in regard to the functions $r_1, \ldots, r_n$. First, we shall let S be a subset of R^n such that $\Pr[(X_1, \ldots, X_n) \in S] = 1$. The subset S could be the entire space R^n. However, if the probability distribution of $X_1, \ldots, X_n$ is concentrated in some proper subset of R^n, then this smaller set could be chosen for S. We shall also let T denote the subset of R^n that is the image of S under the transformation specified by the n equations in (2). In other words, as the values of $(X_1, \ldots, X_n)$ vary over the set S, the values of $(Y_1, \ldots, Y_n)$ vary over the set T. It is further assumed that the transformation from S to T is a one-to-one transformation. In other words, corresponding to each value of $(Y_1, \ldots, Y_n)$ in the set T, there is a *unique* value of $(X_1, \ldots, X_n)$ in the set S such that the n equations in (2) are satisfied.

It follows from this last assumption that there is a one-to-one correspondence between the points $(y_1, \ldots, y_n)$ in T and the points $(x_1, \ldots, x_n)$ in S. Therefore, for $(y_1, \ldots, y_n) \in T$ we can invert the equations in (2) and obtain new equations of the following form:

$$
\begin{aligned}
x_1 &= s_1(y, \ldots, y_n), \\
x_2 &= s_2(y_1, \ldots, y_n), \\
&\ \vdots \qquad \vdots \\
x_n &= s_n(y_1, \ldots, y_n).
\end{aligned}
\tag{3}
$$

Next, we shall assume that for $i = 1, \ldots, n$ and $j = 1, \ldots, n$, each partial derivative $\partial s_i / \partial y_j$ exists at every point $(y_1, \ldots, y_n) \in T$. Under this assumption, the following determinant J can be constructed:

$$
J = \det \begin{bmatrix}
\dfrac{\partial s_1}{\partial y_1} & \cdots & \dfrac{\partial s_1}{\partial y_n} \\
\vdots & & \vdots \\
\dfrac{\partial s_n}{\partial y_1} & \cdots & \dfrac{\partial s_n}{\partial y_n}
\end{bmatrix}.
$$

This determinant is called the *Jacobian* of the transformation specified by the equations in (3).

The joint p.d.f. g of the n random variables $Y_1, \ldots, Y_n$ can now be derived by using the methods of advanced calculus for changing variables in a multiple integral. The derivation will not be presented here, but the result is as follows:

$$
g(y_1, \ldots, y_n) = \begin{cases} f(s_1, \ldots, s_n) |J| & \text{for } (y_1, \ldots, y_n) \in T, \\ 0 & \text{otherwise.} \end{cases}
\tag{4}
$$

In Eq. (4), $|J|$ denotes the absolute value of the determinant J. Thus, the joint p.d.f. $g(y_1, \ldots, y_n)$ is obtained by starting with the joint p.d.f. $f(x_1, \ldots, x_n)$, replacing each value x_i by its expression $s_i(y_1, \ldots, y_n)$ in terms of $y_1, \ldots, y_n$, and then multiplying the result by $|J|$.

Example 1: The Joint p.d.f. of the Quotient and the Product of Two Random Variables. Suppose that two random variables X_1 and X_2 have a continuous joint distribution for which the joint p.d.f. is as follows:

$$f(x_1, x_2) = \begin{cases} 4x_1 x_2 & \text{for } 0 < x_1 < 1 \text{ and } 0 < x_2 < 1, \\ 0 & \text{otherwise.} \end{cases}$$

We shall determine the joint p.d.f. of two new random variables Y_1 and Y_2 which are defined by the relations

$$Y_1 = \frac{X_1}{X_2} \quad \text{and} \quad Y_2 = X_1 X_2. \tag{5}$$

When the equations in (5) are solved for X_1 and X_2 in terms of Y_1 and Y_2, we obtain the relations

$$X_1 = (Y_1 Y_2)^{1/2} \quad \text{and} \quad X_2 = \left(\frac{Y_2}{Y_1}\right)^{1/2}. \tag{6}$$

If S denotes the set of points (x_1, x_2) such that $0 < x_1 < 1$ and $0 < x_2 < 1$, then $\Pr[(X_1, X_2) \in S] = 1$. Furthermore, it can be seen from the relations in (5) and (6) that the conditions that $0 < X_1 < 1$ and $0 < X_2 < 1$ are equivalent to the conditions that $Y_1 > 0$, $Y_2 > 0$, $Y_1 Y_2 < 1$, and $(Y_2/Y_1) < 1$. Therefore, as the values (x_1, x_2) of X_1 and X_2 vary over the set S, the values (y_1, y_2) of Y_1 and Y_2 will vary over the set T containing all points such that $y_1 > 0$, $y_2 > 0$, $(y_1 y_2)^{1/2} < 1$, and $(y_2/y_1)^{1/2} < 1$. The sets S and T are sketched in Fig. 3.18.

The transformation defined by the equations in (5) or, equivalently, by the equations in (6) specifies a one-to-one relation between the points in S and the points in T. For $(y_1, y_2) \in T$, this transformation is given by the following relations:

$$x_1 = s_1(y_1, y_2) = (y_1 y_2)^{1/2},$$

$$x_2 = s_2(y_1, y_2) = \left(\frac{y_2}{y_1}\right)^{1/2}.$$

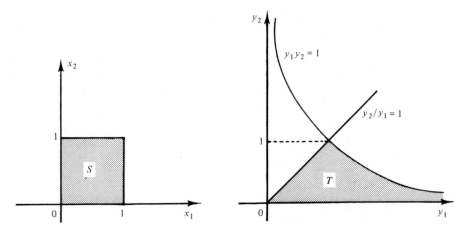

Figure 3.18 The sets S and T in Example 1.

For these relations,

$$\frac{\partial s_1}{\partial y_1} = \frac{1}{2}\left(\frac{y_2}{y_1}\right)^{1/2}, \qquad \frac{\partial s_1}{\partial y_2} = \frac{1}{2}\left(\frac{y_1}{y_2}\right)^{1/2},$$

$$\frac{\partial s_2}{\partial y_1} = -\frac{1}{2}\left(\frac{y_2}{y_1^3}\right)^{1/2}, \qquad \frac{\partial s_2}{\partial y_2} = \frac{1}{2}\left(\frac{1}{y_1 y_2}\right)^{1/2}.$$

Hence,

$$J = \det \begin{bmatrix} \frac{1}{2}\left(\frac{y_2}{y_1}\right)^{1/2} & \frac{1}{2}\left(\frac{y_1}{y_2}\right)^{1/2} \\ -\frac{1}{2}\left(\frac{y_2}{y_1^3}\right)^{1/2} & \frac{1}{2}\left(\frac{1}{y_1 y_2}\right)^{1/2} \end{bmatrix} = \frac{1}{2 y_1}.$$

Since $y_1 > 0$ throughout the set T, $|J| = 1/(2 y_1)$.

The joint p.d.f. $g(y_1, y_2)$ can now be obtained directly from Eq. (4) in the following way: In the expression for $f(x_1, x_2)$, replace x_1 with $(y_1 y_2)^{1/2}$, replace x_2 with $(y_2/y_1)^{1/2}$, and multiply the result by $|J| = 1/(2 y_1)$. Therefore,

$$g(y_1, y_2) = \begin{cases} 2\left(\frac{y_2}{y_1}\right) & \text{for } (y_1, y_2) \in T, \\ 0 & \text{otherwise.} \quad \square \end{cases}$$

Linear Transformations

As a further illustration of the results that have been presented here, we shall continue to assume that the n random variables $X_1, \ldots, X_n$ have a continuous joint distribution for which the joint p.d.f. is f, and we shall now consider a problem in which n new random variables $Y_1, \ldots, Y_n$ are derived from $X_1, \ldots, X_n$ by means of a linear transformation.

We shall let A be a given $n \times n$ matrix, and we shall suppose that the n random variables $Y_1, \ldots, Y_n$ are defined by the following equation:

$$\begin{bmatrix} Y_1 \\ \vdots \\ Y_n \end{bmatrix} = A \begin{bmatrix} X_1 \\ \vdots \\ X_n \end{bmatrix}. \tag{7}$$

Thus, each variable Y_i is a linear combination of the variables $X_1, \ldots, X_n$. We shall also suppose that the matrix A is nonsingular and, therefore, that the matrix A^{-1} exists. It then follows that the transformation in Eq. (7) is a one-to-one transformation of the entire space R^n onto itself. At any point $(y_1, \ldots, y_n) \in R^n$, the inverse transformation can be represented by the equation

$$\begin{bmatrix} x_1 \\ \vdots \\ x_n \end{bmatrix} = A^{-1} \begin{bmatrix} y_1 \\ \vdots \\ y_n \end{bmatrix}. \tag{8}$$

The Jacobian J of the transformation that is defined by Eq. (8) is simply $J = \det A^{-1}$. Also, it is known from the theory of determinants that

$$\det A^{-1} = \frac{1}{\det A}.$$

Therefore, at any point $(y_1, \ldots, y_n) \in R^n$, the joint p.d.f. $g(y_1, \ldots, y_n)$ of $Y_1, \ldots, Y_n$ can be evaluated in the following way: First, for $i = 1, \ldots, n$, the component x_i in $f(x_1, \ldots, x_n)$ is replaced with the ith component of the vector

$$A^{-1} \begin{bmatrix} y_1 \\ \vdots \\ y_n \end{bmatrix}.$$

Then the result is divided by $|\det A|$.

In vector notation, we let

$$x = \begin{bmatrix} x_1 \\ \vdots \\ x_n \end{bmatrix} \quad \text{and} \quad y = \begin{bmatrix} y_1 \\ \vdots \\ y_n \end{bmatrix}.$$

Also, we let $f(x)$ and $g(y)$ denote the joint p.d.f.'s of $X_1, \ldots, X_n$ and $Y_1, \ldots, Y_n$. Then

$$g(y) = \frac{1}{|\det A|} f(A^{-1}y) \qquad \text{for } y \in R^n. \tag{9}$$

The Sum of Two Random Variables

Suppose that two random variables X_1 and X_2 have a given joint p.d.f. f and that it is desired to find the p.d.f. of the random variable $Y = X_1 + X_2$.

As a convenient device, we shall let $Z = X_2$. Then the transformation from X_1 and X_2 to Y and Z will be a one-to-one linear transformation. The inverse transformation is specified by the following equations:

$$X_1 = Y - Z,$$

$$X_2 = Z.$$

The matrix A^{-1} of coefficients of this transformation is

$$A^{-1} = \begin{pmatrix} 1 & -1 \\ 0 & 1 \end{pmatrix}.$$

Therefore,

$$\det A^{-1} = \frac{1}{\det A} = 1.$$

It follows from Eq. (9) that the joint p.d.f. g_0 of Y and Z at any given point (y, z) will be

$$g_0(y, z) = f(y - z, z).$$

Therefore, the marginal p.d.f. g of Y can be obtained from the relation

$$g(y) = \int_{-\infty}^{\infty} f(y - z, z) \, dz \qquad \text{for } -\infty < y < \infty. \tag{10}$$

If we had initially defined Z by the relation $Z = X_1$, then we would have obtained the following alternative and equivalent form for the p.d.f. g:

$$g(y) = \int_{-\infty}^{\infty} f(z, y - z) \, dz \qquad \text{for } -\infty < y < \infty. \tag{11}$$

If X_1 and X_2 are independent random variables with marginal p.d.f.'s f_1 and f_2, then for all values of x_1 and x_2,

$$f(x_1, x_2) = f_1(x_1)f_2(x_2).$$

It therefore follows from Eqs. (10) and (11) that the p.d.f. g of $Y = X_1 + X_2$ is given by either of the following relations:

$$g(y) = \int_{-\infty}^{\infty} f_1(y - z)f_2(z)\, dz \qquad \text{for } -\infty < y < \infty, \tag{12}$$

or

$$g(y) = \int_{-\infty}^{\infty} f_1(z)f_2(y - z)\, dz \qquad \text{for } -\infty < y < \infty. \tag{13}$$

The p.d.f. g determined by either Eq. (12) or Eq. (13) is called the *convolution* of the p.d.f.'s f_1 and f_2.

Example 2: Determining the p.d.f. for a Convolution. Suppose that X_1 and X_2 are i.i.d. random variables and that the p.d.f. of each of these two variables is as follows:

$$f(x) = \begin{cases} e^{-x} & \text{for } x \geqslant 0, \\ 0 & \text{otherwise.} \end{cases}$$

We shall determine the p.d.f. g of the random variable $Y = X_1 + X_2$.
By Eq. (12) or Eq. (13),

$$g(y) = \int_{-\infty}^{\infty} f(y - z)f(z)\, dz \qquad \text{for } -\infty < y < \infty. \tag{14}$$

Since $f(x) = 0$ for $x < 0$, it follows that the integrand in Eq. (14) will be 0 if $z < 0$ or if $z > y$. Therefore, for $y \geqslant 0$,

$$g(y) = \int_0^y f(y - z)f(z)\, dz = \int_0^y e^{-(y-z)}e^{-z}\, dz$$

$$= \int_0^y e^{-y}\, dz = ye^{-y}.$$

Also, $g(y) = 0$ for $y < 0$. $\square$

The Range

Suppose now that the n random variables $X_1, \ldots, X_n$ form a random sample from a continuous distribution for which the p.d.f. is f and the d.f. is F. As earlier in this section, let the random variables Y_1 and Y_n be defined as follows:

$$Y_1 = \min\{X_1, \ldots, X_n\} \quad \text{and} \quad Y_n = \max\{X_1, \ldots, X_n\}. \tag{15}$$

The random variable $W = Y_n - Y_1$ is called the *range* of the sample. In other words, the range W is the difference between the largest value and the smallest value in the sample. We shall determine the p.d.f. of W.

The joint p.d.f. $g(y_1, y_n)$ of Y_1 and Y_n was presented in Eq. (1). If we let $Z = Y_1$, then the transformation from Y_1 and Y_n to W and Z will be a one-to-one linear transformation. The inverse transformation is specified by the following equations:

$$Y_1 = Z,$$

$$Y_n = W + Z.$$

For this transformation, $|J| = 1$. Therefore, the joint p.d.f. $h(w, z)$ of W and Z can be obtained by replacing y_1 with z and replacing y_n with $w + z$ in Eq. (1). The result, for $w > 0$ and $-\infty < z < \infty$, is

$$h(w, z) = n(n-1)[F(w+z) - F(z)]^{n-2} f(z) f(w+z). \tag{16}$$

Otherwise, $h(w, z) = 0$.

The marginal p.d.f. $h_1(w)$ of the range W can be obtained from the relation

$$h_1(w) = \int_{-\infty}^{\infty} h(w, z)\, dz. \tag{17}$$

Example 3: The Range of a Random Sample from a Uniform Distribution. Suppose that the n variables $X_1, \ldots, X_n$ form a random sample from a uniform distribution on the interval $(0, 1)$. We shall determine the p.d.f. of the range of the sample.

In this example,

$$f(x) = \begin{cases} 1 & \text{for } 0 < x < 1, \\ 0 & \text{otherwise,} \end{cases}$$

Also, $F(x) = x$ for $0 < x < 1$. Therefore, in Eq. (16), $h(w, z) = 0$ unless $0 < w < 1$ and $0 < z < 1 - w$. For values of w and z satisfying these conditions, it follows from Eq. (16) that

$$h(w, z) = n(n - 1)w^{n-2}.$$

It follows from Eq. (17) that for $0 < w < 1$, the p.d.f. of W is

$$h_1(w) = \int_0^{1-w} n(n - 1)w^{n-2} dz = n(n - 1)w^{n-2}(1 - w).$$

Otherwise, $h_1(w) = 0$. □

EXERCISES

1. Suppose that three random variables X_1, X_2, and X_3 have a continuous joint distribution for which the joint p.d.f. is as follows:

$$f(x_1, x_2, x_3) = \begin{cases} 8x_1x_2x_3 & \text{for } 0 < x_i < 1 \; (i = 1, 2, 3), \\ 0 & \text{otherwise.} \end{cases}$$

   Suppose also that $Y_1 = X_1$, $Y_2 = X_1X_2$, and $Y_3 = X_1X_2X_3$. Find the joint p.d.f. of Y_1, Y_2, and Y_3.

2. Suppose that X_1 and X_2 are i.i.d. random variables and that each of them has a uniform distribution on the interval $(0, 1)$. Find the p.d.f. of $Y = X_1 + X_2$.

3. Suppose that X_1 and X_2 have a continuous joint distribution for which the joint p.d.f. is as follows:

$$f(x_1, x_2) = \begin{cases} x_1 + x_2 & \text{for } 0 < x_1 < 1 \text{ and } 0 < x_2 < 1, \\ 0 & \text{otherwise.} \end{cases}$$

   Find the p.d.f. of $Y = X_1X_2$.

4. Suppose that the joint p.d.f. of X_1 and X_2 is as given in Exercise 3. Find the p.d.f. of $Z = X_1/X_2$.

5. Let X and Y be random variables for which the joint p.d.f. is as follows:

$$f(x, y) = \begin{cases} 2(x + y) & \text{for } 0 \leqslant x \leqslant y \leqslant 1, \\ 0 & \text{otherwise.} \end{cases}$$

   Find the p.d.f. of $Z = X + Y$.

6. Suppose that X_1 and X_2 are i.i.d. random variables and that the p.d.f. of each of them is as follows:

$$f(x) = \begin{cases} e^{-x} & \text{for } x > 0, \\ 0 & \text{otherwise.} \end{cases}$$

Find the p.d.f. of $Y = X_1 - X_2$.

7. Suppose that $X_1, \ldots, X_n$ form a random sample of size n from the uniform distribution on the interval $(0, 1)$ and that $Y_n = \max\{X_1, \ldots, X_n\}$. Find the smallest value of n such that

$$\Pr\{Y_n \geqslant 0.99\} \geqslant 0.95.$$

8. Suppose that the n variables $X_1, \ldots, X_n$ form a random sample from a uniform distribution on the interval $(0, 1)$ and that the random variables Y_1 and Y_n are defined as in Eq. (15). Determine the value of $\Pr(Y_1 \leqslant 0.1$ and $Y_n \leqslant 0.8)$.

9. For the conditions of Exercise 8, determine the value of $\Pr(Y_1 \leqslant 0.1$ and $Y_n \geqslant 0.8)$.

10. For the conditions of Exercise 8, determine the probability that the interval from Y_1 to Y_n will not contain the point $1/3$.

11. Let W denote the range of a random sample of n observations from a uniform distribution on the interval $(0, 1)$. Determine the value of $\Pr(W > 0.9)$.

12. Determine the p.d.f. of the range of a random sample of n observations from a uniform distribution on the interval $(-3, 5)$.

13. Suppose that $X_1, \ldots, X_n$ form a random sample of n observations from a uniform distribution on the interval $(0, 1)$, and let Y denote the second largest of the observations. Determine the p.d.f. of Y. *Hint:* First determine the d.f. G of Y by noting that

$$G(y) = \Pr(Y \leqslant y) = \Pr(\text{At least } n - 1 \text{ observations} \leqslant y).$$

14. Show that if $X_1, X_2, \ldots, X_n$ are independent random variables and if $Y_1 = r_1(X_1)$, $Y_2 = r_2(X_2), \ldots, Y_n = r_n(X_n)$, then $Y_1, Y_2, \ldots, Y_n$ are also independent random variables.

15. Suppose that $X_1, X_2, \ldots, X_5$ are five random variables for which the joint p.d.f. can be factored in the following form for all points $(x_1, x_2, \ldots, x_5) \in R^5$:

$$f(x_1, x_2, \ldots, x_5) = g(x_1, x_2)h(x_3, x_4, x_5),$$

where g and h are certain nonnegative functions. Show that if $Y_1 = r_1(X_1, X_2)$ and $Y_2 = r_2(X_3, X_4, X_5)$, then the random variables Y_1 and Y_2 are independent.

*3.10. THE BOREL-KOLMOGOROV PARADOX

In this section we shall describe an ambiguity, known as the *Borel-Kolmogorov paradox*, that can arise in the calculation of a conditional p.d.f. The Borel-Kolmogorov paradox is a result of the fact that the conditional p.d.f. of one random variable given another is defined conditionally on an event for which the probability is zero. We shall present two examples that illustrate the general nature of the ambiguity.

Conditioning on a Particular Value

Suppose that X_1 and X_2 are i.i.d. random variables and that the p.d.f. of each of them is as follows:

$$f_1(x) = \begin{cases} e^{-x} & \text{for } x > 0, \\ 0 & \text{otherwise.} \end{cases} \tag{1}$$

Then the joint p.d.f. of X_1 and X_2 is

$$f(x_1, x_2) = \begin{cases} e^{-(x_1 + x_2)} & \text{for } x_1 > 0 \text{ and } x_2 > 0, \\ 0 & \text{otherwise.} \end{cases} \tag{2}$$

Let the random variable Z be defined by the relation

$$Z = \frac{X_2 - 1}{X_1}. \tag{3}$$

Suppose that we are interested in the conditional distribution of X_1 given that $Z = 0$. This conditional p.d.f. can easily be found by applying the methods presented in this chapter as follows: Let $Y = X_1$. Then the inverse of the transformation from X_1 and X_2 to Y and Z is expressed by the relations

$$\begin{aligned} x_1 &= y, \\ x_2 &= yz + 1. \end{aligned} \tag{4}$$

The Jacobian of this transformation is

$$J = \det \begin{bmatrix} 1 & 0 \\ z & y \end{bmatrix} = y. \tag{5}$$

Since $y = x_1 > 0$, it follows from Eq. (4) of Sec. 3.9 and Eq. (2) of this section that the joint p.d.f. of Y and Z is

$$g(y, z) = \begin{cases} ye^{-(yz+y+1)} & \text{for } y > 0 \text{ and } yz > -1, \\ 0 & \text{otherwise.} \end{cases} \tag{6}$$

The value of this joint p.d.f. at the point where $Y = y > 0$ and $Z = 0$ is

$$g(y, 0) = ye^{-(y+1)}.$$

Therefore, the value of the marginal p.d.f. $g_2(z)$ at $z = 0$ is

$$g_2(0) = \int_0^\infty g(y, 0) \, dy = e^{-1}.$$

It follows that the conditional p.d.f. of Y given that $Z = 0$ is

$$\frac{g(y, 0)}{g_2(0)} = ye^{-y} \quad \text{for } y > 0. \tag{7}$$

Since $X_1 = Y$, the conditional p.d.f. (7) is also the conditional p.d.f. of X_1 given that $Z = 0$. Thus, if we let A denote the event that $Z = 0$, we may write

$$g_1(x_1 | A) = x_1 e^{-x_1} \quad \text{for } x_1 > 0. \tag{8}$$

The derivation leading to Eq. (8) is straightforward and follows the rules presented for conditional p.d.f.'s. However, there apparently is a much simpler way to derive $g_1(x_1 | A)$. Since $X_1 > 0$, it follows from Eq. (3) that the event that $Z = 0$ is equivalent to the event that $X_2 = 1$. Hence, the event A could just as well be described as the event that $X_2 = 1$. From this point of view, the conditional p.d.f. of X_1 given A should be the same as the conditional p.d.f. of X_1 given that $X_2 = 1$. Since X_1 and X_2 are independent, this conditional p.d.f. is simply the marginal p.d.f. of X_1 as given by Eq. (1). Thus, from this point of view,

$$g_1(x_1 | A) = f_1(x_1) = e^{-x_1} \quad \text{for } x_1 > 0. \tag{9}$$

Since we have here obtained two different expressions, (8) and (9), for the same conditional p.d.f. $g_1(x_1 | A)$, we have arrived at the Borel-Kolmogorov paradox. Both expressions are correct, but they have different interpretations. If we regard the event A as one point in the sample space of the random variable Z, then Eq. (8) is correct. If we regard A as one point in the sample space of X_2, then Eq. (9) is correct.

The Borel-Kolmogorov paradox arises because $Pr(A) = 0$. It emphasizes the fact that it is not possible to define a conditional distribution in a meaningful manner for just a single event having probability zero. Thus, a conditional distribution can have meaning only in the context of an entire family of conditional distributions that are defined in a consistent fashion.

Conditioning on the Equality of Two Random Variables

We shall conclude this section with another example which is based on the joint p.d.f. of X_1 and X_2 given in (2). Suppose now that we wish to calculate the conditional p.d.f. of X_1 given that $X_1 = X_2$. One way to do this is to let $Z = X_1 - X_2$ and to determine the conditional p.d.f. of X_1 given that $Z = 0$.

It can be found in a straightforward fashion that the joint p.d.f. of X_1 and Z is

$$g(x_1, z) = \begin{cases} e^{-(2x_1 - z)} & \text{for } x_1 > 0, z < x_1, \\ 0 & \text{otherwise.} \end{cases} \tag{10}$$

Hence, for $Z = 0$,

$$g(x_1, 0) = e^{-2x_1} \qquad \text{for } x_1 > 0,$$

and the value of the marginal p.d.f. $g_2(z)$ at $z = 0$ is

$$g_2(0) = \int_0^\infty e^{-2x_1} \, dx_1 = \frac{1}{2}.$$

Thus, if we let B denote the event that $X_1 = X_2$ or, equivalently, that $Z = 0$, then

$$g_1(x_1 \mid B) = \frac{g(x_1, 0)}{g_2(0)} = 2e^{-2x_1} \qquad \text{for } x_1 > 0. \tag{11}$$

Another way to approach the same problem is to let $W = X_2/X_1$. Then the event B is equivalent to the event that $W = 1$. It can again be found in a straightforward fashion that the joint p.d.f. of X_1 and W is

$$h(x_1, w) = \begin{cases} x_1 e^{-(x_1 + wx_1)} & \text{for } x_1 > 0 \text{ and } w > 0, \\ 0 & \text{otherwise.} \end{cases} \tag{12}$$

Hence, for $W = 1$,

$$h(x_1, 1) = x_1 e^{-2x_1} \qquad \text{for } x_1 > 0$$

and the value of the marginal p.d.f. $h_2(w)$ at $w = 1$ is

$$h_2(1) = \int_0^\infty x_1 e^{-2x_1} \, dx_1 = \frac{1}{4}.$$

Thus, on the basis of this approach we find that

$$g_1(x_1 \mid B) = \frac{h(x_1, 1)}{h_2(1)} = 4x_1 e^{-2x_1} \qquad \text{for } x_1 > 0. \tag{13}$$

We have again obtained two different expressions, (11) and (13), for the same conditional p.d.f. $g_1(x_1 \mid B)$. Again, the ambiguity arises because $\Pr(B) = 0$.

3.11. SUPPLEMENTARY EXERCISES

1. Suppose that X and Y are independent random variables; that X has a discrete uniform distribution on the integers 1, 2, 3, 4, 5; and that Y has a continuous uniform distribution on the interval $(0, 5)$. Let Z be a random variable such that $Z = X$ with probability $1/2$ and $Z = Y$ with probability $1/2$. Sketch the d.f. of Z.

2. Suppose that the random variable X has the following d.f.:

$$F(x) = \begin{cases} 0 & \text{for } x \leqslant 0, \\ \dfrac{2}{5}x & \text{for } 0 < x \leqslant 1, \\ \dfrac{3}{5}x - \dfrac{1}{5} & \text{for } 1 \leqslant x \leqslant 2, \\ 1 & \text{for } x > 2. \end{cases}$$

Verify that X has a continuous distribution, and determine the p.d.f. of X.

3. Suppose that the random variable X has a continuous distribution with the following p.d.f.:

$$f(x) = \frac{1}{2} e^{-|x|} \qquad \text{for } -\infty < x < \infty.$$

Determine the value x_0 such that $F(x_0) = 0.9$, where $F(x)$ is the d.f. of X.

4. Suppose that X_1 and X_2 are i.i.d. random variables, and that each has a uniform distribution on the interval $(0, 1)$. Evaluate $\Pr(X_1^2 + X_2^2 \leq 1)$.

5. For any value of $p > 1$, let

$$c(p) = \sum_{x=1}^{\infty} \frac{1}{x^p}.$$

Suppose that the random variable X has a discrete distribution with the following p.f.:

$$f(x) = \frac{1}{c(p)x^p} \qquad \text{for } x = 1, 2, \ldots.$$

(a) For any fixed positive integer n, determine the probability that X will be divisible by n.

(b) Determine the probability that X will be odd.

6. Suppose that X_1 and X_2 are i.i.d. random variables, each of which has the p.f. $f(x)$ specified in Exercise 5. Determine the probability that $X_1 + X_2$ will be even.

7. Suppose that an electronic system comprises four components, and let X_j denote the time until component j fails to operate ($j = 1, 2, 3, 4$). Suppose that X_1, X_2, X_3, and X_4 are i.i.d. random variables, each of which has a continuous distribution with d.f. $F(x)$. Suppose that the system will operate as long as both component 1 and at least one of the other three components operate. Determine the d.f. of the time until the system fails to operate.

8. Suppose that a box contains a large number of tacks, and that the probability X that a particular tack will land with its point up when it is tossed varies from tack to tack in accordance with the following p.d.f.:

$$f(x) = \begin{cases} 2(1 - x) & \text{for } 0 < x < 1, \\ 0 & \text{otherwise.} \end{cases}$$

Suppose that a tack is selected at random from the box and this tack is then tossed three times independently. Determine the probability that the tack will land with its point up on all three tosses.

9. Suppose that the radius X of a circle is a random variable having the following p.d.f.:

$$f(x) = \begin{cases} \frac{1}{8}(3x + 1) & \text{for } 0 < x < 2, \\ 0 & \text{otherwise.} \end{cases}$$

Determine the p.d.f. of the area of the circle.

10. Suppose that the random variable X has the following p.d.f.:

$$f(x) = \begin{cases} 2e^{-2x} & \text{for } x > 0, \\ 0 & \text{otherwise.} \end{cases}$$

Construct a random variable $Y = r(X)$ that has a uniform distribution on the interval $(0, 5)$.

11. Suppose that the 12 random variables $X_1, \ldots, X_{12}$ are i.i.d. and that each has a uniform distribution on the interval $(0, 20)$. For $j = 0, 1, \ldots, 19$, let I_j denote the interval $(j, j + 1)$. Determine the probability that none of the 20 disjoint intervals I_j will contain more than one of the random variables $X_1, \ldots, X_{12}$.

✕ 12. Suppose that the joint distribution of X and Y is uniform over a set A in the xy-plane. For which of the following sets A are X and Y independent?

(a) A circle with a radius of 1 and with its center at the origin.

(b) A circle with a radius of 1 and with its center at the point $(3, 5)$.

(c) A square with vertices at the four points $(1, 1)$, $(1, -1)$, $(-1, -1)$, and $(-1, 1)$.

(d) A rectangle with vertices at the four points $(0, 0)$, $(0, 3)$, $(1, 3)$, and $(1, 0)$.

(e) A square with vertices at the four points $(0, 0)$, $(1, 1)$, $(0, 2)$, and $(-1, 1)$.

13. Suppose that X and Y are independent random variables with the following p.d.f.'s:

$$f_1(x) = \begin{cases} 1 & \text{for } 0 < x < 1, \\ 0 & \text{otherwise,} \end{cases}$$

$$f_2(y) = \begin{cases} 8y & \text{for } 0 < y < \frac{1}{2}, \\ 0 & \text{otherwise.} \end{cases}$$

Determine the value of $\Pr(X > Y)$.

14. Suppose that on a particular day two persons A and B arrive at a certain store independently of each other. Suppose that A remains in the store for 15 minutes and B remains in the store for 10 minutes. If the time of arrival of each person has a uniform distribution over the hour between 9:00 A.M. and 10:00 A.M., what is the probability that A and B will be in the store at the same time?

15. Suppose that X and Y have the following joint p.d.f.:

$$f(x, y) = \begin{cases} 2(x + y) & \text{for } 0 < x < y < 1, \\ 0 & \text{otherwise.} \end{cases}$$

Determine (a) $\Pr(X < 1/2)$; (b) the marginal p.d.f. of X; and (c) the conditional p.d.f. of Y given that $X = x$.

16. Suppose that X and Y are random variables. The marginal p.d.f. of X is

$$f(x) = \begin{cases} 3x^2 & \text{for } 0 < x < 1, \\ 0 & \text{otherwise.} \end{cases}$$

Also, the conditional p.d.f. of Y given that $X = x$ is

$$g(y \mid x) = \begin{cases} \dfrac{3y^2}{x^3} & \text{for } 0 < y < x, \\ 0 & \text{otherwise.} \end{cases}$$

Determine (a) the marginal p.d.f. of Y and (b) the conditional p.d.f. of X given that $Y = y$.

17. Suppose that the joint distribution of X and Y is uniform over the region in the xy-plane bounded by the four lines $x = -1$, $x = 1$, $y = x + 1$, and $y = x - 1$. Determine (a) $\Pr(XY > 0)$ and (b) the conditional p.d.f. of Y given that $X = x$.

18. Suppose that the random variables X, Y, and Z have the following joint p.d.f.:

$$f(x, y, z) = \begin{cases} 6 & \text{for } 0 < x < y < z < 1, \\ 0 & \text{otherwise.} \end{cases}$$

Determine the univariate marginal p.d.f.'s of X, Y, and Z.

19. Suppose that the random variables X, Y, and Z have the following joint p.d.f.:

$$f(x, y, z) = \begin{cases} 2 & \text{for } 0 < x < y < 1 \text{ and } 0 < z < 1, \\ 0 & \text{otherwise.} \end{cases}$$

Evaluate $\Pr(3X > Y \mid 1 < 4Z < 2)$.

20. Suppose that X and Y are i.i.d. random variables, and that each has the following p.d.f.:

$$f(x) = \begin{cases} e^{-x} & \text{for } x > 0, \\ 0 & \text{otherwise.} \end{cases}$$

Also, let $U = X/(X + Y)$ and $V = X + Y$.

(a) Determine the joint p.d.f. of U and V.

(b) Are U and V independent?

21. Suppose that the random variables X and Y have the following joint p.d.f.:

$$f(x, y) = \begin{cases} 8xy & \text{for } 0 \leqslant x \leqslant y \leqslant 1, \\ 0 & \text{otherwise.} \end{cases}$$

Also, let $U = X/Y$ and $V = Y$.

(a) Determine the joint p.d.f. of U and V.

(b) Are X and Y independent?

(c) Are U and V independent?

22. Suppose that $X_1, \ldots, X_n$ are i.i.d. random variables, each having the following d.f.:

$$F(x) = \begin{cases} 0 & \text{for } x \leqslant 0, \\ 1 - e^{-x} & \text{for } x > 0. \end{cases}$$

Let $Y_1 = \min\{X_1, \ldots, X_n\}$ and $Y_n = \max\{X_1, \ldots, X_n\}$. Determine the conditional p.d.f. of Y_1 given that $Y_n = y_n$.

23. Suppose that X_1, X_2, and X_3 form a random sample of three observations from a distribution having the following p.d.f.:

$$f(x) = \begin{cases} 2x & \text{for } 0 < x < 1, \\ 0 & \text{otherwise.} \end{cases}$$

Determine the p.d.f. of the range of the sample.

Expectation

4

4.1. THE EXPECTATION OF A RANDOM VARIABLE

Expectation for a Discrete Distribution

Suppose that a random variable X has a discrete distribution for which the p.f. is f. The *expectation* of X, denoted by $E(X)$, is a number defined as follows:

$$E(X) = \sum_x x f(x). \tag{1}$$

Example 1: Calculating an Expectation from a p.f. Suppose that a random variable X can have only the four different values -2, 0, 1, and 4, and that $\Pr(X = -2) = 0.1$, $\Pr(X = 0) = 0.4$, $\Pr(X = 1) = 0.3$, and $\Pr(X = 4) = 0.2$. Then

$$E(X) = -2(0.1) + 0(0.4) + 1(0.3) + 4(0.2)$$

$$= 0.9. \quad \square$$

It can be seen from Example 1 that the expectation $E(X)$ is not necessarily equal to one of the possible values of X.

If X can have only a finite number of different values, as in Example 1, then there will be only a finite number of terms in the summation in Eq. (1). However, if there is an infinite sequence of different possible values of X, then the summation in Eq. (1) consists of an infinite series of terms. Such a series may not converge for a given p.f. It is said that the expectation $E(X)$ *exists* if and only if

179

the summation in Eq. (1) is *absolutely convergent*, that is, if and only if

$$\sum_x |x| f(x) < \infty.$$ (2)

In other words, if relation (2) is satisfied, then $E(X)$ exists and its value is given by Eq. (1). If relation (2) is not satisfied, then $E(X)$ does not exist.

Expectation for a Continuous Distribution

If a random variable X has a continuous distribution for which the p.d.f. is f, then the expectation $E(X)$ is defined as follows:

$$E(X) = \int_{-\infty}^{\infty} x f(x)\, dx.$$ (3)

Example 2: Calculating an Expectation from a p.d.f. Suppose that the p.d.f. of a random variable X with a continuous distribution is

$$f(x) = \begin{cases} 2x & \text{for } 0 < x < 1, \\ 0 & \text{otherwise.} \end{cases}$$

Then

$$E(X) = \int_0^1 x(2x)\, dx = \int_0^1 2x^2\, dx = \frac{2}{3}. \quad \square$$

It is said that the expectation $E(X)$ exists for a continuous distribution if and only if the integral in Eq. (3) is absolutely convergent, that is, if and only if

$$\int_{-\infty}^{\infty} |x| f(x)\, dx < \infty.$$

Whenever X is a *bounded* random variable, that is, whenever there are numbers a and $b (-\infty < a < b < \infty)$ such that $\Pr(a \leqslant X \leqslant b) = 1$, as in Example 2, then $E(X)$ must exist.

Interpretation of the Expectation

The number $E(X)$ is also called the *expected value* of X or the *mean* of X; and the terms expectation, expected value, and mean can be used interchangeably. The number $E(X)$ is also called the expectation, expected value, or mean *of the*

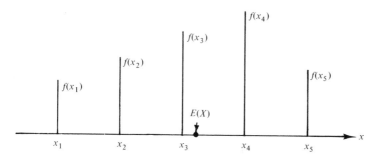

Figure 4.1 The mean of a discrete distribution.

distribution of X. For instance, in Example 2, the number $2/3$ may be called either the expected value of X or the mean of the distribution specified by the p.d.f. f.

Relation of the Mean to the Center of Gravity. The expectation of a random variable or, equivalently, the mean of its distribution can be regarded as being the center of gravity of that distribution. To illustrate this concept, consider, for example, the p.f. sketched in Fig. 4.1. The x-axis may be regarded as a long weightless rod to which weights are attached. If a weight equal to $f(x_j)$ is attached to this rod at each point x_j, then the rod will be balanced if it is supported at the point $E(X)$.

Now consider the p.d.f. sketched in Fig. 4.2. In this case, the x-axis may be regarded as a long rod over which the mass varies continuously. If the density of the rod at each point x is equal to $f(x)$, then the center of gravity of the rod will be located at the point $E(X)$ and the rod will be balanced if it is supported at that point.

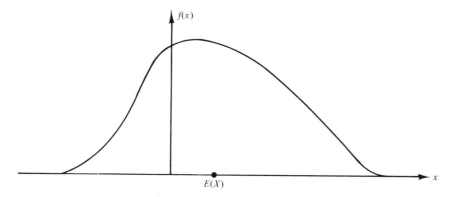

Figure 4.2 The mean of a continuous distribution.

It can be seen from this discussion that the mean of a distribution can be affected greatly by even a very small change in the amount of probability that is assigned to a large value of x. For example, the mean of the distribution represented by the p.f. in Fig. 4.1 can be moved to any specified point on the x-axis, no matter how far from the origin that point may be, by removing an arbitrarily small but positive amount of probability from one of the points x_j and adding this amount of probability at a point far enough from the origin.

Suppose now that the p.f. or p.d.f. f of some distribution is symmetric with respect to a given point x_0 on the x-axis. In other words, suppose that $f(x_0 + \delta) = f(x_0 - \delta)$ for all values of δ. Also assume that the mean $E(X)$ of this distribution exists. In accordance with the interpretation that the mean is at the center of gravity, it follows that $E(X)$ must be equal to x_0, which is the point of symmetry. The following example emphasizes the fact that it is necessary to make certain that the mean $E(X)$ exists before it can be concluded that $E(X) = x_0$.

The Cauchy Distribution. Suppose that a random variable X has a continuous distribution for which the p.d.f. is as follows:

$$f(x) = \frac{1}{\pi(1 + x^2)} \qquad \text{for } -\infty < x < \infty. \tag{4}$$

This distribution is called the *Cauchy distribution*. We can verify the fact that $\int_{-\infty}^{\infty} f(x)\, dx = 1$ by using the following standard result from elementary calculus:

$$\frac{d}{dx} \tan^{-1} x = \frac{1}{1 + x^2} \qquad \text{for } -\infty < x < \infty.$$

The p.d.f. specified by Eq. (4) is sketched in Fig. 4.3. This p.d.f. is symmetric with respect to the point $x = 0$. Therefore, if the mean of the Cauchy distribution existed, its value would have to be 0. However,

$$\int_{-\infty}^{\infty} |x| f(x)\, dx = \frac{2}{\pi} \int_{0}^{\infty} \frac{x}{1 + x^2}\, dx = \infty.$$

Therefore, the mean of the Cauchy distribution does not exist.

The reason for the nonexistence of the mean of the Cauchy distribution is as follows: When the curve $y = f(x)$ is sketched as in Fig. 4.3, its tails approach the x-axis rapidly enough to permit the total area under the curve to be equal to 1. On the other hand, if each value of $f(x)$ is multiplied by x and the curve $y = xf(x)$ is sketched, as in Fig. 4.4, the tails of this curve approach the x-axis so slowly that the total area between the x-axis and each part of the curve is infinite.

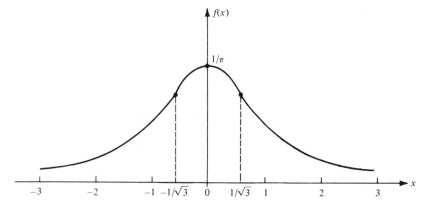

Figure 4.3   The p.d.f. of a Cauchy distribution.

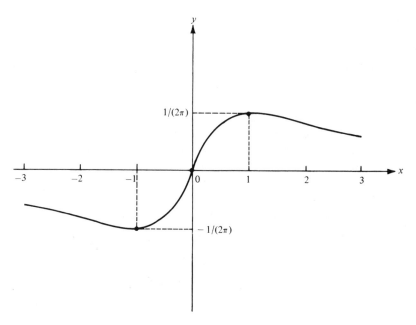

Figure 4.4   The curve $y = xf(x)$ for the Cauchy distribution.

The Expectation of a Function

Functions of a Single Random Variable.   If X is a random variable for which the p.d.f. is f, then the expectation of any function $r(X)$ can be found by applying the definition of expectation to the distribution of $r(X)$ as follows: Let $Y = r(X)$; determine the probability distribution of Y; and then determine $E(Y)$ by applying either Eq. (1) or Eq. (3). For example, suppose that Y has a continuous distribution with the p.d.f. g. Then

$$E[r(X)] = E(Y) = \int_{-\infty}^{\infty} yg(y)\,dy. \tag{5}$$

However, it is not actually necessary to determine the p.d.f. of $r(X)$ in order to calculate the expectation $E[r(X)]$. In fact, it can be shown that the value of $E[r(X)]$ can always be calculated directly from the relation

$$E[r(X)] = \int_{-\infty}^{\infty} r(x)f(x)\,dx. \tag{6}$$

In other words, it can be shown that the two relations (5) and (6) must yield the same value of $E[r(X)]$, and that the expectation $E[r(X)]$ will exist if and only if

$$\int_{-\infty}^{\infty} |r(x)|f(x)\,dx < \infty.$$

If the distribution of Y is discrete, the integral in Eq. (5) should be replaced by a sum. If the distribution of X is discrete, the integral in Eq. (6) should be replaced by a sum.

A general proof of the equality of the values of $E[r(X)]$ determined from Eqs. (5) and (6) will not be given here. However, we shall demonstrate this equality for two special cases. First, suppose that the distribution of X is discrete. Then the distribution of Y must also be discrete. For this case,

$$\sum_y yg(y) = \sum_y y \Pr[r(X) = y]$$

$$= \sum_y y \sum_{x:r(x)=y} f(x)$$

$$= \sum_y \sum_{x:r(x)=y} r(x)f(x) = \sum_x r(x)f(x).$$

Hence, Eqs. (5) and (6) yield the same value.

Second, suppose that the distribution of X is continuous. Suppose also, as in Sec. 3.8, that $r(x)$ is either strictly increasing or strictly decreasing, and that the inverse function $s(y)$ can be differentiated. Then, if we change variables in Eq. (6) from x to $y = r(x)$,

$$\int_{-\infty}^{\infty} r(x)f(x)\,dx = \int_{-\infty}^{\infty} yf[s(y)]\left|\frac{ds(y)}{dy}\right|dy.$$

It now follows from Eq. (3) of Sec. 3.8 that the right side of this equation is equal to

$$\int_{-\infty}^{\infty} y g(y)\, dy.$$

Hence, Eqs. (5) and (6) again yield the same value.

Example 3: Determining the Expectation of $X^{1/2}$. Suppose that the p.d.f. of X is as given in Example 2 and that $Y = X^{1/2}$. Then, by Eq. (6),

$$E(Y) = \int_0^1 x^{1/2}(2x)\, dx = \frac{4}{5}. \quad \Box$$

Functions of Several Random Variables. Suppose that the joint p.d.f. $f(x_1, \ldots, x_n)$ of n random variables $X_1, \ldots, X_n$ is given, and that $Y = r(X_1, \ldots, X_n)$. It can be shown that the expected value $E(Y)$ can be determined directly from the relation

$$E(Y) = \int_{R^n} \cdots \int r(x_1, \ldots, x_n) f(x_1, \ldots, x_n)\, dx_1 \cdots dx_n.$$

Hence, in this case also, it is unnecessary to determine the distribution of Y in order to calculate the value of $E(Y)$.

Example 4: Determining the Expectation of a Function of Two Variables. Suppose that a point (X, Y) is chosen at random from the square S containing all points (x, y) such that $0 \leqslant x \leqslant 1$ and $0 \leqslant y \leqslant 1$. We shall determine the expected value of $X^2 + Y^2$.

Since X and Y have a uniform distribution over the square S, and since the area of S is 1, the joint p.d.f. of X and Y is

$$f(x, y) = \begin{cases} 1 & \text{for } (x, y) \in S, \\ 0 & \text{otherwise.} \end{cases}$$

Therefore,

$$E(X^2 + Y^2) = \int_{-\infty}^{\infty} \int_{-\infty}^{\infty} (x^2 + y^2) f(x, y)\, dx\, dy$$

$$= \int_0^1 \int_0^1 (x^2 + y^2)\, dx\, dy = \frac{2}{3}. \quad \Box$$

EXERCISES

1. If an integer between 1 and 100 is to be chosen at random, what is the expected value?

2. In a class of 50 students, the number of students n_i of each age i is shown in the following table:

Age i	n_i
18	20
19	22
20	4
21	3
25	1

   If a student is to be selected at random from the class, what is the expected value of his age?

3. Suppose that one word is to be selected at random from the sentence THE GIRL PUT ON HER BEAUTIFUL RED HAT. If X denotes the number of letters in the word that is selected, what is the value of $E(X)$?

4. Suppose that one letter is to be selected at random from the 30 letters in the sentence given in Exercise 3. If Y denotes the number of letters in the word in which the selected letter appears, what is the value of $E(Y)$?

5. Suppose that a random variable X has a continuous distribution with the p.d.f. f given in Example 2 of this section. Find the expectation of $1/X$.

6. Suppose that a random variable X has a uniform distribution on the interval $0 < x < 1$. Show that the expectation of $1/X$ does not exist.

7. Suppose that X and Y have a continuous joint distribution for which the joint p.d.f. is as follows:

$$f(x, y) = \begin{cases} 12y^2 & \text{for } 0 \leqslant y \leqslant x \leqslant 1, \\ 0 & \text{otherwise.} \end{cases}$$

   Find the value of $E(XY)$.

8. Suppose that a point is chosen at random on a stick of unit length and that the stick is broken into two pieces at that point. Find the expected value of the length of the longer piece.

9. Suppose that a particle is released at the origin of the xy-plane and travels into the half-plane where $x > 0$. Suppose that the particle travels in a straight line and that the angle between the positive half of the x-axis and this line is

α, which can be either positive or negative. Suppose, finally, that the angle α has a uniform distribution on the interval $(-\pi/2, \pi/2)$. Let Y be the ordinate of the point at which the particle hits the vertical line $x = 1$. Show that the distribution of Y is a Cauchy distribution.

10. Suppose that the random variables $X_1, \ldots, X_n$ form a random sample of size n from a uniform distribution on the interval $(0, 1)$. Let $Y_1 = \min\{X_1, \ldots, X_n\}$, and $Y_n = \max\{X_1, \ldots, X_n\}$. Find $E(Y_1)$ and $E(Y_n)$.

11. Suppose that the random variables $X_1, \ldots, X_n$ form a random sample of size n from a continuous distribution for which the d.f. is F; and let the random variables Y_1 and Y_n be defined as in Exercise 10. Find $E[F(Y_1)]$ and $E[F(Y_n)]$.

4.2. PROPERTIES OF EXPECTATIONS

Basic Theorems

Suppose that X is a random variable for which the expectation $E(X)$ exists. We shall present several results pertaining to the basic properties of expectations.

Theorem 1. *If $Y = aX + b$, where a and b are constants, then*

$$E(Y) = aE(X) + b.$$

Proof. First we shall assume, for convenience, that X has a continuous distribution for which the p.d.f. is f. Then

$$E(Y) = E(aX + b) = \int_{-\infty}^{\infty} (ax + b)f(x)\, dx$$

$$= a\int_{-\infty}^{\infty} xf(x)\, dx + b\int_{-\infty}^{\infty} f(x)\, dx$$

$$= aE(X) + b.$$

A similar proof can be given for a discrete distribution or for a more general type of distribution. $\square$

Example 1: Calculating the Expectation of a Linear Function. Suppose that $E(X) = 5$. Then

$$E(3X - 5) = 3E(X) - 5 = 10$$

and

$$E(-3X + 15) = -3E(X) + 15 = 0. \quad \square$$

It follows from Theorem 1 that for any constant c, $E(c) = c$. In other words, if $X = c$ with probability 1, then $E(X) = c$.

Theorem 2. *If there exists a constant a such that* $\Pr(X \geq a) = 1$, *then* $E(X) \geq a$. *If there exists a constant b such that* $\Pr(X \leq b) = 1$, *then* $E(X) \leq b$.

Proof. We shall assume again, for convenience, that X has a continuous distribution for which the p.d.f. is f, and we shall suppose first that $\Pr(X \geq a) = 1$. Then

$$E(X) = \int_{-\infty}^{\infty} xf(x)\, dx = \int_{a}^{\infty} xf(x)\, dx$$

$$\geq \int_{a}^{\infty} af(x)\, dx = a\Pr(X \geq a) = a.$$

The proof of the other part of the theorem and the proof for a discrete distribution or a more general type of distribution are similar. $\square$

It follows from Theorem 2 that if $\Pr(a \leq X \leq b) = 1$, then $a \leq E(X) \leq b$. Also, it can be shown that if $\Pr(X \geq a) = 1$ and if $E(X) = a$, then it must be true that $\Pr(X > a) = 0$ and $\Pr(X = a) = 1$.

Theorem 3. *If $X_1, \ldots, X_n$ are n random variables such that each expectation $E(X_i)$ exists $(i = 1, \ldots, n)$, then*

$$E(X_1 + \cdots + X_n) = E(X_1) + \cdots + E(X_n).$$

Proof. We shall first assume that $n = 2$ and also, for convenience, that X_1 and X_2 have a continuous joint distribution for which the joint p.d.f. is f. Then

$$E(X_1 + X_2) = \int_{-\infty}^{\infty} \int_{-\infty}^{\infty} (x_1 + x_2) f(x_1, x_2)\, dx_1\, dx_2$$

$$= \int_{-\infty}^{\infty} \int_{-\infty}^{\infty} x_1 f(x_1, x_2)\, dx_1\, dx_2 + \int_{-\infty}^{\infty} \int_{-\infty}^{\infty} x_2 f(x_1, x_2)\, dx_1\, dx_2$$

$$= E(X_1) + E(X_2).$$

The proof for a discrete distribution or a more general joint distribution is similar. Finally, the theorem can be established for any positive integer n by an induction argument. $\square$

It should be emphasized that, in accordance with Theorem 3, the expectation of the sum of several random variables must always be equal to the sum of their individual expectations, regardless of whether or not the random variables are independent.

It follows from Theorems 1 and 3 that for any constants $a_1, \ldots, a_n$ and b,

$$E(a_1 X_1 + \cdots + a_n X_n + b) = a_1 E(X_1) + \cdots + a_n E(X_n) + b.$$

Example 2: Sampling Without Replacement. Suppose that a box contains red balls and blue balls and that the proportion of red balls in the box is $p(0 \leqslant p \leqslant 1)$. Suppose that n balls are selected from the box at random *without replacement*, and let X denote the number of red balls that are selected. We shall determine the value of $E(X)$.

We shall begin by defining n random variables $X_1, \ldots, X_n$ as follows: For $i = 1, \ldots, n$, let $X_i = 1$ if the ith ball that is selected is red, and let $X_i = 0$ if the ith ball is blue. Since the n balls are selected without replacement, the random variables $X_1, \ldots, X_n$ are dependent. However, the marginal distribution of each X_i can be derived easily (see Exercise 9 of Sec. 1.7). We can imagine that all the balls are arranged in the box in some random order, and that the first n balls in this arrangement are selected. Because of randomness, the probability that the ith ball in the arrangement will be red is simply p. Hence, for $i = 1, \ldots, n$,

$$\Pr(X_i = 1) = p \qquad \text{and} \qquad \Pr(X_i = 0) = 1 - p. \tag{1}$$

Therefore, $E(X_i) = 1(p) + 0(1 - p) = p$.

From the definition of $X_1, \ldots, X_n$, it follows that $X_1 + \cdots + X_n$ is equal to the total number of red balls that are selected. Therefore, $X = X_1 + \cdots + X_n$ and, by Theorem 3,

$$E(X) = E(X_1) + \cdots + E(X_n) = np. \quad \square \tag{2}$$

The Mean of a Binomial Distribution

Suppose again that in a box containing red balls and blue balls, the proportion of red balls is p $(0 \leqslant p \leqslant 1)$. Suppose now, however, that a random sample of n balls is selected from the box *with replacement*. If X denotes the number of red balls in the sample, then X has a binomial distribution with parameters n and p, as described in Sec. 3.1. We shall now determine the value of $E(X)$.

As before, for $i = 1, \ldots, n$, we shall let $X_i = 1$ if the ith ball that is selected is red and shall let $X_i = 0$ otherwise. Then, as before, $X = X_1 + \cdots + X_n$. In this problem, the random variables $X_1, \ldots, X_n$ are independent, and the marginal distribution of each X_i is again given by Eq. (1). Therefore, $E(X_i) = p$ for $i = 1, \ldots, n$, and it follows from Theorem 3 that

$$E(X) = np. \tag{3}$$

Thus, the mean of a binomial distribution with parameters n and p is np. The p.f. $f(x)$ of this binomial distribution is given by Eq. (1) of Sec. 3.1, and the mean can be computed directly from the p.f. as follows:

$$E(X) = \sum_{x=0}^{n} x \binom{n}{x} p^x q^{n-x}. \tag{4}$$

Hence, by Eq. (3), the value of the sum in Eq. (4) must be np.

It is seen from Eqs. (2) and (3) that the expected number of red balls in a sample of n balls is np, regardless of whether the sample is selected with or without replacement.

Expected Number of Matches

Consider again the matching problem described in Sec. 1.10. In this problem, a person types n letters, types the addresses on n envelopes, and then places each letter in an envelope in a random manner. We shall now determine the expected value of the number X of letters that are placed in the correct envelopes.

For $i = 1, \ldots, n$, we shall let $X_i = 1$ if the ith letter is placed in the correct envelope, and shall let $X_i = 0$ otherwise. Then, for $i = 1, \ldots, n$,

$$\Pr(X_i = 1) = \frac{1}{n} \quad \text{and} \quad \Pr(X_i = 0) = 1 - \frac{1}{n}.$$

Therefore,

$$E(X_i) = \frac{1}{n} \quad \text{for } i = 1, \ldots, n.$$

Since $X = X_1 + \cdots + X_n$, it follows that

$$E(X) = E(X_1) + \cdots + E(X_n)$$

$$= \frac{1}{n} + \cdots + \frac{1}{n} = 1.$$

Thus, the expected value of the number of correct matches of letters and envelopes is 1, regardless of the value of n.

Expectation of a Product

Theorem 4. *If $X_1, \ldots, X_n$ are n independent random variables such that each expectation $E(X_i)$ exists ($i = 1, \ldots, n$), then*

$$E\left(\prod_{i=1}^{n} X_i\right) = \prod_{i=1}^{n} E(X_i).$$

Proof. We shall again assume, for convenience, that $X_1, \ldots, X_n$ have a continuous joint distribution for which the joint p.d.f. is f. Also, we shall let f_i denote the marginal p.d.f. of X_i ($i = 1, \ldots, n$). Then, since the variables $X_1, \ldots, X_n$ are independent, it follows that at any point $(x_1, \ldots, x_n) \in R^n$,

$$f(x_1, \ldots, x_n) = \prod_{i=1}^{n} f_i(x_i).$$

Therefore,

$$E\left(\prod_{i=1}^{n} X_i\right) = \int_{-\infty}^{\infty} \cdots \int_{-\infty}^{\infty} \left(\prod_{i=1}^{n} x_i\right) f(x_1, \ldots, x_n)\, dx_1 \cdots dx_n$$

$$= \int_{-\infty}^{\infty} \cdots \int_{-\infty}^{\infty} \left[\prod_{i=1}^{n} x_i f_i(x_i)\right] dx_1 \cdots dx_n$$

$$= \prod_{i=1}^{n} \int_{-\infty}^{\infty} x_i f_i(x_i)\, dx_i = \prod_{i=1}^{n} E(X_i).$$

The proof for a discrete distribution or a more general type of distribution is similar. □

The difference between Theorem 3 and Theorem 4 should be emphasized. If it is assumed that each expectation exists, the expectation of the sum of a group of random variables is *always* equal to the sum of their individual expectations. However, the expectation of the product of a group of random variables is *not* always equal to the product of their individual expectations. If the random variables are independent, then this equality will also hold.

Example 3: Calculating the Expectation of a Combination of Random Variables. Suppose that X_1, X_2, and X_3 are independent random variables such that $E(X_i) = 0$ and $E(X_i^2) = 1$ for $i = 1, 2, 3$. We shall determine the value of $E[X_1^2(X_2 - 4X_3)^2]$.

Since X_1, X_2, and X_3 are independent, it follows that the two random variables X_1^2 and $(X_2 - 4X_3)^2$ are also independent. Therefore,

$$E\left[X_1^2(X_2 - 4X_3)^2\right] = E(X_1^2)E\left[(X_2 - 4X_3)^2\right]$$

$$= E(X_2^2 - 8X_2X_3 + 16X_3^2)$$

$$= E(X_2^2) - 8E(X_2X_3) + 16E(X_3^2)$$

$$= 1 - 8E(X_2)E(X_3) + 16$$

$$= 17. \quad \square$$

Expectation for Nonnegative Discrete Distributions

Another Expression for the Expectation. Let X be a random variable that can take only the values $0, 1, 2, \ldots$. Then

$$E(X) = \sum_{n=0}^{\infty} n \Pr(X = n) = \sum_{n=1}^{\infty} n \Pr(X = n). \tag{5}$$

Consider now the following triangular array of probabilities:

$$
\begin{array}{llll}
\Pr(X = 1) & \Pr(X = 2) & \Pr(X = 3) & \cdots \\
 & \Pr(X = 2) & \Pr(X = 3) & \cdots \\
 & & \Pr(X = 3) & \cdots \\
 & & & \cdots
\end{array}
$$

We can compute the sum of all the elements in this array in two different ways. First, we can add the elements in each column of the array and then add these column totals. Thus, we obtain the value $\sum_{n=1}^{\infty} n \Pr(X = n)$. Second, we can add the elements in each row of the array and then add these row totals. In this way we obtain the value $\sum_{n=1}^{\infty} \Pr(X \geqslant n)$. Therefore,

$$\sum_{n=1}^{\infty} n \Pr(X = n) = \sum_{n=1}^{\infty} \Pr(X \geqslant n).$$

It now follows from Eq. (5) that

$$E(X) = \sum_{n=1}^{\infty} \Pr(X \geqslant n). \tag{6}$$

Expected Number of Trials. Suppose that a person repeatedly tries to perform a certain task until he is successful. Suppose also that the probability of success on any given trial is p $(0 < p < 1)$, that the probability of failure is $q = 1 - p$, and that all trials are independent. If X denotes the number of the trial on which the first success is obtained, then $E(X)$ can be determined as follows:

Since at least one trial is always required, $\Pr(X \geqslant 1) = 1$. Also, for $n = 2, 3, \ldots$, at least n trials will be required if and only if every one of the first $n - 1$ trials results in failure. Therefore,

$$\Pr(X \geqslant n) = q^{n-1}.$$

By Eq. (6), it follows that

$$E(X) = 1 + q + q^2 + \cdots = \frac{1}{1-q} = \frac{1}{p}.$$

EXERCISES

1. Suppose that three random variables X_1, X_2, X_3 form a random sample from a distribution for which the mean is 5. Determine the value of

$$E(2X_1 - 3X_2 + X_3 - 4).$$

2. Suppose that three random variables X_1, X_2, X_3 form a random sample from the uniform distribution on the interval $0 < x < 1$. Determine the value of

$$E\left[(X_1 - 2X_2 + X_3)^2\right].$$

3. Suppose that the random variable X has a uniform distribution on the interval $(0, 1)$; that the random variable Y has a uniform distribution on the interval $(5, 9)$; and that X and Y are independent. Suppose also that a rectangle is to be constructed for which the lengths of two adjacent sides are X and Y. Determine the expected value of the area of the rectangle.

4. Suppose that the variables $X_1, \ldots, X_n$ form a random sample of size n from a given continuous distribution on the real line for which the p.d.f. is f. Find the expectation of the number of observations in the sample that fall within a specified interval $a \leqslant x \leqslant b$.

5. Suppose that a particle starts at the origin of the real line and moves along the line in jumps of one unit. For each jump the probability is p $(0 \leqslant p \leqslant 1)$ that the particle will jump one unit to the left and the probability is $1 - p$

that the particle will jump one unit to the right. Find the expected value of the position of the particle after n jumps.

6. Suppose that on each play of a certain game a gambler is equally likely to win or to lose. Suppose that when he wins, his fortune is doubled; and when he loses, his fortune is cut in half. If he begins playing with a given fortune c, what is the expected value of his fortune after n independent plays of the game?

7. Suppose that a class contains 10 boys and 15 girls, and suppose that 8 students are to be selected at random from the class without replacement. Let X denote the number of boys that are selected, and let Y denote the number of girls that are selected. Find $E(X - Y)$.

8. Suppose that the proportion of defective items in a large lot is p, and suppose that a random sample of n items is selected from the lot. Let X denote the number of defective items in the sample, and let Y denote the number of nondefective items. Find $E(X - Y)$.

9. Suppose that a fair coin is tossed repeatedly until a head is obtained for the first time. (a) What is the expected number of tosses that will be required? (b) What is the expected number of tails that will be obtained before the first head is obtained?

10. Suppose that a fair coin is tossed repeatedly until exactly k heads have been obtained. Determine the expected number of tosses that will be required. *Hint:* Represent the total number of tosses X in the form $X = X_1 + \cdots + X_k$, where X_i is the number of tosses required to obtain the ith head after $i - 1$ heads have been obtained.

4.3. VARIANCE

Definitions of the Variance and the Standard Deviation

Suppose that X is a random variable with mean $\mu = E(X)$. The *variance of X*, denoted by $\mathrm{Var}(X)$, is defined as follows:

$$\mathrm{Var}(X) = E\left[(X - \mu)^2\right]. \tag{1}$$

Since $\mathrm{Var}(X)$ is the expected value of the nonnegative random variable $(X - \mu)^2$, it follows that $\mathrm{Var}(X) \geq 0$. It should be kept in mind that the expectation in Eq. (1) may or may not be finite. If the expectation is not finite, it is said that $\mathrm{Var}(X)$ does not exist. However, if the possible values of X are bounded, then $\mathrm{Var}(X)$ must exist.

The value of $\text{Var}(X)$ is also called the *variance of the distribution of X*. Thus, in some cases we speak of the variance of X, and in other cases we speak of the variance of a probability distribution.

The variance of a distribution provides a measure of the spread or dispersion of the distribution around its mean μ. A small value of the variance indicates that the probability distribution is tightly concentrated around μ; and a large value of the variance typically indicates that the probability distribution has a wide spread around μ. However, the variance of any distribution, as well as its mean, can be made arbitrarily large by placing even a very small but positive amount of probability far enough from the origin on the real line.

The *standard deviation* of a random variable or of a distribution is defined as the nonnegative square root of the variance. The standard deviation of a given random variable is commonly denoted by the symbol σ, and the variance is denoted by σ^2.

Properties of the Variance

We shall now present several theorems pertaining to the basic properties of the variance. In these theorems we shall assume that the variances of all the random variables exist. The first theorem shows that the variance of a random variable X cannot be 0 unless the entire probability distribution of X is concentrated at a single point.

Theorem 1. $\text{Var}(X) = 0$ *if and only if there exists a constant c such that* $\Pr(X = c) = 1$.

Proof. Suppose first that there exists a constant c such that $\Pr(X = c) = 1$. Then $E(X) = c$, and $\Pr[(X - c)^2 = 0] = 1$. Therefore,

$$\text{Var}(X) = E\left[(X - c)^2\right] = 0.$$

Conversely, suppose that $\text{Var}(X) = 0$. Then $\Pr[(X - \mu)^2 \geqslant 0] = 1$ but $E[(X - \mu)^2] = 0$. Therefore, in accordance with the comment given immediately following the proof of Theorem 2 in Sec. 4.2, it can be seen that

$$\Pr\left[(X - \mu)^2 = 0\right] = 1.$$

Hence, $\Pr(X = \mu) = 1$. $\square$

Theorem 2. *For any constants a and b,*

$$\text{Var}(aX + b) = a^2\text{Var}(X).$$

Proof. If $E(X) = \mu$, then $E(aX + b) = a\mu + b$. Therefore,

$$\text{Var}(aX + b) = E\left[(aX + b - a\mu - b)^2\right] = E\left[(aX - a\mu)^2\right]$$

$$= a^2 E\left[(X - \mu)^2\right] = a^2 \text{Var}(X). \quad \square$$

It follows from Theorem 2 that $\text{Var}(X + b) = \text{Var}(X)$ for any constant b. This result is intuitively plausible, since shifting the entire distribution of X a distance of b units along the real line will change the mean of the distribution by b units but the shift will not affect the dispersion of the distribution around its mean.

Similarly, it follows from Theorem 2 that $\text{Var}(-X) = \text{Var}(X)$. This result also is intuitively plausible, since reflecting the entire distribution of X with respect to the origin of the real line will result in a new distribution that is the mirror image of the original one. The mean will be changed from μ to $-\mu$, but the total dispersion of the distribution around its mean will not be affected.

The next theorem provides an alternative method for calculating the value of $\text{Var}(X)$.

Theorem 3. *For any random variable X, $\text{Var}(X) = E(X^2) - [E(X)]^2$.*

Proof. Let $E(X) = \mu$. Then

$$\text{Var}(X) = E\left[(X - \mu)^2\right]$$

$$= E(X^2 - 2\mu X + \mu^2)$$

$$= E(X^2) - 2\mu E(X) + \mu^2$$

$$= E(X^2) - \mu^2. \quad \square$$

Example 1: Calculating a Standard Deviation. Suppose that a random variable X can take each of the five values $-2, 0, 1, 3,$ and 4 with equal probability. We shall determine the standard deviation of $Y = 4X - 7$.

In this example,

$$E(X) = \frac{1}{5}(-2 + 0 + 1 + 3 + 4) = 1.2$$

and

$$E(X^2) = \frac{1}{5}\left[(-2)^2 + 0^2 + 1^2 + 3^2 + 4^2\right] = 6.$$

By Theorem 3,

$$\text{Var}(X) = 6 - (1.2)^2 = 4.56.$$

By Theorem 2,

$$\text{Var}(Y) = 16 \text{Var}(X) = 72.96.$$

Therefore, the standard deviation σ of Y is

$$\sigma = (72.96)^{1/2} = 8.54. \quad \square$$

Theorem 4. *If $X_1, \ldots, X_n$ are independent random variables, then*

$$\text{Var}(X_1 + \cdots + X_n) = \text{Var}(X_1) + \cdots + \text{Var}(X_n).$$

Proof. Suppose first that $n = 2$. If $E(X_1) = \mu_1$ and $E(X_2) = \mu_2$, then

$$E(X_1 + X_2) = \mu_1 + \mu_2.$$

Therefore,

$$\begin{aligned}
\text{Var}(X_1 + X_2) &= E\left[(X_1 + X_2 - \mu_1 - \mu_2)^2\right] \\
&= E\left[(X_1 - \mu_1)^2 + (X_2 - \mu_2)^2 + 2(X_1 - \mu_1)(X_2 - \mu_2)\right] \\
&= \text{Var}(X_1) + \text{Var}(X_2) + 2E\left[(X_1 - \mu_1)(X_2 - \mu_2)\right].
\end{aligned}$$

Since X_1 and X_2 are independent,

$$\begin{aligned}
E\left[(X_1 - \mu_1)(X_2 - \mu_2)\right] &= E(X_1 - \mu_1)E(X_2 - \mu_2) \\
&= (\mu_1 - \mu_1)(\mu_2 - \mu_2) \\
&= 0.
\end{aligned}$$

It follows, therefore, that

$$\text{Var}(X_1 + X_2) = \text{Var}(X_1) + \text{Var}(X_2).$$

The theorem can now be established for any positive integer n by an induction argument. $\square$

It should be emphasized that the random variables in Theorem 4 must be independent. The variance of the sum of random variables that are not independent will be discussed in Sec. 4.6. By combining Theorems 2 and 4, we can now obtain the following corollary.

Corollary 1. *If $X_1, \ldots, X_n$ are independent random variables and if $a_1, \ldots, a_n$ and b are arbitrary constants, then*

$$\text{Var}(a_1 X_1 + \cdots + a_n X_n + b) = a_1^2 \text{Var}(X_1) + \cdots + a_n^2 \text{Var}(X_n).$$

The Variance of the Binomial Distribution

We shall now consider again the method of generating a binomial distribution presented in Sec. 4.2. Suppose that a box contains red balls and blue balls, and that the proportion of red balls is p $(0 \leqslant p \leqslant 1)$. Suppose also that a random sample of n balls is selected from the box with replacement. For $i = 1, \ldots, n$, let $X_i = 1$ if the ith ball that is selected is red, and let $X_i = 0$ otherwise. If X denotes the total number of red balls in the sample, then $X = X_1 + \cdots + X_n$ and X will have a binomial distribution with parameters n and p.

Since $X_1, \ldots, X_n$ are independent, it follows from Theorem 4 that

$$\text{Var}(X) = \sum_{i=1}^{n} \text{Var}(X_i).$$

Also, for $i = 1, \ldots, n$,

$$E(X_i) = 1(p) + 0(1 - p) = p$$

and

$$E(X_i^2) = 1^2(p) + 0^2(1 - p) = p.$$

Therefore, by Theorem 3,

$$\text{Var}(X_i) = E(X_i^2) - [E(X_i)]^2$$

$$= p - p^2 = p(1 - p).$$

It now follows that

$$\text{Var}(X) = np(1 - p). \tag{2}$$

EXERCISES

1. Suppose that one word is selected at random from the sentence THE GIRL PUT ON HER BEAUTIFUL RED HAT. If X denotes the number of letters in the word that is selected, what is the value of Var(X)?

2. For any given numbers a and b such that $a < b$, find the variance of the uniform distribution on the interval (a, b).

3. Suppose that X is a random variable for which $E(X) = \mu$ and Var(X) $= \sigma^2$. Show that $E[X(X - 1)] = \mu(\mu - 1) + \sigma^2$.

4. Let X be a random variable for which $E(X) = \mu$ and Var(X) $= \sigma^2$, and let c be any given constant. Show that

$$E\left[(X - c)^2\right] = (\mu - c)^2 + \sigma^2.$$

5. Suppose that X and Y are independent random variables with finite variances such that $E(X) = E(Y)$. Show that

$$E\left[(X - Y)^2\right] = \text{Var}(X) + \text{Var}(Y).$$

6. Suppose that X and Y are independent random variables for which Var(X) = Var(Y) = 3. Find the values of (a) Var($X - Y$) and (b) Var($2X - 3Y + 1$).

7. Construct an example of a distribution for which the mean exists but the variance does not exist.

4.4.   MOMENTS

Existence of Moments

For any random variable X and any positive integer k, the expectation $E(X^k)$ is called the kth *moment of X*. In particular, in accordance with this terminology, the mean of X is the first moment of X.

It is said that the kth moment exists if and only if $E(|X|^k) < \infty$. If the random variable X is bounded, that is, if there are finite numbers a and b such that $\Pr(a \leqslant X \leqslant b) = 1$, then all moments of X must necessarily exist. It is possible, however, that all moments of X exist even though X is not bounded. It is shown in the next theorem that if the kth moment of X exists, then all moments of lower order must also exist.

Theorem 1.   *If $E(|X|^k) < \infty$ for some positive integer k, then $E(|X|^j) < \infty$ for any positive integer j such that $j < k$.*

Proof. We shall assume, for convenience, that the distribution of X is continuous and the p.d.f. is f. Then

$$E(|X|^j) = \int_{-\infty}^{\infty} |x|^j f(x)\, dx$$

$$= \int_{|x| \leqslant 1} |x|^j f(x)\, dx + \int_{|x| > 1} |x|^j f(x)\, dx$$

$$\leqslant \int_{|x| \leqslant 1} 1 \cdot f(x)\, dx + \int_{|x| > 1} |x|^k f(x)\, dx$$

$$\leqslant \Pr(|X| \leqslant 1) + E(|X|^k).$$

By hypothesis, $E(|X|^k) < \infty$. It therefore follows that $E(|X|^j) < \infty$. A similar proof holds for a discrete or a more general type of distribution. □

In particular, it follows from Theorem 1 that if $E(X^2) < \infty$, then both the mean of X and the variance of X exist.

Central Moments.   Suppose that X is a random variable for which $E(X) = \mu$. For any positive integer k, the expectation $E[(X - \mu)^k]$ is called the kth *central moment of X* or the kth *moment of X about the mean*. In particular, in accordance with this terminology, the variance of X is the second central moment of X.

For any distribution, the first central moment must be 0 because

$$E(X - \mu) = \mu - \mu = 0.$$

Furthermore, if the distribution of X is symmetric with respect to its mean μ, and if the central moment $E[(X - \mu)^k]$ exists for a given odd integer k, then the value of $E[(X - \mu)^k]$ will be 0 because the positive and negative terms in this expectation will cancel one another.

Example 1: A Symmetric p.d.f. Suppose that X has a continuous distribution for which the p.d.f. has the following form:

$$f(x) = ce^{-(x-3)^2} \qquad \text{for } -\infty < x < \infty.$$

We shall determine the mean of X and all the odd central moments.

It can be shown that for every positive integer k,

$$\int_{-\infty}^{\infty} |x|^k e^{-(x-3)^2}\, dx < \infty.$$

Hence, all the moments of X exist. Furthermore, since $f(x)$ is symmetric with respect to the point $x = 3$, then $E(X) = 3$. Because of this symmetry, it also follows that $E[(X - 3)^k] = 0$ for every odd positive integer k. □

Moment Generating Functions

We shall now consider a given random variable X; and for each real number t we shall let

$$\psi(t) = E(e^{tX}). \tag{1}$$

The function ψ is called the moment generating function (abbreviated m.g.f.) of X. If the random variable X is bounded, then the expectation in Eq. (1) must exist for all values of t. In this case, therefore, the m.g.f. of X will exist for all values of t. On the other hand, if X is not bounded, then the m.g.f. might exist for some values of t and might not exist for others. It can be seen from Eq. (1), however, that for any random variable X, the m.g.f. $\psi(t)$ must exist at the point $t = 0$ and at that point its value must be $\psi(0) = E(1) = 1$.

Suppose that the m.g.f. of a random variable X exists for all values of t in some interval around the point $t = 0$. It can be shown that the derivative $\psi'(t)$ then exists at the point $t = 0$; and that at $t = 0$ the derivative of the expectation in Eq. (1) must be equal to the expectation of the derivative. Thus,

$$\psi'(0) = \left[\frac{d}{dt} E(e^{tX}) \right]_{t=0} = E\left[\left(\frac{d}{dt} e^{tX} \right)_{t=0} \right].$$

But

$$\left(\frac{d}{dt} e^{tX} \right)_{t=0} = (Xe^{tX})_{t=0} = X.$$

It follows that

$$\psi'(0) = E(X).$$

In other words, the derivative of the m.g.f. $\psi(t)$ at $t = 0$ is the mean of X.

More generally, if the m.g.f. $\psi(t)$ of X exists for all values of t in an interval around the point $t = 0$, then it can be shown that all moments $E(X^k)$ of X must exist $(k = 1, 2, \dots)$. Furthermore, it can be shown that it is possible to differentiate $\psi(t)$ any arbitrary number of times at the point $t = 0$. For $n = 1, 2, \dots$, the

nth derivative $\psi^{(n)}(0)$ at $t = 0$ will satisfy the following relation:

$$\psi^{(n)}(0) = \left[\frac{d^n}{dt^n}E(e^{tX})\right]_{t=0} = E\left[\left(\frac{d^n}{dt^n}e^{tX}\right)_{t=0}\right]$$

$$= E\left[(X^n e^{tX})_{t=0}\right] = E(X^n).$$

Thus, $\psi'(0) = E(X), \psi''(0) = E(X^2), \psi'''(0) = E(X^3)$, and so on.

Example 2: Calculating an m.g.f. Suppose that X is a random variable for which the p.d.f. is as follows:

$$f(x) = \begin{cases} e^{-x} & \text{for } x > 0, \\ 0 & \text{otherwise.} \end{cases}$$

We shall determine the m.g.f. of X and also Var(X).

For any real number t,

$$\psi(t) = E(e^{tX}) = \int_0^\infty e^{tx} e^{-x} \, dx$$

$$= \int_0^\infty e^{(t-1)x} \, dx.$$

The final integral in this equation will be finite if and only if $t < 1$. Therefore, $\psi(t)$ exists only for $t < 1$. For any such value of t,

$$\psi(t) = \frac{1}{1-t}.$$

Since $\psi(t)$ is finite for all values of t in an interval around the point $t = 0$, all moments of X exist. The first two derivatives of ψ are

$$\psi'(t) = \frac{1}{(1-t)^2} \quad \text{and} \quad \psi''(t) = \frac{2}{(1-t)^3}.$$

Therefore, $E(X) = \psi'(0) = 1$ and $E(X^2) = \psi''(0) = 2$. It now follows that

$$\text{Var}(X) = \psi''(0) - [\psi'(0)]^2 = 1. \quad \square$$

Properties of Moment Generating Functions

We shall now present three basic theorems pertaining to moment generating functions.

Theorem 2.   *Let X be a random variable for which the m.g.f. is ψ_1; let $Y = aX + b$, where a and b are given constants; and let ψ_2 denote the m.g.f. of Y. Then for any value of t such that $\psi_1(at)$ exists,*

$$\psi_2(t) = e^{bt}\psi_1(at).$$

Proof. By the definition of an m.g.f.,

$$\psi_2(t) = E(e^{tY}) = E[e^{t(aX+b)}] = e^{bt}E(e^{atX}) = e^{bt}\psi_1(at). \quad \square$$

Example 3: Calculating the m.g.f. of a Linear Function. Suppose that the distribution of X is as specified in Example 2. Then the m.g.f. of X for $t < 1$ is

$$\psi_1(t) = \frac{1}{1-t}.$$

If $Y = 3 - 2X$, then the m.g.f. of Y will exist for $t > -1/2$ and will have the value

$$\psi_2(t) = e^{3t}\psi_1(-2t) = \frac{e^{3t}}{1+2t}. \quad \square$$

The next theorem shows that the m.g.f. of the sum of any number of independent random variables has a very simple form. Because of this property, the m.g.f. is an important tool in the study of such sums.

Theorem 3.   *Suppose that $X_1, \ldots, X_n$ are n independent random variables; and for $i = 1, \ldots, n$, let ψ_i denote the m.g.f. of X_i. Let $Y = X_1 + \cdots + X_n$, and let the m.g.f. of Y be denoted by ψ. Then for any value of t such that $\psi_i(t)$ exists for $i = 1, \ldots, n$,*

$$\psi(t) = \prod_{i=1}^{n}\psi_i(t). \tag{2}$$

Proof. By definition,

$$\psi(t) = E(e^{tY}) = E[e^{t(X_1 + \cdots + X_n)}] = E\left(\prod_{i=1}^{n} e^{tX_i}\right).$$

Since the random variables $X_1, \ldots, X_n$ are independent, it follows from Theorem 4 in Sec. 4.2 that

$$E\left(\prod_{i=1}^{n} e^{tX_i}\right) = \prod_{i=1}^{n} E(e^{tX_i}).$$

Hence,

$$\psi(t) = \prod_{i=1}^{n} \psi_i(t). \quad \square$$

The Moment Generating Function for the Binomial Distribution. Suppose that a random variable X has a binomial distribution with parameters n and p. In Secs. 4.2 and 4.3, the mean and the variance of X were determined by representing X as the sum of n independent random variables $X_1, \ldots, X_n$. In this representation, the distribution of each variable X_i is as follows:

$$\Pr(X_i = 1) = p \quad \text{and} \quad \Pr(X_i = 0) = q = 1 - p.$$

We shall now use this representation to determine the m.g.f. of $X = X_1 + \cdots + X_n$.

Since each of the random variables $X_1, \ldots, X_n$ has the same distribution, the m.g.f. of each variable will be the same. For $i = 1, \ldots, n$, the m.g.f. of X_i is

$$\psi_i(t) = E(e^{tX_i}) = (e^t)\Pr(X_i = 1) + (1)\Pr(X_i = 0)$$

$$= pe^t + q.$$

It follows from Theorem 3 that the m.g.f. of X in this case is

$$\psi(t) = (pe^t + q)^n. \tag{3}$$

Uniqueness of Moment Generating Functions. We shall now state one more important property of the m.g.f. The proof of this property is beyond the scope of this book and is omitted.

Theorem 4. *If the m.g.f.'s of two random variables X_1 and X_2 are identical for all values of t in an interval around the point $t = 0$, then the probability distributions of X_1 and X_2 must be identical.*

The Additive Property of the Binomial Distribution.   Suppose that X_1 and X_2 are independent random variables; that X_1 has a binomial distribution with parameters n_1 and p; and that X_2 has a binomial distribution with parameters n_2 and p. Here, the value of p must be the same for both distributions, but it is not necessarily true that $n_1 = n_2$. We shall determine the distribution of $X_1 + X_2$.

If ψ_i denotes the m.g.f. of X_i for $i = 1, 2$, then it follows from Eq. (3) that

$$\psi_i(t) = (pe^t + q)^{n_i}.$$

If the m.g.f. of $X_1 + X_2$ is denoted by ψ, then by Theorem 3,

$$\psi(t) = (pe^t + q)^{n_1 + n_2}.$$

It can be seen from Eq. (3) that this function ψ is the m.g.f. of a binomial distribution with parameters $n_1 + n_2$ and p. Hence, by Theorem 4, the distribution of $X_1 + X_2$ must be that binomial distribution. Thus, we have established the following result:

> If X_1 and X_2 are independent random variables and if X_i has a binomial distribution with parameters n_i and p ($i = 1, 2$), then $X_1 + X_2$ has a binomial distribution with parameters $n_1 + n_2$ and p.

EXERCISES

1. Suppose that X is a random variable for which $E(X) = 1$, $E(X^2) = 2$, and $E(X^3) = 5$. Find the value of the third central moment of X.

2. If X has a uniform distribution on the interval (a, b), what is the value of the fifth central moment of X?

3. Suppose that X is any random variable such that $E(X^2)$ exists. (a) Show that $E(X^2) \geqslant [E(X)]^2$. (b) Show that $E(X^2) = [E(X)]^2$ if and only if there exists a constant c such that $\Pr(X = c) = 1$. *Hint:* $\mathrm{Var}(X) \geqslant 0$.

4. Suppose that X is a random variable with mean μ and variance σ^2, and that the fourth moment of X exists. Show that

$$E\left[(X - \mu)^4\right] \geqslant \sigma^4.$$

5. Suppose that X has a uniform distribution on the interval (a, b). Determine the m.g.f. of X.

6. Suppose that X is a random variable for which the m.g.f. is as follows:

$$\psi(t) = \frac{1}{4}(3e^t + e^{-t}) \qquad \text{for } -\infty < t < \infty.$$

Find the mean and the variance of X.

7. Suppose that X is a random variable for which the m.g.f. is as follows:

$$\psi(t) = e^{t^2 + 3t} \qquad \text{for } -\infty < t < \infty.$$

Find the mean and the variance of X.

8. Let X be a random variable with mean μ and variance σ^2, and let $\psi_1(t)$ denote the m.g.f. of X for $-\infty < t < \infty$. Let c be a given positive constant, and let Y be a random variable for which the m.g.f. is

$$\psi_2(t) = e^{c[\psi_1(t) - 1]} \qquad \text{for } -\infty < t < \infty.$$

Find expressions for the mean and the variance of Y in terms of the mean and the variance of X.

9. Suppose that the random variables X and Y are i.i.d. and that the m.g.f. of each is

$$\psi(t) = e^{t^2 + 3t} \qquad \text{for } -\infty < t < \infty.$$

Find the m.g.f. of $Z = 2X - 3Y + 4$.

10. Suppose that X is a random variable for which the m.g.f. is as follows:

$$\psi(t) = \frac{1}{5} e^t + \frac{2}{5} e^{4t} + \frac{2}{5} e^{8t} \qquad \text{for } -\infty < t < \infty.$$

Find the probability distribution of X. *Hint:* It is a simple discrete distribution.

11. Suppose that X is a random variable for which the m.g.f. is as follows:

$$\psi(t) = \frac{1}{6}(4 + e^t + e^{-t}) \qquad \text{for } -\infty < t < \infty.$$

Find the probability distribution of X.

4.5. THE MEAN AND THE MEDIAN

The Median

It was mentioned in Sec. 4.1 that the mean of a probability distribution on the real line will be at the center of gravity of that distribution. In this sense, the mean of a distribution can be regarded as the *center* of the distribution. There is another point on the line which might also be regarded as the center of the distribution. This is the point which divides the total probability into two equal parts, that is, the point m_0 such that the probability to the left of m_0 is $1/2$ and the probability to the right of m_0 is also $1/2$. This point is called the median of the distribution. It should be noted, however, that for some discrete distributions there will not be any point at which the total probability is divided into two parts that are exactly equal. Moreover, for other distributions, which may be either discrete or continuous, there will be more than one such point. Therefore, the formal definition of a median, which will now be given, must be general enough to include these possibilities.

For any random variable X, a median of the distribution of X is defined as a point m such that $\Pr(X \leq m) \geq 1/2$ *and* $\Pr(X \geq m) \geq 1/2$.

In other words, a median is a point m that satisfies the following two requirements: First, if m is included with the values of X to the left of m, then

$$\Pr(X \leq m) \geq \Pr(X > m).$$

Second, if m is included with the values of X to the right of m, then

$$\Pr(X \geq m) \geq \Pr(X < m).$$

According to this definition, every distribution must have at least one median, and for some distributions every point in some interval can be a median. If there is a point m such that $\Pr(X < m) = \Pr(X > m)$, that is, if the point m does actually divide the total probability into two equal parts, then m will of course be a median of the distribution of X.

Example 1: The Median of a Discrete Distribution. Suppose that X has the following discrete distribution:

$$\Pr(X = 1) = 0.1, \qquad \Pr(X = 2) = 0.2,$$

$$\Pr(X = 3) = 0.3, \qquad \Pr(X = 4) = 0.4.$$

The value 3 is a median of this distribution because $\Pr(X \leq 3) = 0.6$, which is greater than $1/2$, and $\Pr(X \geq 3) = 0.7$, which is also greater than $1/2$. Furthermore, 3 is the unique median of this distribution. □

Example 2: A Discrete Distribution for Which the Median Is Not Unique. Suppose that X has the following discrete distribution:

$$\Pr(X = 1) = 0.1, \qquad \Pr(X = 2) = 0.4,$$

$$\Pr(X = 3) = 0.3, \qquad \Pr(X = 4) = 0.2.$$

Here, $\Pr(X \leqslant 2) = 1/2$ and $\Pr(X \geqslant 3) = 1/2$. Therefore, every value of m in the closed interval $2 \leqslant m \leqslant 3$ will be a median of this distribution. □

Example 3: The Median of a Continuous Distribution. Suppose that X has a continuous distribution for which the p.d.f. is as follows:

$$f(x) = \begin{cases} 4x^3 & \text{for } 0 < x < 1, \\ 0 & \text{otherwise.} \end{cases}$$

The unique median of this distribution will be the number m such that

$$\int_0^m 4x^3 \, dx = \int_m^1 4x^3 \, dx = \frac{1}{2}.$$

This number is $m = 1/2^{1/4}$. □

Example 4: A Continuous Distribution for Which the Median Is Not Unique. Suppose that X has a continuous distribution for which the p.d.f. is as follows:

$$f(x) = \begin{cases} 1/2 & \text{for } 0 \leqslant x \leqslant 1, \\ 1 & \text{for } 2.5 \leqslant x \leqslant 3, \\ 0 & \text{otherwise.} \end{cases}$$

Here, for any value of m in the closed interval $1 \leqslant m \leqslant 2.5$, $\Pr(X \leqslant m) = \Pr(X \geqslant m) = 1/2$. Therefore, every value of m in the interval $1 \leqslant m \leqslant 2.5$ is a median of this distribution. □

Comparison of the Mean and the Median

Either the mean or the median of a distribution can be used to represent the "average" value of a variable. Some important properties of the mean have already been described in this chapter, and several more properties will be given later in the book. However, for many purposes the median is a more useful measure of the average than is the mean. As mentioned in Sec. 4.1, the mean of a distribution can be made very large by removing a small but positive amount of probability from any part of the distribution and assigning this amount to a sufficiently large value of x. On the other hand, the median may be unaffected by a similar change in probabilities. If any amount of probability is removed from a value of x larger than the median and assigned to an arbitrarily large value of x, the median of the new distribution will be the same as that of the original distribution.

For example, suppose that the mean annual income among the families in a certain community is \$30,000. It is possible that only a few families in the community actually have an income as large as $30,000, but those few families have incomes that are very much larger than $30,000. If, however, the median annual income among the families is $30,000, then at least one-half of the families must have incomes of $30,000 or more.

We shall now consider two specific problems in which the value of a random variable X must be predicted. In the first problem, the optimal prediction that can be made is the mean. In the second problem, the optimal prediction is the median.

Minimizing the Mean Squared Error

Suppose that X is a random variable with mean μ and variance σ^2. Suppose also that the value of X is to be observed in some experiment, but this value must be predicted before the observation can be made. One basis for making the prediction is to select some number d for which the expected value of the square of the error $X - d$ will be a minimum. The number $E[(X - d)^2]$ is called the *mean squared error* of the prediction d. The abbreviation M.S.E. is often used for the term mean squared error. We shall now determine the number d for which the M.S.E. is minimized.

For any value of d,

$$E\left[(X - d)^2\right] = E(X^2 - 2dX + d^2)$$
$$= E(X^2) - 2d\mu + d^2. \tag{1}$$

The final expression in Eq. (1) is simply a quadratic function of d. By elementary differentiation it will be found that the minimum value of this function is attained when $d = \mu$. Hence, in order to minimize the M.S.E., the predicted value of X should be its mean μ. Furthermore, when this prediction is used, the M.S.E. is simply $E[(X - \mu)^2] = \sigma^2$

Minimizing the Mean Absolute Error

Another possible basis for predicting the value of a random variable X is to choose some number d for which $E(|X - d|)$ will be a minimum. The number $E(|X - d|)$ is called the *mean absolute error* of the prediction. We shall use the abbreviation M.A.E. for the term mean absolute error. We shall now show that the M.A.E. is minimized when the chosen value of d is a median of the distribution of X.

Theorem 1. *Let m be a median of the distribution of X, and let d be any other number. Then*

$$E(|X - m|) \leq E(|X - d|). \tag{2}$$

Furthermore, there will be equality in the relation (2) if and only if d is also a median of the distribution of X.

Proof. For convenience, we shall assume that X has a continuous distribution for which the p.d.f. is f. The proof for any other type of distribution is similar. Suppose first that $d > m$. Then

$$E(|X - d|) - E(|X - m|) = \int_{-\infty}^{\infty} (|x - d| - |x - m|) f(x)\, dx$$

$$= \int_{-\infty}^{m} (d - m) f(x)\, dx + \int_{m}^{d} (d + m - 2x) f(x)\, dx + \int_{d}^{\infty} (m - d) f(x)\, dx$$

$$\geq \int_{-\infty}^{m} (d - m) f(x)\, dx + \int_{m}^{d} (m - d) f(x)\, dx + \int_{d}^{\infty} (m - d) f(x)\, dx$$

$$= (d - m)[\Pr(X \leq m) - \Pr(X > m)]. \tag{3}$$

Since m is a median of the distribution of X, it follows that

$$\Pr(X \leq m) \geq 1/2 \geq \Pr(X > m). \tag{4}$$

The final difference in the relation (3) is therefore nonnegative. Hence,

$$E(|X - d|) \geq E(|X - m|). \tag{5}$$

Furthermore, there can be equality in the relation (5) only if the inequalities in relations (3) and (4) are actually equalities. A careful analysis shows that these inequalities will be equalities only if d is also a median of the distribution of X. The proof for any value of d such that $d < m$ is similar. $\square$

Example 5: Predicting the Value of a Discrete Random Variable. Suppose that the probability is $1/6$ that a random variable X will take each of the following six values: $0, 1, 2, 3, 5, 7$. We shall determine the prediction for which the M.S.E. is minimum and the prediction for which the M.A.E. is minimum.

In this example,

$$E(X) = \frac{1}{6}(0 + 1 + 2 + 3 + 5 + 7) = 3.$$

Therefore the M.S.E. will be minimized by the unique value $d = 3$.

Also, any number m in the closed interval $2 \leqslant m \leqslant 3$ is a median of the given distribution. Therefore, the M.A.E. will be minimized by any value of d such that $2 \leqslant d \leqslant 3$ and only by such a value of d. $\square$

EXERCISES

1. Suppose that a random variable X has a discrete distribution for which the p.f. is as follows:

$$f(x) = \begin{cases} cx & \text{for } x = 1, 2, 3, 4, 5, 6, \\ 0 & \text{otherwise.} \end{cases}$$

Determine all the medians of this distribution.

2. Suppose that a random variable X has a continuous distribution for which the p.d.f. is as follows:

$$f(x) = \begin{cases} e^{-x} & \text{for } x > 0, \\ 0 & \text{otherwise.} \end{cases}$$

Determine all the medians of this distribution.

3. In a small community consisting of 153 families, the number of families that have k children ($k = 0, 1, 2, \ldots$) is given in the following table:

Number of children	Number of families
0	21
1	40
2	42
3	27
4 or more	23

Determine the mean and the median of the number of children per family.

4. Suppose that an observed value of X is equally likely to come from a continuous distribution for which the p.d.f. is f or from one for which the p.d.f. is g. Suppose that $f(x) > 0$ for $0 < x < 1$ and $f(x) = 0$ otherwise, and suppose also that $g(x) > 0$ for $2 < x < 4$ and $g(x) = 0$ otherwise. Determine: (a) the mean and (b) the median of the distribution of X.

5. Suppose that a random variable X has a continuous distribution for which the p.d.f. f is as follows:

$$f(x) = \begin{cases} 2x & \text{for } 0 < x < 1, \\ 0 & \text{otherwise.} \end{cases}$$

Determine the value of d that minimizes (a) $E[(X - d)^2]$ and (b) $E(|X - d|)$.

6. Suppose that a person's score X on a certain examination will be a number in the interval $0 \leqslant X \leqslant 1$ and that X has a continuous distribution for which

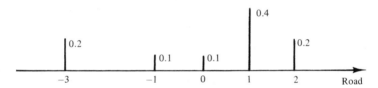

Figure 4.5 Probabilities for Exercise 8.

the p.d.f. is as follows:

$$f(x) = \begin{cases} x + \dfrac{1}{2} & \text{for } 0 \leqslant x \leqslant 1, \\ 0 & \text{otherwise.} \end{cases}$$

Determine the prediction of X which minimizes (a) the M.S.E. and (b) the M.A.E.

7. Suppose that the distribution of a random variable X is symmetric with respect to the point $x = 0$ and that $E(X^4) < \infty$. Show that $E[(X - d)^4]$ is minimized by the value $d = 0$.

8. Suppose that a fire can occur at any one of five points along a road. These points are located at -3, -1, 0, 1, and 2 in Fig. 4.5. Suppose also that the probability that each of these points will be the location of the next fire that occurs along the road is as specified in Fig. 4.5.

   (a) At what point along the road should a fire engine wait in order to minimize the expected value of the square of the distance that it must travel to the next fire?

   (b) Where should the fire engine wait to minimize the expected value of the distance that it must travel to the next fire?

9. If n houses are located at various points along a straight road, at what point along the road should a store be located in order to minimize the sum of the distances from the n houses to the store?

10. Let X be a random variable having a binomial distribution with parameters $n = 7$ and $p = 1/4$, and let Y be a random variable having a binomial distribution with parameters $n = 5$ and $p = 1/2$. Which of these two random variables can be predicted with the smaller M.S.E.?

11. Consider a coin for which the probability of obtaining a head on any given toss is 0.3. Suppose that the coin is to be tossed 15 times, and let X denote the number of heads that will be obtained.

    (a) What prediction of X has the smallest M.S.E.?

    (b) What prediction of X has the smallest M.A.E.?

4.6.   COVARIANCE AND CORRELATION

Covariance

When we consider the joint distribution of two random variables, the means, the medians, and the variances of the variables provide useful information about their marginal distributions. However, these values do not provide any information about the relationship between the two variables or about their tendency to vary together rather than independently. In this section and the next one, we shall introduce new quantities which enable us to measure the association between two random variables, to determine the variance of the sum of any number of dependent random variables, and to predict the value of one random variable by using the observed value of some other related variable.

Let X and Y be random variables having a specified joint distribution; and let $E(X) = \mu_X$, $E(Y) = \mu_Y$, $\mathrm{Var}(X) = \sigma_X^2$, and $\mathrm{Var}(Y) = \sigma_Y^2$. The *covariance of X and Y*, which is denoted by $\mathrm{Cov}(X, Y)$, is defined as follows:

$$\mathrm{Cov}(X, Y) = E[(X - \mu_X)(Y - \mu_Y)]. \tag{1}$$

It can be shown (see Exercise 1 at the end of this section) that if $\sigma_X^2 < \infty$ and $\sigma_Y^2 < \infty$, then the expectation in Eq. (1) will exist and $\mathrm{Cov}(X, Y)$ will be finite. However, the value of $\mathrm{Cov}(X, Y)$ can be positive, negative, or zero.

Correlation

If $0 < \sigma_X^2 < \infty$ and $0 < \sigma_Y^2 < \infty$, then the *correlation of X and Y*, which is denoted by $\rho(X, Y)$, is defined as follows:

$$\rho(X, Y) = \frac{\mathrm{Cov}(X, Y)}{\sigma_X \sigma_Y}. \tag{2}$$

In order to determine the range of possible values of the correlation $\rho(X, Y)$, we shall need the following result:

The Schwarz Inequality.   *For any random variables U and V,*

$$[E(UV)]^2 \leqslant E(U^2)E(V^2). \tag{3}$$

Proof. If $E(U^2) = 0$, then $\mathrm{Pr}(U = 0) = 1$. Therefore, it must also be true that $\mathrm{Pr}(UV = 0) = 1$. Hence, $E(UV) = 0$ and the relation (3) is satisfied. Similarly, if

$E(V^2) = 0$, then the relation (3) will be satisfied. We can assume, therefore, that $E(U^2) > 0$ and $E(V^2) > 0$. Moreover, if either $E(U^2)$ or $E(V^2)$ is infinite, then the right side of the relation (3) will be infinite. In this case, the relation (3) will surely be satisfied.

Thus, we can assume that $0 < E(U^2) < \infty$ and $0 < E(V^2) < \infty$. For any numbers a and b,

$$0 \leqslant E\left[(aU + bV)^2\right] = a^2E(U^2) + b^2E(V^2) + 2abE(UV) \tag{4}$$

and

$$0 \leqslant E\left[(aU - bV)^2\right] = a^2E(U^2) + b^2E(V^2) - 2abE(UV). \tag{5}$$

If we let $a = [E(V^2)]^{1/2}$ and $b = [E(U^2)]^{1/2}$, then it follows from the relation (4) that

$$E(UV) \geqslant -[E(U^2)E(V^2)]^{1/2}.$$

It also follows from the relation (5) that

$$E(UV) \leqslant [E(U^2)E(V^2)]^{1/2}.$$

These two relations together imply that the relation (3) is satisfied. $\square$

If we let $U = X - \mu_X$ and $V = Y - \mu_Y$, then it follows from the Schwarz inequality that

$$[\text{Cov}(X, Y)]^2 \leqslant \sigma_X^2 \sigma_Y^2.$$

In turn, it follows from Eq. (2) that $[\rho(X, Y)]^2 \leqslant 1$ or, equivalently, that

$$-1 \leqslant \rho(X, Y) \leqslant 1.$$

It is said that X and Y are *positively correlated* if $\rho(X, Y) > 0$, that X and Y are *negatively correlated* if $\rho(X, Y) < 0$, and that X and Y are *uncorrelated* if $\rho(X, Y) = 0$. It can be seen from Eq. (2) that $\text{Cov}(X, Y)$ and $\rho(X, Y)$ must have the same sign; that is, both are positive, or both are negative, or both are zero.

Properties of Covariance and Correlation

We shall now present five theorems pertaining to the basic properties of covariance and correlation. The first theorem provides an alternative method for calculating the value of $\text{Cov}(X, Y)$.

Theorem 1. *For any random variables X and Y such that $\sigma_X^2 < \infty$ and $\sigma_Y^2 < \infty$,*

$$\text{Cov}(X, Y) = E(XY) - E(X)E(Y). \tag{6}$$

Proof. It follows from Eq. (1) that

$$\text{Cov}(X, Y) = E(XY - \mu_X Y - \mu_Y X + \mu_X \mu_Y)$$

$$= E(XY) - \mu_X E(Y) - \mu_Y E(X) + \mu_X \mu_Y.$$

Since $E(X) = \mu_X$ and $E(Y) = \mu_Y$, Eq. (6) is obtained. □

The next result shows that independent random variables must be uncorrelated.

Theorem 2. *If X and Y are independent random variables with $0 < \sigma_X^2 < \infty$ and $0 < \sigma_Y^2 < \infty$, then*

$$\text{Cov}(X, Y) = \rho(X, Y) = 0.$$

Proof. If X and Y are independent, then $E(XY) = E(X)E(Y)$. Therefore, by Eq. (6), $\text{Cov}(X, Y) = 0$. Also, it follows that $\rho(X, Y) = 0$. □

The converse of Theorem 2 is not true as a general rule. Two dependent random variables can be uncorrelated. Indeed, even though Y is an explicit function of X, it is possible that $\rho(X, Y) = 0$ as in the following example.

Example 1: Dependent but Uncorrelated Random Variables. Suppose that the random variable X can take only the three values -1, 0, and 1, and that each of these three values has the same probability. Also, let the random variable Y be defined by the relation $Y = X^2$. We shall show that X and Y are dependent but uncorrelated.

In this example X and Y are clearly dependent, since the value of Y is completely determined by the value of X. However,

$$E(XY) = E(X^3) = E(X) = 0.$$

Since $E(XY) = 0$ and $E(X)E(Y) = 0$, it follows from Theorem 1 that $\text{Cov}(X, Y) = 0$ and that X and Y are uncorrelated. □

The next result shows that if Y is a *linear* function of X, then X and Y must be correlated and, in fact, $|\rho(X, Y)| = 1$.

Theorem 3. *Suppose that X is a random variable such that $0 < \sigma_X^2 < \infty$; and that $Y = aX + b$ for some constants a and b, where $a \neq 0$. If $a > 0$, then $\rho(X, Y) = 1$. If $a < 0$, then $\rho(X, Y) = -1$.*

Proof. If $Y = aX + b$, then $\mu_Y = a\mu_X + b$ and $Y - \mu_Y = a(X - \mu_X)$. Therefore, by Eq. (1),

$$\text{Cov}(X, Y) = aE\left[(X - \mu_X)^2\right] = a\sigma_X^2.$$

Since $\sigma_Y = |a|\sigma_X$, the theorem follows from Eq. (2). $\square$

The value of $\rho(X, Y)$ provides a measure of the extent to which two random variables X and Y are linearly related. If the joint distribution of X and Y is relatively concentrated around a straight line in the xy-plane that has a positive slope, then $\rho(X, Y)$ will typically be close to 1. If the joint distribution is relatively concentrated around a straight line that has a negative slope, then $\rho(X, Y)$ will typically be close to -1. We shall not discuss these concepts further here, but we shall consider them again when the bivariate normal distribution is introduced and studied in Sec. 5.12.

We shall now determine the variance of the sum of random variables that are not necessarily independent.

Theorem 4. *If X and Y are random variables such that $\text{Var}(X) < \infty$ and $\text{Var}(Y) < \infty$, then*

$$\text{Var}(X + Y) = \text{Var}(X) + \text{Var}(Y) + 2\,\text{Cov}(X, Y). \tag{7}$$

Proof. Since $E(X + Y) = \mu_X + \mu_Y$, then

$$\text{Var}(X + Y) = E\left[(X + Y - \mu_X - \mu_Y)^2\right]$$

$$= E\left[(X - \mu_X)^2 + (Y - \mu_Y)^2 + 2(X - \mu_X)(Y - \mu_Y)\right]$$

$$= \text{Var}(X) + \text{Var}(Y) + 2\,\text{Cov}(X, Y). \quad \square$$

For any constants a and b, it can be shown that $\text{Cov}(aX, bY) = ab\,\text{Cov}(X, Y)$ (see Exercise 4 at the end of this section). It therefore follows from Theorem 4 that

$$\text{Var}(aX + bY + c) = a^2\text{Var}(X) + b^2\text{Var}(Y) + 2ab\,\text{Cov}(X, Y). \tag{8}$$

In particular,

$$\text{Var}(X - Y) = \text{Var}(X) + \text{Var}(Y) - 2\,\text{Cov}(X, Y). \tag{9}$$

Theorem 4 can also be extended easily to the variance of the sum of n random variables, as follows:

Theorem 5.   *If* $X_1, \ldots, X_n$ *are random variables such that* $\text{Var}(X_i) < \infty$ *for* $i = 1, \ldots, n$, *then*

$$\text{Var}\left(\sum_{i=1}^{n} X_i\right) = \sum_{i=1}^{n} \text{Var}(X_i) + 2\sum\sum_{i<j}\text{Cov}(X_i, X_j). \tag{10}$$

Proof. For any random variable Y, $\text{Cov}(Y, Y) = \text{Var}(Y)$. Therefore, by using the result in Exercise 7 at the end of this section, we can obtain the following relation:

$$\text{Var}\left(\sum_{i=1}^{n} X_i\right) = \text{Cov}\left(\sum_{i=1}^{n} X_i, \sum_{j=1}^{n} X_j\right) = \sum_{i=1}^{n}\sum_{j=1}^{n} \text{Cov}(X_i, X_j).$$

We shall separate the final sum in this relation into two sums: (i) the sum of those terms for which $i = j$ and (ii) the sum of those terms for which $i \neq j$. Then, if we use the fact that $\text{Cov}(X_i, X_j) = \text{Cov}(X_j, X_i)$, we obtain the relation

$$\text{Var}\left(\sum_{i=1}^{n} X_i\right) = \sum_{i=1}^{n} \text{Var}(X_i) + \sum\sum_{i \neq j}\text{Cov}(X_i, X_j)$$

$$= \sum_{i=1}^{n} \text{Var}(X_i) + 2\sum\sum_{i<j}\text{Cov}(X_i, X_j). \quad \square$$

In Theorem 4 of Sec. 4.3 it was shown that if $X_1, \ldots, X_n$ are independent random variables, then

$$\text{Var}\left(\sum_{i=1}^{n} X_i\right) = \sum_{i=1}^{n} \text{Var}(X_i). \tag{11}$$

This result can be extended as follows:

If $X_1, \ldots, X_n$ *are uncorrelated random variables (that is, if* X_i *and* X_j *are uncorrelated whenever* $i \neq j$), *then Eq. (11) is satisfied.*

EXERCISES

1. Prove that if $\text{Var}(X) < \infty$ and $\text{Var}(Y) < \infty$, then $\text{Cov}(X, Y)$ is finite. *Hint:* By considering the relation $[(X - \mu_X) \pm (Y - \mu_Y)]^2 \geqslant 0$, show that

$$|(X - \mu_X)(Y - \mu_Y)| \leqslant \frac{1}{2}\left[(X - \mu_X)^2 + (Y - \mu_Y)^2\right].$$

2. Suppose that X has a uniform distribution on the interval $(-2, 2)$ and that $Y = X^6$. Show that X and Y are uncorrelated.

3. Suppose that the distribution of a random variable X is symmetric with respect to the point $x = 0$, that $0 < E(X^4) < \infty$, and that $Y = X^2$. Show that X and Y are uncorrelated.

4. For any random variables X and Y and any constants a, b, c, and d, show that

$$\text{Cov}(aX + b, cY + d) = ac \, \text{Cov}(X, Y).$$

5. Let X and Y be random variables such that $0 < \sigma_X^2 < \infty$ and $0 < \sigma_Y^2 < \infty$. Suppose that $U = aX + b$ and $V = cY + d$, where $a \neq 0$ and $c \neq 0$. Show that $\rho(U, V) = \rho(X, Y)$ if $ac > 0$ and that $\rho(U, V) = -\rho(X, Y)$ if $ac < 0$.

6. Let X, Y, and Z be three random variables such that $\text{Cov}(X, Z)$ and $\text{Cov}(Y, Z)$ exist, and let a, b, and c be any given constants. Show that

$$\text{Cov}(aX + bY + c, Z) = a \, \text{Cov}(X, Z) + b \, \text{Cov}(Y, Z).$$

7. Suppose that $X_1, \ldots, X_m$ and $Y_1, \ldots, Y_n$ are random variables such that $\text{Cov}(X_i, Y_j)$ exists for $i = 1, \ldots, m$ and $j = 1, \ldots, n$; and suppose that $a_1, \ldots, a_m$ and $b_1, \ldots, b_n$ are constants. Show that

$$\text{Cov}\left(\sum_{i=1}^{m} a_i X_i, \sum_{j=1}^{n} b_j Y_j \right) = \sum_{i=1}^{m} \sum_{j=1}^{n} a_i b_j \text{Cov}(X_i, Y_j).$$

8. Suppose that X and Y are two random variables, which may be dependent, and that $\text{Var}(X) = \text{Var}(Y)$. Assuming that $0 < \text{Var}(X + Y) < \infty$ and $0 < \text{Var}(X - Y) < \infty$, show that the random variables $X + Y$ and $X - Y$ are uncorrelated.

9. Suppose that X and Y are negatively correlated. Is $\text{Var}(X + Y)$ larger or smaller than $\text{Var}(X - Y)$?

10. Show that two random variables X and Y cannot possibly have the following properties: $E(X) = 3$, $E(Y) = 2$, $E(X^2) = 10$, $E(Y^2) = 29$, and $E(XY) = 0$.

11. Suppose that X and Y have a continuous joint distribution for which the joint p.d.f. is as follows:

$$f(x, y) = \begin{cases} \dfrac{1}{3}(x + y) & \text{for } 0 \leqslant x \leqslant 1 \text{ and } 0 \leqslant y \leqslant 2, \\ 0 & \text{otherwise.} \end{cases}$$

Determine the value of $\text{Var}(2X - 3Y + 8)$.

12. Suppose that X and Y are random variables such that $\text{Var}(X) = 9$, $\text{Var}(Y) = 4$, and $\rho(X, Y) = -1/6$. Determine (a) $\text{Var}(X + Y)$ and (b) $\text{Var}(X - 3Y + 4)$.

13. Suppose that X, Y, and Z are three random variables such that $\text{Var}(X) = 1$, $\text{Var}(Y) = 4$, $\text{Var}(Z) = 8$, $\text{Cov}(X, Y) = 1$, $\text{Cov}(X, Z) = -1$, and $\text{Cov}(Y, Z) = 2$. Determine (a) $\text{Var}(X + Y + Z)$ and (b) $\text{Var}(3X - Y - 2Z + 1)$.

14. Suppose that $X_1, \ldots, X_n$ are random variables such that the variance of each variable is 1 and the correlation between each pair of different variables is $1/4$. Determine $\text{Var}(X_1 + \cdots + X_n)$.

4.7. CONDITIONAL EXPECTATION

Definition and Basic Properties

Suppose that X and Y are random variables with a continuous joint distribution. Let $f(x, y)$ denote their joint p.d.f.; let $f_1(x)$ denote the marginal p.d.f. of X; and for any value of x such that $f_1(x) > 0$, let $g(y \mid x)$ denote the conditional p.d.f. of Y given that $X = x$.

The conditional expectation of Y given X is denoted by $E(Y \mid X)$ and is defined as a function of the random variable X whose value $E(Y \mid x)$, when $X = x$, is specified as follows:

$$E(Y \mid x) = \int_{-\infty}^{\infty} y g(y \mid x) \, dy.$$

In other words, $E(Y \mid x)$ is the mean of the conditional distribution of Y given that $X = x$. The value of $E(Y \mid x)$ will not be defined for any value of x such that $f_1(x) = 0$. However, since these values of x form a set of points whose probability is 0, the definition of $E(Y \mid x)$ at such a point is irrelevant.

Similarly, if X and Y have a discrete joint distribution and $g(y \mid x)$ is the conditional p.f. of Y given that $X = x$, then the conditional expectation $E(Y \mid X)$ is defined as the function of X whose value $E(Y \mid x)$, when $X = x$, is

$$E(Y \mid x) = \sum_y y g(y \mid x).$$

Since $E(Y \mid X)$ is a function of the random variable X, it is itself a random variable with its own probability distribution, which can be derived from the distribution of X. We shall now show that the mean of $E(Y \mid X)$ must be $E(Y)$.

Theorem 1. *For any random variables X and Y,*

$$E[E(Y|X)] = E(Y). \tag{1}$$

Proof. We shall assume, for convenience, that X and Y have a continuous joint distribution. Then

$$E[E(Y|X)] = \int_{-\infty}^{\infty} E(Y|x)f_1(x)\,dx$$

$$= \int_{-\infty}^{\infty}\int_{-\infty}^{\infty} yg(y|x)f_1(x)\,dy\,dx.$$

Since $g(y|x) = f(x, y)/f_1(x)$, it follows that

$$E[E(Y|X)] = \int_{-\infty}^{\infty}\int_{-\infty}^{\infty} yf(x, y)\,dy\,dx = E(Y).$$

The proof for a discrete distribution or a more general type of distribution is similar. $\square$

Example 1: Choosing Points from Uniform Distributions. Suppose that a point X is chosen in accordance with a uniform distribution on the interval $(0, 1)$. Also, suppose that after the value $X = x$ has been observed $(0 < x < 1)$, a point Y is chosen in accordance with a uniform distribution on the interval $(x, 1)$. We shall determine the value of $E(Y)$.

For any given value of x $(0 < x < 1)$, $E(Y|x)$ will be equal to the midpoint $(1/2)(x + 1)$ of the interval $(x, 1)$. Therefore, $E(Y|X) = (1/2)(X + 1)$ and

$$E(Y) = E[E(Y|X)] = \frac{1}{2}[E(X) + 1] = \frac{1}{2}\left(\frac{1}{2} + 1\right) = \frac{3}{4}. \quad \square$$

More generally, suppose that X and Y have a continuous joint distribution and that $r(X, Y)$ is any function of X and Y. Then the conditional expectation $E[r(X, Y)|X]$ is defined as the function of X whose value $E[r(X, Y)|x]$, when $X = x$, is

$$E[r(X, Y)|x] = \int_{-\infty}^{\infty} r(x, y)g(y|x)\,dy.$$

For two arbitrary random variables X and Y, it can be shown that

$$E\{E[r(X, Y)|X]\} = E[r(X, Y)]. \tag{2}$$

We can define in a similar manner the conditional expectation of $r(X, Y)$ given Y and the conditional expectation of a function $r(X_1, \ldots, X_n)$ of several random variables given one or more of the variables $X_1, \ldots, X_n$.

Example 2: Linear Conditional Expectation. Suppose that $E(Y|X) = aX + b$ for some constants a and b. We shall determine the value of $E(XY)$ in terms of $E(X)$ and $E(X^2)$.

By Eq. (2), $E(XY) = E[E(XY|X)]$. Furthermore, since X is considered to be given and fixed in the conditional expectation,

$$E(XY|X) = XE(Y|X) = X(aX + b) = aX^2 + bX.$$

Therefore,

$$E(XY) = E(aX^2 + bX) = aE(X^2) + bE(X). \quad \square$$

Prediction

Consider two random variables X and Y that have a specified joint distribution and suppose that after the value of X has been observed, the value of Y must be predicted. In other words, the predicted value of Y can depend on the value of X. We shall assume that this predicted value $d(X)$ must be chosen so as to minimize the mean squared error $E\{[Y - d(X)]^2\}$.

It follows from Eq. (2) that

$$E\{[Y - d(X)]^2\} = E\big(E\{[Y - d(X)]^2|X\}\big). \tag{3}$$

Therefore, the M.S.E. in Eq. (3) will be minimized if $d(X)$ is chosen, for each given value of X, so as to minimize the conditional expectation $E\{[Y - d(X)]^2|X\}$ on the right side of Eq. (3). After the value x of X has been observed, the distribution of Y is specified by the conditonal distribution of Y given that $X = x$. As discussed in Sec. 4.5, when $X = x$ the conditional expectation in Eq. (3) will be minimized when $d(x)$ is chosen to be equal to the mean $E(Y|x)$ of the conditional distribution of Y. It follows that the function $d(X)$ for which the M.S.E. in Eq. (3) is minimized is $d(X) = E(Y|X)$.

For any given value x, we shall let $\text{Var}(Y|x)$ denote the variance of the conditional distribution of Y given that $X = x$. That is,

$$\text{Var}(Y|x) = E\{[Y - E(Y|x)]^2|x\}. \; = \; E(Y^2/x) - E(Y/x)^2$$

Therefore, if the value $X = x$ is observed and the value $E(Y|x)$ is predicted for Y, then the M.S.E. of this predicted value will be $\text{Var}(Y|x)$. It follows from

Eq. (3) that if the prediction is to be made by using the function $d(X) = E(Y | X)$, then the overall M.S.E., averaged over all the possible values of X, will be $E[\text{Var}(Y | X)]$.

If the value of Y must be predicted without any information about the value of X, then as shown in Sec. 4.5, the best prediction is the mean $E(Y)$ and the M.S.E. is $\text{Var}(Y)$. However, if X can be observed before the prediction is made, the best prediction is $d(X) = E(Y | X)$ and the M.S.E. is $E[\text{Var}(Y | X)]$. Thus, the reduction in the M.S.E. that can be achieved by using the observation X is

$$\text{Var}(Y) - E[\text{Var}(Y | X)]. \tag{4}$$

This reduction provides a measure of the usefulness of X in predicting Y. It is shown in Exercise 10 at the end of this section that this reduction can also be expressed as $\text{Var}[E(Y | X)]$.

It is important to distinguish carefully between the overall M.S.E., which is $E[\text{Var}(Y | X)]$, and the M.S.E. of the particular prediction to be made when $X = x$, which is $\text{Var}(Y | x)$. *Before* the value of X has been observed, the appropriate value for the M.S.E. of the complete process of observing X and then predicting Y is $E[\text{Var}(Y | X)]$. *After* a particular value x of X has been observed and the prediction $E(Y | x)$ has been made, the appropriate measure of the M.S.E. of this prediction is $\text{Var}(Y | x)$.

Example 3: Predicting the Value of an Observation. Suppose that an observation Y is equally likely to be taken from either of two populations Π_1 and Π_2. Suppose also that the following assumptions are made: If the observation is taken from population Π_1, then the p.d.f. of Y is

$$g_1(y) = \begin{cases} 1 & \text{for } 0 \leqslant y \leqslant 1, \\ 0 & \text{otherwise.} \end{cases}$$

If the observation is taken from population Π_2, then the p.d.f. of Y is

$$g_2(y) = \begin{cases} 2y & \text{for } 0 \leqslant y \leqslant 1, \\ 0 & \text{otherwise.} \end{cases}$$

First, we shall suppose that the population from which Y is to be taken is not known; and we shall determine both the predicted value of Y that minimizes the M.S.E. and the minimum value of the M.S.E. Since the population from which Y is to be taken is not known, the marginal p.d.f. of Y is $g(y) = \left(\frac{1}{2}\right)[g_1(y) + g_2(y)]$ for $-\infty < y < \infty$. The predicted value of Y with the smallest M.S.E. will be $E(Y)$, and the M.S.E. of this predicted value will be $\text{Var}(Y)$. It is found that

$$E(Y) = \int_0^1 y g(y)\, dy = \frac{1}{2} \int_0^1 (y + 2y^2)\, dy = \frac{7}{12}$$

and

$$E(Y^2) = \int_0^1 y^2 g(y)\, dy = \frac{1}{2}\int_0^1 (y^2 + 2y^3)\, dy = \frac{5}{12}.$$

Hence,

$$\mathrm{Var}(Y) = \frac{5}{12} - \left(\frac{7}{12}\right)^2 = \frac{11}{144}.$$

Next, we shall suppose that before the value of Y must be predicted, it is possible to learn from which of the two populations Y is to be taken. For each of the two populations, we shall again determine the predicted value of Y that minimizes the M.S.E. and also the M.S.E. of this predicted value. Finally, we shall determine the overall M.S.E. for this process of learning from which population the observation Y is to be taken and then predicting the value of Y.

For convenience, we shall define a new random variable X such that $X = 1$ if Y is to be taken from population Π_1 and $X = 2$ if Y is to be taken from population Π_2. For $i = 1, 2$, we know that if Y is to be taken from the population Π_i, then the predicted value of Y with the smallest M.S.E. will be $E(Y \mid X = i)$ and the M.S.E. of this prediction will be $\mathrm{Var}(Y \mid X = i)$.

The conditional p.d.f. of Y given that $X = 1$ is g_1. Since g_1 is the p.d.f. of a uniform distribution on the interval $(0, 1)$, it is known that $E(Y \mid X = 1) = 1/2$ and $\mathrm{Var}(Y \mid X = 1) = 1/12$. Therefore, if it is known that Y is to be taken from population Π_1, the predicted value with the smallest M.S.E. is $Y = 1/2$ and the M.S.E. of this predicted value is $1/12$.

For $X = 2$,

$$E(Y \mid X = 2) = \int_0^1 y g_2(y)\, dy = \int_0^1 2y^2\, dy = \frac{2}{3}$$

and

$$E(Y^2 \mid X = 2) = \int_0^1 y^2 g_2(y)\, dy = \int_0^1 2y^3\, dy = \frac{1}{2}.$$

Hence,

$$\mathrm{Var}(Y \mid X = 2) = \frac{1}{2} - \left(\frac{2}{3}\right)^2 = \frac{1}{18}.$$

Therefore, if it is known that Y is to be taken from population Π_2, the predicted value with the smallest M.S.E. is $Y = 2/3$ and the M.S.E. of this predicted value is $1/18$.

Finally, since $\Pr(X = 1) = \Pr(X = 2) = 1/2$, the overall M.S.E. of predicting Y from X in this way is

$$E[\text{Var}(Y \mid X)] = \frac{1}{2}\text{Var}(Y \mid X = 1) + \frac{1}{2}\text{Var}(Y \mid X = 2) = \frac{5}{72}.$$

It follows from these results that if a person is able to observe the value of X before predicting the value of Y, the M.S.E. can be reduced from $\text{Var}(Y) = 11/144$ to $E[\text{Var}(Y \mid X)] = 5/72 = 10/144$. □

It should be emphasized that for the conditions of Example 3, $10/144$ is the appropriate value of the overall M.S.E. when it is known that the value of X will be available for predicting Y but before the explicit value of X has been determined. After the value of X has been determined, the appropriate value of the M.S.E. is either $\text{Var}(Y \mid X = 1) = \dfrac{1}{12} = \dfrac{12}{144}$ or $\text{Var}(Y \mid X = 2) = \dfrac{1}{18} = \dfrac{8}{144}$. Therefore, if $X = 1$, the M.S.E. of the predicted value is actually larger than $\text{Var}(Y) = 11/144$. However, if $X = 2$, the M.S.E. of the predicted value is relatively small.

EXERCISES

1. Suppose that 20 percent of the students who took a certain test were from school A and that the arithmetic average of their scores on the test was 80. Suppose also that 30 percent of the students were from school B and that the arithmetic average of their scores was 76. Suppose finally that the other 50 percent of the students were from school C and that the arithmetic average of their scores was 84. If a student is selected at random from the entire group that took the test, what is the expected value of his score?

2. Suppose that $0 < \text{Var}(X) < \infty$ and $0 < \text{Var}(Y) < \infty$. Show that if $E(X \mid Y)$ is constant for all values of Y, then X and Y are uncorrelated.

3. Suppose that the distribution of X is symmetric with respect to the point $x = 0$; that all moments of X exist; and that $E(Y \mid X) = aX + b$, where a and b are given constants. Show that X^{2m} and Y are uncorrelated for $m = 1, 2, \ldots$.

4. Suppose that a point X_1 is chosen from a uniform distribution on the interval $(0, 1)$; and that after the value $X_1 = x_1$ is observed, a point X_2 is chosen from a uniform distribution on the interval $(x_1, 1)$. Suppose further that additional variables $X_3, X_4, \ldots$ are generated in the same way. In general, for $j = 1, 2, \ldots$, after the value $X_j = x_j$ has been observed, X_{j+1} is chosen from a uniform distribution on the interval $(x_j, 1)$. Find the value of $E(X_n)$.

5. Suppose that the joint distribution of X and Y is a uniform distribution on the circle $x^2 + y^2 < 1$. Find $E(X \mid Y)$.

6. Suppose that X and Y have a continuous joint distribution for which the joint p.d.f. is as follows:

$$f(x, y) = \begin{cases} x + y & \text{for } 0 \leqslant x \leqslant 1 \text{ and } 0 \leqslant y \leqslant 1, \\ 0 & \text{otherwise.} \end{cases}$$

Find $E(Y \mid X)$ and $\text{Var}(Y \mid X)$.

7. Consider again the conditions of Exercise 6. (a) If it is observed that $X = 1/2$, what predicted value of Y will have the smallest M.S.E.? (b) What will be the value of this M.S.E.?

8. Consider again the conditions of Exercise 6. If the value of Y is to be predicted from the value of X, what will be the minimum value of the overall M.S.E.?

9. Suppose that, for the conditions in Exercises 6 and 8, a person either can pay a cost c for the opportunity of observing the value of X before predicting the value of Y or can simply predict the value of Y without first observing the value of X. If the person considers his total loss to be the cost c plus the M.S.E. of his predicted value, what is the maximum value of c that he should be willing to pay?

10. If X and Y are any random variables for which the necessary expectations and variances exist, prove that

$$\text{Var}(Y) = E\left[\text{Var}(Y \mid X)\right] + \text{Var}\left[E(Y \mid X)\right].$$

11. Suppose that X and Y are random variables such that $E(Y \mid X) = aX + b$. Assuming that $\text{Cov}(X, Y)$ exists and that $0 < \text{Var}(X) < \infty$, determine expressions for a and b in terms of $E(X)$, $E(Y)$, $\text{Var}(X)$, and $\text{Cov}(X, Y)$.

12. Suppose that a person's score X on a mathematics aptitude test is a number in the interval $(0, 1)$ and that his score Y on a music aptitude test is also a number in the interval $(0, 1)$. Suppose also that in the population of all college students in the United States, the scores X and Y are distributed in accordance with the following joint p.d.f.:

$$f(x, y) = \begin{cases} \dfrac{2}{5}(2x + 3y) & \text{for } 0 \leqslant x \leqslant 1 \text{ and } 0 \leqslant y \leqslant 1, \\ 0 & \text{otherwise.} \end{cases}$$

(a) If a college student is selected at random, what predicted value of his score on the music test has the smallest M.S.E.?

(b) What predicted value of his score on the mathematics test has the smallest M.A.E.?

13. Consider again the conditions of Exercise 12. Are the scores of college students on the mathematics test and the music test positively correlated, negatively correlated, or uncorrelated?

14. Consider again the conditions of Exercise 12. (a) If a student's score on the mathematics test is 0.8, what predicted value of his score on the music test has the smallest M.S.E.? (b) If a student's score on the music test is $1/3$, what predicted value of his score on the mathematics test has the smallest M.A.E.?

4.8. THE SAMPLE MEAN

The Markov and Chebyshev Inequalities

We shall begin this section by presenting two simple and general results, known as the Markov inequality and the Chebyshev inequality. We shall then apply these inequalities to random samples.

The Markov Inequality. *Suppose that X is a random variable such that $\Pr(X \geq 0) = 1$. Then for any given number $t > 0$,*

$$\Pr(X \geq t) \leq \frac{E(X)}{t}. \tag{1}$$

Proof. For convenience, we shall assume that X has a discrete distribution for which the p.f. is f. The proof for a continuous distribution or a more general type of distribution is similar. For a discrete distribution,

$$E(X) = \sum_x xf(x) = \sum_{x<t} xf(x) + \sum_{x \geq t} xf(x).$$

Since X can have only nonnegative values, all the terms in the summations are nonnegative. Therefore,

$$E(X) \geq \sum_{x \geq t} xf(x) \geq \sum_{x \geq t} tf(x) = t \Pr(X \geq t). \quad \square$$

The Markov inequality is primarily of interest for large values of t. In fact, when $t \leq E(X)$, the inequality is of no interest whatsoever, since it is known that $\Pr(X \leq t) \leq 1$. However, it is found from the Markov inequality that for any nonnegative random variable X whose mean is 1, the maximum possible value of $\Pr(X \geq 100)$ is 0.01. Furthermore, it can be verified that this maximum value is actually attained by the random variable X for which $\Pr(X = 0) = 0.99$ and $\Pr(X = 100) = 0.01$.

The Chebyshev Inequality.   *Let X be a random variable for which* $\operatorname{Var}(X)$ *exists. Then for any given number $t > 0$,*

$$\Pr\left(|X - E(X)| \geqslant t\right) \leqslant \frac{\operatorname{Var}(X)}{t^2}. \tag{2}$$

Proof. Let $Y = [X - E(X)]^2$. Then $\Pr(Y \geqslant 0) = 1$ and $E(Y) = \operatorname{Var}(X)$. By applying the Markov inequality to Y, we obtain the following result:

$$\Pr\left(|X - E(X)| \geqslant t\right) = \Pr(Y \geqslant t^2) \leqslant \frac{\operatorname{Var}(X)}{t^2}. \quad \square$$

It can be seen from this proof that the Chebyshev inequality is simply a special case of the Markov inequality. Therefore, the comments that were given following the proof of the Markov inequality can be applied as well to the Chebyshev inequality. Because of their generality, these inequalities are very useful. For example, if $\operatorname{Var}(X) = \sigma^2$ and we let $t = 3\sigma$, then the Chebyshev inequality yields the result that

$$\Pr\left(|X - E(X)| \geqslant 3\sigma\right) \leqslant \frac{1}{9}.$$

In words, this result states that the probability that any given random variable will differ from its mean by more than 3 standard deviations *cannot* exceed $1/9$. This probability will actually be much smaller than $1/9$ for many of the random variables and distributions that will be discussed in this book. The Chebyshev inequality is useful because of the fact that this probability must be $1/9$ or less for *every* distribution. It can also be shown (see Exercise 3 at the end of this section) that the upper bound in (2) is sharp in the sense that it cannot be made any smaller and still hold for *all* distributions.

Properties of the Sample Mean

Suppose that the random variables $X_1, \ldots, X_n$ form a random sample of size n from some distribution for which the mean is μ and the variance is σ^2. In other words, suppose that the random variables $X_1, \ldots, X_n$ are i.i.d. and that each has mean μ and variance σ^2. We shall let $\overline{X}_n$ represent the arithmetic average of the n observations in the sample. Thus,

$$\overline{X}_n = \frac{1}{n}(X_1 + \cdots + X_n).$$

This random variable $\overline{X}_n$ is called the *sample mean*.

The mean and the variance of $\overline{X}_n$ can easily be computed. It follows directly from the definition of $\overline{X}_n$ that

$$E(\overline{X}_n) = \frac{1}{n} \sum_{i=1}^{n} E(X_i) = \frac{1}{n} \cdot n\mu = \mu.$$

Furthermore, since the variables $X_1, \ldots, X_n$ are independent,

$$\mathrm{Var}(\overline{X}_n) = \frac{1}{n^2} \mathrm{Var}\left(\sum_{i=1}^{n} X_i \right)$$

$$= \frac{1}{n^2} \sum_{i=1}^{n} \mathrm{Var}(X_i) = \frac{1}{n^2} \cdot n\sigma^2 = \frac{\sigma^2}{n}.$$

In words, the mean of $\overline{X}_n$ is equal to the mean of the distribution from which the random sample was drawn, but the variance of $\overline{X}_n$ is only $1/n$ times the variance of that distribution. It follows that the probability distribution of $\overline{X}_n$ will be more concentrated around the mean value μ than was the original distribution. In other words, the sample mean $\overline{X}_n$ is more likely to be close to μ than is the value of just a single observation X_i from the given distribution.

These statements can be made more precise by applying the Chebyshev inequality to $\overline{X}_n$. Since $E(\overline{X}_n) = \mu$ and $\mathrm{Var}(\overline{X}_n) = \sigma^2/n$, it follows from the relation (2) that for any given number $t > 0$,

$$\Pr(|\overline{X}_n - \mu| \geqslant t) \leqslant \frac{\sigma^2}{nt^2}. \tag{3}$$

Example 1: Determining the Required Number of Observations. Suppose that a random sample is to be taken from a distribution for which the value of the mean μ is not known, but for which it is known that the standard deviation σ is 2 units. We shall determine how large the sample size must be in order to make the probability at least 0.99 that $|\overline{X}_n - \mu|$ will be less than 1 unit.

Since $\sigma^2 = 4$, it follows from the relation (3) that for any given sample size n,

$$\Pr(|\overline{X}_n - \mu| \geqslant 1) \leqslant \frac{4}{n}.$$

Since n must be chosen so that $\Pr(|\overline{X}_n - \mu| < 1) \geqslant 0.99$, it follows that n must be chosen so that $4/n \leqslant 0.01$. Hence, it is required that $n \geqslant 400$. $\square$

It should be emphasized that the use of the Chebyshev inequality in Example 1 guarantees that a sample for which $n = 400$ will be large enough to meet the specified probability requirements, regardless of the particular type of distribution from which the sample is to be taken. If further information about this distribution is available, then it can often be shown that a smaller value for n will be sufficient. This property is illustrated in the next example.

Example 2: Tossing a Coin. Suppose that a fair coin is to be tossed n times independently. For $i = 1, \ldots, n$, let $X_i = 1$ if a head is obtained on the ith toss and let $X_i = 0$ if a tail is obtained on the ith toss. Then the sample mean $\overline{X}_n$ will simply be equal to the proportion of heads that are obtained on the n tosses. We shall determine the number of times the coin must be tossed in order to make $\Pr(0.4 \leqslant \overline{X}_n \leqslant 0.6) \geqslant 0.7$. We shall determine this number in two ways: first, by using the Chebyshev inequality; second, by using the exact probabilities for the binomial distribution of the total number of heads.

Let $T = \sum_{i=1}^{n} X_i$ denote the total number of heads that are obtained when n tosses are made. Then T has a binomial distribution with parameters n and $p = 1/2$. Therefore, it follows from Eq. (3) of Sec. 4.2 that $E(T) = n/2$ and it follows from Eq. (2) of Sec. 4.3 that $\mathrm{Var}(T) = n/4$. Since $\overline{X}_n = T/n$, we can obtain the following relation from the Chebyshev inequality:

$$\Pr(0.4 \leqslant \overline{X}_n \leqslant 0.6) = \Pr(0.4n \leqslant T \leqslant 0.6n)$$

$$= \Pr\left(\left|T - \frac{n}{2}\right| \leqslant 0.1n\right)$$

$$\geqslant 1 - \frac{n}{4(0.1n)^2} = 1 - \frac{25}{n}.$$

Hence, if $n \geqslant 84$, this probability will be at least 0.7, as required.

However, from the tables of the binomial distribution given at the end of this book, it is found that for $n = 15$,

$$\Pr(0.4 \leqslant \overline{X}_n \leqslant 0.6) = \Pr(6 \leqslant T \leqslant 9) = 0.70.$$

Hence, 15 tosses would actually be sufficient to satisfy the specified probability requirement. □

The Law of Large Numbers

The discussion in Example 2 indicates that the Chebyshev inequality may not be a practical tool for determining the appropriate sample size in a particular problem, because it may specify a much greater sample size than is actually needed for the particular distribution from which the sample is being taken. However, the Chebyshev inequality is a valuable theoretical tool, and it will be used here to prove an important result known as the *law of large numbers*.

Convergence in Probability. Suppose that $Z_1, Z_2, \ldots$ is a sequence of random variables. Roughly speaking, it is said that this sequence converges to a given number b if the probability distribution of Z_n becomes more and more concentrated around b as $n \to \infty$. Formally, it is said that the sequence $Z_1, Z_2, \ldots$ *converges to b in probability* if for any given number $\varepsilon > 0$,

$$\lim_{n \to \infty} \Pr(|Z_n - b| < \varepsilon) = 1.$$

In other words, the sequence converges to b in probability if the probability that Z_n lies in any given interval around b, no matter how small this interval may be, approaches 1 as $n \to \infty$.

The statement that the sequence $Z_1, Z_2, \ldots$ converges to b in probability is represented either by the notation

$$\plim_{n \to \infty} Z_n = b$$

or by the notation

$$Z_n \xrightarrow{p} b.$$

Sometimes it is simply said that Z_n converges to b in probability.

We shall now show that the sample mean always converges in probability to the mean of the distribution from which the random sample was taken.

Law of Large Numbers. *Suppose that $X_1, \ldots, X_n$ form a random sample from a distribution for which the mean is μ, and let $\overline{X}_n$ denote the sample mean. Then*

$$\plim_{n \to \infty} \overline{X}_n = \mu. \tag{4}$$

Proof. For the purposes of this proof we shall assume that the distribution from which the random sample is taken has both a finite mean μ and a finite variance σ^2. It then follows from the Chebyshev inequality that for any given number $\varepsilon > 0$,

$$\Pr(|\overline{X}_n - \mu| < \varepsilon) \geqslant 1 - \frac{\sigma^2}{n\varepsilon^2}.$$

Hence,

$$\lim_{n \to \infty} \Pr(|\overline{X}_n - \mu| < \varepsilon) = 1,$$

which means that $\plim_{n \to \infty} \overline{X}_n = \mu$.

It can also be shown that Eq. (4) is satisfied if the distribution from which the random sample is taken has a finite mean μ but an infinite variance. However, the proof for this case is beyond the scope of this book.   □

Since $\overline{X}_n$ converges to μ in probability, it follows that there is high probability that $\overline{X}_n$ will be close to μ if the sample size n is large. Hence, if a large random sample is taken from a distribution for which the mean is unknown, then the arithmetic average of the values in the sample will usually be a close estimate of the unknown mean. This topic will be discussed again in Chapter 5 after the central limit theorem has been derived. It will then be possible to present a more precise probability distribution for the difference between $\overline{X}_n$ and μ.

Weak Laws and Strong Laws.   There are other concepts of the convergence of a sequence of random variables, in addition to the concept of convergence in probability that has been presented here. For example, it is said that a sequence $Z_1, Z_2, \ldots$ *converges to a constant b with probability 1 if*

$$\Pr\left(\lim_{n \to \infty} Z_n = b \right) = 1.$$

A careful investigation of the concept of convergence with probability 1 is beyond the scope of this book. It can be shown that if a sequence $Z_1, Z_2, \ldots$ converges to b with probability 1, then the sequence will also converge to b in probability. For this reason, convergence with probability 1 is often called *strong convergence*, whereas convergence in probability is called *weak convergence*. In order to emphasize the distinction between these two concepts of convergence, the result which here has been called simply the law of large numbers is often called the *weak law of large numbers*. The *strong law of large numbers* can then be stated as follows: If $\overline{X}_n$ is the sample mean of a random sample of size n from a distribution with mean μ, then

$$\Pr\left(\lim_{n \to \infty} \overline{X}_n = \mu \right) = 1.$$

The proof of this result will not be given here.

EXERCISES

1. Suppose that X is a random variable for which

   $$\Pr(X \geqslant 0) = 1 \text{ and } \Pr(X \geqslant 10) = 1/5.$$

   Prove that $E(X) \geqslant 2$.

2. Suppose that X is a random variable for which $E(X) = 10$, $\Pr(X \leqslant 7) = 0.2$, and $\Pr(X \geqslant 13) = 0.3$. Prove that $\text{Var}(X) \geqslant 9/2$.

3. Let X be a random variable for which $E(X) = \mu$ and $\text{Var}(X) = \sigma^2$. Construct a probability distribution for X such that

$$\Pr(|X - \mu| \geqslant 3\sigma) = 1/9.$$

4. How large a random sample must be taken from a given distribution in order for the probability to be at least 0.99 that the sample mean will be within 2 standard deviations of the mean of the distribution?

5. Suppose that $X_1, \ldots, X_n$ form a random sample of size n from a distribution for which the mean is 6.5 and the variance is 4. Determine how large the value of n must be in order for the following relation to be satisfied:

$$\Pr(6 \leqslant \overline{X}_n \leqslant 7) \geqslant 0.8.$$

6. Suppose that X is a random variable for which $E(X) = \mu$ and $E[(X - \mu)^4] = \beta_4$. Prove that

$$\Pr(|X - \mu| \geqslant t) \leqslant \frac{\beta_4}{t^4}.$$

7. Suppose that 30 percent of the items in a large manufactured lot are of poor quality. Suppose also that a random sample of n items is to be taken from the lot, and let Q_n denote the proportion of the items in the sample that are of poor quality. Find a value of n such that $\Pr(0.2 \leqslant Q_n \leqslant 0.4) \geqslant 0.75$ by using (a) the Chebyshev inequality and (b) the tables of the binomial distribution at the end of this book.

8. Let $Z_1, Z_2, \ldots$ be a sequence of random variables; and suppose that, for $n = 1, 2, \ldots$, the distribution of Z_n is as follows:

$$\Pr(Z_n = n^2) = \frac{1}{n} \quad \text{and} \quad \Pr(Z_n = 0) = 1 - \frac{1}{n}.$$

Show that

$$\lim_{n \to \infty} E(Z_n) = \infty \text{ but } \plim_{n \to \infty} Z_n = 0.$$

9. It is said that a sequence of random variables $Z_1, Z_2, \ldots$ *converges to a constant b in the quadratic mean* if

$$\lim_{n \to \infty} E\left[(Z_n - b)^2\right] = 0. \tag{5}$$

Show that Eq. (5) is satisfied if and only if

$$\lim_{n \to \infty} E(Z_n) = b \quad \text{and} \quad \lim_{n \to \infty} \text{Var}(Z_n) = 0.$$

Hint: Use Exercise 4 of Sec. 4.3.

10. Prove that if a sequence $Z_1, Z_2, \ldots$ converges to a constant b in the quadratic mean, then the sequence also converges to b in probability.

11. Let $\overline{X}_n$ be the sample mean of a random sample of size n from a distribution for which the mean is μ and the variance is σ^2, where $\sigma^2 < \infty$. Show that $\overline{X}_n$ converges to μ in the quadratic mean as $n \to \infty$.

12. Let $Z_1, Z_2, \ldots$ be a sequence of random variables; and suppose that for $n = 2, 3, \ldots$, the distribution of Z_n is as follows:

$$\Pr\left(Z_n = \frac{1}{n}\right) = 1 - \frac{1}{n^2} \quad \text{and} \quad \Pr(Z_n = n) = \frac{1}{n^2}.$$

(a) Does there exist a constant c to which the sequence converges in probability?

(b) Does there exist a constant c to which the sequence converges in the quadratic mean?

*4.9. UTILITY

Utility Functions

Consider a gamble in which one of the three following outcomes will occur: A person will win 100 dollars with probability $1/5$, will win 0 dollars with probability $2/5$, or will lose 40 dollars with probability $2/5$. The expected *gain* from this gamble is

$$\frac{1}{5}(100) + \frac{2}{5}(0) + \frac{2}{5}(-40) = 4.$$

In general, any gamble of this type, in which the possible gains or losses are different amounts of money, can be represented as a random variable X with a specified probability distribution. It is to be understood that a positive value of X represents an actual monetary gain to the person from the gamble, and that a negative value of X represents a loss (which is considered a negative gain). The expected gain from a gamble X is then simply $E(X)$.

Although two different gambles X and Y may have the same expected gain, a person who is forced to accept one of the two gambles would typically prefer one of them to the other. For example, consider two gambles X and Y for which the gains have the following probability distributions:

$$\Pr(X = 500) = \Pr(X = -400) = 1/2 \tag{1}$$

and

$$\Pr(Y = 60) = \Pr(Y = 50) = \Pr(Y = 40) = 1/3. \tag{2}$$

Here, $E(X) = E(Y) = 50$. However, a person who does not desire to risk losing 400 dollars for the chance of winning 500 dollars would typically prefer Y, which yields a certain gain of at least 40 dollars.

The *theory of utility* was developed during the 1930's and 1940's to describe a person's preference among gambles like those just described. According to that theory, a person will prefer a gamble X for which the expectation of a certain function $U(X)$ is a maximum, rather than a gamble for which simply the expected gain $E(X)$ is a maximum. The function U is called the person's *utility function*. Roughly speaking, a person's utility function is a function that assigns to each possible amount x ($-\infty < x < \infty$) a number $U(x)$ representing the actual worth to the person of gaining the amount x.

For example, suppose that a person's utility function is U and he must choose between the gambles X and Y defined by Eqs. (1) and (2). Then

$$E[U(X)] = \frac{1}{2}U(500) + \frac{1}{2}U(-400) \tag{3}$$

and

$$E[U(Y)] = \frac{1}{3}U(60) + \frac{1}{3}U(50) + \frac{1}{3}U(40). \tag{4}$$

The person would prefer the gamble for which the expected utility of the gain, as specified by Eq. (3) or Eq. (4), is larger.

Formally, a person's utility function is defined as a function U having the following property: When the person must choose between any two gambles X and Y, he will prefer X to Y if $E[U(X)] > E[U(Y)]$, and will be indifferent between X and Y if $E[U(X)] = E[U(Y)]$. When the person is choosing from more than two gambles, he will choose a gamble X for which $E[U(X)]$ is a maximum.

We shall not consider here the problem of determining the conditions that must be satisfied by a person's preferences among all possible gambles in order to be certain that these preferences can be represented by a utility function. This problem and other aspects of the theory of utility are discussed by DeGroot (1970).

Examples of Utility Functions

Since it is reasonable to assume that every person prefers a larger gain to a smaller gain, we shall assume that every utility function $U(x)$ is an increasing function of the gain x. However, the shape of the function $U(x)$ will vary from person to person and will depend on each person's willingness to risk losses of various amounts in attempting to increase his gains.

For example, consider two gambles X and Y for which the gains have the following probability distributions:

$$\Pr(X = -3) = 0.5, \qquad \Pr(X = 2.5) = 0.4, \qquad \Pr(X = 6) = 0.1 \qquad (5)$$

and

$$\Pr(Y = -2) = 0.3, \qquad \Pr(Y = 1) = 0.4, \qquad \Pr(Y = 3) = 0.3. \qquad (6)$$

We shall assume that a person must choose one of the following three decisions: (i) accept gamble X, (ii) accept gamble Y, or (iii) do not accept either gamble. We shall now determine the decision that a person would choose for three different utility functions.

Example 1: Linear Utility Function. Suppose that $U(x) = ax + b$ for some constants a and b, where $a > 0$. In this case, for any gamble X, $E[U(X)] = aE(X) + b$. Hence, for any two gambles X and Y, $E[U(X)] > E[U(Y)]$ if and only if $E(X) > E(Y)$. In other words, a person who has a linear utility function will always choose a gamble for which the expected gain is a maximum.

When the gambles X and Y are defined by Eqs. (5) and (6),

$$E(X) = (0.5)(-3) + (0.4)(2.5) + (0.1)(6) = 0.1$$

and

$$E(Y) = (0.3)(-2) + (0.4)(1) + (0.3)(3) = 0.7.$$

Furthermore, since the gain from not accepting either of these gambles is 0, the expected gain from choosing not to accept either gamble is clearly 0. Since $E(Y) > E(X) > 0$, it follows that a person who has a linear utility function would choose to accept gamble Y. If gamble Y were not available, then the person would prefer to accept gamble X rather than not to gamble at all. □

Example 2: Cubic Utility Function. Suppose that a person's utility function is $U(x) = x^3$ for $-\infty < x < \infty$. Then for the gambles defined by Eqs. (5) and (6),

$$E[U(X)] = (0.5)(-3)^3 + (0.4)(2.5)^3 + (0.1)(6)^3 = 14.35$$

and

$$E[U(Y)] = (0.3)(-2)^3 + (0.4)(1)^3 + (0.3)(3)^3 = 6.1.$$

Furthermore, the utility of not accepting either gamble is $U(0) = 0^3 = 0$. Since $E[U(X)] > E[U(Y)] > 0$, it follows that the person would choose to accept gamble X. If gamble X were not available, the person would prefer to accept gamble Y rather than not to gamble at all.   □

Example 3: Logarithmic Utility Function. Suppose that a person's utility function is $U(x) = \log(x + 4)$ for $x > -4$. Since $\lim_{x \to -4} \log(x + 4) = -\infty$, a person who has this utility function cannot choose a gamble in which there is any possibility of his gain being -4 or less. For the gambles X and Y defined by Eqs. (5) and (6),

$$E[U(X)] = (0.5)(\log 1) + (0.4)(\log 6.5) + (0.1)(\log 10) = 0.9790$$

and

$$E[U(Y)] = (0.3)(\log 2) + (0.4)(\log 5) + (0.3)(\log 7) = 1.4355.$$

Furthermore, the utility of not accepting either gamble is $U(0) = \log 4 = 1.3863$. Since $E[U(Y)] > U(0) > E[U(X)]$, it follows that the person would choose to accept gamble Y. If gamble Y were not available, the person would prefer not to gamble at all rather than to accept gamble X.   □

Selling a Lottery Ticket

Suppose that a person has a lottery ticket from which he will receive a random gain of X dollars, where X has a specified probability distribution. We shall determine the number of dollars for which the person would be willing to sell this lottery ticket.

Let U denote the person's utility function. Then the expected utility of his gain from the lottery ticket is $E[U(X)]$. If he sells the lottery ticket for x_0 dollars, then his gain is x_0 dollars and the utility of this gain is $U(x_0)$. The person will prefer to accept x_0 dollars as a certain gain rather than to accept the random gain X from the lottery ticket if and only if $U(x_0) > E[U(X)]$. Hence, the person would be willing to sell the lottery ticket for any amount x_0 such that $U(x_0) > E[U(X)]$. If $U(x_0) = E[U(X)]$, he would be indifferent between selling the lottery ticket and accepting the random gain X.

Example 4: Quadratic Utility Function. Suppose that $U(x) = x^2$ for $x \geqslant 0$, and suppose that the person has a lottery ticket from which he will win either 36 dollars with probability 1/4 or 0 dollars with probability 3/4. For how many dollars x_0 would he be willing to sell this lottery ticket?

The expected utility of the gain from the lottery ticket is

$$E[U(X)] = \frac{1}{4}U(36) + \frac{3}{4}U(0)$$

$$= \frac{1}{4}(36^2) + \frac{3}{4}(0) = 324.$$

Therefore, the person would be willing to sell the lottery ticket for any amount x_0 such that $U(x_0) = x_0^2 > 324$. Hence, $x_0 > 18$. In other words, although the expected gain from the lottery ticket in this example is only 9 dollars, the person would not sell the ticket for less than 18 dollars. □

Example 5: Square-Root Utility Function. Suppose now that $U(x) = x^{1/2}$ for $x \geq 0$, and consider again the lottery ticket described in Example 4. The expected utility of the gain from the lottery ticket in this case is

$$E[U(X)] = \frac{1}{4}U(36) + \frac{3}{4}U(0)$$

$$= \frac{1}{4}(6) + \frac{3}{4}(0) = 1.5.$$

Therefore, the person would be willing to sell the lottery ticket for any amount x_0 such that $U(x_0) = x_0^{1/2} > 1.5$. Hence, $x_0 > 2.25$. In other words, although the expected gain from the lottery ticket in this example is 9 dollars, the person would be willing to sell the ticket for as little as 2.25 dollars. □

EXERCISES

1. Consider three gambles X, Y, and Z for which the probability distributions of the gains are as follows:

$$\Pr(X = 5) = \Pr(X = 25) = 1/2,$$

$$\Pr(Y = 10) = \Pr(Y = 20) = 1/2,$$

$$\Pr(Z = 15) = 1.$$

Suppose that a person's utility function has the form $U(x) = x^2$ for $x > 0$. Which of the three gambles would he prefer?

2. Determine which of the three gambles in Exercise 1 would be preferred by a person whose utility function is $U(x) = x^{1/2}$ for $x > 0$.

3. Determine which of the three gambles in Exercise 1 would be preferred by a person whose utility function has the form $U(x) = ax + b$, where a and b are constants $(a > 0)$.

4. Consider a utility function U for which $U(0) = 0$ and $U(100) = 1$. Suppose that a person who has this utility function is indifferent between accepting a gamble from which his gain will be 0 dollars with probability $1/3$ or 100 dollars with probability $2/3$ and accepting 50 dollars as a sure thing. What is the value of $U(50)$?

5. Consider a utility function U for which $U(0) = 5$, $U(1) = 8$, and $U(2) = 10$. Suppose that a person who has this utility function is indifferent between two gambles X and Y for which the probability distributions of the gains are as follows:

$\Pr(X = -1) = 0.6$, $\Pr(X = 0) = 0.2$, $\Pr(X = 2) = 0.2$;

$\Pr(Y = 0) = 0.9$, $\Pr(Y = 1) = 0.1$.

What is the value of $U(-1)$?

6. Suppose that a person must accept a gamble X of the following form:

$\Pr(X = a) = p$ and $\Pr(X = 1 - a) = 1 - p$,

where p is a given number such that $0 < p < 1$. Suppose also that the person can choose and fix the value of a $(0 \leqslant a \leqslant 1)$ to be used in this gamble. Determine the value of a that the person would choose if his utility function is $U(x) = \log x$ for $x > 0$.

7. Determine the value of a that a person would choose in Exercise 6 if his utility function is $U(x) = x^{1/2}$ for $x \geqslant 0$.

8. Determine the value of a that a person would choose in Exercise 6 if his utility function is $U(x) = x$ for $x \geqslant 0$.

9. Consider four gambles X_1, X_2, X_3, and X_4 for which the probability distributions of the gains are as follows:

$\Pr(X_1 = 0) = 0.2$, $\Pr(X_1 = 1) = 0.5$, $\Pr(X_1 = 2) = 0.3$;

$\Pr(X_2 = 0) = 0.4$, $\Pr(X_2 = 1) = 0.2$, $\Pr(X_2 = 2) = 0.4$;

$\Pr(X_3 = 0) = 0.3$, $\Pr(X_3 = 1) = 0.3$, $\Pr(X_3 = 2) = 0.4$;

$\Pr(X_4 = 0) = \Pr(X_4 = 2) = 0.5$.

Suppose that a person's utility function is such that he prefers X_1 to X_2. If the person were forced to accept either X_3 or X_4, which one would he choose?

10. Suppose that a person has a given fortune $A > 0$ and can bet any amount b of this fortune in a certain game $(0 \leqslant b \leqslant A)$. If he wins the bet, then his fortune becomes $A + b$; if he loses the bet, then his fortune becomes $A - b$. In general, let X denote his fortune after he has won or lost. Assume that the probability of his winning is p $(0 < p < 1)$ and the probability of his losing is $1 - p$. Assume also that his utility function, as a function of his final fortune x, is $U(x) = \log x$ for $x > 0$. If the person wishes to bet an amount b for which the expected utility of his fortune $E[U(X)]$ will be a maximum, what amount b should he bet?

11. Determine the amount b that the person should bet in Exercise 10 if his utility function is $U(x) = x^{1/2}$ for $x \geqslant 0$.

12. Determine the amount b that the person should bet in Exercise 10 if his utility function is $U(x) = x$ for $x \geqslant 0$.

13. Determine the amount b that the person should bet in Exercise 10 if his utility function is $U(x) = x^2$ for $x \geqslant 0$.

14. Suppose that a person has a lottery ticket from which he will win X dollars, where X has a uniform distribution on the interval $(0, 4)$. Suppose also that the person's utility function is $U(x) = x^\alpha$ for $x \geqslant 0$, where α is a given positive constant. For how many dollars x_0 would the person be willing to sell this lottery ticket?

4.10. SUPPLEMENTARY EXERCISES

1. Suppose that the random variable X has a continuous distribution with d.f. $F(x)$. Suppose also that $\Pr(X \geqslant 0) = 1$ and that $E(X)$ exists. Show that

$$E(X) = \int_0^\infty [1 - F(x)] \, dx.$$

Hint: You may use the fact that if $E(X)$ exists, then

$$\lim_{x \to \infty} x[1 - F(x)] = 0.$$

2. Consider again the conditions of Exercise 1, but suppose now that X has a discrete distribution with d.f. $F(x)$, rather than a continuous distribution. Show that the conclusion of Exercise 1 still holds.

3. Suppose that X, Y, and Z are nonnegative random variables such that $\Pr(X + Y + Z \leqslant 1.3) = 1$. Show that X, Y, and Z cannot possibly have a joint distribution under which each of their marginal distributions is a uniform distribution on the interval $(0, 1)$.

4. Suppose that the random variable X has mean μ and variance σ^2, and that $Y = aX + b$. Determine the values of a and b for which $E(Y) = 0$ and $\text{Var}(Y) = 1$.

5. Determine the expectation of the range of a random sample of size n from a uniform distribution on the interval $(0, 1)$.

6. Suppose that an automobile dealer pays an amount X (in thousands of dollars) for a used car and then sells it for an amount Y. Suppose that the random variables X and Y have the following joint p.d.f.:

$$f(x, y) = \begin{cases} \dfrac{1}{36}x & \text{for } 0 < x < y < 6, \\ 0 & \text{otherwise.} \end{cases}$$

Determine the dealer's expected gain from the sale.

7. Suppose that $X_1, \ldots, X_n$ form a random sample of size n from a continuous distribution with the following p.d.f.:

$$f(x) = \begin{cases} 2x & \text{for } 0 < x < 1, \\ 0 & \text{otherwise.} \end{cases}$$

Let $Y_n = \max\{X_1, \ldots, X_n\}$. Evaluate $E(Y_n)$.

8. If m is a median of the distribution of X and if $Y = r(X)$ is either a nondecreasing or a nonincreasing function of X, show that $r(m)$ is a median of the distribution of Y.

9. Suppose that $X_1, \ldots, X_n$ are i.i.d. random variables, each of which has a continuous distribution with median m. Let $Y_n = \max\{X_1, \ldots, X_n\}$. Determine the value of $\Pr(Y_n > m)$.

10. Suppose that you are going to sell cola at a football game and must decide in advance how much to order. Suppose that the demand for cola at the game, in liters, has a continuous distribution with p.d.f. $f(x)$. Suppose that you make a profit of g cents on each liter that you sell at the game and suffer a loss of c cents on each liter that you order but do not sell. What is the optimal amount of cola for you to order so as to maximize your expected net gain?

11. Suppose that the number of hours X for which a machine will operate before it fails has a continuous distribution with p.d.f. $f(x)$. Suppose that at the time at which the machine begins operating you must decide when you will return to inspect it. If you return before the machine has failed, you incur a cost of b dollars for having wasted an inspection. If you return after the machine has failed, you incur a cost of c dollars per hour for the length of time during which the machine was not operating after its failure. What is the

optimal number of hours to wait before you return for inspection in order to minimize your expected cost?

12. Suppose that X and Y are random variables for which $E(X) = 3$, $E(Y) = 1$, $\text{Var}(X) = 4$, and $\text{Var}(Y) = 9$. Let $Z = 5X - Y + 15$. Find $E(Z)$ and $\text{Var}(Z)$ under each of the following conditions: (a) X and Y are independent; (b) X and Y are uncorrelated; (c) the correlation of X and Y is 0.25.

13. Suppose that $X_0, X_1, \ldots, X_n$ are independent random variables, each having the same variance σ^2. Let $Y_j = X_j - X_{j-1}$ for $j = 1, \ldots, n$, and let $\overline{Y}_n = \frac{1}{n}\sum_{j=1}^{n} Y_j$. Determine the value of $\text{Var}(\overline{Y}_n)$.

14. Suppose that $X_1, \ldots, X_n$ are random variables for which $\text{Var}(X_i)$ has the same value σ^2 for $i = 1, \ldots, n$ and $\rho(X_i, X_j)$ has the same value ρ for every pair of values i and j such that $i \neq j$. Prove that $\rho \geq -\dfrac{1}{n-1}$.

15. Suppose that the joint distribution of X and Y is a uniform distribution over a circle in the xy-plane. Determine the correlation of X and Y.

16. Suppose that n letters are put at random into n envelopes, as in the matching problem described in Sec. 1.10. Determine the variance of the number of letters that are placed in the correct envelopes.

17. Suppose that the random variable X has mean μ and variance σ^2. Show that the third central moment of X can be expressed as $E(X^3) - 3\mu\sigma^2 - \mu^3$.

18. Suppose that X is a random variable with m.g.f. $\psi(t)$, mean μ, and variance σ^2; and let $c(t) = \log[\psi(t)]$. Prove that $c'(0) = \mu$ and $c''(0) = \sigma^2$.

19. Suppose that X and Y have a joint distribution with means μ_X and μ_Y, standard deviations σ_X and σ_Y, and correlation ρ. Show that if $E(Y \mid X)$ is a linear function of X, then

$$E(Y \mid X) = \mu_Y + \rho\frac{\sigma_Y}{\sigma_X}(X - \mu_X).$$

20. Suppose that X and Y are random variables such that $E(Y \mid X) = 7 - (1/4)X$ and $E(X \mid Y) = 10 - Y$. Determine the correlation of X and Y.

21. Suppose that a stick having a length of 3 feet is broken into two pieces, and that the point at which the stick is broken is chosen in accordance with the p.d.f. $f(x)$. What is the correlation between the length of the longer piece and the length of the shorter piece?

22. Suppose that X and Y have a joint distribution with correlation $\rho > 1/2$ and that $\text{Var}(X) = \text{Var}(Y) = 1$. Show that $b = -\dfrac{1}{2\rho}$ is the unique value of b such that the correlation of X and $X + bY$ is also ρ.

23. Suppose that four apartment buildings A, B, C, and D are located along a highway at the points 0, 1, 3, and 5, as shown in the following figure. Suppose

also that 10 percent of the employees of a certain company live in building A, 20 percent live in B, 30 percent live in C, and 40 percent live in D.

```
  A   B     C     D
──•───•──┼──•──┼──•──┼──┼──
  0   1   2   3   4   5   6   7
```

(a) Where should the company build its new office in order to minimize the total distance that its employees must travel?

(b) Where should the company build its new office in order to minimize the sum of the squared distances that its employees must travel?

24. Suppose that X and Y have the following joint p.d.f.:

$$f(x, y) = \begin{cases} 8xy & \text{for } 0 < y < x < 1, \\ 0 & \text{otherwise.} \end{cases}$$

Suppose also that the observed value of X is 0.2.

(a) What predicted value of Y has the smallest M.S.E.?

(b) What predicted value of Y has the smallest M.A.E.?

25. For any random variables X, Y, and Z, let $\text{Cov}(X, Y \mid z)$ denote the covariance of X and Y in their conditional joint distribution given $Z = z$. Prove that

$$\text{Cov}(X, Y) = E[\text{Cov}(X, Y \mid Z)] + \text{Cov}[E(X \mid Z), E(Y \mid Z)].$$

26. Suppose that X is a random variable such that $E(X^k)$ exists and $\Pr(X \geq 0) = 1$. Prove that for $k > 0$ and $t > 0$,

$$\Pr(X \geq t) \leq \frac{E(X^k)}{t^k}.$$

27. Suppose that a person's utility function is $U(x) = x^2$ for $x \geq 0$. Show that the person will always prefer to take a gamble in which he will receive a random gain of X dollars rather than receive the amount $E(X)$ with certainty, where $\Pr(X \geq 0) = 1$ and $E(X) < \infty$.

28. A person is given m dollars, which he must allocate between an event A and its complement A^c. Suppose that he allocates a dollars to A and $m - a$ dollars to A^c. The person's gain is then determined as follows: If A occurs, his gain is $g_1 a$; if A^c occurs, his gain is $g_2(m - a)$. Here, g_1 and g_2 are given positive constants. Suppose also that $\Pr(A) = p$ and the person's utility function is $U(x) = \log x$ for $x > 0$. Determine the amount a that will maximize the person's expected utility and show that this amount does not depend on the values of g_1 and g_2.

Special
Distributions

5

5.1. INTRODUCTION

In this chapter we shall define and discuss several special distributions that are widely used in applications of probability and statistics. The distributions that will be presented here include discrete and continuous distributions of univariate, bivariate, and multivariate types. The discrete univariate distributions are the Bernoulli, binomial, hypergeometric, Poisson, negative binomial, and geometric distributions. The continuous univariate distributions are the normal, gamma, exponential, and beta distributions. Other continuous univariate distributions (introduced in exercises) are the lognormal, Weibull, and Pareto distributions. Also discussed are the multivariate discrete distribution called the multinomial distribution and the bivariate continuous distribution called the bivariate normal distribution.

We shall describe briefly how each of these distributions arises in applied problems and shall show why each might be an appropriate probability model for some experiment. For each distribution we shall present the p.f. or the p.d.f., and discuss some of the basic properties of the distribution.

5.2. THE BERNOULLI AND BINOMIAL DISTRIBUTIONS

The Bernoulli Distribution

An experiment of a particularly simple type is one in which there are only two possible outcomes, such as head or tail, success or failure, or defective or nondefective. It is convenient to designate the two possible outcomes of such an experiment as 0 and 1. The following definition can then be applied to any experiment of this type.

It is said that a random variable X has a *Bernoulli distribution with parameter* p $(0 \leqslant p \leqslant 1)$ if X can take only the values 0 and 1 and the probabilities are

$$\Pr(X = 1) = p \qquad \text{and} \qquad \Pr(X = 0) = 1 - p. \tag{1}$$

If we let $q = 1 - p$, then the p.f. of X can be written as follows:

$$f(x \mid p) = \begin{cases} p^x q^{1-x} & \text{for } x = 0, 1, \\ 0 & \text{otherwise.} \end{cases} \tag{2}$$

To verify that this p.f. $f(x \mid p)$ actually does represent the Bernoulli distribution specified by the probabilities (1), it is simply necessary to note that $f(1 \mid p) = p$ and $f(0 \mid p) = q$.

If X has a Bernoulli distribution with parameter p then, as derived at the end of Sec. 4.3,

$$E(X) = 1 \cdot p + 0 \cdot q = p,$$
$$E(X^2) = 1^2 \cdot p + 0^2 \cdot q = p, \tag{3}$$

and

$$\text{Var}(X) = E(X^2) - [E(X)]^2 = pq. \tag{4}$$

Furthermore, the m.g.f. of X is

$$\psi(t) = E(e^{tX}) = pe^t + q \qquad \text{for } -\infty < t < \infty. \tag{5}$$

Bernoulli Trials

If the random variables in an infinite sequence $X_1, X_2, \ldots$ are i.i.d., and if each random variable X_i has a Bernoulli distribution with parameter p, then it is said that the random variables $X_1, X_2, \ldots$ form an infinite sequence of *Bernoulli trials with parameter p*. Similarly, if n random variables $X_1, \ldots, X_n$ are i.i.d. and each has a Bernoulli distribution with parameter p, then it is said that the variables $X_1, \ldots, X_n$ form n *Bernoulli trials with parameter p*.

For example, suppose that a fair coin is tossed repeatedly. Let $X_i = 1$ if a head is obtained on the ith toss, and let $X_i = 0$ if a tail is obtained $(i = 1, 2, \ldots)$. Then the random variables $X_1, X_2, \ldots$ form an infinite sequence of Bernoulli trials with parameter $p = 1/2$. Similarly, suppose that 10 percent of the items produced by a certain machine are defective, and that n items are selected at random and inspected. In this case, let $X_i = 1$ if the ith item is defective, and let $X_i = 0$ if it is nondefective ($i = 1, \ldots, n$). Then the variables $X_1, \ldots, X_n$ form n Bernoulli trials with parameter $p = 1/10$.

The Binomial Distribution

As stated in Sec. 3.1, a random variable X has a *binomial distribution with parameters n and p* if X has a discrete distribution for which the p.f. is as follows:

$$f(x \mid n, p) = \begin{cases} \binom{n}{x} p^x q^{n-x} & \text{for } x = 0, 1, 2, \ldots, n, \\ 0 & \text{otherwise.} \end{cases} \tag{6}$$

In this distribution n must be a positive integer and p must lie in the interval $0 \leqslant p \leqslant 1$.

The binomial distribution is of fundamental importance in probability and statistics because of the following result, which was derived in Sec. 3.1: Suppose that the outcome of an experiment can be either success or failure; that the experiment is performed n times independently; and that the probability of the success of any given performance is p. If X denotes the total number of successes in the n performances, then X has a binomial distribution with parameters n and p. This result can be restated as follows:

If the random variables $X_1, \ldots, X_n$ form n Bernoulli trials with parameter p and if $X = X_1 + \cdots + X_n$, then X has a binomial distribution with parameters n and p.

When X is represented as the sum of n Bernoulli trials in this way, the values of the mean, variance, and m.g.f. of X can be derived very easily. These values, which were already obtained in Secs. 4.2, 4.3, and 4.4, are

$$E(X) = \sum_{i=1}^{n} E(X_i) = np, \tag{7}$$

$$\text{Var}(X) = \sum_{i=1}^{n} \text{Var}(X_i) = npq, \tag{8}$$

and

$$\psi(t) = E(e^{tX}) = \prod_{i=1}^{n} E(e^{tX_i}) = (pe^t + q)^n. \tag{9}$$

Probabilities for the binomial distribution for various values of n and p can be obtained from the table given at the end of this book.

We shall now use the m.g.f. in Eq. (9) to establish the following simple extension of a result which was derived in Sec. 4.4.

If $X_1, \ldots, X_k$ *are independent random variables and if* X_i *has a binomial distribution with parameters* n_i *and* p *(* $i = 1, \ldots, k$ *), then the sum* $X_1 + \cdots + X_k$ *has a binomial distribution with parameters* $n = n_1 + \cdots + n_k$ *and* p.

This result also follows easily if we represent each X_i as the sum of n_i Bernoulli trials with parameter p. If $n = n_1 + \cdots + n_k$ and if all n trials are independent, then the sum $X_1 + \cdots + X_k$ will simply be the sum of n Bernoulli trials with parameter p. Hence, this sum must have a binomial distribution with parameters n and p.

EXERCISES

1. Suppose that X is a random variable such that $E(X^k) = 1/3$ for $k = 1, 2, \ldots$. Assuming that there cannot be more than one distribution with this same sequence of moments, determine the distribution of X.

2. Suppose that a random variable X can take only the two values a and b with the following probabilities:

   $$\Pr(X = a) = p \quad \text{and} \quad \Pr(X = b) = q.$$

   Express the p.f. of X in a form similar to that given in Eq. (2).

3. Suppose that the probability that a certain experiment will be successful is 0.4, and let X denote the number of successes that are obtained in 15 independent performances of the experiment. Use the table of the binomial distribution given at the end of this book to determine the value of $\Pr(6 \leqslant X \leqslant 9)$.

4. A coin for which the probability of a head is 0.6 is tossed nine times. Use the table of the binomial distribution given at the end of this book to find the probability of obtaining an even number of heads.

5. Three men A, B, and C shoot at a target. Suppose that A shoots three times and the probability that he will hit the target on any given shot is $1/8$; that B shoots five times and the probability that he will hit the target on any given shot is $1/4$; and that C shoots twice and the probability that he will hit the target on any given shot is $1/2$. What is the expected number of times that the target will be hit?

6. For the conditions of Exercise 5, what is the variance of the number of times that the target will be hit?

7. A certain electronic system contains ten components. Suppose that the probability that any individual component will fail is 0.2 and that the

components fail independently of each other. Given that at least one of the components has failed, what is the probability that at least two of the components have failed?

8. Suppose that the random variables $X_1, \ldots, X_n$ form n Bernoulli trials with parameter p. Determine the conditional probability that $X_1 = 1$, given that

$$\sum_{i=1}^{n} X_i = k \qquad (k = 1, \ldots, n).$$

9. The probability that any specific child in a given family will inherit a certain disease is p. If it is known that at least one child in a family of n children has inherited the disease, what is the expected number of children in the family who have inherited the disease?

10. For $0 \le p \le 1$, $q = 1 - p$, and $n = 2, 3, \ldots$, determine the value of

$$\sum_{x=2}^{n} x(x-1)\binom{n}{x}p^x q^{n-x}.$$

11. If a random variable X has a discrete distribution for which the p.f. is $f(x)$, then the value of x for which $f(x)$ is maximum is called the *mode* of the distribution. If this same maximum $f(x)$ is attained at more than one value of x, then all such values of x are called *modes* of the distribution. Find the mode or modes of the binomial distribution with parameters n and p. *Hint:* Study the ratio $f(x + 1 \mid n, p)/f(x \mid n, p)$.

5.3.   THE HYPERGEOMETRIC DISTRIBUTION

Definition of the Hypergeometric Distribution

Suppose that a box contains A red balls and B blue balls. Suppose also that n balls are selected at random from the box without replacement, and let X denote the number of red balls that are obtained. Clearly, the value of X can neither exceed n nor exceed A. Therefore, it must be true that $X \le \min\{n, A\}$. Similarly, since the number of blue balls $n - X$ that are obtained cannot exceed B, the value of X must be at least $n - B$. Since the value of X cannot be less than 0, it must be true that $X \ge \max\{0, n - B\}$. Hence, the value of X must be an integer in the interval

$$\max\{0, n - B\} \le x \le \min\{n, A\}. \tag{1}$$

Let $f(x \mid A, B, n)$ denote the p.f. of X. Then for any integer x in the interval (1), the probability of obtaining exactly x red balls, as explained in Example 3 of

Sec. 1.8, is

$$f(x \mid A, B, n) = \frac{\binom{A}{x}\binom{B}{n - x}}{\binom{A + B}{n}}. \tag{2}$$

Furthermore, $f(x \mid A, B, n) = 0$ for all other values of x. If a random variable X has a discrete distribution with this p.f., then it is said that X has a *hypergeometric distribution with parameters A, B, and n.*

Extending the Definition of Binomial Coefficients

In order to simplify the expression for the p.f. of the hypergeometric distribution, it is convenient to extend the definition of a binomial coefficient given in Sec. 1.8. For any positive integers r and m, where $r \leqslant m$, the binomial coefficient $\binom{m}{r}$ was defined to be

$$\binom{m}{r} = \frac{m!}{r!(m - r)!}. \tag{3}$$

It can be seen that the value of $\binom{m}{r}$ specified by Eq. (3) can also be written in the form

$$\binom{m}{r} = \frac{m(m - 1) \cdots (m - r + 1)}{r!}. \tag{4}$$

For any real number m, which is not necessarily a positive integer, and for any positive integer r, the value of the right side of Eq. (4) is a well-defined number. Therefore, for any real number m and any positive integer r, we can extend the definition of the binomial coefficient $\binom{m}{r}$ by defining its value as that given by Eq. (4).

The value of the binomial coefficient $\binom{m}{r}$ can be obtained from this definition for any positive integers r and m. If $r \leqslant m$, the value of $\binom{m}{r}$ is given by Eq. (3). If $r > m$, it follows that $\binom{m}{r} = 0$. Finally, for any real number m, we shall define the value of $\binom{m}{0}$ to be $\binom{m}{0} = 1$.

When this extended definition of a binomial coefficient is used, it can be seen that the value of $\binom{A}{x}\binom{B}{n - x}$ is 0 for any integer x such that either $x > A$ or $n - x > B$. Therefore, we can write the p.f. of a hypergeometric distribution with parameters A, B, and n as follows:

$$f(x \mid A, B, n) = \begin{cases} \dfrac{\dbinom{A}{x}\dbinom{B}{n-x}}{\dbinom{A+B}{n}} & \text{for } x = 0, 1, \ldots, n, \\ 0 & \text{otherwise.} \end{cases} \tag{5}$$

It then follows from Eq. (4) that $f(x \mid A, B, n) > 0$ if and only if x is an integer in the interval (1).

The Mean and Variance for a Hypergeometric Distribution

We shall continue to assume that n balls are selected at random without replacement from a box containing A red balls and B blue balls. For $i = 1, \ldots, n$, we shall let $X_i = 1$ if the ith ball that is selected is red and shall let $X_i = 0$ if the ith ball is blue. As explained in Example 2 of Sec. 4.2, we can imagine that the n balls are selected from the box by first arranging all the balls in the box in some random order and then selecting the first n balls from this arrangement. It can be seen from this interpretation that, for $i = 1, \ldots, n$,

$$\Pr(X_i = 1) = \frac{A}{A + B} \qquad \text{and} \qquad \Pr(X_i = 0) = \frac{B}{A + B}.$$

Therefore, for $i = 1, \ldots, n$,

$$E(X_i) = \frac{A}{A + B} \qquad \text{and} \qquad \text{Var}(X_i) = \frac{AB}{(A + B)^2}. \tag{6}$$

Since $X = X_1 + \cdots + X_n$, then as shown in Example 2 of Sec. 4.2,

$$E(X) = \sum_{i=1}^{n} E(X_i) = \frac{nA}{A + B}. \tag{7}$$

In other words, the mean of a hypergeometric distribution with parameters A, B, and n is $nA/(A + B)$.

Furthermore, by Theorem 5 of Sec. 4.6,

$$\text{Var}(X) = \sum_{i=1}^{n} \text{Var}(X_i) + 2\sum\sum_{i<j} \text{Cov}(X_i, X_j). \tag{8}$$

Because of the symmetry among the variables $X_1, \ldots, X_n$, every term $\text{Cov}(X_i, X_j)$ in the final summation in Eq. (8) will have the same value as $\text{Cov}(X_1, X_2)$. Since there are $\binom{n}{2}$ terms in this summation, it follows from Eqs. (6) and (8) that

$$\text{Var}(X) = \frac{nAB}{(A + B)^2} + n(n - 1)\text{Cov}(X_1, X_2). \tag{9}$$

If $n = A + B$, then it must be true that $X = A$ because *all* the balls in the box will be selected without replacement. Thus, for $n = A + B$, $\text{Var}(X) = 0$ and it follows from Eq. (9) that

$$\text{Cov}(X_1, X_2) = -\frac{AB}{(A + B)^2(A + B - 1)}.$$

Hence, by Eq. (9),

$$\text{Var}(X) = \frac{nAB}{(A + B)^2} \cdot \frac{A + B - n}{A + B - 1}. \tag{10}$$

We shall use the following notation: $T = A + B$ denotes the total number of balls in the box; $p = A/T$ denotes the proportion of red balls in the box; and $q = 1 - p$ denotes the proportion of blue balls in the box. Then $\text{Var}(X)$ can be rewritten as follows:

$$\text{Var}(X) = npq\frac{T - n}{T - 1}. \tag{11}$$

Comparison of Sampling Methods

It is interesting to compare the variance of the hypergeometric distribution given in Eq. (11) with the variance npq of the binomial distribution. If n balls are selected from the box *with replacement*, then the number X of red balls that are obtained will have a binomial distribution with variance npq. The factor $\alpha = (T - n)/(T - 1)$ in Eq. (11) therefore represents the reduction in $\text{Var}(X)$ caused by sampling without replacement from a finite population.

If $n = 1$, the value of this factor α is 1, because there is no distinction between sampling with replacement and sampling without replacement when only one ball is being selected. If $n = T$, then (as previously mentioned) $\alpha = 0$ and $\text{Var}(X) = 0$. For values of n between 1 and T, the value of α will be between 0 and 1.

For any fixed sample size n, it can be seen that $\alpha \to 1$ as $T \to \infty$. This limit reflects the fact that when the population size T is very large in comparison with the sample size n, there is very little difference between sampling with replacement and sampling without replacement. In other words, if the sample size n represents a negligible fraction of the total population $A + B$, then the hypergeometric distribution with parameters A, B, and n will be very nearly the same as the binomial distribution with parameters n and $p = A/(A + B)$.

EXERCISES

1. Suppose that a box contains five red balls and ten blue balls. If seven balls are selected at random without replacement, what is the probability that at least three red balls will be obtained?

2. Suppose that seven balls are selected at random without replacement from a box containing five red balls and ten blue balls. If $\overline{X}$ denotes the proportion of red balls in the sample, what are the mean and the variance of $\overline{X}$?

3. Find the value of $\begin{pmatrix} 3/2 \\ 4 \end{pmatrix}$.

4. Show that for any positive integers n and k,

$$\binom{-n}{k} = (-1)^k \binom{n + k - 1}{k}.$$

5. If a random variable X has a hypergeometric distribution with parameters $A = 8$, $B = 20$, and n, for what value of n will $\mathrm{Var}(X)$ be a maximum?

6. Suppose that n students are selected at random without replacement from a class containing T students, of whom A are boys and $T - A$ are girls. Let X denote the number of boys that are obtained. For what sample size n will $\mathrm{Var}(X)$ be a maximum?

7. Suppose that X_1 and X_2 are independent random variables; that X_1 has a binomial distribution with parameters n_1 and p; and that X_2 has a binomial distribution with parameters n_2 and p, where p is the same for both X_1 and X_2. For any fixed value of k ($k = 1, 2, \ldots, n_1 + n_2$), determine the conditional distribution of X_1 given that $X_1 + X_2 = k$.

8. Suppose that in a large lot containing T manufactured items, 30 percent of the items are defective and 70 percent are nondefective. Also, suppose that ten items are selected at random without replacement from the lot. Determine (a) an exact expression for the probability that not more than one defective item will be obtained and (b) an approximate expression for this probability based on the binomial distribution.

9. Consider a group of T persons, and let $a_1, \ldots, a_T$ denote the heights of these T persons. Suppose that n persons are selected from this group at random without replacement, and let X denote the sum of the heights of these n persons. Determine the mean and variance of X.

5.4.   THE POISSON DISTRIBUTION

Definition and Properties of the Poisson Distribution

The Probability Function.   Let X be a random variable with a discrete distribution, and suppose that the value of X must be a nonnegative integer. It is said that X has a *Poisson distribution with mean* λ $(\lambda > 0)$ if the p.f. of X is as follows:

$$f(x \mid \lambda) = \begin{cases} \dfrac{e^{-\lambda}\lambda^x}{x!} & \text{for } x = 0, 1, 2, \ldots, \\ 0 & \text{otherwise.} \end{cases} \tag{1}$$

It is clear that $f(x \mid \lambda) \geqslant 0$ for each value of x. In order to verify that the function $f(x \mid \lambda)$ defined by Eq. (1) satisfies the requirements of every p.f., it must be shown that $\sum_{x=0}^{\infty} f(x \mid \lambda) = 1$. It is known from elementary calculus that for any real number λ,

$$e^\lambda = \sum_{x=0}^{\infty} \frac{\lambda^x}{x!}. \tag{2}$$

Therefore,

$$\sum_{x=0}^{\infty} f(x \mid \lambda) = e^{-\lambda} \sum_{x=0}^{\infty} \frac{\lambda^x}{x!} = e^{-\lambda}e^\lambda = 1. \tag{3}$$

The Mean and Variance.   We have stated that the distribution for which the p.f. is given by Eq. (1) is defined to be the Poisson distribution with mean λ. In order to justify this definition, we must show that λ is, in fact, the mean of this distribution. The mean $E(X)$ is specified by the following infinite series:

$$E(X) = \sum_{x=0}^{\infty} xf(x \mid \lambda).$$

Since the term corresponding to $x = 0$ in this series is 0, we can omit this term and can begin the summation with the term for $x = 1$. Therefore,

$$E(X) = \sum_{x=1}^{\infty} xf(x \mid \lambda) = \sum_{x=1}^{\infty} x \frac{e^{-\lambda}\lambda^x}{x!} = \lambda \sum_{x=1}^{\infty} \frac{e^{-\lambda}\lambda^{x-1}}{(x-1)!}.$$

If we now let $y = x - 1$ in this summation, we obtain

$$E(X) = \lambda \sum_{y=0}^{\infty} \frac{e^{-\lambda}\lambda^y}{y!}.$$

By Eq. (3) the sum of the series in this equation is 1. Hence, $E(X) = \lambda$.

The variance of a Poisson distribution can be found by a technique similar to the one that has just been given. We begin by considering the following expectation:

$$E[X(X-1)] = \sum_{x=0}^{\infty} x(x-1)f(x \mid \lambda) = \sum_{x=2}^{\infty} x(x-1)f(x \mid \lambda)$$

$$= \sum_{x=2}^{\infty} x(x-1)\frac{e^{-\lambda}\lambda^x}{x!} = \lambda^2 \sum_{x=2}^{\infty} \frac{e^{-\lambda}\lambda^{x-2}}{(x-2)!}.$$

If we let $y = x - 2$, we obtain

$$E[X(X-1)] = \lambda^2 \sum_{y=0}^{\infty} \frac{e^{-\lambda}\lambda^y}{y!} = \lambda^2. \tag{4}$$

Since $E[X(X-1)] = E(X^2) - E(X) = E(X^2) - \lambda$, it follows from Eq. (4) that $E(X^2) = \lambda^2 + \lambda$. Therefore,

$$\text{Var}(X) = E(X^2) - [E(X)]^2 = \lambda. \tag{5}$$

Thus, for the Poisson distribution for which the p.f. is defined by Eq. (1), we have established the fact that both the mean and the variance are equal to λ.

The Moment Generating Function.   We shall now determine the m.g.f. $\psi(t)$ of the Poisson distribution for which the p.f. is defined by Eq. (1). For any value of t $(-\infty < t < \infty)$,

$$\psi(t) = E(e^{tX}) = \sum_{x=0}^{\infty} \frac{e^{tx}e^{-\lambda}\lambda^x}{x!} = e^{-\lambda} \sum_{x=0}^{\infty} \frac{(\lambda e^t)^x}{x!}.$$

It follows from Eq. (2) that, for $-\infty < t < \infty$,

$$\psi(t) = e^{-\lambda}e^{\lambda e^{t}} = e^{\lambda(e^{t}-1)}. \tag{6}$$

The mean and the variance, as well as all other moments, can be determined from the m.g.f. given in Eq. (6). We shall not derive the values of any other moments here, but we shall use the m.g.f. to derive the following property of Poisson distributions.

> **Theorem 1.** *If the random variables $X_1, \ldots, X_k$ are independent and if X_i has a Poisson distribution with mean λ_i $(i = 1, \ldots, k)$, then the sum $X_1 + \cdots + X_k$ has a Poisson distribution with mean $\lambda_1 + \cdots + \lambda_k$.*

Proof. Let $\psi_i(t)$ denote the m.g.f. of X_i for $i = 1, \ldots, k$, and let $\psi(t)$ denote the m.g.f. of the sum $X_1 + \cdots + X_k$. Since $X_1, \ldots, X_k$ are independent, it follows that, for $-\infty < t < \infty$,

$$\psi(t) = \prod_{i=1}^{k} \psi_i(t) = \prod_{i=1}^{k} e^{\lambda_i(e^{t}-1)} = e^{(\lambda_1 + \cdots + \lambda_k)(e^{t}-1)}.$$

It can be seen from Eq. (6) that this m.g.f. $\psi(t)$ is the m.g.f. of a Poisson distribution with mean $\lambda_1 + \cdots + \lambda_k$. Hence, the distribution of $X_1 + \cdots + X_k$ must be that Poisson distribution. $\square$

A table of probabilities for Poisson distributions with various values of the mean λ is given at the end of this book.

The Poisson Process

The Poisson distribution will often serve as an appropriate probability distribution for random variables such as the number of telephone calls received at a switchboard during a fixed period of time, the number of atomic particles emitted from a radioactive source which strike a certain target during a fixed period of time, or the number of defects on a specified length of magnetic recording tape. Each of these random variables represents the total number X of occurrences of some phenomenon during a fixed period of time or within a fixed region of space. It can be shown that if the physical process which generates these occurrences satisfies three specific mathematical conditions, then the distribution of X must be a Poisson distribution. We shall now present a complete description of the three conditions that are needed. In the following discussion, we shall assume that we are observing the number of occurrences of a particular phenomenon during a fixed period of time.

The first condition is that the numbers of occurrences in any two *disjoint* intervals of time must be independent of each other. For example, even though an unusually large number of telephone calls are received at a switchboard during a particular interval, the probability that at least one call will be received during a forthcoming interval remains unchanged. Similarly, even though no call has been received at the switchboard for an unusually long interval, the probability that a call will be received during the next short interval remains unchanged.

The second condition is that the probability of an occurrence during any particular very short interval of time must be approximately proportional to the length of that interval. To express this condition more formally, we shall use the standard mathematical notation in which $o(t)$ denotes any function of t having the property that

$$\lim_{t \to 0} \frac{o(t)}{t} = 0. \tag{7}$$

According to (7), $o(t)$ must be a function that approaches 0 as $t \to 0$ and, furthermore, this function must approach 0 at a rate faster than t itself. An example of such a function is $o(t) = t^{\alpha}$, where $\alpha > 1$. It can be verified that this function satisfies (7). The second condition can now be expressed as follows: There exists a constant $\lambda > 0$ such that for any time interval of length t, the probability of at least one occurrence during that interval has the form $\lambda t + o(t)$. Thus, for any very small value of t, the probability of at least one occurrence during an interval of length t is equal to λt plus a quantity having a smaller order of magnitude.

One of the consequences of the second condition is that the process being observed must be *stationary* over the entire period of observation; that is, the probability of an occurrence must be the same over the entire period. There can be neither busy intervals, during which we know in advance that occurrences are likely to be more frequent, nor quiet intervals, during which we know in advance that occurrences are likely to be less frequent. This condition is reflected in the fact that the same constant λ expresses the probability of an occurrence in any interval over the entire period of observation.

The third condition that must be satisfied is that the probability that there will be two or more occurrences in any particular very short interval of time must have a smaller order of magnitude than the probability that there will be just one occurrence. In symbols, the probability of two or more occurrences in any time interval of length t must be $o(t)$. Thus, the probability of two or more occurrences in any small interval must be negligible in comparison with the probability of one occurrence. Of course, it follows from the second condition that the probability of one occurrence in that same interval will itself be negligible in comparison with the probability of no occurrences.

If the preceding three conditions are met, then it can be shown by the methods of elementary differential equations that the process will satisfy the following two properties: (i) The number of occurrences in any fixed interval of time of length t will have a Poisson distribution for which the mean is λt. (ii) As assumed in the first condition, the numbers of occurrences in any two disjoint intervals will be independent. A process for which these two properties are satisfied is called a *Poisson process*. The positive constant λ is the expected number of occurrences per unit time.

Example 1: Radioactive Particles. Suppose that radioactive particles strike a certain target in accordance with a Poisson process at an average rate of 3 particles per minute. We shall determine the probability that 10 or more particles will strike the target in a particular 2-minute period.

In a Poisson process, the number of particles striking the target in any particular one-minute period has a Poisson distribution with mean λ. Since the mean number of strikes in any one-minute period is 3, it follows that $\lambda = 3$ in this example. Therefore, the number of strikes X in any 2-minute period will have a Poisson distribution with mean 6. It can be found from the table of the Poisson distribution at the end of this book that $\Pr(X \geq 10) = 0.0838$. □

The Poisson Approximation to the Binomial Distribution

We shall now show that when the value of n is large and the value of p is close to 0, the binomial distribution with parameters n and p can be approximated by a Poisson distribution with mean np. Suppose that a random variable X has a binomial distribution with parameters n and p, and let $\Pr(X = x) = f(x \mid n, p)$ for any given value of x. Then by Eq. (6) of Sec. 5.2, for $x = 1, 2, \ldots, n$,

$$f(x \mid n, p) = \frac{n(n-1) \cdots (n-x+1)}{x!} p^x (1-p)^{n-x}.$$

If we let $\lambda = np$, then $f(x \mid n, p)$ can be rewritten in the following form:

$$f(x \mid n, p) = \frac{\lambda^x}{x!} \cdot \frac{n}{n} \cdot \frac{n-1}{n} \cdots \frac{n-x+1}{n} \left(1 - \frac{\lambda}{n}\right)^n \left(1 - \frac{\lambda}{n}\right)^{-x}. \tag{8}$$

We shall now let $n \to \infty$ and $p \to 0$ in such a way that the value of the product np remains equal to the fixed value λ throughout this limiting process. Since the values of λ and x are held fixed as $n \to \infty$, then

$$\lim_{n \to \infty} \frac{n}{n} \cdot \frac{n-1}{n} \cdots \frac{n-x+1}{n} \left(1 - \frac{\lambda}{n}\right)^{-x} = 1.$$

Furthermore, it is known from elementary calculus that

$$\lim_{n \to \infty} \left(1 - \frac{\lambda}{n}\right)^n = e^{-\lambda}. \tag{9}$$

It now follows from Eq. (8) that for any fixed positive integer x,

$$f(x \mid n, p) \to \frac{e^{-\lambda} \lambda^x}{x!}. \tag{10}$$

Finally, for $x = 0$,

$$f(x \mid n, p) = (1 - p)^n = \left(1 - \frac{\lambda}{n}\right)^n.$$

It therefore follows from Eq. (9) that the relationship (10) is also satisfied for $x = 0$. Hence, the relationship (10) is satisfied for every nonnegative integer x.

The expression on the right side of the relation (10) is the p.f. $f(x \mid \lambda)$ of the Poisson distribution with mean λ. Therefore, when n is large and p is close to 0, the value of the p.f. $f(x \mid n, p)$ of the binomial distribution can be approximated, for $x = 0, 1, \ldots$, by the value of the p.f. $f(x \mid \lambda)$ of the Poisson distribution for which $\lambda = np$.

Example 2: Approximating a Probability. Suppose that in a large population the proportion of people that have a certain disease is 0.01. We shall determine the probability that in a random group of 200 people at least four people will have the disease.

In this example, we can assume that the exact distribution of the number of people having the disease among the 200 people in the random group is a binomial distribution with parameters $n = 200$ and $p = 0.01$. Therefore, this distribution can be approximated by a Poisson distribution for which the mean is $\lambda = np = 2$. If X denotes a random variable having this Poisson distribution, then it can be found from the table of the Poisson distribution at the end of this book that $\Pr(X \geqslant 4) = 0.1428$. Hence, the probability that at least four people will have the disease is approximately 0.1428. □

EXERCISES

1. Suppose that on a given weekend the number of accidents at a certain intersection has a Poisson distribution with mean 0.7. What is the probability that there will be at least three accidents at the intersection during the weekend?

2. Suppose that the number of defects on a bolt of cloth produced by a certain process has a Poisson distribution with mean 0.4. If a random sample of five bolts of cloth is inspected, what is the probability that the total number of defects on the five bolts will be at least 6?

3. Suppose that in a certain book there are, on the average, λ misprints per page. What is the probability that a particular page will contain no misprints?

4. Suppose that a book with n pages contains, on the average, λ misprints per page. What is the probability that there will be at least m pages which contain more than k misprints?

5. Suppose that a certain type of magnetic tape contains, on the average, 3 defects per 1000 feet. What is the probability that a roll of tape 1200 feet long contains no defects?

6. Suppose that, on the average, a certain store serves 15 customers per hour. What is the probability that the store will serve more than 20 customers in a particular 2-hour period?

7. Suppose that X_1 and X_2 are independent random variables and that X_i has a Poisson distribution with mean λ_i $(i = 1, 2)$. For any fixed value of k $(k = 1, 2, \dots)$, determine the conditional distribution of X_1 given that $X_1 + X_2 = k$.

8. Suppose that the total number of items produced by a certain machine has a Poisson distribution with mean λ; that all items are produced independently of one another; and that the probability that any given item produced by the machine will be defective is p. Determine the marginal distribution of the number of defective items produced by the machine.

9. For the problem described in Exercise 8, let X denote the number of defective items produced by the machine and let Y denote the number of nondefective items produced by the machine. Show that X and Y are independent random variables.

10. The mode of a discrete distribution was defined in Exercise 11 of Sec. 5.2. Determine the mode or modes of the Poisson distribution with mean λ.

11. Suppose that the proportion of colorblind people in a certain population is 0.005. What is the probability that there will not be more than one colorblind person in a randomly chosen group of 600 people?

12. The probability of triplets in human births is approximately 0.001. What is the probability that there will be exactly one set of triplets among 700 births in a large hospital?

13. An airline sells 200 tickets for a certain flight on an airplane which has only 198 seats because, on the average, 1 percent of purchasers of airline tickets do not appear for the departure of their flight. Determine the probability that everyone who appears for the departure of this flight will have a seat.

5.5. THE NEGATIVE BINOMIAL DISTRIBUTION

Definition of the Negative Binomial Distribution

Suppose that in an infinite sequence of independent experiments the outcome of each experiment must be either a success or a failure. Suppose also that the probability of a success in any particular experiment is p $(0 < p < 1)$ and the probability of a failure is $q = 1 - p$. Then these experiments form an infinite sequence of Bernoulli trials with parameter p. In this section we shall study the distribution of the total number of failures that will occur before exactly r successes have been obtained, where r is a fixed positive integer.

For $n = r, r + 1, \ldots$, we shall let A_n denote the event that the total number of trials required to obtain exactly r successes is n. As explained in Example 7 of Sec. 1.11, the event A_n will occur if and only if exactly $r - 1$ successes occur among the first $n - 1$ trials and the rth success is obtained on the nth trial. Since all trials are independent, it follows that

$$
\Pr(A_n) = \binom{n-1}{r-1} p^{r-1} q^{(n-1)-(r-1)} \cdot p = \binom{n-1}{r-1} p^r q^{n-r}. \tag{1}
$$

For any value of x ($x = 0, 1, 2, \ldots$), the event that exactly x failures are obtained before the rth success is obtained is equivalent to the event that the total number of trials required to obtain r successes is $r + x$. In other words, if X denotes the number of failures that will occur before the rth success is obtained, then $\Pr(X = x) = \Pr(A_{r+x})$. If we denote $\Pr(X = x)$ by $f(x \mid r, p)$, it follows from Eq. (1) that

$$
f(x \mid r, p) = \begin{cases} \binom{r+x-1}{x} p^r q^x & \text{for } x = 0, 1, 2, \ldots, \\ 0 & \text{otherwise.} \end{cases} \tag{2}
$$

It is said that a random variable X has a *negative binomial distribution with parameters r and p* ($r = 1, 2, \ldots$ and $0 < p < 1$) if X has a discrete distribution for which the p.f. $f(x \mid r, p)$ is as specified by Eq. (2).

By using the definition of binomial coefficients given in Eq. (4) of Sec. 5.3, the function $f(x \mid r, p)$ can be regarded as the p.f. of a discrete distribution for any number $r > 0$ (not necessarily an integer) and any number p in the interval $0 < p < 1$. In other words, it can be verified that for $r > 0$ and $0 < p < 1$,

$$
\sum_{x=0}^{\infty} \binom{r+x-1}{x} p^r q^x = 1. \tag{3}
$$

In this section, however, we shall restrict our attention to negative binomial distributions for which the parameter r is a positive integer.

It follows from the results given in Exercise 4 at the end of Sec. 5.3 that the p.f. of the negative binomial distribution can be written in the following alternative form:

$$
f(x \mid r, p) = \begin{cases} \binom{-r}{x} p^r (-q)^x & \text{for } x = 0, 1, 2, \ldots, \\ 0 & \text{otherwise.} \end{cases} \tag{4}
$$

The Geometric Distribution

A negative binomial distribution for which $r = 1$ is called a geometric distribution. In other words, it is said that a random variable X has a *geometric distribution with parameter p* $(0 < p < 1)$ if X has a discrete distribution for which the p.f. $f(x \mid 1, p)$ is as follows:

$$f(x \mid 1, p) = \begin{cases} pq^x & \text{for } x = 0, 1, 2, \ldots, \\ 0 & \text{otherwise.} \end{cases} \tag{5}$$

Consider again an infinite sequence of Bernoulli trials in which the outcome of any trial is either a success or a failure and the probability of a success on any trial is p. If we let X_1 denote the number of failures that occur before the first success is obtained, then X_1 will have a geometric distribution with parameter p.

More generally, for $j = 2, 3, \ldots$, we shall let X_j denote the number of failures that occur after $j - 1$ successes have been obtained but before the jth success is obtained. Since all the trials are independent and the probability of obtaining a success on any given trial is p, it follows that each random variable X_j will have a geometric distribution with parameter p and that the random variables $X_1, X_2, \ldots$ will be independent. Furthermore, for $r = 1, 2, \ldots$, the sum $X_1 + \cdots + X_r$ will be equal to the total number of failures that occur before exactly r successes have been obtained. Therefore, this sum will have a negative binomial distribution with parameters r and p. We have thus derived the following result:

If $X_1, \ldots, X_r$ are i.i.d. random variables and if each X_i has a geometric distribution with parameter p, then the sum $X_1 + \cdots + X_r$ has a negative binomial distribution with parameters r and p.

Other Properties of Negative Binomial and Geometric Distributions

The Moment Generating Function. If X_1 has a geometric distribution with parameter p, then the m.g.f. $\psi_1(t)$ of X_1 is as follows:

$$\psi_1(t) = E(e^{tX_1}) = p \sum_{x=0}^{\infty} (qe^t)^x. \tag{6}$$

The infinite series in Eq. (6) will have a finite sum for any value of t such that $0 < qe^t < 1$, that is, for $t < \log(1/q)$. It is known from elementary calculus that for any number α $(0 < \alpha < 1)$,

$$\sum_{x=0}^{\infty} \alpha^x = \frac{1}{1 - \alpha}.$$

Therefore, for $t < \log(1/q)$,

$$\psi_1(t) = \frac{p}{1 - qe^t}. \tag{7}$$

It is known from Theorem 3 of Sec. 4.4 that if the random variables $X_1, \ldots, X_r$ are i.i.d. and if the m.g.f. of each of them is $\psi_1(t)$, then the m.g.f. of the sum $X_1 + \cdots + X_r$ is $[\psi_1(t)]^r$. Since the distribution of the sum $X_1 + \cdots + X_r$ is a negative binomial distribution with parameters r and p, we have established the following result:

If X has a negative binomial distribution with parameters r and p, then the m.g.f. of X is as follows:

$$\psi(t) = \left(\frac{p}{1 - qe^t}\right)^r \quad \text{for } t < \log\left(\frac{1}{q}\right). \tag{8}$$

The Mean and Variance. If X_1 has a geometric distribution with parameter p, then the mean and the variance of X_1 can be found by differentiating the m.g.f. given in Eq. (6). The results are as follows:

$$E(X_1) = \psi_1'(0) = \frac{q}{p} \tag{9}$$

and

$$\text{Var}(X_1) = \psi_1''(0) - [\psi_1'(0)]^2 = \frac{q}{p^2}. \tag{10}$$

Now suppose that X has a negative binomial distribution with parameters r and p. If X is represented as the sum $X_1 + \cdots + X_r$ of r independent random variables, each having the same distribution as X_1, it follows from Eqs. (9) and (10) that the mean and the variance of X must be

$$E(X) = \frac{rq}{p} \quad \text{and} \quad \text{Var}(X) = \frac{rq}{p^2}. \tag{11}$$

The Memoryless Property of the Geometric Distribution. We shall continue to consider an infinite sequence of Bernoulli trials in which the outcome of any trial is either a success or a failure and the probability of a success on any trial is p. Then the distribution of the number of failures that will occur before the first success occurs is a geometric distribution with parameter p. Suppose now that a failure occurred on each of the first 20 trials. Then, since all trials are indepen-dent, the distribution of the *additional* failures that will occur before the first success is obtained will again be a geometric distribution with parameter p. In effect, the process begins anew with the twenty-first trial, and the long sequence of failures that were obtained on the first 20 trials can have no effect on the future outcomes of the process. This property is often called the *memoryless property* of the geometric distribution.

At the beginning of the experiment, the expected number of failures that will occur before the first success is obtained is q/p, as given by Eq. (9). If it is known that failures were obtained on the first 20 trials, then the expected total number of failures before the first success is simply $20 + (q/p)$. In purely mathematical terms, the memoryless property can be stated as follows:

If X has a geometric distribution with parameter p, then for any nonnegative integers k and t,

$$\Pr(X = k + t \mid X \geqslant k) = \Pr(X = t). \tag{12}$$

A simple mathematical proof of Eq. (12) could be given by using the p.f. $f(x \mid 1, p)$ specified in Eq. (5). This proof is required in Exercise 7 at the end of this section.

EXERCISES

1. Suppose that a sequence of independent tosses are made with a coin for which the probability of obtaining a head on any given toss is $1/30$.
   (a) What is the expected number of tails that will be obtained before five heads have been obtained?
   (b) What is the variance of the number of tails that will be obtained before five heads have been obtained?

2. Consider the sequence of coin tosses described in Exercise 1.
   (a) What is the expected number of tosses that will be required in order to obtain five heads?
   (b) What is the variance of the number of tosses that will be required in order to obtain five heads?

3. Suppose that two players A and B are trying to throw a basketball through a hoop. The probability that player A will succeed on any given throw is p, and he throws until he has succeeded r times. The probability that player B will succeed on any given throw is mp, where m is a given integer ($m = 2, 3, \ldots$) such that $mp < 1$; and he throws until he has succeeded mr times.
   (a) For which player is the expected number of throws smaller?
   (b) For which player is the variance of the number of throws smaller?

4. Suppose that the random variables $X_1, \ldots, X_k$ are independent and that X_i has a negative binomial distribution with parameters r_i and p $(i = 1, \ldots, k)$. Prove that the sum $X_1 + \cdots + X_k$ has a negative binomial distribution with parameters $r = r_1 + \cdots + r_k$ and p.

5. Suppose that X has a geometric distribution with parameter p. Determine the probability that the value of X will be one of the even integers $0, 2, 4, \ldots$.

6. Suppose that X has a geometric distribution with parameter p. Show that for any nonnegative integer k, $\Pr(X \geq k) = q^k$.

7. Prove Eq. (12).

8. Suppose that an electronic system contains n components which function independently of each other, and suppose that these components are connected in series, as defined in Exercise 4 of Sec. 3.7. Suppose also that each component will function properly for a certain number of periods and then will fail. Finally, suppose that for $i = 1, \ldots, n$, the number of periods for which component i will function properly is a discrete random variable having a geometric distribution with parameter p_i. Determine the distribution of the number of periods for which the system will function properly.

9. Let $f(x \mid r, p)$ denote the p.f. of the negative binomial distribution with parameters r and p; and let $f(x \mid \lambda)$ denote the p.f. of the Poisson distribution with mean λ, as defined by Eq. (1) of Sec. 5.4. Suppose $r \to \infty$ and $q \to 0$ in such a way that the value of rq remains constant and is equal to λ throughout the process. Show that for each fixed nonnegative integer x,

$$f(x \mid r, p) \to f(x \mid \lambda).$$

5.6. THE NORMAL DISTRIBUTION

Importance of the Normal Distribution

The normal distribution, which will be defined and discussed in this section, is by far the single most important probability distribution in statistics. There are three main reasons for this preeminent position of the normal distribution.

The first reason is directly related to the mathematical properties of the normal distribution. We shall demonstrate in this section and in several later sections of this book that if a random sample is taken from a normal distribution, then the distributions of various important functions of the observations in the sample can be derived explicitly and will themselves have simple forms. Therefore, it is a mathematical convenience to be able to assume that the distribution from which a random sample is drawn is a normal distribution.

The second reason is that many scientists have observed that the random variables studied in various physical experiments often have distributions which are approximately normal. For example, a normal distribution will usually be a close approximation to the distribution of the heights or weights of individuals in a homogeneous population of people, of corn stalks, or of mice, or to the distribution of the tensile strength of pieces of steel produced by a certain process.

The third reason for the preeminence of the normal distribution is the central limit theorem, which will be stated and proved in the next section. If a large random sample is taken from some distribution, then even though this distribution is not itself approximately normal, a consequence of the central limit theorem is that many important functions of the observations in the sample will have distributions which are approximately normal. In particular, for a large random sample from any distribution that has a finite variance, the distribution of the sample mean will be approximately normal. We shall return to this topic in the next section.

Properties of the Normal Distribution

Definition of the Distribution. It is said that a random variable X has a *normal distribution with mean μ and variance σ^2* ($-\infty < \mu < \infty$ and $\sigma > 0$) if X has a continuous distribution for which the p.d.f. $f(x \mid \mu, \sigma^2)$ is as follows:

$$f(x \mid \mu, \sigma^2) = \frac{1}{(2\pi)^{1/2}\sigma} \exp\left[-\frac{1}{2}\left(\frac{x-\mu}{\sigma}\right)^2\right] \qquad \text{for } -\infty < x < \infty. \qquad (1)$$

We shall now verify that the nonnegative function defined in Eq. (1) is a proper p.d.f. by showing that

$$\int_{-\infty}^{\infty} f(x \mid \mu, \sigma^2)\, dx = 1. \qquad (2)$$

If we let $y = (x - \mu)/\sigma$, then

$$\int_{-\infty}^{\infty} f(x \mid \mu, \sigma^2)\, dx = \int_{-\infty}^{\infty} \frac{1}{(2\pi)^{1/2}} \exp\left(-\frac{1}{2}y^2\right) dy.$$

We shall now let

$$I = \int_{-\infty}^{\infty} \exp\left(-\frac{1}{2}y^2\right) dy. \qquad (3)$$

Then we must show that $I = (2\pi)^{1/2}$.

From Eq. (3), it follows that

$$I^2 = I \cdot I = \int_{-\infty}^{\infty} \exp\left(-\frac{1}{2}y^2\right) dy \int_{-\infty}^{\infty} \exp\left(-\frac{1}{2}z^2\right) dz$$

$$= \int_{-\infty}^{\infty} \int_{-\infty}^{\infty} \exp\left[-\frac{1}{2}(y^2 + z^2)\right] dy\, dz.$$

We shall now change the variables in this integral from y and z to the polar coordinates r and θ by letting $y = r\cos\theta$ and $z = r\sin\theta$. Then, since $y^2 + z^2 = r^2$,

$$I^2 = \int_0^{2\pi} \int_0^{\infty} \exp\left(-\frac{1}{2}r^2\right) r\, dr\, d\theta = 2\pi.$$

Therefore, $I = (2\pi)^{1/2}$ and the correctness of Eq. (2) has been established.

The Moment Generating Function. In the definition of the normal distribution, it is stated that the parameters μ and σ^2 are the mean and the variance of the distribution. In order to justify the use of these terms, we must verify that μ is actually the mean and σ^2 is actually the variance for the p.d.f. specified by Eq. (1). We shall do this after we have derived the m.g.f. $\psi(t)$ of this normal distribution.

By the definition of an m.g.f.,

$$\psi(t) = E(e^{tX}) = \int_{-\infty}^{\infty} \frac{1}{(2\pi)^{1/2}\sigma} \exp\left[tx - \frac{(x-\mu)^2}{2\sigma^2}\right] dx.$$

By completing the square inside the brackets, we obtain the relation

$$tx - \frac{(x-\mu)^2}{2\sigma^2} = \mu t + \frac{1}{2}\sigma^2 t^2 - \frac{[x-(\mu + \sigma^2 t)]^2}{2\sigma^2}.$$

Therefore,

$$\psi(t) = C\exp\left(\mu t + \frac{1}{2}\sigma^2 t^2\right),$$

where

$$C = \int_{-\infty}^{\infty} \frac{1}{(2\pi)^{1/2}\sigma} \exp\left\{-\frac{[x-(\mu + \sigma^2 t)]^2}{2\sigma^2}\right\} dx.$$

If we now replace μ with $\mu + \sigma^2 t$ in Eq. (1), it follows from Eq. (2) that $C = 1$. Hence, the m.g.f. of the normal distribution is as follows:

$$\psi(t) = \exp\left(\mu t + \frac{1}{2}\sigma^2 t^2\right) \qquad \text{for } -\infty < t < \infty. \tag{4}$$

If a random variable X has a normal distribution for which the p.d.f. is as given in Eq. (1), it follows from Eq. (4) that

$$E(X) = \psi'(0) = \mu$$

and

$$\text{Var}(X) = \psi''(0) - \left[\psi'(0)\right]^2 = \sigma^2.$$

Thus we have shown that the parameters μ and σ^2 are indeed the mean and the variance of the normal distribution defined by Eq. (1).

Since the m.g.f. $\psi(t)$ is finite for all values of t, all the moments $E(X^k)$ ($k = 1, 2, \ldots$) will also be finite.

The Shape of the Normal Distribution. It can be seen from Eq. (1) that the p.d.f. $f(x \mid \mu, \sigma^2)$ of a normal distribution with mean μ and variance σ^2 is symmetric with respect to the point $x = \mu$. Therefore, μ is both the mean and the median of the distribution. Furthermore, μ is also the mode of the distribution. In other words, the p.d.f. $f(x \mid \mu, \sigma^2)$ attains its maximum value at the point $x = \mu$. Finally, by differentiating $f(x \mid \mu, \sigma^2)$ twice, it can be found that there are points of inflection at $x = \mu + \sigma$ and at $x = \mu - \sigma$.

The p.d.f. $f(x \mid \mu, \sigma^2)$ is sketched in Fig. 5.1. It is seen that the curve is "bell-shaped." However, it is not necessarily true that any arbitrary bell-shaped p.d.f. can be approximated by the p.d.f. of a normal distribution. For example, the p.d.f. of a Cauchy distribution, as sketched in Fig. 4.3, is a symmetric bell-shaped curve which apparently resembles the p.d.f. sketched in Fig. 5.1. However, since no moment of the Cauchy distribution—not even the mean—exists, the tails of the Cauchy p.d.f. must be quite different from the tails of the normal p.d.f.

Linear Transformations. We shall now show that if a random variable X has a normal distribution, then any linear function of X will also have a normal distribution.

Theorem 1.   *If X has a normal distribution with mean μ and variance σ^2 and if $Y = aX + b$, where a and b are given constants and $a \neq 0$, then Y has a normal distribution with mean $a\mu + b$ and variance $a^2\sigma^2$.*

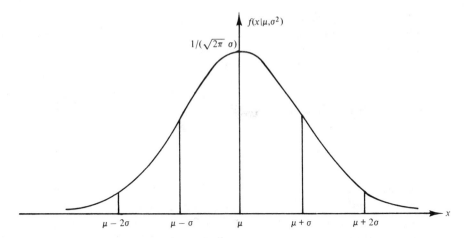

Figure 5.1 The p.d.f. of a normal distribution.

Proof. The m.g.f. ψ of X is given by Eq. (4). If ψ_Y denotes the m.g.f. of Y, then

$$\psi_Y(t) = e^{bt}\psi(at) = \exp\left[(a\mu + b)t + \frac{1}{2}a^2\sigma^2t^2\right] \qquad \text{for } -\infty < t < \infty.$$

By comparing this expression for ψ_Y with the m.g.f. of a normal distribution given in Eq. (4), we see that ψ_Y is the m.g.f. of a normal distribution with mean $a\mu + b$ and variance $a^2\sigma^2$. Hence, Y must have this normal distribution. $\square$

The Standard Normal Distribution

The normal distribution with mean 0 and variance 1 is called the *standard normal distribution*. The p.d.f. of the standard normal distribution is usually denoted by the symbol ϕ, and the d.f. is denoted by the symbol Φ. Thus,

$$\phi(x) = f(x|0, 1) = \frac{1}{(2\pi)^{1/2}}\exp\left(-\frac{1}{2}x^2\right) \qquad \text{for } -\infty < x < \infty \qquad (5)$$

and

$$\Phi(x) = \int_{-\infty}^{x} \phi(u)\, du \qquad \text{for } -\infty < x < \infty, \qquad (6)$$

where the symbol u is used in Eq. (6) as a dummy variable of integration.

The d.f. $\Phi(x)$ cannot be expressed in closed form in terms of elementary functions. Therefore, probabilities for the standard normal distribution or any other normal distribution can be found only by numerical approximations or by using a table of values of $\Phi(x)$ such as the one given at the end of this book. In that table, the values of $\Phi(x)$ are given only for $x \geq 0$. Since the p.d.f. of the standard normal distribution is symmetric with respect to the point $x = 0$, it follows that $\Pr(X \leq x) = \Pr(X \geq -x)$ for any number x $(-\infty < x < \infty)$. Since $\Pr(X \leq x) = \Phi(x)$ and $\Pr(X \geq -x) = 1 - \Phi(-x)$, we have

$$\Phi(x) + \Phi(-x) = 1 \qquad \text{for } -\infty < x < \infty. \tag{7}$$

It follows from Theorem 1 that if a random variable X has a normal distribution with mean μ and variance σ^2, then the variable $Z = (X - \mu)/\sigma$ will have a standard normal distribution. Therefore, probabilities for a normal distribution with any specified mean and variance can be found from a table of the standard normal distribution.

Example 1: Determining Probabilities for a Normal Distribution. Suppose that X has a normal distribution with mean 5 and standard deviation 2. We shall determine the value of $\Pr(1 < X < 8)$.

If we let $Z = (X - 5)/2$, then Z will have a standard normal distribution and

$$\Pr(1 < X < 8) = \Pr\left(\frac{1-5}{2} < \frac{X-5}{2} < \frac{8-5}{2}\right) = \Pr(-2 < Z < 1.5).$$

Furthermore,

$$\Pr(-2 < Z < 1.5) = \Pr(Z < 1.5) - \Pr(Z \leq -2)$$

$$= \Phi(1.5) - \Phi(-2)$$

$$= \Phi(1.5) - [1 - \Phi(2)].$$

From the table at the end of this book, it is found that $\Phi(1.5) = 0.9332$ and $\Phi(2) = 0.9773$. Therefore,

$$\Pr(1 < X < 8) = 0.9105. \quad \square$$

Comparisons of Normal Distributions

The p.d.f. of the normal distribution is sketched in Fig. 5.2 for a fixed value of μ and three different values of σ ($\sigma = 1/2$, 1, and 2). It can be seen from this figure

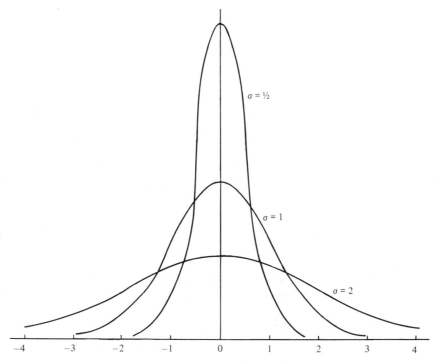

Figure 5.2 The normal p.d.f. for $\mu = 0$ and $\sigma = 1/2, 1, 2$.

that the p.d.f. of a normal distribution with a small value of σ has a high peak and is very concentrated around the mean μ, whereas the p.d.f. of a normal distribution with a larger value of σ is relatively flat and is spread out more widely over the real line.

An important fact is that every normal distribution contains the same total amount of probability within one standard deviation of its mean, the same amount within two standard deviations of its mean, and the same amount within any other fixed number of standard deviations of its mean. In general, if X has a normal distribution with mean μ and variance σ^2 and if Z has a standard normal distribution, then for $k > 0$,

$$p_k = \Pr(|X - \mu| \leqslant k\sigma) = \Pr(|Z| \leqslant k).$$

In Table 5.1 the values of this probability p_k are given for various values of k. Although the p.d.f. of a normal distribution is positive over the entire real line, it can be seen from this table that the total amount of probability outside an interval of four standard deviations on each side of the mean is only 0.00006.

Table 5.1

k	p_k
1	0.6826
2	0.9544
3	0.9974
4	0.99994
5	$1 - 6 \times 10^{-7}$
10	$1 - 2 \times 10^{-23}$

Linear Combinations of Normally Distributed Variables

In the next theorem and corollary we shall prove the following important result: Any linear combination of random variables that are independent and normally distributed will also have a normal distribution.

Theorem 2. *If the random variables $X_1, \ldots, X_k$ are independent and if X_i has a normal distribution with mean μ_i and variance σ_i^2 ($i = 1, \ldots, k$), then the sum $X_1 + \cdots + X_k$ has a normal distribution with mean $\mu_1 + \cdots + \mu_k$ and variance $\sigma_1^2 + \cdots + \sigma_k^2$.*

Proof. Let $\psi_i(t)$ denote the m.g.f. of X_i for $i = 1, \ldots, k$, and let $\psi(t)$ denote the m.g.f. of $X_1 + \cdots + X_k$. Since the variables $X_1, \ldots, X_k$ are independent, then

$$\psi(t) = \prod_{i=1}^{k} \psi_i(t) = \prod_{i=1}^{k} \exp\left(\mu_i t + \frac{1}{2}\sigma_i^2 t^2 \right)$$

$$= \exp\left[\left(\sum_{i=1}^{k} \mu_i \right) t + \frac{1}{2}\left(\sum_{i=1}^{k} \sigma_i^2 \right) t^2 \right] \qquad \text{for } -\infty < t < \infty.$$

From Eq. (4) the m.g.f. $\psi(t)$ can be identified as the m.g.f. of a normal distribution for which the mean is $\sum_{i=1}^{k} \mu_i$ and the variance is $\sum_{i=1}^{k} \sigma_i^2$. Hence, the distribution of $X_1 + \cdots + X_k$ must be that normal distribution. $\square$

The following corollary is now obtained by combining Theorem 1 and Theorem 2.

Corollary 1. *If the random variables $X_1, \ldots, X_k$ are independent, if X_i has a normal distribution with mean μ_i and variance σ_i^2 ($i = 1, \ldots, k$), and if $a_1, \ldots, a_k$ and b are constants for which at least one of the values $a_1, \ldots, a_k$ is different from 0, then the variable $a_1 X_1 + \cdots + a_k X_k + b$ has a normal distribution with mean $a_1\mu_1 + \cdots + a_k\mu_k + b$ and variance $a_1^2\sigma_1^2 + \cdots + a_k^2\sigma_k^2$.*

The distribution of the sample mean for a random sample from a normal distribution can now be derived easily.

Corollary 2. *Suppose that the random variables $X_1, \ldots, X_n$ form a random sample from a normal distribution with mean μ and variance σ^2, and let $\overline{X}_n$ denote the sample mean. Then $\overline{X}_n$ has a normal distribution with mean μ and variance σ^2/n.*

Proof. Since $\overline{X}_n = (1/n)\sum_{i=1}^{n} X_i$, it follows from Corollary 1 that the distribution of $\overline{X}_n$ is normal. It is known from Sec. 4.8 that $E(\overline{X}_n) = \mu$ and $\text{Var}(\overline{X}_n) = \sigma^2/n$.   □

Example 2: Determining a Sample Size. Suppose that a random sample of size n is to be taken from a normal distribution with mean μ and variance 9. We shall determine the minimum value of n for which

$$\Pr(|\overline{X}_n - \mu| \leqslant 1) \geqslant 0.95.$$

It is known from Corollary 2 that the sample mean $\overline{X}_n$ will have a normal distribution for which the mean is μ and the standard deviation is $3/n^{1/2}$. Therefore, if we let

$$Z = \frac{n^{1/2}}{3}(\overline{X}_n - \mu),$$

then Z will have a standard normal distribution. In this example, n must be chosen so that

$$\Pr(|\overline{X}_n - \mu| \leqslant 1) = \Pr\left(|Z| \leqslant \frac{n^{1/2}}{3}\right) \geqslant 0.95. \tag{8}$$

For any positive number x, it will be true that $\Pr(|Z| \leqslant x) \geqslant 0.95$ if and only if $1 - \Phi(x) = \Pr(Z > x) \leqslant 0.025$. From the table of the standard normal distribution at the end of this book, it is found that $1 - \Phi(x) \leqslant 0.025$ if and only if $x \geqslant 1.96$. Therefore, the inequality in relation (8) will be satisfied if and only if

$$\frac{n^{1/2}}{3} \geqslant 1.96.$$

Since the smallest permissible value of n is 34.6, the sample size must be at least 35 in order that the specified relation will be satisfied.   □

Example 3: Heights of Men and Women. Suppose that the heights, in inches, of the women in a certain population follow a normal distribution with mean 65 and standard deviation 1, and that the heights of the men follow a normal distribution with mean 68 and standard deviation 2. Suppose also that one woman is selected at random and, independently, one man is selected at random. We shall determine the probability that the woman will be taller than the man.

Let W denote the height of the selected woman, and let M denote the height of the selected man. Then the difference $W - M$ has a normal distribution with mean $65 - 68 = -3$ and variance $1^2 + 2^2 = 5$. Therefore, if we let

$$Z = \frac{1}{5^{1/2}}(W - M + 3),$$

then Z has a standard normal distribution. It follows that

$$\Pr(W > M) = \Pr(W - M > 0)$$

$$= \Pr\left(Z > \frac{3}{5^{1/2}}\right) = \Pr(Z > 1.342)$$

$$= 1 - \Phi(1.342) = 0.090.$$

Thus, the probability that the woman will be taller than the man is 0.090. □

EXERCISES

1. Suppose that X has a normal distribution for which the mean is 1 and the variance is 4. Find the value of each of the following probabilities:
   (a) $\Pr(X \leq 3)$ (b) $\Pr(X > 1.5)$
   (c) $\Pr(X = 1)$ (d) $\Pr(2 < X < 5)$
   (e) $\Pr(X \geq 0)$ (f) $\Pr(-1 < X < 0.5)$
   (g) $\Pr(|X| \leq 2)$ (h) $\Pr(1 \leq -2X + 3 \leq 8)$.

2. If the temperature in degrees Fahrenheit at a certain location is normally distributed with a mean of 68 degrees and a standard deviation of 4 degrees, what is the distribution of the temperature in degrees Centigrade at the same location?

3. If the m.g.f. of a random variable X is $\psi(t) = e^{t^2}$ for $-\infty < t < \infty$, what is the distribution of X?

4. Suppose that the measured voltage in a certain electric circuit has a normal distribution with mean 120 and standard deviation 2. If three independent measurements of the voltage are made, what is the probability that all three measurements will lie between 116 and 118?

5. Evaluate the integral $\int_0^\infty e^{-3x^2}\, dx$.

6. A straight rod is formed by connecting three sections A, B, and C, each of which is manufactured on a different machine. The length of section A, in

inches, has a normal distribution with mean 20 and variance 0.04. The length of section B has a normal distribution with mean 14 and variance 0.01. The length of section C has a normal distribution with mean 26 and variance 0.04. As indicated in Fig. 5.3, the three sections are joined so that there is an overlap of 2 inches at each connection. Suppose that the rod can be used in the construction of an airplane wing if its total length in inches is between 55.7 and 56.3. What is the probability that the rod can be used?

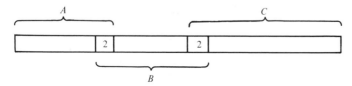

Figure 5.3   Sections of the rod in Exercise 6.

7. If a random sample of 25 observations is taken from a normal distribution with mean μ and standard deviation 2, what is the probability that the sample mean will lie within one unit of μ?

8. Suppose that a random sample of size n is to be taken from a normal distribution with mean μ and standard deviation 2. Determine the smallest value of n such that

$$\Pr(|\overline{X}_n - \mu| < 0.1) \geq 0.9.$$

9. (a) Sketch the d.f. Φ of the standard normal distribution from the values given in the table at the end of this book.

   (b) From the sketch given in part (a) of this exercise, sketch the d.f. of a normal distribution for which the mean is -2 and the standard deviation is 3.

10. Suppose that the diameters of the bolts in a large box follow a normal distribution with a mean of 2 centimeters and a standard deviation of 0.03 centimeter. Also, suppose that the diameters of the holes in the nuts in another large box follow a normal distribution with a mean of 2.02 centimeters and a standard deviation of 0.04 centimeter. A bolt and a nut will fit together if the diameter of the hole in the nut is greater than the diameter of the bolt and the difference between these diameters is not greater than 0.05 centimeter. If a bolt and a nut are selected at random, what is the probability that they will fit together?

11. Suppose that on a certain examination in advanced mathematics, students from university A achieve scores which are normally distributed with a mean of 625 and a variance of 100, and that students from university B achieve scores which are normally distributed with a mean of 600 and a variance of

150. If two students from university A and three students from university B take this examination, what is the probability that the average of the scores of the two students from university A will be greater than the average of the scores of the three students from university B? *Hint:* Determine the distribution of the difference between the two averages.

12. Suppose that 10 percent of the people in a certain population have the eye disease glaucoma. For persons who have glaucoma, measurements of eye pressure X will be normally distributed with a mean of 25 and a variance of 1. For persons who do not have glaucoma, the pressure X will be normally distributed with a mean of 20 and a variance of 1. Suppose that a person is selected at random from the population and his eye pressure X is measured.

   (a) Determine the conditional probability that the person has glaucoma given that $X = x$.

   (b) For what values of x is the conditional probability in part (a) greater than $1/2$?

13. Suppose that the joint p.d.f. of two random variables X and Y is

$$f(x, y) = \frac{1}{2\pi} e^{-(1/2)(x^2 + y^2)} \qquad \text{for } -\infty < x < \infty \text{ and } -\infty < y < \infty.$$

   Find $\Pr(-\sqrt{2} < X + Y < 2\sqrt{2})$.

14. Consider a random variable X for which $\Pr(X > 0) = 1$. It is said that X has a lognormal distribution if the random variable $\log X$ has a normal distribution. Suppose that X has a lognormal distribution and that $E(\log X) = \mu$ and $\text{Var}(\log X) = \sigma^2$. Determine the p.d.f. of X.

15. Suppose that the random variables X and Y are independent and that each has a standard normal distribution. Show that the quotient X/Y has a Cauchy distribution.

5.7. THE CENTRAL LIMIT THEOREM

Statement of the Theorem

It was shown in Sec. 5.6 that if a random sample of size n is taken from a normal distribution with mean μ and variance σ^2, then the sample mean $\overline{X}_n$ has a normal distribution with mean μ and variance σ^2/n. In this section we shall discuss the central limit theorem, a simple version of which can be stated roughly as follows: Whenever a random sample of size n is taken from *any* distribution with mean μ and variance σ^2, the sample mean $\overline{X}_n$ will have a distribution which is *approximately* normal with mean μ and variance σ^2/n.

This result was established for a random sample from a Bernoulli distribution by A. de Moivre in the early part of the eighteenth century. The proof for a random sample from an arbitrary distribution was given independently by J. W. Lindeberg and P. Lévy in the early 1920's. A precise statement of their theorem will be given now, and an outline of the proof of that theorem will be given later in this section. We shall also give here another central limit theorem pertaining to the sum of independent random variables that are not necessarily identically distributed, and shall present some examples illustrating both theorems.

The Central Limit Theorem (Lindeberg and Lévy) for the Sample Mean. As in Sec. 5.6, we shall let Φ denote the d.f. of the standard normal distribution.

Theorem 1. *If the random variables $X_1, \ldots, X_n$ form a random sample of size n from a given distribution with mean μ and variance σ^2 $(0 < \sigma^2 < \infty)$, then for any fixed number x,*

$$\lim_{n \to \infty} \Pr\left[\frac{n^{1/2}(\overline{X}_n - \mu)}{\sigma} \leqslant x\right] = \Phi(x). \tag{1}$$

The interpretation of Eq. (1) is as follows: If a large random sample is taken from any distribution with mean μ and variance σ^2, regardless of whether this distribution is discrete or continuous, then the distribution of the random variable $n^{1/2}(\overline{X}_n - \mu)/\sigma$ will be approximately a standard normal distribution. Therefore, the distribution of $\overline{X}_n$ will be approximately a normal distribution with mean μ and variance σ^2/n; or, equivalently, the distribution of the sum $\sum_{i=1}^{n} X_i$ will be approximately a normal distribution with mean $n\mu$ and variance $n\sigma^2$.

Example 1: Tossing a Coin. Suppose that a fair coin is tossed 900 times. We shall determine the probability of obtaining more than 495 heads.

For $i = 1, \ldots, 900$, let $X_i = 1$ if a head is obtained on the ith toss and let $X_i = 0$ otherwise. Then $E(X_i) = 1/2$ and $\text{Var}(X_i) = 1/4$. Therefore, the values $X_1, \ldots, X_{900}$ form a random sample of size $n = 900$ from a distribution with mean $1/2$ and variance $1/4$. It follows from the central limit theorem that the distribution of the total number of heads $H = \sum_{i=1}^{900} X_i$ will be approximately a normal distribution for which the mean is $(900)(1/2) = 450$, the variance is $(900)(1/4) = 225$, and the standard deviation is $(225)^{1/2} = 15$. Therefore, the variable $Z = (H - 450)/15$ will have approximately a standard normal distribution. Thus,

$$\Pr(H > 495) = \Pr\left(\frac{H - 450}{15} > \frac{495 - 450}{15}\right)$$

$$= \Pr(Z > 3) = 1 - \Phi(3) = 0.0013. \quad \square$$

Example 2: Sampling from a Uniform Distribution. Suppose that a random sample of size $n = 12$ is taken from the uniform distribution on the interval $(0, 1)$. We shall determine the value of $\Pr\left(\left|\bar{X}_n - \dfrac{1}{2}\right| \leqslant 0.1\right)$.

The mean of the uniform distribution on the interval $(0, 1)$ is $1/2$ and the variance is $1/12$ (see Exercise 2 of Sec. 4.3). Since $n = 12$ in this example, it follows from the central limit theorem that the distribution of $\bar{X}_n$ will be approximately a normal distribution with mean $1/2$ and variance $1/144$. Therefore, the distribution of the variable $Z = 12\left(\bar{X}_n - \dfrac{1}{2}\right)$ will be approximately a standard normal distribution. Hence,

$$\Pr\left(\left|\bar{X}_n - \frac{1}{2}\right| \leqslant 0.1\right) = \Pr\left[12\left|\bar{X}_n - \frac{1}{2}\right| \leqslant 1.2\right]$$

$$= \Pr(|Z| \leqslant 1.2) = 2\Phi(1.2) - 1 = 0.7698. \quad \square$$

The Central Limit Theorem (Liapounov) for the Sum of Independent Random Variables. We shall now state a central limit theorem which applies to a sequence of random variables $X_1, X_2, \ldots$ that are independent but not necessarily identically distributed. This theorem was first proved by A. Liapounov in 1901. We shall assume that $E(X_i) = \mu_i$ and $\text{Var}(X_i) = \sigma_i^2$ for $i = 1, \ldots, n$. Also, we shall let

$$Y_n = \frac{\sum_{i=1}^n X_i - \sum_{i=1}^n \mu_i}{\left(\sum_{i=1}^n \sigma_i^2\right)^{1/2}}. \tag{2}$$

Then $E(Y_n) = 0$ and $\text{Var}(Y_n) = 1$. The theorem that is stated next gives a sufficient condition for the distribution of this random variable Y_n to be approximately a standard normal distribution.

Theorem 2. *Suppose that the random variables $X_1, X_2, \ldots$ are independent and that $E(|X_i - \mu_i|^3) < \infty$ for $i = 1, 2, \ldots$. Also, suppose that*

$$\lim_{n \to \infty} \frac{\sum_{i=1}^n E\left(|X_i - \mu_i|^3\right)}{\left(\sum_{i=1}^n \sigma_i^2\right)^{3/2}} = 0. \tag{3}$$

Finally, let the random variable Y_n be as defined in Eq. (2). Then, for any fixed number x,

$$\lim_{n \to \infty} \Pr(Y_n \leqslant x) = \Phi(x). \tag{4}$$

The interpretation of this theorem is as follows: If Eq. (3) is satisfied, then for any large value of n, the distribution of $\sum_{i=1}^n X_i$ will be approximately a normal distribution with mean $\sum_{i=1}^n \mu_i$ and variance $\sum_{i=1}^n \sigma_i^2$. It should be noted that when the random variables $X_1, X_2, \ldots$ are identically distributed and the third moments of the variables exist, Eq. (3) will automatically be satisfied and Eq. (4) then reduces to Eq. (1).

The distinction between the theorem of Lindeberg and Lévy and the theorem of Liapounov should be emphasized. The theorem of Lindeberg and Lévy applies to a sequence of i.i.d. random variables. In order for this theorem to be applicable, it is sufficient to assume only that the variance of each random variable is finite. The theorem of Liapounov applies to a sequence of independent random variables that are not necessarily identically distributed. In order for this theorem to be applicable, it must be assumed that the third moment of each random variable is finite and satisfies Eq. (3).

The Central Limit Theorem for Bernoulli Random Variables. By applying the theorem of Liapounov, we can establish the following result:

Theorem 3. *Suppose that the random variables $X_1, \ldots, X_n$ are independent and that X_i has a Bernoulli distribution with parameter p_i $(i = 1, 2, \ldots)$. Suppose also that the infinite series $\sum_{i=1}^{\infty} p_i q_i$ is divergent, and let*

$$Y_n = \frac{\sum_{i=1}^{n} X_i - \sum_{i=1}^{n} p_i}{\left(\sum_{i=1}^{n} p_i q_i\right)^{1/2}}. \tag{5}$$

Then for any fixed number x,

$$\lim_{n \to \infty} \Pr(Y_n \leqslant x) = \Phi(x). \tag{6}$$

Proof. Here $\Pr(X_i = 1) = p_i$ and $\Pr(X_i = 0) = q_i$. Therefore, $E(X_i) = p_i$, $\mathrm{Var}(X_i) = p_i q_i$, and

$$E\left(|X_i - p_i|^3\right) = p_i q_i^3 + q_i p_i^3 = p_i q_i\left(p_i^2 + q_i^2\right) \leqslant p_i q_i.$$

It follows that

$$\frac{\sum_{i=1}^{n} E\left(|X_i - p_i|^3\right)}{\left(\sum_{i=1}^{n} p_i q_i\right)^{3/2}} \leqslant \frac{1}{\left(\sum_{i=1}^{n} p_i q_i\right)^{1/2}}. \tag{7}$$

Since the infinite series $\sum_{i=1}^{\infty} p_i q_i$ is divergent, then $\sum_{i=1}^{n} p_i q_i \to \infty$ as $n \to \infty$ and it can be seen from the relation (7) that Eq. (3) will be satisfied. In turn, it follows from Theorem 2 that Eq. (4) will be satisfied. Since Eq. (6) is simply a restatement of Eq. (4) for the particular random variables being considered here, the proof of the theorem is complete. $\square$

Theorem 3 implies that if the infinite series $\sum_{i=1}^{\infty} p_i q_i$ is divergent, then the distribution of the sum $\sum_{i=1}^{n} X_i$ of a large number of independent Bernoulli random variables will be approximately a normal distribution with mean $\sum_{i=1}^{n} p_i$ and variance $\sum_{i=1}^{n} p_i q_i$. It should be kept in mind, however, that a typical practical problem will involve only a finite number of random variables $X_1, \ldots, X_n$ rather than an infinite sequence of random variables. In such a problem, it is not meaningful to consider whether or not the infinite series $\sum_{i=1}^{\infty} p_i q_i$ is divergent, because only a finite number of values $p_1, \ldots, p_n$ will be specified in the problem. In a certain sense, therefore, the distribution of the sum $\sum_{i=1}^{n} X_i$ can *always* be approximated by a normal distribution. The critical question is whether or not this normal distribution provides a *good* approximation to the actual distribution of $\sum_{i=1}^{n} X_i$. The answer depends, of course, on the values of $p_1, \ldots, p_n$.

Since the normal distribution will be attained more and more closely as $\sum_{i=1}^{n} p_i q_i \to \infty$, the normal distribution provides a good approximation when the value of $\sum_{i=1}^{n} p_i q_i$ is large. Furthermore, since the value of each term $p_i q_i$ is a maximum when $p_i = 1/2$, the approximation will be best when n is large and the values of $p_1, \ldots, p_n$ are close to $1/2$.

Example 3: Examination Questions. Suppose that an examination contains 99 questions arranged in a sequence from the easiest to the most difficult. Suppose that the probability that a particular student will answer the first question correctly is 0.99; the probability that he will answer the second question correctly is 0.98; and, in general, the probability that he will answer the ith question correctly is $1 - i/100$ for $i = 1, \ldots, 99$. It is assumed that all questions will be answered independently and that the student must answer at least 60 questions correctly to pass the examination. We shall determine the probability that the student will pass.

Let $X_i = 1$ if the ith question is answered correctly and let $X_i = 0$ otherwise. Then $E(X_i) = p_i = 1 - (i/100)$ and $\text{Var}(X_i) = p_i q_i = (i/100)[1 - (i/100)]$. Also,

$$\sum_{i=1}^{99} p_i = 99 - \frac{1}{100} \sum_{i=1}^{99} i = 99 - \frac{1}{100} \cdot \frac{(99)(100)}{2} = 49.5$$

and

$$\sum_{i=1}^{99} p_i q_i = \frac{1}{100} \sum_{i=1}^{99} i - \frac{1}{(100)^2} \sum_{i=1}^{99} i^2$$

$$= 49.5 - \frac{1}{(100)^2} \cdot \frac{(99)(100)(199)}{6} = 16.665.$$

It follows from the central limit theorem that the distribution of the total number of questions that are answered correctly, which is $\sum_{i=1}^{99} X_i$, will be approximately a normal distribution with mean 49.5 and standard deviation

$(16.665)^{1/2} = 4.08$. Therefore, the distribution of the variable

$$Z = \frac{\sum_{i=1}^{n} X_i - 49.5}{4.08}$$

will be approximately a standard normal distribution. It follows that

$$\Pr\left(\sum_{i=1}^{n} X_i \geq 60 \right) = \Pr(Z \geq 2.5735) \simeq 1 - \Phi(2.5735) = 0.0050. \quad \square$$

Effect of the Central Limit Theorem. The central limit theorem provides a plausible explanation for the fact that the distributions of many random variables studied in physical experiments are approximately normal. For example, a person's height is influenced by many random factors. If the height of each person is determined by adding the values of these individual factors, then the distribution of the heights of a large number of persons will be approximately normal. In general, the central limit theorem indicates that the distribution of the sum of many random variables can be approximately normal, even though the distribution of each random variable in the sum differs from the normal.

Convergence in Distribution

Let $X_1, X_2, \ldots$ be a sequence of random variables; and for $n = 1, 2, \ldots,$ let F_n denote the d.f. of X_n. Also, let X^* denote another random variable for which the d.f. is F^*. We shall assume that F^* is a continuous function over the entire real line. Then it is said that the sequence $X_1, X_2, \ldots$ *converges in distribution* to the random variable X^* if

$$\lim_{n \to \infty} F_n(x) = F^*(x) \qquad \text{for } -\infty < x < \infty. \tag{8}$$

Sometimes, it is simply said that X_n converges in distribution to X^*, and the distribution of X^* is called the *asymptotic distribution* of X_n. Thus, according to the central limit theorem of Lindeberg and Lévy, as indicated in Eq. (1), the random variable $n^{1/2}(\overline{X}_n - \mu)/\sigma$ converges in distribution to a random variable having a standard normal distribution; or, equivalently, the asymptotic distribution of $n^{1/2}(\overline{X}_n - \mu)/\sigma$ is a standard normal distribution.

Convergence of the Moment Generating Functions. Moment generating functions are important in the study of convergence in distribution because of the following theorem, the proof of which is too advanced to be presented here.

Theorem 4. *Let $X_1, X_2, \ldots$ be a sequence of random variables; and for $n = 1, 2, \ldots,$ let F_n denote the d.f. of X_n and let ψ_n denote the m.g.f. of X_n.*

Also, let X^* denote another random variable with d.f. F^* and m.g.f. ψ^*. Suppose that the m.g.f.'s ψ_n and ψ^* exist $(n = 1, 2, \ldots)$. If $\lim_{n \to \infty} \psi_n(t) = \psi^*(t)$ for all values of t in some interval around the point $t = 0$, then the sequence $X_1, X_2, \ldots$ converges in distribution to X^*.

In other words, the sequence of d.f.'s $F_1, F_2, \ldots$ must converge to the d.f. F^* if the corresponding sequence of m.g.f.'s $\psi_1, \psi_2, \ldots$ converges to the m.g.f. ψ^*.

Outline of the Proof of the Central Limit Theorem.   We are now ready to outline a proof of Theorem 1, which is the central limit theorem of Lindeberg and Lévy. We shall assume that the variables $X_1, \ldots, X_n$ form a random sample of size n from a distribution with mean μ and variance σ^2. We shall also assume, for convenience, that the m.g.f. of this distribution exists, although the central limit theorem is true even without this assumption.

For $i = 1, \ldots, n$, let $Y_i = (X_i - \mu)/\sigma$. Then the random variables $Y_1, \ldots, Y_n$ are i.i.d., and each has mean 0 and variance 1. Furthermore, let

$$Z_n = \frac{n^{1/2}(\overline{X}_n - \mu)}{\sigma} = \frac{1}{n^{1/2}} \sum_{i=1}^{n} Y_i.$$

We shall show that Z_n converges in distribution to a random variable having a standard normal distribution, as indicated in Eq. (1), by showing that the m.g.f. of Z_n converges to the m.g.f. of the standard normal distribution.

If $\psi(t)$ denotes the m.g.f. of each random variable Y_i $(i = 1, \ldots, n)$, then it follows from Theorem 3 of Sec. 4.4 that the m.g.f. of the sum $\sum_{i=1}^{n} Y_i$ will be $[\psi(t)]^n$. Also, it follows from Theorem 2 of Sec. 4.4 that the m.g.f. $\zeta_n(t)$ of Z_n will be

$$\zeta_n(t) = \left[\psi\left(\frac{t}{n^{1/2}}\right)\right]^n.$$

In this problem, $\psi'(0) = E(Y_i) = 0$ and $\psi''(0) = E(Y_i^2) = 1$. Therefore, the Taylor series expansion of $\psi(t)$ about the point $t = 0$ has the following form:

$$\psi(t) = \psi(0) + t\psi'(0) + \frac{t^2}{2!}\psi''(0) + \frac{t^3}{3!}\psi'''(0) + \cdots$$

$$= 1 + \frac{t^2}{2} + \frac{t^3}{3!}\psi'''(0) + \cdots .$$

Also,

$$\zeta_n(t) = \left[1 + \frac{t^2}{2n} + \frac{t^3\psi'''(0)}{3!n^{3/2}} + \cdots\right]^n.$$

It is shown in advanced calculus that if $\lim_{n \to \infty} a_n = b$ for some numbers a_n and b, then

$$\lim_{n \to \infty} \left(1 + \frac{a_n}{n} \right)^n = e^b.$$

But

$$\lim_{n \to \infty} \left[\frac{t^2}{2} + \frac{t^3 \psi'''(0)}{3! n^{1/2}} + \cdots \right] = \frac{t^2}{2}.$$

Hence,

$$\lim_{n \to \infty} \zeta_n(t) = \exp\left(\frac{1}{2} t^2 \right). \tag{9}$$

Since the right side of Eq. (9) is the m.g.f. of the standard normal distribution, it follows from Theorem 4 that the asymptotic distribution of Z_n must be the standard normal distribution.

An outline of the proof of the central limit theorem of Liapounov can also be given by proceeding along similar lines, but we shall not consider this problem further here.

EXERCISES

1. Suppose that 75 percent of the people in a certain metropolitan area live in the city and 25 percent of the people live in the suburbs. If 1200 people attending a certain concert represent a random sample from the metropolitan area, what is the probability that the number of people from the suburbs attending the concert will be fewer than 270?

2. Suppose that the distribution of the number of defects on any given bolt of cloth is a Poisson distribution with mean 5, and that the number of defects on each bolt is counted for a random sample of 125 bolts. Determine the probability that the average number of defects per bolt in the sample will be less than 5.5.

3. Suppose that a random sample of size n is to be taken from a distribution for which the mean is μ and the standard deviation is 3. Use the central limit theorem to determine approximately the smallest value of n for which the following relation will be satisfied:

$$\Pr(|\bar{X}_n - \mu| < 0.3) \geq 0.95.$$

4. Suppose that the proportion of defective items in a large manufactured lot is 0.1. What is the smallest random sample of items that must be taken from the lot in order for the probability to be at least 0.99 that the proportion of defective items in the sample will be less than 0.13?

5. Suppose that three boys A, B, and C throw snowballs at a target. Suppose also that boy A throws 10 times and the probability that he will hit the target on any given throw is 0.3; boy B throws 15 times and the probability that he will hit the target on any given throw is 0.2; and boy C throws 20 times and the probability that he will hit the target on any given throw is 0.1. Determine the probability that the target will be hit at least 12 times.

6. If 16 digits are chosen from a table of random digits, what is the probability that their average will lie between 4 and 6?

7. Suppose that people attending a party pour drinks from a bottle containing 63 ounces of a certain liquid. Suppose also that the expected size of each drink is 2 ounces, that the standard deviation of each drink is 1/2 ounce, and that all drinks are poured independently. Determine the probability that the bottle will not be empty after 36 drinks have been poured.

8. A physicist makes 25 independent measurements of the specific gravity of a certain body. He knows that the limitations of his equipment are such that the standard deviation of each measurement is σ units.

   (a) By using the Chebyshev inequality, find a lower bound for the probability that the average of his measurements will differ from the actual specific gravity of the body by less than $\sigma/4$ units.

   (b) By using the central limit theorem, find an approximate value for the probability in part (a).

9. A random sample of n items is to be taken from a distribution with mean μ and standard deviation σ.

   (a) Use the Chebyshev inequality to determine the smallest number of items n that must be taken in order to satisfy the following relation:

   $$\Pr\left(|\bar{X}_n - \mu| \leq \frac{\sigma}{4}\right) \geq 0.99.$$

   (b) Use the central limit theorem to determine the smallest number of items n that must be taken in order to satisfy the relation in part (a) approximately.

10. Suppose that, on the average, one third of the graduating seniors at a certain college have two parents attend the graduation ceremony, another third of these seniors have one parent attend the ceremony, and the remaining third of these seniors have no parents attend. If there are 600 graduating seniors in a particular class, what is the probability that not more than 650 parents will attend the graduation ceremony?

5.8.   THE CORRECTION FOR CONTINUITY

Approximating a Discrete Distribution by a Continuous Distribution

Suppose that $X_1, \ldots, X_n$ form a random sample from a discrete distribution, and let $X = X_1 + \cdots + X_n$. It was shown in the previous section that even though the distribution of X will be discrete, this distribution can be approximated by the normal distribution, which is continuous. In this section, we shall describe a standard method for improving the quality of the approximation that is obtained when a probability based on a discrete distribution is approximated by one based on a continuous distribution.

Suppose, therefore, that the random variable X has a discrete distribution with p.f. $f(x)$ and it is desired to approximate this distribution by a continuous distribution with p.d.f. $g(x)$. For simplicity, we shall consider only a discrete distribution for which all possible values of X are integers. This condition is satisfied for the binomial, hypergeometric, Poisson, and negative binomial distributions described in this chapter.

If the p.d.f. $g(x)$ provides a good approximation to the distribution of X, then for any integers a and b, we can simply approximate the probability

$$\Pr(a \leqslant X \leqslant b) = \sum_{x=a}^{b} f(x) \tag{1}$$

by the integral

$$\int_a^b g(x)\, dx. \tag{2}$$

Indeed, this approximation was used in Examples 1 and 3 of Sec. 5.7, where $g(x)$ was the appropriate normal p.d.f. specified by the central limit theorem.

This simple approximation has the following shortcoming: Although $\Pr(X \geqslant a)$ and $\Pr(X > a)$ will typically have different values for the discrete distribution, these probabilities will always be equal for the continuous distribution. Another way of expressing this shortcoming is as follows: Although $\Pr(X = x) > 0$ for any integer x that is a possible value of X, this probability is necessarily 0 under the approximating p.d.f.

Approximating a Histogram

The p.f. $f(x)$ of X can be represented by a *histogram*, or *bar chart*, as sketched in Fig. 5.4. For each integer x, the probability of x is represented by the area of a rectangle with a base that extends from $x - \dfrac{1}{2}$ to $x + \dfrac{1}{2}$ and with a height $f(x)$.

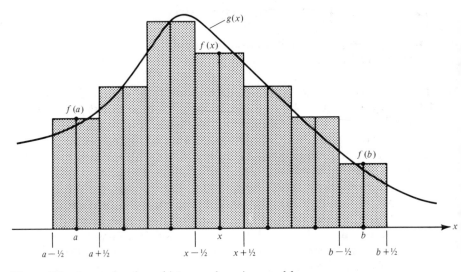

Figure 5.4   Approximating a histogram by using a p.d.f.

Thus, the area of the rectangle for which the center of the base is at the integer x is simply $f(x)$. An approximating p.d.f. $g(x)$ is also sketched in Fig. 5.4.

From this point of view it can be seen that $\Pr(a \leqslant X \leqslant b)$, as specified in Eq. (1), is the sum of the areas of the rectangles in Fig. 5.4 that are centered at $a, a + 1, \ldots, b$. It can also be seen from Fig. 5.4 that the sum of these areas is approximated by the integral

$$\int_{a-(1/2)}^{b+(1/2)} g(x)\, dx. \tag{3}$$

The adjustment from the integral (2) to the integral (3) is called the *correction for continuity*.

If we use the correction for continuity, we find that the probability $f(a)$ of the single integer a can be approximated as follows:

$$\Pr(X = a) = \Pr\left(a - \frac{1}{2} \leqslant X \leqslant a + \frac{1}{2}\right) \tag{4}$$

$$\approx \int_{a-(1/2)}^{a+(1/2)} g(x)\, dx.$$

Similarly,

$$Pr(X > a) = Pr(X \geq a + 1) = Pr\left(X \geq a + \frac{1}{2}\right)$$

$$\approx \int_{a+(1/2)}^{\infty} g(x)\, dx. \tag{5}$$

Example 1: Examination Questions. To illustrate the use of the correction for continuity, we shall again consider Example 3 of Sec. 5.7. In that example, an examination contains 99 questions of varying difficulty and it is desired to determine $Pr(X \geq 60)$, where X denotes the total number of questions that a particular student answers correctly. Then under the conditions of the example, it is found from the central limit theorem that the discrete distribution of X could be approximated by a normal distribution with mean 49.5 and standard deviation 4.08.

If we use the correction for continuity, we obtain

$$Pr(X \geq 60) = Pr(X \geq 59.5) = Pr\left(Z \geq \frac{59.5 - 49.5}{4.08}\right)$$

$$\approx 1 - \Phi(2.4510) = 0.007.$$

This value can be compared with the value 0.005 that was obtained in Sec. 5.7 without the correction. □

Example 2: Coin Tossing. Suppose that a fair coin is tossed 20 times, and that all tosses are independent. What is the probability of obtaining exactly 10 heads?

Let X denote the total number of heads obtained in the 20 tosses. According to the central limit theorem, the distribution of X will be approximately a normal distribution with mean 10 and standard deviation $[(20)(1/2)(1/2)]^{1/2} = 2.236$. If we use the correction for continuity,

$$Pr(X = 10) = Pr(9.5 \leq X \leq 10.5)$$

$$= Pr\left(-\frac{0.5}{2.236} \leq Z \leq \frac{0.5}{2.236}\right)$$

$$\approx \Phi(0.2236) - \Phi(-0.2236) = 0.177.$$

The exact value of $Pr(X = 10)$ found from the table of binomial probabilities given at the back of this book is 0.1762. Thus, the normal approximation with the correction for continuity is quite good. □

EXERCISES

1. Let X denote the total number of successes in 15 Bernoulli trials, with probability of success $p = 0.3$ on each trial.

(a) Determine approximately the value of $\Pr(X = 4)$ by using the central limit theorem with the correction for continuity.

(b) Compare the answer obtained in part (a) with the exact value of this probability.

2. Using the correction for continuity, determine the probability required in Example 1 of Sec. 5.7.

3. Using the correction for continuity, determine the probability required in Exercise 1 of Sec. 5.7.

4. Using the correction for continuity, determine the probability required in Exercise 2 of Sec. 5.7.

5. Using the correction for continuity, determine the probability required in Exercise 5 of Sec. 5.7.

6. Using the correction for continuity, determine the probability required in Exercise 6 of Sec. 5.7.

5.9. THE GAMMA DISTRIBUTION

The Gamma Function

For any positive number α, let the value $\Gamma(\alpha)$ be defined by the following integral:

$$\Gamma(\alpha) = \int_0^\infty x^{\alpha - 1} e^{-x} \, dx. \tag{1}$$

It can be shown that the value of this integral will be finite for any value of $\alpha > 0$. The function Γ whose values are defined by Eq. (1) for $\alpha > 0$ is called the *gamma function*. We shall now derive some properties of the gamma function.

Theorem 1. *If $\alpha > 1$, then*

$$\Gamma(\alpha) = (\alpha - 1)\Gamma(\alpha - 1). \tag{2}$$

Proof. We shall apply the method of integration by parts to the integral in Eq. (1). If we let $u = x^{\alpha - 1}$ and $dv = e^{-x} \, dx$, then $du = (\alpha - 1)x^{\alpha - 2} \, dx$ and $v = -e^{-x}$. Therefore,

$$\Gamma(\alpha) = \int_0^\infty u \, dv = [uv]_0^\infty - \int_0^\infty v \, du$$

$$= [-x^{\alpha - 1} e^{-x}]_0^\infty + (\alpha - 1) \int_0^\infty x^{\alpha - 2} e^{-x} \, dx$$

$$= 0 + (\alpha - 1)\Gamma(\alpha - 1). \quad \square$$

It follows from Theorem 1 that for any integer $n \geqslant 2$,

$$\Gamma(n) = (n-1)\Gamma(n-1) = (n-1)(n-2)\Gamma(n-2)$$
$$= (n-1)(n-2)\cdots 1 \cdot \Gamma(1)$$
$$= (n-1)!\Gamma(1).$$

Furthermore, by Eq. (1),

$$\Gamma(1) = \int_0^\infty e^{-x} dx = 1.$$

Hence, $\Gamma(n) = (n-1)!$ for $n = 2, 3, \ldots$. Moreover, since $\Gamma(1) = 1 = 0!$, we have established the following result:

Theorem 2. *For any positive integer n,*

$$\Gamma(n) = (n-1)!. \tag{3}$$

In many statistical applications, $\Gamma(\alpha)$ must be evaluated when α is either a positive integer or of the form $\alpha = n + (1/2)$ for some positive integer n. It follows from Eq. (2) that for any positive integer n,

$$\Gamma\left(n + \frac{1}{2}\right) = \left(n - \frac{1}{2}\right)\left(n - \frac{3}{2}\right)\cdots \left(\frac{1}{2}\right)\Gamma\left(\frac{1}{2}\right). \tag{4}$$

Hence, it will be possible to determine the value of $\Gamma\left(n + \frac{1}{2}\right)$ if we can evaluate $\Gamma\left(\frac{1}{2}\right)$.

From Eq. (1),

$$\Gamma\left(\frac{1}{2}\right) = \int_0^\infty x^{-1/2} e^{-x} dx.$$

If we let $x = (1/2)y^2$ in this integral, then $dx = y\, dy$ and

$$\Gamma\left(\frac{1}{2}\right) = 2^{1/2} \int_0^\infty \exp\left(-\frac{1}{2}y^2\right) dy. \tag{5}$$

Since the integral of the p.d.f. of the standard normal distribution is equal to 1, it follows that

$$\int_{-\infty}^\infty \exp\left(-\frac{1}{2}y^2\right) dy = (2\pi)^{1/2}.$$

Therefore,

$$\int_0^\infty \exp\left(-\frac{1}{2}y^2\right) dy = \frac{1}{2}(2\pi)^{1/2} = \left(\frac{\pi}{2}\right)^{1/2}.$$

It now follows from Eq. (5) that

$$\Gamma\left(\frac{1}{2}\right) = \pi^{1/2}. \tag{6}$$

For example, it is found from Eqs. (4) and (6) that

$$\Gamma\left(\frac{7}{2}\right) = \left(\frac{5}{2}\right)\left(\frac{3}{2}\right)\left(\frac{1}{2}\right)\pi^{1/2} = \frac{15}{8}\pi^{1/2}.$$

The Gamma Distribution

It is said that a random variable X has a *gamma distribution with parameters α and β ($\alpha > 0$ and $\beta > 0$)* if X has a continuous distribution for which the p.d.f. $f(x \mid \alpha, \beta)$ is specified as follows:

$$f(x \mid \alpha, \beta) = \begin{cases} \dfrac{\beta^\alpha}{\Gamma(\alpha)} x^{\alpha-1} e^{-\beta x} & \text{for } x > 0, \\ 0 & \text{for } x \leqslant 0. \end{cases} \tag{7}$$

The integral of this p.d.f. is 1, since it follows from the definition of the gamma function that

$$\int_0^\infty x^{\alpha-1} e^{-\beta x} dx = \frac{\Gamma(\alpha)}{\beta^\alpha}. \tag{8}$$

If X has a gamma distribution with parameters α and β, then the moments of X are easily found from Eqs. (7) and (8). For $k = 1, 2, \ldots$, we have

$$E(X^k) = \int_0^\infty x^k f(x \mid \alpha, \beta) \, dx = \frac{\beta^\alpha}{\Gamma(\alpha)} \int_0^\infty x^{\alpha+k-1} e^{-\beta x} \, dx$$

$$= \frac{\beta^\alpha}{\Gamma(\alpha)} \cdot \frac{\Gamma(\alpha + k)}{\beta^{\alpha+k}} = \frac{\Gamma(\alpha + k)}{\beta^k \Gamma(\alpha)}$$

$$= \frac{\alpha(\alpha + 1) \cdots (\alpha + k - 1)}{\beta^k}.$$

In particular, therefore,

$$E(X) = \frac{\alpha}{\beta}$$

and

$$\text{Var}(X) = \frac{\alpha(\alpha + 1)}{\beta^2} - \left(\frac{\alpha}{\beta}\right)^2 = \frac{\alpha}{\beta^2}.$$

The m.g.f. ψ of X can be obtained similarly, as follows:

$$\psi(t) = \int_0^\infty e^{tx} f(x \mid \alpha, \beta) \, dx = \frac{\beta^\alpha}{\Gamma(\alpha)} \int_0^\infty x^{\alpha-1} e^{-(\beta-t)x} \, dx.$$

This integral will be finite for any value of t such that $t < \beta$. Therefore, it follows from Eq. (8) that

$$\psi(t) = \frac{\beta^\alpha}{\Gamma(\alpha)} \cdot \frac{\Gamma(\alpha)}{(\beta - t)^\alpha} = \left(\frac{\beta}{\beta - t}\right)^\alpha \qquad \text{for } t < \beta. \tag{9}$$

We can now show that the sum of independent random variables which have gamma distributions with a common value of the parameter β will also have a gamma distribution.

Theorem 3. *If the random variables $X_1, \ldots, X_k$ are independent and if X_i has a gamma distribution with parameters α_i and β $(i = 1, \ldots, k)$, then the sum $X_1 + \cdots + X_k$ has a gamma distribution with parameters $\alpha_1 + \cdots + \alpha_k$ and β.*

Proof. If ψ_i denotes the m.g.f. of X_i, then it follows from Eq. (9) that for $i = 1, \ldots, k$,

$$\psi_i(t) = \left(\frac{\beta}{\beta - t}\right)^{\alpha_i} \qquad \text{for } t < \beta.$$

If ψ denotes the m.g.f. of the sum $X_1 + \cdots + X_k$, then by Theorem 3 of Sec. 4.4,

$$\psi(t) = \prod_{i=1}^k \psi_i(t) = \left(\frac{\beta}{\beta - t}\right)^{\alpha_1 + \cdots + \alpha_k} \qquad \text{for } t < \beta.$$

The m.g.f. ψ can now be recognized as the m.g.f. of a gamma distribution with parameters $\alpha_1 + \cdots + \alpha_k$ and β. Hence, the sum $X_1 + \cdots + X_k$ must have this gamma distribution. $\square$

The Exponential Distribution

It is said that a random variable X has an *exponential distribution with parameter β* $(\beta > 0)$ if X has a continuous distribution for which the p.d.f. $f(x \mid \beta)$ is specified as follows:

$$f(x \mid \beta) = \begin{cases} \beta e^{-\beta x} & \text{for } x > 0, \\ 0 & \text{for } x \leqslant 0. \end{cases} \tag{10}$$

It can be seen from Eq. (10) that an exponential distribution with parameter β is the same as a gamma distribution with parameters $\alpha = 1$ and β. Therefore, if X has an exponential distribution with parameter β, it follows from the expressions derived for the gamma distribution that

$$E(X) = \frac{1}{\beta} \quad \text{and} \quad \text{Var}(X) = \frac{1}{\beta^2}.$$

Similarly, it follows from Eq. (9) that the m.g.f. of X is

$$\psi(t) = \frac{\beta}{\beta - t} \quad \text{for } t < \beta.$$

An exponential distribution is often used in a practical problem to represent the distribution of the time that elapses before the occurrence of some event. For example, this distribution has been used to represent such periods of time as the period for which a machine or an electronic component will operate without breaking down, the period required to take care of a customer at some service facility, and the period between the arrivals of two successive customers at a facility.

If the events being considered occur in accordance with a Poisson process, as defined in Sec. 5.4, then both the waiting time until an event occurs and the period of time between any two successive events will have exponential distributions. This fact provides theoretical support for the use of the exponential distribution in many types of problems.

The exponential distribution has a memoryless property similar to that described in Sec. 5.5 for the geometric distribution. This property can be derived as follows: If X has an exponential distribution with parameter β, then for any number $t > 0$,

$$\Pr(X \geqslant t) = \int_t^\infty \beta e^{-\beta x} \, dx = e^{-\beta t}. \tag{11}$$

Hence, for $t > 0$ and any other number $h > 0$,

$$\Pr(X \geqslant t + h \mid X \geqslant t) = \frac{\Pr(X \geqslant t + h)}{\Pr(X \geqslant t)} \tag{12}$$

$$= \frac{e^{-\beta(t+h)}}{e^{-\beta t}} = e^{-\beta h} = \Pr(X \geqslant h).$$

To illustrate the memoryless property, we shall suppose that X represents the number of minutes that elapse before some event occurs. According to Eq. (12), if the event has not occurred in t minutes, then the probability that the event will not occur during the next h minutes is simply $e^{-\beta h}$. This is the same as the probability that the event would not occur during an interval of h minutes starting from time 0. In other words, regardless of the length of time that has elapsed without the occurrence of the event, the probability that the event will occur during the next h minutes always has the same value. Theoretically, therefore, it is not necessary to consider past occurrences of an event in order to calculate probabilities for future occurrences of the event.

This memoryless property will not strictly be satisfied in all practical problems. For example, suppose that X is the length of time for which a light bulb will burn before it fails. The length of time for which the bulb can be expected to continue to burn in the future will depend on the length of time for which it has been burning in the past. Nevertheless, the exponential distribution has been used effectively as an approximate distribution for such variables as the lengths of the lives of various products.

Life Tests

Suppose that n light bulbs are burning simultaneously in a test to determine the lengths of their lives. We shall assume that the n bulbs burn independently of one another and that the lifetime of each bulb has an exponential distribution with parameter β. In other words, if X_i denotes the lifetime of bulb i, for $i = 1, \ldots, n$, then it is assumed that the random variables $X_1, \ldots, X_n$ are i.i.d. and that each has an exponential distribution with parameter β. We shall now determine the distribution of the length of time Y_1 until one of the n bulbs fails.

Since the time Y_1 at which the first bulb fails will be equal to the smallest of the n lifetimes $X_1, \ldots, X_n$, we can write $Y_1 = \min\{X_1, \ldots, X_n\}$. For any number $t > 0$,

$$\Pr(Y_1 > t) = \Pr(X_1 > t, \ldots, X_n > t)$$

$$= \Pr(X_1 > t) \cdots \Pr(X_n > t)$$

$$= e^{-\beta t} \cdots e^{-\beta t} = e^{-n\beta t}.$$

By comparing this result with Eq. (11), we see that the distribution of Y_1 must be an exponential distribution with parameter $n\beta$.

In summary, we have established the following result:

Theorem 4. *Suppose that the variables $X_1, \ldots, X_n$ form a random sample from an exponential distribution with parameter β. Then the distribution of $Y_1 = \min\{X_1, \ldots, X_n\}$ will be an exponential distribution with parameter $n\beta$.*

Next, we shall determine the distribution of the interval of time Y_2 between the failure of the first bulb and the failure of a second bulb.

After one bulb has failed, $n-1$ bulbs are still burning. Furthermore, regardless of the time at which the first bulb failed, it follows from the memoryless property of the exponential distribution that the distribution of the remaining lifetime of each of the other $n-1$ bulbs is still an exponential distribution with parameter β. In other words, the situation is the same as it would be if we were starting the test over again from time $t=0$ with $n-1$ new bulbs. Therefore, Y_2 will be equal to the smallest of $n-1$ i.i.d. random variables, each of which has an exponential distribution with parameter β. It follows from Theorem 4 that Y_2 will have an exponential distribution with parameter $(n-1)\beta$.

By continuing in this way, we find that the distribution of the interval of time Y_3 between the failure of the second bulb and the failure of a third bulb will be an exponential distribution with parameter $(n-2)\beta$. Finally, after all but one of the bulbs have failed, the distribution of the additional interval of time until the final bulb fails will be an exponential distribution with parameter β.

EXERCISES

1. Suppose that X has a gamma distribution with parameters α and β, and that c is a positive constant. Show that cX has a gamma distribution with parameters α and β/c.

2. Sketch the p.d.f. of the gamma distribution for each of the following pairs of values of the parameters α and β: (a) $\alpha = 1/2$ and $\beta = 1$; (b) $\alpha = 1$ and $\beta = 1$; (c) $\alpha = 2$ and $\beta = 1$.

3. Determine the mode of the gamma distribution with parameters α and β.

4. Sketch the p.d.f. of the exponential distribution for each of the following values of the parameter β: (a) $\beta = 1/2$, (b) $\beta = 1$, and (c) $\beta = 2$.

5. Suppose that $X_1, \ldots, X_n$ form a random sample of size n from an exponential distribution with parameter β. Determine the distribution of the sample mean $\overline{X}_n$.

6. Suppose that the number of minutes required to serve a customer at the checkout counter of a supermarket has an exponential distribution for which the mean is 3. Using the central limit theorem, determine the probability that the total time required to serve a random sample of 16 customers will exceed 1 hour.

7. Suppose that the random variables $X_1, \ldots, X_k$ are independent and that X_i has an exponential distribution with parameter β_i $(i = 1, \ldots, k)$. Let $Y = \min\{X_1, \ldots, X_k\}$. Show that Y has an exponential distribution with parameter $\beta_1 + \cdots + \beta_k$.

8. Suppose that a certain system contains three components which function independently of each other and which are connected in series, as defined in

Exercise 4 of Sec. 3.7, so that the system fails as soon as one of the components fails. Suppose that the length of life of the first component, measured in hours, has an exponential distribution with parameter $\beta = 0.001$; the length of life of the second component has an exponential distribution with parameter $\beta = 0.003$; and the length of life of the third component has an exponential distribution with parameter $\beta = 0.006$. Determine the probability that the system will not fail before 100 hours.

9. Suppose that an electronic system contains n similar components which function independently of each other and which are connected in series, so that the system fails as soon as one of the components fails. Suppose also that the length of life of each component, measured in hours, has an exponential distribution with mean μ. Determine the mean and the variance of the length of time until the system fails.

10. Suppose that n items are being tested simultaneously; that the items are independent; and that the length of life of each item has an exponential distribution with parameter β. Determine the expected length of time until three items have failed. *Hint:* The required value is $E(Y_1 + Y_2 + Y_3)$.

11. Consider again the electronic system described in Exercise 9; but suppose now that the system will continue to operate until two components have failed. Determine the mean and the variance of the length of time until the system fails.

12. Suppose that a certain examination is to be taken by five students independently of one another, and that the number of minutes required by any particular student to complete the examination has an exponential distribution for which the mean is 80. Suppose that the examination begins at 9:00 A.M. Determine the probability that at least one of the students will complete the examination before 9:40 A.M.

13. Suppose again that the examination considered in Exercise 12 is taken by five students, and that the first student to complete the examination finishes at 9:25 A.M. Determine the probability that at least one other student will complete the examination before 10:00 A.M.

14. Suppose again that the examination considered in Exercise 12 is taken by five students. Determine the probability that no two students will complete the examination within 10 minutes of each other.

15. It is said that a random variable X has a *Pareto distribution with parameters x_0 and α* ($x_0 > 0$ and $\alpha > 0$) if X has a continuous distribution for which the p.d.f. $f(x \mid x_0, \alpha)$ is as follows:

$$
f(x \mid x_0, \alpha) = \begin{cases} \dfrac{\alpha x_0^{\alpha}}{x^{\alpha+1}} & \text{for } x \geq x_0, \\ 0 & \text{for } x < x_0. \end{cases}
$$

Show that if X has this Pareto distribution, then the random variable $\log(X/x_0)$ has an exponential distribution with parameter α.

16. Suppose that a random variable X has a normal distribution with mean μ and variance σ^2. Determine the value of $E[(X - \mu)^{2n}]$ for $n = 1, 2, \ldots$.

17. Consider a random variable X for which $\Pr(X > 0) = 1$, and for which the p.d.f. is f and the d.f. is F. Consider also the function h defined as follows:

$$h(x) = \frac{f(x)}{1 - F(x)} \qquad \text{for } x > 0.$$

The function h is called the *failure rate* or the *hazard function* of X. Show that if X has an exponential distribution, then the failure rate $h(x)$ is constant for $x > 0$.

18. It is said that a random variable has a *Weibull distribution with parameters a and b* $(a > 0$ and $b > 0)$ if X has a continuous distribution for which the p.d.f. $f(x \mid a, b)$ is as follows:

$$f(x \mid a, b) = \begin{cases} \dfrac{b}{a^b} x^{b-1} e^{-(x/a)^b} & \text{for } x > 0, \\ 0 & \text{for } x \leqslant 0. \end{cases}$$

Show that if X has this Weibull distribution, then the random variable X^b has an exponential distribution with parameter $\beta = a^{-b}$.

19. It is said that a random variable X has an *increasing failure rate* if the failure rate $h(x)$ defined in Exercise 17 is an increasing function of x for $x > 0$; and it is said that X has a *decreasing failure rate* if $h(x)$ is a decreasing function of x for $x > 0$. Suppose that X has a Weibull distribution with parameters a and b, as defined in Exercise 18. Show that X has an increasing failure rate if $b > 1$ and that X has a decreasing failure rate if $b < 1$.

5.10. THE BETA DISTRIBUTION

Definition of the Beta Distribution

It is said that a random variable X has a *beta distribution with parameters α and β* $(\alpha > 0$ and $\beta > 0)$ if X has a continuous distribution for which the p.d.f. $f(x \mid \alpha, \beta)$ is as follows:

$$f(x \mid \alpha, \beta) = \begin{cases} \dfrac{\Gamma(\alpha + \beta)}{\Gamma(\alpha)\Gamma(\beta)} x^{\alpha-1}(1 - x)^{\beta-1} & \text{for } 0 < x < 1, \\ 0 & \text{otherwise.} \end{cases} \qquad (1)$$

In order to verify that the integral of this p.d.f. over the real line has the value 1, we must show that for $\alpha > 0$ and $\beta > 0$,

$$\int_0^1 x^{\alpha-1}(1-x)^{\beta-1}\,dx = \frac{\Gamma(\alpha)\Gamma(\beta)}{\Gamma(\alpha+\beta)}. \qquad (2)$$

From the definition of the gamma function, it follows that

$$\Gamma(\alpha)\Gamma(\beta) = \int_0^\infty u^{\alpha-1}e^{-u}\,du \int_0^\infty v^{\beta-1}e^{-v}\,dv$$
$$= \int_0^\infty \int_0^\infty u^{\alpha-1}v^{\beta-1}e^{-(u+v)}\,du\,dv. \qquad (3)$$

Now we shall let

$$x = \frac{u}{u+v} \qquad \text{and} \qquad y = u + v.$$

Then $u = xy$ and $v = (1-x)y$, and it can be found that the value of the Jacobian of this inverse transformation is y. Furthermore, as u and v vary over all positive values, x will vary over the interval $(0,1)$ and y will vary over all positive values. From Eq. (3), we now obtain the relation

$$\Gamma(\alpha)\Gamma(\beta) = \int_0^1 \int_0^\infty x^{\alpha-1}(1-x)^{\beta-1}y^{\alpha+\beta-1}e^{-y}\,dy\,dx$$
$$= \Gamma(\alpha+\beta)\int_0^1 x^{\alpha-1}(1-x)^{\beta-1}\,dx.$$

Therefore, Eq. (2) has been established.

It can be seen from Eq. (1) that the beta distribution with parameters $\alpha = 1$ and $\beta = 1$ is simply the uniform distribution on the interval $(0,1)$.

Moments of the Beta Distribution

When the p.d.f. of a random variable X is given by Eq. (1), the moments of X are easily calculated. For $k = 1, 2, \ldots$,

$$E(X^k) = \int_0^1 x^k f(x \mid \alpha, \beta)\,dx$$
$$= \frac{\Gamma(\alpha+\beta)}{\Gamma(\alpha)\Gamma(\beta)} \int_0^1 x^{\alpha+k-1}(1-x)^{\beta-1}\,dx.$$

Therefore, by Eq. (2),

$$E(X^k) = \frac{\Gamma(\alpha + \beta)}{\Gamma(\alpha)\Gamma(\beta)} \cdot \frac{\Gamma(\alpha + k)\Gamma(\beta)}{\Gamma(\alpha + k + \beta)}$$

$$= \frac{\alpha(\alpha + 1) \cdots (\alpha + k - 1)}{(\alpha + \beta)(\alpha + \beta + 1) \cdots (\alpha + \beta + k - 1)}.$$

It follows that

$$E(X) = \frac{\alpha}{\alpha + \beta}$$

and

$$\text{Var}(X) = \frac{\alpha(\alpha + 1)}{(\alpha + \beta)(\alpha + \beta + 1)} - \left(\frac{\alpha}{\alpha + \beta}\right)^2$$

$$= \frac{\alpha\beta}{(\alpha + \beta)^2(\alpha + \beta + 1)}.$$

EXERCISES

1. Determine the mode of the beta distribution with parameters α and β, assuming that $\alpha > 1$ and $\beta > 1$.

2. Sketch the p.d.f. of the beta distribution for each of the following pairs of values of the parameters:
   (a) $\alpha = 1/2$ and $\beta = 1/2$,   (b) $\alpha = 1/2$ and $\beta = 1$,
   (c) $\alpha = 1/2$ and $\beta = 2$,   (d) $\alpha = 1$ and $\beta = 1$,
   (e) $\alpha = 1$ and $\beta = 2$,   (f) $\alpha = 2$ and $\beta = 2$,
   (g) $\alpha = 25$ and $\beta = 100$,   (h) $\alpha = 100$ and $\beta = 25$.

3. Suppose that X has a beta distribution with parameters α and β. Show that $1 - X$ has a beta distribution with parameters β and α.

4. Suppose that X has a beta distribution with parameters α and β, and let r and s be given positive integers. Determine the value of $E[X^r(1 - X)^s]$.

5. Suppose that X and Y are independent random variables; that X has a gamma distribution with parameters α_1 and β; and that Y has a gamma distribution with parameters α_2 and β. Let $U = X/(X + Y)$ and $V = X + Y$. Show (a) that U has a beta distribution with parameters α_1 and α_2; and (b) that U and V are independent.

6. Suppose that X_1 and X_2 form a random sample of two observed values from an exponential distribution with parameter β. Show that $X_1/(X_1 + X_2)$ has a uniform distribution on the interval $(0, 1)$.

7. Suppose that the proportion X of defective items in a large lot is unknown, and that X has a beta distribution with parameters α and β.

   (a) If one item is selected at random from the lot, what is the probability that it will be defective?

   (b) If two items are selected at random from the lot, what is the probability that both will be defective?

5.11. THE MULTINOMIAL DISTRIBUTION

Definition of the Multinomial Distribution

Suppose that a population contains items of k different types ($k \geqslant 2$) and that the proportion of the items in the population that are of type i is p_i ($i = 1, \ldots, k$). It is assumed that $p_i > 0$ for $i = 1, \ldots, k$ and that $\sum_{i=1}^{k} p_i = 1$. Furthermore, suppose that n items are selected at random from the population, with replacement; and let X_i denote the number of selected items that are of type i ($i = 1, \ldots, k$). Then it is said that the random vector $X = (X_1, \ldots, X_k)$ has a *multinomial distribution with parameters n and $p = (p_1, \ldots, p_k)$*. We shall now derive the p.f. of X.

We can imagine that the n items are selected from the population one at a time, with replacement. Since the n selections are made independently of each other, the probability that the first item will be of type i_1, the second item of type i_2, and so on is simply $p_{i_1} p_{i_2} \cdots p_{i_n}$. Therefore, the probability that the sequence of n outcomes will consist of exactly x_1 items of type 1, x_2 items of type 2, and so on, selected in a *particular prespecified order*, is $p_1^{x_1} p_2^{x_2} \cdots p_k^{x_k}$. It follows that the probability of obtaining exactly x_i items of type i ($i = 1, \ldots, k$) is equal to the probability $p_1^{x_1} p_2^{x_2} \cdots p_k^{x_k}$ multiplied by the total number of different ways in which the order of the n items can be specified.

From the discussion given in Sec. 1.9, it follows that the total number of different ways in which n items can be arranged when there are x_i items of type i ($i = 1, \ldots, k$) is given by the multinomial coefficient

$$\frac{n!}{x_1! x_2! \cdots x_k!}.$$

Hence,

$$\Pr(X_1 = x_1, \ldots, X_k = x_k) = \frac{n!}{x_1! \cdots x_k!} p_1^{x_1} \cdots p_k^{x_k}. \tag{1}$$

For any vector $x = (x_1, \ldots, x_k)$, the p.f. $f(x \mid n, p)$ of X is defined by the following relation:

$$f(x \mid n, p) = \Pr(X = x) = \Pr(X_1 = x_1, \ldots, X_k = x_k).$$

If $x_1, \ldots, x_k$ are nonnegative integers such that $x_1 + \cdots + x_k = n$, then it follows from Eq. (1) that

$$f(x \mid n, p) = \frac{n!}{x_1! \cdots x_k!} p_1^{x_1} \cdots p_k^{x_k}. \tag{2}$$

Furthermore, $f(x \mid n, p) = 0$ for any other vector x.

Example 1: Attendance at a Baseball Game. Suppose that 23 percent of the people attending a certain baseball game live within 10 miles of the stadium; 59 percent live between 10 and 50 miles from the stadium; and 18 percent live more than 50 miles from the stadium. Suppose also that 20 people are selected at random from the crowd attending the game. We shall determine the probability that seven of the people selected live within 10 miles of the stadium, eight of them live between 10 and 50 miles from the stadium, and five of them live more than 50 miles from the stadium.

We shall assume that the crowd attending the game is so large that it is irrelevant whether the 20 people are selected with or without replacement. We can therefore assume that they were selected with replacement. It then follows from Eq. (1) or Eq. (2) that the required probability is

$$\frac{20!}{7! \, 8! \, 5!} (0.23)^7 (0.59)^8 (0.18)^5 = 0.0094. \quad \square$$

Relation Between the Multinomial and Binomial Distributions

When the population being sampled contains only two different types of items, that is, when $k = 2$, the multinomial distribution reduces to the binomial distribution. The truth of this relationship can be demonstrated as follows: Suppose that, for $k = 2$, the random vector $X = (X_1, X_2)$ has a multinomial distribution with parameters n and $p = (p_1, p_2)$. Then it must be true that $X_2 = n - X_1$ and $p_2 = 1 - p_1$. Therefore, the random vector X is actually determined by the single random variable X_1, and the distribution of X_1 depends only on the parameters n and p_1. Furthermore, since X_1 denotes the total number of items of type 1 that are selected in n Bernoulli trials, when the probability of selection on each trial is p_1, it follows that X_1 has a binomial distribution with parameters n and p_1.

More generally, for any given value of k ($k = 2, 3, \ldots$), suppose that the random vector $X = (X_1, \ldots, X_k)$ has a multinomial distribution with parameters

n and $p = (p_1, \ldots, p_k)$. Since X_i can be regarded as the total number of items of type i that are selected in n Bernoulli trials, when the probability of selection on each trial is p_i, it follows that the *marginal distribution* of each variable X_i $(i = 1, \ldots, k)$ must be a binomial distribution with parameters n and p_i.

Means, Variances, and Covariances

Suppose that a random vector X has a multinomial distribution with parameters n and p. Since the marginal distribution of each component X_i is a binomial distribution with parameters n and p_i, it follows that

$$E(X_i) = np_i \quad \text{and} \quad \text{Var}(X_i) = np_i(1 - p_i) \quad \text{for } i = 1, \ldots, k. \quad (3)$$

A similar argument can be used to derive the value of the covariance of any two different components X_i and X_j. Since the sum $X_i + X_j$ can be regarded as the total number of items of either type i or type j that are selected in n Bernoulli trials, when the probability of selection on each trial is $p_i + p_j$, it follows that $X_i + X_j$ has a binomial distribution with parameters n and $p_i + p_j$. Hence,

$$\text{Var}(X_i + X_j) = n(p_i + p_j)(1 - p_i - p_j). \quad (4)$$

However, it is also true that

$$\text{Var}(X_i + X_j) = \text{Var}(X_i) + \text{Var}(X_j) + 2\,\text{Cov}(X_i, X_j)$$
$$= np_i(1 - p_i) + np_j(1 - p_j) + 2\,\text{Cov}(X_i, X_j). \quad (5)$$

By equating the right sides of (4) and (5), we obtain the following result:

$$\text{Cov}(X_i, X_j) = -np_i p_j. \quad (6)$$

Together, Eqs. (3) and (6) specify the values of the means, the variances, and the covariances for the multinomial distribution with parameters n and p.

EXERCISES

1. Suppose that F is a continuous d.f. on the real line; and let α_1 and α_2 be numbers such that $F(\alpha_1) = 0.3$ and $F(\alpha_2) = 0.8$. If 25 observations are selected at random from the distribution for which the d.f. is F, what is the probability that six of the observed values will be less than α_1, ten of the observed values will be between α_1 and α_2, and nine of the observed values will be greater than α_2?

2. If five balanced dice are rolled, what is the probability that the number 1 and the number 4 will appear the same number of times?

3. Suppose that a die is loaded so that each of the numbers 1, 2, 3, 4, 5, and 6 has a different probability of appearing when the die is rolled. For $i = 1, \ldots, 6$, let p_i denote the probability that the number i will be obtained; and suppose that $p_1 = 0.11$, $p_2 = 0.30$, $p_3 = 0.22$, $p_4 = 0.05$, $p_5 = 0.25$, and $p_6 = 0.07$. Suppose also that the die is to be rolled 40 times. Let X_1 denote the number of rolls for which an even number appears, and let X_2 denote the number of rolls for which either the number 1 or the number 3 appears. Find the value of $\Pr(X_1 = 20 \text{ and } X_2 = 15)$.

4. Suppose that 16 percent of the students in a certain high school are freshmen, 14 percent are sophomores, 38 percent are juniors, and 32 percent are seniors. If 15 students are selected at random from the school, what is the probability that at least 8 will be either freshmen or sophomores?

5. In Exercise 4, let X_3 denote the number of juniors in the random sample of 15 students, and let X_4 denote the number of seniors in the sample. Find the value of $E(X_3 - X_4)$ and the value of $\text{Var}(X_3 - X_4)$.

6. Suppose that the random variables $X_1, \ldots, X_k$ are independent, and that X_i has a Poisson distribution with mean λ_i ($i = 1, \ldots, k$). Show that for any fixed positive integer n, the conditional distribution of the random vector $X = (X_1, \ldots, X_k)$, given that $\sum_{i=1}^{k} X_i = n$, is a multinomial distribution with parameters n and $p = (p_1, \ldots, p_k)$, where

$$p_i = \frac{\lambda_i}{\sum_{j=1}^{k} \lambda_j} \qquad \text{for } i = 1, \ldots, k.$$

5.12. THE BIVARIATE NORMAL DISTRIBUTION

Definition of the Bivariate Normal Distribution

Suppose that Z_1 and Z_2 are independent random variables each of which has a standard normal distribution. Then the joint p.d.f. $g(z_1, z_2)$ of Z_1 and Z_2 is specified for any values of z_1 and z_2 by the equation

$$g(z_1, z_2) = \frac{1}{2\pi} \exp\left[-\frac{1}{2}(z_1^2 + z_2^2) \right]. \tag{1}$$

For any constants μ_1, μ_2, σ_1, σ_2, and ρ such that $-\infty < \mu_i < \infty$ ($i = 1, 2$), $\sigma_i > 0$ ($i = 1, 2$), and $-1 < \rho < 1$, we shall now define two new random variables

X_1 and X_2 as follows:

$$X_1 = \sigma_1 Z_1 + \mu_1,$$

$$X_2 = \sigma_2 \left[\rho Z_1 + (1 - \rho^2)^{1/2} Z_2 \right] + \mu_2. \tag{2}$$

We shall derive the joint p.d.f. $f(x_1, x_2)$ of X_1 and X_2.

The transformation from Z_1 and Z_2 to X_1 and X_2 is a linear transformation; and it will be found that the determinant Δ of the matrix of coefficients of Z_1 and Z_2 has the value $\Delta = (1 - \rho^2)^{1/2} \sigma_1 \sigma_2$. Therefore, as discussed in Sec. 3.9, the Jacobian J of the inverse transformation from X_1 and X_2 to Z_1 and Z_2 is

$$J = \frac{1}{\Delta} = \frac{1}{(1 - \rho^2)^{1/2} \sigma_1 \sigma_2}. \tag{3}$$

Since $J > 0$, the value of $|J|$ is equal to the value of J itself. If the relations (2) are solved for Z_1 and Z_2 in terms of X_1 and X_2, then the joint p.d.f. $f(x_1, x_2)$ can be obtained by replacing z_1 and z_2 in Eq. (1) by their expressions in terms of x_1 and x_2, and then multiplying by $|J|$. It can be shown that the result is, for $-\infty < x_1 < \infty$ and $-\infty < x_2 < \infty$,

$$f(x_1, x_2) = \frac{1}{2\pi(1 - \rho^2)^{1/2} \sigma_1 \sigma_2} \exp \left\{ - \frac{1}{2(1 - \rho^2)} \left[\left(\frac{x_1 - \mu_1}{\sigma_1} \right)^2 \right. \right.$$

$$- 2\rho \left(\frac{x_1 - \mu_1}{\sigma_1} \right) \left(\frac{x_2 - \mu_2}{\sigma_2} \right) \tag{4}$$

$$\left. \left. + \left(\frac{x_2 - \mu_2}{\sigma_2} \right)^2 \right] \right\}.$$

When the joint p.d.f. of two random variables X_1 and X_2 is of the form in Eq. (4), it is said that X_1 and X_2 have a *bivariate normal distribution*. The means and the variances of the bivariate normal distribution specified by Eq. (4) are easily derived from the definitions in Eq. (2). Since Z_1 and Z_2 are independent and each has mean 0 and variance 1, it follows that $E(X_1) = \mu_1$, $E(X_2) = \mu_2$, $\text{Var}(X_1) = \sigma_1^2$, and $\text{Var}(X_2) = \sigma_2^2$. Furthermore, it can be shown by using Eq. (2) that $\text{Cov}(X_1, X_2) = \rho \sigma_1 \sigma_2$. Therefore, the correlation of X_1 and X_2 is simply ρ. In summary, if X_1 and X_2 have a bivariate normal distribution for which the p.d.f. is specified by Eq. (4), then

$$E(X_i) = \mu_i \quad \text{and} \quad \text{Var}(X_i) = \sigma_i^2 \quad \text{for } i = 1, 2.$$

Also,

$$\rho(X_1, X_2) = \rho.$$

It has been convenient for us to introduce the bivariate normal distribution as the joint distribution of certain linear combinations of independent random variables having standard normal distributions. It should be emphasized, however, that the bivariate normal distribution arises directly and naturally in many practical problems. For example, for many populations, the joint distribution of two physical characteristics such as the heights and the weights of the individuals in the population will be approximately a bivariate normal distribution. For other populations, the joint distribution of the scores of the individuals in the population on two related tests will be approximately a bivariate normal distribution.

Marginal and Conditional Distributions

Marginal Distributions. We shall continue to assume that the random variables X_1 and X_2 have a bivariate normal distribution and their joint p.d.f. is specified by Eq. (4). In the study of the properties of this distribution, it will be convenient to represent X_1 and X_2 as in Eq. (2), where Z_1 and Z_2 are independent random variables with standard normal distributions. In particular, since both X_1 and X_2 are linear combinations of Z_1 and Z_2, it follows from this representation and from Corollary 1 of Sec. 5.6 that the marginal distributions of both X_1 and X_2 are also normal distributions. Thus, for $i = 1, 2$, the marginal distribution of X_i is a normal distribution with mean μ_i and variance σ_i^2.

Independence and Correlation. If X_1 and X_2 are uncorrelated, then $\rho = 0$. In this case, it can be seen from Eq. (4) that the joint p.d.f. $f(x_1, x_2)$ factors into the product of the marginal p.d.f. of X_1 and the marginal p.d.f. of X_2. Hence, X_1 and X_2 are independent, and the following result has been established:

   Two random variables X_1 and X_2 that have a bivariate normal distribution are independent if and only if they are uncorrelated.

   We have already seen in Sec. 4.6 that two random variables X_1 and X_2 with an arbitrary joint distribution can be uncorrelated without being independent.

Conditional Distributions. The conditional distribution of X_2 given that $X_1 = x_1$ can also be derived from the representation in Eq. (2). If $X_1 = x_1$, then $Z_1 = (x_1 - \mu_1)/\sigma_1$. Therefore, the conditional distribution of X_2 given that $X_1 = x_1$ is the same as the conditional distribution of

$$\left(1 - \rho^2\right)^{1/2} \sigma_2 Z_2 + \mu_2 + \rho\sigma_2\left(\frac{x_1 - \mu_1}{\sigma_1}\right). \tag{5}$$

Since Z_2 has a standard normal distribution and is independent of X_1, it follows from (5) that the conditional distribution of X_2 given that $X_1 = x_1$ is a normal distribution for which the mean is

$$E(X_2 \mid x_1) = \mu_2 + \rho\sigma_2\left(\frac{x_1 - \mu_1}{\sigma_1}\right) \tag{6}$$

and the variance is $(1 - \rho^2)\sigma_2^2$.

The conditional distribution of X_1 given that $X_2 = x_2$ cannot be derived so easily from Eq. (2) because of the different ways in which Z_1 and Z_2 enter Eq. (2). However, it is seen from Eq. (4) that the joint p.d.f. $f(x_1, x_2)$ is symmetric in the two variables $(x_1 - \mu_1)/\sigma_1$ and $(x_2 - \mu_2)/\sigma_2$. Therefore, it follows that the conditional distribution of X_1 given that $X_2 = x_2$ can be found from the conditional distribution of X_2 given that $X_1 = x_1$ (this distribution has just been derived) simply by interchanging x_1 and x_2, interchanging μ_1 and μ_2, and interchanging σ_1 and σ_2. Thus, the conditional distribution of X_1 given that $X_2 = x_2$ must be a normal distribution for which the mean is

$$E(X_1 \mid x_2) = \mu_1 + \rho\sigma_1\left(\frac{x_2 - \mu_2}{\sigma_2}\right) \tag{7}$$

and the variance is $(1 - \rho^2)\sigma_1^2$.

We have now shown that each marginal distribution and each conditional distribution of a bivariate normal distribution is a univariate normal distribution.

Some particular features of the conditional distribution of X_2 given that $X_1 = x_1$ should be noted. If $\rho \neq 0$, then $E(X_2 \mid x_1)$ is a linear function of the given value x_1. If $\rho > 0$, the slope of this linear function is positive. If $\rho < 0$, the slope of the function is negative. However, the variance of the conditional distribution of X_2 given that $X_1 = x_1$ is $(1 - \rho^2)\sigma_2^2$, and its value does not depend on the given value x_1. Furthermore, this variance of the conditional distribution of X_2 is smaller than the variance σ_2^2 of the marginal distribution of X_2.

Example 1: Predicting a Person's Weight. Let X_1 denote the height of a person selected at random from a certain population, and let X_2 denote the weight of the person. Suppose that these random variables have a bivariate normal distribution for which the p.d.f. is specified by Eq. (4) and that the person's weight X_2 must be predicted. We shall compare the smallest M.S.E. that can be attained if the person's height X_1 is known when his weight must be predicted with the smallest M.S.E. that can be attained if his height is not known.

If the person's height is not known, then the best prediction of his weight is the mean $E(X_2) = \mu_2$; and the M.S.E. of this prediction is the variance σ_2^2. If it is known that the person's height is x_1, then the best prediction is the mean $E(X_2 \mid x_1)$ of the conditional distribution of X_2 given that $X_1 = x_1$; and the M.S.E. of this prediction is the variance $(1 - \rho^2)\sigma_2^2$ of that conditional distribution. Hence, when the value of X_1 is known, the M.S.E. is reduced from σ_2^2 to $(1 - \rho^2)\sigma_2^2$. $\square$

Since the variance of the conditional distribution in Example 1 is $(1 - \rho^2)\sigma_2^2$, regardless of the known height x_1 of the person, it follows that the difficulty of predicting the person's weight is the same for a tall person, a short person, or a person of medium height. Furthermore, since the variance $(1 - \rho^2)\sigma_2^2$ decreases as $|\rho|$ increases, it follows that it is easier to predict a person's weight from his height when the person is selected from a population in which height and weight are highly correlated.

Example 2: Determining a Marginal Distribution. Suppose that a random variable X has a normal distribution with mean μ and variance σ^2; and that for any number x, the conditional distribution of another random variable Y given that $X = x$ is a normal distribution with mean x and variance τ^2. We shall determine the marginal distribution of Y.

We know that the marginal distribution of X is a normal distribution and that the conditional distribution of Y given that $X = x$ is a normal distribution for which the mean is a linear function of x and the variance is constant. It follows that the joint distribution of X and Y must be a bivariate normal distribution. Hence, the marginal distribution of Y is also a normal distribution. The mean and the variance of Y must be determined.

The mean of Y is

$$E(Y) = E[E(Y|X)] = E(X) = \mu.$$

Furthermore, by Exercise 10 at the end of Sec. 4.7,

$$\text{Var}(Y) = E[\text{Var}(Y|X)] + \text{Var}[E(Y|X)]$$

$$= E(\tau^2) + \text{Var}(X)$$

$$= \tau^2 + \sigma^2.$$

Hence, the distribution of Y is a normal distribution with mean μ and variance $\tau^2 + \sigma^2$. $\square$

Linear Combinations

Suppose again that two random variables X_1 and X_2 have a bivariate normal distribution for which the p.d.f. is specified by Eq. (4). Now consider the random variable $Y = a_1 X_1 + a_2 X_2 + b$, where a_1, a_2, and b are arbitrary given constants. Both X_1 and X_2 can be represented, as in Eq. (2), as linear combinations of independent and normally distributed random variables Z_1 and Z_2. Since Y is a linear combination of X_1 and X_2, it follows that Y can also be represented as a linear combination of Z_1 and Z_2. Therefore, by Corollary 1 of Sec. 5.6, the distribution of Y will also be a normal distribution. Thus, the following important property has been established.

If two random variables X_1 and X_2 have a bivariate normal distribution, then any linear combination $Y = a_1 X_1 + a_2 X_2 + b$ will have a normal distribution.

The mean and variance of Y are as follows:

$$E(Y) = a_1 E(X_1) + a_2 E(X_2) + b$$

$$= a_1 \mu_1 + a_2 \mu_2 + b$$

and

$$\text{Var}(Y) = a_1^2 \text{Var}(X_1) + a_2^2 \text{Var}(X_2) + 2a_1 a_2 \text{Cov}(X_1, X_2)$$

$$= a_1^2 \sigma_1^2 + a_2^2 \sigma_2^2 + 2a_1 a_2 \rho \sigma_1 \sigma_2.$$

Example 3: Heights of Husbands and Wives. Suppose that a married couple is selected at random from a certain population of married couples, and that the joint distribution of the height of the wife and the height of her husband is a bivariate normal distribution. Suppose that the heights of the wives have a mean of 66.8 inches and a standard deviation of 2 inches; that the heights of the husbands have a mean of 70 inches and a standard deviation of 2 inches; and that the correlation between these two heights is 0.68. We shall determine the probability that the wife will be taller than her husband.

If we let X denote the height of the wife and let Y denote the height of her husband, then we must determine the value of $\text{Pr}(X - Y > 0)$. Since X and Y have a bivariate normal distribution, it follows that the distribution of $X - Y$ will be a normal distribution for which the mean is

$$E(X - Y) = 66.8 - 70 = -3.2$$

and the variance is

$$\text{Var}(X - Y) = \text{Var}(X) + \text{Var}(Y) - 2\,\text{Cov}(X, Y)$$

$$= 4 + 4 - 2(0.68)(2)(2) = 2.56.$$

Hence the standard deviation of $X - Y$ is 1.6.

The random variable $Z = (X - Y + 3.2)/(1.6)$ will have a standard normal distribution. It can be found from the table given at the end of this book that

$$\text{Pr}(X - Y > 0) = \text{Pr}(Z > 2) = 1 - \Phi(2)$$

$$= 0.0227.$$

Therefore, the probability that the wife will be taller than her husband is 0.0227.

□

EXERCISES

1. Suppose that two different tests A and B are to be given to a student chosen at random from a certain population. Suppose also that the mean score on test A is 85 and the standard deviation is 10; that the mean score on test B is 90 and the standard deviation is 16; that the scores on the two tests have a bivariate normal distribution; and that the correlation of the two scores is 0.8. If the student's score on test A is 80, what is the probability that his score on test B will be higher than 90?

2. Consider again the two tests A and B described in Exercise 1. If a student is chosen at random, what is the probability that the sum of his scores on the two tests will be greater than 200?

3. Consider again the two tests A and B described in Exercise 1. If a student is chosen at random, what is the probability that his score on test A will be higher than his score on test B?

4. Consider again the two tests A and B described in Exercise 1. If a student is chosen at random and his score on test B is 100 what predicted value of his score on test A has the smallest M.S.E. and what is the value of this minimum M.S.E.?

5. Suppose that the random variables X_1 and X_2 have a bivariate normal distribution for which the joint p.d.f. is specified by Eq. (4). Determine the value of the constant b for which $\text{Var}(X_1 + bX_2)$ will be a minimum.

6. Suppose that X_1 and X_2 have a bivariate normal distribution for which $E(X_1 | X_2) = 3.7 - 0.15X_2$, $E(X_2 | X_1) = 0.4 - 0.6X_1$, and $\text{Var}(X_2 | X_1) = 3.64$. Find the mean and the variance of X_1, the mean and the variance of X_2, and the correlation of X_1 and X_2.

7. Let $f(x_1, x_2)$ denote the p.d.f. of the bivariate normal distribution specified by Eq. (4). Show that the maximum value of $f(x_1, x_2)$ is attained at the point at which $x_1 = \mu_1$ and $x_2 = \mu_2$.

8. Let $f(x_1, x_2)$ denote the p.d.f. of the bivariate normal distribution specified by Eq. (4), and let k be a constant such that

$$0 < k < \frac{1}{2\pi(1 - \rho^2)^{1/2}\sigma_1\sigma_2}.$$

Show that the points (x_1, x_2) such that $f(x_1, x_2) = k$ lie on a circle if $\rho = 0$ and $\sigma_1 = \sigma_2$, and that these points lie on an ellipse otherwise.

9. Suppose that two random variables X_1 and X_2 have a bivariate normal distribution, and that two other random variables Y_1 and Y_2 are defined as follows:

$$Y_1 = a_{11}X_1 + a_{12}X_2 + b_1,$$

$$Y_2 = a_{21}X_1 + a_{22}X_2 + b_2,$$

where

$$\begin{vmatrix} a_{11} & a_{12} \\ a_{21} & a_{22} \end{vmatrix} \neq 0.$$

Show that Y_1 and Y_2 also have a bivariate normal distribution.

10. Suppose that two random variables X_1 and X_2 have a bivariate normal distribution and that $\text{Var}(X_1) = \text{Var}(X_2)$. Show that the sum $X_1 + X_2$ and the difference $X_1 - X_2$ are independent random variables.

5.13. SUPPLEMENTARY EXERCISES

1. Suppose that X, Y, and Z are i.i.d. random variables, and that each has a standard normal distribution. Evaluate $\text{Pr}(3X + 2Y < 6Z - 7)$.

2. Suppose that X and Y are independent Poisson random variables such that $\text{Var}(X) + \text{Var}(Y) = 5$. Evaluate $\text{Pr}(X + Y < 2)$.

3. Suppose that X has a normal distribution such that $\text{Pr}(X < 116) = 0.20$ and $\text{Pr}(X < 328) = 0.90$. Determine the mean and the variance of X.

4. Suppose that a random sample of four observations is drawn from a Poisson distribution with mean λ, and let $\overline{X}$ denote the sample mean. Show that

$$\text{Pr}\left(\overline{X} < \frac{1}{2}\right) = (4\lambda + 1)e^{-4\lambda}.$$

5. The lifetime X of an electronic component has an exponential distribution such that $\text{Pr}(X \leqslant 1000) = 0.75$. What is the expected lifetime of the component?

6. Suppose that X has a normal distribution with mean μ and variance σ^2. Express $E(X^3)$ in terms of μ and σ^2.

7. Suppose that a random sample of 16 observations is drawn from a normal distribution with mean μ and standard deviation 12; and that independently another random sample of 25 observations is drawn from a normal distribu-

tion with the same mean μ and standard deviation 20. Let $\overline{X}$ and $\overline{Y}$ denote the sample means of the two samples. Evaluate $\Pr(|\overline{X} - \overline{Y}| < 5)$.

8. Suppose that men arrive at a ticket counter according to a Poisson process at the rate of 120 per hour, and that women arrive according to an independent Poisson process at the rate of 60 per hour. Determine the probability that four or fewer people arrive in a one-minute period.

9. Suppose that $X_1, X_2, \ldots$ are i.i.d. random variables, each of which has m.g.f. $\psi(t)$. Let $Y = X_1 + \cdots + X_N$, where the number of terms N in this sum is a random variable having a Poisson distribution with mean λ. Assume that N and $X_1, X_2, \ldots$ are independent, and that $Y = 0$ if $N = 0$. Determine the m.g.f. of Y.

10. Every Sunday morning two children, Craig and Jill, independently try to launch their model airplanes. On each Sunday Craig has probability $1/3$ of a successful launch, and Jill has probability $1/5$ of a successful launch. Determine the expected number of Sundays required until at least one of the two children has a successful launch.

11. Suppose that a fair coin is tossed until at least one head and at least one tail have been obtained. Let X denote the number of tosses that are required. Find the p.f. of X.

12. Suppose that a pair of balanced dice are rolled 120 times; and let X denote the number of rolls on which the sum of the two numbers is 7. Use the central limit theorem to determine a value of k such that $\Pr(|X - 20| \leqslant k)$ is approximately 0.95.

13. Suppose that $X_1, \ldots, X_n$ form a random sample from the uniform distribution on the interval $(0, 1)$. Let $Y_1 = \min\{X_1, \ldots, X_n\}$, $Y_n = \max\{X_1, \ldots, X_n\}$, and $W = Y_n - Y_1$. Show that each of the random variables Y_1, Y_n, and W has a beta distribution.

14. Suppose that events occur in accordance with a Poisson process at the rate of five events per hour.

    (a) Determine the distribution of the waiting time T_1 until the first event occurs.

    (b) Determine the distribution of the total waiting time T_k until k events have occurred.

    (c) Determine the probability that none of the first k events will occur within 20 minutes of one another.

15. Suppose that five components are functioning simultaneously; that the lifetimes of the components are i.i.d.; and that each lifetime has an exponential distribution with parameter β. Let T_1 denote the time from the beginning of the process until one of the components fails; and let T_5 denote the total time until all five components have failed. Evaluate $\text{Cov}(T_1, T_5)$.

16. Suppose that X_1 and X_2 are independent random variables, and that X_i has an exponential distribution with parameter β_i $(i = 1, 2)$. Show that for any constant $k > 0$,

$$\Pr(X_1 > kX_2) = \frac{\beta_2}{k\beta_1 + \beta_2}.$$

17. Suppose that 15,000 people in a city with a population of 500,000 are watching a certain television program. If 200 people in the city are contacted at random, what is the probability, approximately, that fewer than four of them are watching the program?

18. Suppose that it is desired to estimate the proportion of persons in a large population who have a certain characteristic. A random sample of 100 persons is selected from the population without replacement, and the proportion $\bar{X}$ of persons in the sample who have the characteristic is observed. Show that, no matter how large the population is, the standard deviation of $\bar{X}$ is at most 0.05.

19. Suppose that X has a binomial distribution with parameters n and p; and that Y has a negative binomial distribution with parameters r and p, where r is a positive integer. Show that $\Pr(X < r) = \Pr(Y > n - r)$ by showing that both the left side and the right side of this equation can be regarded as the probability of the same event in a sequence of Bernoulli trials with probability p of success.

20. Suppose that X has a Poisson distribution with mean λt; and that Y has a gamma distribution with parameters $\alpha = k$ and $\beta = \lambda$, where k is a positive integer. Show that $\Pr(X \geq k) = \Pr(Y \leq t)$ by showing that both the left side and the right side of this equation can be regarded as the probability of the same event in a Poisson process in which the expected number of occurrences per unit time is λ.

21. Suppose that X has a Poisson distribution with a very large mean λ. Explain why the distribution of X can be approximated by a normal distribution with mean λ and variance λ. In other words, explain why $(X - \lambda)/\lambda^{1/2}$ converges in distribution, as $\lambda \to \infty$, to a random variable having a standard normal distribution.

22. Suppose that X has a Poisson distribution with mean 10. Use the central limit theorem, both without and with the correction for continuity, to determine an approximate value for $\Pr(8 \leq X \leq 12)$. Use the table of Poisson probabilities given in the back of this book to assess the quality of these approximations.

23. Suppose that X is a random variable having a continuous distribution with p.d.f. $f(x)$ and d.f. $F(x)$, and for which $\Pr(X > 0) = 1$. Let the failure rate

$h(x)$ be as defined in Exercise 17 of Sec. 5.9. Show that

$$\exp\left[-\int_0^x h(t)\,dt\right] = 1 - F(x).$$

24. Suppose that 40 percent of the students in a large population are freshmen, 30 percent are sophomores, 20 percent are juniors, and 10 percent are seniors. Suppose that 10 students are selected at random from the population; and let X_1, X_2, X_3, X_4 denote, respectively, the numbers of freshmen, sophomores, juniors, and seniors that are obtained.

    (a) Determine $\rho(X_i, X_j)$ for each pair of values i and $j(i < j)$.

    (b) For what values of i and $j(i < j)$ is $\rho(X_i, X_j)$ most negative?

    (c) For what values of i and $j(i < j)$ is $\rho(X_i, X_j)$ closest to 0?

25. Suppose that X_1 and X_2 have a bivariate normal distribution with means μ_1 and μ_2, variances σ_1^2 and σ_2^2, and correlation ρ. Determine the distribution of $X_1 - 3X_2$.

26. Suppose that X has a standard normal distribution, and that the conditional distribution of Y given X is a normal distribution with mean $2X - 3$ and variance 12. Determine the marginal distribution of Y and the value of $\rho(X, Y)$.

27. Suppose that X_1 and X_2 have a bivariate normal distribution with $E(X_2) = 0$. Evaluate $E(X_1^2 X_2)$.

Estimation 6

6.1. STATISTICAL INFERENCE

Nature of Statistical Inference

In the first five chapters of this book we discussed the theory and methods of probability. In the last five chapters we shall discuss the theory and methods of *statistical inference*. A problem of statistical inference or, more simply, a statistics problem is a problem in which data that have been generated in accordance with some unknown probability distribution must be analyzed and some type of inference about the unknown distribution must be made. In other words, in a statistics problem there are two or more probability distributions which might have generated some experimental data. In most real problems, there are an infinite number of different possible distributions which might have generated the data. By analyzing the data, we attempt to learn about the unknown distribution, to make some inferences about certain properties of the distribution, and to determine the relative likelihood that each possible distribution is actually the correct one.

Parameters

In many statistics problems, the probability distribution that generated the experimental data is completely known except for the values of one or more parameters. For example, it might be known that the length of life of a certain type of nuclear pacemaker has an exponential distribution with parameter β, as defined in Sec. 5.9, but the exact value of β might be unknown. If the lifetimes of several pacemakers of this type can be observed, then from these observed values and any other relevant information that might be available, it is possible to make an inference about the unknown value of the parameter β. For example, we might wish to give our best estimate of the value of β, or to specify an interval in which we think the value of β is likely to lie, or to decide whether or not β is smaller than some specified value. It is typically not possible to determine the value of β exactly.

As another example, suppose that the distribution of the heights of the individuals in a certain population is known to be a normal distribution with mean μ and variance σ^2, but that the exact values of μ and σ^2 are unknown. If we can observe the heights of the individuals in a random sample selected from the given population, then from these observed heights and any other information we might have about the distribution of heights, we can make an inference about the values of μ and σ^2.

In a problem of statistical inference, any characteristic of the distribution generating the experimental data which has an unknown value, such as the mean μ or the variance σ^2 in the example just presented, is called a *parameter* of the distribution. The set Ω of all possible values of a parameter θ or of a vector of parameters $(\theta_1, \ldots, \theta_k)$ is called the *parameter space*.

In the first example we presented, the parameter β of the exponential distribution must be positive. Therefore, unless certain positive values of β can be explicitly ruled out as possible values of β, the parameter space Ω will be the set of all positive numbers. In the second example we presented, the mean μ and the variance σ^2 of the normal distribution can be regarded as a pair of parameters. Here the value of μ can be any real number and σ^2 must be positive. Therefore, the parameter space Ω can be taken as the set of all pairs (μ, σ^2) such that $-\infty < \mu < \infty$ and $\sigma^2 > 0$. More specifically, if the normal distribution in this example represents the distribution of the heights in inches of the individuals in some particular population, we might be certain that $30 < \mu < 100$ and $\sigma^2 < 50$. In this case, the parameter space Ω could be taken as the smaller set of all pairs (μ, σ^2) such that $30 < \mu < 100$ and $0 < \sigma^2 < 50$.

The important feature of the parameter space Ω is that it must contain all possible values of the parameters in a given problem, in order that we can be certain that the actual value of the vector of parameters is a point in Ω.

Statistical Decision Problems

In many statistics problems, after the experimental data have been analyzed, we must choose a decision from some available class of decisions with the property that the consequences of each available decision depend on the unknown value of some parameter. For example, we might have to estimate the unknown value of a parameter θ when the consequences depend on how close our estimate is to the correct value θ. As another example, we might have to decide whether the unknown value of θ is larger or smaller than some specified constant when the consequences depend on whether the decision is right or wrong.

Experimental Design

In some statistics problems, we have some control over the type or the amount of experimental data that will be collected. For example, consider an experiment to determine the mean tensile strength of a certain type of alloy as a function of the pressure and temperature at which the alloy is produced. Within the limits of certain budgetary and time constraints, it may be possible for the experimenter to choose the levels of pressure and temperature at which experimental specimens of the alloy are to be produced, and also to specify the number of specimens to be produced at each of these levels.

Such a problem, in which the experimenter can choose (at least to some extent) the particular experiment that is to be carried out, is called a problem of *experimental design*. Of course, the design of an experiment and the statistical analysis of the experimental data are closely related. One cannot design an effective experiment without considering the subsequent statistical analysis that is to be carried out on the data that will be obtained; and one cannot carry out a meaningful statistical analysis of experimental data without considering the particular type of experiment from which the data were derived.

References

In the remainder of this book we shall consider many different problems of statistical inference, statistical decision, and experimental design. Some books that discuss statistical theory and methods at about the same level as they will be discussed in this book were mentioned at the end of Sec. 1.1. Some statistics books which are written at a more advanced level are Cramér (1946), Rao (1973), Zacks (1971) and (1981), DeGroot (1970), Ferguson (1967), Lehmann (1959, 1983), Bickel and Doksum (1977), and Rohatgi (1976).

6.2. PRIOR AND POSTERIOR DISTRIBUTIONS

The Prior Distribution

Specifying a Prior Distribution. Consider a problem of statistical inference in which observations are to be taken from a distribution for which the p.d.f. or the p.f. is $f(x \mid \theta)$, where θ is a parameter having an unknown value. It is assumed that the unknown value of θ must lie in a specified parameter space Ω. The problem of statistical inference introduced in Sec. 6.1 can be roughly described as the problem of trying to determine, on the basis of observations from the p.d.f. or p.f. $f(x | \theta)$, where in the parameter space Ω the actual value of θ is likely to lie.

In many problems, before any observations from $f(x | \theta)$ are available, the experimenter or statistician will be able to summarize his previous information and knowledge about where in Ω the value of θ is likely to lie by constructing a probability distribution for θ on the set Ω. In other words, before any experimental data have been collected or observed, the experimenter's past experience and knowledge will lead him to believe that θ is more likely to lie in certain regions of Ω than in others. We shall assume that the relative likelihoods of the different regions can be expressed in terms of a probability distribution on Ω. This distribution is called the *prior distribution* of θ because it represents the relative likelihood that the true value of θ lies in each of various regions of Ω prior to the observation of any values from $f(x | \theta)$.

Controversial Nature of Prior Distributions.   The concept of a prior distribution is very controversial in statistics. This controversy is closely related to the controversy in regard to the meaning of probability, which was discussed in Sec. 1.2. Some statisticians believe that a prior distribution can be chosen for the parameter θ in every statistics problem. They believe that this distribution is a subjective probability distribution in the sense that it represents an individual experimenter's information and subjective beliefs about where the true value of θ is likely to lie. They also believe, however, that a prior distribution is no different from any other probability distribution used in the field of statistics, and that all the rules of probability theory apply to a prior distribution. It is said that these statisticians adhere to the Bayesian philosophy of statistics.

Other statisticians believe that in many problems it is not appropriate to speak of a probability distribution of θ because the true value of θ is not a random variable at all but is rather a certain fixed number whose value happens to be unknown to the experimenter. These statisticians believe that a prior distribution can be assigned to a parameter θ only when there is extensive previous information about the relative frequencies with which θ has taken each of its possible values in the past. It would then be possible for two different scientists to agree on the correct prior distribution to be used. For example, suppose that the proportion θ of defective items in a certain large manufactured lot is unknown. Suppose also that the same manufacturer has produced many such lots of items in the past and that detailed records have been kept about the proportions of defective items in past lots. The relative frequencies for past lots could then be used to construct a prior distribution for θ.

Both groups of statisticians agree that whenever a meaningful prior distribution can be chosen, the theory and methods to be described in this section are applicable and useful. In this section and Secs. 6.3 and 6.4, we shall proceed under the assumption that we can assign to θ a prior distribution which represents the probabilities that the unknown value of θ lies in various subsets of the parameter space. Beginning in Sec. 6.5, we shall consider techniques of estimation that are not based on the assignment of a prior distribution.

Discrete and Continuous Prior Distributions. In some problems, the parameter θ can take only a finite number of different values or, at most, an infinite sequence of different values. The prior distribution of θ will therefore be a discrete distribution. The p.f. $\xi(\theta)$ of this distribution is called the *prior p.f.* of θ. In other problems, the parameter θ can take any value on the real line or in some interval of the real line, and a continuous prior distribution is assigned to θ. The p.d.f. $\xi(\theta)$ of this distribution is called the *prior p.d.f.* of θ.

Example 1: Fair or Two-Headed Coin. Let θ denote the probability of obtaining a head when a certain coin is tossed; and suppose that it is known that the coin either is fair or has a head on each side. Therefore, the only possible values of θ are $\theta = 1/2$ and $\theta = 1$. If the prior probability that the coin is fair is p, then the prior p.f. of θ is $\xi(1/2) = p$ and $\xi(1) = 1 - p$. $\square$

Example 2: Proportion of Defective Items. Suppose that the proportion θ of defective items in a large manufactured lot is unknown, and that the prior distribution assigned to θ is a uniform distribution on the interval $(0, 1)$. Then the prior p.d.f. of θ is

$$\xi(\theta) = \begin{cases} 1 & \text{for } 0 < \theta < 1, \\ 0 & \text{otherwise.} \end{cases} \qquad \square \tag{1}$$

Example 3: Parameter of an Exponential Distribution. Suppose that the lifetimes of fluorescent lamps of a certain type are to be observed, and that the distribution of the lifetime of any particular lamp is an exponential distribution with parameter β, as defined in Sec. 5.9. Suppose also that the exact value of β is unknown and on the basis of previous experience the prior distribution of β is taken as a gamma distribution for which the mean is 0.0002 and the standard deviation is 0.0001. We shall determine the prior p.d.f. of β.

Suppose that the prior distribution of β is a gamma distribution with parameters α_0 and β_0. It was shown in Sec. 5.9 that the mean of this distribution is α_0/β_0 and the variance is α_0/β_0^2. Therefore, $\alpha_0/\beta_0 = 0.0002$ and $\alpha_0^{1/2}/\beta_0 = 0.0001$. It can now be found that $\alpha_0 = 4$ and $\beta_0 = 20,000$. It follows from Eq. (7) of Sec. 5.9 that the prior p.d.f. of β for $\beta > 0$ is as follows:

$$\xi(\beta) = \frac{(20,000)^4}{3!}\beta^3 e^{-20,000\beta}. \tag{2}$$

Also, $\xi(\beta) = 0$ for $\beta \leqslant 0$. $\square$

The Posterior Distribution

Suppose now that the n random variables $X_1, \ldots, X_n$ form a random sample from a distribution for which the p.d.f. or the p.f. is $f(x \mid \theta)$. Suppose also that the value of the parameter θ is unknown and the prior p.d.f. or prior p.f. of θ is $\xi(\theta)$. For simplicity, we shall assume that the parameter space Ω is either an interval of the real line or the entire real line; that $\xi(\theta)$ is a prior p.d.f. on Ω, rather than a prior p.f.; and that $f(x \mid \theta)$ is a p.d.f., rather than a p.f. However, the discussion that will be given here can be easily adapted for a problem in which $\xi(\theta)$ or $f(x \mid \theta)$ is a p.f.

Since the random variables $X_1, \ldots, X_n$ form a random sample from the distribution for which the p.d.f. is $f(x \mid \theta)$, it follows from Sec. 3.7 that their joint p.d.f. $f_n(x_1, \ldots, x_n \mid \theta)$ will be given by the equation

$$f_n(x_1, \ldots, x_n \mid \theta) = f(x_1 \mid \theta) \cdots f(x_n \mid \theta). \tag{3}$$

If we use the vector notation $x = (x_1, \ldots, x_n)$, then the joint p.d.f. in Eq. (3) can be written simply as $f_n(x \mid \theta)$.

Since the parameter θ itself is now regarded as having a distribution for which the p.d.f. is $\xi(\theta)$, the joint p.d.f. $f_n(x \mid \theta)$ should properly be regarded as the conditional joint p.d.f. of $X_1, \ldots, X_n$ for a given value of θ. If we multiply this conditional joint p.d.f. by the p.d.f. $\xi(\theta)$, we obtain the $(n + 1)$-dimensional joint p.d.f. of $X_1, \ldots, X_n$ and θ in the form $f_n(x \mid \theta)\xi(\theta)$. The marginal joint p.d.f. of $X_1, \ldots, X_n$ can now be obtained by integrating this joint p.d.f. over all values of θ. Therefore, the n-dimensional marginal joint p.d.f. $g_n(x)$ of $X_1, \ldots, X_n$ can be written in the form

$$g_n(x) = \int_\Omega f_n(x \mid \theta)\xi(\theta)\,d\theta. \tag{4}$$

Furthermore, the conditional p.d.f. of θ given that $X_1 = x_1, \ldots, X_n = x_n$, which we shall denote by $\xi(\theta \mid x)$, must be equal to the joint p.d.f. of $X_1, \ldots, X_n$ and θ divided by the marginal joint p.d.f. of $X_1, \ldots, X_n$. Thus, we have

$$\xi(\theta \mid x) = \frac{f_n(x \mid \theta)\xi(\theta)}{g_n(x)} \qquad \text{for } \theta \in \Omega. \tag{5}$$

The probability distribution over Ω represented by the conditional p.d.f. in Eq. (5) is called the *posterior distribution* of θ because it is the distribution of θ after the values of $X_1, \ldots, X_n$ have been observed. Similarly, the conditional p.d.f. of θ in Eq. (5) is called the *posterior p.d.f.* of θ. We may say that a prior p.d.f. $\xi(\theta)$ represents the relative likelihood, before the values of $X_1, \ldots, X_n$ have been observed, that the true value of θ lies in each of various regions of Ω; and that

Posterior p.d.f. $\xi(\theta|x)$

the posterior p.d.f. $\xi(\theta \mid x)$ represents this relative likelihood after the values $X_1 = x_1, \ldots, X_n = x_n$ have been observed.

The Likelihood Function

The denominator on the right side of Eq. (5) is simply the integral of the numerator over all possible values of θ. Although the value of this integral depends on the observed values $x_1, \ldots, x_n$, it does not depend on θ and it may be treated as a constant when the right side of Eq. (5) is regarded as a p.d.f. of θ. We may therefore replace Eq. (5) with the following relation:

$$\xi(\theta \mid x) \propto f_n(x \mid \theta) \xi(\theta). \tag{6}$$

The proportionality symbol $\propto$ is used here to indicate that the left side is equal to the right side except possibly for a constant factor, the value of which may depend on the observed values $x_1, \ldots, x_n$ but does not depend on θ. The appropriate constant factor which will establish the equality of the two sides in the relation (6) can be determined at any time by using the fact that $\int_\Omega \xi(\theta \mid x) \, d\theta = 1$, because $\xi(\theta \mid x)$ is a p.d.f. of θ.

When the joint p.d.f. or the joint p.f. $f_n(x \mid \theta)$ of the observations in a random sample is regarded as a function of θ for given values of $x_1, \ldots, x_n$, it is called the *likelihood function*. In this terminology, the relation (6) states that the posterior p.d.f. of θ is proportional to the product of the likelihood function and the prior p.d.f. of θ.

By using the proportionality relation (6), it is often possible to determine the posterior p.d.f. of θ without explicitly performing the integration in Eq. (4). If we can recognize the right side of the relation (6) as being equal to one of the standard p.d.f.'s introduced in Chapter 5 or elsewhere in this book, except possibly for a constant factor, then we can easily determine the appropriate factor which will convert the right side of (6) into a proper p.d.f. of θ. We shall illustrate these ideas by considering again Examples 2 and 3.

Example 4: Proportion of Defective Items. Suppose again, as in Example 2, that the proportion θ of defective items in a large manufactured lot is unknown and that the prior distribution of θ is a uniform distribution on the interval $(0, 1)$. Suppose also that a random sample of n items is taken from the lot; and for $i = 1, \ldots, n$, let $X_i = 1$ if the ith item is defective and let $X_i = 0$ otherwise. Then $X_1, \ldots, X_n$ form n Bernoulli trials with parameter θ. We shall determine the posterior p.d.f. of θ.

It follows from Eq. (2) of Sec. 5.2 that the p.f. of each observation X_i is

$$f(x \mid \theta) = \begin{cases} \theta^x (1 - \theta)^{1-x} & \text{for } x = 0, 1, \\ 0 & \text{otherwise.} \end{cases} \tag{7}$$

UNDERSTAND USE OF BINOMIAL DISTRIBUTION

Hence, if we let $y = \sum_{i=1}^{n} x_i$, then the joint p.f. of $X_1, \ldots, X_n$ can be written in the following form for $x_i = 0$ or 1 ($i = 1, \ldots, n$):

$$f_n(x \mid \theta) = \theta^y (1 - \theta)^{n-y}. \tag{8}$$

Since the prior p.d.f. $\xi(\theta)$ is given by Eq. (1), it follows that for $0 < \theta < 1$,

$$f_n(x \mid \theta)\xi(\theta) = \theta^y (1 - \theta)^{n-y}. \tag{9}$$

UNDERSTAND THE TRICK

When we compare this expression with Eq. (1) of Sec. 5.10, we can see that, except for a constant factor, the right side of Eq. (9) has the same form as the p.d.f. of a beta distribution with parameters $\alpha = y + 1$ and $\beta = n - y + 1$. Since the posterior p.d.f. $\xi(\theta \mid x)$ is proportional to the right side of Eq. (9), it follows that $\xi(\theta \mid x)$ must be the p.d.f. of a beta distribution with parameters $\alpha = y + 1$ and $\beta = n - y + 1$. Therefore, for $0 < \theta < 1$,

$$\xi(\theta \mid x) = \frac{\Gamma(n + 2)}{\Gamma(y + 1)\Gamma(n - y + 1)} \theta^y (1 - \theta)^{n-y}. \quad \square \tag{10}$$

Example 5: Parameter of an Exponential Distribution. Suppose again, as in Example 3, that the distribution of the lifetimes of fluorescent lamps of a certain type is an exponential distribution with parameter β, and that the prior distribution of β is a particular gamma distribution for which the p.d.f. $\xi(\beta)$ is given by Eq. (2). Suppose also that the lifetimes $X_1, \ldots, X_n$ of a random sample of n lamps of this type are observed. We shall determine the posterior p.d.f. of β given that $X_1 = x_1, \ldots, X_n = x_n$.

By Eq. (10) of Sec. 5.9, the p.d.f. of each observation X_i is

$$f(x \mid \beta) = \begin{cases} \beta e^{-\beta x} & \text{for } x > 0, \\ 0 & \text{otherwise.} \end{cases} \tag{11}$$

Hence, if we let $y = \sum_{i=1}^{n} x_i$, then the joint p.d.f. of $X_1, \ldots, X_n$ can be written in the following form, for $x_i > 0$ ($i = 1, \ldots, n$):

$$f_n(x \mid \beta) = \beta^n e^{-\beta y}. \tag{12}$$

Since the prior p.d.f. $\xi(\beta)$ is given by Eq. (2), it follows that for $\beta > 0$,

$$f_n(x \mid \beta)\xi(\beta) \propto \beta^{n+3} e^{-(y + 20,000)\beta}. \tag{13}$$

A constant factor which does not involve β has been omitted from the right side of the relation (13).

When we compare this expression with Eq. (7) of Sec. 5.9, we can see that, except for a constant factor, it has the same form as the p.d.f. of a gamma distribution with parameters $n + 4$ and $y + 20,000$. Since the posterior p.d.f. $\xi(\beta \mid x)$ is proportional to $f_n(x \mid \beta)\xi(\beta)$, it follows that $\xi(\beta \mid x)$ must be the p.d.f. of a gamma distribution with parameters $n + 4$ and $y + 20,000$. Therefore, for $\beta > 0$,

$$\xi(\beta \mid x) = \frac{(y + 20,000)^{n+4}}{(n + 3)!} \beta^{n+3} e^{-(y+20,000)\beta}. \quad \Box \tag{14}$$

Sequential Observations

In many experiments, the observations $X_1, \ldots, X_n$ which form the random sample must be obtained sequentially, that is, one at a time. In such an experiment, the value of X_1 is observed first, the value of X_2 is observed next, the value of X_3 is then observed, and so on. Suppose that the prior p.d.f. of the parameter θ is $\xi(\theta)$. After the value x_1 of X_1 has been observed, the posterior p.d.f. $\xi(\theta \mid x_1)$ can be calculated in the usual way from the relation

$$\xi(\theta \mid x_1) \propto f(x_1 \mid \theta)\xi(\theta). \tag{15}$$

This p.d.f., in turn, serves as the prior p.d.f. of θ when the value of X_2 is to be observed. Thus, after the value x_2 of X_2 has been observed, the posterior p.d.f. $\xi(\theta \mid x_1, x_2)$ can be calculated from the relation

$$\xi(\theta \mid x_1, x_2) \propto f(x_2 \mid \theta)\xi(\theta \mid x_1). \tag{16}$$

We can continue in this way, calculating an updated posterior p.d.f. of θ after each observation and using that p.d.f. as the prior p.d.f. of θ for the next observation. The posterior p.d.f. $\xi(\theta \mid x_1, \ldots, x_{n-1})$ after the values $x_1, \ldots, x_{n-1}$ have been observed will ultimately be the prior p.d.f. of θ for the final observed value of X_n. The posterior p.d.f. after all n values $x_1, \ldots, x_n$ have been observed will therefore be specified by the relation

$$\xi(\theta \mid x) \propto f(x_n \mid \theta)\xi(\theta \mid x_1, \ldots, x_{n-1}). \tag{17}$$

Alternatively, after all n values $x_1, \ldots, x_n$ have been observed, we could calculate the posterior p.d.f. $\xi(\theta \mid x)$ in the usual way by combining the joint p.d.f. $f_n(x \mid \theta)$ with the original prior p.d.f. $\xi(\theta)$, as indicated in Eq. (5). It can be shown (see Exercise 7) that the posterior p.d.f. $\xi(\theta \mid x)$ will be the same regardless of whether it is calculated directly by using Eq. (5) or it is calculated sequentially by using Eqs. (15), (16), and (17). This property was illustrated in Sec. 2.2 for a coin that is known either to be fair or to have a head on each side. After each toss of the coin, the posterior probability that the coin is fair is updated.

EXERCISES

1. Suppose that the proportion θ of defective items in a large manufactured lot is known to be either 0.1 or 0.2, and that the prior p.f. of θ is as follows:

   $$\xi(0.1) = 0.7 \quad \text{and} \quad \xi(0.2) = 0.3.$$

   Suppose also that when eight items are selected at random from the lot, it is found that exactly two of them are defective. Determine the posterior p.f. of θ.

2. Suppose that the number of defects on a roll of magnetic recording tape has a Poisson distribution for which the mean λ is either 1.0 or 1.5, and that the prior p.f. of λ is as follows:

   $$\xi(1.0) = 0.4 \quad \text{and} \quad \xi(1.5) = 0.6.$$

   If a roll of tape selected at random is found to have three defects, what is the posterior p.f. of λ?

3. Suppose that the prior distribution of some parameter θ is a gamma distribution for which the mean is 10 and the variance is 5. Determine the prior p.d.f. of θ.

4. Suppose that the prior distribution of some parameter θ is a beta distribution for which the mean is $1/3$ and the variance is $1/45$. Determine the prior p.d.f. of θ.

5. Suppose that the proportion θ of defective items in a large manufactured lot is unknown, and that the prior distribution of θ is a uniform distribution on the interval $(0, 1)$. When eight items are selected at random from the lot, it is found that exactly three of them are defective. Determine the posterior distribution of θ.

6. Consider again the problem described in Exercise 5, but suppose now that the prior p.d.f. of θ is as follows:

   $$\xi(\theta) = \begin{cases} 2(1 - \theta) & \text{for } 0 < \theta < 1, \\ 0 & \text{otherwise.} \end{cases}$$

   As in Exercise 5, suppose that in a random sample of eight items exactly three are found to be defective. Determine the posterior distribution of θ.

7. Suppose that $X_1, \ldots, X_n$ form a random sample from a distribution for which the p.d.f. is $f(x \mid \theta)$, that the value of θ is unknown, and that the prior p.d.f. of θ is $\xi(\theta)$. Show that the posterior p.d.f. $\xi(\theta \mid x)$ is the same regardless of whether it is calculated directly by using Eq. (5) or it is calculated sequentially by using Eqs. (15), (16), and (17).

8. Consider again the problem described in Exercise 5, and assume the same prior distribution of θ. Suppose now, however, that instead of selecting a random sample of eight items from the lot, we perform the following experiment: Items from the lot are selected at random one by one until exactly three defectives have been found. If we find that we must select a total of eight items in this experiment, what is the posterior distribution of θ at the end of the experiment?

9. Suppose that a single observation X is to be taken from a uniform distribution on the interval $\left(\theta - \dfrac{1}{2}, \theta + \dfrac{1}{2} \right)$; that the value of θ is unknown; and that the prior distribution of θ is a uniform distribution on the interval $(10, 20)$. If the observed value of X is 12, what is the posterior distribution of θ?

10. Consider again the conditions of Exercise 9, and assume the same prior distribution of θ. Suppose now, however, that six observations are selected at random from the uniform distribution on the interval $\left(\theta - \dfrac{1}{2}, \theta + \dfrac{1}{2} \right)$; and that their values are 11.0, 11.5, 11.7, 11.1, 11.4, and 10.9. Determine the posterior distribution of θ.

6.3. CONJUGATE PRIOR DISTRIBUTIONS

Sampling from a Bernoulli Distribution

The Basic Theorem.   Certain prior distributions are particularly convenient for use with samples from certain other distributions. For example, suppose that a random sample is taken from a Bernoulli distribution for which the value of the parameter θ is unknown. If the prior distribution of θ is a beta distribution, then for any possible set of observed sample values, the posterior distribution of θ will again be a beta distribution. Specifically, the following result can be established:

Theorem 1.   *Suppose that $X_1, \ldots, X_n$ form a random sample from a Bernoulli distribution for which the value of the parameter θ is unknown $(0 < \theta < 1)$. Suppose also that the prior distribution of θ is a beta distribution with given parameters α and β ($\alpha > 0$ and $\beta > 0$). Then the posterior distribution of θ given that $X_i = x_i$ $(i = 1, \ldots, n)$ is a beta distribution with parameters $\alpha + \sum_{i=1}^n x_i$ and $\beta + n - \sum_{i=1}^n x_i$.*

Proof. Let $y = \sum_{i=1}^{n} x_i$. Then the likelihood function, that is, the joint p.f. $f_n(x \mid \theta)$ of $X_1, \ldots, X_n$, is given by Eq. (7) of Sec. 6.2. Also, the prior p.d.f. $\xi(\theta)$ satisfies the following relation:

$$\xi(\theta) \propto \theta^{\alpha-1}(1 - \theta)^{\beta-1} \quad \text{for } 0 < \theta < 1.$$

Since the posterior p.d.f. $\xi(\theta \mid x)$ is proportional to the product $f_n(x \mid \theta)\xi(\theta)$, it follows that

$$\xi(\theta \mid x) \propto \theta^{\alpha+y-1}(1 - \theta)^{\beta+n-y-1} \quad \text{for } 0 < \theta < 1.$$

The right side of this relation can be recognized as being, except for a constant factor, equal to the p.d.f. of a beta distribution with parameters $\alpha + y$ and $\beta + n - y$. Therefore, the posterior distribution of θ is as specified in the theorem. $\square$

Updating the Posterior Distribution. One implication of Theorem 1 is the following: Suppose that the proportion θ of defective items in a large shipment is unknown; that the prior distribution of θ is a beta distribution with parameters α and β; and that n items are selected one at a time at random from the shipment and inspected. If the first item inspected is defective, the posterior distribution of θ will be a beta distribution with parameters $\alpha + 1$ and β. If the first item is nondefective, the posterior distribution will be a beta distribution with parameters α and $\beta + 1$. The process can be continued in the following way: Each time an item is inspected, the current posterior beta distribution of θ is changed to a new beta distribution in which the value of either the parameter α or the parameter β is increased by one unit. The value of α is increased by one unit each time a defective item is found, and the value of β is increased by one unit each time a nondefective item is found.

The family of beta distributions is called a *conjugate family of prior distributions* for samples from a Bernoulli distribution. If the prior distribution of θ is a beta distribution, then the posterior distribution at each stage of sampling will also be a beta distribution, regardless of the observed values in the sample. It is also said that the family of beta distributions is *closed under sampling* from a Bernoulli distribution.

Example 1: The Variance of the Posterior Beta Distribution. Suppose that the proportion θ of defective items in a large shipment is unknown; that the prior distribution of θ is a uniform distribution on the interval $(0, 1)$; and that items are to be selected at random from the shipment and inspected until the variance of the posterior distribution of θ has been reduced to the value 0.01 or less. We shall determine the total number of defective and nondefective items that must be obtained before the sampling process is stopped.

As stated in Sec. 5.10, the uniform distribution on the interval $(0, 1)$ is a beta distribution for which $\alpha = 1$ and $\beta = 1$. Therefore, after y defective items and z nondefective items have been obtained, the posterior distribution of θ will be a beta distribution with $\alpha = y + 1$ and $\beta = z + 1$. It was shown in Sec. 5.10 that the variance of a beta distribution with parameters α and β is $\alpha\beta/[(\alpha + \beta)^2(\alpha + \beta + 1)]$. Therefore, the variance V of the posterior distribution of θ will be

$$V = \frac{(y + 1)(z + 1)}{(y + z + 2)^2(y + z + 3)}.$$

Sampling is to stop as soon as the number of defectives y and the number of nondefectives z that have been obtained are such that $V \leqslant 0.01$. It can be shown (see Exercise 1) that it will not be necessary to select more than 22 items.   □

Sampling from a Poisson Distribution

When samples are taken from a Poisson distribution, the family of gamma distributions is a conjugate family of prior distributions. This relationship is shown in the next theorem.

Theorem 2.   *Suppose that $X_1, \ldots, X_n$ form a random sample from a Poisson distribution for which the value of the mean θ is unknown ($\theta > 0$). Suppose also that the prior distribution of θ is a gamma distribution with given parameters α and β ($\alpha > 0$ and $\beta > 0$). Then the posterior distribution of θ, given that $X_i = x_i$ ($i = 1, \ldots, n$), is a gamma distribution with parameters $\alpha + \sum_{i=1}^{n} x_i$ and $\beta + n$.*

Proof. Let $y = \sum_{i=1}^{n} x_i$. Then the likelihood function $f_n(x \mid \theta)$ satisfies the relation

$$f_n(x \mid \theta) \propto e^{-n\theta}\theta^y.$$

In this relation, a factor that involves x but does not depend on θ has been dropped from the right side. Furthermore, the prior p.d.f. of θ has the form

$$\xi(\theta) \propto \theta^{\alpha-1}e^{-\beta\theta} \qquad \text{for } \theta > 0.$$

Since the posterior p.d.f. $\xi(\theta \mid x)$ is proportional to $f_n(x \mid \theta)\xi(\theta)$, it follows that

$$\xi(\theta \mid x) \propto \theta^{\alpha+y-1}e^{-(\beta+n)\theta} \qquad \text{for } \theta > 0.$$

The right side of this relation can be recognized as being, except for a constant factor, the p.d.f. of a gamma distribution with parameters $\alpha + y$ and $\beta + n$. Therefore, the posterior distribution of θ is as specified in the theorem. □

Example 2: The Variance of the Posterior Gamma Distribution. Consider a Poisson distribution for which the mean θ is unknown, and suppose that the prior p.d.f. of θ is as follows:

$$\xi(\theta) = \begin{cases} 2e^{-2\theta} & \text{for } \theta > 0, \\ 0 & \text{for } \theta \leqslant 0. \end{cases}$$

Suppose also that observations are to be taken at random from the given Poisson distribution until the variance of the posterior distribution of θ has been reduced to the value 0.01 or less. We shall determine the number of observations that must be taken before the sampling process is stopped.

The given prior p.d.f. $\xi(\theta)$ is the p.d.f. of a gamma distribution for which $\alpha = 1$ and $\beta = 2$. Therefore, after we have obtained n observed values $x_1, \ldots, x_n$, the sum of which is $y = \sum_{i=1}^{n} x_i$, the posterior distribution of θ will be a gamma distribution with $\alpha = y + 1$ and $\beta = n + 2$. It was shown in Sec. 5.9 that the variance of a gamma distribution with parameters α and β is α/β^2. Therefore, the variance V of the posterior distribution of θ will be

$$V = \frac{y + 1}{(n + 2)^2}.$$

Sampling is to stop as soon as the sequence of observed values $x_1, \ldots, x_n$ that has been obtained is such that $V \leqslant 0.01$. □

Sampling from a Normal Distribution

When samples are taken from a normal distribution for which the value of the mean θ is unknown but the value of the variance σ^2 is known, the family of normal distributions is itself a conjugate family of prior distributions, as is shown in the next theorem.

Theorem 3. *Suppose that $X_1, \ldots, X_n$ form a random sample from a normal distribution for which the value of the mean θ is unknown ($-\infty < \theta < \infty$) and the value of the variance σ^2 is known ($\sigma^2 > 0$). Suppose also that the prior distribution of θ is a normal distribution with given values of the mean μ and the variance v^2. Then the posterior distribution of θ, given that $X_i = x_i$ ($i = 1, \ldots, n$), is a normal distribution for which the mean μ_1 and the variance v_1^2 are as follows:*

$$\mu_1 = \frac{\sigma^2 \mu + n v^2 \bar{x}_n}{\sigma^2 + n v^2} \tag{1}$$

and

$$v_1^2 = \frac{\sigma^2 v^2}{\sigma^2 + nv^2}. \tag{2}$$

Proof. The likelihood function $f_n(x \mid \theta)$ has the form

$$f_n(x \mid \theta) \propto \exp\left[-\frac{1}{2\sigma^2} \sum_{i=1}^{n} (x_i - \theta)^2\right].$$

Here a constant factor has been dropped from the right side. To transform this expression, we shall use the identity

$$\sum_{i=1}^{n} (x_i - \theta)^2 = n(\theta - \bar{x}_n)^2 + \sum_{i=1}^{n} (x_i - \bar{x}_n)^2,$$

and we shall omit a factor that involves $x_1, \ldots, x_n$ but does not depend on θ. As a result, we may rewrite $f_n(x \mid \theta)$ in the following form:

$$f_n(x \mid \theta) \propto \exp\left[-\frac{n}{2\sigma^2}(\theta - \bar{x}_n)^2\right].$$

Since the prior p.d.f. $\xi(\theta)$ has the form

$$\xi(\theta) \propto \exp\left[-\frac{1}{2v^2}(\theta - \mu)^2\right],$$

it follows that the posterior p.d.f. $\xi(\theta \mid x)$ satisfies the relation

$$\xi(\theta \mid x) \propto \exp\left\{-\frac{1}{2}\left[\frac{n}{\sigma^2}(\theta - \bar{x}_n)^2 + \frac{1}{v^2}(\theta - \mu)^2\right]\right\}.$$

If μ_1 and v_1^2 are as specified in Eqs. (1) and (2), we can now verify the following identity:

$$\frac{n}{\sigma^2}(\theta - \bar{x}_n)^2 + \frac{1}{v^2}(\theta - \mu)^2 = \frac{1}{v_1^2}(\theta - \mu_1)^2 + \frac{n}{\sigma^2 + nv^2}(\bar{x}_n - \mu)^2.$$

Since the final term on the right side of this equation does not involve θ, it can be absorbed in the proportionality factor, and we obtain the relation

$$\xi(\theta \mid x) \propto \exp\left[-\frac{1}{2v_1^2}(\theta - \mu_1)^2\right].$$

The right side of this relation can be recognized as being, except for a constant factor, the p.d.f. of a normal distribution with mean μ_1 and variance v_1^2. Therefore, the posterior distribution of θ is as specified in the theorem. □

The mean μ_1 of the posterior distribution of θ, as given in Eq. (1), can be rewritten as follows:

$$
\mu_1 = \frac{\sigma^2}{\sigma^2 + nv^2}\mu + \frac{nv^2}{\sigma^2 + nv^2}\bar{x}_n. \tag{3}
$$

It can be seen from Eq. (3) that μ_1 is a weighted average of the mean μ of the prior distribution and the sample mean $\bar{x}_n$. Furthermore, it can be seen that the relative weight given to $\bar{x}_n$ satisfies the following three properties: (i) For fixed values of v^2 and σ^2, the larger the sample size n, the greater will be the relative weight that is given to $\bar{x}_n$. (ii) For fixed values of v^2 and n, the larger the variance σ^2 of each observation in the sample, the smaller will be the relative weight that is given to $\bar{x}_n$. (iii) For fixed values of σ^2 and n, the larger the variance v^2 of the prior distribution, the larger will be the relative weight that is given to $\bar{x}_n$.

Moreover, it can be seen from Eq. (2) that the variance v_1^2 of the posterior distribution of θ depends on the number n of observations that have been taken, but does not depend on the magnitudes of the observed values. Suppose, therefore, that a random sample of n observations is to be taken from a normal distribution for which the value of the mean θ is unknown, the value of the variance is known, and the prior distribution of θ is a specified normal distribution. Then, before any observations have been taken, we can use Eq. (2) to calculate the actual value of the variance v_1^2 of the posterior distribution. However, the value of the mean μ_1 of the posterior distribution will depend on the observed values that are obtained in the sample.

Example 3: The Variance of the Posterior Normal Distribution. Suppose that observations are to be taken at random from a normal distribution for which the value of the mean θ is unknown and the variance is 1, and that the prior distribution of θ is a normal distribution for which the variance is 4. Also, observations are to be taken until the variance of the posterior distribution of θ has been reduced to the value 0.01 or less. We shall determine the number of observations that must be taken before the sampling process is stopped.

It follows from Eq. (2) that after n observations have been taken, the variance v_1^2 of the posterior distribution of θ will be

$$
v_1^2 = \frac{4}{4n + 1}.
$$

Therefore, the relation $v_1^2 \leqslant 0.01$ will be satisfied if and only if $n \geqslant 99.75$. Hence, the relation $v_1^2 \leqslant 0.01$ will be satisfied after 100 observations have been taken and not before then. □

Sampling from an Exponential Distribution

We shall conclude this section by considering a random sample from an exponential distribution for which the value of the parameter θ is unknown. For this problem, the family of gamma distributions serves as a conjugate family of prior distributions, as shown in the next theorem. It should be remarked that we have changed the notation here from that used earlier in this chapter. To avoid confusion that might result from different uses of the symbol β in Theorem 4, the parameter of the exponential distribution is denoted by the symbol θ rather than by β.

> **Theorem 4.** *Suppose that $X_1, \ldots, X_n$ form a random sample from an exponential distribution for which the value of the parameter θ is unknown ($\theta > 0$). Suppose also that the prior distribution of θ is a gamma distribution with given parameters α and β ($\alpha > 0$ and $\beta > 0$). Then the posterior distribution of θ, given that $X_i = x_i$ ($i = 1, \ldots, n$), is a gamma distribution with parameters $\alpha + n$ and $\beta + \sum_{i=1}^{n} x_i$.*

Proof. Again, let $y = \sum_{i=1}^{n} x_i$. Then the likelihood function $f_n(x \mid \theta)$ is

$$f_n(x \mid \theta) = \theta^n e^{-\theta y}.$$

Also, the prior p.d.f. $\xi(\theta)$ has the form

$$\xi(\theta) \propto \theta^{\alpha-1} e^{-\beta\theta} \qquad \text{for } \theta > 0.$$

It follows, therefore, that the posterior p.d.f. $\xi(\theta \mid x)$ has the form

$$\xi(\theta \mid x) \propto \theta^{\alpha+n-1} e^{-(\beta+y)\theta} \qquad \text{for } \theta > 0.$$

The right side of this relation can be recognized as being, except for a constant factor, the p.d.f. of a gamma distribution with parameters $\alpha + n$ and $\beta + y$. Therefore, the posterior distribution of θ is as specified in the theorem. $\square$

EXERCISES

1. Show that in Example 1 it must be true that $V \leqslant 0.01$ after 22 items have been selected.

2. Suppose that the proportion θ of defective items in a large shipment is unknown, and that the prior distribution of θ is a beta distribution for which the parameters are $\alpha = 2$ and $\beta = 200$. If 100 items are selected at random from the shipment and if three of these items are found to be defective, what is the posterior distribution of θ?

3. Consider again the conditions of Exercise 2. Suppose that after a certain statistician has observed that there were three defective items among the 100 items selected at random, the posterior distribution which he assigns to θ is a beta distribution for which the mean is $2/51$ and the variance is $98/[(51)^2(103)]$. What prior distribution had the statistician assigned to θ?

4. Suppose that the number of defects in a 1200-foot roll of magnetic recording tape has a Poisson distribution for which the value of the mean θ is unknown, and that the prior distribution of θ is a gamma distribution with parameters $\alpha = 3$ and $\beta = 1$. When five rolls of this tape are selected at random and inspected, the number of defects found on the rolls are 2, 2, 6, 0, and 3. Determine the posterior distribution of θ.

5. Let θ denote the average number of defects per 100 feet of a certain type of magnetic tape. Suppose that the value of θ is unknown, and that the prior distribution of θ is a gamma distribution with parameters $\alpha = 2$ and $\beta = 10$. When a 1200-foot roll of this tape is inspected, exactly four defects are found. Determine the posterior distribution of θ.

6. Suppose that the heights of the individuals in a certain population have a normal distribution for which the value of the mean θ is unknown and the standard deviation is 2 inches. Suppose also that the prior distribution of θ is a normal distribution for which the mean is 68 inches and the standard deviation is 1 inch. If ten people are selected at random from the population and their average height is found to be 69.5 inches, what is the posterior distribution of θ?

7. Consider again the problem described in Exercise 6.

   (a) Which interval 1 inch long had the highest prior probability of containing the value of θ?

   (b) Which interval 1 inch long has the highest posterior probability of containing the value of θ?

   (c) Find the values of the probabilities in parts (a) and (b).

8. Suppose that a random sample of 20 observations is taken from a normal distribution for which the value of the mean θ is unknown and the variance is 1. After the sample values have been observed, it is found that $\overline{X}_n = 10$ and that the posterior distribution of θ is a normal distribution for which the mean is 8 and the variance is $1/25$. What was the prior distribution of θ?

9. Suppose that a random sample is to be taken from a normal distribution for which the value of the mean θ is unknown and the standard deviation is 2, and that the prior distribution of θ is a normal distribution for which the standard deviation is 1. What is the smallest number of observations that must be included in the sample in order to reduce the standard deviation of the posterior distribution of θ to the value 0.1?

10. Suppose that a random sample of 100 observations is to be taken from a normal distribution for which the value of the mean θ is unknown and the standard deviation is 2, and that the prior distribution of θ is a normal distribution. Show that no matter how large the standard deviation of the prior distribution is, the standard deviation of the posterior distribution will be less than $1/5$.

11. Suppose that the time in minutes required to serve a customer at a certain facility has an exponential distribution for which the value of the parameter θ is unknown, and that the prior distribution of θ is a gamma distribution for which the mean is 0.2 and the standard deviation is 1. If the average time required to serve a random sample of 20 customers is observed to be 3.8 minutes, what is the posterior distribution of θ?

12. For a distribution with mean $\mu \neq 0$ and standard deviation $\sigma > 0$, the *coefficient of variation* of the distribution is defined as $\sigma/|\mu|$. Consider again the problem described in Exercise 11, and suppose that the coefficient of variation of the prior gamma distribution of θ is 2. What is the smallest number of customers that must be observed in order to reduce the coefficient of variation of the posterior distribution to 0.1?

13. Show that the family of beta distributions is a conjugate family of prior distributions for samples from a negative binomial distribution with a known value of the parameter r and an unknown value of the parameter p $(0 < p < 1)$.

14. Let $\xi(\theta)$ be a p.d.f. which is defined as follows for constants $\alpha > 0$ and $\beta > 0$:

$$\xi(\theta) = \begin{cases} \dfrac{\beta^\alpha}{\Gamma(\alpha)} \theta^{-(\alpha+1)} e^{-\beta/\theta} & \text{for } \theta > 0, \\ 0 & \text{for } \theta \leqslant 0. \end{cases}$$

(a) Verify that $\xi(\theta)$ is actually a p.d.f. by verifying that $\int_0^\infty \xi(\theta)\, d\theta = 1$.

(b) Consider the family of probability distributions that can be represented by a p.d.f. $\xi(\theta)$ having the given form for all possible pairs of constants $\alpha > 0$ and $\beta > 0$. Show that this family is a conjugate family of prior distributions for samples from a normal distribution with a known value of the mean μ and an unknown value of the variance θ.

15. Suppose that in Exercise 14 the parameter is taken as the standard deviation of the normal distribution, rather than the variance. Determine a conjugate family of prior distributions for samples from a normal distribution with a known value of the mean μ and an unknown value of the standard deviation σ.

16. Suppose that the number of minutes a person must wait for a bus each morning has a uniform distribution on the interval $(0, \theta)$, where the value of the endpoint θ is unknown. Suppose also that the prior p.d.f. of θ is as follows:

$$\xi(\theta) = \begin{cases} \dfrac{192}{\theta^4} & \text{for } \theta \geqslant 4, \\ 0 & \text{otherwise.} \end{cases}$$

If the observed waiting times on three successive mornings are 5, 3, and 8 minutes, what is the posterior p.d.f. of θ?

17. The Pareto distribution with parameters x_0 and α $(x_0 > 0$ and $\alpha > 0)$ is defined in Exercise 15 of Sec. 5.9. Show that the family of Pareto distributions is a conjugate family of prior distributions for samples from a uniform distribution on the interval $(0, \theta)$, where the value of the endpoint θ is unknown.

18. Suppose that $X_1, \dots, X_n$ form a random sample from a distribution for which the p.d.f. $f(x \mid \theta)$ is as follows:

$$f(x \mid \theta) = \begin{cases} \theta x^{\theta - 1} & \text{for } 0 < x < 1, \\ 0 & \text{otherwise.} \end{cases}$$

Suppose also that the value of the parameter θ is unknown $(\theta > 0)$ and that the prior distribution of θ is a gamma distribution with parameters α and β $(\alpha > 0$ and $\beta > 0)$. Determine the mean and the variance of the posterior distribution of θ.

6.4. BAYES ESTIMATORS

Nature of an Estimation Problem

Suppose that a random sample $X_1, \dots, X_n$ is to be taken from a distribution for which the p.f. or p.d.f. is $f(x \mid \theta)$, where the value of the parameter θ is unknown. Suppose also that the value of θ must lie in a given interval Ω of the real line. The interval Ω could be either bounded or unbounded; in particular, it could be the entire real line. Finally, suppose that the value of θ must be estimated from the observed values in the sample.

An *estimator* of the parameter θ, based on the random variables $X_1, \dots, X_n$, is a real-valued function $\delta(X_1, \dots, X_n)$ which specifies the estimated value of θ for each possible set of values of $X_1, \dots, X_n$. In other words, if the observed values of $X_1, \dots, X_n$ turn out to be $x_1, \dots, x_n$, then the estimated value of θ is $\delta(x_1, \dots, x_n)$. Since the value of θ must belong to the interval Ω, it is reasonable to require that every possible value of an estimator $\delta(X_1, \ldots, X_n)$ must also belong to Ω.

It is convenient to distinguish between the terms *estimator* and *estimate*. Since an estimator $\delta(X_1, \ldots, X_n)$ is a function of the random variables $X_1, \ldots, X_n$, the estimator itself is a random variable; and its probability distribution can be derived from the joint distribution of $X_1, \ldots, X_n$. On the other hand, an *estimate* is a specific value $\delta(x_1, \ldots, x_n)$ of the estimator that is determined by using specific observed values $x_1, \ldots, x_n$. It will often be convenient to use vector notation and to let $X = (X_1, \ldots, X_n)$ and $x = (x_1, \ldots, x_n)$. In this notation, an estimator is a function $\delta(X)$ of the random vector X, and an estimate is a specific value $\delta(x)$. It will often be convenient to denote an estimator $\delta(X)$ simply by the symbol δ.

Loss Functions

The foremost requirement of a good estimator δ is that it yield an estimate of θ which is close to the actual value of θ. In other words, a good estimator is one for which it is highly probable that the error $\delta(X) - \theta$ will be close to 0. We shall assume that for each possible value of $\theta \in \Omega$ and each possible estimate $a \in \Omega$, there is a number $L(\theta, a)$ which measures the loss or cost to the statistician when the true value of the parameter is θ and his estimate is a. Typically, the greater the distance between a and θ, the larger will be the value of $L(\theta, a)$.

As before, let $\xi(\theta)$ denote the prior p.d.f. of θ on the interval Ω; and consider a problem in which the statistician must estimate the value of θ without being able to observe the values in a random sample. If the statistician chooses a particular estimate a, then his expected loss will be

$$E[L(\theta, a)] = \int_\Omega L(\theta, a)\xi(\theta)\, d\theta. \tag{1}$$

We shall assume that the statistician wishes to choose an estimate a for which the expected loss in Eq. (1) is a minimum. In any estimation problem, a function L for which the expectation $E[L(\theta, a)]$ is to be minimized is called a loss function.

Definition of a Bayes Estimator

Suppose now that the statistician can observe the value x of the random vector X before estimating θ, and let $\xi(\theta \mid x)$ denote the posterior p.d.f. of θ on the interval Ω. For any estimate a that the statistician might use, his expected loss in this case will be

$$E[L(\theta, a) \mid x] = \int_\Omega L(\theta, a)\xi(\theta \mid x)\, d\theta. \tag{2}$$

Hence, the statistician should now choose an estimate a for which the expectation in Eq. (2) is a minimum.

For each possible value x of the random vector X, let $\delta^*(x)$ denote a value of the estimate a for which the expected loss in Eq. (2) is a minimum. Then the function $\delta^*(X)$ for which the values are specified in this way will be an estimator of θ. This estimator is called a *Bayes estimator* of θ. In other words, for each possible value x of X, the value $\delta^*(x)$ of the Bayes estimator is chosen so that

$$E\left[L\left(\theta, \delta^*(x)\right)\mid x\right] = \min_{a \in \Omega} E\left[L(\theta, a)\mid x\right]. \tag{3}$$

In summary, we have considered an estimation problem in which a random sample $X = (X_1, \ldots, X_n)$ is to be taken from a distribution involving a parameter θ that has an unknown value in some specified interval Ω. For any given loss function $L(\theta, a)$ and any prior p.d.f. $\xi(\theta)$, the Bayes estimator of θ is the estimator $\delta^*(X)$ for which Eq. (3) is satisfied for every possible value x of X. It should be emphasized that the form of the Bayes estimator will depend on both the loss function that is used in the problem and the prior distribution that is assigned to θ.

Different Loss Functions

The Squared Error Loss Function.   By far the most commonly used loss function in estimation problems is the squared error loss function. This function is defined as follows:

MINIMUM WHEN a
CHOSEN TO BE MEAN
OF SAMPLE

$$L(\theta, a) = (\theta - a)^2. \tag{4}$$

When the squared error loss function is used, the Bayes estimate $\delta^*(x)$ for any observed value of x will be the value of a for which the expectation $E[(\theta - a)^2 \mid x]$ is a minimum.

It was shown in Sec. 4.5 that for any given probability distribution of θ, the expectation of $(\theta - a)^2$ will be a minimum when a is chosen to be equal to the mean of the distribution of θ. Therefore, when the expectation of $(\theta - a)^2$ is calculated with respect to the posterior distribution of θ, this expectation will be a minimum when a is chosen to be equal to the mean $E(\theta \mid x)$ of the posterior distribution. This discussion shows that when the squared error loss function (4) is used, the Bayes estimator is $\delta^*(X) = E(\theta \mid X)$.

Example 1: Estimating the Parameter of a Bernoulli Distribution.   Suppose that a random sample $X_1, \ldots, X_n$ is to be taken from a Bernoulli distribution for which the value of the parameter θ is unknown and must be estimated, and that the prior distribution of θ is a beta distribution with given parameters α and β

($\alpha > 0$ and $\beta > 0$). Suppose also that the squared error loss function is used, as specified by Eq. (4), for $0 < \theta < 1$ and $0 < a < 1$. We shall determine the Bayes estimator of θ.

For any observed values $x_1, \ldots, x_n$, let $y = \sum_{i=1}^{n} x_i$. Then it follows from Theorem 1 of Sec. 6.3 that the posterior distribution of θ will be a beta distribution with parameters $\alpha + y$ and $\beta + n - y$. Since the mean of a beta distribution with parameters α_1 and β_1 is $\alpha_1/(\alpha_1 + \beta_1)$, the mean of this posterior distribution of θ will be $(\alpha + y)/(\alpha + \beta + n)$. The Bayes estimate $\delta(x)$ will be equal to this value for any observed vector x. Therefore, the Bayes estimator $\delta^*(X)$ is specified as follows:

$$\delta^*(X) = \frac{\alpha + \sum_{i=1}^{n} X_i}{\alpha + \beta + n}. \quad \square \tag{5}$$

Example 2: Estimating the Mean of a Normal Distribution. Suppose that a random sample $X_1, \ldots, X_n$ is to be taken from a normal distribution for which the value of the mean θ is unknown and the value of the variance σ^2 is known. Suppose also that the prior distribution of θ is a normal distribution with given values of the mean μ and the variance v^2. Suppose, finally, that the squared error loss function is to be used, as specified in Eq. (4), for $-\infty < \theta < \infty$ and $-\infty < a < \infty$. We shall determine the Bayes estimator of θ.

It follows from Theorem 3 of Sec. 6.3 that for any observed values $x_1, \ldots, x_n$, the posterior distribution of θ will be a normal distribution for which the mean μ_1 is specified by Eq. (1) of Sec. 6.3. Therefore, the Bayes estimator $\delta^*(X)$ is specified as follows:

$$\delta^*(X) = \frac{\sigma^2 \mu + n v^2 \overline{X}_n}{\sigma^2 + n v^2}. \quad \square \tag{6}$$

The Absolute Error Loss Function. Another commonly used loss function in estimation problems is the absolute error loss function. This function is defined as follows:

$$L(\theta, a) = |\theta - a|. \tag{7}$$

MINIMUM WHEN a CHOSEN TO BE MEDIAN

For any observed value of x, the Bayes estimate $\delta(x)$ will now be the value of a for which the expectation $E(|\theta - a||x)$ is a minimum.

It was shown in Sec. 4.5 that for any given probability distribution of θ, the expectation of $|\theta - a|$ will be a minimum when a is chosen to be equal to a median of the distribution of θ. Therefore, when the expectation of $|\theta - a|$ is calculated with respect to the posterior distribution of θ, this expectation will be a minimum when a is chosen to be equal to a median of the posterior distribution of θ. It follows that when the absolute error loss function (7) is used, the Bayes estimator $\delta^*(X)$ is an estimator for which the value is always equal to a median of the posterior distribution of θ. We shall now consider Examples 1 and 2 again, but we shall use the absolute error loss function instead of the squared error loss function.

Example 3: Estimating the Parameter of a Bernoulli Distribution. Consider again the conditions of Example 1, but suppose now that the absolute error loss function is used, as specified by Eq. (7). For any observed values $x_1, \ldots, x_n$, the Bayes estimate $\delta^*(x)$ will be equal to the median of the posterior distribution of θ, which is a beta distribution with parameters $\alpha + y$ and $\beta + n - y$. There is no simple expression for this median. It must be determined by numerical approximations for each given set of observed values. □

Example 4: Estimating the Mean of a Normal Distribution. Consider again the conditions of Example 2, but suppose now that the absolute error loss function is used, as specified by Eq. (7). For any observed values $x_1, \ldots, x_n$, the Bayes estimate $\delta^*(x)$ will be equal to the median of the posterior normal distribution of θ. However, since the mean and the median of any normal distribution are equal, $\delta^*(x)$ is also equal to the mean of the posterior distribution. Therefore, the Bayes estimator with respect to the absolute error loss function is the same as the Bayes estimator with respect to the squared error loss function; and it is again given by Eq. (6). □

Other Loss Functions.   Although the squared error loss function and, to a lesser extent, the absolute error loss function are the ones most commonly used in estimation problems, neither of these loss functions might be appropriate in a particular problem. In some problems, it might be appropriate to use a loss function having the form $L(\theta, a) = |\theta - a|^k$, where k is some positive number other than 1 or 2. In other problems, the loss that results when the error $|\theta - a|$ has a given magnitude might depend on the actual value of θ. In such a problem, it might be appropriate to use a loss function having the form $L(\theta, a) = \lambda(\theta)(\theta - a)^2$ or $L(\theta, a) = \lambda(\theta)|\theta - a|$, where $\lambda(\theta)$ is a given positive function of θ. In still other problems, it might be more costly to overestimate the value of θ by a certain amount than to underestimate the value of θ by the same amount. One specific loss function which reflects this property is as follows:

$$L(\theta, a) = \begin{cases} 3(\theta - a)^2 & \text{for } \theta \leq a, \\ (\theta - a)^2 & \text{for } \theta > a. \end{cases}$$

Various other types of loss functions might be relevant in specific estimation problems. However, in this book we shall consider only the squared error and absolute error loss functions.

The Bayes Estimate for Large Samples

Effect of Different Prior Distributions.   Suppose that the proportion θ of defective items in a large shipment is unknown, and that the prior distribution of θ is a uniform distribution on the interval $(0, 1)$. Suppose also that the value of θ must be estimated, and that the squared error loss function is used. Suppose, finally, that in a random sample of 100 items from the shipment, exactly ten items are found to be defective. Since the uniform distribution is a beta distribution with parameters $\alpha = 1$ and $\beta = 1$, and since $n = 100$ and $y = 10$ for the given sample, it follows from Eq. (5) that the Bayes estimate is $\delta(x) = 11/102 = 0.108$.

Next, suppose that the prior p.d.f. of θ has the form $\xi(\theta) = 2(1 - \theta)$ for $0 < \theta < 1$, instead of being a uniform distribution; and that again in a random sample of 100 items, exactly ten items are found to be defective. Since $\xi(\theta)$ is the p.d.f. of a beta distribution with parameters $\alpha = 1$ and $\beta = 2$, it follows from Eq. (5) that in this case the Bayes estimate of θ is $\delta(x) = 11/103 = 0.107$.

The two prior distributions considered here are quite different. The mean of the uniform prior distribution is $1/2$, and the mean of the other beta prior distribution is $1/3$. Nevertheless, because the number of observations in the sample is so large ($n = 100$), the Bayes estimates with respect to the two different prior distributions are almost the same. Furthermore, the values of both estimates are very close to the observed proportion of defective items in the sample, which is $\bar{x}_n = 0.1$.

Consistency of the Bayes Estimator.   Since the unknown value of θ is the mean of the Bernoulli distribution from which the observations are being taken, it follows from the law of large numbers discussed in Sec. 4.8 that $\bar{X}_n$ converges in probability to this unknown value as $n \to \infty$. Since the difference between the Bayes estimator $\delta^*(X)$ and $\bar{X}_n$ converges in probability to 0 as $n \to \infty$, it can also be concluded that $\delta^*(X)$ converges in probability to the unknown value of θ as $n \to \infty$.

A sequence of estimators which converges to the unknown value of the parameter being estimated, as $n \to \infty$, is called a *consistent sequence of estimators*. Thus, we have shown that the Bayes estimators $\delta^*(X)$ form a consistent sequence of estimators in the problem considered here. The practical interpretation of this result is as follows: When large numbers of observations are taken, there is high probability that the Bayes estimator will be very close to the unknown value of θ.

The results which have just been presented for estimating the parameter of a Bernoulli distribution are also true for other estimation problems. Under fairly general conditions and for a wide class of loss functions, the Bayes estimators of some parameter θ will form a consistent sequence of estimators as the sample size $n \to \infty$. In particular, for random samples from any one of the various families of distributions discussed in Sec. 6.3, if a conjugate prior distribution is assigned to the parameter and the squared error loss function is used, the Bayes estimators will form a consistent sequence of estimators.

For example, consider again the conditions of Example 2. In that example, a random sample is taken from a normal distribution for which the value of the mean θ is unknown, and the Bayes estimator $\delta^*(X)$ is specified by Eq. (6). By the law of large numbers, $\overline{X}_n$ will converge to the unknown value of the mean θ as $n \to \infty$. It can now be seen from Eq. (6) that $\delta^*(X)$ will also converge to θ as $n \to \infty$. Thus, the Bayes estimators again form a consistent sequence of estimators.

Other examples are given in Exercises 6 and 10 at the end of this section.

EXERCISES

1. Suppose that the proportion θ of defective items in a large shipment is unknown, and that the prior distribution of θ is a beta distribution for which the parameters are $\alpha = 5$ and $\beta = 10$. Suppose also that 20 items are selected at random from the shipment, and that exactly one of these items is found to be defective. If the squared error loss function is used, what is the Bayes estimate of θ?

2. Consider again the conditions of Exercise 1. Suppose that the prior distribution of θ is as given in Exercise 1, and suppose again that 20 items are selected at random from the shipment.

    (a) For what number of defective items in the sample will the mean squared error of the Bayes estimate be a maximum?

    (b) For what number will the mean squared error of the Bayes estimate be a minimum?

3. Suppose that a random sample of size n is taken from a Bernoulli distribution for which the value of the parameter θ is unknown, and that the prior distribution of θ is a beta distribution for which the mean is μ_0. Show that the mean of the posterior distribution of θ will be a weighted average having the form $\gamma_n \overline{X}_n + (1 - \gamma_n)\mu_0$, and show that $\gamma_n \to 1$ as $n \to \infty$.

4. Suppose that the number of defects in a 1200-foot roll of magnetic recording tape has a Poisson distribution for which the value of the mean θ is unknown, and that the prior distribution of θ is a gamma distribution with parameters $\alpha = 3$ and $\beta = 1$. When five rolls of this tape are selected at random and inspected, the numbers of defects found on the rolls are 2, 2, 6, 0, and 3. If the squared error loss function is used, what is the Bayes estimate of θ? (See Exercise 4 of Sec. 6.3.)

5. Suppose that a random sample of size n is taken from a Poisson distribution for which the value of the mean θ is unknown, and that the prior distribution of θ is a gamma distribution for which the mean is μ_0. Show that the mean of the posterior distribution of θ will be a weighted average having the form $\gamma_n \overline{X}_n + (1 - \gamma_n)\mu_0$, and show that $\gamma_n \to 1$ as $n \to \infty$.

6. Consider again the conditions of Exercise 5, and suppose that the value of θ must be estimated by using the squared error loss function. Show that the Bayes estimators, for $n = 1, 2, \ldots$, form a consistent sequence of estimators of θ.

7. Suppose that the heights of the individuals in a certain population have a normal distribution for which the value of the mean θ is unknown and the standard deviation is 2 inches. Suppose also that the prior distribution of θ is a normal distribution for which the mean is 68 inches and the standard deviation is 1 inch. Suppose finally that ten people are selected at random from the population, and that their average height is found to be 69.5 inches.

   (a) If the squared error loss function is used, what is the Bayes estimate of θ?

   (b) If the absolute error loss function is used, what is the Bayes estimate of θ? (See Exercise 6 of Sec. 6.3.)

8. Suppose that a random sample is to be taken from a normal distribution for which the value of the mean θ is unknown and the standard deviation is 2; that the prior distribution of θ is a normal distribution for which the standard deviation is 1; and that the value of θ must be estimated by using the squared error loss function. What is the smallest random sample that must be taken in order for the mean squared error of the Bayes estimator of θ to be 0.01 or less? (See Exercise 9 of Sec. 6.3.)

9. Suppose that the time in minutes required to serve a customer at a certain facility has an exponential distribution for which the value of the parameter θ is unknown; that the prior distribution of θ is a gamma distribution for which the mean is 0.2 and the standard deviation is 1; and that the average time required to serve a random sample of 20 customers is observed to be 3.8 minutes. If the squared error loss function is used, what is the Bayes estimate of θ? (See Exercise 11 of Sec. 6.3.)

10. Suppose that a random sample of size n is taken from an exponential distribution for which the value of the parameter θ is unknown; that the prior distribution of θ is a specified gamma distribution; and that the value of θ must be estimated by using the squared error loss function. Show that the Bayes estimators, for $n = 1, 2, \ldots$, form a consistent sequence of estimators of θ.

11. Let θ denote the proportion of registered voters in a large city who are in favor of a certain proposition. Suppose that the value of θ is unknown, and that two statisticians A and B assign to θ the following different prior p.d.f.'s $\xi_A(\theta)$ and $\xi_B(\theta)$, respectively:

$$\xi_A(\theta) = 2\theta \qquad \text{for } 0 < \theta < 1,$$
$$\xi_B(\theta) = 4\theta^3 \qquad \text{for } 0 < \theta < 1.$$

In a random sample of 1000 registered voters from the city, it is found that 710 are in favor of the proposition.

(a) Find the posterior distribution that each statistician assigns to θ.

(b) Find the Bayes estimate for each statistician based on the squared error loss function.

(c) Show that after the opinions of the 1000 registered voters in the random sample had been obtained, the Bayes estimates for the two statisticians could possibly differ by more than 0.002, regardless of the number in the sample who were in favor of the proposition.

12. Suppose that $X_1, \ldots, X_n$ form a random sample from a uniform distribution on the interval $(0, \theta)$, where the value of the parameter θ is unknown. Suppose also that the prior distribution of θ is a Pareto distribution with parameters x_0 and α ($x_0 > 0$ and $\alpha > 0$), as defined in Exercise 15 of Sec. 5.9. If the value of θ is to be estimated by using the squared error loss function, what is the Bayes estimator of θ? (See Exercise 17 of Sec. 6.3.)

13. Suppose that $X_1, \ldots, X_n$ form a random sample from an exponential distribution for which the value of the parameter θ is unknown ($\theta > 0$). Let $\xi(\theta)$ denote the prior p.d.f. of θ; and let $\hat{\theta}$ denote the Bayes estimator of θ with respect to the prior p.d.f. $\xi(\theta)$ when the squared error loss function is used. Let $\psi = \theta^2$ and suppose that instead of estimating θ, it is desired to estimate the value of ψ subject to the following squared error loss function:

$$L(\psi, a) = (\psi - a)^2 \quad \text{for } \psi > 0 \text{ and } a > 0.$$

Let $\hat{\psi}$ denote the Bayes estimator of ψ. Explain why $\hat{\psi} > \hat{\theta}^2$. *Hint:* Use the fact that for any random variable Z which can have two or more values, $E(Z^2) > [E(Z)]^2$.

6.5. MAXIMUM LIKELIHOOD ESTIMATORS

Limitations of Bayes Estimators

The theory of Bayes estimators, as described in the preceding sections, provides a satisfactory and coherent theory for the estimation of parameters. Indeed, according to statisticians who adhere to the Bayesian philosophy, it provides the only coherent theory of estimation that can possibly be developed. Nevertheless, there are certain limitations to the applicability of this theory in practical statistical problems. To apply the theory, it is necessary to specify a particular loss function, such as the squared error or absolute error function, and also a prior distribution for the parameter. Meaningful specifications may exist, in principle, but it may be very difficult and very time-consuming to determine them. In some problems, the statistician must determine the specifications that would be appropriate for clients or employers who are unavailable or otherwise unable to communicate their preferences and knowledge. In other problems, it may be necessary for an estimate to be made jointly by members of a group or committee; and it may be difficult for the members of the group to reach agreement about an appropriate loss function and prior distribution.

Another possible difficulty is that in a particular problem the parameter θ may actually be a vector of real-valued parameters for which all the values are unknown. The theory of Bayes estimation which has been developed in the preceding sections can easily be generalized to include the estimation of a vector parameter θ. However, to apply this theory in such a problem it is necessary to specify a multivariate prior distribution for the vector θ and also to specify a loss function $L(\theta, a)$ that is a function of the vector θ and the vector a which will be used to estimate θ. Even though, in a given problem, the statistician may be interested in estimating only one or two components of the vector θ he must still assign a multivariate prior distribution to the entire vector θ. In many important statistical problems, some of which will be discussed later in this book, θ may have a large number of components. In such a problem, it is especially difficult to specify a meaningful prior distribution on the multidimensional parameter space Ω.

It should be emphasized that there is no simple way to resolve these difficulties. Other methods of estimation that are not based on prior distributions and loss functions typically have not only serious defects in their theoretical structure but also severe practical limitations. Nevertheless, it is useful to be able to apply a relatively simple method of constructing an estimator without having to specify a loss function and a prior distribution. In this section we shall describe such a method, which is called the method of *maximum likelihood*. This method, which was introduced by R. A. Fisher in 1912, can be applied in most problems, has a strong intuitive appeal, and will often yield a reasonable estimator of θ. Furthermore, if the sample is large, the method will typically yield an excellent estimator of θ. For these reasons, the method of maximum likelihood is probably the most widely used method of estimation in statistics.

Definition of a Maximum Likelihood Estimator

Suppose that the random variables $X_1, \ldots, X_n$ form a random sample from a discrete distribution or a continuous distribution for which the p.f. or the p.d.f. is $f(x \mid \theta)$, where the parameter θ belongs to some parameter space Ω. Here, θ can be either a real-valued parameter or a vector. For any observed vector $x = (x_1, \ldots, x_n)$ in the sample, the value of the joint p.f. or joint p.d.f. will, as usual, be denoted by $f_n(x \mid \theta)$. As before, when $f_n(x \mid \theta)$ is regarded as a function of θ for a given vector x, it is called the *likelihood function*.

Suppose, for the moment, that the observed vector x came from a discrete distribution. If an estimate of θ must be selected, we would certainly not consider any value of $\theta \in \Omega$ for which it would be impossible to obtain the vector x that was actually observed. Furthermore, suppose that the probability $f_n(x \mid \theta)$ of obtaining the actual observed vector x is very high when θ has a particular value, say $\theta = \theta_0$, and is very small for every other value of $\theta \in \Omega$. Then we would naturally estimate the value of θ to be θ_0 (unless we had strong prior information which outweighed the evidence in the sample and pointed toward some other value). When the sample comes from a continuous distribution, it would again be natural to try to find a value of θ for which the probability density $f_n(x \mid \theta)$ is large, and to use this value as an estimate of θ. For any given observed vector x, we are led by this reasoning to consider a value of θ for which the likelihood function $f_n(x \mid \theta)$ is a maximum and to use this value as an estimate of θ. This concept is formalized in the following definition:

For each possible observed vector x, let $\delta(x) \in \Omega$ denote a value of $\theta \in \Omega$ for which the likelihood function $f_n(x \mid \theta)$ is a maximum, and let $\hat{\theta} = \delta(X)$ be the estimator of θ defined in this way. The estimator $\hat{\theta}$ is called the *maximum likelihood estimator of θ*. The expression *maximum likelihood estimator* or *maximum likelihood estimate* is abbreviated M.L.E.

Examples of Maximum Likelihood Estimators

It should be noted that in some problems, for certain observed vectors x, the maximum value of $f_n(x \mid \theta)$ may not actually be attained for any point $\theta \in \Omega$. In such a case, an M.L.E. of θ does not exist. For certain other observed vectors x, the maximum value of $f_n(x \mid \theta)$ may actually be attained at more than one point in the space Ω. In such a case, the M.L.E. is not uniquely defined, and any one of these points can be chosen as the estimate $\hat{\theta}$. In many practical problems, however, the M.L.E. exists and is uniquely defined.

We shall now illustrate the method of maximum likelihood and these various possibilities by considering seven examples. In each example, we shall attempt to determine an M.L.E.

Example 1: Sampling from a Bernoulli Distribution. Suppose that the random variables $X_1, \ldots, X_n$ form a random sample from a Bernoulli distribution for which the parameter θ is unknown ($0 \leqslant \theta \leqslant 1$). For any observed values $x_1, \ldots, x_n$, where each x_i is either 0 or 1, the likelihood function is

$$f_n(x \mid \theta) = \prod_{i=1}^{n} \theta^{x_i}(1 - \theta)^{1-x_i}. \tag{1}$$

The value of θ which maximizes the likelihood function $f_n(x \mid \theta)$ will be the same as the value of θ which maximizes $\log f_n(x \mid \theta)$. Therefore, it will be convenient to determine the M.L.E. by finding the value of θ which maximizes

$$
\begin{aligned}
L(\theta) = \log f_n(x \mid \theta) &= \sum_{i=1}^{n} \left[x_i \log \theta + (1 - x_i)\log(1 - \theta) \right] \\
&= \left(\sum_{i=1}^{n} x_i \right) \log \theta + \left(n - \sum_{i=1}^{n} x_i \right) \log(1 - \theta).
\end{aligned}
\tag{2}
$$

If we now calculate the derivative $dL(\theta)/d\theta$, set this derivative equal to 0, and solve the resulting equation for θ, we find that $\theta = \bar{x}_n$. It can be verified that this value does indeed maximize $L(\theta)$. Hence, it also maximizes the likelihood function defined by Eq. (1). It follows therefore that the M.L.E. of θ is $\hat{\theta} = \bar{X}_n$. $\quad\square$

It follows from Example 1 that if $X_1, \ldots, X_n$ are regarded as n Bernoulli trials, then the M.L.E. of the unknown probability of success on any given trial is simply the proportion of successes observed in the n trials.

Example 2: Sampling from a Normal Distribution. Suppose that $X_1, \ldots, X_n$ form a random sample from a normal distribution for which the mean μ is unknown and the variance σ^2 is known. For any observed values $x_1, \ldots, x_n$, the likelihood function $f_n(x \mid \mu)$ will be

$$
f_n(x \mid \mu) = \frac{1}{(2\pi\sigma^2)^{n/2}} \exp\left[-\frac{1}{2\sigma^2} \sum_{i=1}^{n} (x_i - \mu)^2 \right].
\tag{3}
$$

It can be seen from Eq. (3) that $f_n(x \mid \mu)$ will be maximized by the value of μ which minimizes

$$
Q(\mu) = \sum_{i=1}^{n} (x_i - \mu)^2 = \sum_{i=1}^{n} x_i^2 - 2\mu \sum_{i=1}^{n} x_i + n\mu^2.
$$

If we now calculate the derivative $dQ(\mu)/d\mu$, set this derivative equal to 0, and solve the resulting equation for μ, we find that $\mu = \bar{x}_n$. It follows, therefore, that the M.L.E. of μ is $\hat{\mu} = \bar{X}_n$. $\quad\square$

It can be seen in Example 2 that the estimator $\hat{\mu}$ is not affected by the value of the variance σ^2, which we assumed was known. The M.L.E. of the unknown mean μ is simply the sample mean $\bar{X}_n$, regardless of the value of σ^2. We shall see this again in the next example, in which both μ and σ^2 must be estimated.

Example 3: Sampling from a Normal Distribution with Unknown Variance. Suppose again that $X_1, \ldots, X_n$ form a random sample from a normal distribution, but suppose now that both the mean μ and the variance σ^2 are unknown. For any observed values $x_1, \ldots, x_n$, the likelihood function $f_n(x \mid \mu, \sigma^2)$ will again be given by the right side of Eq. (3). This function must now be maximized over all possible values of μ and σ^2, where $-\infty < \mu < \infty$ and $\sigma^2 > 0$. Instead of maximizing the likelihood function $f_n(x \mid \mu, \sigma^2)$ directly, it is again easier to maximize $\log f_n(x \mid \mu, \sigma^2)$. We have

$$L(\mu, \sigma^2) = \log f_n(x \mid \mu, \sigma^2)$$

$$= -\frac{n}{2}\log(2\pi) - \frac{n}{2}\log \sigma^2 - \frac{1}{2\sigma^2}\sum_{i=1}^{n}(x_i - \mu)^2. \tag{4}$$

We shall find the values of μ and σ^2 for which $L(\mu, \sigma^2)$ is maximum by finding the values of μ and σ^2 that satisfy the following two equations:

$$\frac{\partial L(\mu, \sigma^2)}{\partial \mu} = 0, \tag{5a}$$

$$\frac{\partial L(\mu, \sigma^2)}{\partial \sigma^2} = 0. \tag{5b}$$

From Eq. (4), we obtain the relation

$$\frac{\partial L(\mu, \sigma^2)}{\partial \mu} = \frac{1}{\sigma^2}\sum_{i=1}^{n}(x_i - \mu) = \frac{1}{\sigma^2}\left(\sum_{i=1}^{n}x_i - n\mu\right).$$

Therefore, from Eq. (5a) we find that $\mu = \bar{x}_n$.
Furthermore, from Eq. (4),

$$\frac{\partial L(\mu, \sigma^2)}{\partial \sigma^2} = -\frac{n}{2\sigma^2} + \frac{1}{2\sigma^4}\sum_{i=1}^{n}(x_i - \mu)^2.$$

When μ is replaced with the value $\bar{x}_n$ which we have just obtained, we find from Eq. (5b) that

$$\sigma^2 = \frac{1}{n}\sum_{i=1}^{n}(x_i - \bar{x}_n)^2. \tag{6}$$

Just as $\bar{x}_n$ is called the sample mean, the statistic on the right side of Eq. (6) is called the *sample variance*. It is the variance of a distribution that assigns probability $1/n$ to each of the n observed values $x_1, \ldots, x_n$ in the sample.

It can be verified that the values of μ and σ^2 which satisfy Eqs. (5a) and (5b), and which have just been derived, actually yield the maximum value of $L(\mu, \sigma^2)$. Therefore, the M.L.E.'s of μ and σ^2 are

$$\hat{\mu} = \bar{X}_n \qquad \text{and} \qquad \widehat{\sigma^2} = \frac{1}{n} \sum_{i=1}^{n} (X_i - \bar{X}_n)^2.$$

In other words, the M.L.E.'s of the mean and the variance of a normal distribution are the sample mean and the sample variance. □

Example 4: Sampling from a Uniform Distribution. Suppose that $X_1, \ldots, X_n$ form a random sample from a uniform distribution on the interval $(0, \theta)$, where the value of the parameter θ is unknown $(\theta > 0)$. The p.d.f. $f(x \mid \theta)$ of each observation has the following form:

$$f(x \mid \theta) = \begin{cases} \dfrac{1}{\theta} & \text{for } 0 \leqslant x \leqslant \theta, \\ 0 & \text{otherwise.} \end{cases} \tag{7}$$

Therefore, the joint p.d.f. $f_n(x \mid \theta)$ of $X_1, \ldots, X_n$ has the form

$$f_n(x \mid \theta) = \begin{cases} \dfrac{1}{\theta^n} & \text{for } 0 \leqslant x_i \leqslant \theta \ (i = 1, \ldots, n), \\ 0 & \text{otherwise.} \end{cases} \tag{8}$$

It can be seen from Eq. (8) that the M.L.E. of θ must be a value of θ for which $\theta \geqslant x_i$ for $i = 1, \ldots, n$ and which maximizes $1/\theta^n$ among all such values. Since $1/\theta^n$ is a decreasing function of θ, the estimate will be the smallest value of θ such that $\theta \geqslant x_i$ for $i = 1, \ldots, n$. Since this value is $\theta = \max(x_1, \ldots, x_n)$, the M.L.E. of θ is $\hat{\theta} = \max(X_1, \ldots, X_n)$. □

It should be remarked that in Example 4, the M.L.E. $\hat{\theta}$ does not seem to be a suitable estimator of θ. Since $\max(X_1, \ldots, X_n) < \theta$ with probability 1, it follows that $\hat{\theta}$ surely underestimates the value of θ. Indeed, if any prior distribution is assigned to θ, then the Bayes estimator of θ will surely be greater than $\hat{\theta}$. The actual amount by which the Bayes estimator exceeds $\hat{\theta}$ will, of course, depend on the particular prior distribution that is used and on the observed values of $X_1, \ldots, X_n$.

Example 5: Nonexistence of an M.L.E. Suppose again that $X_1, \ldots, X_n$ form a random sample from a uniform distribution on the interval $(0, \theta)$. However, suppose now that instead of writing the p.d.f. $f(x \mid \theta)$ of the uniform distribution in the form given in Eq. (7), we write it in the following form:

$$f(x \mid \theta) = \begin{cases} \dfrac{1}{\theta} & \text{for } 0 < x < \theta, \\ 0 & \text{otherwise.} \end{cases} \tag{9}$$

The only difference between Eq. (7) and Eq. (9) is that the value of the p.d.f. at each of the two endpoints 0 and θ has been changed by replacing the weak inequalities in Eq. (7) with strict inequalities in Eq. (9). Therefore, either equation could be used as the p.d.f. of the uniform distribution. However, if Eq. (9) is used as the p.d.f, then an M.L.E. of θ will be a value of θ for which $\theta > x_i$ for $i = 1, \ldots, n$ and which maximizes $1/\theta^n$ among all such values. It should be noted that the possible values of θ no longer include the value $\theta = \max(x_1, \ldots, x_n)$, since θ must be *strictly* greater than each observed value x_i $(i = 1, \ldots, n)$. Since θ can be chosen arbitrarily close to the value $\max(x_1, \ldots, x_n)$ but cannot be chosen equal to this value, it follows that the M.L.E. of θ does not exist. □

Examples 4 and 5 illustrate one shortcoming of the concept of an M.L.E. In all our previous discussions about p.d.f.'s, we emphasized the fact that it is irrelevant whether the p.d.f. of the uniform distribution is chosen to be equal to $1/\theta$ over the open interval $0 < x < \theta$ or over the closed interval $0 \leqslant x \leqslant \theta$. Now, however, we see that the existence of an M.L.E. depends on this irrelevant and unimportant choice. This difficulty is easily avoided in Example 5 by using the p.d.f. given by Eq. (7) rather than that given by Eq. (9). In many other problems as well, a difficulty of this type in regard to the existence of an M.L.E. can be avoided simply by choosing one particular appropriate version of the p.d.f. to represent the given distribution. However, as we shall see in Example 7, the difficulty cannot always be avoided.

Example 6: Nonuniqueness of an M.L.E. Suppose that $X_1, \ldots, X_n$ form a random sample from a uniform distribution on the interval $(\theta, \theta + 1)$, where the value of the parameter θ is unknown $(-\infty < \theta < \infty)$. In this example, the joint p.d.f. $f_n(x \mid \theta)$ has the form

$$f_n(x \mid \theta) = \begin{cases} 1 & \text{for } \theta \leqslant x_i \leqslant \theta + 1 \ (i = 1, \ldots, n), \\ 0 & \text{otherwise.} \end{cases} \tag{10}$$

The condition that $\theta \leqslant x_i$ for $i = 1, \ldots, n$ is equivalent to the condition that $\theta \leqslant \min(x_1, \ldots, x_n)$. Similarly, the condition that $x_i \leqslant \theta + 1$ for $i = 1, \ldots, n$ is equivalent to the condition that $\theta \geqslant \max(x_1, \ldots, x_n) - 1$. Therefore, instead of writing $f_n(x \mid \theta)$ in the form given in Eq. (10), we can use the following form:

$$f_n(x \mid \theta) = \begin{cases} 1 & \text{for } \max(x_1, \ldots, x_n) - 1 \leqslant \theta \leqslant \min(x_1, \ldots, x_n), \\ 0 & \text{otherwise.} \end{cases} \tag{11}$$

Thus, it is possible to select as an M.L.E. any value of θ in the interval

$$\max(x_1, \ldots, x_n) - 1 \leqslant \theta \leqslant \min(x_1, \ldots, x_n). \tag{12}$$

In this example, the M.L.E. is not uniquely specified. In fact, the method of maximum likelihood provides no help at all in choosing an estimate of θ. The likelihood of any value of θ outside the interval (12) is actually 0. Therefore, no value θ outside this interval would ever be estimated, and all values inside the interval are M.L.E.'s. $\square$

Example 7: Sampling from a Mixture of Two Distributions. Consider a random variable X that can come with equal probability either from a normal distribution with mean 0 and variance 1 or from another normal distribution with mean μ and variance σ^2, where both μ and σ^2 are unknown. Under these conditions, the p.d.f. $f(x \mid \mu, \sigma^2)$ of X will be the average of the p.d.f.'s of the two different normal distributions. Thus,

$$f(x \mid \mu, \sigma^2) = \frac{1}{2}\left\{ \frac{1}{(2\pi)^{1/2}} \exp\left(-\frac{x^2}{2}\right) + \frac{1}{(2\pi)^{1/2}\sigma} \exp\left[-\frac{(x-\mu)^2}{2\sigma^2}\right] \right\}. \tag{13}$$

Suppose now that $X_1, \ldots, X_n$ form a random sample from the distribution for which the p.d.f. is given by Eq. (13). As usual, the likelihood function $f_n(x \mid \mu, \sigma^2)$ has the form

$$f_n(x \mid \mu, \sigma^2) = \prod_{i=1}^{n} f(x_i \mid \mu, \sigma^2). \tag{14}$$

To find the M.L.E.'s of μ and σ^2, we must find values of μ and σ^2 for which $f_n(x \mid \mu, \sigma^2)$ is maximized.

Let x_k denote any one of the observed values $x_1, \ldots, x_n$. If we let $\mu = x_k$ and let $\sigma^2 \to 0$, then the factor $f(x_k \mid \mu, \sigma^2)$ on the right side of Eq. (14) will grow large without bound, while each factor $f(x_i \mid \mu, \sigma^2)$ for $x_i \neq x_k$ will approach the value

$$\frac{1}{2(2\pi)^{1/2}} \exp\left(-\frac{x_i^2}{2}\right).$$

Hence, when $\mu = x_k$ and $\sigma^2 \to 0$, we find that $f_n(x \mid \mu, \sigma^2) \to \infty$.

The value 0 is not a permissible estimate of σ^2 because we know in advance that $\sigma^2 > 0$. Since the likelihood function can be made arbitrarily large by choosing $\mu = x_k$ and choosing σ^2 arbitrarily close to 0, it follows that M.L.E.'s do not exist.

If we try to correct this difficulty by allowing the value 0 to be a permissible estimate of σ^2, then we find that there are n different pairs of M.L.E.'s of μ and σ^2; namely, $\hat{\mu} = x_k$ and $\widehat{\sigma^2} = 0$ for $k = 1, \ldots, n$. All of these estimates appear silly. Consider again the description, given at the beginning of this example, of the two normal distributions from which each observation might come. Suppose, for example, that $n = 1000$ and that we use the estimates $\hat{\mu} = x_3$ and $\widehat{\sigma^2} = 0$. Then, we would be estimating the value of the unknown variance to be 0; and also, in effect, we would be concluding that exactly one observed value x_3 came from the given unknown normal distribution whereas all the other 999 observed values came from the normal distribution with mean 0 and variance 1. In fact, however, since each observation was equally likely to come from either of the two distributions, it is much more probable that hundreds of observed values, rather than just one, came from the unknown normal distribution. In this example, the method of maximum likelihood is obviously unsatisfactory. $\square$

EXERCISES

1. It is not known what proportion p of the purchases of a certain brand of breakfast cereal are made by women and what proportion are made by men. In a random sample of 70 purchases of this cereal, it was found that 58 were made by women and 12 were made by men. Find the M.L.E. of p.

2. Consider again the conditions in Exercise 1, but suppose also that it is known that $\frac{1}{2} \leqslant p \leqslant \frac{2}{3}$. If the observations in the random sample of 70 purchases are as given in Exercise 1, what is the M.L.E. of p?

3. Suppose that $X_1, \ldots, X_n$ form a random sample from a Bernoulli distribution for which the parameter θ is unknown, but it is known that θ lies in the open interval $0 < \theta < 1$. Show that the M.L.E. of θ does not exist if every observed value is 0 or if every observed value is 1.

4. Suppose that $X_1, \ldots, X_n$ form a random sample from a Poisson distribution for which the mean θ is unknown ($\theta > 0$).

   (a) Determine the M.L.E. of θ, assuming that at least one of the observed values is different from 0.

   (b) Show that the M.L.E. of θ does not exist if every observed value is 0.

5. Suppose that $X_1, \ldots, X_n$ form a random sample from a normal distribution for which the mean μ is known but the variance σ^2 is unknown. Find the M.L.E. of σ^2.

6. Suppose that $X_1, \ldots, X_n$ form a random sample from an exponential distribution for which the value of the parameter β is unknown ($\beta > 0$). Find the M.L.E. of β.

7. Suppose that $X_1, \ldots, X_n$ form a random sample from a distribution for which the p.d.f. $f(x \mid \theta)$ is as follows:

$$f(x \mid \theta) = \begin{cases} e^{\theta - x} & \text{for } x > \theta, \\ 0 & \text{for } x \leqslant \theta. \end{cases}$$

Also, suppose that the value of θ is unknown ($-\infty < \theta < \infty$).

(a) Show that the M.L.E. of θ does not exist.

(b) Determine another version of the p.d.f. of this same distribution for which the M.L.E. of θ will exist, and find this estimator.

8. Suppose that $X_1, \ldots, X_n$ form a random sample from a distribution for which the p.d.f. $f(x \mid \theta)$ is as follows:

$$f(x \mid \theta) = \begin{cases} \theta x^{\theta - 1} & \text{for } 0 < x < 1, \\ 0 & \text{otherwise.} \end{cases}$$

Also, suppose that the value of θ is unknown ($\theta > 0$). Find the M.L.E. of θ.

9. Suppose that $X_1, \ldots, X_n$ form a random sample from a distribution for which the p.d.f. $f(x \mid \theta)$ is as follows:

$$f(x \mid \theta) = \frac{1}{2} e^{-|x - \theta|} \qquad \text{for } -\infty < x < \infty.$$

Also, suppose that the value of θ is unknown ($-\infty < \theta < \infty$). Find the M.L.E. of θ.

10. Suppose that $X_1, \ldots, X_n$ form a random sample from a uniform distribution on the interval (θ_1, θ_2), where both θ_1 and θ_2 are unknown ($-\infty < \theta_1 < \theta_2 < \infty$). Find the M.L.E.'s of θ_1 and θ_2.

11. Suppose that a certain large population contains k different types of individuals ($k \geqslant 2$); and let θ_i denote the proportion of individuals of type i, for $i = 1, \ldots, k$. Here, $0 \leqslant \theta_i \leqslant 1$ and $\theta_1 + \cdots + \theta_k = 1$. Suppose also that in a random sample of n individuals from this population, exactly n_i individuals are of type i, where $n_1 + \cdots + n_k = n$. Find the M.L.E.'s of $\theta_1, \ldots, \theta_k$.

12. Suppose that the two-dimensional vectors $(X_1, Y_1), (X_2, Y_2), \ldots, (X_n, Y_n)$ form a random sample from a bivariate normal distribution for which the means of X and Y are unknown but the variances of X and Y and the correlation between X and Y are known. Find the M.L.E.'s of the means.

13. Suppose that the two-dimensional vectors $(X_1, Y_1), (X_2, Y_2), \ldots, (X_n, Y_n)$ form a random sample from a bivariate normal distribution for which the

means of X and Y, the variances of X and Y, and the correlation between X and Y are unknown. Show that the M.L.E.'s of these five parameters are as follows:

$$\hat{\mu}_1 = \overline{X}_n \quad \text{and} \quad \hat{\mu}_2 = \overline{Y}_n,$$

$$\widehat{\sigma_1^2} = \frac{1}{n} \sum_{i=1}^{n} (X_i - \overline{X}_n)^2 \quad \text{and} \quad \widehat{\sigma_2^2} = \frac{1}{n} \sum_{i=1}^{n} (Y_i - \overline{Y}_n)^2,$$

$$\hat{\rho} = \frac{\sum_{i=1}^{n}(X_i - \overline{X}_n)(Y_i - \overline{Y}_n)}{\left[\sum_{i=1}^{n}(X_i - \overline{X}_n)^2\right]^{1/2}\left[\sum_{i=1}^{n}(Y_i - \overline{Y}_n)^2\right]^{1/2}}.$$

6.6. PROPERTIES OF MAXIMUM LIKELIHOOD ESTIMATORS

Invariance

Suppose that the variables $X_1, \ldots, X_n$ form a random sample from a distribution for which either the p.f. or the p.d.f. is $f(x \mid \theta)$, where the value of the parameter θ is unknown, and let $\hat{\theta}$ denote the M.L.E. of θ. Thus, for any observed values $x_1, \ldots, x_n$, the likelihood function $f_n(x \mid \theta)$ is maximized when $\theta = \hat{\theta}$.

Suppose now that we change the parameter in the distribution as follows: Instead of expressing the p.f. or the p.d.f. $f(x \mid \theta)$ in terms of the parameter θ, we shall express it in terms of a new parameter $\tau = g(\theta)$, where g is a one-to-one function of θ. We shall let $\theta = h(\tau)$ denote the inverse function. Then, expressed in terms of the new parameter τ, the p.f. or p.d.f. of each observed value will be $f[x \mid h(\tau)]$ and the likelihood function will be $f_n[x \mid h(\tau)]$.

The M.L.E. $\hat{\tau}$ of τ will be equal to the value of τ for which $f_n[x \mid h(\tau)]$ is maximized. Since $f_n(x \mid \theta)$ is maximized when $\theta = \hat{\theta}$, it follows that $f_n[x \mid h(\tau)]$ will be maximized when $h(\tau) = \hat{\theta}$. Hence, the M.L.E. $\hat{\tau}$ must satisfy the relation $h(\hat{\tau}) = \hat{\theta}$ or, equivalently, $\hat{\tau} = g(\hat{\theta})$. We have therefore established the following property, which is called the *invariance* property of maximum likelihood estimators:

> If $\hat{\theta}$ is the maximum likelihood estimator of θ, then $g(\hat{\theta})$ is the maximum likelihood estimator of $g(\theta)$.

The invariance property can be extended to functions of a vector parameter $\boldsymbol{\theta}$. Suppose that $\boldsymbol{\theta} = (\theta_1, \ldots, \theta_k)$ is a vector of k real-valued parameters. If $\tau = g(\theta_1, \ldots, \theta_k)$ is a real-valued function of $\theta_1, \ldots, \theta_k$, then τ can be regarded as a single component of a one-to-one transformation from the set of parameters

$\theta_1, \ldots, \theta_k$ to a new set of k real-valued parameters. Therefore, if $\hat{\theta}_1, \ldots, \hat{\theta}_k$ are the M.L.E.'s of $\theta_1, \ldots, \theta_k$, it follows from the invariance property that the M.L.E. of τ is $\hat{\tau} = g(\hat{\theta}_1, \ldots, \hat{\theta}_k)$.

Example 1: Estimating the Standard Deviation and the Second Moment. Suppose that the variables $X_1, \ldots, X_n$ form a random sample from a normal distribution for which both the mean μ and the variance σ^2 are unknown. We shall determine the M.L.E. of the standard deviation σ and the M.L.E. of the second moment of the normal distribution $E(X^2)$. It was found in Example 3 of Sec. 6.5 that the M.L.E.'s $\hat{\mu}$ and $\widehat{\sigma^2}$ of the mean and the variance are the sample mean and the sample variance, respectively. From the invariance property, we can conclude that the M.L.E. $\hat{\sigma}$ of the standard deviation is simply the square root of the sample variance. In symbols $\hat{\sigma}^2 = \widehat{\sigma^2}$. Also, since $E(X^2) = \sigma^2 + \mu^2$, the M.L.E. of $E(X^2)$ will be $\hat{\sigma}^2 + \hat{\mu}^2$. $\square$

Numerical Computation

In many problems there exists a unique M.L.E. $\hat{\theta}$ of a given parameter θ, but this M.L.E. cannot be expressed as an explicit algebraic function of the observations in the sample. For a given set of observed values, it is necessary to determine the value of $\hat{\theta}$ by numerical computation. We shall illustrate this condition by two examples.

Example 2: Sampling from a Gamma Distribution. Suppose that the variables $X_1, \ldots, X_n$ form a random sample from a gamma distribution for which the p.d.f. is as follows:

$$f(x \mid \alpha) = \frac{1}{\Gamma(\alpha)} x^{\alpha-1} e^{-x} \qquad \text{for } x > 0. \tag{1}$$

Suppose also that the value of α is unknown ($\alpha > 0$) and is to be estimated. The likelihood function is

$$f_n(x \mid \alpha) = \frac{1}{\Gamma^n(\alpha)} \left(\prod_{i=1}^{n} x_i \right)^{\alpha-1} \exp\left(-\sum_{i=1}^{n} x_i \right). \tag{2}$$

The M.L.E. of α will be the value of α which satisfies the equation

$$\frac{\partial \log f_n(x \mid \alpha)}{\partial \alpha} = 0. \tag{3}$$

When we apply Eq. (3) in this example, we obtain the following equation:

$$\frac{\Gamma'(\alpha)}{\Gamma(\alpha)} = \frac{1}{n} \sum_{i=1}^{n} \log x_i. \tag{4}$$

Tables of the function $\Gamma'(\alpha)/\Gamma(\alpha)$, which is called the *digamma function*, are included in various published collections of mathematical tables. For any given values of $x_1, \ldots, x_n$, the unique value of α that satisfies Eq. (4) must be determined either by referring to these tables or by carrying out a numerical analysis of the digamma function. This value will be the M.L.E. of α. $\square$

Example 3: Sampling from a Cauchy Distribution. Suppose that the variables $X_1, \ldots, X_n$ form a random sample from a Cauchy distribution centered at an unknown point θ $(-\infty < \theta < \infty)$, for which the p.d.f. is as follows:

$$f(x \mid \theta) = \frac{1}{\pi \left[1 + (x - \theta)^2 \right]} \qquad \text{for } -\infty < x < \infty. \tag{5}$$

Suppose also that the value of θ is to be estimated.

The likelihood function is

$$f_n(x \mid \theta) = \frac{1}{\pi^n \prod_{i=1}^{n} \left[1 + (x_i - \theta)^2 \right]}. \tag{6}$$

Therefore, the M.L.E. of θ will be the value which minimizes

$$\prod_{i=1}^{n} \left[1 + (x_i - \theta)^2 \right]. \tag{7}$$

For any given values of $x_1, \ldots, x_n$, the value of θ which minimizes the expression (7) must be determined by a numerical computation. $\square$

Consistency

Consider an estimation problem in which a random sample is to be taken from a distribution involving a parameter θ. Suppose that for every sufficiently large sample size n, that is, for every value of n greater than some given minimum number, there exists a unique M.L.E. of θ. Then, under certain conditions which are typically satisfied in practical problems, the sequence of M.L.E.'s is a consistent sequence of estimators of θ. In other words, in most problems the sequence of M.L.E.'s converges in probability to the unknown value of θ as $n \to \infty$.

We have remarked in Sec. 6.4 that under certain general conditions the sequence of Bayes estimators of a parameter θ is also a consistent sequence of estimators. Therefore, for a given prior distribution and a sufficiently large sample size n, the Bayes estimator and the M.L.E. of θ will typically be very close to each other, and both will be very close to the unknown value of θ.

We shall not present any formal details of the conditions that are needed to prove this result. We shall, however, illustrate the result by considering again a random sample $X_1, \ldots, X_n$ from a Bernoulli distribution for which the parameter θ is unknown ($0 \leqslant \theta \leqslant 1$). It was shown in Sec. 6.4 that if the given prior distribution of θ is a beta distribution, then the difference between the Bayes estimator of θ and the sample mean $\overline{X}_n$ converges to 0 as $n \to \infty$. Furthermore, it was shown in Example 1 of Sec. 6.5 that the M.L.E. of θ is $\overline{X}_n$. Thus, as $n \to \infty$, the difference between the Bayes estimator and the M.L.E. will converge to 0. Finally, as remarked in Sec. 6.4, the sample mean $\overline{X}_n$ converges in probability to θ as $n \to \infty$. Therefore both the sequence of Bayes estimators and the sequence of M.L.E.'s are consistent sequences.

Sampling Plans

Suppose that an experimenter wishes to take observations from a distribution for which the p.f. or the p.d.f. is $f(x \mid \theta)$ in order to gain information about the value of the parameter θ. The experimenter could simply take a random sample of a predetermined size from the distribution. Instead, however, he may begin by first observing a few values at random from the distribution, and noting the cost and the time spent in taking these observations. He may then decide to observe a few more values at random from the distribution, and to study all the values thus far obtained. At some point the experimenter will decide to stop taking observations and will estimate the value of θ from all the observed values that have been obtained up to that point. He might decide to stop either because he feels that he has enough information to be able to make a good estimate of θ or because he feels that he cannot afford to spend any more money or time on sampling.

In this experiment, the number n of observations in the sample is not fixed beforehand. It is a random variable whose value may very well depend on the magnitudes of the observations as they are obtained.

Regardless of whether an experimenter decides to fix the value of n before any observations are taken or prefers to use some other sampling plan, such as the one just described, it can be shown that the likelihood function $L(\theta)$ based on the observed values can be taken to be

$$L(\theta) = f(x_1 \mid \theta) \cdots f(x_n \mid \theta).$$

It follows, therefore, that the M.L.E. of θ will be the same, no matter what type of sampling plan is used. In other words, the value of θ depends only on the values $x_1, \ldots, x_n$ that are actually observed and does not depend on the plan (if there was one) that was used by the experimenter to decide when to stop sampling.

To illustrate this property, suppose that the intervals of time, in minutes, between arrivals of successive customers at a certain service facility are i.i.d. random variables. Suppose also that each interval has an exponential distribution with parameter β, and that a set of observed intervals $X_1, \ldots, X_n$ form a random sample from this distribution. It follows from Exercise 6 of Sec. 6.5 that the M.L.E. of β will be $\hat{\beta} = 1/\bar{X}_n$. Also, since the mean μ of the exponential distribution is $1/\beta$, it follows from the invariance property of M.L.E.'s that $\hat{\mu} = \bar{X}_n$. In other words, the M.L.E. of the mean is the average of the observations in the sample.

Consider now the following three sampling plans:

(i) An experimenter decides in advance to take exactly 20 observations, and the average of these 20 observations turns out to be 6. Then the M.L.E. of μ is $\hat{\mu} = 6$.

(ii) An experimenter decides to take observations $X_1, X_2 \ldots$ until he obtains a value greater than 10. He finds that $X_i < 10$ for $i = 1, \ldots, 19$ and that $X_{20} > 10$. Hence, sampling terminates after 20 observations. If the average of these 20 observations is 6, then the M.L.E. is again $\hat{\mu} = 6$.

(iii) An experimenter takes observations one at a time, with no particular plan in mind, either until he is forced to stop sampling or until he feels that he has gained enough information to make a good estimate of μ. If for either reason he stops after he has taken 20 observations and if the average of the 20 observations is 6, then the M.L.E. is again $\hat{\mu} = 6$.

Sometimes, an experiment of this type must be terminated during an interval when the experimenter is waiting for the next customer to arrive. If a certain amount of time has elapsed since the arrival of the last customer, this time should not be omitted from the sample data, even though the full interval to the arrival of the next customer has not been observed. Suppose, for example, that the average of the first 20 observations is 6; that the experimenter waits another 15 minutes but no other customer arrives; and that he then terminates the experiment. In this case, we know that the M.L.E. of μ would have to be greater than 6, since the value of the 21st observation must be greater than 15, even though its exact value is unknown. The new M.L.E. can be obtained by multiplying the likelihood function for the first 20 observations by the probability that the 21st observation is greater than 15, and finding the value of θ which maximizes this new likelihood function (see Exercise 14).

Other properties of M.L.E.'s will be discussed later in this chapter and in Chapter 7.

The Likelihood Principle

The values of both the M.L.E. of a parameter θ and the Bayes estimator of θ depend on the observed values in the sample only through the likelihood function $f_n(x \mid \theta)$ determined by these observed values. Therefore, if two different sets of observed values determine the same likelihood function, then the same value of the M.L.E. of θ will be obtained for both sets of observed values. Likewise, for a given prior p.d.f. of θ, the same value of the Bayes estimate of θ will be obtained for both sets of observed values.

Now suppose that two different sets of observed values x and y determine likelihood functions which are proportional to each other. In other words, the likelihood functions differ only by a factor which may depend on x and y but does not depend on θ. In this case, it can be verified that the M.L.E. of θ will again be the same regardless of whether x or y is considered. Also, for any given prior p.d.f. of θ, the Bayes estimate of θ will be the same. Thus, both M.L.E.'s and Bayes estimators are compatible with the following principle of statistical inference, which is known as the *likelihood principle*:

> *Suppose that two different sets of observed values x and y that might be obtained either from the same experiment or from two different experiments have the property that they determine the same likelihood function for a certain parameter θ or determine likelihood functions which are proportional to each other. Then x and y furnish the same information about the unknown value of θ, and a statistician should obtain the same estimate of θ from either x or y.*

For example, suppose that a statistician must estimate the unknown proportion θ of defective items in a large manufactured lot. Suppose also that the statistician is informed that ten items were selected at random from the lot, and that exactly two were defective and eight were nondefective. Suppose, however, that the statistician does not know which one of the following two experiments had been performed: (i) A fixed sample of ten items had been selected from the lot and it was found that two of the items were defective. (ii) Items had been selected at random from the lot, one at a time, until two defective items had been obtained and it was found that a total of ten items had to be selected.

For each of these two possible experiments, the observed values determine a likelihood function that is proportional to $\theta^2(1 - \theta)^8$ for $0 \leqslant \theta \leqslant 1$. Therefore, if the statistician uses a method of estimation that is compatible with the likelihood principle, he does not need to know which one of the two possible experiments was actually performed. His estimate of θ would be the same for either case.

We have already remarked that for any given prior distribution of θ and any loss function, the Bayes estimate of θ would be the same for either experiment. We have also remarked that the M.L.E. of θ would be the same for either experiment. However, in Chapter 7 we shall discuss a method of estimation called *unbiased estimation*. Although this method is widely used in statistical problems, it violates the likelihood principle and specifies that a different estimate of θ should be used in each of the two experiments.

EXERCISES

1. Suppose that the variables $X_1, \ldots, X_n$ form a random sample from a Poisson distribution for which the mean is unknown. Determine the M.L.E. of the standard deviation of the distribution.

2. Suppose that $X_1, \ldots, X_n$ form a random sample from an exponential distribution for which the value of the parameter β is unknown. Determine the M.L.E. of the median of the distribution.

3. Suppose that the lifetime of a certain type of lamp has an exponential distribution for which the value of the parameter β is unknown. A random sample of n lamps of this type are tested for a period of T hours and the number X of lamps which fail during this period is observed; but the times at which the failures occurred are not noted. Determine the M.L.E. of β based on the observed value of X.

4. Suppose that $X_1, \ldots, X_n$ form a random sample from a uniform distribution on the interval (a, b), where both endpoints a and b are unknown. Find the M.L.E. of the mean of the distribution.

5. Suppose that $X_1, \ldots, X_n$ form a random sample from a normal distribution for which both the mean and the variance are unknown. Find the M.L.E. of the 0.95 quantile of the distribution, that is, of the point θ such that $\Pr(X < \theta) = 0.95$.

6. For the conditions of Exercise 5, find the M.L.E. of $\nu = \Pr(X > 2)$.

7. Suppose that $X_1, \ldots, X_n$ form a random sample from a gamma distribution for which the p.d.f. is given by Eq. (1) in this section. Find the M.L.E. of $\Gamma'(\alpha)/\Gamma(\alpha)$.

8. Suppose that $X_1, \ldots, X_n$ form a random sample from a gamma distribution for which both parameters α and β are unknown. Find the M.L.E. of α/β.

9. Suppose that $X_1, \ldots, X_n$ form a random sample from a beta distribution for which both parameters α and β are unknown. Show that the M.L.E.'s of α and β satisfy the following equation:

$$\frac{\Gamma'(\hat{\alpha})}{\Gamma(\hat{\alpha})} - \frac{\Gamma'(\hat{\beta})}{\Gamma(\hat{\beta})} = \frac{1}{n} \sum_{i=1}^{n} \log \frac{X_i}{1 - X_i}.$$

10. Suppose that $X_1, \ldots, X_n$ form a random sample of size n from a uniform

distribution on the interval $(0, \theta)$, where the value of θ is unknown. Show that the sequence of M.L.E.'s of θ is a consistent sequence.

11. Suppose that $X_1, \ldots, X_n$ form a random sample from an exponential distribution for which the value of the parameter β is unknown. Show that the sequence of M.L.E.'s of β is a consistent sequence.

12. Suppose that $X_1, \ldots, X_n$ form a random sample from a distribution for which the p.d.f. is as specified in Exercise 8 of Sec. 6.5. Show that the sequence of M.L.E.'s of θ is a consistent sequence.

13. Suppose that a scientist desires to estimate the proportion p of monarch butterflies that have a special type of marking on their wings.

    (a) Suppose that he captures monarch butterflies one at a time until he has found five that have this special marking. If he must capture a total of 43 butterflies, what is the M.L.E. of p?

    (b) Suppose that at the end of a day the scientist had captured 58 monarch butterflies and had found only three with the special marking. What is the M.L.E. of p?

14. Suppose that 21 observations are taken at random from an exponential distribution for which the mean μ is unknown ($\mu > 0$); that the average of 20 of these observations is 6; and that although the exact value of the other observation could not be determined, it was known to be greater than 15. Determine the M.L.E. of μ.

15. Suppose that each of two statisticians A and B must estimate a certain parameter θ whose value is unknown ($\theta > 0$). Statistician A can observe the value of a random variable X which has a gamma distribution with parameters α and β, where $\alpha = 3$ and $\beta = \theta$; and statistician B can observe the value of a random variable Y which has a Poisson distribution with mean 2θ. Suppose that the value observed by statistician A is $X = 2$ and the value observed by statistician B is $Y = 3$. Show that the likelihood functions determined by these observed values are proportional, and find the common value of the M.L.E. of θ obtained by each statistician.

16. Suppose that each of two statisticians A and B must estimate a certain parameter p whose value is unknown ($0 < p < 1$). Statistician A can observe the value of a random variable X which has a binomial distribution with parameters $n = 10$ and p; and statistician B can observe the value of a random variable Y which has a negative binomial distribution with parameters $r = 4$ and p. Suppose that the value observed by statistician A is $X = 4$ and the value observed by statistician B is $Y = 6$. Show that the likelihood functions determined by these observed values are proportional, and find the common value of the M.L.E. of p obtained by each statistician.

6.7. SUFFICIENT STATISTICS

Definition of a Statistic

In many problems in which a parameter θ must be estimated, it is possible to find either an M.L.E. or a Bayes estimator that will be suitable. In some problems, however, neither of these estimators may be suitable. There may not be any M.L.E., or there may be more than one. Even when an M.L.E. is unique, it may not be a suitable estimator, as in Example 4 of Sec. 6.5, where the M.L.E. always underestimates the value of θ. Reasons why there may not be a suitable Bayes estimator were presented at the beginning of Sec. 6.5. In such problems, the search for a good estimator must be extended beyond the methods that have been introduced thus far.

In this section, we shall define the concept of a sufficient statistic, which was introduced by R. A. Fisher in 1922; and we shall show how this concept can be used to expedite the search for a good estimator in many problems. We shall assume, as usual, that the random variables $X_1, \ldots, X_n$ form a random sample from either a discrete distribution or a continuous distribution, and we shall let $f(x \mid \theta)$ denote the p.f. or the p.d.f. of this distribution. For simplicity, it will be convenient in most of the following discussion to assume that the distribution is continuous and that $f(x \mid \theta)$ is a p.d.f. However, it should be kept in mind that the discussion applies equally well, except perhaps with some obvious changes, to a discrete distribution for which the p.f. is $f(x \mid \theta)$. We shall also assume that the unknown value of θ must belong to some specified parameter space Ω.

Since the random variables $X_1, \ldots, X_n$ form a random sample, it is known that their joint p.d.f. $f_n(x \mid \theta)$ has the following form for some particular value of $\theta \in \Omega$:

$$f_n(x \mid \theta) = f(x_1 \mid \theta) \cdots f(x_n \mid \theta). \tag{1}$$

In other words, it is known that the joint p.d.f. of $X_1, \ldots, X_n$ is a member of the family containing all p.d.f.'s having the form (1) for all possible values of $\theta \in \Omega$. The problem of estimating the value of θ can therefore be viewed as the problem of selecting by inference the particular distribution in this family which generated the observations $X_1, \ldots, X_n$.

Any real-valued function $T = r(X_1, \ldots, X_n)$ of the observations in the random sample is called a *statistic*. Three examples of statistics are the sample mean $\overline{X}_n$, the maximum Y_n of the values of $X_1, \ldots, X_n$, and the function $r(X_1, \ldots, X_n)$ which has the constant value 3 for all values of $X_1, \ldots, X_n$. In any estimation problem, we can say that an estimator of θ is a statistic whose value can be regarded as an estimate of the value of θ.

For any fixed value of $\theta \in \Omega$, the distribution of any given statistic T can be derived from the joint p.d.f. of $X_1, \ldots, X_n$ given in Eq. (1). In general, this distribution will depend on the value of θ. Hence, there will be a family of possible distributions of T corresponding to the different possible values of $\theta \in \Omega$.

Definition of a Sufficient Statistic

Suppose that in a specific estimation problem, two statisticians A and B must estimate the value of the parameter θ; that statistician A can observe the values of the observations $X_1, \ldots, X_n$ in a random sample; and that statistician B cannot observe the individual values of $X_1, \ldots, X_n$ but can learn the value of a certain statistic $T = r(X_1, \ldots, X_n)$. In this case statistician A can choose any function of the observations $X_1, \ldots, X_n$ as an estimator of θ, whereas statistician B can use only a function of T. Hence, it follows that A will generally be able to find a better estimator than will B.

In some problems, however, B will be able to do just as well as A. In such a problem, the single function $T = r(X_1, \ldots, X_n)$ will in some sense summarize all the information contained in the random sample, and knowledge of the individual values of $X_1, \ldots, X_n$ will be irrelevant in the search for a good estimator of θ. A statistic T having this property is called a *sufficient statistic*. We shall now present the formal definition of a sufficient statistic.

If T is a statistic and t is any particular value of T, then the conditional joint distribution of $X_1, \ldots, X_n$, given that $T = t$, can be calculated from Eq. (1). In general, this conditional joint distribution will depend on the value of θ. Therefore, for each value of t, there will be a family of possible conditional distributions corresponding to the different possible values of $\theta \in \Omega$. It may happen, however, that for each possible value of t, the conditional joint distribution of $X_1, \ldots, X_n$, given that $T = t$, is the same for all values of $\theta \in \Omega$ and therefore does not actually depend on the value of θ. In this case it is said that T is a *sufficient statistic for the parameter θ*.

Before we describe a simple method for finding a sufficient statistic, and before we consider an example of a sufficient statistic, we shall indicate why a sufficient statistic T that satisfies the definition just given is regarded as summarizing all the relevant information about θ contained in the sample $X_1, \ldots, X_n$. Let us return to the case of the statistician B who can learn only the value of the statistic T and cannot observe the individual values of $X_1, \ldots, X_n$. If T is a sufficient statistic, then the conditional joint distribution of $X_1, \ldots, X_n$, given that $T = t$, is completely known for any observed value t and does not depend on the unknown value of θ. Therefore, for any value t that might be observed, statistician B could, in principle, generate n random variables $X_1', \ldots, X_n'$ in accordance with this conditional joint distribution. The process of generating random variables $X'_1, \ldots, X'_n$ having a specified joint probability distribution is called an *auxiliary randomization*.

When we use this process of first observing T and then generating $X'_1, \ldots, X'_n$ in accordance with the specified conditional joint distribution, it follows that for any given value of $\theta \in \Omega$, the marginal joint distribution of $X'_1, \ldots, X'_n$ will be the same as the joint distribution of $X_1, \ldots, X_n$. Hence, if statistician B can observe the value of a sufficient statistic T, then he can generate n random variables $X'_1, \ldots, X'_n$ which have the same joint distribution as the original random sample $X_1, \ldots, X_n$. The property that distinguishes a sufficient statistic T from a statistic which is not sufficient may be described as follows: The auxiliary randomization used to generate the random variables $X'_1, \ldots, X'_n$ after the sufficient statistic T has been observed does not require any knowledge about the value of θ, since the conditional joint distribution of $X_1, \ldots, X_n$ when T is given does not depend on the value of θ. If the statistic T were not sufficient, this auxiliary randomization could not be carried out because the conditional joint distribution of $X_1, \ldots, X_n$ for a given value of T would involve the value of θ and this value is unknown.

We can now show why statistician B, who observes only the value of a sufficient statistic T, can nevertheless estimate θ just as well as can statistician A, who observes the values of $X_1, \ldots, X_n$. Suppose that A plans to use a particular estimator $\delta(X_1, \ldots, X_n)$ to estimate θ, and that B observes the value of T and generates $X'_1, \ldots, X'_n$, which have the same joint distribution as $X_1, \ldots, X_n$. If B uses the estimator $\delta(X'_1, \ldots, X'_n)$, then it follows that the probability distribution of B's estimator will be the same as the probability distribution of A's estimator. This discussion illustrates why, when searching for a good estimator, a statistician can restrict the search to estimators which are functions of a sufficient statistic T. We shall return to this point in Sec. 6.9.

The Factorization Criterion

We shall now present a simple method for finding a sufficient statistic which can be applied in many problems. This method is based on the following result, which was developed with increasing generality by R. A. Fisher in 1922, J. Neyman in 1935, and P. R. Halmos and L. J. Savage in 1949.

The Factorization Criterion. *Let $X_1, \ldots, X_n$ form a random sample from either a continuous distribution or a discrete distribution for which the p.d.f. or the p.f. is $f(x \mid \theta)$, where the value of θ is unknown and belongs to a given parameter space Ω. A statistic $T = r(X_1, \ldots, X_n)$ is a sufficient statistic for θ if and only if the joint p.d.f. or the joint p.f. $f_n(x \mid \theta)$ of $X_1, \ldots, X_n$ can be factored as follows for all values of $x = (x_1, \ldots, x_n) \in R^n$ and all values of $\theta \in \Omega$:*

$$f_n(x \mid \theta) = u(x)v[r(x), \theta].$$ (2)

Here, the functions u and v are nonnegative; the function u may depend on x but does not depend on θ; and the function v will depend on θ but depends on the observed value x only through the value of the statistic r(x).

Proof. We shall give the proof only when the random vector $X = (X_1, \ldots, X_n)$ has a discrete distribution, in which case

$$f_n(x \mid \theta) = \Pr(X = x \mid \theta).$$

Suppose first that $f_n(x \mid \theta)$ can be factored as in Eq. (2) for all values of $x \in R^n$ and $\theta \in \Omega$. For each possible value t of T, let $A(t)$ denote the set of all points $x \in R^n$ such that $r(x) = t$. For any given value of $\theta \in \Omega$, we shall determine the conditional distribution of X given that $T = t$. For any point $x \in A(t)$,

$$\Pr(X = x \mid T = t, \theta) = \frac{\Pr(X = x \mid \theta)}{\Pr(T = t \mid \theta)} = \frac{f_n(x \mid \theta)}{\sum_{y \in A(t)} f_n(y \mid \theta)}.$$

Since $r(y) = t$ for every point $y \in A(t)$, and since $x \in A(t)$, it follows from Eq. (2) that

$$\Pr(X = x \mid T = t, \theta) = \frac{u(x)}{\sum_{y \in A(t)} u(y)}. \tag{3}$$

Finally, for any point x that does not belong to $A(t)$,

$$\Pr(X = x \mid T = t, \theta) = 0. \tag{4}$$

It can be seen from Eqs. (3) and (4) that the conditional distribution of X does not depend on θ. Therefore, T is a sufficient statistic.

Conversely, suppose that T is a sufficient statistic. Then, for any given value t of T, any point $x \in A(t)$, and any value of $\theta \in \Omega$, the conditional probability $\Pr(X = x \mid T = t, \theta)$ will not depend on θ and will therefore have the form

$$\Pr(X = x \mid T = t, \theta) = u(x).$$

If we let $v(t, \theta) = \Pr(T = t \mid \theta)$, it follows that

$$f_n(x \mid \theta) = \Pr(X = x \mid \theta) = \Pr(X = x \mid T = t, \theta)\Pr(T = t \mid \theta)$$

$$= u(x)v(t, \theta).$$

Hence, $f_n(x \mid \theta)$ has been factored in the form specified in Eq. (2).

The proof for a random sample $X_1, \ldots, X_n$ from a continuous distribution requires somewhat different methods and will not be given here. $\square$

For any value of x for which $f_n(x \mid \theta) = 0$ for all values of $\theta \in \Omega$, the value of the function $u(x)$ in Eq. (2) can be chosen to be 0. Therefore, when the factorization criterion is being applied, it is sufficient to verify that a factorization of the form given in Eq. (2) is satisfied for every value of x such that $f_n(x \mid \theta) > 0$ for at least one value of $\theta \in \Omega$.

We shall now illustrate the use of the factorization criterion by giving four examples.

Example 1: Sampling from a Poisson Distribution. Suppose that $X_1, \ldots, X_n$ form a random sample from a Poisson distribution for which the value of the mean θ is unknown ($\theta > 0$). We shall show that $T = \sum_{i=1}^{n} X_i$ is a sufficient statistic for θ.

For any set of nonnegative integers $x_1, \ldots, x_n$, the joint p.f. $f_n(x \mid \theta)$ of $X_1, \ldots, X_n$ is as follows:

$$f_n(x \mid \theta) = \prod_{i=1}^{n} \frac{e^{-\theta} \theta^{x_i}}{x_i!} = \left(\prod_{i=1}^{n} \frac{1}{x_i!} \right) e^{-n\theta} \theta^{y},$$

where $y = \sum_{i=1}^{n} x_i$. It can be seen that $f_n(x \mid \theta)$ has been expressed, as in Eq. (2), as the product of a function that does not depend on θ and a function that depends on θ but depends on the observed vector x only through the value of y. It follows that $T = \sum_{i=1}^{n} X_i$ is a sufficient statistic for θ. □

Example 2: Applying the Factorization Criterion to a Continuous Distribution. Suppose that $X_1, \ldots, X_n$ form a random sample from a continuous distribution with the following p.d.f.:

$$f(x \mid \theta) = \begin{cases} \theta x^{\theta - 1} & \text{for } 0 < x < 1, \\ 0 & \text{otherwise.} \end{cases}$$

It is assumed that the value of the parameter θ is unknown ($\theta > 0$). We shall show that $T = \prod_{i=1}^{n} X_i$ is a sufficient statistic for θ.

For $0 < x_i < 1$ ($i = 1, \ldots, n$), the joint p.d.f. $f_n(x \mid \theta)$ of $X_1, \ldots, X_n$ is as follows:

$$f_n(x \mid \theta) = \theta^n \left(\prod_{i=1}^{n} x_i \right)^{\theta - 1}. \tag{5}$$

Furthermore, if at least one value of x_i is outside the interval $0 < x_i < 1$, then $f_n(x \mid \theta) = 0$ for every value of $\theta \in \Omega$. The right side of Eq. (5) depends on x only through the value of the product $\prod_{i=1}^{n} x_i$. Therefore, if we let $u(x) = 1$ and $r(x) = \prod_{i=1}^{n} x_i$, then $f_n(x \mid \theta)$ in Eq. (5) can be considered to be factored in the form specified in Eq. (2). It follows from the factorization criterion that the statistic $T = \prod_{i=1}^{n} X_i$ is a sufficient statistic for θ. □

Example 3: Sampling from a Normal Distribution. Suppose that $X_1, \ldots, X_n$ form a random sample from a normal distribution for which the mean μ is unknown and the variance σ^2 is known. We shall show that $T = \sum_{i=1}^{n} X_i$ is a sufficient statistic for μ.

For $-\infty < x_i < \infty$ $(i = 1,\ldots, n)$, the joint p.d.f. of $X_1,\ldots, X_n$ is as follows:

$$f_n(x \mid \mu) = \prod_{i=1}^{n} \frac{1}{(2\pi)^{1/2}\sigma} \exp\left[-\frac{(x_i - \mu)^2}{2\sigma^2}\right]. \tag{6}$$

This equation can be rewritten in the form

$$f_n(x \mid \mu) = \frac{1}{(2\pi)^{n/2}\sigma^n} \exp\left(-\frac{1}{2\sigma^2}\sum_{i=1}^{n} x_i^2\right) \exp\left(\frac{\mu}{\sigma^2}\sum_{i=1}^{n} x_i - \frac{n\mu^2}{2\sigma^2}\right). \tag{7}$$

It can be seen that $f_n(x \mid \mu)$ has now been expressed as the product of a function that does not depend on μ and a function that depends on x only through the value of $\sum_{i=1}^{n} x_i$. It follows from the factorization criterion that $T = \sum_{i=1}^{n} X_i$ is a sufficient statistic for μ. $\square$

Since $\sum_{i=1}^{n} x_i = n\bar{x}_n$, we can state equivalently that the final factor in Eq. (7) depends on $x_1,\ldots, x_n$ only through the value of $\bar{x}_n$. Therefore, in Example 3 the statistic $\bar{X}_n$ is also a sufficient statistic for μ. More generally (see Exercise 13 at the end of this section), any one-to-one function of $\bar{X}_n$ will be a sufficient statistic for μ.

Example 4: Sampling from a Uniform Distribution. Suppose that $X_1,\ldots, X_n$ form a random sample from a uniform distribution on the interval $(0, \theta)$, where the value of the parameter θ is unknown $(\theta > 0)$. We shall show that $T = \max(X_1,\ldots, X_n)$ is a sufficient statistic for θ.

The p.d.f. $f(x \mid \theta)$ of each individual observation X_i is

$$f(x \mid \theta) = \begin{cases} \dfrac{1}{\theta} & \text{for } 0 \leqslant x \leqslant \theta, \\ 0 & \text{otherwise.} \end{cases}$$

Therefore, the joint p.d.f. $f_n(x \mid \theta)$ of $X_1,\ldots, X_n$ is

$$f_n(x \mid \theta) = \begin{cases} \dfrac{1}{\theta^n} & \text{for } 0 \leqslant x_i \leqslant \theta \ (i = 1,\ldots, n), \\ 0 & \text{otherwise.} \end{cases}$$

It can be seen that if $x_i < 0$ for at least one value of i $(i = 1,\ldots, n)$, then $f_n(x \mid \theta) = 0$ for every value of $\theta > 0$. Therefore, it is only necessary to consider the factorization of $f_n(x \mid \theta)$ for values of $x_i \geqslant 0$ $(i = 1,\ldots, n)$.

Let $h[\max(x_1,\ldots, x_n), \theta]$ be defined as follows:

$$h[\max(x_1,\ldots, x_n), \theta] = \begin{cases} 1 & \text{if } \max(x_1,\ldots, x_n) \leqslant \theta, \\ 0 & \text{if } \max(x_1,\ldots, x_n) > \theta. \end{cases}$$

Also, $x_i \leqslant \theta$ for $i = 1,\ldots, n$ if and only if $\max(x_1,\ldots, x_n) \leqslant \theta$. Therefore, for

$x_i \geq 0$ $(i = 1, \ldots, n)$, we can rewrite $f_n(x \mid \theta)$ as follows:

$$f_n(x \mid \theta) = \frac{1}{\theta^n} h\left[\max(x_1, \ldots, x_n), \theta\right].$$ (8)

Since the right side of Eq. (8) depends on x only through the value of $\max(x_1, \ldots, x_n)$, it follows that $T = \max(X_1, \ldots, X_n)$ is a sufficient statistic for θ. $\square$

EXERCISES

Instructions for Exercises 1 to 10: In each of these exercises, assume that the random variables $X_1, \ldots, X_n$ form a random sample of size n from the distribution specified in that exercise, and show that the statistic T specified in the exercise is a sufficient statistic for the parameter.

1. A Bernoulli distribution for which the value of the parameter p is unknown $(0 \leq p \leq 1)$; $T = \sum_{i=1}^{n} X_i$.

2. A geometric distribution for which the value of the parameter p is unknown $(0 < p < 1)$; $T = \sum_{i=1}^{n} X_i$.

3. A negative binomial distribution with parameters r and p, where the value of r is known and the value of p is unknown $(0 < p < 1)$; $T = \sum_{i=1}^{n} X_i$.

4. A normal distribution for which the mean μ is known and the variance σ^2 is unknown; $T = \sum_{i=1}^{n} (X_i - \mu)^2$.

5. A gamma distribution with parameters α and β, where the value of α is known and the value of β is unknown $(\beta > 0)$; $T = \overline{X}_n$.

6. A gamma distribution with parameters α and β, where the value of β is known and the value of α is unknown $(\alpha > 0)$; $T = \prod_{i=1}^{n} X_i$.

7. A beta distribution with parameters α and β, where the value of β is known and the value of α is unknown $(\alpha > 0)$; $T = \prod_{i=1}^{n} X_i$.

8. A uniform distribution on the integers $1, 2, \ldots, \theta$, as defined in Sec. 3.1, where the value of θ is unknown $(\theta = 1, 2, \ldots)$; $T = \max(X_1, \ldots, X_n)$.

9. A uniform distribution on the interval (a, b), where the value of a is known and the value of b is unknown $(b > a)$; $T = \max(X_1, \ldots, X_n)$.

10. A uniform distribution on the interval (a, b), where the value of b is known and the value of a is unknown $(a < b)$; $T = \min(X_1, \ldots, X_n)$.

11. Consider a distribution for which the p.d.f. or the p.f. is $f(x \mid \theta)$, where θ belongs to some parameter space Ω. It is said that the family of distributions obtained by letting θ vary over all values in Ω is an *exponential family*, or a

Koopman-Darmois family, if $f(x \mid \theta)$ can be written as follows for $\theta \in \Omega$ and all values of x:

$$f(x \mid \theta) = a(\theta)b(x)\exp[c(\theta)d(x)].$$

Here $a(\theta)$ and $c(\theta)$ are arbitrary functions of θ, and $b(x)$ and $d(x)$ are arbitrary functions of x. Assuming that the random variables $X_1, \ldots, X_n$ form a random sample from a distribution which belongs to an exponential family of this type, show that $T = \sum_{i=1}^{n} d(X_i)$ is a sufficient statistic for θ.

12. Show that each of the following families of distributions is an exponential family, as defined in Exercise 11:

   (a) The family of Bernoulli distributions with an unknown value of the parameter p.

   (b) The family of Poisson distributions with an unknown mean.

   (c) The family of negative binomial distributions for which the value of r is known and the value of p is unknown.

   (d) The family of normal distributions with an unknown mean and a known variance.

   (e) The family of normal distributions with an unknown variance and a known mean.

   (f) The family of gamma distributions for which the value of α is unknown and the value of β is known.

   (g) The family of gamma distributions for which the value of α is known and the value of β is unknown.

   (h) The family of beta distributions for which the value of α is unknown and the value of β is known.

   (i) The family of beta distributions for which the value of α is known and the value of β is unknown.

13. Suppose that $X_1, \ldots, X_n$ form a random sample from a distribution for which the p.d.f. is $f(x \mid \theta)$, where the value of the parameter θ belongs to a given parameter space Ω. Suppose that $T = r(X_1, \ldots, X_n)$ and $T' = r'(X_1, \ldots, X_n)$ are two statistics such that T' is a one-to-one function of T; that is, the value of T' can be determined from the value of T without knowing the values of $X_1, \ldots, X_n$, and the value of T can be determined from the value of T' without knowing the values of $X_1, \ldots, X_n$. Show that T' is a sufficient statistic for θ if and only if T is a sufficient statistic for θ.

14. Suppose that $X_1, \ldots, X_n$ form a random sample from the gamma distribution specified in Exercise 6. Show that the statistic $T = \sum_{i=1}^{n} \log X_i$ is a sufficient statistic for the parameter α.

15. Suppose that $X_1, \ldots, X_n$ form a random sample from a beta distribution with parameters α and β, where the value of α is known and the value of β is

unknown ($\beta > 0$). Show that the following statistic T is a sufficient statistic for β:

$$T = \frac{1}{n} \left(\sum_{i=1}^{n} \log \frac{1}{1 - X_i} \right)^4.$$

6.8. JOINTLY SUFFICIENT STATISTICS

Definition of Jointly Sufficient Statistics

We shall continue to suppose that the variables $X_1, \ldots, X_n$ form a random sample from a distribution for which the p.d.f. or the p.f. is $f(x \mid \theta)$, where the parameter θ must belong to some parameter space Ω. However, we shall now explicitly consider the possibility that θ may be a vector of real-valued parameters. For example, if the sample comes from a normal distribution for which both the mean μ and the variance σ^2 are unknown, then θ would be a two-dimensional vector whose components are μ and σ^2. Similarly, if the sample comes from a uniform distribution on some interval (a, b) for which both endpoints a and b are unknown, then θ would be a two-dimensional vector whose components are a and b. We shall, of course, continue to include the possibility that θ is a one-dimensional parameter.

In almost every problem in which θ is a vector, as well as in many problems in which θ is one-dimensional, there does not exist a single statistic T which is sufficient. In such a problem it is necessary to find two or more statistics $T_1, \ldots, T_k$ which together are *jointly sufficient statistics* in a sense that will now be described.

Suppose that in a given problem the statistics $T_1, \ldots, T_k$ are defined by k different functions of the vector of observations $X = (X_1, \ldots, X_n)$. Specifically, let $T_i = r_i(X)$ for $i = 1, \ldots, k$. Loosely speaking, the statistics $T_1, \ldots, T_k$ are jointly sufficient statistics for θ if a statistician who learns only the values of the k functions $r_1(X), \ldots, r_k(X)$ can estimate any component of θ, or any function of the components of θ, as well as can a statistician who observes the n individual values of $X_1, \ldots, X_n$. In terms of the factorization criterion, the following version can now be stated:

The statistics $T_1, \ldots, T_k$ are jointly sufficient statistics for θ if and only if the joint p.d.f. or the joint p.f. $f_n(x \mid \theta)$ can be factored as follows for all values of $x \in R^n$ and all values of $\theta \in \Omega$:

$$f_n(x \mid \theta) = u(x) v[r_1(x), \ldots, r_k(x), \theta]. \tag{1}$$

*Here the functions u and v are nonnegative, the function u may depend on **x** but does not depend on θ, and the function v will depend on θ but depends on **x** only through the k functions $r_1(x), \ldots, r_k(x)$.*

Example 1: Jointly Sufficient Statistics for the Parameters of a Normal Distribution. Suppose that $X_1, \ldots, X_n$ form a random sample from a normal distribution for which both the mean μ and the variance σ^2 are unknown. The joint p.d.f. of $X_1, \ldots, X_n$ is given by Eq. (7) of Sec. 6.7, and it can be seen that this joint p.d.f. depends on **x** only through the values of $\sum_{i=1}^n x_i$ and $\sum_{i=1}^n x_i^2$. Therefore, by the factorization criterion, the statistics $T_1 = \sum_{i=1}^n X_i$ and $T_2 = \sum_{i=1}^n X_i^2$ are jointly sufficient statistics for μ and σ^2. □

Suppose now that in a given problem the statistics $T_1, \ldots, T_k$ are jointly sufficient statistics for some parameter vector θ. If k other statistics $T_1', \ldots, T_k'$ are obtained from $T_1, \ldots, T_k$ by a one-to-one transformation, then it can be shown that $T_1', \ldots, T_k'$ will also be jointly sufficient statistics for θ.

Example 2: Another Pair of Jointly Sufficient Statistics for the Parameters of a Normal Distribution. Suppose again that $X_1, \ldots, X_n$ form a random sample from a normal distribution for which both the mean μ and the variance σ^2 are unknown; and let T_1' and T_2' denote the sample mean and the sample variance, respectively. Thus,

$$T_1' = \overline{X}_n \quad \text{and} \quad T_2' = \frac{1}{n} \sum_{i=1}^n (X_i - \overline{X}_n)^2.$$

We shall show that T_1' and T_2' are jointly sufficient statistics for μ and σ^2.

Let T_1 and T_2 be the jointly sufficient statistics for μ and σ^2 derived in Example 1. Then

$$T_1' = \frac{1}{n} T_1 \quad \text{and} \quad T_2' = \frac{1}{n} T_2 - \frac{1}{n^2} T_1^2.$$

Also, equivalently,

$$T_1 = n T_1' \quad \text{and} \quad T_2 = n(T_2' + T_1'^2).$$

Hence, the statistics T_1' and T_2' are obtained from the jointly sufficient statistics T_1 and T_2 by a one-to-one transformation. It follows, therefore, that T_1' and T_2' themselves are jointly sufficient statistics for μ and σ^2. □

We have now shown that the jointly sufficient statistics for the unknown mean and variance of a normal distribution can be chosen to be either T_1 and T_2, as given in Example 1, or T_1' and T_2', as given in Example 2.

Example 3: Jointly Sufficient Statistics for the Parameters of a Uniform Distribution. Suppose that $X_1, \ldots, X_n$ form a random sample from a uniform distribution on the interval (a, b), where the values of both endpoints a and b are unknown $(a < b)$. The joint p.d.f. $f_n(x \mid a, b)$ of $X_1, \ldots, X_n$ will be 0 unless all the observed values $x_1, \ldots, x_n$ lie between a and b; that is, $f_n(x \mid a, b) = 0$ unless $\min(x_1, \ldots, x_n) \geq a$ and $\max(x_1, \ldots, x_n) \leq b$. Furthermore, for any vector x such that $\min(x_1, \ldots, x_n) \geq a$ and $\max(x_1, \ldots, x_n) \leq b$, we have

$$f_n(x \mid a, b) = \frac{1}{(b - a)^n}.$$

For any two numbers y and z, we shall let $h(y, z)$ be defined as follows:

$$h(y, z) = \begin{cases} 1 & \text{for } y \leq z, \\ 0 & \text{for } y > z. \end{cases}$$

For any value of $x \in R^n$, we can then write

$$f_n(x \mid a, b) = \frac{h[a, \min(x_1, \ldots, x_n)] \, h[\max(x_1, \ldots, x_n), b]}{(b - a)^n}.$$

Since this expression depends on x only through the values of $\min(x_1, \ldots, x_n)$ and $\max(x_1, \ldots, x_n)$, it follows that the statistics $T_1 = \min(X_1, \ldots, X_n)$ and $T_2 = \max(X_1, \ldots, X_n)$ are jointly sufficient statistics for a and b. $\square$

Minimal Sufficient Statistics

In a given problem we want to try to find a sufficient statistic, or a set of jointly sufficient statistics for θ, because the values of such statistics summarize all the relevant information about θ contained in the random sample. When a set of jointly sufficient statistics are known, the search for a good estimator of θ is simplified because we need only consider functions of these statistics as possible estimators. Therefore, in a given problem it is desirable to find, not merely any set of jointly sufficient statistics, but the *simplest* set of jointly sufficient statistics. For example, it is correct but completely useless to say that in every problem the n observations $X_1, \ldots, X_n$ are jointly sufficient statistics.

We shall now describe another set of jointly sufficient statistics which exist in every problem and are slightly more useful. Suppose that $X_1, \ldots, X_n$ form a random sample from some distribution. Let Y_1 denote the smallest value in the random sample, let Y_2 denote the next smallest value, let Y_3 denote the third smallest value, and so on. In this way, Y_n denotes the largest value in the sample, and Y_{n-1} denotes the next largest value. The random variables $Y_1, \ldots, Y_n$ are called the *order statistics* of the sample.

Now let $y_1 \leqslant y_2 \leqslant \cdots \leqslant y_n$ denote the values of the order statistics for a given sample. If we are told the values of $y_1, \ldots, y_n$, then we know that these n values were obtained in the sample. However, we do not know which one of the observations $X_1, \ldots, X_n$ actually yielded the value y_1, which one actually yielded the value y_2, and so on. All we know is that the smallest of the values of $X_1, \ldots, X_n$ was y_1, the next smallest value was y_2, and so on.

If the variables $X_1, \ldots, X_n$ form a random sample from a distribution for which the p.d.f. or the p.f. is $f(x \mid \theta)$, then the joint p.d.f. or joint p.f. of $X_1, \ldots, X_n$ has the following form:

$$f_n(x \mid \theta) = \prod_{i=1}^{n} f(x_i \mid \theta). \tag{2}$$

Since the order of the factors in the product on the right side of Eq. (2) is irrelevant, Eq. (2) could just as well be rewritten in the form

$$f_n(x \mid \theta) = \prod_{i=1}^{n} f(y_i \mid \theta).$$

Hence, $f_n(x \mid \theta)$ depends on x only through the values of $y_1, \ldots, y_n$. It follows, therefore, that the order statistics $Y_1, \ldots, Y_n$ are always jointly sufficient statistics for θ. In other words, it is sufficient to know the set of n numbers that were obtained in the sample, and it is not necessary to know which particular one of these numbers was, for example, the value of X_3.

In each of the examples that have been given in this section and in Sec. 6.7, we considered a distribution for which either there was a single sufficient statistic or there were two statistics which were jointly sufficient. For some distributions, however, the order statistics $Y_1, \ldots, Y_n$ are the simplest set of jointly sufficient statistics that exist, and no further reduction in terms of sufficient statistics is possible.

Example 4: Sufficient Statistics for the Parameter of a Cauchy Distribution.
Suppose that $X_1, \ldots, X_n$ form a random sample from a Cauchy distribution centered at an unknown point θ $(-\infty < \theta < \infty)$. The p.d.f. $f(x \mid \theta)$ of this distribution is given by Eq. (5) of Sec. 6.6, and the joint p.d.f. $f_n(x \mid \theta)$ of $X_1, \ldots, X_n$ is given by Eq. (6) of Sec. 6.6. It can be shown that the only jointly sufficient statistics that exist in this problem are the order statistics $Y_1, \ldots, Y_n$ or some other set of n statistics $T_1, \ldots, T_n$ that can be derived from the order statistics by a one-to-one transformation. The details of the argument will not be given here. $\square$

These considerations lead us to the concepts of a minimal sufficient statistic and a minimal set of jointly sufficient statistics. Roughly speaking, a sufficient statistic T is a minimal sufficient statistic if it cannot be reduced further without destroying the property of sufficiency. Alternatively, a sufficient statistic T is a minimal sufficient statistic if every function of T which itself is a sufficient statistic is a one-to-one function of T. Formally, we shall use the following definition, which is equivalent to either of the informal definitions just given:

A statistic T is a *minimal sufficient statistic* if T is a sufficient statistic and is a function of every other sufficient statistic.

In any problem in which there is no single sufficient statistic, *minimal jointly sufficient statistics* are defined in a similar manner. Thus, in Example 4, the order statistics $Y_1, \ldots, Y_n$ are minimal jointly sufficient statistics.

Maximum Likelihood Estimators and Bayes Estimators as Sufficient Statistics

Suppose again that $X_1, \ldots, X_n$ form a random sample from a distribution for which the p.f. or the p.d.f. is $f(x \mid \theta)$, where the value of the parameter θ is unknown, and also that $T = r(X_1, \ldots, X_n)$ is a sufficient statistic for θ. We shall now show that the M.L.E. $\hat{\theta}$ depends on the observations $X_1, \ldots, X_n$ only through the statistic T.

It follows from the factorization criterion presented in Sec. 6.7 that the likelihood function $f_n(x \mid \theta)$ can be written in the form

$$f_n(x \mid \theta) = u(x)v[r(x), \theta].$$

The M.L.E. $\hat{\theta}$ is the value of θ for which $f_n(x \mid \theta)$ is a maximum. It follows, therefore, that $\hat{\theta}$ will be the value of θ for which $v[r(x), \theta]$ is a maximum. Since $v[r(x), \theta]$ depends on the observed vector x only through the function $r(x)$, it follows that $\hat{\theta}$ will also depend on x only through the function $r(x)$. Thus, the estimator $\hat{\theta}$ is a function of $T = r(X_1, \ldots, X_n)$.

Since the estimator $\hat{\theta}$ is a function of the observations $X_1, \ldots, X_n$ and is not a function of the parameter θ, the estimator is itself a statistic. In many problems $\hat{\theta}$ is actually a sufficient statistic. Since $\hat{\theta}$ will always be a function of any other sufficient statistic, we may now state the following result:

If the M.L.E. $\hat{\theta}$ is a sufficient statistic, then it is a minimal sufficient statistic.

These properties can be extended to a vector parameter $\boldsymbol{\theta}$. If $\boldsymbol{\theta} = (\theta_1, \ldots, \theta_k)$ is a vector of k real-valued parameters, then the M.L.E.'s $\hat{\theta}_1, \ldots, \hat{\theta}_k$ will depend on the observations $X_1, \ldots, X_n$ only through the functions in any set of jointly sufficient statistics. In many problems, the estimators $\hat{\theta}_1, \ldots, \hat{\theta}_k$ themselves will form a set of jointly sufficient statistics. When they are jointly sufficient statistics, they will be *minimal jointly sufficient statistics*.

Example 5: Minimal Jointly Sufficient Statistics for the Parameters of a Normal Distribution. Suppose that $X_1, \ldots, X_n$ form a random sample from a normal distribution for which both the mean μ and the variance σ^2 are unknown. It was shown in Example 3 of Sec. 6.5 that the M.L.E.'s $\hat{\mu}$ and $\hat{\sigma}^2$ are the sample mean and the sample variance. Also, it was shown in Example 2 of this section that $\hat{\mu}$ and $\hat{\sigma}^2$ are jointly sufficient statistics. Hence, $\hat{\mu}$ and $\hat{\sigma}^2$ are minimal jointly sufficient statistics. □

The statistician can restrict the search for good estimators of μ and σ^2 to functions of minimal jointly sufficient statistics. It follows, therefore, from Example 5 that if the M.L.E.'s $\hat{\mu}$ and $\hat{\sigma}^2$ themselves are not used as estimators of μ and σ^2, the only other estimators that need to be considered are functions of $\hat{\mu}$ and $\hat{\sigma}^2$.

The discussion just presented for M.L.E.'s pertains also to Bayes estimators. Suppose that a parameter θ is to be estimated and that a prior p.d.f. $\xi(\theta)$ is assigned to θ. If a statistic $T = r(X_1, \ldots, X_n)$ is a sufficient statistic, then it follows from relation (6) of Sec. 6.2 and the factorization criterion that the posterior p.d.f. $\xi(\theta \mid x)$ will satisfy the following relation:

$$\xi(\theta \mid x) \propto v[r(x), \theta] \xi(\theta).$$

It can be seen from this relation that the posterior p.d.f. of θ will depend on the observed vector x only through the value of $r(x)$. Since the Bayes estimator of θ with respect to any specified loss function is calculated from this posterior p.d.f., the estimator also will depend on the observed vector x only through the value of $r(x)$. In other words, the Bayes estimator is a function of $T = r(X_1, \ldots, X_n)$.

Since the Bayes estimator is itself a statistic and is a function of any sufficient statistic T, we may state the following result:

If the Bayes estimator is a sufficient statistic, then it is a minimal sufficient statistic.

EXERCISES

Instructions for Exercises 1 to 4: In each exercise assume that the random variables $X_1, \ldots, X_n$ form a random sample of size n from the distribution specified in the exercise, and show that the statistics T_1 and T_2 specified in the exercise are jointly sufficient statistics.

1. A gamma distribution for which both parameters α and β are unknown ($\alpha > 0$ and $\beta > 0$); $T_1 = \prod_{i=1}^{n} X_i$ and $T_2 = \sum_{i=1}^{n} X_i$.

2. A beta distribution for which both parameters α and β are unknown ($\alpha > 0$ and $\beta > 0$); $T_1 = \prod_{i=1}^{n} X_i$ and $T_2 = \prod_{i=1}^{n} (1 - X_i)$.

3. A Pareto distribution (see Exercise 15 of Sec. 5.9) for which both parameters x_0 and α are unknown ($x_0 > 0$ and $\alpha > 0$); $T_1 = \min(X_1, \ldots, X_n)$ and $T_2 = \prod_{i=1}^{n} X_i$.

4. A uniform distribution on the interval $(\theta, \theta + 3)$, where the value of θ is unknown ($-\infty < \theta < \infty$); $T_1 = \min(X_1, \ldots, X_n)$ and $T_2 = \max(X_1, \ldots, X_n)$.

5. Suppose that the vectors $(X_1, Y_1), (X_2, Y_2), \ldots, (X_n, Y_n)$ form a random sample of two-dimensional vectors from a bivariate normal distribution for which the means, the variances, and the correlation are unknown. Show that the following five statistics are jointly sufficient: $\sum_{i=1}^{n} X_i$, $\sum_{i=1}^{n} Y_i$, $\sum_{i=1}^{n} X_i^2$, $\sum_{i=1}^{n} Y_i^2$, and $\sum_{i=1}^{n} X_i Y_i$.

6. Consider a distribution for which the p.d.f. or the p.f. is $f(x \mid \theta)$, where the parameter θ is a k-dimensional vector belonging to some parameter space Ω. It is said that the family of distributions indexed by the values of θ in Ω is a *k-parameter exponential family*, or a *k-parameter Koopman-Darmois family*, if $f(x \mid \theta)$ can be written as follows for $\theta \in \Omega$ and all values of x:

$$f(x \mid \theta) = a(\theta)b(x)\exp\left[\sum_{i=1}^{k} c_i(\theta)d_i(x) \right].$$

Here a and $c_1, \ldots, c_k$ are arbitrary functions of θ, and b and $d_1, \ldots, d_k$ are arbitrary functions of x. Suppose now that $X_1, \ldots, X_n$ form a random sample from a distribution which belongs to a k-parameter exponential family of this type, and define the k statistics $T_1, \ldots, T_k$ as follows:

$$T_i = \sum_{j=1}^{n} d_i(X_j) \qquad \text{for } i = 1, \ldots, k.$$

Show that the statistics $T_1, \ldots, T_k$ are jointly sufficient statistics for θ.

7. Show that each of the following families of distributions is a two-parameter exponential family as defined in Exercise 6:

(a) The family of all normal distributions for which both the mean and the variance are unknown.

(b) The family of all gamma distributions for which both α and β are unknown.

(c) The family of all beta distributions for which both α and β are unknown.

8. Suppose that $X_1, \ldots, X_n$ form a random sample from a Bernoulli distribution for which the value of the parameter p is unknown ($0 \le p \le 1$). Is the M.L.E. of p a minimal sufficient statistic?

9. Suppose that $X_1, \ldots, X_n$ form a random sample from a uniform distribution on the interval $(0, \theta)$, where the value of θ is unknown ($\theta > 0$). Is the M.L.E. of θ a minimal sufficient statistic?

10. Suppose that $X_1, \ldots, X_n$ form a random sample from a Cauchy distribution centered at an unknown point θ ($-\infty < \theta < \infty$). Is the M.L.E. of θ a minimal sufficient statistic?

11. Suppose that $X_1, \ldots, X_n$ form a random sample from a distribution for which the p.d.f. is as follows:

$$f(x \mid \theta) = \begin{cases} \dfrac{2x}{\theta^2} & \text{for } 0 \leqslant x \leqslant \theta, \\ 0 & \text{otherwise.} \end{cases}$$

Here, the value of the parameter θ is unknown ($\theta > 0$). Determine the M.L.E. of the median of this distribution, and show that this estimator is a minimal sufficient statistic for θ.

12. Suppose that $X_1, \ldots, X_n$ form a random sample from a uniform distribution on the interval (a, b), where both endpoints a and b are unknown. Are the M.L.E.'s of a and b minimal jointly sufficient statistics?

13. For the conditions of Exercise 5, the M.L.E.'s of the means, the variances, and the correlation are given in Exercise 13 of Sec. 6.5. Are these five estimators minimal jointly sufficient statistics?

14. Suppose that $X_1, \ldots, X_n$ form a random sample from a Bernoulli distribution for which the value of the parameter p is unknown, and that the prior distribution of p is a certain specified beta distribution. Is the Bayes estimator of p with respect to the squared error loss function a minimal sufficient statistic?

15. Suppose that $X_1, \ldots, X_n$ form a random sample from a Poisson distribution for which the value of the mean λ is unknown, and that the prior distribution of λ is a certain specified gamma distribution. Is the Bayes estimator of λ with respect to the squared error loss function a minimal sufficient statistic?

16. Suppose that $X_1, \ldots, X_n$ form a random sample from a normal distribution for which the value of the mean μ is unknown and the value of the variance is known, and that the prior distribution of μ is a certain specified normal distribution. Is the Bayes estimator of μ with respect to the squared error loss function a minimal sufficient statistic?

6.9. IMPROVING AN ESTIMATOR

The Mean Squared Error of an Estimator

Suppose again that the random variables $X_1, \ldots, X_n$ form a random sample from a distribution for which the p.d.f. or the p.f. is $f(x \mid \theta)$, where the parameter θ

must belong to some parameter space Ω. For any random variable $Z = g(X_1, \ldots, X_n)$, we shall let $E_\theta(Z)$ denote the expectation of Z calculated with respect to the joint p.d.f. or joint p.f. $f_n(x \mid \theta)$. Thus, if $f_n(x \mid \theta)$ is a p.d.f.,

$$E_\theta(Z) = \int_{-\infty}^{\infty} \cdots \int_{-\infty}^{\infty} g(x) f_n(x \mid \theta) \, dx_1 \cdots dx_n.$$

In other words, $E_\theta(Z)$ is the expectation of Z for a given value of $\theta \in \Omega$.

We shall suppose, as usual, that the value of θ is unknown and must be estimated, and we shall assume that the squared error loss function is to be used. Also, for any given estimator $\delta(X_1, \ldots, X_n)$ and any given value of $\theta \in \Omega$, we shall let $R(\theta, \delta)$ denote the M.S.E. of δ calculated with respect to the given value of θ. Thus,

$$R(\theta, \delta) = E_\theta\big[(\delta - \theta)^2\big]. \tag{1}$$

If we do not assign a prior distribution to θ, then it is desired to find an estimator δ for which the M.S.E. $R(\theta, \delta)$ is small for every value of $\theta \in \Omega$ or, at least, for a wide range of values of θ.

Suppose now that $T = r(X_1, \ldots, X_n)$ is a sufficient statistic for θ, and consider a statistician A who plans to use a particular estimator $\delta(X_1, \ldots, X_n)$. In Sec. 6.7 we remarked that another statistician B who learns only the value of the sufficient statistic T can generate, by means of an auxiliary randomization, an estimator that will have exactly the same properties as δ and, in particular, will have the same mean squared error as δ for every value of $\theta \in \Omega$. We shall now show that even without using an auxiliary randomization, statistician B can find an estimator δ_0 which depends on the observations $X_1, \ldots, X_n$ only through the sufficient statistic T and which is at least as good an estimator as δ in the sense that $R(\theta, \delta_0) \leqslant R(\theta, \delta)$ for every value of $\theta \in \Omega$.

Conditional Expectation When a Sufficient Statistic Is Known

We shall define the estimator $\delta_0(T)$ by the following conditional expectation:

$$\delta_0(T) = E_\theta\big[\delta(X_1, \ldots, X_n) \mid T\big]. \tag{2}$$

Since T is a sufficient statistic, the conditional joint distribution of $X_1, \ldots, X_n$ for any given value of T is the same for every value of $\theta \in \Omega$. Therefore, for any given value of T, the conditional expectation of the function $\delta(X_1, \ldots, X_n)$ will be the same for every value of $\theta \in \Omega$. It follows that the conditional expectation in Eq. (2) will depend on the value of T but will not actually depend on the value of

θ. In other words, the function $\delta_0(T)$ is indeed an estimator of θ because it depends only on the observations $X_1, \ldots, X_n$ and does not depend on the unknown value of θ. For this reason, we can omit the subscript θ on the expectation symbol E in Eq. (2), and can write the relation as follows:

$$\delta_0(T) = E[\delta(X_1, \ldots, X_n) \mid T]. \tag{3}$$

It will be convenient to write this relation simply as $\delta_0 = E(\delta \mid T)$. It should be emphasized that the estimator δ_0 will depend on the observations $X_1, \ldots, X_n$ only through the statistic T.

We can now prove the following theorem, which was established independently by D. Blackwell and C. R. Rao in the late 1940's.

Theorem 1. *Let $\delta(X_1, \ldots, X_n)$ be any given estimator, let T be a sufficient statistic for θ, and let the estimator $\delta_0(T)$ be defined as in Eq. (3). Then for every value of $\theta \in \Omega$,*

$$R(\theta, \delta_0) \leqslant R(\theta, \delta). \tag{4}$$

Proof. If the risk $R(\theta, \delta)$ is infinite for a given value of $\theta \in \Omega$, then the relation (3) is automatically satisfied. We shall assume, therefore, that $R(\theta, \delta) < \infty$. It follows from Exercise 3 of Sec. 4.4 that $E_\theta[(\delta - \theta)^2] \geqslant [E_\theta(\delta) - \theta]^2$, and it can be shown that this same relationship must also hold if the expectations are replaced by conditional expectations given T. Therefore,

$$E_\theta\left[(\delta - \theta)^2 \mid T\right] \geqslant \left[E_\theta(\delta \mid T) - \theta\right]^2 = (\delta_0 - \theta)^2. \tag{5}$$

It now follows from relation (5) that

$$\begin{aligned}
R(\theta, \delta_0) &= E_\theta\left[(\delta_0 - \theta)^2\right] \leqslant E_\theta\left\{E_\theta\left[(\delta - \theta)^2 \mid T\right]\right\} \\
&= E_\theta\left[(\delta - \theta)^2\right] = R(\theta, \delta).
\end{aligned} \tag{6}$$

Hence, $R(\theta, \delta_0) \leqslant R(\theta, \delta)$ for every value of $\theta \in \Omega$. $\square$

If $R(\theta, \delta) < \infty$, then it can also be shown that there will be strict inequality in relation (4) unless $\delta = \delta_0$, that is, unless δ itself depends on $X_1, \ldots, X_n$ only through the statistic T.

A result similar to Theorem 1 holds when θ is a vector of real-valued parameters and the statistics $T_1, \ldots, T_k$ are jointly sufficient statistics for θ. Suppose that it is desired to estimate a particular function $v(\theta)$ of the vector θ, such as one particular component of θ or the sum of all the components of θ. For any value of $\theta \in \Omega$, the M.S.E. of an estimator $\delta(X_1, \ldots, X_n)$ would now be defined as follows:

$$R(\theta, \delta) = E_\theta \big\{ [\delta(X_1, \ldots, X_n) - \nu(\theta)]^2 \big\}. \tag{7}$$

Since the statistics $T_1, \ldots, T_k$ are jointly sufficient, the conditional expectation $E[\delta(X_1, \ldots, X_n) \mid T_1, \ldots, T_k]$ will not depend on θ. Therefore, another estimator $\delta_0(T_1, \ldots, T_k)$ can be defined by the relation

$$\delta_0(T_1, \ldots, T_k) = E\big[\delta(X_1, \ldots, X_n) \mid T_1, \ldots, T_k\big]. \tag{8}$$

It now follows by a proof similar to the one given for Theorem 1 that

$$R(\theta, \delta_0) \leqslant R(\theta, \delta) \qquad \text{for } \theta \in \Omega. \tag{9}$$

If $R(\theta, \delta) < \infty$, there is strict inequality in relation (9) unless $\delta = \delta_0$.

Furthermore, a result similar to Theorem 1 holds if $R(\theta, \delta)$ is defined as the M.A.E. of an estimator for a given value of $\theta \in \Omega$ instead of the M.S.E. of δ. In other words, suppose that $R(\theta, \delta)$ is defined as follows:

$$R(\theta, \delta) = E_\theta(\mid \delta - \theta \mid). \tag{10}$$

Then it can be shown (see Exercise 9 at the end of this section) that Theorem 1 is still true.

Suppose now that $R(\theta, \delta)$ is defined by either Eq. (1) or Eq. (10). It is said that an estimator δ is *inadmissible* if there exists another estimator δ_0 such that $R(\theta, \delta_0) \leqslant R(\theta, \delta)$ for every value of $\theta \in \Omega$ and there is strict inequality in this relation for at least one value of $\theta \in \Omega$. Under these conditions, it is also said that the estimator δ_0 *dominates* the estimator δ. An estimator is *admissible* if it is not dominated by any other estimator.

In this terminology, Theorem 1 can be summarized as follows: An estimator δ that is not a function of the sufficient statistic T alone must be inadmissible. Theorem 1 also explicitly identifies an estimator $\delta_0 = E(\delta \mid T)$ that dominates δ. However, this part of the theorem is somewhat less useful in a practical problem, because it is usually very difficult to calculate the conditional expectation $E(\delta \mid T)$. Theorem 1 is valuable mainly because it provides further strong evidence that we can restrict our search for a good estimator of θ to those estimators which depend on the observations only through a sufficient statistic.

Example 1: Estimating the Mean of a Normal Distribution. Suppose that $X_1, \ldots, X_n$ form a random sample from a normal distribution for which the mean μ is unknown and the variance is known, and let $Y_1 \leqslant \cdots \leqslant Y_n$ denote the order statistics of the sample, as defined in Sec. 6.8. If n is an odd number, then the middle observation $Y_{(n+1)/2}$ is called the *sample median*. If n is an even number, then any value between the two middle observations $Y_{n/2}$ and $Y_{(n/2)+1}$ is a *sample median*, but the particular value $\frac{1}{2}[Y_{n/2} + Y_{(n/2)+1}]$ is often referred to as *the sample median*.

Since the normal distribution from which the sample is drawn is symmetric with respect to the point μ, the median of the normal distribution is μ. Therefore, we might consider the use of the sample median, or a simple function of the sample median, as an estimator of μ. However, it was shown in Example 3 of Sec. 6.7 that the sample mean $\overline{X}_n$ is a sufficient statistic for μ. It follows from Theorem 1 that any function of the sample median which might be used as an estimator of μ will be dominated by some other function of $\overline{X}_n$. In searching for an estimator of μ, we need consider only functions of $\overline{X}_n$. □

Example 2: Estimating the Standard Deviation of a Normal Distribution. Suppose that $X_1, \ldots, X_n$ form a random sample from a normal distribution for which both the mean μ and the variance σ^2 are unknown, and again let $Y_1 \leqslant \cdots \leqslant Y_n$ denote the order statistics of the sample. The difference $Y_n - Y_1$ is called the *range* of the sample; and we might consider using some simple function of the range as an estimator of the standard deviation σ. However, it was shown in Example 1 of Sec. 6.8 that the statistics $\sum_{i=1}^n X_i$ and $\sum_{i=1}^n X_i^2$ are jointly sufficient for the parameters μ and σ^2. Therefore, any function of the range that might be used as an estimator of σ will be dominated by a function of $\sum_{i=1}^n X_i$ and $\sum_{i=1}^n X_i^2$. □

Limitation of the Use of Sufficient Statistics

When the foregoing theory of sufficient statistics is applied in a statistical problem, it is important to keep in mind the following limitation. The existence and the form of a sufficient statistic in a particular problem depend critically on the form of the function assumed for the p.d.f. or the p.f. A statistic which is a sufficient statistic when it is assumed that the p.d.f. is $f(x \mid \theta)$ may not be a sufficient statistic when it is assumed that the p.d.f. is $g(x \mid \theta)$, even though $g(x \mid \theta)$ may be quite similar to $f(x \mid \theta)$ for every value of $\theta \in \Omega$. Suppose that a statistician is in doubt about the exact form of the p.d.f. in a specific problem but assumes for convenience that the p.d.f. is $f(x \mid \theta)$; and suppose also that the statistic T is a sufficient statistic under this assumption. Because of the statistician's uncertainty about the exact form of the p.d.f., he may wish to use an estimator of θ that performs reasonably well for a wide variety of possible p.d.f.'s, even though the selected estimator may not meet the requirement that it should depend on the observations only through the statistic T.

An estimator that performs reasonably well for a wide variety of possible p.d.f.'s, even though it may not necessarily be the best available estimator for any particular family of p.d.f.'s, is often called a *robust estimator*. We shall consider robust estimators further in Chapter 9.

EXERCISES

1. Suppose that the random variables $X_1, \ldots, X_n$ form a random sample of size n $(n \geqslant 2)$ from a uniform distribution on the interval $(0, \theta)$, where the value of the parameter θ is unknown $(\theta > 0)$ and must be estimated. Suppose also that for any estimator $\delta(X_1, \ldots, X_n)$, the M.S.E. $R(\theta, \delta)$ is defined by Eq. (1). Explain why the estimator $\delta_1(X_1, \ldots, X_n) = 2\overline{X}_n$ is inadmissible.

2. Consider again the conditions of Exercise 1, and let the estimator δ_1 be as defined in that exercise. Determine the value of the M.S.E. $R(\theta, \delta_1)$ for $\theta > 0$.

3. Consider again the conditions of Exercise 1. Let $Y_n = \max(X_1, \ldots, X_n)$ and consider the estimator $\delta_2(X_1, \ldots, X_n) = Y_n$.

   (a) Determine the M.S.E. $R(\theta, \delta_2)$ for $\theta > 0$.

   (b) Show that for $n = 2$, $R(\theta, \delta_2) = R(\theta, \delta_1)$ for $\theta > 0$.

   (c) Show that for $n \geqslant 3$, the estimator δ_2 dominates the estimator δ_1.

4. Consider again the conditions of Exercises 1 and 3. Show that there exists a constant c^* such that the estimator $c^* Y_n$ dominates every other estimator having the form $c Y_n$ for $c \neq c^*$.

5. Suppose that $X_1, \ldots, X_n$ form a random sample of size n $(n \geqslant 2)$ from a gamma distribution with parameters α and β, where the value of α is unknown $(\alpha > 0)$ and the value of β is known. Explain why $\overline{X}_n$ is an inadmissible estimator of the mean of this distribution when the squared error loss function is used.

6. Suppose that $X_1, \ldots, X_n$ form a random sample from an exponential distribution for which the value of the parameter β is unknown $(\beta > 0)$ and must be estimated by using the squared error loss function. Let δ be an estimator such that $\delta(X_1, \ldots, X_n) = 3$ for all possible values of $X_1, \ldots, X_n$.

   (a) Determine the value of the M.S.E. $R(\beta, \delta)$ for $\beta > 0$.

   (b) Explain why the estimator δ must be admissible.

7. Suppose that a random sample of n observations is taken from a Poisson distribution for which the value of the mean θ is unknown $(\theta > 0)$, and that the value of $\beta = e^{-\theta}$ must be estimated by using the squared error loss function. Since β is equal to the probability that an observation from this Poisson distribution will have the value 0, a natural estimator of β is the proportion $\hat{\beta}$ of observations in the random sample that have the value 0. Explain why $\hat{\beta}$ is an inadmissible estimator of β.

8. For any random variable X, show that $|E(X)| \leqslant E(|X|)$.

9. Let $X_1, \ldots, X_n$ form a random sample from a distribution for which the p.d.f. or the p.f. is $f(x | \theta)$, where $\theta \in \Omega$. Suppose that the value of θ must be estimated, and that T is a sufficient statistic for θ. Let δ be any given estimator of θ, and let δ_0 be another estimator defined by the relation $\delta_0 = E(\delta | T)$. Show that for every value of $\theta \in \Omega$,

$$E_\theta(|\delta_0 - \theta|) \leqslant E_\theta(|\delta - \theta|).$$

10. Suppose that the variables $X_1, \ldots, X_n$ form a random sample from a distribution for which the p.d.f. or the p.f. is $f(x | \theta)$, where $\theta \in \Omega$; and let $\hat{\theta}$ denote the M.L.E. of θ. Suppose also that the statistic T is a sufficient statistic for θ; and let the estimator δ_0 be defined by the relation $\delta_0 = E(\hat{\theta} | T)$. Compare the estimators $\hat{\theta}$ and δ_0.

11. Suppose that $X_1, \ldots, X_n$ form a sequence of n Bernoulli trials for which the probability p of success on any given trial is unknown $(0 \leqslant p \leqslant 1)$, and let $T = \sum_{i=1}^n X_i$. Determine the form of the estimator $E(X_1 | T)$.

12. Suppose that $X_1, \ldots, X_n$ form a random sample from a Poisson distribution for which the value of the mean θ is unknown $(\theta > 0)$. Let $T = \sum_{i=1}^n X_i$; and for $i = 1, \ldots, n$, let the statistic Y_i be defined as follows:

$$Y_i = 1 \quad \text{if} \quad X_i = 0,$$
$$Y_i = 0 \quad \text{if} \quad X_i > 0.$$

Determine the form of the estimator $E(Y_i | T)$.

13. Consider again the conditions of Exercises 7 and 12. Determine the form of the estimator $E(\hat{\beta} | T)$.

6.10. SUPPLEMENTARY EXERCISES

1. Suppose that $X_1, \ldots, X_n$ are i.i.d. with $\Pr(X_i = 1) = \theta$ and $\Pr(X_i = 0) = 1 - \theta$, where θ is unknown $(0 \leqslant \theta \leqslant 1)$. Find the M.L.E. of θ^2.

2. Suppose that the proportion θ of bad apples in a large lot is unknown, and has the following prior p.d.f.:

$$\xi(\theta) = \begin{cases} 600\theta^2(1-\theta)^3 & \text{for } 0 < \theta < 1, \\ 0 & \text{otherwise.} \end{cases}$$

Suppose that a random sample of 10 apples is drawn from the lot, and it is found that 3 are bad. Find the Bayes estimate of θ with respect to the squared error loss function.

3. Suppose that $X_1, \ldots, X_n$ form a random sample from a uniform distribution with the following p.d.f.:

$$f(x \mid \theta) = \begin{cases} \dfrac{1}{\theta} & \text{for } \theta \leqslant x \leqslant 2\theta, \\ 0 & \text{otherwise.} \end{cases}$$

Assuming that the value of θ is unknown ($\theta > 0$), determine the M.L.E. of θ.

4. Suppose that X_1 and X_2 are independent random variables, and that X_i has a normal distribution with mean $b_i \mu$ and variance σ_i^2 for $i = 1, 2$. Suppose also that b_1, b_2, σ_1^2, and σ_2^2 are known positive constants, and that μ is an unknown parameter. Determine the M.L.E. of μ based on X_1 and X_2.

5. Let $\psi(\alpha) = \Gamma'(\alpha)/\Gamma(\alpha)$ for $\alpha > 0$ (the digamma function). Show that

$$\psi(\alpha + 1) = \psi(\alpha) + \frac{1}{\alpha}.$$

6. Suppose that a regular light bulb, a long-life light bulb, and an extra-long-life light bulb are being tested. The lifetime X_1 of the regular bulb has an exponential distribution with mean θ, the lifetime X_2 of the long-life bulb has an exponential distribution with mean 2θ, and the lifetime X_3 of the extra-long-life bulb has an exponential distribution with mean 3θ.

(a) Determine the M.L.E. of θ based on the observations X_1, X_2, and X_3.

(b) Let $\psi = 1/\theta$, and suppose that the prior distribution of ψ is a gamma distribution with parameters α and β. Determine the posterior distribution of ψ given X_1, X_2, and X_3.

7. Consider a Markov chain with two possible states s_1 and s_2 and with stationary transition probabilities as given in the following transition matrix P:

$$P = \begin{array}{c} \\ s_1 \\ s_2 \end{array} \begin{array}{c} \overset{s_1}{} \qquad \overset{s_2}{} \\ \begin{bmatrix} \theta & 1 - \theta \\ 3/4 & 1/4 \end{bmatrix}, \end{array}$$

where the value of θ is unknown ($0 \leqslant \theta \leqslant 1$). Suppose that the initial state X_1 of the chain is s_1, and let $X_2, \ldots, X_{n+1}$ denote the state of the chain at each of the next n successive periods. Determine the M.L.E. of θ based on the observations $X_2, \ldots, X_{n+1}$.

8. Suppose that an observation X is drawn from a distribution with the following p.d.f.:

$$f(x \mid \theta) = \begin{cases} \dfrac{1}{\theta} & \text{for } 0 < x < \theta, \\ 0 & \text{otherwise.} \end{cases}$$

Also, suppose that the prior p.d.f. of θ is

$$\xi(\theta) = \begin{cases} \theta e^{-\theta} & \text{for } \theta > 0 \\ 0 & \text{otherwise.} \end{cases}$$

Determine the Bayes estimator of θ with respect to (a) the mean squared error loss function and (b) the absolute error loss function.

9. Suppose that $X_1, \ldots, X_n$ form n Bernoulli trials with parameter $\theta = (1/3)(1 + \beta)$, where the value of β is unknown $(0 \le \beta \le 1)$. Determine the M.L.E. of β.

10. The method of *randomized response* is sometimes used to conduct surveys on sensitive topics. A simple version of the method can be described as follows: A random sample of n persons is drawn from a large population. For each person in the sample there is probability $1/2$ that the person will be asked a standard question and probability $1/2$ that the person will be asked a sensitive question. Furthermore, this selection of the standard or the sensitive question is made independently from person to person. If a person is asked the standard question, then there is probability $1/2$ that he will give a positive response; but if he is asked the sensitive question, then there is an unknown probability p that he will give a positive response. The statistician can observe only the total number X of positive responses that were given by the n persons in the sample. He cannot observe which of these persons were asked the sensitive question or how many persons in the sample were asked the sensitive question. Determine the M.L.E. of p based on the observation X.

11. Suppose that a random sample of four observations is to be drawn from a uniform distribution on the interval $(0, \theta)$, and that the prior distribution of θ has the following p.d.f.:

$$\xi(\theta) = \begin{cases} \dfrac{1}{\theta^2} & \text{for } \theta \ge 1, \\ 0 & \text{otherwise.} \end{cases}$$

Suppose that the values of the observations in the sample are found to be 0.6, 0.4, 0.8, and 0.9. Determine the Bayes estimate of θ with respect to the squared error loss function.

12. For the conditions of Exercise 11, determine the Bayes estimate of θ with respect to the absolute error loss function.

13. Suppose that $X_1, \ldots, X_n$ form a random sample from a distribution with the following p.d.f.:

$$f(x \mid \beta, \theta) = \begin{cases} \beta e^{-\beta(x-\theta)} & \text{for } x \ge \theta, \\ 0 & \text{otherwise,} \end{cases}$$

where β and θ are unknown ($\beta > 0$, $-\infty < \theta < \infty$). Determine a pair of jointly sufficient statistics.

14. Suppose that $X_1, \ldots, X_n$ form a random sample from a Pareto distribution with parameters x_0 and α (see Exercise 15 of Sec. 5.9), where x_0 is unknown and α is known. Determine the M.L.E. of x_0.

15. Determine whether the estimator found in Exercise 14 is a minimal sufficient statistic.

16. Consider again the conditions of Exercise 14, but suppose now that both parameters x_0 and α are unknown. Determine the M.L.E.'s of x_0 and α.

17. Determine whether the estimators found in Exercise 16 are minimal jointly sufficient statistics.

18. Suppose that the random variable X has a binomial distribution with an unknown value of n and a known value of p ($0 < p < 1$). Determine the M.L.E. of n based on the observation X. *Hint:* Consider the ratio

$$\frac{f(x \mid n+1, p)}{f(x \mid n, p)}.$$

19. Suppose that two observations X_1 and X_2 are drawn at random from a uniform distribution with the following p.d.f.:

$$f(x \mid \theta) = \begin{cases} \dfrac{1}{2\theta} & \text{for } 0 \leqslant x \leqslant \theta \text{ or } 2\theta \leqslant x \leqslant 3\theta, \\ 0 & \text{otherwise,} \end{cases}$$

where the value of θ is unknown ($\theta > 0$). Determine the M.L.E. of θ for each of the following pairs of observed values of X_1 and X_2:

(a) $X_1 = 7$ and $X_2 = 9$.

(b) $X_1 = 4$ and $X_2 = 9$.

(c) $X_1 = 5$ and $X_2 = 9$.

20. Suppose that a random sample $X_1, \ldots, X_n$ is to be taken from a normal distribution with unknown mean θ and variance 100; and that the prior distribution of θ is a normal distribution with specified mean μ and variance 25. Suppose that θ is to be estimated using the squared error loss function, and that the sampling cost of each observation is 0.25 (in appropriate units). If the total cost of the estimation procedure is equal to the expected loss of the Bayes estimator plus the sampling cost $(0.25)n$, what is the sample size n for which the total cost will be a minimum?

21. Suppose that $X_1, \ldots, X_n$ form a random sample from a Poisson distribution with unknown mean θ, and that the variance of this distribution is to be estimated using the squared error loss function. Determine whether or not the sample variance is an admissible estimator.

Sampling Distributions of Estimators

7

7.1. THE SAMPLING DISTRIBUTION OF A STATISTIC

Statistics and Estimators

Suppose that the random variables $X_1, \ldots, X_n$ form a random sample from a distribution involving a parameter θ whose value is unknown. In Sec. 6.7, a *statistic* was defined as any real-valued function $T = r(X_1, \ldots, X_n)$ of the variables $X_1, \ldots, X_n$. Since a statistic T is a function of random variables, it follows that T is itself a random variable and its distribution can, in principle, be derived from the joint distribution of $X_1, \ldots, X_n$. This distribution is often called the sampling distribution of the statistic T because it is derived from the joint distribution of the observations in a random sample.

As mentioned in Sec. 6.8, an estimator of θ is a statistic since it is a function of the observations $X_1, \ldots, X_n$. Therefore, in principle, it is possible to derive the sampling distribution of any estimator of θ. In fact, the distributions of many estimators and statistics have already been found in previous chapters of this book. For example, if $X_1, \ldots, X_n$ form a random sample from a normal distribution with mean μ and variance σ^2, then it is known from Sec. 6.5 that the sample mean $\overline{X}_n$ is the M.L.E. of μ. Furthermore, it was found in Corollary 2 of Sec. 5.6 that the distribution of $\overline{X}_n$ is a normal distribution with mean μ and variance σ^2/n.

In this chapter we shall derive, for random samples from a normal distribution, the distribution of the sample variance and the distributions of various functions of the sample mean and the sample variance. These derivations will lead us to the definitions of some new distributions that play an important part in

381

problems of statistical inference. In addition, we shall study certain general properties of estimators and their sampling distributions.

Purpose of the Sampling Distribution

It can be seen from the discussion given in Chapter 6 that a Bayes estimator or an M.L.E. in a given problem can be determined without calculating the sampling distribution of the estimator. Indeed, after the values in the sample have been observed and the Bayes estimate of the parameter θ has been found, the relevant properties of this estimate can be determined from the posterior distribution of θ. For example, the probability that the estimate does not differ from the unknown value of θ by more than a specified number of units or the M.S.E. of the estimate can be determined from the posterior distribution of θ.

However, before the sample is taken, we may wish to calculate the probability that the estimator will not differ from θ by more than a specified number of units or to calculate the M.S.E. of the estimator. It is then typically necessary to determine the sampling distribution of the estimator for each possible value of θ. In particular, if the statistician must decide which one of two or more available experiments should be performed in order to obtain the best estimator of θ, or if he must choose the best sample size in a given experiment, then he will typically base his decision on the sampling distributions of the different estimators that might be used.

In addition, as mentioned in Secs. 6.2 and 6.5, many statisticians believe that it is either inappropriate or too difficult to assign a prior distribution to a parameter θ in certain problems. Therefore, in such a problem, it is not possible to assign a posterior distribution to the parameter θ. Hence, after the sample values have been observed and the numerical value of the estimate of θ has been calculated, it would not be possible or would not be appropriate to consider the posterior probability that this estimate is close to θ. Before the sample has been taken, the statistician can use the sampling distribution of the estimator to calculate the probability that the estimator will be close to θ. If this probability is high for every possible value of θ, then the statistician can conclude that the particular estimate which he has obtained from the observed sample values is likely to be close to θ, even though explicit posterior probabilities cannot be given.

EXERCISES

1. Suppose that a random sample is to be taken from a normal distribution for which the value of the mean θ is unknown and the standard deviation is 2.

How large a random sample must be taken in order that $E_\theta(|\bar{X}_n - \theta|^2) \leqslant 0.1$ for every possible value of θ?

2. For the conditions of Exercise 1, how large a random sample must be taken in order that $E_\theta(|\bar{X}_n - \theta|) \leqslant 0.1$ for every possible value of θ?

3. For the conditions of Exercise 1, how large a random sample must be taken in order that $\Pr(|\bar{X}_n - \theta| \leqslant 0.1) \geqslant 0.95$ for every possible value of θ?

4. Suppose that a random sample is to be taken from a Bernoulli distribution for which the value of the parameter p is unknown. Suppose also that it is believed that the value of p is in the neighborhood of 0.2. How large a random sample must be taken in order that $\Pr(|\bar{X}_n - p| \leqslant 0.1) \geqslant 0.75$ when $p = 0.2$?

5. For the conditions of Exercise 4, use the central limit theorem to find approximately the size of a random sample that must be taken in order that $\Pr(|\bar{X}_n - p| \leqslant 0.1) \geqslant 0.95$ when $p = 0.2$.

6. For the conditions of Exercise 4, how large a random sample must be taken in order that $E_p(|\bar{X}_n - p|^2) \leqslant 0.01$ when $p = 0.2$?

7. For the conditions of Exercise 4, how large a random sample must be taken in order that $E_p(|\bar{X}_n - p|^2) \leqslant 0.01$ for every possible value of p $(0 \leqslant p \leqslant 1)$?

7.2. THE CHI-SQUARE DISTRIBUTION

Definition of the Distribution

In this section we shall introduce and discuss a particular type of gamma distribution known as the chi-square (χ^2) distribution. This distribution, which is closely related to random samples from a normal distribution, is widely applied in the field of statistics; and in the remainder of this book we shall see how it is applied in many important problems of statistical inference. In this section we shall present the definition of the χ^2 distribution and some of its basic mathematical properties.

The gamma distribution with parameters α and β was defined in Sec. 5.9. For any given positive integer n, the gamma distribution for which $\alpha = n/2$ and $\beta = 1/2$ is called the χ^2 *distribution with n degrees of freedom.* If a random variable X has a χ^2 distribution with n degrees of freedom, it follows from Eq. (7) of Sec. 5.9 that the p.d.f. of X for $x > 0$ is

$$f(x) = \frac{1}{2^{n/2}\Gamma(n/2)} x^{(n/2)-1} e^{-x/2}. \tag{1}$$

Also, $f(x) = 0$ for $x \leqslant 0$.

A short table of probabilities for the χ^2 distribution for various values of n is given at the end of this book.

It follows from the definition of the χ^2 distribution, and it can be seen from Eq. (1), that the χ^2 distribution with two degrees of freedom is an exponential distribution with parameter $1/2$ or, equivalently, an exponential distribution for which the mean is 2. Thus, the following three distributions are all the same: the gamma distribution with parameters $\alpha = 1$ and $\beta = 1/2$, the χ^2 distribution with two degrees of freedom, and the exponential distribution for which the mean is 2.

Properties of the Distribution

If a random variable X has a χ^2 distribution with n degrees of freedom, it follows from the expressions for the mean and the variance of the gamma distribution, as given in Sec. 5.9, that

$$E(X) = n \quad \text{and} \quad \text{Var}(X) = 2n. \tag{2}$$

Furthermore, it follows from the moment generating function given in Sec. 5.9 that the m.g.f. of X is

$$\psi(t) = \left(\frac{1}{1 - 2t} \right)^{n/2} \quad \text{for } t < \frac{1}{2}. \tag{3}$$

The additivity property of the χ^2 distribution which is presented in the next theorem follows directly from Theorem 3 of Sec. 5.9.

Theorem 1. *If the random variables $X_1, \ldots, X_k$ are independent and if X_i has a χ^2 distribution with n_i degrees of freedom $(i = 1, \ldots, k)$, then the sum $X_1 + \cdots + X_k$ has a χ^2 distribution with $n_1 + \cdots + n_k$ degrees of freedom.*

We shall now establish the basic relation between the χ^2 distribution and the normal distribution. We begin by showing that if a random variable X has a standard normal distribution, then the random variable $Y = X^2$ will have a χ^2 distribution with one degree of freedom. For this purpose, we shall let $f(y)$ and $F(y)$ denote the p.d.f. and the d.f. of Y. Also, since X has a standard normal distribution, we shall let $\phi(x)$ and $\Phi(x)$ denote the p.d.f. and the d.f. of X. Then for $y > 0$,

$$F(y) = \Pr(Y \leqslant y) = \Pr(X^2 \leqslant y) = \Pr(-y^{1/2} \leqslant X \leqslant y^{1/2})$$

$$= \Phi(y^{1/2}) - \Phi(-y^{1/2}).$$

Since $f(y) = F'(y)$ and $\phi(x) = \Phi'(x)$, it follows from the chain rule for derivatives that

$$f(y) = \phi(y^{1/2})\left(\frac{1}{2}y^{-1/2}\right) + \phi(-y^{1/2})\left(\frac{1}{2}y^{-1/2}\right).$$

Furthermore, since $\phi(y^{1/2}) = \phi(-y^{1/2}) = (2\pi)^{-1/2}e^{-y/2}$, it now follows that

$$f(y) = \frac{1}{(2\pi)^{1/2}}y^{-1/2}e^{-y/2} \qquad \text{for } y > 0.$$

By comparing this equation with Eq. (1), it is seen that the p.d.f. of Y is indeed the p.d.f. of a χ^2 distribution with one degree of freedom.

We can now combine this result with Theorem 1 to obtain the following theorem, which provides the main reason that the χ^2 distribution is important in statistics.

Theorem 2.   *If the random variables $X_1, \ldots, X_k$ are i.i.d. and if each of these variables has a standard normal distribution, then the sum of squares $X_1^2 + \cdots + X_k^2$ has a χ^2 distribution with k degrees of freedom.*

EXERCISES

1. Find the mode of the χ^2 distribution with n degrees of freedom ($n = 1, 2, \ldots$).

2. Sketch the p.d.f. of the χ^2 distribution with n degrees of freedom for each of the following values of n. Locate the mean, the median, and the mode on each sketch. (a) $n = 1$; (b) $n = 2$; (c) $n = 3$; (d) $n = 4$.

3. Suppose that a point (X, Y) is to be chosen at random in the xy-plane, where X and Y are independent random variables and each has a standard normal distribution. If a circle is drawn in the xy-plane with its center at the origin, what is the radius of the smallest circle that can be chosen in order for there to be probability 0.99 that the point (X, Y) will lie inside the circle?

4. Suppose that a point (X, Y, Z) is to be chosen at random in three-dimensional space, where X, Y, and Z are independent random variables and each has a standard normal distribution. What is the probability that the distance from the origin to the point will be less than 1 unit?

5. When the motion of a microscopic particle in a liquid or a gas is observed, it is seen that the motion is irregular because the particle collides frequently with other particles. The probability model for this motion, which is called Brownian motion, is as follows: A coordinate system is chosen in the liquid or gas. Suppose that the particle is at the origin of this coordinate system at time $t = 0$; and let (X, Y, Z) denote the coordinates of the particle at any time $t > 0$. The random variables X, Y, and Z are i.i.d. and each of them has

a normal distribution with mean 0 and variance $\sigma^2 t$. Find the probability that at time $t = 2$ the particle will lie within a sphere whose center is at the origin and whose radius is 4σ.

6. Suppose that the random variables $X_1, \ldots, X_n$ are independent, and that each random variable X_i has a continuous d.f. F_i. Also, let the random variable Y be defined by the relation $Y = -2\sum_{i=1}^{n} \log F_i(X_i)$. Show that Y has a χ^2 distribution with $2n$ degrees of freedom.

7. Suppose that $X_1, \ldots, X_n$ form a random sample from a uniform distribution on the interval $(0, 1)$, and let W denote the range of the sample, as defined in Sec. 3.9. Also, let $g_n(x)$ denote the p.d.f. of the random variable $2n(1 - W)$, and let $g(x)$ denote the p.d.f. of the χ^2 distribution with four degrees of freedom. Show that

$$\lim_{n \to \infty} g_n(x) = g(x) \qquad \text{for } x > 0.$$

8. Suppose that $X_1, \ldots, X_n$ form a random sample from a normal distribution with mean μ and variance σ^2. Find the distribution of

$$\frac{n(\overline{X}_n - \mu)^2}{\sigma^2}.$$

9. Suppose that six random variables $X_1, \ldots, X_6$ form a random sample from a standard normal distribution, and let

$$Y = (X_1 + X_2 + X_3)^2 + (X_4 + X_5 + X_6)^2.$$

Determine a value of c such that the random variable cY will have a χ^2 distribution.

10. If a random variable X has a χ^2 distribution with n degrees of freedom, then the distribution of $X^{1/2}$ is called a *chi* (χ) *distribution with n degrees of freedom*. Determine the mean of this distribution.

7.3. JOINT DISTRIBUTION OF THE SAMPLE MEAN AND SAMPLE VARIANCE

Independence of the Sample Mean and Sample Variance

Suppose that the variables $X_1, \ldots, X_n$ form a random sample from a normal distribution with an unknown mean μ and an unknown variance σ^2. Then, as was shown in Sec. 6.5, the M.L.E.'s of μ and σ^2 are the sample mean $\overline{X}_n$ and the sample variance $(1/n)\sum_{i=1}^{n}(X_i - \overline{X}_n)^2$. In this section we shall derive the joint distribution of these two estimators.

We already know that the sample mean itself has a normal distribution with mean μ and variance σ^2/n. We shall establish the noteworthy property that the sample mean and the sample variance are independent random variables, even though both are functions of the same variables $X_1, \ldots, X_n$. Furthermore, we shall show that, except for a scale factor, the sample variance has a χ^2 distribution with $n - 1$ degrees of freedom. More precisely, we shall show that the random variable $\sum_{i=1}^{n}(X_i - \overline{X}_n)^2/\sigma^2$ has a χ^2 distribution with $n - 1$ degrees of freedom. This result is also a rather striking property of random samples from a normal distribution, as the following discussion indicates.

Since the random variables $X_1, \ldots, X_n$ are independent and since each has a normal distribution with mean μ and variance σ^2, then the random variables $(X_1 - \mu)/\sigma, \ldots, (X_n - \mu)/\sigma$ will also be independent and each of these variables will have a standard normal distribution. It follows from Theorem 2 of Sec. 7.2 that the sum of their squares $\sum_{i=1}^{n}(X_i - \mu)^2/\sigma^2$ will have a χ^2 distribution with n degrees of freedom. Hence, the striking property just mentioned is that if the population mean μ is replaced by the sample mean $\overline{X}_n$ in this sum of squares, the effect is simply to reduce the degrees of freedom in the χ^2 distribution from n to $n - 1$. In summary, we shall establish the following theorem:

Theorem 1. *Suppose that $X_1, \ldots, X_n$ form a random sample from a normal distribution with mean μ and variance σ^2. Then the sample mean $\overline{X}_n$ and the sample variance $(1/n)\sum_{i=1}^{n}(X_i - \overline{X}_n)^2$ are independent random variables; $\overline{X}_n$ has a normal distribution with mean μ and variance σ^2/n; and $\sum_{i=1}^{n}(X_i - \overline{X}_n)^2/\sigma^2$ has a χ^2 distribution with $n - 1$ degrees of freedom.*

Furthermore, it can be shown that the sample mean and the sample variance are independent *only* when the random sample is drawn from a normal distribution. We shall not consider this result further in this book. However, it does emphasize the fact that the independence of the sample mean and the sample variance is indeed a noteworthy property of samples from a normal distribution.

The proof of Theorem 1 makes use of the properties of orthogonal matrices, which we shall now introduce.

Orthogonal Matrices

Definition. It is said that an $n \times n$ matrix A is *orthogonal* if $A^{-1} = A'$, where A' is the transpose of A. Thus, a matrix A is orthogonal if and only if $AA' = A'A = I$, where I is the $n \times n$ identity matrix. It follows from this definition that a matrix is orthogonal if and only if the sum of the squares of the elements in each row is 1 and the sum of the products of the corresponding elements in any two different rows is 0. Alternatively, a matrix is orthogonal if and only if the sum of the squares of the elements in each column is 1 and the sum of the products of the corresponding elements in any two different columns is 0.

Properties of Orthogonal Matrices. We shall now derive two important properties of orthogonal matrices.

Property 1: *If A is orthogonal, then $|\det A| = 1$.*

Proof. To prove this result, it should be recalled that $\det A = \det A'$ for any square matrix A. Therefore,

$$\det(AA') = (\det A)(\det A') = (\det A)^2.$$

Also, if A is orthogonal, then $AA' = I$, and it follows that

$$\det(AA') = \det I = 1.$$

Hence, $(\det A)^2 = 1$ or, equivalently, $|\det A| = 1$. □

Property 2: *Consider two n-dimensional random vectors*

$$X = \begin{bmatrix} X_1 \\ \vdots \\ X_n \end{bmatrix} \quad and \quad Y = \begin{bmatrix} Y_1 \\ \vdots \\ Y_n \end{bmatrix}, \tag{1}$$

and suppose that $Y = AX$, where A is an orthogonal matrix. Then

$$\sum_{i=1}^{n} Y_i^2 = \sum_{i=1}^{n} X_i^2. \tag{2}$$

Proof. This result follows from the fact that $A'A = I$, because

$$\sum_{i=1}^{n} Y_i^2 = Y'Y = X'A'AX = X'X = \sum_{i=1}^{n} X_i^2. \quad □$$

Together, these two properties of orthogonal matrices imply that if a random vector Y is obtained from a random vector X by an *orthogonal* linear transformation $Y = AX$, then the absolute value of the Jacobian of the transformation is 1 and $\sum_{i=1}^{n} Y_i^2 = \sum_{i=1}^{n} X_i^2$.

Orthogonal Linear Transformations of a Random Sample. Suppose that $X_1, \ldots, X_n$ form a random sample from a standard normal distribution. Then the joint p.d.f. of $X_1, \ldots, X_n$ is as follows, for $-\infty < x_i < \infty$ $(i = 1, \ldots, n)$:

$$f_n(x) = \frac{1}{(2\pi)^{n/2}} \exp\left(-\frac{1}{2} \sum_{i=1}^{n} x_i^2\right). \tag{3}$$

Suppose also that A is an orthogonal $n \times n$ matrix, and that the random variables $Y_1, \ldots, Y_n$ are defined by the relation $Y = AX$, where the vectors X and Y are as specified in Eq. (1). It now follows from Eq. (3) and the properties of orthogonal matrices previously derived that the joint p.d.f. of $Y_1, \ldots, Y_n$ is as follows, for $-\infty < y_i < \infty$ $(i = 1, \ldots, n)$:

$$g_n(y) = \frac{1}{(2\pi)^{n/2}} \exp\left(-\frac{1}{2} \sum_{i=1}^{n} y_i^2\right). \tag{4}$$

It can be seen from Eq. (4) that the joint p.d.f. of $Y_1, \ldots, Y_n$ is exactly the same as the joint p.d.f. of $X_1, \ldots, X_n$. Thus, we have established the following result:

Theorem 2. *Suppose that the random variables, $X_1, \ldots, X_n$ are i.i.d. and that each has a standard normal distribution. Suppose also that A is an orthogonal $n \times n$ matrix, and $Y = AX$. Then the random variables $Y_1, \ldots, Y_n$ are also i.i.d.; each also has a standard normal distribution; and $\sum_{i=1}^{n} X_i^2 = \sum_{i=1}^{n} Y_i^2$.*

Proof of the Independence of the Sample Mean and Sample Variance

Random Samples from a Standard Normal Distribution. We shall begin by proving Theorem 1 under the assumption that $X_1, \ldots, X_n$ form a random sample from a standard normal distribution. Consider the n-dimensional vector u, in which each of the n components has the value $1/\sqrt{n}$:

$$u = \left[\frac{1}{\sqrt{n}} \cdots \frac{1}{\sqrt{n}}\right]. \tag{5}$$

Since the sum of the squares of the n components of the vector u is 1, it is possible to construct an orthogonal matrix A such that the components of the vector u form the first row of A. This construction is described in textbooks on linear algebra and will not be discussed here. We shall assume that such a matrix A has been constructed, and we shall again define the random variables $Y_1, \ldots, Y_n$ by the transformation $Y = AX$.

Since the components of u form the first row of A, it follows that

$$Y_1 = uX = \sum_{i=1}^{n} \frac{1}{\sqrt{n}} X_i = \sqrt{n}\, \bar{X}_n. \tag{6}$$

Furthermore, by Theorem 2, $\sum_{i=1}^{n} X_i^2 = \sum_{i=1}^{n} Y_i^2$. Therefore,

$$\sum_{i=2}^{n} Y_i^2 = \sum_{i=1}^{n} Y_i^2 - Y_1^2 = \sum_{i=1}^{n} X_i^2 - n\bar{X}_n^2 = \sum_{i=1}^{n} (X_i - \bar{X}_n)^2.$$

We have thus obtained the relation

$$\sum_{i=2}^{n} Y_i^2 = \sum_{i=1}^{n} (X_i - \bar{X}_n)^2. \tag{7}$$

It is known from Theorem 2 that the random variables $Y_1, \ldots, Y_n$ are independent. Therefore, the two random variables Y_1 and $\sum_{i=2}^{n} Y_i^2$ will be independent, and it follows from Eqs. (6) and (7) that $\bar{X}_n$ and $\sum_{i=1}^{n} (X_i - \bar{X}_n)^2$ are independent. Furthermore, it is known from Theorem 2 that the $n - 1$ random variables $Y_2, \ldots, Y_n$ are i.i.d., and that each of these random variables has a standard normal distribution. Hence, by Theorem 2 of Sec. 7.2, the random variable $\sum_{i=2}^{n} Y_i^2$ has a χ^2 distribution with $n - 1$ degrees of freedom. It follows from Eq. (7) that $\sum_{i=1}^{n} (X_i - \bar{X}_n)^2$ also has a χ^2 distribution with $n - 1$ degrees of freedom.

Random Samples from an Arbitrary Normal Distribution. Thus far, in proving Theorem 1, we have considered only random samples from a standard normal distribution. Suppose now that the random variables $X_1, \ldots, X_n$ form a random sample from an arbitrary normal distribution with mean μ and variance σ^2.

If we let $Z_i = (X_i - \mu)/\sigma$ for $i = 1, \ldots, n$, then the random variables $Z_1, \ldots, Z_n$ will be independent and each will have a standard normal distribution. In other words, the joint distribution of $Z_1, \ldots, Z_n$ will be the same as the joint distribution of a random sample from a standard normal distribution. It follows from the results that have just been obtained that $\bar{Z}_n$ and $\sum_{i=1}^{n} (Z_i - \bar{Z}_n)^2$ are independent, and that $\sum_{i=1}^{n} (Z_i - \bar{Z}_n)^2$ has a χ^2 distribution with $n - 1$ degrees of freedom. However, $\bar{Z}_n = (\bar{X}_n - \mu)/\sigma$ and

$$\sum_{i=1}^{n} (Z_i - \bar{Z}_n)^2 = \frac{1}{\sigma^2} \sum_{i=1}^{n} (X_i - \bar{X}_n)^2. \tag{8}$$

It can be concluded that the sample mean $\bar{X}_n$ and the sample variance $(1/n)\sum_{i=1}^{n} (X_i - \bar{X}_n)^2$ are independent, and that the random variable on the right side of Eq. (8) has a χ^2 distribution with $n - 1$ degrees of freedom. All the results stated in Theorem 1 have now been established.

Estimation of the Mean and Variance

We shall assume that $X_1, \ldots, X_n$ form a random sample from a normal distribution for which both the mean μ and the variance σ^2 are unknown. Also, as usual, we shall denote the M.L.E.'s of μ and σ^2 by $\hat{\mu}$ and $\hat{\sigma}^2$. Thus,

$$\hat{\mu} = \overline{X}_n \quad \text{and} \quad \hat{\sigma}^2 = \frac{1}{n} \sum_{i=1}^{n} (X_i - \overline{X}_n)^2.$$

As an illustration of the application of Theorem 1, we shall now determine the smallest possible sample size n such that the following relation will be satisfied:

$$\Pr\left(|\hat{\mu} - \mu| \leqslant \frac{1}{5}\sigma \quad \text{and} \quad |\hat{\sigma} - \sigma| \leqslant \frac{1}{5}\sigma\right) \geqslant \frac{1}{2}. \tag{9}$$

In other words, we shall determine the minimum sample size n for which the probability will be at least $1/2$ that neither $\hat{\mu}$ nor $\hat{\sigma}$ will differ from the unknown value it is estimating by more than $(1/5)\sigma$.

Because of the independence of $\hat{\mu}$ and $\hat{\sigma}^2$, the relation (9) can be rewritten as follows:

$$\Pr\left(|\hat{\mu} - \mu| < \frac{1}{5}\sigma\right)\Pr\left(|\hat{\sigma} - \sigma| < \frac{1}{5}\sigma\right) \geqslant \frac{1}{2}. \tag{10}$$

If we let p_1 denote the first probability on the left side of the relation (10), and let U be a random variable that has a standard normal distribution, this probability can be written in the following form:

$$p_1 = \Pr\left(\frac{\sqrt{n}|\hat{\mu} - \mu|}{\sigma} < \frac{1}{5}\sqrt{n}\right) = \Pr\left(|U| < \frac{1}{5}\sqrt{n}\right).$$

Similarly, if we let p_2 denote the second probability on the left side of the relation (10), and let $V = n\hat{\sigma}^2/\sigma^2$, this probability can be written in the following form:

$$p_2 = \Pr\left(0.8 < \frac{\hat{\sigma}}{\sigma} < 1.2\right) = \Pr\left(0.64n < \frac{n\hat{\sigma}^2}{\sigma^2} < 1.44n\right)$$

$$= \Pr(0.64n < V < 1.44n).$$

$$V = \frac{n\hat{\sigma}^2}{\sigma^2} = \frac{n}{\sigma^2}\frac{1}{n}\sum_{i=1}^{n}(x_i - \overline{x}_n)^2$$

$$= \frac{1}{\sigma^2}\sum_{i=1}^{n}(x_i - \overline{x}_n)^2$$

$$= \chi^2(n-1)$$

By Theorem 1, the random variable V has a χ^2 distribution with $n - 1$ degrees of freedom.

For any given value of n, the values of p_1 and p_2 can be found, at least approximately, from the table of the standard normal distribution and the table of the χ^2 distribution given at the end of this book. In particular, after various values of n have been tried, it will be found that for $n = 21$ the values of p_1 and p_2 are $p_1 = 0.64$ and $p_2 = 0.78$. Hence, $p_1 p_2 = 0.50$ and it follows that the relation (9) will be satisfied for $n = 21$.

EXERCISES

1. Determine whether or not each of the five following matrices is orthogonal:

(a) $\begin{bmatrix} 0 & 1 & 0 \\ 0 & 0 & 1 \\ 1 & 0 & 0 \end{bmatrix}$
(b) $\begin{bmatrix} 0.8 & 0 & 0.6 \\ -0.6 & 0 & 0.8 \\ 0 & -1 & 0 \end{bmatrix}$
(c) $\begin{bmatrix} 0.8 & 0 & 0.6 \\ -0.6 & 0 & 0.8 \\ 0 & 0.5 & 0 \end{bmatrix}$

(d) $\begin{bmatrix} -\dfrac{1}{\sqrt{3}} & \dfrac{1}{\sqrt{3}} & \dfrac{1}{\sqrt{3}} \\ \dfrac{1}{\sqrt{3}} & -\dfrac{1}{\sqrt{3}} & \dfrac{1}{\sqrt{3}} \\ \dfrac{1}{\sqrt{3}} & \dfrac{1}{\sqrt{3}} & -\dfrac{1}{\sqrt{3}} \end{bmatrix}$
(e) $\begin{bmatrix} \dfrac{1}{2} & \dfrac{1}{2} & \dfrac{1}{2} & \dfrac{1}{2} \\ -\dfrac{1}{2} & -\dfrac{1}{2} & \dfrac{1}{2} & \dfrac{1}{2} \\ -\dfrac{1}{2} & \dfrac{1}{2} & -\dfrac{1}{2} & \dfrac{1}{2} \\ -\dfrac{1}{2} & \dfrac{1}{2} & \dfrac{1}{2} & -\dfrac{1}{2} \end{bmatrix}$

2. (a) Construct a 2×2 orthogonal matrix for which the first row is as follows:

$$\begin{bmatrix} \dfrac{1}{\sqrt{2}} & \dfrac{1}{\sqrt{2}} \end{bmatrix}.$$

(b) Construct a 3×3 orthogonal matrix for which the first row is as follows:

$$\begin{bmatrix} \dfrac{1}{\sqrt{3}} & \dfrac{1}{\sqrt{3}} & \dfrac{1}{\sqrt{3}} \end{bmatrix}.$$

3. Suppose that the random variables X_1, X_2, and X_3 are i.i.d., and that each has a standard normal distribution. Also, suppose that

$$Y_1 = 0.8X_1 + 0.6X_2,$$

$$Y_2 = \sqrt{2}\,(0.3X_1 - 0.4X_2 - 0.5X_3),$$

$$Y_3 = \sqrt{2}\,(0.3X_1 - 0.4X_2 + 0.5X_3).$$

Find the joint distribution of Y_1, Y_2, and Y_3.

4. Suppose that the random variables X_1 and X_2 are independent, and that each has a normal distribution with mean μ and variance σ^2. Prove that the random variables $X_1 + X_2$ and $X_1 - X_2$ are independent.

5. Suppose that $X_1, \ldots, X_n$ form a random sample from a normal distribution with mean μ and variance σ^2. Assuming that the sample size n is 16, determine the values of the following probabilities:

(a) $\Pr\left[\dfrac{1}{2}\sigma^2 \leqslant \dfrac{1}{n}\sum_{i=1}^{n}(X_i - \mu)^2 \leqslant 2\sigma^2\right]$,

(b) $\Pr\left[\dfrac{1}{2}\sigma^2 \leqslant \dfrac{1}{n}\sum_{i=1}^{n}(X_i - \overline{X}_n)^2 \leqslant 2\sigma^2\right]$.

6. Suppose that $X_1, \ldots, X_n$ form a random sample from a normal distribution with mean μ and variance σ^2, and let $\hat{\sigma}^2$ denote the sample variance. Determine the smallest values of n for which the following relations are satisfied:

(a) $\Pr\left(\dfrac{\hat{\sigma}^2}{\sigma^2} \leqslant 1.5\right) \geqslant 0.95$;

(b) $\Pr\left(|\hat{\sigma}^2 - \sigma^2| \leqslant \dfrac{1}{2}\sigma^2\right) \geqslant 0.8$.

7. Suppose that X has a χ^2 distribution with 200 degrees of freedom. Explain why the central limit theorem can be used to determine the approximate value of $\Pr(160 < X < 240)$, and find this approximate value.

8. Suppose that each of two statisticians A and B independently takes a random sample of 20 observations from a normal distribution for which the mean μ is unknown and the known value of the variance is 4. Suppose also that statistician A finds the sample variance in his random sample to be 3.8, and that statistician B finds the sample variance in his random sample to be 9.4. For which random sample is the sample mean likely to be closer to the unknown value of μ?

7.4. THE *t* DISTRIBUTION

Definition of the Distribution

In this section we shall introduce and discuss another distribution, called the *t* distribution, which is closely related to random samples from a normal distribution. The t distribution, like the χ^2 distribution, has been widely applied in important problems of statistical inference. The t distribution is also known as Student's distribution in honor of W. S. Gosset, who published his studies of this distribution in 1908 under the pen-name "Student." The distribution is defined as follows:

Consider two independent random variables Y and Z, such that Y has a standard normal distribution and Z has a χ^2 distribution with n degrees of freedom. Suppose that a random variable X is defined by the equation

$$X = \frac{Y}{\left(\dfrac{Z}{n}\right)^{1/2}}. \tag{1}$$

Then the distribution of X is called the *t distribution with n degrees of freedom*.

Derivation of the p.d.f.   We shall now derive the p.d.f. of the t distribution with n degrees of freedom. Suppose that the joint distribution of Y and Z is as specified in the definition of the t distribution. Then, since Y and Z are independent, their joint p.d.f. is equal to the product $f_1(y)f_2(z)$, where $f_1(y)$ is the p.d.f. of the standard normal distribution and $f_2(z)$ is the p.d.f. of the χ^2 distribution with n degrees of freedom. Let X be defined by Eq. (1) and, as a convenient device, let $W = Z$. We shall determine first the joint p.d.f. of X and W.

From the definitions of X and W,

$$Y = \frac{1}{n^{1/2}} X W^{1/2} \quad \text{and} \quad Z = W. \tag{2}$$

The Jacobian of the transformation (2) from X and W to Y and Z is $(W/n)^{1/2}$. The joint p.d.f. $f(x,w)$ of X and W can be obtained from the joint p.d.f. $f_1(y)f_2(z)$ by replacing y and z by the expressions given in (2) and then multiplying the result by $(w/n)^{1/2}$. It is then found that the value of $f(x,w)$ is as follows, for $-\infty < x < \infty$ and $w > 0$:

$$f(x,w) = c w^{(n-1)/2} \exp\left[-\frac{1}{2}\left(1 + \frac{x^2}{n}\right) w \right], \tag{3}$$

where

$$c = \left[2^{(n+1)/2}(n\pi)^{1/2} \Gamma\left(\frac{n}{2}\right) \right]^{-1}.$$

The marginal p.d.f. $g(x)$ of X can be obtained from Eq. (3) by using the relation

$$g(x) = \int_0^\infty f(x,w)\, dw.$$

It is thus found that

$$g(x) = \frac{\Gamma\left(\dfrac{n+1}{2}\right)}{(n\pi)^{1/2}\Gamma\left(\dfrac{n}{2}\right)}\left(1 + \frac{x^2}{n}\right)^{-(n+1)/2} \qquad \text{for} -\infty < x < \infty. \qquad (4)$$

Therefore, if X has a t distribution with n degrees of freedom, then the p.d.f. of X is specified by Eq. (4).

Relation to the Cauchy Distribution and to the Normal Distribution. It can be seen from Eq. (4) that the p.d.f. $g(x)$ is a symmetric, bell-shaped function with its maximum value at $x = 0$. Thus, its general shape is similar to that of the p.d.f. of a normal distribution with mean 0. However, as $x \to \infty$ or $x \to -\infty$, the tails of the p.d.f. $g(x)$ approach 0 much more slowly than do the tails of the p.d.f. of a normal distribution. In fact, it can be seen from Eq. (4) that when $n = 1$, the t distribution is the Cauchy distribution which was described in Sec. 4.1. The p.d.f. of the Cauchy distribution was sketched in Fig. 4.3. It was shown in Sec. 4.1 that the mean of the Cauchy distribution does not exist, because the integral which specifies the value of the mean is not absolutely convergent. It follows that, although the p.d.f. of the t distribution with one degree of freedom is symmetric with respect to the point $x = 0$, the mean of this distribution does not exist.

It can also be shown from Eq. (4) that, as $n \to \infty$, the p.d.f. $g(x)$ converges to the p.d.f. $\phi(x)$ of the standard normal distribution for every value of x $(-\infty < x < \infty)$. Hence, when n is large, the t distribution with n degrees of freedom can be approximated by the standard normal distribution.

A short table of probabilities for the t distribution for various values of n is given at the end of this book. The probabilities in the first line of the table, corresponding to $n = 1$, are those for the Cauchy distribution. The probabilities in the bottom line of the table, corresponding to $n = \infty$, are those for the standard normal distribution.

Moments of the *t* Distribution

Although the mean of the t distribution does not exist when $n = 1$, the mean does exist for any value of $n > 1$. Of course, whenever the mean does exist, its value is 0 because of the symmetry of the t distribution.

In general, if a random variable X has a t distribution with n degrees of freedom $(n > 1)$, then it can be shown that $E(|X|^k) < \infty$ for $k < n$ and that $E(|X|^k) = \infty$ for $k \geq n$. In other words, the first $n - 1$ moments of X exist, but no moments of higher order exist. It follows, therefore, that the m.g.f. of X does not exist.

It can be shown (see Exercise 2 at the end of this section) that if X has a t distribution with n degrees of freedom ($n > 2$), then $\text{Var}(X) = n/(n - 2)$.

Relation to Random Samples from a Normal Distribution

Suppose again that $X_1, \ldots, X_n$ form a random sample from a normal distribution with mean μ and variance σ^2. As usual, we shall let $\overline{X}_n$ denote the sample mean and we shall let

$$S_n^2 = \sum_{i=1}^{n} (X_i - \overline{X}_n)^2.$$

If we define random variables Y and Z by the relations $Y = n^{1/2}(\overline{X}_n - \mu)/\sigma$ and $Z = S_n^2/\sigma^2$, then it follows from Theorem 1 of Sec. 7.3 that Y and Z are independent; that Y has a standard normal distribution; and that Z has a χ^2 distribution with $n - 1$ degrees of freedom. Now consider another random variable U, which is defined by the relation

$$U = \frac{Y}{\left(\dfrac{Z}{n-1}\right)^{1/2}}.$$

It follows from the definition of the t distribution that U has a t distribution with $n - 1$ degrees of freedom. It is easily seen that the expression for U can be rewritten in the following form:

$$U = \frac{n^{1/2}(\overline{X}_n - \mu)}{\left(\dfrac{S_n^2}{n-1}\right)^{1/2}}. \tag{5}$$

The first rigorous proof that the distribution of U is a t distribution with $n - 1$ degrees of freedom was given by R. A. Fisher in 1923.

One important aspect of Eq. (5) is that neither the value of U nor the distribution of U depends on the value of the variance σ^2. We have remarked previously that the random variable $Y = n^{1/2}(\overline{X}_n - \mu)/\sigma$ has a standard normal distribution. If the standard deviation σ is now replaced in this expression by an estimator of σ calculated from the sample $X_1, \ldots, X_n$, then the distribution of Y will be changed. In particular, if we replace σ by its M.L.E. $\hat{\sigma} = (S_n^2/n)^{1/2}$, then we obtain the random variable Y', which is defined as follows:

$$Y' = \frac{n^{1/2}(\overline{X}_n - \mu)}{\hat{\sigma}} = \left(\frac{n}{n-1}\right)^{1/2} U. \tag{6}$$

Since Y' differs from U only because of a constant factor, the distribution of Y' can easily be derived from the distribution of U, which has already been found to be a t distribution with $n - 1$ degrees of freedom. Thus, if the standard deviation σ is replaced by its M.L.E. $\hat{\sigma}$, the resulting random variable Y' and its distribution no longer depend on the variance σ^2 of the underlying normal distribution. The practical importance of this fact will be demonstrated later in this section and in other sections of this book.

A similar statement can be made in regard to the random variable U itself. In the expression for Y, we shall replace the standard deviation σ by the estimator

$$\sigma' = \left[\frac{S_n^2}{n-1} \right]^{1/2} = \left(\frac{n}{n-1} \right)^{1/2} \hat{\sigma}. \tag{7}$$

Then the random variable Y will be replaced by the random variable U. It can be seen from Eq. (7) that for large values of n the estimators σ' and $\hat{\sigma}$ will be very close to each other. The estimator σ' will be discussed further in Sec. 7.7.

If the sample size n is large, the probability that the estimator σ' will be close to σ is high. Hence, replacing σ by σ' in the random variable Y will not greatly change the standard normal distribution of Y. For this reason, as we have already mentioned, the p.d.f. of the t distribution converges to the p.d.f. of the standard normal distribution as $n \to \infty$.

EXERCISES

1. Sketch, on a single graph, the p.d.f. of the t distribution with one degree of freedom, the p.d.f. of the t distribution with five degrees of freedom, and the p.d.f. of the standard normal distribution.

2. Suppose that X has a t distribution with n degrees of freedom ($n > 2$). Show that $\mathrm{Var}(X) = n/(n-2)$. *Hint:* To evaluate $E(X^2)$, restrict the integral to the positive half of the real line and change the variable from x to

$$y = \frac{\dfrac{x^2}{n}}{1 + \dfrac{x^2}{n}}.$$

Then compare the integral with the p.d.f. of a beta distribution.

3. Suppose that $X_1, \ldots, X_n$ form a random sample from a normal distribution for which the mean μ and the standard deviation σ are unknown, and let $\hat{\mu}$ and $\hat{\sigma}$ denote the M.L.E.'s of μ and σ. For the sample size $n = 17$, find a value of k such that

$$\Pr(\hat{\mu} > \mu + k\hat{\sigma}) = 0.95.$$

4. Suppose that the five random variables $X_1, \ldots, X_5$ are i.i.d. and that each has a standard normal distribution. Determine a constant c such that the random variable

$$\frac{c(X_1 + X_2)}{\left(X_3^2 + X_4^2 + X_5^2\right)^{1/2}}$$

will have a t distribution.

5. By using the table of the t distribution given in the back of this book, determine the value of the integral

$$\int_{-\infty}^{2.5} \frac{dx}{(12 + x^2)^2}.$$

6. Suppose that the random variables X_1 and X_2 are independent and that each has a normal distribution with mean 0 and variance σ^2. Determine the value of

$$\Pr\left[\frac{(X_1 + X_2)^2}{(X_1 - X_2)^2} < 4\right].$$

Hint: $(X_1 - X_2)^2 = 2\left[\left(X_1 - \dfrac{X_1 + X_2}{2}\right)^2 + \left(X_2 - \dfrac{X_1 + X_2}{2}\right)^2\right].$

7.5. CONFIDENCE INTERVALS

Confidence Intervals for the Mean of a Normal Distribution

We shall continue to assume that $X_1, \ldots, X_n$ form a random sample from a normal distribution with unknown mean μ and unknown variance σ^2. Let $g_{n-1}(x)$ denote the p.d.f. of the t distribution with $n - 1$ degrees of freedom, and let c be a constant such that

$$\int_{-c}^{c} g_{n-1}(x)\, dx = 0.95. \tag{1}$$

For any given value of n, the value of c can be found from the table of the t distribution given at the end of this book. For example, if $n = 12$ and if $G_{11}(x)$

denotes the d.f. of the t distribution with 11 degrees of freedom, then

$$\int_{-c}^{c} g_{11}(x)\, dx = G_{11}(c) - G_{11}(-c) = G_{11}(c) - \left[1 - G_{11}(c)\right]$$

$$= 2G_{11}(c) - 1.$$

Hence, it follows from Eq. (1) that $G_{11}(c) = 0.975$. It is found from the table that $c = 2.201$.

Since the random variable U defined by Eq. (5) of Sec. 7.4 has a t distribution with $n - 1$ degrees of freedom, Eq. (1) implies that $\Pr(-c < U < c) = 0.95$. Furthermore, it follows from Eqs. (5) and (7) of Sec. 7.4 that this relation can be rewritten as follows:

$$\Pr\left(\bar{X}_n - \frac{c\sigma'}{n^{1/2}} < \mu < \bar{X}_n + \frac{c\sigma'}{n^{1/2}} \right) = 0.95. \tag{2}$$

Thus, Eq. (2) states that regardless of the unknown value of σ, the probability is 0.95 that μ will lie between the random variables $A = \bar{X}_n - (c\sigma'/n^{1/2})$ and $B = \bar{X}_n + (c\sigma'/n^{1/2})$.

In a practical problem, Eq. (2) is applied as follows: After the values of the variables $X_1, \ldots, X_n$ in the random sample have been observed, the values of A and B are computed. If these values are $A = a$ and $B = b$, then the interval (a, b) is called a *confidence interval for μ with confidence coefficient* 0.95. We can then make the statement that the unknown value of μ lies in the interval (a, b) with *confidence* 0.95.

For a given confidence coefficient, it is possible to construct many different confidence intervals for μ. For example, if the confidence coefficient is 0.95, any pair of constants c_1 and c_2 such that $\Pr(c_1 < U < c_2) = 0.95$ leads to a confidence interval for μ with endpoints

$$a = \bar{x}_n - \frac{c_2 \sigma'}{n^{1/2}} \quad \text{and} \quad b = \bar{x}_n - \frac{c_1 \sigma'}{n^{1/2}}.$$

However, it can be shown that among all such confidence intervals with confidence coefficient 0.95, the interval defined by Eq. (2), which is symmetric with respect to the value $\bar{x}_n$, is the shortest one.

Confidence Intervals for an Arbitrary Parameter

As a general case, suppose that $X_1, \ldots, X_n$ form a random sample from a distribution which involves a parameter θ whose value is unknown. Suppose also that two statistics $A(X_1, \ldots, X_n)$ and $B(X_1, \ldots, X_n)$ can be found such that, no matter what the true value of θ may be,

$$\Pr[A(X_1,\ldots,X_n) < \theta < B(X_1,\ldots,X_n)] = \gamma, \tag{3}$$

where γ is a fixed probability $(0 < \gamma < 1)$. If the observed values of $A(X_1,\ldots,X_n)$ and $B(X_1,\ldots,X_n)$ are a and b, then it is said that the interval (a,b) is a *confidence interval for θ with confidence coefficient γ* or, in other words, that θ lies in the interval (a,b) with *confidence γ*.

It should be emphasized that it is *not* correct to state that θ lies in the interval (a,b) with *probability γ*. We shall explain this point further here. *Before* the values of the statistics $A(X_1,\ldots,X_n)$ and $B(X_1,\ldots,X_n)$ are observed, these statistics are random variables. It follows, therefore, from Eq. (3) that θ will lie in the random interval having endpoints $A(X_1,\ldots,X_n)$ and $B(X_1,\ldots,X_n)$ with probability γ. *After* the specific values $A(X_1,\ldots,X_n) = a$ and $B(X_1,\ldots,X_n) = b$ have been observed, it is not possible to assign a probability to the event that θ lies in the specific interval (a,b) without regarding θ as a random variable which itself has a probability distribution. In other words, it is necessary first to assign a prior distribution to θ and then to use the resulting posterior distribution to calculate the probability that θ lies in the interval (a,b). Instead of assigning a prior distribution to the parameter θ, many statisticians prefer to state that there is confidence γ, rather than probability γ, that θ lies in the interval (a,b). Because of this distinction between confidence and probability, the meaning and the relevance of confidence intervals in statistical practice is a somewhat controversial topic.

Shortcoming of Confidence Intervals

In accordance with the preceding explanation, the interpretation of a confidence coefficient γ for a confidence interval is as follows: Before a sample is taken, there is probability γ that the interval that will be constructed from the sample will include the unknown value of θ. This concept has the following shortcoming: Although the particular sample values that are observed may give the experimenter additional information about whether or not the interval formed from these particular values actually does include θ, there is no standard way to adjust the confidence coefficient γ in the light of this information.

For example, suppose that two observations X_1 and X_2 are taken at random from a uniform distribution on the interval $\left(\theta - \dfrac{1}{2}, \theta + \dfrac{1}{2}\right)$, where the value of θ is unknown $(-\infty < \theta < \infty)$. If we let $Y_1 = \min(X_1, X_2)$ and $Y_2 = \max(X_1, X_2)$, then

$$
\begin{aligned}
\Pr(Y_1 < \theta < Y_2) &= \Pr(X_1 < \theta < X_2) + \Pr(X_2 < \theta < X_1) \\
&= \Pr(X_1 < \theta)\Pr(X_2 > \theta) + \Pr(X_2 < \theta)\Pr(X_1 > \theta) \\
&= (1/2)(1/2) + (1/2)(1/2) = 1/2.
\end{aligned}
\tag{4}
$$

It follows from Eq. (4) that if the values $Y_1 = y_1$ and $Y_2 = y_2$ are observed, then the interval (y_1, y_2) will be a confidence interval for θ with confidence coefficient $1/2$. However, the analysis can be carried further.

Since both observations X_1 and X_2 must be greater than $\theta - (1/2)$ and both must be less than $\theta + (1/2)$, we know with certainty that $y_1 > \theta - (1/2)$ and $y_2 < \theta + (1/2)$. In other words, we know with certainty that

$$y_2 - (1/2) < \theta < y_1 + (1/2). \tag{5}$$

Suppose now that $(y_2 - y_1) \geqslant 1/2$. Then $y_1 \leqslant y_2 - (1/2)$ and it follows from Eq. (5) that $y_1 < \theta$. Moreover, since $y_1 + (1/2) \leqslant y_2$, it also follows from Eq. (5) that $\theta < y_2$. Thus, if $(y_2 - y_1) \geqslant 1/2$, then $y_1 < \theta < y_2$. In other words, if $(y_2 - y_1) \geqslant 1/2$, then we know with certainty that the confidence interval (y_1, y_2) includes the unknown value of θ, even though the confidence coefficient of this interval is only $1/2$.

Indeed, even when $(y_2 - y_1) < 1/2$, the closer the value of $(y_2 - y_1)$ is to $1/2$, the more certain we feel that the interval (y_1, y_2) includes θ. Also, the closer the value of $(y_2 - y_1)$ is to 0, the more certain we feel that the interval (y_1, y_2) does not include θ. However, the confidence coefficient necessarily remains $1/2$ and does not depend on the observed values y_1 and y_2.

In the next section we shall discuss Bayesian methods for analyzing a random sample from a normal distribution for which both the mean μ and the variance σ^2 are unknown. We shall assign a joint prior distribution to μ and σ^2, and shall then calculate the posterior probability that μ belongs to any given interval (a, b). It can be shown [see, e.g., DeGroot (1970)] that if the joint prior p.d.f. of μ and σ^2 is fairly smooth and does not assign high probability to any particular small set of values of μ and σ^2, and if the sample size n is large, then the confidence coefficient assigned to a particular confidence interval (a, b) for the mean μ will be approximately equal to the posterior probability that μ lies in the interval (a, b). An example of this approximate equality is included in the next section. Therefore, under these conditions, the differences between the results obtained by the practical application of methods based on confidence intervals and methods based on prior probabilities will be small. Nevertheless the philosophical differences between these methods will persist.

EXERCISES

1. Suppose that a random sample of eight observations is taken from a normal distribution for which both the mean μ and the variance σ^2 are unknown; and that the observed values are 3.1, 3.5, 2.6, 3.4, 3.8, 3.0, 2.9, and 2.2. Find the shortest confidence interval for μ with each of the following three confidence coefficients: (a) 0.90, (b) 0.95, and (c) 0.99.

2. Suppose that $X_1, \ldots, X_n$ form a random sample from a normal distribution for which both the mean μ and variance σ^2 are unknown; and let the random variable L denote the length of the shortest confidence interval for μ that can be constructed from the observed values in the sample. Find the value of $E(L^2)$ for the following values of the sample size n and the confidence coefficient γ:

(a) $n = 5$, $\gamma = 0.95$.   (d) $n = 8$, $\gamma = 0.90$.

(b) $n = 10$, $\gamma = 0.95$.   (e) $n = 8$, $\gamma = 0.95$.

(c) $n = 30$, $\gamma = 0.95$.   (f) $n = 8$, $\gamma = 0.99$.

3. Suppose that $X_1, \ldots, X_n$ form a random sample from a normal distribution for which the mean μ is unknown and the variance σ^2 is known. How large a random sample must be taken in order that there will be a confidence interval for μ with confidence coefficient 0.95 and length less than 0.01σ?

4. Suppose that $X_1, \ldots, X_n$ form a random sample from a normal distribution for which both the mean μ and the variance σ^2 are unknown. Describe a method for constructing a confidence interval for σ^2 with a specified confidence coefficient γ $(0 < \gamma < 1)$. *Hint:* Determine constants c_1 and c_2 such that

$$\Pr\left[c_1 < \frac{\sum_{i=1}^{n}(X_i - \overline{X}_n)^2}{\sigma^2} < c_2 \right] = \gamma.$$

5. Suppose that $X_1, \ldots, X_n$ form a random sample from an exponential distribution with unknown mean μ. Describe a method for constructing a confidence interval for μ with a specified confidence coefficient γ $(0 < \gamma < 1)$. *Hint:* Determine constants c_1 and c_2 such that $\Pr[c_1 < (1/\mu)\sum_{i=1}^{n}X_i < c_2] = \gamma$.

*7.6. BAYESIAN ANALYSIS OF SAMPLES FROM A NORMAL DISTRIBUTION

The Precision of a Normal Distribution

Suppose that the variables $X_1, \ldots, X_n$ form a random sample from a normal distribution for which both the mean μ and the variance σ^2 are unknown. In this section we shall consider the assignment of a joint prior distribution to the parameters μ and σ^2, and shall study the posterior distribution that is then derived from the observed values in the sample.

The *precision* τ of a normal distribution is defined as the reciprocal of the variance; that is, $\tau = 1/\sigma^2$. In a Bayesian analysis of the type to be discussed in this section, it is convenient to specify a normal distribution by its mean μ and its precision τ, rather than by its mean and its variance. Thus, if a random variable has a normal distribution with mean μ and precision τ, then its p.d.f. $f(x \mid \mu, \tau)$ is specified as follows, for $-\infty < x < \infty$:

$$f(x \mid \mu, \tau) = \left(\frac{\tau}{2\pi}\right)^{1/2} \exp\left[-\frac{1}{2}\tau(x - \mu)^2\right]. \tag{1}$$

Similarly, if $X_1, \ldots, X_n$ form a random sample from a normal distribution with mean μ and precision τ, then their joint p.d.f. $f_n(x \mid \mu, \tau)$ is as follows, for $-\infty < x_i < \infty$ $(i = 1, \ldots, n)$:

$$f_n(x \mid \mu, \tau) = \left(\frac{\tau}{2\pi}\right)^{n/2} \exp\left[-\frac{1}{2}\tau \sum_{i=1}^{n}(x_i - \mu)^2\right]. \tag{2}$$

A Conjugate Family of Prior Distributions

We shall now describe a conjugate family of joint prior distributions of μ and τ. We shall specify the joint distribution of μ and τ by specifying both the conditional distribution of μ given τ and the marginal distribution of τ. In particular, we shall assume that the conditional distribution of μ for any given value of τ is a normal distribution for which the precision is proportional to the given value of τ, and also that the marginal distribution of τ is a gamma distribution. The family of all joint distributions of this type is a conjugate family of joint prior distributions. If the joint prior distribution of μ and τ belongs to this family, then for any possible set of observed values in the random sample, the joint posterior distribution of μ and τ will also belong to the family. This result is established in the next theorem.

Theorem 1. *Suppose that $X_1, \ldots, X_n$ form a random sample from a normal distribution for which both the mean μ and the precision τ are unknown $(-\infty < \mu < \infty$ and $\tau > 0)$. Suppose also that the joint prior distribution of μ and τ is as follows: The conditional distribution of μ given τ is a normal distribution with mean μ_0 and precision $\lambda_0\tau$ $(-\infty < \mu_0 < \infty$ and $\lambda_0 > 0)$; and the marginal distribution of τ is a gamma distribution with parameters α_0 and β_0 $(\alpha_0 > 0$ and $\beta_0 > 0)$. Then the joint posterior distribution of μ and τ, given that $X_i = x_i$ $(i = 1, \ldots, n)$, is as follows: The conditional distribution of μ given τ is a normal distribution with mean μ_1 and precision $\lambda_1\tau$, where*

$$\mu_1 = \frac{\lambda_0\mu_0 + n\bar{x}_n}{\lambda_0 + n} \quad and \quad \lambda_1 = \lambda_0 + n; \tag{3}$$

and the marginal distribution of τ is a gamma distribution with parameters α_1 and β_1, where

$$\alpha_1 = \alpha_0 + \frac{n}{2} \quad and \quad \beta_1 = \beta_0 + \frac{1}{2}\sum_{i=1}^{n}(x_i - \bar{x}_n)^2 + \frac{n\lambda_0(\bar{x}_n - \mu_0)^2}{2(\lambda_0 + n)}. \tag{4}$$

Proof. The joint prior p.d.f. $\xi(\mu, \tau)$ of μ and τ can be found by multiplying the conditional p.d.f. $\xi_1(\mu \mid \tau)$ of μ given τ by the marginal p.d.f. $\xi_2(\tau)$ of τ. By the conditions of the theorem we have, for $-\infty < \mu < \infty$ and $\tau > 0$,

$$\xi_1(\mu \mid \tau) \propto \tau^{1/2}\exp\left[-\frac{1}{2}\lambda_0\tau(\mu - \mu_0)^2\right] \tag{5}$$

and

$$\xi_2(\tau) \propto \tau^{\alpha_0 - 1}e^{-\beta_0\tau}. \tag{6}$$

A constant factor involving neither μ nor τ has been dropped from the right side of each of these relations.

The joint posterior p.d.f. $\xi(\mu, \tau \mid x)$ of μ and τ satisfies the relation

$$\xi(\mu, \tau \mid x) \propto f_n(x \mid \mu, \tau)\xi_1(\mu \mid \tau)\xi_2(\tau). \tag{7}$$

If we use the definition of μ_1 given in Eq. (3), then the following identity can be established:

$$\sum_{i=1}^{n}(x_i - \mu)^2 + \lambda_0(\mu - \mu_0)^2 = (\lambda_0 + n)(\mu - \mu_1)^2 + \sum_{i=1}^{n}(x_i - \bar{x}_n)^2$$

$$+ \frac{n\lambda_0(\bar{x}_n - \mu_0)^2}{\lambda_0 + n}. \tag{8}$$

It now follows from (2), (5), and (6) that the posterior p.d.f. $\xi(\mu, \tau \mid x)$ can be written in the form

$$\xi(\mu, \tau \mid x) \propto \left\{\tau^{1/2}\exp\left[-\frac{1}{2}\lambda_1\tau(\mu - \mu_1)^2\right]\right\}(\tau^{\alpha_1 - 1}e^{-\beta_1\tau}), \tag{9}$$

where λ_1, α_1, and β_1 are defined by Eqs. (3) and (4).

When the expression inside the braces on the right side of Eq. (9) is regarded as a function of μ for a fixed value of τ, this expression can be recognized as being (except for a constant factor) the p.d.f. of a normal distribution with mean μ_1 and precision $\lambda_1\tau$. Since the variable μ does not appear elsewhere on the right side of Eq. (9), it follows that this p.d.f. must be the conditional posterior p.d.f. of μ given τ. It now follows in turn that the expression outside the braces on the right side of Eq. (9) must be proportional to the marginal posterior p.d.f. of τ. This expression can be recognized as being (except for a constant factor) the p.d.f. of a gamma distribution with parameters α_1 and β_1. Hence, the joint posterior distribution of μ and τ is as specified in the theorem.   □

If the joint distribution of μ and τ belongs to the conjugate family described in Theorem 1, then it is said that μ and τ have a joint normal-gamma distribution. By choosing appropriate values of μ_0, λ_0, α_0, and β_0, it is usually possible in a particular problem to find a normal-gamma distribution which approximates an experimenter's actual prior distribution of μ and τ sufficiently well. It should be emphasized, however, that if the joint distribution of μ and τ is a normal-gamma distribution, then μ and τ are not independent. Thus, it is not possible to use a normal-gamma distribution as a joint prior distribution of μ and τ in a problem in which the experimenter's prior information about μ and his prior information about τ are independent and he wishes to assign a joint prior distribution under which μ and τ will be independent. Although this characteristic of the family of normal-gamma distributions is a deficiency, it is not an important deficiency because of the following fact: Even if a joint prior distribution under which μ and τ are independent is chosen from outside the conjugate family, it will be found that after just a single value of X has been observed, μ and τ will have a posterior distribution under which they are dependent. In other words, it is not possible for μ and τ to remain independent in the light of even one observation from the underlying normal distribution.

The Marginal Distribution of the Mean

When the joint distribution of μ and τ is a normal-gamma distribution of the type described in Theorem 1, then the conditional distribution of μ for a given value of τ is a certain normal distribution and the marginal distribution of τ is a gamma distribution. It is not clear from this specification, however, what the marginal distribution of μ will be. We shall now derive this marginal distribution.

Suppose that the joint distribution of μ and τ is the prior normal-gamma distribution described in Theorem 1 and specified by the constants μ_0, λ_0, α_0, and β_0. We shall again let $\xi(\mu, \tau)$ denote the joint p.d.f. of μ and τ, let $\xi_1(\mu \mid \tau)$ denote the conditional p.d.f. of μ given τ, and let $\xi_2(\tau)$ denote the marginal p.d.f. of τ. Then, if $\xi_3(\mu)$ denotes the marginal p.d.f. of μ for $-\infty < \mu < \infty$,

$$\xi_3(\mu) = \int_0^\infty \xi(\mu, \tau) \, d\tau = \int_0^\infty \xi_1(\mu \mid \tau)\xi_2(\tau) \, d\tau. \tag{10}$$

If we make use of the proportionality symbol, then it follows from Eqs. (5) and (6) that, for $-\infty < \mu < \infty$,

$$\xi_3(\mu) \propto \int_0^\infty \tau^{\alpha_0 - \frac{1}{2}} e^{-[\beta_0 + \frac{1}{2}\lambda_0(\mu - \mu_0)^2]\tau} \, d\tau. \tag{11}$$

This integral was evaluated in Eq. (8) of Sec. 5.9. By dropping a factor that does not involve μ, we obtain the relation

$$\xi_3(\mu) \propto \left[\beta_0 + \frac{1}{2}\lambda_0(\mu - \mu_0)^2\right]^{-(\alpha_0 + \frac{1}{2})} \tag{12}$$

or

$$\xi_3(\mu) \propto \left[1 + \frac{1}{2\alpha_0}\frac{\lambda_0\alpha_0}{\beta_0}(\mu - \mu_0)^2\right]^{-(2\alpha_0 + 1)/2}. \tag{13}$$

We shall now define a new random variable Y by the relation

$$Y = \left(\frac{\lambda_0\alpha_0}{\beta_0}\right)^{1/2}(\mu - \mu_0). \tag{14}$$

From this relation,

$$\mu = \left(\frac{\beta_0}{\lambda_0\alpha_0}\right)^{1/2}Y + \mu_0. \tag{15}$$

Then the p.d.f. $g(y)$ of Y will be specified by the following equation:

$$g(y) = \left(\frac{\beta_0}{\lambda_0\alpha_0}\right)^{1/2}\xi_3\left[\left(\frac{\beta_0}{\lambda_0\alpha_0}\right)^{1/2}y + \mu_0\right] \quad \text{for } -\infty < y < \infty. \tag{16}$$

It follows, therefore, from Eq. (13) that

$$g(y) \propto \left(1 + \frac{y^2}{2\alpha_0}\right)^{-(2\alpha_0 + 1)/2}. \tag{17}$$

The p.d.f. of the t distribution with n degrees of freedom was given in Eq. (4) of Sec. 7.4. The p.d.f. in Eq. (17) can now be recognized as being (except for a constant factor) the p.d.f. of a t distribution with $2\alpha_0$ degrees of freedom. Thus, we have established the fact that the distribution of Y, which is the linear function of μ given by Eq. (14), is a t distribution with $2\alpha_0$ degrees of freedom. In other words, the distribution of μ can be obtained from a t distribution by translating the t distribution so that it is centered at μ_0 rather than at 0, and also changing the scale factor.

The mean and the variance of the marginal distribution of μ can easily be obtained from the mean and the variance of the t distribution that are given in Sec. 7.4. Since Y has a t distribution with $2\alpha_0$ degrees of freedom, it follows from Sec. 7.4 that $E(Y) = 0$ if $\alpha_0 > 1/2$ and that $\text{Var}(Y) = \alpha_0/(\alpha_0 - 1)$ if $\alpha_0 > 1$. Therefore, we can obtain the following results from Eq. (15). If $\alpha_0 > 1/2$, then $E(\mu) = \mu_0$. Also, if $\alpha_0 > 1$, then

$$\text{Var}(\mu) = \frac{\beta_0}{\lambda_0(\alpha_0 - 1)}. \tag{18}$$

Furthermore, the probability that μ lies in any specified interval can, in principle, be obtained from a table of the t distribution. It should be noted, however, that when the t distribution was introduced in Sec. 7.4, the number of degrees of freedom had to be a positive integer. In the present problem, the number of degrees of freedom is $2\alpha_0$. Since α_0 can have any positive value, $2\alpha_0$ can be equal to any positive number and need not necessarily be an integer.

When the joint prior distribution of μ and τ is a normal-gamma distribution in which the constants are μ_0, λ_0, α_0, and β_0, it was established in Theorem 1 that the joint posterior distribution of μ and τ is also a normal-gamma distribution in which the constants μ_1, λ_1, α_1, and β_1, are given by Eqs. (3) and (4). It follows, therefore, that the marginal posterior distribution of μ can also be reduced to a t distribution by using a linear transformation, like that in Eq. (14), with the constants of the posterior distribution. Hence, the mean and the variance of this marginal posterior distribution, and also the probability that μ lies in any specified interval, can again be obtained from the corresponding t distribution.

A Numerical Example

General Procedure.   To illustrate the concepts that have been developed in this section, we shall consider again a normal distribution for which both the mean μ and the precision τ are unknown. We shall now suppose that we wish to assign a joint normal-gamma prior distribution to μ and τ such that $E(\mu) = 10$, $\text{Var}(\mu) = 8$, $E(\tau) = 2$, and $\text{Var}(\tau) = 2$.

We shall first determine the values of the constants μ_0, λ_0, α_0, and β_0 for which the prior distribution will satisfy these conditions. Since the marginal distribution of τ is a gamma distribution with parameters α_0 and β_0, we know that $E(\tau) = \alpha_0/\beta_0$ and $\text{Var}(\tau) = \alpha_0/\beta_0^2$. It follows from the given conditions that $\alpha_0 = 2$ and $\beta_0 = 1$. Furthermore, we know that $E(\mu) = \mu_0$ and that the value of $Var(\mu)$ is given by Eq. (18). Therefore, it follows from the given conditions that $\mu_0 = 10$ and $\lambda_0 = 1/8$. The joint prior normal-gamma distribution of μ and τ is completely specified by these values of μ_0, λ_0, α_0, and β_0.

Next, we shall determine an interval for μ centered at the point $\mu_0 = 10$ such that the probability that μ lies in this interval is 0.95. Since the random variable Y defined by Eq. (14) has a t distribution with $2\alpha_0$ degrees of freedom, it follows that for the numerical values just obtained, the random variable $(1/2)(\mu - 10)$ has a t distribution with 4 degrees of freedom. From the table of the t distribution, it is found that

$$\Pr\left[-2.776 < \frac{1}{2}(\mu - 10) < 2.776\right] = 0.95. \tag{19}$$

An equivalent statement is that

$$\Pr(4.448 < \mu < 15.552) = 0.95. \tag{20}$$

Thus, under the prior distribution assigned to μ and τ, there is probability 0.95 that μ lies in the interval (4.448, 15.552).

Suppose now that a random sample of 20 observations is taken from the given normal distribution and that for the 20 observed values, $\bar{x}_n = 7.5$ and $\sum_{i=1}^{n}(x_i - \bar{x}_n)^2 = 28$. Then it follows from Theorem 1 that the joint posterior distribution of μ and τ is a normal-gamma distribution specified by the following values:

$$\mu_1 = 7.516, \qquad \lambda_1 = 20.125, \qquad \alpha_1 = 12, \qquad \beta_1 = 15.388. \tag{21}$$

Hence, the values of the means and the variances of μ and τ, as found from this joint posterior distribution, are:

$$E(\mu) = \mu_1 = 7.516, \qquad Var(\mu) = \frac{\beta_1}{\lambda_1(\alpha_1 - 1)} = 0.070,$$

$$E(\tau) = \frac{\alpha_1}{\beta_1} = 0.780, \qquad Var(\tau) = \frac{\alpha_1}{\beta_1^2} = 0.051. \tag{22}$$

It follows from Eq. (3) that the mean μ_1 of the posterior distribution of μ is a weighted average of μ_0 and $\bar{x}_n$. In this numerical example, it is seen that μ_1 is quite close to $\bar{x}_n$.

Next, we shall determine the marginal posterior distribution of μ. By substituting the values given in (21) into Eq. (14), we find that $Y = (3.962)(\mu - 7.516)$. Also, the posterior distribution of Y will be a t distribution with $2\alpha_1 = 24$ degrees of freedom. From the table of the t distribution it is found that

$$\Pr(-2.064 < Y < 2.064) = 0.95. \tag{23}$$

An equivalent statement is that

$$\Pr(6.995 < \mu < 8.037) = 0.95. \tag{24}$$

In other words, under the posterior distribution of μ and τ, the probability that μ lies in the interval (6.995, 8.037) is 0.95.

It should be noted that the interval in Eq. (24) determined from the posterior distribution of μ is much shorter than the interval in Eq. (20) determined from the prior distribution. This result reflects the fact that the posterior distribution of μ is much more concentrated around its mean than was the prior distribution. The variance of the prior distribution of μ was 8 and the variance of the posterior distribution is 0.07.

Comparison with Confidence Intervals. Using the sample data which we have assumed in this example, we shall now construct a confidence interval for μ with confidence coefficient 0.95; and we shall compare this interval with the interval in Eq. (24) for which the posterior probability is 0.95. Since the sample size n in this example is 20, the random variable U defined by Eq. (5) of Sec. 7.4 has a t distribution with 19 degrees of freedom. It is found from the table of the t distribution that

$$\Pr(-2.093 < U < 2.093) = 0.95.$$

It now follows from Eqs. (1) and (2) of Sec. 7.5 that the endpoints of a confidence interval for μ with confidence coefficient 0.95 will be

$$\bar{x}_n - 2.093 \left[\frac{\sum_{i=1}^{n}(x_i - \bar{x}_n)^2}{n(n-1)} \right]^{1/2}$$

and

$$\bar{x}_n + 2.093 \left[\frac{\sum_{i=1}^{n}(x_i - \bar{x}_n)^2}{n(n-1)} \right]^{1/2}.$$

When the numerical values of $\bar{x}_n$ and $\sum_{i=1}^{n}(x_i - \bar{x}_n)^2$ are used here, we find that this confidence interval for μ is the interval (6.932, 8.068).

This interval is close to the interval (6.995, 8.037) in Eq. (24), for which the posterior probability is 0.95. The similarity of the two intervals illustrates the truth of the statement made at the end of Sec. 7.4. In this problem, and in many other problems involving the normal distribution, the method of confidence intervals and the method of using prior probabilities yield similar results, even though the philosophical bases of the two methods are quite different.

EXERCISES

1. Suppose that a random variable X has a normal distribution with mean μ and precision τ. Show that the random variable $Y = aX + b$ $(a \neq 0)$ has a normal distribution with mean $a\mu + b$ and precision τ/a^2.

2. Suppose that $X_1, \ldots, X_n$ form a random sample from a normal distribution for which the value of the mean μ is unknown $(-\infty < \mu < \infty)$ and the value of the precision τ is known $(\tau > 0)$. Suppose also that the prior distribution of μ is a normal distribution with mean μ_0 and precision λ_0. Show that the posterior distribution of μ, given that $X_i = x_i$ $(i = 1, \ldots, n)$, is a normal distribution for which the mean is

$$\frac{\lambda_0 \mu_0 + n\tau \bar{x}_n}{\lambda_0 + n\tau}$$

and the precision is $\lambda_0 + n\tau$.

3. Suppose that $X_1, \ldots, X_n$ form a random sample from a normal distribution for which the value of the mean μ is known and the value of the precision τ is unknown $(\tau > 0)$. Suppose also that the prior distribution of τ is a gamma distribution with parameters α_0 and β_0 $(\alpha_0 > 0$ and $\beta_0 > 0)$. Show that the posterior distribution of τ, given that $X_i = x_i$ $(i = 1, \ldots, n)$, is a gamma distribution with parameters $\alpha_0 + (n/2)$ and

$$\beta_0 + \frac{1}{2} \sum_{i=1}^{n} (x_i - \mu)^2.$$

4. Suppose that two random variables μ and τ have a joint normal-gamma distribution such that $E(\mu) = -5$, $\text{Var}(\mu) = 1$, $E(\tau) = 1/2$, and $\text{Var}(\tau) = 1/8$. Find the values of μ_0, λ_0, α_0, and β_0 which specify the normal-gamma distribution.

5. Show that two random variables μ and τ cannot have a joint normal-gamma distribution such that $E(\mu) = 0$, $\text{Var}(\mu) = 1$, $E(\tau) = 1/2$, and $\text{Var}(\tau) = 1/4$.

6. Show that two random variables μ and τ cannot have a joint normal-gamma distribution such that $E(\mu) = 0$, $E(\tau) = 1$, and $\text{Var}(\tau) = 4$.

7. Suppose that two random variables μ and τ have a joint normal-gamma distribution specified by the values $\mu_0 = 4$, $\lambda_0 = 0.5$, $\alpha_0 = 1$, and $\beta_0 = 8$. Find the values of (a) $\Pr(\mu > 0)$ and (b) $\Pr(0.736 < \mu < 15.680)$.

8. Using the data in the numerical example presented at the end of this section, find (a) the shortest possible interval such that the posterior probability that μ lies in the interval is 0.90 and (b) the shortest possible confidence interval for μ for which the confidence coefficient is 0.90.

9. Suppose that $X_1, \ldots, X_n$ form a random sample from a normal distribution for which both the mean μ and the precision τ are unknown; and also that the joint prior distribution of μ and τ is a normal-gamma distribution satisfying the following conditions: $E(\mu) = 0$, $E(\tau) = 2$, $E(\tau^2) = 5$, and $\Pr(|\mu| < 1.412) = 0.5$. Determine the values of μ_0, λ_0, α_0, and β_0 which specify the normal-gamma distribution.

10. Consider again the conditions of Exercise 9. Suppose also that in a random sample of size $n = 10$, it is found that $\bar{x}_n = 1$ and $\sum_{i=1}^{n}(x_i - \bar{x}_n)^2 = 8$. Find the shortest possible interval such that the posterior probability that μ lies in the interval is 0.95.

11. Suppose that $X_1, \ldots, X_n$ form a random sample from a normal distribution for which both the mean μ and the precision τ are unknown; and also that the joint prior distribution of μ and τ is a normal-gamma distribution satisfying the following conditions: $E(\tau) = 1$, $\mathrm{Var}(\tau) = 1/3$, $\Pr(\mu > 3) = 0.5$, and $\Pr(\mu > 0.12) = 0.9$. Determine the values of μ_0, λ_0, α_0, and β_0 which specify the normal-gamma distribution.

12. Consider again the conditions of Exercise 11. Suppose also that in a random sample of size $n = 8$, it is found that $\sum_{i=1}^{n} x_i = 16$ and $\sum_{i=1}^{n} x_i^2 = 48$. Find the shortest possible interval such that the posterior probability that μ lies in the interval is 0.99.

7.7. UNBIASED ESTIMATORS

Definition of an Unbiased Estimator

Consider again a problem in which the variables $X_1, \ldots, X_n$ form a random sample from a distribution that involves a parameter θ whose value is unknown and must be estimated. In a problem of this type it is desirable to use an estimator $\delta(X_1, \ldots, X_n)$ that, with high probability, will be close to θ. In other words, it is desirable to use an estimator δ whose value changes with the value of θ in such a way that no matter what the true value of θ is, the probability distribution of δ is concentrated around this value.

For example, suppose that $X_1, \ldots, X_n$ form a random sample from a normal distribution for which the mean θ is unknown and the variance is 1. In this case the M.L.E. of θ is the sample mean $\bar{X}_n$. The estimator $\bar{X}_n$ is a reasonably good estimator of θ because its probability distribution is a normal distribution with mean θ and variance $1/n$; and this distribution is concentrated around the unknown value of θ, no matter how large or how small that value is.

These considerations lead to the following definition: An estimator $\delta(X_1, \ldots, X_n)$ is an *unbiased estimator* of a parameter θ if $E_\theta[\delta(X_1, \ldots, X_n)] = \theta$ for every possible value of θ. In other words, an estimator of a parameter θ is unbiased if its expectation is equal to the unknown true value of θ.

In the example just mentioned, $\overline{X}_n$ is an unbiased estimator of the unknown mean θ of a normal distribution because $E_\theta(\overline{X}_n) = \theta$ for $-\infty < \theta < \infty$. In fact, if $X_1, \ldots, X_n$ form a random sample from any arbitrary distribution for which the mean μ is unknown, the sample mean $\overline{X}_n$ will always be an unbiased estimator of μ because it is always true that $E(\overline{X}_n) = \mu$.

If an estimator δ of some parameter θ is unbiased, then the distribution of δ must indeed change with the value of θ, since the mean of this distribution is θ. It should be emphasized, however, that this distribution might be either closely concentrated around θ or widely spread out. For example, an estimator that is equally likely to underestimate θ by 1,000,000 units or to overestimate θ by 1,000,000 units would be an unbiased estimator, but it would never yield an estimate close to θ. Therefore, the mere fact that an estimator of θ is unbiased does not necessarily imply that the estimator is good, or even reasonable. However, if an unbiased estimator of θ also has a small variance, it follows that the distribution of the estimator will necessarily be concentrated around its mean θ and there will be high probability that the estimator will be close to θ.

For the reasons just mentioned, the study of unbiased estimators in a particular problem is largely devoted to the search for an unbiased estimator that has a small variance. However, if an estimator δ is unbiased, then its M.S.E. $E_\theta[(\delta - \theta)^2]$ is equal to its variance $\mathrm{Var}_\theta(\delta)$. Therefore, the search for an unbiased estimator with a small variance is equivalent to the search for an unbiased estimator with a small M.S.E.

Unbiased Estimation of the Variance

Sampling from an Arbitrary Distribution. We have already remarked that if $X_1, \ldots, X_n$ form a random sample from any distribution for which the mean μ is unknown, then the sample mean $\overline{X}_n$ is an unbiased estimator of μ. We shall now suppose that the variance σ^2 of the underlying distribution is also unknown, and we shall determine an unbiased estimator of σ^2.

Since the sample mean is an unbiased estimator of the mean μ, it is more or less natural to consider first the sample variance $S_0^2 = (1/n)\sum_{i=1}^n(X_i - \overline{X}_n)^2$ and to attempt to determine if it is an unbiased estimator of the variance σ^2. We shall use the identity

$$\sum_{i=1}^n (X_i - \mu)^2 = \sum_{i=1}^n (X_i - \overline{X}_n)^2 + n(\overline{X}_n - \mu)^2. \tag{1}$$

Then it follows that

$$E(S_0^2) = E\left[\frac{1}{n}\sum_{i=1}^n (X_i - \overline{X}_n)^2\right]$$

$$= E\left[\frac{1}{n}\sum_{i=1}^n (X_i - \mu)^2\right] - E\left[(\overline{X}_n - \mu)^2\right]. \tag{2}$$

Since each observation X_i has mean μ and variance σ^2, then $E[(X_i - \mu)^2] = \sigma^2$ for $i = 1, \ldots, n$. Therefore,

$$E\left[\frac{1}{n}\sum_{i=1}^{n}(X_i - \mu)^2\right] = \frac{1}{n}\sum_{i=1}^{n}E\left[(X_i - \mu)^2\right] = \frac{1}{n}n\sigma^2 = \sigma^2. \tag{3}$$

Furthermore, the sample mean $\overline{X}_n$ has mean μ and variance σ^2/n. Therefore,

$$E\left[(\overline{X}_n - \mu)^2\right] = \text{Var}(\overline{X}_n) = \frac{\sigma^2}{n}. \tag{4}$$

It now follows from Eqs. (2), (3), and (4) that

$$E(S_0^2) = \sigma^2 - \frac{1}{n}\sigma^2 = \frac{n-1}{n}\sigma^2. \tag{5}$$

It can be seen from Eq. (5) that the sample variance S_0^2 is not an unbiased estimator of σ^2 because its expectation is $[(n-1)/n]\sigma^2$, rather than σ^2. However, if S_0^2 is multiplied by the factor $n/(n-1)$ to obtain the statistic S_1^2, then the expectation of S_1^2 will indeed be σ^2. Therefore, S_1^2 will be an unbiased estimator of σ^2, and we have established the following result:

If the random variables $X_1, \ldots, X_n$ form a random sample from any distribution for which the variance σ^2 is unknown, then the estimator

$$S_1^2 = \frac{1}{n-1}\sum_{i=1}^{n}(X_i - \overline{X}_n)^2$$

will be an unbiased estimator of σ^2.

In other words, although S_0^2 is not an unbiased estimator of σ^2, the estimator S_1^2 obtained by dividing the sum of squared deviations $\sum_{i=1}^{n}(X_i - \overline{X}_n)^2$ by $n - 1$, rather than by n, will always be unbiased. For this reason, in many textbooks the sample variance is defined initially as S_1^2, rather than as S_0^2.

Sampling from a Specific Family of Distributions. When it can be assumed that $X_1, \ldots, X_n$ form a random sample from a specific family of distributions, such as the family of Poisson distributions, it will generally be desirable to consider not only S_1^2 but also other unbiased estimators of the variance. For example, suppose that the sample does indeed come from a Poisson distribution for which the mean θ is unknown. We have already seen that $\overline{X}_n$ will be an unbiased estimator of the mean θ. Moreover, since the variance of a Poisson distribution is also equal to θ, it follows that $\overline{X}_n$ is also an unbiased estimator of the variance. In this example, therefore, both $\overline{X}_n$ and S_1^2 are unbiased estimators of the unknown variance θ. Furthermore, any combination of $\overline{X}_n$ and S_1^2 having the form $\alpha\overline{X}_n + (1 - \alpha)S_1^2$, where α is a given constant $(-\infty < \alpha < \infty)$, will also be an unbiased estimator of θ because its expectation will be

$$E[\alpha \bar{X}_n + (1 - \alpha) S_1^2] = \alpha E(\bar{X}_n) + (1 - \alpha) E(S_1^2) = \alpha \theta + (1 - \alpha) \theta = \theta.$$

Other unbiased estimators of θ can also be constructed.

If an unbiased estimator is to be used, the problem is to determine which one of the possible unbiased estimators has the smallest variance or, equivalently, has the smallest M.S.E. We shall not derive the solution to this problem right now. However, it will be shown in Sec. 7.8 that in this example, for every possible value of θ, the estimator $\bar{X}_n$ has the smallest variance among all unbiased estimators of θ. This result is not surprising. We know from Example 1 of Sec. 6.7 that $\bar{X}_n$ is a sufficient statistic for θ; and it was argued in Sec. 6.9 that we can restrict our attention to estimators which are functions of the sufficient statistic alone. (See also Exercise 12 at the end of this section.)

Estimation of the Variance of a Normal Distribution

We shall suppose now that $X_1, \ldots, X_n$ form a random sample from a normal distribution for which both the mean μ and the variance σ^2 are unknown, and we shall consider the problem of estimating σ^2. We know from the discussion in this section that the estimator S_1^2 is an unbiased estimator of σ^2. Moreover, we know from Sec. 6.5 that the sample variance S_0^2 is the M.L.E. of σ^2. We want to determine whether the M.S.E. $E[(S_i^2 - \sigma^2)^2]$ is smaller for the estimator S_0^2 or for the estimator S_1^2, and also whether or not there is some other estimator of σ^2 which has a smaller M.S.E. than both S_0^2 and S_1^2.

Both the estimator S_0^2 and the estimator S_1^2 have the following form:

$$T^2 = c \sum_{i=1}^{n} (X_i - \bar{X}_n)^2, \tag{6}$$

where $c = 1/n$ for S_0^2 and $c = 1/(n - 1)$ for S_1^2. We shall now determine the M.S.E. for an arbitrary estimator having the form in Eq. (6) and shall then determine the value of c for which this M.S.E. is minimum. We shall demonstrate the striking property that the same value of c minimizes the M.S.E. for all possible values of the parameters μ and σ^2. Therefore, among all estimators having the form in Eq. (6), there is a single one that has the smallest M.S.E. for all possible values of μ and σ^2.

It was shown in Sec. 7.3 that when $X_1, \ldots, X_n$ form a random sample from a normal distribution, the random variable $\sum_{i=1}^{n}(X_i - \bar{X}_n)^2/\sigma^2$ has a χ^2 distribution with $n - 1$ degrees of freedom. By Eq. (2) of Sec. 7.2, the mean of this variable is $n - 1$ and the variance is $2(n - 1)$. Therefore, if T^2 is defined by

Eq. (6), then

$$E(T^2) = (n - 1)c\sigma^2 \quad \text{and} \quad \text{Var}(T^2) = 2(n - 1)c^2\sigma^4. \tag{7}$$

Thus, by Exercise 4 of Sec. 4.3, the M.S.E. of T^2 can be found as follows:

$$
\begin{aligned}
E\left[(T^2 - \sigma^2)^2\right] &= \left[E(T^2) - \sigma^2\right]^2 + \text{Var}(T^2) \\
&= \left[(n - 1)c - 1\right]^2\sigma^4 + 2(n - 1)c^2\sigma^4 \\
&= \left[(n^2 - 1)c^2 - 2(n - 1)c + 1\right]\sigma^4.
\end{aligned}
\tag{8}
$$

The coefficient of σ^4 in Eq. (8) is simply a quadratic function of c. Hence, for any given value of σ^4, the minimizing value of c is found by elementary differentiation to be $c = 1/(n + 1)$.

In summary, we have established the following fact: Among all estimators of σ^2 having the form in Eq. (6), the estimator that has the smallest M.S.E. for all possible values of μ and σ^2 is $T_0^2 = [1/(n + 1)]\sum_{i=1}^{n}(X_i - \bar{X}_n)^2$. In particular, T_0^2 has a smaller M.S.E. than both the M.L.E. S_0^2 and the unbiased estimator S_1^2. Therefore, the estimators S_0^2 and S_1^2, as well as all other estimators having the form in Eq. (6) with $c \neq 1/(n + 1)$, are inadmissible. Furthermore, it was shown by C. Stein in 1964 that even the estimator T_0^2 is dominated by other estimators, and that T_0^2 itself is therefore inadmissible.

The estimators S_0^2 and S_1^2 are compared in Exercise 5 at the end of this section. Of course, when the sample size n is large, it makes little difference whether n, $n - 1$, or $n + 1$ is used as the divisor in the estimate of σ^2; all three estimators S_0^2, S_1^2, and T_0^2 will be approximately equal.

Discussion of the Concept of Unbiased Estimation

Any estimator that is not unbiased is called a *biased estimator*. The difference between the expectation of a biased estimator and the parameter θ which is being estimated is called the *bias* of the estimator. Since no scientist wishes to be biased or to be accused of being biased, merely the terminology of the theory of unbiased estimation seems to make the use of unbiased estimators highly desirable. Indeed, the concept of unbiased estimation has played an important part in the historical development of statistics; and the feeling that an unbiased estimator should be preferred to a biased estimator is prevalent in current statistical practice.

However, as explained in this section, the quality of an unbiased estimator must be evaluated in terms of its variance or its M.S.E. It has just been shown that when the variance of a normal distribution is to be estimated, and also in many other problems, there will exist a biased estimator which has a smaller M.S.E. than any unbiased estimator for every possible value of the parameter. Furthermore, it can be shown that a Bayes estimator, which makes use of all relevant prior information about the parameter and which minimizes the overall M.S.E., is unbiased only in trivial problems in which the parameter can be estimated perfectly.

Some other undesirable features of the theory of unbiased estimation will now be described.

Nonexistence of an Unbiased Estimator.   In many problems there does not exist any unbiased estimator of the parameter, or of some particular function of the parameter that must be estimated. For example, suppose that $X_1, \ldots, X_n$ form n Bernoulli trials for which the parameter p is unknown ($0 \leqslant p \leqslant 1$). Then the sample mean $\overline{X}_n$ will be an unbiased estimator of p; but it can be shown that there will be no unbiased estimator of $p^{1/2}$. (See Exercise 6 at the end of this section.) Furthermore, if it is known in this example that p must lie in the interval $\frac{1}{3} \leqslant p \leqslant \frac{2}{3}$, then there is no unbiased estimator of p whose possible values are confined to that same interval.

Inappropriate Unbiased Estimators.   Consider an infinite sequence of Bernoulli trials for which the parameter p is unknown ($0 < p < 1$), and let X denote the number of failures that occur before the first success is obtained. Then X will have a geometric distribution for which the p.f. is given by Eq. (5) of Sec. 5.5. If it is desired to estimate the value of p from the observation X, then it can be shown (see Exercise 7) that the *only* unbiased estimator of p yields the estimate 1 if $X = 0$ and yields the estimate 0 if $X > 0$. This estimator seems inappropriate. For example, if the first success is obtained on the second trial, that is, if $X = 1$, then it is silly to estimate that the probability of success p is 0.

As another example of an inappropriate unbiased estimator, suppose that the random variable X has a Poisson distribution for which the mean λ is unknown ($\lambda > 0$), and suppose also that it is desired to estimate the value of $e^{-2\lambda}$. It can be shown (see Exercise 8) that the *only* unbiased estimator of $e^{-2\lambda}$ yields the estimate 1 if X is an even integer and the estimate -1 if X is an odd integer. This estimator is inappropriate for two reasons. First, it yields the estimate 1 or -1 for a parameter $e^{-2\lambda}$ which must lie between 0 and 1. Second, the value of the estimate depends on whether X is odd or even, rather than on whether X is large or small.

Violation of the Likelihood Principle.   A final criticism of the concept of unbiased estimation is that the principle of always using an unbiased estimator for a parameter θ violates the likelihood principle introduced in Sec. 6.6. For example, suppose that in a sequence of n Bernoulli trials for which the parameter p is unknown $(0 < p < 1)$, there are x successes $(x \geq 2)$ and $n - x$ failures. If the number of trials n had been fixed in advance, then an unbiased estimate of p would be the sample mean x/n. On the other hand, if the number of successes x had been fixed in advance and sampling had continued until exactly x successes were obtained, then it can be shown (see Exercise 9) that x/n is not an unbiased estimate of p and that $(x - 1)/(n - 1)$ is unbiased. However, as explained in Sec. 6.6, any method of estimation that is consistent with the likelihood principle should yield the same estimate of p regardless of which method of sampling was used to generate the Bernoulli trials.

As another example of how the concept of unbiased estimation violates the likelihood principle, suppose that the average voltage θ in a certain electric circuit is unknown; that this voltage is to be measured by a voltmeter for which the reading X has a normal distribution with mean θ and known variance σ^2; and that the observed reading on the voltmeter is 205 volts. Since X is an unbiased estimator of θ in this example, a scientist who wished to use an unbiased estimator would estimate the value of θ to be 205 volts.

However, suppose also that after the scientist reported the value 205 as his estimate of θ, he discovered that the voltmeter was not functioning properly at the time at which the reading was made. It was verified that the reading of the voltmeter was accurate for any voltage less than 280 volts, but a voltage greater than 280 volts would not have been recorded accurately. Since the actual reading was 205 volts, this reading was unaffected by the defect in the voltmeter. Nevertheless, the observed reading would no longer be an unbiased estimator of θ because the distribution of X when the reading was made was not a normal distribution with mean θ. Therefore, if the scientist still wished to use an unbiased estimator, he would have to change his estimate of θ from 205 volts to a different value.

This violation of the likelihood principle seems unacceptable. Since the actual observed reading was only 205 volts, the scientist's likelihood function remained unchanged when he learned that the voltmeter would not have functioned properly if the reading had been larger. Since the observed reading is correct, it would seem that the information that there might have been erroneous readings is irrelevant to the estimation of θ. However, since this information does change the sample space of X and its probability distribution, this information will also change the form of the unbiased estimator of θ.

EXERCISES

1. Suppose that X is a random variable whose distribution is completely unknown; but it is known that all the moments $E(X^k)$, for $k = 1, 2, \ldots$, are finite. Suppose also that $X_1, \ldots, X_n$ form a random sample from this distribu-

tion. Show that for $k = 1, 2, \ldots$, the kth sample moment $(1/n)\sum_{i=1}^{n} X_i^k$ is an unbiased estimator of $E(X^k)$.

2. For the conditions of Exercise 1, find an unbiased estimator of $[E(X)]^2$. *Hint:* $[E(X)]^2 = E(X^2) - \text{Var}(X)$.

3. Suppose that a random variable X has a geometric distribution, as defined in Sec. 5.5, for which the parameter p is unknown ($0 < p < 1$). Find a statistic $\delta(X)$ that will be an unbiased estimator of $1/p$.

4. Suppose that a random variable X has a Poisson distribution for which the mean λ is unknown ($\lambda > 0$). Find a statistic $\delta(X)$ that will be an unbiased estimator of e^λ. *Hint:* If $E[\delta(X)] = e^\lambda$, then

$$\sum_{x=0}^{\infty} \frac{\delta(x)e^{-\lambda}\lambda^x}{x!} = e^\lambda.$$

Multiply both sides of this equation by e^λ; expand the right side in a power series in λ; and then equate the coefficients of λ^x on both sides of the equation for $x = 0, 1, 2, \ldots$.

5. Suppose that $X_1, \ldots, X_n$ form a random sample from a normal distribution for which both the mean μ and the variance σ^2 are unknown. Let S_0^2 and S_1^2 be the two estimators of σ^2 which are defined as follows:

$$S_0^2 = \frac{1}{n} \sum_{i=1}^{n} (X_i - \bar{X}_n)^2 \quad \text{and} \quad S_1^2 = \frac{1}{n-1} \sum_{i=1}^{n} (X_i - \bar{X}_n)^2.$$

Show that the M.S.E. of S_0^2 is smaller than the M.S.E. of S_1^2 for all possible values of μ and σ^2.

6. Suppose that $X_1, \ldots, X_n$ form n Bernoulli trials for which the parameter p is unknown ($0 \leq p \leq 1$). Show that the expectation of any function $\delta(X_1, \ldots, X_n)$ is a polynomial in p whose degree does not exceed n.

7. Suppose that a random variable X has a geometric distribution for which the parameter p is unknown ($0 < p < 1$). Show that the only unbiased estimator of p is the estimator $\delta(X)$ such that $\delta(0) = 1$ and $\delta(X) = 0$ for $X > 0$.

8. Suppose that a random variable X has a Poisson distribution for which the mean λ is unknown ($\lambda > 0$). Show that the only unbiased estimator of $e^{-2\lambda}$ is the estimator $\delta(X)$ such that $\delta(X) = 1$ if X is an even integer and $\delta(X) = -1$ if X is an odd integer.

9. Consider an infinite sequence of Bernoulli trials for which the parameter p is unknown ($0 < p < 1$), and suppose that sampling is continued until exactly k successes have been obtained, where k is a fixed integer ($k \geq 2$). Let N denote the total number of trials that are needed to obtain the k successes. Show that the estimator $(k - 1)/(N - 1)$ is an unbiased estimator of p.

10. Suppose that a certain drug is to be administered to two different types of animals A and B. It is known that the mean response of animals of type A is the same as the mean response of animals of type B; but the common value θ of this mean is unknown and must be estimated. It is also known that the variance of the response of animals of type A is four times as large as the variance of the response of animals of type B. Let $X_1, \ldots, X_m$ denote the responses of a random sample of m animals of type A, and let $Y_1, \ldots, Y_n$ denote the responses of an independent random sample of n animals of type B. Finally, consider the estimator $\hat{\theta} = \alpha \overline{X}_m + (1 - \alpha)\overline{Y}_n$.

    (a) For what values of α, m, and n is $\hat{\theta}$ an unbiased estimator of θ?

    (b) For fixed values of m and n, what value of α yields an unbiased estimator with minimum variance?

11. Suppose that a certain population of individuals is composed of k different strata $(k \geqslant 2)$; and that for $i = 1, \ldots, k$, the proportion of individuals in the total population who belong to stratum i is p_i, where $p_i > 0$ and $\sum_{i=1}^{k} p_i = 1$. We are interested in estimating the mean value μ of a certain characteristic among the total population. Among the individuals in stratum i this characteristic has mean μ_i and variance σ_i^2, where the value of μ_i is unknown and the value of σ_i^2 is known. Suppose that a *stratified sample* is taken from the population as follows: From each stratum i, a random sample of n_i individuals is taken, and the characteristic is measured for each of these individuals. The samples from the k strata are taken independently of each other. Let $\overline{X}_i$ denote the average of the n_i measurements in the sample from stratum i.

    (a) Show that $\mu = \sum_{i=1}^{k} p_i \mu_i$, and show also that $\hat{\mu} = \sum_{i=1}^{k} p_i \overline{X}_i$ is an unbiased estimator of μ.

    (b) Let $n = \sum_{i=1}^{k} n_i$ denote the total number of observations in the k samples. For a fixed value of n, find the values of $n_1, \ldots, n_k$ for which the variance of $\hat{\mu}$ will be a minimum.

12. Suppose that $X_1, \ldots, X_n$ form a random sample from a distribution for which the p.d.f. or the p.f. is $f(x \mid \theta)$, where the value of the parameter θ is unknown; and that the statistic $T = r(X_1, \ldots, X_n)$ is a sufficient statistic for θ. Let $\delta(X_1, \ldots, X_n)$ denote an unbiased estimator of θ, and let $\delta_0(T)$ be another estimator which is defined by Eq. (3) of Sec. 6.9.

    (a) Show that $\delta_0(T)$ is also an unbiased estimator of θ.

    (b) Show that $\text{Var}_\theta(\delta_0) \leqslant \text{Var}_\theta(\delta)$ for every possible value of θ.

    Hint: Use Theorem 1 of Sec. 6.9.

13. Suppose that $X_1, \ldots, X_n$ form a random sample from a uniform distribution on the interval $(0, \theta)$, where the value of the parameter θ is unknown; and let $Y_n = \max(X_1, \ldots, X_n)$. Show that $[(n + 1)/n]Y_n$ is an unbiased estimator of θ.

14. Suppose that a random variable X can take only the five values $x = 1, 2, 3, 4, 5$ with the following probabilities:

$$f(1 \mid \theta) = \theta^3, \qquad f(2 \mid \theta) = \theta^2(1 - \theta), \qquad f(3 \mid \theta) = 2\theta(1 - \theta),$$

$$f(4 \mid \theta) = \theta(1 - \theta)^2, \qquad f(5 \mid \theta) = (1 - \theta)^3.$$

Here the value of the parameter θ is unknown $(0 \leqslant \theta \leqslant 1)$.

(a) Verify that the sum of the five given probabilities is 1 for any given value of θ.

(b) Consider an estimator $\delta_c(X)$ that has the following form:

$$\delta_c(1) = 1, \quad \delta_c(2) = 2 - 2c, \quad \delta_c(3) = c, \quad \delta_c(4) = 1 - 2c, \quad \delta_c(5) = 0.$$

Show that for any given constant c, $\delta_c(X)$ is an unbiased estimator of θ.

(c) Let θ_0 be a number such that $0 < \theta_0 < 1$. Determine a constant c_0 such that when $\theta = \theta_0$, the variance of $\delta_{c_0}(X)$ is smaller than the variance of $\delta_c(X)$ for any other value of c.

---

*7.8.   FISHER INFORMATION

Definition and Properties of Fisher Information

The Fisher Information in a Single Random Variable.   In this section we shall introduce a concept, called the Fisher information, which enters various aspects of the theory of statistical inference; and we shall describe a few uses of this concept.

Consider a random variable X for which the p.f. or the p.d.f. is $f(x \mid \theta)$. It is assumed that $f(x \mid \theta)$ involves a parameter θ whose value is unknown but must lie in a given open interval Ω of the real line. Furthermore, it is assumed that X takes values in a specified sample space S and that $f(x \mid \theta) > 0$ for each value of $x \in S$ and each value of $\theta \in \Omega$. This assumption eliminates from consideration the uniform distribution on the interval $(0, \theta)$, where the value of θ is unknown, because for that distribution $f(x \mid \theta) > 0$ only when $x < \theta$ and $f(x \mid \theta) = 0$ when $x > \theta$. The assumption does not eliminate any distribution where the set of values of x for which $f(x \mid \theta) > 0$ is a fixed set that does not depend on θ.

Next, we define $\lambda(x \mid \theta)$ as follows:

$$\lambda(x \mid \theta) = \log f(x \mid \theta). \tag{1}$$

It is assumed that for each value of $x \in S$, the p.f. or p.d.f. $f(x \mid \theta)$ is a twice differentiable function of θ, and we let

$$\lambda'(x \mid \theta) = \frac{\partial}{\partial \theta} \lambda(x \mid \theta) \qquad \text{and} \qquad \lambda''(x \mid \theta) = \frac{\partial^2}{\partial \theta^2} \lambda(x \mid \theta). \tag{2}$$

The *Fisher information* $I(\theta)$ *in the random variable* X is defined as follows:

$$I(\theta) = E_\theta \left\{ [\lambda'(X \mid \theta)]^2 \right\}. \tag{3}$$

Thus, if $f(x \mid \theta)$ is a p.d.f., then

$$I(\theta) = \int_S [\lambda'(x \mid \theta)]^2 f(x \mid \theta) \, dx. \tag{4}$$

If $f(x \mid \theta)$ is a p.f., the integral in Eq. (4) is replaced by a sum over the points in S. In the discussion that follows, we shall assume for convenience that $f(x \mid \theta)$ is a p.d.f. However, all the results hold also when $f(x \mid \theta)$ is a p.f.

We know that $\int_S f(x \mid \theta) \, dx = 1$ for every value of $\theta \in \Omega$. Therefore, if the integral on the left side of this equation is differentiated with respect to θ, the result will be 0. We shall assume that we can reverse the order in which we perform the integration with respect to x and the differentiation with respect to θ, and will still obtain the value 0. In other words, we shall assume that we can take the derivative "inside the integral sign" and obtain

$$\int_S f'(x \mid \theta) \, dx = 0 \qquad \text{for } \theta \in \Omega. \tag{5}$$

Furthermore, we shall assume that we can take a second derivative with respect to θ "inside the integral sign" and obtain

$$\int_S f''(x \mid \theta) \, dx = 0 \qquad \text{for } \theta \in \Omega. \tag{6}$$

We can now give two additional forms for the Fisher information $I(\theta)$. First, since $\lambda'(x \mid \theta) = f'(x \mid \theta)/f(x \mid \theta)$, then

$$E_\theta[\lambda'(X \mid \theta)] = \int_S \lambda'(x \mid \theta) f(x \mid \theta) \, dx = \int_S f'(x \mid \theta) \, dx.$$

Hence, it follows from Eq. (5) that

$$E_\theta[\lambda'(X \mid \theta)] = 0. \tag{7}$$

Since the mean of $\lambda'(X|\theta)$ is 0, it follows from Eq. (3) that

$$I(\theta) = \text{Var}_\theta[\lambda'(X|\theta)]. \tag{8}$$

Also, it should be noted that

$$\lambda''(x|\theta) = \frac{f(x|\theta)f''(x|\theta) - [f'(x|\theta)]^2}{[f(x|\theta)]^2}$$

$$= \frac{f''(x|\theta)}{f(x|\theta)} - [\lambda'(x|\theta)]^2. \tag{9}$$

Therefore,

$$E_\theta[\lambda''(X|\theta)] = \int_S f''(x|\theta)\,dx - I(\theta). \tag{10}$$

It follows from Eqs. (10) and (6) that

$$I(\theta) = -E_\theta[\lambda''(X|\theta)]. \tag{11}$$

In many problems, it is easier to determine the value of $I(\theta)$ from Eq. (11) than from Eq. (3).

Example 1: The Bernoulli Distribution. Suppose that X has a Bernoulli distribution for which the parameter p is unknown $(0 < p < 1)$. We shall determine the Fisher information $I(p)$ in X.

In this example X can have only the two values 0 and 1. For $x = 0$ or 1,

$$\lambda(x|p) = \log f(x|p) = x \log p + (1 - x)\log(1 - p).$$

Hence,

$$\lambda'(x|p) = \frac{x}{p} - \frac{1 - x}{1 - p}$$

and

$$\lambda''(x|p) = -\left[\frac{x}{p^2} + \frac{1 - x}{(1 - p)^2}\right].$$

Since $E(X) = p$, the Fisher information is

$$I(p) = -E[\lambda''(X|p)] = \frac{1}{p} + \frac{1}{1 - p} = \frac{1}{p(1 - p)}.$$

In this example it can be readily verified that the assumptions represented by Eqs. (5) and (6) are satisfied. Indeed, since X can take only the two values 0 and 1, the integrals in Eqs. (5) and (6) reduce to summations over the two values $x = 0$ and $x = 1$. Since it is always possible to take a derivative "inside a finite summation" and to differentiate the sum term by term, Eqs. (5) and (6) must be satisfied. □

Example 2: The Normal Distribution. Suppose that X has a normal distribution for which the mean μ is unknown $(-\infty < \mu < \infty)$ and the variance σ^2 is known. We shall determine the Fisher information $I(\mu)$ in X.
     For $-\infty < x < \infty$,

$$\lambda(x \mid \mu) = -\frac{1}{2}\log(2\pi\sigma^2) - \frac{(x-\mu)^2}{2\sigma^2}.$$

Hence,

$$\lambda'(x \mid \mu) = \frac{x-\mu}{\sigma^2} \quad \text{and} \quad \lambda''(x \mid \mu) = -\frac{1}{\sigma^2}.$$

It now follows from Eq. (11) that the Fisher information is

$$I(\mu) = \frac{1}{\sigma^2}.$$

In this example, it can be verified directly (see Exercise 1 at the end of this section) that Eqs. (5) and (6) are satisfied. □

It should be emphasized that the concept of Fisher information cannot be applied to a distribution, such as the uniform distribution on the interval $(0, \theta)$, for which the necessary assumptions are not satisfied.

The Fisher Information in a Random Sample.   Suppose that the random variables $X_1, \ldots, X_n$ form a random sample from a distribution for which the p.d.f. is $f(x \mid \theta)$, where the value of the parameter θ must lie in an open interval Ω of the real line. As usual, we shall let $f_n(x \mid \theta)$ denote the joint p.d.f. of $X_1, \ldots, X_n$. Furthermore, we shall let

$$\lambda_n(x \mid \theta) = \log f_n(x \mid \theta). \tag{12}$$

Then, analogously to Eq. (3), the *Fisher information* $I_n(\theta)$ *in the random sample* $X_1, \ldots, X_n$ is defined as follows:

$$I_n(\theta) = E_\theta\left\{\left[\lambda_n'(X \mid \theta)\right]^2\right\}. \tag{13}$$

Thus, the Fisher information $I_n(\theta)$ in the entire sample is given by the following n-dimensional integral:

$$I_n(\theta) = \int_S \cdots \int_S [\lambda'_n(x \mid \theta)]^2 f_n(x \mid \theta) \, dx_1 \cdots dx_n. \tag{14}$$

Furthermore, if we again assume that Eqs. (5) and (6) are satisfied when $f(x \mid \theta)$ is replaced by $f_n(x \mid \theta)$, then we may express $I_n(\theta)$ in either of the following two ways:

$$I_n(\theta) = \mathrm{Var}_\theta [\lambda'_n(X \mid \theta)] \tag{15}$$

or

$$I_n(\theta) = -E_\theta [\lambda''_n(X \mid \theta)]. \tag{16}$$

We shall now show that there is a simple relation between the Fisher information $I_n(\theta)$ in the entire sample and the Fisher information $I(\theta)$ in a single observation X_i. Since $f_n(x \mid \theta) = f(x_1 \mid \theta) \cdots f(x_n \mid \theta)$, it follows that

$$\lambda_n(x \mid \theta) = \sum_{i=1}^n \lambda(x_i \mid \theta).$$

Hence,

$$\lambda''_n(x \mid \theta) = \sum_{i=1}^n \lambda''(x_i \mid \theta). \tag{17}$$

Since each observation X_i has the p.d.f. $f(x \mid \theta)$, the Fisher information in each X_i is $I(\theta)$. It follows from Eqs. (11) and (16) that by taking expectations on both sides of Eq. (17) we obtain the result

$$I_n(\theta) = nI(\theta). \tag{18}$$

In other words, the Fisher information in a random sample of n observations is simply n times the Fisher information in a single observation.

The Information Inequality

As one application of the results that have just been derived, we shall show how the Fisher information can be used to determine a lower bound for the variance of an arbitrary estimator of the parameter θ in a given problem. We shall suppose again that $X_1, \ldots, X_n$ form a random sample from a distribution for which the p.d.f. is $f(x \mid \theta)$, and also that all the assumptions which have been made about $f(x \mid \theta)$ thus far in this section continue to hold.

Now let $T = r(X_1, \ldots, X_n) = r(X)$ denote an arbitrary estimator of θ for which the variance is finite, and consider the covariance between T and the random variable $\lambda'_n(X \mid \theta)$. Since $\lambda'_n(x \mid \theta) = f'_n(x \mid \theta)/f_n(x \mid \theta)$, it follows just as for a single observation that

$$E_\theta[\lambda'_n(X \mid \theta)] = \int_S \cdots \int_S f'_n(x \mid \theta)\, dx_1 \cdots dx_n = 0.$$

Therefore,

$$\mathrm{Cov}_\theta[T, \lambda'_n(X \mid \theta)] = E_\theta[T\lambda'_n(X \mid \theta)]$$

$$= \int_S \cdots \int_S r(x)\lambda'_n(x \mid \theta) f_n(x \mid \theta)\, dx_1 \cdots dx_n \qquad (19)$$

$$= \int_S \cdots \int_S r(x) f'_n(x \mid \theta)\, dx_1 \cdots dx_n.$$

Next, let $E_\theta(T) = m(\theta)$ for $\theta \in \Omega$, so

$$\int_S \cdots \int_S r(x) f_n(x \mid \theta)\, dx_1 \cdots dx_n = m(\theta) \qquad \text{for } \theta \in \Omega. \qquad (20)$$

Finally, suppose that when both sides of Eq. (20) are differentiated with respect to θ, the derivative can be taken "inside the integrals" on the left side. Then

$$\int_S \cdots \int_S r(x) f'_n(x \mid \theta)\, dx_1 \cdots dx_n = m'(\theta) \qquad \text{for } \theta \in \Omega. \qquad (21)$$

It follows from Eqs. (19) and (21) that

$$\mathrm{Cov}_\theta[T, \lambda'_n(X \mid \theta)] = m'(\theta) \qquad \text{for } \theta \in \Omega. \qquad (22)$$

It was shown in Sec. 4.6 that

$$\{\mathrm{Cov}_\theta[T, \lambda'_n(X \mid \theta)]\}^2 \leqslant \mathrm{Var}_\theta(T)\mathrm{Var}_\theta[\lambda'_n(X \mid \theta)]. \qquad (23)$$

Therefore, it follows from Eqs. (15), (18), (22), and (23) that

$$\mathrm{Var}_\theta(T) \geqslant \frac{[m'(\theta)]^2}{nI(\theta)}. \qquad (24)$$

The inequality (24) is called the *information inequality*. It is also known as the *Cramér-Rao inequality*, in honor of the statisticians H. Cramér and C. R. Rao who independently developed this inequality during the 1940's.

As an example of the use of the information inequality, suppose that T is an unbiased estimator of θ. Then $m(\theta) = \theta$ and $m'(\theta) = 1$ for every value of $\theta \in \Omega$. Hence, by the inequality (24), $\text{Var}_\theta(T) \geq 1/[nI(\theta)]$. In words, the variance of any unbiased estimator of θ cannot be smaller than the reciprocal of the Fisher information in the sample.

Efficient Estimators

It is said that an estimator T is an *efficient estimator of its expectation* $m(\theta)$ if there is equality in the information inequality (24) for every value of $\theta \in \Omega$. One difficulty with this definition is that, in a given problem, there may be no estimator of a particular function $m(\theta)$ whose variance actually attains the lower bound given in the information inequality. For example, if the random variable X has a normal distribution for which the mean is 0 and the standard deviation σ is unknown ($\sigma > 0$), then it can be shown that the variance of any unbiased estimator of σ based on the single observation X is strictly greater than $1/I(\sigma)$ for every value of $\sigma > 0$ (see Exercise 8).

On the other hand, in many standard estimation problems there do exist efficient estimators. Of course, the estimator which is identically equal to a constant is an efficient estimator of that constant, since the variance of this estimator is 0. However, as we shall now show, there are often efficient estimators of more interesting functions of θ as well.

There will be equality in the inequality (23) and, hence, there will be equality in the information inequality (24) if and only if the estimator T is a linear function of $\lambda'_n(X \mid \theta)$. In other words, T will be an efficient estimator if and only if there exist functions $u(\theta)$ and $v(\theta)$ which may depend on θ but do not depend on the observations $X_1, \ldots, X_n$ and which satisfy the relation

$$T = u(\theta)\lambda'_n(X \mid \theta) + v(\theta). \tag{25}$$

It is possible that the only efficient estimators in a given problem will be constants. The reason is as follows: Since T is an estimator, it cannot involve the parameter θ. Therefore, in order for T to be efficient, it must be possible to find functions $u(\theta)$ and $v(\theta)$ such that the parameter θ will actually be canceled from the right side of Eq. (25) and the value of T will depend only on the observations $X_1, \ldots, X_n$ and not on θ.

Example 3: Sampling from a Poisson Distribution. Suppose that $X_1, \ldots, X_n$ form a random sample from a Poisson distribution for which the mean θ is unknown ($\theta > 0$). We shall show that $\overline{X}_n$ is an efficient estimator of θ.

The joint p.f. of $X_1, \ldots, X_n$ can be written in the form

$$f_n(x \mid \theta) = \frac{e^{-n\theta}\theta^{n\overline{x}_n}}{\prod_{i=1}^n (x_i!)}.$$

Therefore,

$$\lambda_n(X \mid \theta) = -n\theta + n\overline{X}_n \log \theta - \sum_{i=1}^{n} \log(X_i!)$$

and

$$\lambda'_n(X \mid \theta) = -n + \frac{n\overline{X}_n}{\theta}. \tag{26}$$

If we now let $u(\theta) = \theta/n$ and $v(\theta) = \theta$, then it is found from Eq. (26) that

$$\overline{X}_n = u(\theta)\lambda'_n(X \mid \theta) + v(\theta).$$

Since the statistic $\overline{X}_n$ has been represented as a linear function of $\lambda'_n(X \mid \theta)$, it follows that $\overline{X}_n$ will be an efficient estimator of its expectation θ. In other words, the variance of $\overline{X}_n$ will attain the lower bound given by the information inequality, which in this example will be θ/n (see Exercise 2). This fact can also be verified directly. $\square$

Unbiased Estimators with Minimum Variance

Suppose that in a given problem a particular estimator T is an efficient estimator of its expectation $m(\theta)$, and let T_1 denote any other unbiased estimator of $m(\theta)$. Then for every value of $\theta \in \Omega$, $\mathrm{Var}_\theta(T)$ will be equal to the lower bound provided by the information inequality and $\mathrm{Var}_\theta(T_1)$ will be at least as large as that lower bound. Hence, $\mathrm{Var}_\theta(T) \leqslant \mathrm{Var}_\theta(T_1)$ for $\theta \in \Omega$. In other words, if T is an efficient estimator of $m(\theta)$, then among all unbiased estimators of $m(\theta)$, T will have the smallest variance for every possible value of θ.

In particular, it was shown in Example 3 that $\overline{X}_n$ is an efficient estimator of the mean θ of a Poisson distribution. Therefore, for every value of $\theta > 0$, $\overline{X}_n$ has the smallest variance among all unbiased estimators of θ. This discussion establishes a result which had been stated without proof in Sec. 7.7.

Properties of Maximum Likelihood Estimators for Large Samples

Suppose that $X_1, \ldots, X_n$ form a random sample from a distribution for which the p.d.f. or the p.f. is $f(x \mid \theta)$, and suppose also that $f(x \mid \theta)$ satisfies conditions similar to those which were needed to derive the information inequality. For any given sample size n, let $\hat{\theta}_n$ denote the M.L.E. of θ. We shall show that if n is large, then the distribution of $\hat{\theta}_n$ will be approximately a normal distribution with mean θ and variance $1/[nI(\theta)]$.

Asymptotic Distribution of an Efficient Estimator.   Consider first the random variable $\lambda'_n(X \mid \theta)$. Since $\lambda_n(X \mid \theta) = \sum_{i=1}^n \lambda(X_i \mid \theta)$, then

$$\lambda'_n(X \mid \theta) = \sum_{i=1}^n \lambda'(X_i \mid \theta). \tag{27}$$

Furthermore, since the n random variables $X_1, \ldots, X_n$ are i.i.d., the n random variables $\lambda'(X_1 \mid \theta), \ldots, \lambda'(X_n \mid \theta)$ will also be i.i.d. We know from Eqs. (7) and (8) that the mean of each of these variables is 0 and the variance of each is $I(\theta)$. Hence, it follows from the central limit theorem of Lindeberg and Lévy that the asymptotic distribution of the random variable $\lambda'_n(X \mid \theta)/[nI(\theta)]^{1/2}$ will be a standard normal distribution.

Now suppose that an estimator T is an efficient estimator of θ. Then $E_\theta(T) = \theta$ and $\text{Var}_\theta(T) = 1/[nI(\theta)]$. Furthermore, there must exist functions $u(\theta)$ and $v(\theta)$ which satisfy Eq. (25). Since the random variable $\lambda'_n(X \mid \theta)$ has mean 0 and variance $nI(\theta)$, it follows from Eq. (25) that $E_\theta(T) = v(\theta)$ and $\text{Var}_\theta(T) = [u(\theta)]^2 nI(\theta)$. When these values for the mean and the variance of T are compared with the values we gave first, it is found that $v(\theta) = \theta$ and $|u(\theta)| = 1/[nI(\theta)]$. To be specific, we shall assume that $u(\theta) = 1/[nI(\theta)]$, although the same conclusions would be obtained if $u(\theta) = -1/[nI(\theta)]$.

When the values $u(\theta) = 1/[nI(\theta)]$ and $v(\theta) = \theta$ are substituted into Eq. (25), we obtain

$$[nI(\theta)]^{1/2}(T - \theta) = \frac{\lambda'_n(X \mid \theta)}{[nI(\theta)]^{1/2}}. \tag{28}$$

We have already shown that the asymptotic distribution of the random variable on the right side of Eq. (28) is a standard normal distribution. Therefore, the asymptotic distribution of the random variable on the left side of Eq. (28) is also a standard normal distribution.

Asymptotic Distribution of an M.L.E.   It follows from the result just presented that if the M.L.E. $\hat{\theta}_n$ is an efficient estimator of θ for each value of n, then the asymptotic distribution of $[nI(\theta)]^{1/2}(\hat{\theta}_n - \theta)$ will be a standard normal distribution. However, it can be shown that even in an arbitrary problem in which $\hat{\theta}_n$ is not an efficient estimator, $[nI(\theta)]^{1/2}(\hat{\theta}_n - \theta)$ will have this same asymptotic distribution under certain conditions. Without presenting all the required conditions in full detail, we can state the following result.

Suppose that in an arbitrary problem the M.L.E. $\hat{\theta}_n$ is determined by solving the equation $\lambda'_n(x \mid \theta) = 0$; and in addition both the second and third derivatives $\lambda''_n(x \mid \theta)$ and $\lambda'''_n(x \mid \theta)$ exist and satisfy certain regularity conditions. Then the asymptotic distribution of $[nI(\theta)]^{1/2}(\hat{\theta}_n - \theta)$ will be a standard normal distribution. The proof of this result is beyond the scope of this book and will not be given here.

In practical terms, the foregoing result states that in most problems in which the sample size n is large and the M.L.E. $\hat{\theta}_n$ is found by differentiating the likelihood function $f_n(x \mid \theta)$ or its logarithm, the distribution of $[nI(\theta)]^{1/2}(\hat{\theta}_n - \theta)$ will be approximately a standard normal distribution. Equivalently, the distribution of $\hat{\theta}_n$ will be approximately a normal distribution with mean θ and variance $1/[nI(\theta)]$. Under these conditions it is said that $\hat{\theta}_n$ is an *asymptotically efficient estimator*.

Example 4: Estimating the Standard Deviation of a Normal Distribution. Suppose that $X_1, \ldots, X_n$ form a random sample from a normal distribution for which the mean is 0 and the standard deviation σ is unknown ($\sigma > 0$). It can be shown that the M.L.E. of σ is

$$
\hat{\sigma} = \left[\frac{1}{n} \sum_{i=1}^{n} X_i^2 \right]^{1/2}.
$$

Also, it can be shown (see Exercise 3) that the Fisher information in a single observation is $I(\sigma) = 2/\sigma^2$. Therefore, if the sample size n is large, the distribution of $\hat{\sigma}$ will be approximately a normal distribution with mean σ and variance $\sigma^2/(2n)$. □

The Bayesian Point of View. Another general property of the M.L.E. $\hat{\theta}_n$ pertains to making inferences about a parameter θ from the Bayesian point of view. Suppose that the prior distribution of θ is represented by a positive and differentiable p.d.f. over the interval Ω, and that the sample size n is large. Then under conditions similar to the regularity conditions that are needed to assure the asymptotic normality of the distribution of $\hat{\theta}_n$, it can be shown that the posterior distribution of θ, after the values of $X_1, \ldots, X_n$ have been observed, will be approximately a normal distribution with mean $\hat{\theta}_n$ and variance $1/[nI(\hat{\theta}_n)]$.

Example 5: The Posterior Distribution of the Standard Deviation. Suppose again that $X_1, \ldots, X_n$ form a random sample from a normal distribution for which the mean is 0 and the standard deviation σ is unknown. Suppose also that the prior p.d.f. of σ is a positive and differentiable function for $\sigma > 0$ and that the sample size n is large. Since $I(\sigma) = 2/\sigma^2$, it follows that the posterior distribution of σ will be approximately a normal distribution with mean $\hat{\sigma}$ and variance $\hat{\sigma}^2/(2n)$, where $\hat{\sigma}$ is the M.L.E. of σ calculated from the observed values in the sample. □

The Delta Method

Suppose that $X_1, \ldots, X_n$ form a random sample from a distribution for which the p.d.f. or the p.f. is $f(x \mid \theta)$, where again the parameter θ must lie in a given interval Ω of the real line. Suppose that T_n is an estimator of θ which is based on the observations $X_1, \ldots, X_n$ and has the following property: For some positive function $b(\theta)$, the asymptotic distribution of $[nb(\theta)]^{1/2}(T_n - \theta)$ is a standard normal distribution. In other words, suppose that for a large sample, the distribution of T_n is approximately normal with mean θ and variance $[nb(\theta)]^{-1}$.

Now suppose that it is desired to estimate the function $\alpha(\theta)$, where α is a differentiable function of θ such that $\alpha'(\theta) \neq 0$ for every $\theta \in \Omega$. It is natural to consider using the estimator $\alpha(T_n)$. We shall determine the asymptotic distribution of this estimator by a method known in statistics as the *delta method*.

It follows from the asymptotic distribution of T_n that, for a large sample, T_n will be close to θ with high probability. Hence, $T_n - \theta$ will be small with high probability. Therefore, we shall represent the function $\alpha(T_n)$ as a Taylor series in $T_n - \theta$, and we shall ignore all the terms involving $(T_n - \theta)^2$ and higher powers. Thus,

$$\alpha(T_n) \approx \alpha(\theta) + \alpha'(\theta)(T_n - \theta) \tag{29}$$

and

$$\frac{[nb(\theta)]^{1/2}}{\alpha'(\theta)}[\alpha(T_n) - \alpha(\theta)] \approx [nb(\theta)]^{1/2}(T_n - \theta). \tag{30}$$

We can conclude that the asymptotic distribution of the left side of Eq. (30) will be a standard normal distribution, since that is the asymptotic distribution of the right side. In other words, for a large sample, the distribution of $\alpha(T_n)$ will be approximately normal with mean $\alpha(\theta)$ and variance $[\alpha'(\theta)]^2/[nb(\theta)]$.

EXERCISES

1. Suppose that a random variable X has a normal distribution for which the mean μ is unknown $(-\infty < \mu < \infty)$ and the variance σ^2 is known. Let $f(x \mid \mu)$ denote the p.d.f. of X, and let $f'(x \mid \mu)$ and $f''(x \mid \mu)$ denote the first and second partial derivatives with respect to μ. Show that

$$\int_{-\infty}^{\infty} f'(x \mid \mu)\, dx = 0 \quad \text{and} \quad \int_{-\infty}^{\infty} f''(x \mid \mu)\, dx = 0.$$

2. Suppose that a random variable X has a Poisson distribution for which the mean θ is unknown $(\theta > 0)$. Find the Fisher information $I(\theta)$ in X.

3. Suppose that a random variable has a normal distribution for which the mean is 0 and the standard deviation σ is unknown $(\sigma > 0)$. Find the Fisher information $I(\sigma)$ in X.

4. Suppose that a random variable X has a normal distribution for which the mean is 0 and the variance σ^2 is unknown ($\sigma^2 > 0$). Find the Fisher information $I(\sigma^2)$ in X. Note that in this exercise the variance σ^2 is regarded as the parameter, whereas in Exercise 3 the standard deviation σ is regarded as the parameter.

5. Suppose that X is a random variable for which the p.d.f. or the p.f. is $f(x|\theta)$, where the value of the parameter θ is unknown but must lie in an open interval Ω. Let $I_0(\theta)$ denote the Fisher information in X. Suppose now that the parameter θ is replaced by a new parameter μ, where $\theta = \psi(\mu)$ and ψ is a differentiable function. Let $I_1(\mu)$ denote the Fisher information in X when the parameter is regarded as μ. Show that

$$I_1(\mu) = [\psi'(\mu)]^2 I_0[\psi(\mu)].$$

6. Suppose that $X_1, \ldots, X_n$ form a random sample from a Bernoulli distribution for which the parameter p is unknown. Show that $\overline{X}_n$ is an efficient estimator of p.

7. Suppose that $X_1, \ldots, X_n$ form a random sample from a normal distribution for which the mean μ is unknown and the variance σ^2 is known. Show that $\overline{X}_n$ is an efficient estimator of μ.

8. Suppose that a single observation X is taken from a normal distribution for which the mean is 0 and the standard deviation σ is unknown. Find an unbiased estimator of σ, determine its variance, and show that this variance is greater than $1/I(\sigma)$ for every value of $\sigma > 0$. Note that the value of $I(\sigma)$ was found in Exercise 3.

9. Suppose that $X_1, \ldots, X_n$ form a random sample from a normal distribution for which the mean is 0 and the standard deviation σ is unknown ($\sigma > 0$). Find the lower bound specified by the information inequality for the variance of any unbiased estimator of $\log \sigma$.

10. Suppose that $X_1, \ldots, X_n$ form a random sample from an exponential family for which the p.d.f. or the p.f. $f(x|\theta)$ is as specified in Exercise 11 of Sec. 6.7. Suppose also that the unknown value of θ must belong to an open interval Ω of the real line. Show that the estimator $T = \sum_{i=1}^{n} d(X_i)$ is an efficient estimator. *Hint:* Show that T can be represented in the form given in Eq. (25).

11. Suppose that $X_1, \ldots, X_n$ form a random sample from a normal distribution for which the mean is known and the variance is unknown. Construct an efficient estimator that is not identically equal to a constant, and determine the expectation and the variance of this estimator.

12. Determine what is wrong with the following argument: Suppose that the random variable X has a uniform distribution on the interval $(0, \theta)$, where

the value of θ is unknown ($\theta > 0$). Then $f(x \mid \theta) = 1/\theta$, $\lambda(x \mid \theta) = -\log \theta$, and $\lambda'(x \mid \theta) = -(1/\theta)$. Therefore,

$$I(\theta) = E_\theta\{[\lambda'(X \mid \theta)]^2\} = \frac{1}{\theta^2}.$$

Since $2X$ is an unbiased estimator of θ, the information inequality states that

$$\mathrm{Var}(2X) \geqslant \frac{1}{I(\theta)} = \theta^2.$$

But

$$\mathrm{Var}(2X) = 4\,\mathrm{Var}(X) = 4 \cdot \frac{\theta^2}{12} = \frac{\theta^2}{3} < \theta^2.$$

Hence, the information inequality is not correct.

13. Suppose that $X_1, \ldots, X_n$ form a random sample from a gamma distribution for which the value of the parameter α is unknown and the value of β is known. Show that if n is large the distribution of the M.L.E. of α will be approximately a normal distribution with mean α and variance

$$\frac{[\Gamma(\alpha)]^2}{n\{\Gamma(\alpha)\Gamma''(\alpha) - [\Gamma'(\alpha)]^2\}}.$$

14. Suppose that $X_1, \ldots, X_n$ form a random sample from a normal distribution for which the mean μ is unknown and the variance σ^2 is known; and that the prior p.d.f. of μ is a positive and differentiable function over the entire real line. Show that if n is large the posterior distribution of μ given that $X_i = x_i$ ($i = 1, \ldots, n$) will be approximately a normal distribution with mean $\bar{x}_n$ and variance σ^2/n.

15. Suppose that $X_1, \ldots, X_n$ form a random sample from a Bernoulli distribution for which the parameter p is unknown; and that the prior p.d.f. of p is a positive and differentiable function over the interval $0 < p < 1$. Suppose, furthermore, that n is large; that the observed values of $X_1, \ldots, X_n$ are $x_1, \ldots, x_n$; and that $0 < \bar{x}_n < 1$. Show that the posterior distribution of p will be approximately a normal distribution with mean $\bar{x}_n$ and variance $\bar{x}_n(1 - \bar{x}_n)/n$.

16. Suppose that $X_1, \ldots, X_n$ form a random sample from a normal distribution with unknown mean θ and variance σ^2. Assuming that $\theta \neq 0$, determine the asymptotic distribution of $\bar{X}_n^3$.

17. Consider again the conditions of Exercise 3. Determine the asymptotic distribution of the statistic $\left(\frac{1}{n}\sum_{i=1}^{n}X_i^2\right)^{-1}$.

7.9. SUPPLEMENTARY EXERCISES

1. Prove that if X has a t distribution with one degree of freedom, then $1/X$ also has a t distribution with one degree of freedom.

2. Suppose that U and V are independent random variables, and that each has a standard normal distribution. Show that U/V, $U/|V|$, and $|U|/V$ each has a t distribution with one degree of freedom.

3. Suppose that X_1 and X_2 are independent random variables, and that each has a normal distribution with mean 0 and variance σ^2. Show that $(X_1 + X_2)/(X_1 - X_2)$ has a t distribution with one degree of freedom.

4. Suppose that $X_1, \ldots, X_n$ form a random sample from an exponential distribution with parameter β. Show that $2\beta\sum_{i=1}^{n}X_i$ has a χ^2 distribution with $2n$ degrees of freedom.

5. Suppose that $X_1, \ldots, X_n$ form a random sample from an unknown probability distribution P on the real line. Let A be a given subset of the real line, and let $\theta = P(A)$. Construct an unbiased estimator of θ, and specify its variance.

6. Suppose that $X_1, \ldots, X_m$ form a random sample from a normal distribution with mean μ_1 and variance σ^2; and that $Y_1, \ldots, Y_n$ form an independent random sample from a normal distribution with mean μ_2 and variance $2\sigma^2$. Let $S_X^2 = \sum_{i=1}^{m}(X_i - \overline{X}_m)^2$ and $S_Y^2 = \sum_{i=1}^{n}(Y_i - \overline{Y}_n)^2$.

   (a) For what pairs of values of α and β is $\alpha S_X^2 + \beta S_Y^2$ an unbiased estimator of σ^2?

   (b) Determine the values of α and β for which $\alpha S_X^2 + \beta S_Y^2$ will be an unbiased estimator with minimum variance.

7. Suppose that $X_1, \ldots, X_{n+1}$ form a random sample from a normal distribution, and let $\overline{X}_n = \frac{1}{n}\sum_{i=1}^{n}X_i$ and $T_n = \left[\frac{1}{n}\sum_{i=1}^{n}(X_i - \overline{X}_n)^2\right]^{1/2}$. Determine the value of a constant k such that the random variable $k(X_{n+1} - \overline{X}_n)/T_n$ will have a t distribution.

8. Suppose that $X_1, \ldots, X_n$ form a random sample from a normal distribution with mean μ and variance σ^2, and that Y is an independent random variable having a normal distribution with mean 0 and variance $4\sigma^2$. Determine a function of $X_1, \ldots, X_n$ and Y that does not involve μ or σ^2 but has a t distribution with $n - 1$ degrees of freedom.

9. Suppose that $X_1, \ldots, X_n$ form a random sample from a normal distribution for which both the mean μ and the variance σ^2 are unknown. A confidence interval for μ is to be constructed with confidence coefficient 0.90. Determine the smallest value of n such that the expected squared length of this interval will be less than $\sigma^2/2$.

10. Suppose that $X_1, \ldots, X_n$ form a random sample from a normal distribution for which both the mean μ and the variance σ^2 are unknown. Construct a lower confidence limit $L(X_1, \ldots, X_n)$ for μ such that

$$\Pr[\mu > L(X_1, \ldots, X_n)] = 0.99.$$

11. Consider again the conditions of Exercise 10. Construct an upper confidence limit $U(X_1, \ldots, X_n)$ for σ^2 such that

$$\Pr[\sigma^2 < U(X_1, \ldots, X_n)] = 0.99.$$

12. Suppose that $X_1, \ldots, X_n$ form a random sample from a normal distribution with unknown mean θ and known variance σ^2. Suppose also that the prior distribution of θ is normal with mean μ and variance ν^2.

   (a) Determine the shortest interval I such that $\Pr(\theta \in I \mid x_1, \ldots, x_n) = 0.95$, where the probability is calculated with respect to the posterior distribution of θ, as indicated.

   (b) Show that as $\nu^2 \to \infty$, the interval I converges to an interval I^* that is a confidence interval for θ with confidence coefficient 0.95.

13. Suppose that $X_1, \ldots, X_n$ form a random sample from a Poisson distribution with unknown mean θ, and let $Y = \sum_{i=1}^{n} X_i$.

   (a) Determine the value of a constant c such that the estimator e^{-cY} is an unbiased estimator of $e^{-\theta}$.

   (b) Use the information inequality to obtain a lower bound for the variance of the unbiased estimator found in part (a).

14. Suppose that $X_1, \ldots, X_n$ form a random sample from a distribution for which the p.d.f. is as follows:

$$f(x \mid \theta) = \begin{cases} \theta x^{\theta - 1} & \text{for } 0 < x < 1, \\ 0 & \text{otherwise,} \end{cases}$$

   where the value of θ is unknown $(\theta > 0)$. Determine the asymptotic distribution of the M.L.E. of θ. (*Note:* The M.L.E. was found in Exercise 8 of Sec. 6.5.)

15. Suppose that a random variable X has an exponential distribution for which the parameter β is unknown $(\beta > 0)$. Find the Fisher information $I(\beta)$ in X.

16. Suppose that $X_1, \ldots, X_n$ form a random sample from a Bernoulli distribution for which the parameter p is unknown. Show that the variance of any unbiased estimator of $(1 - p)^2$ must be at least $4p(1 - p)^3/n$.

17. Suppose that $X_1, \ldots, X_n$ form a random sample from an exponential distribution for which the parameter β is unknown. Construct an efficient estimator that is not identically equal to a constant, and determine the expectation and the variance of this estimator.

18. Suppose that $X_1, \ldots, X_n$ form a random sample from an exponential distribution for which the parameter β is unknown. Show that if n is large, the distribution of the M.L.E. of β will be approximately a normal distribution with mean β and variance β^2/n.

19. Consider again the conditions of Exercise 18, and let $\hat{\beta}_n$ denote the M.L.E. of β.

    (a) Use the delta method to determine the asymptotic distribution of $1/\hat{\beta}_n$.

    (b) Show that $1/\hat{\beta}_n = \bar{X}_n$, and use the central limit theorem to determine the asymptotic distribution of $1/\hat{\beta}_n$.

Testing Hypotheses

8

8.1. PROBLEMS OF TESTING HYPOTHESES

The Null and Alternative Hypotheses

In this chapter we shall again consider statistical problems involving a parameter θ whose value is unknown but must lie in a certain parameter space Ω. We shall suppose now, however, that Ω can be partitioned into two disjoint subsets Ω_0 and Ω_1, and that the statistician must decide whether the unknown value of θ lies in Ω_0 or in Ω_1.

We shall let H_0 denote the hypothesis that $\theta \in \Omega_0$, and shall let H_1 denote the hypothesis that $\theta \in \Omega_1$. Since the subsets Ω_0 and Ω_1 are disjoint and $\Omega_0 \cup \Omega_1 = \Omega$, exactly one of the hypotheses H_0 and H_1 must be true. The statistician must decide whether to accept the hypothesis H_0 or to accept the hypothesis H_1. A problem of this type, in which there are only two possible decisions, is called a problem of *testing hypotheses*. If the statistician makes the wrong decision, he typically must suffer a certain loss or pay a certain cost. In many problems, he will have an opportunity to take some observations before he has to make his decision, and the observed values will provide him with information about the value of θ. A procedure for deciding whether to accept the hypothesis H_0 or to accept the hypothesis H_1 is called a *test procedure* or simply a *test*.

In our discussion up to this point, we have treated the hypotheses H_0 and H_1 on an equal basis. In most problems, however, the two hypotheses are treated quite differently. To distinguish between them, the hypothesis H_0 is called the *null hypothesis* and the hypothesis H_1 is called the *alternative hypothesis*. We shall use this terminology in all the problems of testing hypotheses that will be discussed in the later sections of this chapter and in the rest of the book.

One way of describing the decisions available to the statistician is that he may accept either H_0 or H_1. However, since there are only two possible decisions, accepting H_0 is equivalent to rejecting H_1, and accepting H_1 is equivalent to rejecting H_0. We shall use all these descriptions in our discussions of testing hypotheses.

The Critical Region

Consider now a problem in which hypotheses having the following form are to be tested:

$$H_0: \quad \theta \in \Omega_0,$$
$$H_1: \quad \theta \in \Omega_1.$$

Suppose that before the statistician has to decide which hypothesis to accept, he can observe a random sample $X_1, \ldots, X_n$ drawn from a distribution which involves the unknown parameter θ. We shall let S denote the sample space of the n-dimensional random vector $X = (X_1, \ldots, X_n)$. In other words, S is the set of all possible outcomes of the random sample.

In a problem of this type, the statistician specifies a test procedure by partitioning the sample space S into two subsets. One subset contains the values of X for which he will accept H_0, and the other subset contains the values of X for which he will reject H_0 and therefore accept H_1. The subset for which H_0 will be rejected is called the *critical region* of the test. In summary, a test procedure is determined by specifying the critical region of the test. The complement of the critical region must then contain all the outcomes for which H_0 will be accepted.

The Power Function

The characteristics of a test procedure can be described by specifying, for each value of $\theta \in \Omega$, either the probability $\pi(\theta)$ that the procedure will lead to the rejection of H_0 or the probability $1 - \pi(\theta)$ that it will lead to the acceptance of H_0. The function $\pi(\theta)$ is called the *power function* of the test. Thus, if C denotes the critical region of the test, then the power function $\pi(\theta)$ is determined by the relation

$$\pi(\theta) = \Pr(X \in C \mid \theta) \qquad \text{for } \theta \in \Omega. \tag{1}$$

Since the power function $\pi(\theta)$ specifies for each possible value of the parameter θ the probability that H_0 will be rejected, it follows that the ideal power function would be one for which $\pi(\theta) = 0$ for every value of $\theta \in \Omega_0$ and $\pi(\theta) = 1$ for every value of $\theta \in \Omega_1$. If the power function of a test actually had these values, then regardless of the actual value of θ, the test would lead to the correct decision with probability 1. In a practical problem, however, there would seldom exist any test procedure having this ideal power function.

For any value of $\theta \in \Omega_0$, the decision to reject H_0 is an incorrect decision. Therefore, if $\theta \in \Omega_0$, $\pi(\theta)$ is the probability that the statistician will make an incorrect decision. In many problems, a statistician will specify an upper bound α_0 ($0 < \alpha_0 < 1$) and will consider only tests for which $\pi(\theta) \leq \alpha_0$ for every value of $\theta \in \Omega_0$. An upper bound α_0 that is specified in this way is called the *level of significance* of the tests to be considered.

The *size* α of a given test is defined as follows:

$$\alpha = \sup_{\theta \in \Omega_0} \pi(\theta). \tag{2}$$

In words, the size of a test is the maximum probability, among all the values of θ which satisfy the null hypothesis, of making an incorrect decision.

The relationship between the level of significance and the size can be summarized as follows: If a statistician specifies a certain level of significance α_0 in a given problem of testing hypotheses, then he will consider only tests for which the size α is such that $\alpha \leq \alpha_0$.

Example 1: Testing Hypotheses About a Uniform Distribution. Suppose that a random sample $X_1, \ldots, X_n$ is taken from a uniform distribution on the interval $(0, \theta)$, where the value of θ is unknown ($\theta > 0$); and suppose also that it is desired to test the following hypotheses:

$$\begin{aligned} H_0 &: \quad 3 \leq \theta \leq 4, \\ H_1 &: \quad \theta < 3 \quad \text{or} \quad \theta > 4. \end{aligned} \tag{3}$$

We know from Sec. 6.5 that the M.L.E. of θ is $Y_n = \max(X_1, \ldots, X_n)$. Although Y_n must be less than θ, there is a high probability that Y_n will be close to θ if the sample size n is fairly large. For illustrative purposes, we shall suppose that the hypothesis H_0 will be accepted if the observed value of Y_n lies in the interval $2.9 \leq Y_n \leq 4$ and that H_0 will be rejected if Y_n does not lie in this interval. Thus, the critical region of the test contains all the values of $X_1, \ldots, X_n$ for which either $Y_n < 2.9$ or $Y_n > 4$.

The power function of the test is specified by the relation

$$\pi(\theta) = \Pr(Y_n < 2.9 \mid \theta) + \Pr(Y_n > 4 \mid \theta).$$

If $\theta \leqslant 2.9$, then $\Pr(Y_n < 2.9 \,|\, \theta) = 1$ and $\Pr(Y_n > 4 \,|\, \theta) = 0$. Therefore, $\pi(\theta) = 1$. If $2.9 < \theta \leqslant 4$, then $\Pr(Y_n < 2.9 \,|\, \theta) = (2.9/\theta)^n$ and $\Pr(Y_n > 4 \,|\, \theta) = 0$. In this case, $\pi(\theta) = (2.9/\theta)^n$. Finally, if $\theta > 4$, then $\Pr(Y_n < 2.9 \,|\, \theta) = (2.9/\theta)^n$ and $\Pr(Y_n > 4 \,|\, \theta) = 1 - (4/\theta)^n$. In this case, $\pi(\theta) = (2.9/\theta)^n + 1 - (4/\theta)^n$. The power function $\pi(\theta)$ is sketched in Fig. 8.1.

By Eq. (2), the size of the test is $\alpha = \sup_{3 \leqslant \theta \leqslant 4} \pi(\theta)$. It can be seen from Fig. 8.1 and the calculations just given that $\alpha = \pi(3) = (29/30)^n$. In particular, if the sample size is $n = 68$, then the size of the test is $(29/30)^{68} = 0.100$. $\square$

Simple and Composite Hypotheses

Suppose that $X_1, \ldots, X_n$ form a random sample from a distribution for which the p.d.f. or the p.f. is $f(x \,|\, \theta)$, where the value of the parameter θ must lie in the parameter space Ω; that Ω_0 and Ω_1 are disjoint sets with $\Omega_0 \cup \Omega_1 = \Omega$; and that it is desired to test the following hypotheses:

$$H_0: \quad \theta \in \Omega_0,$$
$$H_1: \quad \theta \in \Omega_1.$$

For $i = 0$ or $i = 1$, the set Ω_i may contain just a single value of θ. It is then said that the hypothesis H_i is a *simple hypothesis*. If the set Ω_i contains more than one value of θ, then it is said that the hypothesis H_i is a *composite hypothesis*. Under a simple hypothesis, the distribution of the observations is completely specified. Under a composite hypothesis, it is specified only that the distribution of the observations belongs to a certain class. For example, suppose that in a given problem the null hypothesis H_0 has the form

$$H_0: \quad \theta = \theta_0.$$

Since this hypothesis is simple, it follows from Eq. (2) that the size of any test procedure will just be $\pi(\theta_0)$.

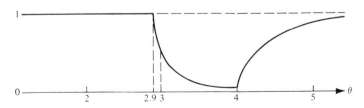

Figure 8.1   The power function $\pi(\theta)$.

EXERCISES

1. Suppose that $X_1, \ldots, X_n$ form a random sample from a uniform distribution on the interval $(0, \theta)$, and that the following hypotheses are to be tested:

$$H_0: \quad \theta \geqslant 2,$$
$$H_1: \quad \theta < 2.$$

Let $Y_n = \max(X_1, \ldots, X_n)$, and consider a test procedure such that the critical region contains all the outcomes for which $Y_n \leqslant 1.5$.

(a) Determine the power function of the test.

(b) Determine the size of the test.

2. Suppose that the proportion p of defective items in a large population of items is unknown, and that it is desired to test the following hypotheses:

$$H_0: \quad p = 0.2,$$
$$H_1: \quad p \neq 0.2.$$

Suppose also that a random sample of 20 items is drawn from the population. Let Y denote the number of defective items in the sample, and consider a test procedure such that the critical region contains all the outcomes for which either $Y \geqslant 7$ or $Y \leqslant 1$.

(a) Determine the value of the power function $\pi(p)$ at the points $p = 0, 0.1$, $0.2, 0.3, 0.4, 0.5, 0.6, 0.7, 0.8, 0.9$, and 1; and sketch the power function.

(b) Determine the size of the test.

3. Suppose that $X_1, \ldots, X_n$ form a random sample from a normal distribution for which the mean μ is unknown and the variance is 1. Suppose also that μ_0 is a certain specified number, and that the following hypotheses are to be tested:

$$H_0: \quad \mu = \mu_0,$$
$$H_1: \quad \mu \neq \mu_0.$$

Finally, suppose that the sample size n is 25, and consider a test procedure such that H_0 is to be accepted if $|\overline{X}_n - \mu_0| < c$. Determine the value of c such that the size of the test will be 0.05.

4. Suppose that $X_1, \ldots, X_n$ form a random sample from a normal distribution for which both the mean μ and the variance σ^2 are unknown. Classify each

of the following hypotheses as either simple or composite:

(a) H_0: $\mu = 0$ and $\sigma = 1$.

(b) H_0: $\mu > 3$ and $\sigma < 1$.

(c) H_0: $\mu = -2$ and $\sigma^2 < 5$.

(d) H_0: $\mu = 0$.

5. Suppose that a single observation X is to be taken from a uniform distribution on the interval $\left(\theta - \dfrac{1}{2}, \theta + \dfrac{1}{2} \right)$, and suppose that the following hypotheses are to be tested:

H_0: $\theta \leqslant 3$,
H_1: $\theta \geqslant 4$.

Construct a test procedure for which the power function has the following values: $\pi(\theta) = 0$ for $\theta \leqslant 3$ and $\pi(\theta) = 1$ for $\theta \geqslant 4$.

8.2.   TESTING SIMPLE HYPOTHESES

Two Types of Errors

In this section we shall consider problems of testing hypotheses in which a random sample is taken from one of two possible distributions, and the statistician must decide from which distribution the sample actually came. In this type of problem, the parameter space Ω contains exactly two points, and both the null hypothesis and the alternative hypothesis are simple.

Specifically, we shall assume that the variables $X_1, \ldots, X_n$ form a random sample from a distribution for which the p.d.f. or the p.f. is $f(x \mid \theta)$, and we shall also assume that either $\theta = \theta_0$ or $\theta = \theta_1$, where θ_0 and θ_1 are two specific values of θ. The following simple hypotheses are to be tested:

$$H_0: \theta = \theta_0, \tag{1}$$
$$H_1: \theta = \theta_1.$$

For $i = 0$ or $i = 1$, we shall let

$$f_i(x) = f(x_1 \mid \theta_i) f(x_2 \mid \theta_i) \cdots f(x_n \mid \theta_i). \tag{2}$$

Thus, $f_i(x)$ represents the joint p.d.f. or joint p.f. of the observations in the sample if the hypothesis H_i is true ($i = 0, 1$).

When a test of the hypotheses (1) is being carried out, two possible types of errors must be kept in mind. First, the test might result in the rejection of the null hypothesis H_0 when, in fact, H_0 is true. It has become traditional to call this result an error of type 1, or an error of the first kind. Second, the test might result in the acceptance of the null hypothesis H_0 when, in fact, the alternative hypothesis H_1 is true. This result is called an error of type 2, or an error of the second kind. Furthermore, for any given test procedure δ, we shall let $\alpha(\delta)$ denote the probability of an error of type 1 and shall let $\beta(\delta)$ denote the probability of an error of type 2. Thus,

$$\alpha(\delta) = \Pr(\text{Rejecting } H_0 \mid \theta = \theta_0),$$
$$\beta(\delta) = \Pr(\text{Accepting } H_0 \mid \theta = \theta_1).$$

It is desirable to find a test procedure for which the probabilities $\alpha(\delta)$ and $\beta(\delta)$ of the two types of error will be small. It is easy to construct a test procedure for which $\alpha(\delta) = 0$ by using a procedure that always accepts H_0. However, for this procedure $\beta(\delta) = 1$. Similarly, it is easy to construct a test procedure for which $\beta(\delta) = 0$ but $\alpha(\delta) = 1$. For a given sample size, it is typically not possible to find a test procedure for which both $\alpha(\delta)$ and $\beta(\delta)$ will be arbitrarily small. Therefore, we shall now show how to construct a procedure for which the value of a specific linear combination of α and β will be minimized.

Optimal Tests

Minimizing a Linear Combination. Suppose that a and b are specified positive constants, and that it is desired to find a procedure δ for which $a\alpha(\delta) + b\beta(\delta)$ will be a minimum. The following result shows that a procedure which is optimal in this sense has a very simple form.

Theorem 1. *Let δ^* denote a test procedure such that the hypothesis H_0 is accepted if $af_0(x) > bf_1(x)$ and the hypothesis H_1 is accepted if $af_0(x) < bf_1(x)$. Either H_0 or H_1 may be accepted if $af_0(x) = bf_1(x)$. Then for any other test procedure δ,*

$$a\alpha(\delta^*) + b\beta(\delta^*) \leqslant a\alpha(\delta) + b\beta(\delta). \tag{3}$$

Proof. For convenience, we shall present the proof for a problem in which the random sample $X_1, \ldots, X_n$ is drawn from a discrete distribution. In this case, $f_i(x)$ represents the joint p.f. of the observations in the sample when H_i is true $(i = 0, 1)$. If the sample comes from a continuous distribution, in which case $f_i(x)$ is a joint p.d.f., then each of the sums that will appear in this proof should be replaced by an n-dimensional integral.

If we let R denote the critical region of an arbitrary test procedure δ, then R contains every sample outcome x for which δ specifies that H_0 should be rejected, and R^c contains every outcome x for which H_0 should be accepted. Therefore,

$$a\alpha(\delta) + b\beta(\delta) = a \sum_{x \in R} f_0(x) + b \sum_{x \in R^c} f_1(x)$$

$$= a \sum_{x \in R} f_0(x) + b\left[1 - \sum_{x \in R} f_1(x)\right] \tag{4}$$

$$= b + \sum_{x \in R} \left[af_0(x) - bf_1(x)\right].$$

It follows from Eq. (4) that the value of the linear combination $a\alpha(\delta) + b\beta(\delta)$ will be a minimum if the critical region R is chosen so that the value of the final summation in Eq. (4) is a minimum. Furthermore, the value of this summation will be a minimum if the summation includes every point x for which $af_0(x) - bf_1(x) < 0$ and does not include any point x for which $af_0(x) - bf_1(x) > 0$. In other words, $a\alpha(\delta) + b\beta(\delta)$ will be a minimum if the critical region R is chosen to include every point x such that $af_0(x) - bf_1(x) < 0$ and to exclude every point x such that this inequality is reversed. If $af_0(x) - bf_1(x) = 0$ for some point x, then it is irrelevant whether or not x is included in R, since the corresponding term would contribute zero to the final summation in Eq. (4). It can be seen that this description of the critical region corresponds to the description of the test procedure δ^* given in the statement of the theorem. $\square$

The ratio $f_1(x)/f_0(x)$ is called the likelihood ratio of the sample. Thus, Theorem 1 states that a test procedure for which the value of $a\alpha(\delta) + b\beta(\delta)$ is a minimum rejects H_0 when the likelihood ratio exceeds a/b and accepts H_0 when the likelihood ratio is less than a/b.

Minimizing the Probability of an Error of Type 2.   Next, suppose that the probability $\alpha(\delta)$ of an error of type 1 is not permitted to be greater than a specified level of significance, and it is desired to find a procedure δ for which $\beta(\delta)$ will be a minimum. In this problem we can apply the following result, which is closely related to Theorem 1 and is known as the *Neyman-Pearson lemma* in honor of the statisticians J. Neyman and E. S. Pearson who developed these ideas in 1933.

Neyman-Pearson Lemma.   *Suppose that δ^* is a test procedure which has the following form for some constant $k > 0$: The hypothesis H_0 is accepted if $f_0(x) > kf_1(x)$ and the hypothesis H_1 is accepted if $f_0(x) < kf_1(x)$. Either H_0 or H_1 may be accepted if $f_0(x) = kf_1(x)$. If δ is any other test procedure such that $\alpha(\delta) \leqslant \alpha(\delta^*)$, then it follows that $\beta(\delta) \geqslant \beta(\delta^*)$. Furthermore, if $\alpha(\delta) < \alpha(\delta^*)$, then $\beta(\delta) > \beta(\delta^*)$.*

Proof. From the description of the procedure δ^* and from Theorem 1, it follows that for any other test procedure δ,

$$\alpha(\delta^*) + k\beta(\delta^*) \leqslant \alpha(\delta) + k\beta(\delta). \tag{5}$$

If $\alpha(\delta) \leqslant \alpha(\delta^*)$, then it follows from the relation (5) that $\beta(\delta) \geqslant \beta(\delta^*)$. Also, if $\alpha(\delta) < \alpha(\delta^*)$, then it follows that $\beta(\delta) > \beta(\delta^*)$. $\square$

To illustrate the use of the Neyman-Pearson lemma, we shall suppose that a statistician wishes to use a test procedure for which $\alpha(\delta) = 0.05$ and $\beta(\delta)$ is a minimum. According to the lemma, he should try to find a value of k for which $\alpha(\delta^*) = 0.05$. The procedure δ^* will then have the minimum possible value of $\beta(\delta)$. If the distribution from which the random sample is taken is continuous, then it is usually (but not always) possible to find a value of k such that $\alpha(\delta^*)$ is equal to a specified value such as 0.05. However, if the distribution from which the random sample is taken is discrete, then it is typically not possible to choose k so that $\alpha(\delta^*)$ is equal to a specified value. These remarks are considered further in the following examples and in the exercises at the end of this section.

Example 1: Sampling from a Normal Distribution. Suppose that $X_1, \ldots, X_n$ form a random sample from a normal distribution for which the value of the mean θ is unknown and the variance is 1; and that the following hypotheses are to be tested:

$$\begin{array}{ll} H_0: & \theta = 0, \\ H_1: & \theta = 1. \end{array} \tag{6}$$

We shall begin by determining a test procedure for which $\beta(\delta)$ will be a minimum among all test procedures for which $\alpha(\delta) \leqslant 0.05$.

When H_0 is true, the variables $X_1, \ldots, X_n$ form a random sample from a standard normal distribution. When H_1 is true, these variables form a random sample from a normal distribution for which both the mean and the variance are 1. Therefore,

$$f_0(x) = \frac{1}{(2\pi)^{n/2}} \exp\left(-\frac{1}{2} \sum_{i=1}^{n} x_i^2\right) \tag{7}$$

and

$$f_1(x) = \frac{1}{(2\pi)^{n/2}} \exp\left[-\frac{1}{2} \sum_{i=1}^{n} (x_i - 1)^2\right]. \tag{8}$$

After some algebraic simplification, the likelihood ratio $f_1(x)/f_0(x)$ can be written in the form

$$\frac{f_1(x)}{f_0(x)} = \exp\left[n\left(\bar{x}_n - \frac{1}{2}\right)\right]. \tag{9}$$

It now follows from Eq. (9) that rejecting the hypothesis H_0 when the likelihood ratio is greater than a specified positive constant k is equivalent to rejecting H_0 when the sample mean $\bar{x}_n$ is greater than $(1/2) + (1/n)\log k$.

Let $k' = (1/2) + (1/n)\log k$, and suppose that we can find a value of k' such that

$$\Pr(\bar{X}_n > k' \mid \theta = 0) = 0.05. \tag{10}$$

Then the procedure δ^* which rejects H_0 when $\bar{X}_n > k'$ will be such that $\alpha(\delta^*) = 0.05$. Furthermore, by the Neyman-Pearson lemma, δ^* will be an optimal procedure in the sense of minimizing the value of $\beta(\delta)$ among all procedures for which $\alpha(\delta) \leqslant 0.05$.

It is easy to find a value of k' that satisfies Eq. (10). When $\theta = 0$, the distribution of $\bar{X}_n$ will be a normal distribution with mean 0 and variance $1/n$. Therefore, if we let $Z = n^{1/2}\bar{X}_n$, then Z will have a standard normal distribution, and Eq. (10) can be rewritten in the form

$$\Pr(Z > n^{1/2}k') = 0.05. \tag{11}$$

From a table of the standard normal distribution, it is found that Eq. (11) will be satisfied, and therefore Eq. (10) will also be satisfied, when $n^{1/2}k' = 1.645$ or, equivalently, when $k' = 1.645n^{-1/2}$.

In summary, among all test procedures for which $\alpha(\delta) \leqslant 0.05$, the procedure which rejects H_0 when $\bar{X}_n > 1.645n^{-1/2}$ is optimal.

Next, we shall determine the probability $\beta(\delta^*)$ of an error of type 2 for this procedure δ^*. Since $\beta(\delta^*)$ is the probability of accepting H_0 when H_1 is true,

$$\beta(\delta^*) = \Pr(\bar{X}_n < 1.645n^{-1/2} \mid \theta = 1). \tag{12}$$

When $\theta = 1$, the distribution of $\bar{X}_n$ will be a normal distribution with mean 1 and variance $1/n$. If we let $Z' = n^{1/2}(\bar{X}_n - 1)$, then Z' will have a standard normal distribution. Hence,

$$\beta(\delta^*) = \Pr(Z' < 1.645 - n^{1/2}). \tag{13}$$

For instance, when $n = 9$, it is found from a table of the standard normal distribution that

$$\beta(\delta^*) = \Pr(Z' < -1.355) = 1 - \Phi(1.355) = 0.0877. \tag{14}$$

Finally, for this same random sample and the same hypotheses (6), we shall determine the test procedure δ_0 for which the value of $2\alpha(\delta) + \beta(\delta)$ is a minimum, and we shall calculate the value of $2\alpha(\delta_0) + \beta(\delta_0)$ when $n = 9$.

It follows from Theorem 1 that the procedure δ_0 for which $2\alpha(\delta) + \beta(\delta)$ is a minimum rejects H_0 when the likelihood ratio is greater than 2. By Eq. (9), this procedure is equivalent to rejecting H_0 when $\overline{X}_n > (1/2) + (1/n)\log 2$. Thus, when $n = 9$, the optimal procedure δ_0 rejects H_0 when $\overline{X}_n > 0.577$. For this procedure we then have

$$\alpha(\delta_0) = \Pr(\overline{X}_n > 0.577 \mid \theta = 0) \tag{15}$$

and

$$\beta(\delta_0) = \Pr(\overline{X}_n < 0.577 \mid \theta = 1). \tag{16}$$

If Z and Z' are defined as earlier in this example, then it is found that

$$\alpha(\delta_0) = \Pr(Z > 1.731) = 0.0417 \tag{17}$$

and

$$\beta(\delta_0) = \Pr(Z' < -1.269) = 1 - \Phi(1.269) = 0.1022. \tag{18}$$

The minimum value of $2\alpha(\delta) + \beta(\delta)$ is therefore

$$2\alpha(\delta_0) + \beta(\delta_0) = 2(0.0417) + (0.1022) = 0.1856. \quad \square \tag{19}$$

Example 2: Sampling from a Bernoulli Distribution. Suppose that $X_1, \ldots, X_n$ form a random sample from a Bernoulli distribution for which the value of the parameter p is unknown; and that the following hypotheses are to be tested:

$$\begin{aligned} H_0{:} \quad & p = 0.2, \\ H_1{:} \quad & p = 0.4. \end{aligned} \tag{20}$$

It is desired to find a test procedure for which $\alpha(\delta) = 0.05$ and $\beta(\delta)$ is a minimum.

In this example, each observed value x_i must be either 0 or 1. If we let $y = \sum_{i=1}^{n} x_i$, then the joint p.f. of $X_1, \ldots, X_n$ when $p = 0.2$ is

$$f_0(x) = (0.2)^y (0.8)^{n-y} \tag{21}$$

and the joint p.f. when $p = 0.4$ is

$$f_1(x) = (0.4)^y (0.6)^{n-y}. \tag{22}$$

Hence, the likelihood ratio is

$$\frac{f_1(x)}{f_0(x)} = \left(\frac{3}{4}\right)^n \left(\frac{8}{3}\right)^y. \tag{23}$$

It follows that rejecting H_0 when the likelihood ratio is greater than a specified positive constant k is equivalent to rejecting H_0 when y is greater than k', where

$$k' = \frac{\log k + n\log(4/3)}{\log(8/3)}. \tag{24}$$

To find a test procedure for which $\alpha(\delta) = 0.05$ and $\beta(\delta)$ is a minimum, we use the Neyman-Pearson lemma. If we let $Y = \sum_{i=1}^{n} X_i$, we should try to find a value of k' such that

$$\Pr\left(Y > k' \mid p = 0.2\right) = 0.05. \tag{25}$$

When the hypothesis H_0 is true, the random variable Y will have a binomial distribution with parameters n and $p = 0.2$. However, because of the discreteness of this distribution, it generally will not be possible to find a value of k' for which Eq. (25) is satisfied. For example, suppose that $n = 10$. Then it is found from a table of the binomial distribution that $\Pr(Y > 4 \mid p = 0.2) = 0.0328$ and also that $\Pr(Y > 3 \mid p = 0.2) = 0.1209$. Therefore, there is no critical region of the desired form for which $\alpha(\delta) = 0.05$. If it is desired to use a test δ based on the likelihood ratio as specified by the Neyman-Pearson lemma, then $\alpha(\delta)$ must be either 0.0328 or 0.1209. $\square$

Randomized Tests.   It has been emphasized by some statisticians that $\alpha(\delta)$ can be made exactly 0.05 in Example 2 if a *randomized* test procedure is used. Such a procedure is described as follows: When the critical region of the test procedure contains all values of y greater than 4, we found in Example 2 that the size of the test is $\alpha(\delta) = 0.0328$. Also, when the point $y = 4$ is added to this critical region, then the value of $\alpha(\delta)$ jumps to 0.1209. Suppose, however, that instead of either always including the point $y = 4$ in the critical region or always excluding that point, we can use an auxiliary randomization to decide which hypothesis to accept when $y = 4$. For example, we may toss a coin or spin a wheel to arrive at this decision. Then, by choosing appropriate probabilities to be used in this randomization, we can make $\alpha(\delta)$ exactly 0.05.

Specifically, consider the following test procedure: The hypothesis H_0 is always rejected if $y > 4$, and H_0 is always accepted if $y < 4$. However, if $y = 4$, then an auxiliary randomization is carried out in which H_0 will be rejected with probability 0.195 and H_0 will be accepted with probability 0.805. The size $\alpha(\delta)$

of this test will then be

$$\alpha(\delta) = \Pr(Y > 4 \mid p = 0.2) + (0.195)\Pr(Y = 4 \mid p = 0.2)$$
$$= 0.0328 + (0.195)(0.0881) = 0.05.$$

(26)

Randomized tests do not seem to have any place in practical applications of statistics. It does not seem reasonable for a statistician to decide which hypothesis he will accept by tossing a coin or performing some other type of randomization for the sole purpose of obtaining a value of $\alpha(\delta)$ that is equal to some arbitrarily specified value such as 0.05. The main consideration for the statistician is to use a nonrandomized test procedure δ^* having the form specified in the Neyman-Pearson lemma. The proof of Theorem 1 can be extended to show that δ^* will be optimal, in the sense of the Neyman-Pearson lemma, among all test procedures regardless of whether they are randomized or nonrandomized.

Furthermore, rather than fixing a specific size $\alpha(\delta)$ and trying to minimize $\beta(\delta)$, it is more reasonable for the statistician to minimize a linear combination of the form $a\alpha(\delta) + b\beta(\delta)$. As we have seen in Theorem 1, such a minimization can always be achieved without recourse to an auxiliary randomization. We shall now present another argument that indicates why it is more reasonable to minimize a linear combination of the form $a\alpha(\delta) + b\beta(\delta)$ than to specify a value of $\alpha(\delta)$ and then minimize $\beta(\delta)$.

Choosing a Level of Significance

In many statistical applications, it has become standard practice for an experimenter to specify a level of significance α_0, and then to find a test procedure for which $\beta(\delta)$ is a minimum among all procedures for which $\alpha(\delta) \leqslant \alpha_0$. The Neyman-Pearson lemma explicitly describes how to construct such a procedure. Furthermore, it has become traditional in these applications to choose the level of significance α_0 to be 0.10, 0.05, or 0.01. The selected level depends on how serious the consequences of an error of type 1 are judged to be. The value of α_0 most commonly used is 0.05. If the consequences of an error of type 1 are judged to be relatively mild in a particular problem, the experimenter may choose α_0 to be 0.10. On the other hand, if these consequences are judged to be especially serious, the experimenter may choose α_0 to be 0.01.

Because these values of α_0 have become established in statistical practice, the choice of $\alpha_0 = 0.01$ is sometimes made by an experimenter who wishes to use a cautious test procedure, or one that will not reject H_0 unless the sample data provide strong evidence that H_0 is not true. We shall now show, however, that when the sample size n is large, the choice of $\alpha_0 = 0.01$ can actually lead to a test procedure that will reject H_0 for certain samples which, in fact, provide strong evidence that H_0 is true.

To illustrate this property, we shall again suppose, as in Example 1, that a random sample is taken from a normal distribution for which the mean θ is unknown and the variance is 1; and that the hypotheses (6) are to be tested. It follows from the discussion in Example 1 that, among all test procedures for which $\alpha(\delta) \leqslant 0.01$, the value of $\beta(\delta)$ will be a minimum for the procedure δ^* which rejects H_0 when $\overline{X}_n > k'$, where k' is chosen so that $\Pr(\overline{X}_n > k' \mid \theta = 0) = 0.01$. When $\theta = 0$, the random variable $\overline{X}_n$ has a normal distribution with mean 0 and variance $1/n$. Therefore, it can be found from a table of the standard normal distribution that $k' = 2.326n^{-1/2}$.

Furthermore, it follows from Eq. (9) that this test procedure δ^* is equivalent to rejecting H_0 when $f_1(x)/f_0(x) > k$, where $k = \exp(2.326n^{1/2} - 0.5n)$. The probability of an error of type 1 will be $\alpha(\delta^*) = 0.01$. Also, by an argument similar to the one leading to Eq. (13), the probability of an error of type 2 will be $\beta(\delta^*) = \Phi(2.326 - n^{1/2})$, where Φ denotes the d.f. of the standard normal distribution. For $n = 1$, 25, and 100, the values of $\beta(\delta^*)$ and k are as follows:

n	$\alpha(\delta^*)$	$\beta(\delta^*)$	k
1	0.01	0.91	6.21
25	0.01	0.0038	0.42
100	0.01	8×10^{-15}	2.5×10^{-12}

It can be seen from this tabulation that when $n = 1$, the null hypothesis H_0 will be rejected only if the likelihood ratio $f_1(x)/f_0(x)$ exceeds the value $k = 6.21$. In other words, H_0 will not be rejected unless the observed values $x_1, \ldots, x_n$ in the sample are at least 6.21 times as likely under H_1 as they are under H_0. In this case, the procedure δ^* therefore satisfies the experimenter's desire to use a test that is cautious about rejecting H_0.

If $n = 100$, however, the procedure δ^* will reject H_0 whenever the likelihood ratio exceeds the value $k = 2.5 \times 10^{-12}$. Therefore, H_0 will be rejected for certain observed values $x_1, \ldots, x_n$ that are actually millions of times as likely under H_0 as they are under H_1. The reason for this result is that the value of $\beta(\delta^*)$ that can be achieved when $n = 100$, which is 8×10^{-15}, is extremely small relative to the specified value $\alpha_0 = 0.01$. Hence, the procedure δ^* actually turns out to be much more cautious about an error of type 2 than it is about an error of type 1. We can see from this discussion that a value of α_0 that is an appropriate choice for a small value of n might be unnecessarily large for a large value of n.

Suppose now that the experimenter regards an error of type 1 to be much more serious than an error of type 2; and he therefore desires to use a test procedure for which the value of the linear combination $100\alpha(\delta) + \beta(\delta)$ will be a minimum. Then it follows from Theorem 1 that he should reject H_0 if and only if the likelihood ratio exceeds the value $k = 100$, regardless of the sample size n. In other words, the procedure that minimizes the value of $100\alpha(\delta) + \beta(\delta)$ will not reject H_0 unless the observed values $x_1, \ldots, x_n$ are at least 100 times as likely under H_1 as they are under H_0.

From this discussion it seems more reasonable for the experimenter to minimize the value of a linear combination of the form $a\alpha(\delta) + b\beta(\delta)$, rather than to fix a value of $\alpha(\delta)$ and minimize $\beta(\delta)$. Indeed, the following discussion shows that, from the Bayesian point of view, it is natural to try to minimize a linear combination of this form.

Bayes Test Procedures

Consider a general problem in which $X_1, \ldots, X_n$ form a random sample from a distribution for which the p.d.f. or the p.f. is $f(x \mid \theta)$ and it is desired to test the following simple hypotheses:

$$H_0: \quad \theta = \theta_0,$$
$$H_1: \quad \theta = \theta_1.$$

We shall let d_0 denote the decision to accept the hypothesis H_0, and we shall let d_1 denote the decision to accept the hypothesis H_1. Also, we shall assume that the losses resulting from choosing an incorrect decision are as follows: If decision d_1 is chosen when H_0 is actually the true hypothesis, then the loss is w_0 units; and if decision d_0 is chosen when H_1 is actually the true hypothesis, then the loss is w_1 units. If the decision d_0 is chosen when H_0 is the true hypothesis or if the decision d_1 is chosen when H_1 is the true hypothesis, then the correct decision has been made and the loss is 0. Thus, for $i = 0, 1$ and $j = 0, 1$, the loss $L(\theta_i, d_j)$ that occurs when θ_i is the true value of θ and the decision d_j is chosen is given by the following table:

	d_0	d_1
θ_0	0	w_0
θ_1	w_1	0

Next, suppose that the prior probability that H_0 is true is ξ_0 and the prior probability that H_1 is true is $\xi_1 = 1 - \xi_0$. Then the expected loss $r(\delta)$ of any test procedure δ will be

$$r(\delta) = \xi_0 E(\text{Loss} \mid \theta = \theta_0) + \xi_1 E(\text{Loss} \mid \theta = \theta_1). \tag{27}$$

If $\alpha(\delta)$ and $\beta(\delta)$ again denote the probabilities of the two types of errors for the procedure δ, and if the table of losses just given is used, it follows that

$$E(\text{Loss} \mid \theta = \theta_0) = w_0 \Pr(\text{Choosing } d_1 \mid \theta = \theta_0) = w_0 \alpha(\delta),$$
$$E(\text{Loss} \mid \theta = \theta_1) = w_1 \Pr(\text{Choosing } d_0 \mid \theta = \theta_1) = w_1 \beta(\delta). \tag{28}$$

Hence,

$$r(\delta) = \xi_0 w_0 \alpha(\delta) + \xi_1 w_1 \beta(\delta). \tag{29}$$

A procedure for which this expected loss $r(\delta)$ is minimized is called a *Bayes test procedure*.

Since $r(\delta)$ is simply a linear combination of the form $a\alpha(\delta) + b\beta(\delta)$ with $a = \xi_0 w_0$ and $b = \xi_1 w_1$, a Bayes test procedure can immediately be determined from Theorem 1. Thus, a Bayes procedure will accept H_0 whenever $\xi_0 w_0 f_0(x) > \xi_1 w_1 f_1(x)$ and will accept H_1 whenever $\xi_0 w_0 f_0(x) < \xi_1 w_1 f_1(x)$. Either H_0 or H_1 may be accepted if $\xi_0 w_0 f_0(x) = \xi_1 w_1 f_1(x)$.

EXERCISES

1. Consider two p.d.f.'s $f_0(x)$ and $f_1(x)$ which are defined as follows:

$$f_0(x) = \begin{cases} 1 & \text{for } 0 \leqslant x \leqslant 1, \\ 0 & \text{otherwise,} \end{cases}$$

and

$$f_1(x) = \begin{cases} 2x & \text{for } 0 \leqslant x \leqslant 1, \\ 0 & \text{otherwise.} \end{cases}$$

Suppose that a single observation X is taken from a distribution for which the p.d.f. $f(x)$ is either $f_0(x)$ or $f_1(x)$, and that the following simple hypotheses are to be tested:

$H_0: \ f(x) = f_0(x),$
$H_1: \ f(x) = f_1(x).$

(a) Describe a test procedure for which the value of $\alpha(\delta) + 2\beta(\delta)$ is a minimum.

(b) Determine the minimum value of $\alpha(\delta) + 2\beta(\delta)$ attained by that procedure.

2. Consider again the conditions of Exercise 1, but suppose now that it is desired to find a test procedure for which the value of $3\alpha(\delta) + \beta(\delta)$ is a minimum.

(a) Describe the procedure.

(b) Determine the minimum value of $3\alpha(\delta) + \beta(\delta)$ attained by the procedure.

3. Consider again the conditions of Exercise 1, but suppose now that it is desired to find a test procedure for which $\alpha(\delta) \leq 0.1$ and $\beta(\delta)$ is a minimum.

(a) Describe the procedure.

(b) Determine the minimum value of $\beta(\delta)$ attained by the procedure.

4. Suppose that $X_1, \ldots, X_n$ form a random sample from a normal distribution for which the value of the mean θ is unknown and the variance is 1; and that the following hypotheses are to be tested:

$$H_0: \quad \theta = 3.5,$$
$$H_1: \quad \theta = 5.0.$$

(a) Among all test procedures for which $\beta(\delta) \leq 0.05$, describe a procedure for which $\alpha(\delta)$ is a minimum.

(b) For $n = 4$, find the minimum value of $\alpha(\delta)$ attained by the procedure described in part (a).

5. Suppose that $X_1, \ldots, X_n$ form a random sample from a Bernoulli distribution for which the value of the parameter p is unknown. Let p_0 and p_1 be specified values such that $0 < p_1 < p_0 < 1$, and suppose that it is desired to test the following simple hypotheses:

$$H_0: \quad p = p_0,$$
$$H_1: \quad p = p_1.$$

(a) Show that a test procedure for which $\alpha(\delta) + \beta(\delta)$ is a minimum rejects H_0 when $\overline{X}_n < c$.

(b) Find the value of the constant c.

6. Suppose that $X_1, \ldots, X_n$ form a random sample from a normal distribution for which the value of the mean μ is known and the value of the variance σ^2 is unknown; and that the following simple hypotheses are to be tested:

$$H_0: \quad \sigma^2 = 2,$$
$$H_1: \quad \sigma^2 = 3.$$

(a) Show that among all test procedures for which $\alpha(\delta) \leq 0.05$, the value of $\beta(\delta)$ is minimized by a procedure which rejects H_0 when $\sum_{i=1}^{n}(X_i - \mu)^2 > c$.

(b) For $n = 8$, find the value of the constant c which appears in part (a).

7. Suppose that a single observation X is taken from a uniform distribution on the interval $(0, \theta)$, where the value of θ is unknown; and that the following simple hypotheses are to be tested:

$$H_0: \quad \theta = 1,$$
$$H_1: \quad \theta = 2.$$

(a) Show that there exists a test procedure for which $\alpha(\delta) = 0$ and $\beta(\delta) < 1$.

(b) Among all test procedures for which $\alpha(\delta) = 0$, find the one for which $\beta(\delta)$ is a minimum.

8. Suppose that a random sample $X_1, \ldots, X_n$ is drawn from a uniform distribution on the interval $(0, \theta)$; and consider again the problem of testing the simple hypotheses described in Exercise 7. Find the minimum value of $\beta(\delta)$ that can be attained among all test procedures for which $\alpha(\delta) = 0$.

9. Suppose that $X_1, \ldots, X_n$ form a random sample from a Poisson distribution for which the value of the mean λ is unknown. Let λ_0 and λ_1 be specified values such that $\lambda_1 > \lambda_0 > 0$, and suppose that it is desired to test the following simple hypotheses:

$$H_0: \quad \lambda = \lambda_0,$$
$$H_1: \quad \lambda = \lambda_1.$$

(a) Show that the value of $\alpha(\delta) + \beta(\delta)$ is minimized by a test procedure which rejects H_0 when $\overline{X}_n > c$.

(b) Find the value of c.

* (c) For $\lambda_0 = 1/4$, $\lambda_1 = 1/2$, and $n = 20$, determine the minimum value of $\alpha(\delta) + \beta(\delta)$ that can be attained.

10. Suppose that $X_1, \ldots, X_n$ form a random sample from a normal distribution for which the value of the mean μ is unknown and the standard deviation is 2; and that the following simple hypotheses are to be tested:

$$H_0: \quad \mu = -1,$$
$$H_1: \quad \mu = 1.$$

Determine the minimum value of $\alpha(\delta) + \beta(\delta)$ that can be attained for each of the following values of the sample size n:

(a) $n = 1$, (b) $n = 4$, (c) $n = 16$, (d) $n = 36$.

11. A single observation X is to be taken from a continuous distribution for which the p.d.f. is either f_0 or f_1, where

$$f_0(x) = \begin{cases} 1 & \text{for } 0 < x < 1, \\ 0 & \text{otherwise,} \end{cases}$$

and

$$f_1(x) = \begin{cases} 4x^3 & \text{for } 0 < x < 1, \\ 0 & \text{otherwise.} \end{cases}$$

On the basis of the observation X, it must be decided whether f_0 or f_1 is the correct p.d.f. Suppose that the prior probability that f_0 is correct is $2/3$ and the prior probability that f_1 is correct is $1/3$. Suppose also that the loss from choosing the correct decision is 0; the loss from deciding that f_1 is correct when in fact f_0 is correct is 1 unit; and the loss from deciding that f_0 is correct when in fact f_1 is correct is 4 units. If the expected loss is to be minimized, for what values of X should it be decided that f_0 is correct?

12. Suppose that a certain industrial process can be either in control or out of control; and that at any specified time the prior probability that it will be in control is 0.9 and the prior probability that it will be out of control is 0.1. A single observation X of the output of the process is to be taken, and it must be decided immediately whether the process is in control or out of control. If the process is in control, then X will have a normal distribution with mean 50 and variance 1. If the process is out of control, then X will have a normal distribution with mean 52 and variance 1. If it is decided that the process is out of control when in fact it is in control, then the loss from unnecessarily stopping the process will be $1000. If it is decided that the process is in control when in fact it is out of control, then the loss from continuing the process will be $18,000. If a correct decision is made, then the loss will be 0. It is desired to find a test procedure for which the expected loss will be a minimum. For what values of X should it be decided that the process is out of control?

13. Suppose that the proportion p of defective items in a large manufactured lot is unknown, and it is desired to test the following simple hypotheses:

$$H_0: \quad p = 0.3,$$
$$H_1: \quad p = 0.4.$$

Suppose that the prior probability that $p = 0.3$ is $1/4$ and the prior probability that $p = 0.4$ is $3/4$; and also that the loss from choosing an incorrect decision is 1 unit and the loss from choosing a correct decision is 0. Suppose that a random sample of n items is selected from the lot. Show that the Bayes test procedure is to reject H_0 if and only if the proportion of defective items in the sample is greater than

$$\frac{\log\left(\frac{7}{6}\right) + \frac{1}{n}\log\left(\frac{1}{3}\right)}{\log\left(\frac{14}{9}\right)}.$$

14. Suppose that a failure in a certain electronic system can occur because of either a minor defect or a major defect. Suppose also that 80 percent of the failures are caused by minor defects and 20 percent of the failures are caused by major defects. When a failure occurs, n independent soundings $X_1, \ldots, X_n$ are made on the system. If the failure was caused by a minor defect, these

soundings form a random sample from a Poisson distribution for which the mean is 3. If the failure was caused by a major defect, these soundings form a random sample from a Poisson distribution for which the mean is 7. The cost of deciding that the failure was caused by a major defect when it was actually caused by a minor defect is \$400. The cost of deciding that the failure was caused by a minor defect when it was actually caused by a major defect is \$2500. The cost of choosing a correct decision is 0. For a given set of observed values of $X_1, \ldots, X_n$, which decision minimizes the expected cost?

*8.3. MULTIDECISION PROBLEMS

Finite Number of Parameter Values and Finite Number of Decisions

In a problem of testing a pair of simple hypotheses, there are only two possible values of the parameter θ and there are only two possible decisions for the experimenter, namely, either to accept H_0 or to accept H_1. This problem therefore belongs to the class of problems in which there are a finite number of possible values of the parameter θ and a finite number of possible decisions. Such problems are called *multidecision problems*.

For a general multidecision problem we shall let $\theta_1, \ldots, \theta_k$ denote the k possible values of θ, and we shall let $d_1, \ldots, d_m$ denote the m possible decisions that can be chosen. Furthermore, for $i = 1, \ldots, k$ and $j = 1, \ldots, m$, we shall let w_{ij} denote the loss incurred by the experimenter when $\theta = \theta_i$ and decision d_j is chosen. Finally, for $i = 1, \ldots, k$, we shall let ξ_i denote the prior probability that $\theta = \theta_i$. Thus $\xi_i \geq 0$ and $\xi_1 + \cdots + \xi_k = 1$.

If the experimenter must choose one of the decisions $d_1, \ldots, d_m$ without being able to observe any relevant sample data, then the expected loss or *risk* ρ_j from choosing decision d_j will be $\rho_j = \sum_{i=1}^{k} \xi_i w_{ij}$. A decision for which the risk is a minimum is called a *Bayes decision*.

Example 1: Finding a Bayes Decision. Consider a multidecision problem in which $k = 3$ and $m = 4$, and the losses w_{ij} are given by the following table:

	d_1	d_2	d_3	d_4
θ_1	1	2	3	4
θ_2	3	0	1	2
θ_3	4	2	1	0

It follows from this table that the risks of the four possible decisions are as follows:

$$
\begin{aligned}
\rho_1 &= \xi_1 + 3\xi_2 + 4\xi_3, \\
\rho_2 &= 2\xi_1 + 2\xi_3, \\
\rho_3 &= 3\xi_1 + \xi_2 + \xi_3, \\
\rho_4 &= 4\xi_1 + 2\xi_2.
\end{aligned}
\tag{1}
$$

For any given prior probabilities ξ_1, ξ_2, and ξ_3, a Bayes decision is found simply by determining the decision for which the risk is smallest. As an illustration, if $\xi_1 = 0.5$, $\xi_2 = 0.2$, and $\xi_3 = 0.3$, then $\rho_1 = 2.3$, $\rho_2 = 1.6$, $\rho_3 = 2.0$, and $\rho_4 = 2.4$. Therefore, d_2 is the unique Bayes decision. If $\theta = \theta_1$, it can be seen from the first row of the table of losses that d_1 has the smallest loss among the four decisions. Therefore, if the prior probability ξ_1 is sufficiently close to 1, then d_1 will be the Bayes decision. Similarly, if $\theta = \theta_2$, then d_2 has the smallest loss among the four decisions. Therefore, if ξ_2 is sufficiently close to 1, then d_2 will be the Bayes decision. Finally, if $\theta = \theta_3$, then d_4 has the smallest loss among the four decisions. Therefore, if ξ_3 is sufficiently close to 1, then d_4 will be the Bayes decision. We shall now determine whether there are any prior probabilities ξ_1, ξ_2, and ξ_3 for which d_3 will be a Bayes decision.

The following results can be obtained from the relations (1): $\rho_2 < \rho_3$ if and only if $\xi_1 + \xi_2 > \xi_3$; and $\rho_4 < \rho_3$ if and only if $\xi_1 + \xi_2 < \xi_3$. Therefore, the only condition under which d_3 could be a Bayes decision is when $\xi_1 + \xi_2 = \xi_3$. But, if $\xi_1 + \xi_2 = \xi_3$, then it follows that $\xi_1 + \xi_2 = 1/2$ and $\xi_3 = 1/2$, and it can be verified from the relations (1) that $\rho_2 = \rho_3 = \rho_4 = 1 + 2\xi_1$ and $\rho_1 = (5/2) + 2\xi_2 > 1 + 2\xi_1$. We can conclude from this discussion that d_3 is a Bayes decision only if $\xi_1 + \xi_2 = 1/2$ and $\xi_3 = 1/2$. However, in this case d_2 and d_4 are also Bayes decisions, and any one of these three decisions could be chosen. $\square$

Bayes Decision Procedures

Consider a general multidecision problem in which the conditions are as follows: There are k possible values of the parameter θ; there are m possible decisions; the loss that results from choosing decision d_j when $\theta = \theta_i$ is w_{ij} for $i = 1, \ldots, k$ and $j = 1, \ldots, m$; and the prior probability that $\theta = \theta_i$ is ξ_i for $i = 1, \ldots, k$. Suppose now, however, that before the experimenter chooses a decision d_j, he can observe the values in a random sample $X_1, \ldots, X_n$ drawn from some distribution which depends on the parameter θ.

For $i = 1, \ldots, k$, we shall let $f_n(x \mid \theta_i)$ denote the joint p.d.f. or the joint p.f. of the observations $X_1, \ldots, X_n$ when $\theta = \theta_i$. After the vector x of values in the sample has been observed, the posterior probability $\xi_i(x)$ that $\theta = \theta_i$ will be

$$\xi_i(x) = \Pr(\theta = \theta_i \mid x) = \frac{\xi_i f_n(x \mid \theta_i)}{\sum_{t=1}^{k} \xi_t f_n(x \mid \theta_t)} \qquad \text{for } i = 1, \ldots, k. \tag{2}$$

Thus, after the vector x of values in the sample has been observed, the risk $\rho_j(x)$ from choosing decision d_j will be

$$\rho_j(x) = \sum_{i=1}^{k} \xi_i(x) w_{ij} \qquad \text{for } j = 1, \ldots, m. \tag{3}$$

It follows that after x has been observed, a Bayes decision will be a decision for which the risk in Eq. (3) is a minimum. Such a decision is called a *Bayes decision with respect to the posterior distribution of θ.*

In a multidecision problem of this type, a *decision procedure* is defined to be a function δ which specifies, for each possible vector x, the decision $\delta(x)$ that is to be chosen if the observed vector is x. Thus, for each vector x, $\delta(x)$ must be one of the m possible decisions $d_1, \ldots, d_m$.

A decision procedure δ is called a *Bayes decision procedure* if, for each possible vector x, the decision $\delta(x)$ is a Bayes decision with respect to the posterior distribution of θ. In other words, when a Bayes decision procedure is used, the decision that is chosen after the vector x has been observed is always a decision for which the risk $\rho_j(x)$ is a minimum.

Before any observations have been taken, the risk that the experimenter faces by using a specific decision procedure δ can be calculated as follows: For $j = 1, \ldots, m$, let A_j denote the set of all outcomes x for which $\delta(x) = d_j$, that is, for which decision d_j will be chosen. For convenience, we shall assume that the observations $X_1, \ldots, X_n$ have a discrete distribution and that $f_n(x \mid \theta_i)$ represents their joint p.f. when $\theta = \theta_i$. If $f_n(x \mid \theta_i)$ is actually a joint p.d.f., then the summations over values of x that appear in the development to be given here should be replaced by integrals.

If $\theta = \theta_i$, the risk $\rho(\delta \mid \theta = \theta_i)$ from using the procedure δ is

$$\rho(\delta \mid \theta = \theta_i) = \sum_{j=1}^{m} w_{ij} \Pr\left[\delta(x) = d_j \mid \theta = \theta_i\right]$$

$$= \sum_{j=1}^{m} w_{ij} \sum_{x \in A_j} f_n(x \mid \theta_i). \tag{4}$$

Since the prior probability that $\theta = \theta_i$ is ξ_i, the overall risk $\rho(\delta)$ from using the procedure δ will be

$$p(\delta) = \sum_{i=1}^{k} \xi_i \rho(\delta \,|\, \theta = \theta_i) = \sum_{i=1}^{k} \sum_{j=1}^{m} \sum_{x \in A_j} \xi_i w_{ij} f_n(x \,|\, \theta_i). \tag{5}$$

This risk $\rho(\delta)$ will be minimized when δ is a Bayes decision procedure.

Example 2: Determining a Bayes Decision Procedure. Suppose that in a large shipment of fruit, the only three possible values for the proportion θ of bruised pieces are 0.1, 0.3, and 0.5, and that there are three possible decisions d_1, d_2, and d_3. Suppose also that the losses from these decisions are as follows:

	d_1	d_2	d_3
$\theta = 0.1$	0	1	3
$\theta = 0.3$	2	0	2
$\theta = 0.5$	3	1	0

Furthermore, suppose that on the basis of previous shipments from the same supplier, it is believed that the prior probabilities of the three possible values of θ are as follows:

$$\begin{aligned}
\Pr(\theta = 0.1) &= 0.5, \\
\Pr(\theta = 0.3) &= 0.3, \\
\Pr(\theta = 0.5) &= 0.2.
\end{aligned} \tag{6}$$

Finally, suppose that we can observe the number Y of bruised pieces of fruit in a random sample of 20 pieces selected from the shipment. We shall determine a Bayes decision procedure, and shall calculate the risk from that procedure.

When $\theta = 0.1$, the distribution of Y is a binomial distribution with parameters 20 and 0.1. The p.f. $g(y \,|\, \theta = 0.1)$ is as follows:

$$g(y \,|\, \theta = 0.1) = \binom{20}{y}(0.1)^y (0.9)^{20-y} \qquad \text{for } y = 0, 1, \ldots, 20. \tag{7}$$

When $\theta = 0.3$ or $\theta = 0.5$, the distribution of Y is a similar binomial distribution and expressions for $g(y \,|\, \theta = 0.3)$ and $g(y \,|\, \theta = 0.5)$ will have a form similar to Eq. (7). The values of these p.f.'s for specific values of y can be obtained from the table of the binomial distribution given at the end of this book.

It follows from Eq. (2) that after the value $Y = y$ has been observed, the posterior probability that $\theta = 0.1$ will be

$$\Pr(\theta = 0.1 \mid Y = y)$$

$$= \frac{(0.5)g(y \mid \theta = 0.1)}{(0.5)g(y \mid \theta = 0.1) + (0.3)g(y \mid \theta = 0.3) + (0.2)g(y \mid \theta = 0.5)} .$$

$$(8)$$

Similar expressions can be written for the posterior probabilities that $\theta = 0.3$ and that $\theta = 0.5$. These posterior probabilities, for each possible value of y, are given in Table 8.1.

After the value of y has been observed, the risk $\rho_j(y)$ from each possible decision d_j $(j = 1, 2, 3)$ can be calculated by applying Eq. (3) and using these posterior probabilities and the table of losses given at the beginning of this example. Thus, the risk $\rho_1(y)$ from choosing decision d_1 will be

$$\rho_1(y) = 2 \Pr(\theta = 0.3 \mid Y = y) + 3 \Pr(\theta = 0.5 \mid Y = y),$$

Table 8.1

y	$\Pr(\theta = 0.1 \mid Y = y)$	$\Pr(\theta = 0.3 \mid Y = y)$	$\Pr(\theta = 0.5 \mid Y = y)$
0	0.9961	0.0039	0.0000
1	0.9850	0.0150	0.0000
2	0.9444	0.0553	0.0002
3	0.8141	0.1840	0.0019
4	0.5285	0.4606	0.0109
5	0.2199	0.7393	0.0408
6	0.0640	0.8294	0.1066
7	0.0151	0.7575	0.2273
8	0.0031	0.5864	0.4105
9	0.0005	0.3795	0.6200
10	0.0001	0.2078	0.7921
11	0.0000	0.1011	0.8989
12	0.000	0.046	0.954
13	0.000	0.020	0.980
14	0.000	0.009	0.991
15	0.000	0.004	0.996
16	0.000	0.000	1.000
17	0.000	0.000	1.000
18	0.000	0.000	1.000
19	0.000	0.000	1.000
20	0.000	0.000	1.000

the risk from choosing d_2 will be

$$\rho_2(y) = \Pr(\theta = 0.1 \mid Y = y) + \Pr(\theta = 0.5 \mid Y = y),$$

and the risk from choosing d_3 will be

$$\rho_3(y) = 3 \Pr(\theta = 0.1 \mid Y = y) + 2 \Pr(\theta = 0.3 \mid Y = y).$$

The values of these risks for each possible value of y and each possible decision are given in Table 8.2.

The following conclusions can be drawn from the tabulated values: If $y \leqslant 3$, then the Bayes decision is d_1; if $4 \leqslant y \leqslant 9$, then the Bayes decision is d_2; and if $y \geqslant 10$, then the Bayes decision is d_3. In other words, the Bayes decision

Table 8.2

y	$\rho_1(y)$	$\rho_2(y)$	$\rho_3(y)$
0	0.0078	0.9961	2.9961
1	0.0300	0.9850	2.9850
2	0.1124	0.9446	2.9430
3	0.3737	0.8160	2.8103
4	0.9539	0.5394	2.5067
5	1.6010	0.2607	2.1383
6	1.9786	0.1706	1.8508
7	2.1969	0.2428	1.5603
8	2.4043	0.4136	1.1821
9	2.6190	0.6205	0.7605
10	2.7919	0.7922	0.4159
11	2.8989	0.8989	0.2022
12	2.954	0.954	0.092
13	2.980	0.980	0.040
14	2.991	0.991	0.018
15	2.996	0.996	0.008
16	3.00	1.00	0.00
17	3.00	1.00	0.00
18	3.00	1.00	0.00
19	3.00	1.00	0.00
20	3.00	1.00	0.00

procedure δ is defined as follows:

$$\delta(y) = \begin{cases} d_1 & \text{if } y = 0, 1, 2, 3, \\ d_2 & \text{if } y = 4, 5, 6, 7, 8, 9, \\ d_3 & \text{if } y = 10, 11, \dots, 20. \end{cases} \tag{9}$$

It now follows from Eq. (5) that the risk $\rho(\delta)$ from using the procedure δ is

$$\rho(\delta) = (0.5) \sum_{y=4}^{9} g(y \mid \theta = 0.1) + (1.5) \sum_{y=10}^{20} g(y \mid \theta = 0.1)$$

$$+ (0.6) \sum_{y=0}^{3} g(y \mid \theta = 0.3) + (0.6) \sum_{y=10}^{20} g(y \mid \theta = 0.3) \tag{10}$$

$$+ (0.6) \sum_{y=0}^{3} g(y \mid \theta = 0.5) + (0.2) \sum_{y=4}^{9} g(y \mid \theta = 0.5)$$

$$= 0.2423.$$

Thus, the risk of the Bayes decision procedure is $\rho(\delta) = 0.2423$. □

Example 3: The Value of Sample Information. Suppose now that it was necessary to choose one of the three decisions d_1, d_2, or d_3 in Example 2 without being able to observe the number of bruised pieces in a random sample. In this case, it is found from the table of losses w_{ij} and the prior probabilities (6) that the risks ρ_1, ρ_2, and ρ_3 from choosing each of the decisions d_1, d_2, and d_3 are as follows:

$$\begin{aligned} \rho_1 &= 2(0.3) + 3(0.2) = 1.2, \\ \rho_2 &= (0.5) + (0.2) = 0.7, \\ \rho_3 &= 3(0.5) + 2(0.3) = 2.1. \end{aligned} \tag{11}$$

Hence, the Bayes decision without any observations would be d_2 and the risk from that decision would be 0.7.

By being able to observe the number of bruised pieces in a random sample of 20 pieces, we can reduce the risk from 0.7 to 0.2423. □

EXERCISES

1. Suppose that a malfunction which causes a certain system to break down and become inoperative can occur in either of two different parts of the system, part A or part B. Suppose also that when the system does become inoper-

ative, it is not known immediately whether the malfunction causing the breakdown has occurred in part A or in part B. It is assumed that the repair procedures are quite different for the two different parts. Therefore, when a breakdown occurs in the system, one of the following three decisions must be chosen:

Decision d_1: The repair procedure for a breakdown in part A is activated immediately. If the malfunction causing the breakdown actually occurred in part B, then the cost of this decision in terms of unnecessary labor and lost time is \$1000. If the malfunction actually occurred in part A, then this decision leads to the repair of the malfunction in the most efficient manner and the cost is regarded as zero.

Decision d_2: The repair procedure for a breakdown in part B is activated immediately. If the malfunction actually occurred in part A, then the cost of this decision is \$3000. If the malfunction occurred in part B, then the cost is again regarded as zero.

Decision d_3: A test is applied to the system that will determine with certainty whether the malfunction has occurred in part A or in part B. The cost of applying this test is \$300.

(a) If 75 percent of all malfunctions occur in part A and only 25 percent occur in part B, what is the Bayes decision when the system breaks down?

(b) Suppose that the breakdown in the system is always caused by a defect in one of 36 similar components, all of which are equally likely to be defective. If 4 of these components are used in part A and the other 32 components are used in part B, what is the Bayes decision when the system breaks down?

2. Consider a multidecision problem in which θ can take only two values, there are four possible decisions, and the losses are as given in the following table:

	d_1	d_2	d_3	d_4
θ_1	0	10	1	6
θ_2	10	0	8	6

For $i = 1$ or 2, let ξ_i denote the prior probability that $\theta = \theta_i$.

(a) Show that d_4 is never a Bayes decision for any values of ξ_1 and ξ_2.

(b) For what values of ξ_1 and ξ_2 is the Bayes decision not unique?

3. Suppose that an unmanned rocket is being launched, and that at the time of the launching a certain electronic component is either functioning or not functioning. In the control center there is a warning light that is not completely reliable. If the electronic component is not functioning, the

warning light goes on with probability $1/2$; if the component is functioning, the warning light goes on with probability $1/3$. At the time of launching, an observer notes whether the warning light is on or off. It must then be decided immediately whether or not to launch the rocket. Suppose that the losses, in millions of dollars, are as follows:

	Launch rocket	Do not launch rocket
Component functioning	0	2
Component not functioning	5	0

(a) Suppose that the prior probability that the component is not functioning is $\xi = 2/5$. If the warning light does not go on, is the Bayes decision to launch the rocket or not to launch it?

(b) For what values of the prior probability ξ is the Bayes decision to launch the rocket, even if the warning light goes on?

4. Suppose that 10 percent of all the workers in a particular type of factory have a certain lung disease. Suppose also that a diagnostic test is available to help determine whether a particular worker has the disease, and that the outcome of the test is a random variable X with the following distribution: If the worker has the disease, then X has a normal distribution with mean 50 and variance 1. If the worker does not have the disease, then X has a normal distribution with mean 52 and variance 1. As a result of the outcome X, a worker may be required to undergo a complete medical examination. Suppose that the loss from requiring such an examination when the worker does not have the disease is \$100; that the loss from not requiring an examination when the worker has the disease is \$2000; and that otherwise the loss is 0. If the test is administered to a worker selected at random from a factory of this type, for what values of X is the Bayes decision to require a complete medical examination?

5. On any given day, a production system may operate at a low level w_1, at a medium level w_2, or at a high level w_3. The output of the system during the first hour of a given day is measured as a number X between 0 and 2; and on the basis of the observed value of X it must be decided at which of the three levels, w_1, w_2, or w_3, the system is operating on that day. When the system is operating at a low level, the p.d.f. of X is

$$f(x \mid w_1) = \frac{1}{2} \qquad \text{for } 0 \leqslant x \leqslant 2.$$

When the system is operating at a medium level, the p.d.f. is

$$f(x \mid w_2) = \frac{1}{2}x \qquad \text{for } 0 \leqslant x \leqslant 2.$$

When the system is operating at a high level, the p.d.f. is

$$f(x \mid w_3) = \frac{3}{8}x^2 \qquad \text{for } 0 \leqslant x \leqslant 2.$$

Suppose that it is known that the system operates at a low level on 10 percent of the days, at a medium level on 70 percent of the days, and at a high level on 20 percent of the days. Suppose finally that the loss from an incorrect decision is 1 unit, and that the loss from a correct decision is 0.

(a) Determine a Bayes decision procedure as a function of X, and calculate its risk.

(b) Compare this risk with the minimum risk that could be attained if a decision had to be made without the observation X being available.

6. Suppose that it is known that the probability p of a head when a certain coin is tossed is either 0.3 or 0.4; and that an experimenter must decide which value of p is the correct one after observing the outcome, head or tail, of just a single toss of the coin. Suppose also that the prior probabilities are as follows:

$$\Pr(p = 0.3) = 0.8 \qquad \text{and} \qquad \Pr(p = 0.4) = 0.2.$$

Finally, suppose that the loss from an incorrect decision is 1 unit, and the loss from a correct decision is 0. Show that observing the outcome of a single toss is of no value in this problem because the risk of the Bayes decision procedure based on the observation is just as large as the risk of making a Bayes decision without the observation.

7. Suppose that the variables $X_1, \ldots, X_n$ form a random sample from a normal distribution with mean θ and variance 1. Suppose that it is known that either $\theta = 0$ or $\theta = 1$, and that the prior probabilities are $\Pr(\theta = 0) = \Pr(\theta = 1) = 1/2$. Suppose also that one of three decisions d_1, d_2, and d_3 must be chosen, and that the losses from these decisions are as follows:

	d_1	d_2	d_3
$\theta = 0$	0	1	5
$\theta = 1$	5	1	0

(a) Show that a Bayes decision procedure has the following form: Choose decision d_1 if $\bar{X}_n \leqslant c_1$; choose decision d_2 if $c_1 < \bar{X}_n < c_2$; and choose decision d_3 if $\bar{X}_n \geqslant c_2$. Find the values of c_1 and c_2.

(b) Determine the risk of the Bayes decision procedure when the sample size is $n = 4$.

8.4. UNIFORMLY MOST POWERFUL TESTS

Definition of a Uniformly Most Powerful Test

Consider again a general problem of testing hypotheses and suppose that the random variables $X_1, \ldots, X_n$ form a random sample from a distribution for which either the p.d.f. or the p.f. is $f(x \mid \theta)$, where the value of the parameter θ is unknown but must lie in a specified parameter space Ω. In this section, we shall assume that the unknown value of θ is a real number and that the parameter space Ω is therefore a subset of the real line. As usual, we shall suppose that Ω_0 and Ω_1 are disjoint subsets of Ω, and that the hypotheses to be tested are

$$
\begin{aligned}
H_0: \quad & \theta \in \Omega_0, \\
H_1: \quad & \theta \in \Omega_1.
\end{aligned} \tag{1}
$$

We shall assume that the subset Ω_1 contains at least two distinct values of θ, in which case the alternative hypothesis H_1 is composite. The null hypothesis H_0 may be either simple or composite.

We shall also suppose that it is desired to test the hypotheses (1) at a specified level of significance α_0, where α_0 is a given number in the interval $0 < \alpha_0 < 1$. In other words, we shall consider only procedures in which $\Pr(\text{Rejecting } H_0 \mid \theta) \leqslant \alpha_0$ for every value of $\theta \in \Omega_0$. If $\pi(\theta \mid \delta)$ denotes the power function of a given test procedure δ, this requirement can be written simply as

$$
\pi(\theta \mid \delta) \leqslant \alpha_0 \qquad \text{for } \theta \in \Omega_0. \tag{2}
$$

Equivalently, if $\alpha(\delta)$ denotes the size of a test procedure δ, as defined by Eq. (2) of Sec. 8.1, then the requirement (2) can also be expressed by the relation

$$
\alpha(\delta) \leqslant \alpha_0. \tag{3}
$$

We must find a test procedure which satisfies the requirement (3) and for which the value of $\pi(\theta \mid \delta)$ is as large as possible for every value of $\theta \in \Omega_1$.

It may not be possible to satisfy this criterion. If θ_1 and θ_2 are two different values of θ in Ω_1, then the test procedure for which the value of $\pi(\theta_1 \mid \delta)$ is a maximum might be different from the test procedure for which the value of $\pi(\theta_2 \,|\, \delta)$ is a maximum. In other words, there might be no single test procedure δ that maximizes the power function $\pi(\theta \,|\, \delta)$ simultaneously for every value of θ in Ω_1. In some problems, however, there will exist a test procedure that satisfies this criterion. Such a procedure, when it exists, is called a *uniformly most powerful* test or, more briefly, a *UMP test*. The formal definition of a UMP test is as follows:

A test procedure δ^* is a UMP test of the hypotheses (1) at the level of significance α_0 if $\alpha(\delta^*) \leq \alpha_0$ and, for any other test procedure δ such that $\alpha(\delta) \leq \alpha_0$, it is true that

$$\pi(\theta \,|\, \delta) \leq \pi(\theta \,|\, \delta^*) \qquad \text{for every value of } \theta \in \Omega_1. \tag{4}$$

In this section we shall show that a UMP test exists in many problems in which the random sample comes from one of the standard families of distributions that we have been considering in this book.

Monotone Likelihood Ratio

As usual, we shall let $f_n(x \,|\, \theta)$ denote the joint p.d.f. or the joint p.f. of the observations $X_1, \ldots, X_n$. Consider now a statistic $T = r(X)$, which is some particular function of the vector $X = (X_1, \ldots, X_n)$. It is said that $f_n(x \,|\, \theta)$ has a *monotone likelihood ratio in the statistic T* if the following property is satisfied: For any two values $\theta_1 \in \Omega$ and $\theta_2 \in \Omega$, with $\theta_1 < \theta_2$, the ratio $f_n(x \,|\, \theta_2)/f_n(x \,|\, \theta_1)$ depends on the vector x only through the function $r(x)$, and this ratio is an increasing function of $r(x)$ over the range of possible values of $r(x)$.

Example 1: Sampling from a Bernoulli Distribution. Suppose that $X_1, \ldots, X_n$ form a random sample from a Bernoulli distribution for which the parameter p is unknown ($0 < p < 1$). If we let $y = \sum_{i=1}^{n} x_i$, then the joint p.f. $f_n(x \,|\, p)$ is as follows:

$$f_n(x \,|\, p) = p^y (1 - p)^{n-y}. \tag{5}$$

Therefore, for any two values p_1 and p_2 such that $0 < p_1 < p_2 < 1$,

$$\frac{f_n(x \,|\, p_2)}{f_n(x \,|\, p_1)} = \left[\frac{p_2(1 - p_1)}{p_1(1 - p_2)} \right]^y \left(\frac{1 - p_2}{1 - p_1} \right)^n. \tag{6}$$

It can be seen from Eq. (6) that the ratio $f_n(x \,|\, p_2)/f_n(x \,|\, p_1)$ depends on the vector x only through the value of y, and that this ratio is an increasing function of y. Therefore, $f_n(x \,|\, p)$ has a monotone likelihood ratio in the statistic $Y = \sum_{i=1}^{n} X_i$. $\square$

Example 2: Sampling from a Normal Distribution. Suppose that $X_1, \ldots, X_n$ form a random sample from a normal distribution for which the value of the mean μ is unknown $(-\infty < \mu < \infty)$ and the value of the variance σ^2 is known. The joint p.d.f. $f_n(x \mid \mu)$ is as follows:

$$f_n(x \mid \mu) = \frac{1}{(2\pi)^{n/2} \sigma^n} \exp\left[-\frac{1}{2\sigma^2} \sum_{i=1}^{n} (x_i - \mu)^2 \right]. \tag{7}$$

Therefore, for any two values μ_1 and μ_2 such that $\mu_1 < \mu_2$,

$$\frac{f_n(x \mid \mu_2)}{f_n(x \mid \mu_1)} = \exp\left\{ \frac{n(\mu_2 - \mu_1)}{\sigma^2} \left[\bar{x}_n - \frac{1}{2}(\mu_2 + \mu_1) \right] \right\}. \tag{8}$$

It can be seen from Eq. (8) that the ratio $f_n(x \mid \mu_2)/f_n(x \mid \mu_1)$ depends on the vector x only through the value of $\bar{x}_n$, and that this ratio is an increasing function of $\bar{x}_n$. Therefore, $f_n(x \mid \mu)$ has a monotone likelihood ratio in the statistic $\bar{X}_n$. $\quad\square$

One-Sided Alternatives

Suppose that θ_0 is a particular value in the parameter space Ω, and consider the following hypotheses:

$$\begin{aligned} H_0: & \quad \theta \leqslant \theta_0, \\ H_1: & \quad \theta > \theta_0. \end{aligned} \tag{9}$$

We shall now show that if the joint p.d.f. or the joint p.f. $f_n(x \mid \theta)$ of the observations in a random sample has a monotone likelihood ratio in some statistic T, then there will exist UMP tests of the hypotheses (9). Furthermore (see Exercise 11), there will exist UMP tests of the hypotheses obtained by reversing the inequalities in both H_0 and H_1 in (9).

Theorem 1.   *Suppose that $f_n(x \mid \theta)$ has a monotone likelihood ratio in the statistic $T = r(X)$, and let c be a constant such that*

$$\Pr(T \geqslant c \mid \theta = \theta_0) = \alpha_0. \tag{10}$$

Then the test procedure which rejects H_0 if $T \geqslant c$ is a UMP test of the hypotheses (9) at the level of significance α_0.

Proof. Let θ_1 be a specific value of θ such that $\theta_1 > \theta_0$. Also, for any test procedure δ, let

$$\alpha(\delta) = \Pr(\text{Rejecting } H_0 \mid \theta = \theta_0) = \pi(\theta_0 \mid \delta)$$

and let

$$\beta(\delta) = \Pr(\text{Accepting } H_0 \mid \theta = \theta_1) = 1 - \pi(\theta_1 \mid \delta).$$

It follows from the Neyman-Pearson lemma that among all procedures for which $\alpha(\delta) \leqslant \alpha_0$, the value of $\beta(\delta)$ will be minimized by a procedure which rejects H_0 when $f_n(x \mid \theta_1)/f_n(x \mid \theta_0) \geqslant k$. The constant k is to be chosen so that

$$\Pr(\text{Rejecting } H_0 \mid \theta = \theta_0) = \alpha_0.$$

However, it follows from the assumptions presented in the theorem that the ratio $f_n(x \mid \theta_1)/f_n(x \mid \theta_0)$ is an increasing function of $r(x)$. Therefore, a procedure which rejects H_0 when this ratio is at least equal to k will be equivalent to a procedure which rejects H_0 when $r(x)$ is at least equal to some other number c. The value of c is to be chosen so that $\Pr(\text{Rejecting } H_0 \mid \theta = \theta_0) = \alpha_0$ or, in other words, so that Eq. (10) is satisfied.

In summary, we have established the following result: If the constant c is chosen so that Eq. (10) is satisfied, then for any value of $\theta_1 > \theta_0$, the procedure δ^* which rejects H_0 when $T \geqslant c$ will minimize the value of $\beta(\delta) = 1 - \pi(\theta_1 \mid \delta)$ among all procedures for which $\alpha(\delta) = \pi(\theta_0 \mid \delta) \leqslant \alpha_0$. In other words, among all procedures for which $\pi(\theta_0 \mid \delta) \leqslant \alpha_0$, the procedure δ^* maximizes the value of $\pi(\theta \mid \delta)$ at every value of $\theta > \theta_0$. We may, therefore, state that δ^* is a UMP test among all procedures for which $\pi(\theta_0 \mid \delta) \leqslant \alpha_0$. However, in order to complete the proof of the theorem we must establish the result that δ^* is a UMP test, not only in the class $\mathscr{C}$ of all procedures for which $\pi(\theta_0 \mid \delta) \leqslant \alpha_0$ but in the class $\mathscr{C}'$ of all procedures for which $\pi(\theta \mid \delta) \leqslant \alpha_0$ for every value of $\theta \leqslant \theta_0$.

To establish this result, it should be noted first that $\mathscr{C}' \subset \mathscr{C}$. This relation is true because if δ is a procedure for which $\pi(\theta \mid \delta) \leqslant \alpha_0$ for every value of $\theta \leqslant \theta_0$, then, in particular, $\pi(\theta_0 \mid \delta) \leqslant \alpha_0$. Therefore, if a procedure which belongs to the smaller class $\mathscr{C}'$ is actually a UMP test among all procedures in the larger class $\mathscr{C}$, then the procedure certainly will also be a UMP test among all procedures in $\mathscr{C}'$. It follows that to complete the proof of the theorem, we need only show that the procedure δ^* actually belongs to the class $\mathscr{C}'$.

Let θ' and θ'' be any two values of θ such that $\theta' < \theta''$. We shall show that $\pi(\theta' \mid \delta^*) \leqslant \pi(\theta'' \mid \delta^*)$. Let $\alpha = \pi(\theta' \mid \delta^*)$, and consider the class of all procedures, either randomized or nonrandomized, for which $\pi(\theta' \mid \delta) = \alpha$. It follows from the discussion in the first part of this proof that among all such procedures, the value of $\pi(\theta'' \mid \delta)$ will be maximized by the procedure δ^*. In particular, if δ_0 is a procedure for which $\pi(\theta \mid \delta_0) = \alpha$ for every value of $\theta \in \Omega$ (see Exercise 6 at the end of this section), then $\pi(\theta'' \mid \delta^*) \geqslant \pi(\theta'' \mid \delta_0)$. But

$$\pi(\theta'' \mid \delta_0) = \pi(\theta' \mid \delta_0) = \alpha = \pi(\theta' \mid \delta^*).$$

Therefore, $\pi(\theta' \mid \delta^*) \leqslant \pi(\theta'' \mid \delta^*)$.

It follows from this derivation that the power function $\pi(\theta \mid \delta^*)$ of the procedure δ^* is a nondecreasing function of θ. Therefore, $\pi(\theta \mid \delta^*) \leqslant \pi(\theta_0 \mid \delta^*)$ if $\theta \leqslant \theta_0$. Since $\pi(\theta_0 \mid \delta^*) \leqslant \alpha_0$, we may conclude that $\pi(\theta \mid \delta^*) \leqslant \alpha_0$ for every value of $\theta \leqslant \theta_0$. This statement means that δ^* belongs to the class $\mathscr{C}'$. Thus, we have now shown that δ^* is a UMP test among all procedures in the class $\mathscr{C}'$. $\square$

Example 3: Testing Hypotheses About the Proportion of Defective Items. Suppose that the proportion p of defective items in a large manufactured lot is unknown; that 20 items are to be selected at random from the lot and inspected; and that the following hypotheses are to be tested:

$$\begin{aligned} H_0&: \quad p \leqslant 0.1, \\ H_1&: \quad p > 0.1. \end{aligned} \tag{11}$$

We shall show first that there exist UMP tests of the hypotheses (11). We shall then determine the form of these tests, and shall discuss the different levels of significance that can be attained with nonrandomized tests.

Let $X_1, \ldots, X_{20}$ denote the 20 observations in the sample. Then $X_1, \ldots, X_{20}$ form a random sample of size 20 from a Bernoulli distribution with parameter p; and it is known from Example 1 that the joint p.f. of $X_1, \ldots, X_{20}$ has a monotone likelihood ratio in the statistic $Y = \sum_{i=1}^{20} X_i$. Therefore, by Theorem 1, a test procedure which rejects H_0 when $Y \geqslant c$ will be a UMP test of the hypotheses (11).

For any specified choice of the constant c, the level of significance α_0 of the UMP test will be $\alpha_0 = \Pr(Y \geqslant c \mid p = 0.1)$. When $p = 0.1$, the random variable Y has a binomial distribution with parameters $n = 20$ and $p = 0.1$. Since Y has a discrete distribution and can have only a finite number of different possible values, it follows that there are only a finite number of different possible values for α_0. To illustrate this remark, it is found from a table of the binomial distribution that if $c = 7$, then $\alpha_0 = \Pr(Y \geqslant 7 \mid p = 0.1) = 0.0024$; and if $c = 6$, then $\alpha_0 = \Pr(Y \geqslant 6 \mid p = 0.1) = 0.0113$. Therefore, if an experimenter wishes to use a level of significance that is approximately 0.01, he could choose either $c = 7$ and $\alpha_0 = 0.0024$ or $c = 6$ and $\alpha_0 = 0.0113$.

If the experimenter wishes to use a test for which the level of significance is exactly 0.01, then he can use a randomized test procedure of the type described in Sec. 8.2. However, it seems more reasonable for the experimenter to use one of the levels of significance that can be attained with a nonrandomized UMP test than to employ randomization for the sake of attaining some specific level of significance such as 0.01. $\square$

Example 4: Testing Hypotheses About the Mean of a Normal Distribution. Suppose that $X_1, \ldots, X_n$ form a random sample from a normal distribution for which the value of the mean μ is unknown and the value of the variance σ^2 is known. Let μ_0 be a specified number, and suppose that the following hypotheses are to be tested:

$$H_0: \quad \mu \leqslant \mu_0,$$
$$H_1: \quad \mu > \mu_0. \tag{12}$$

We shall show first that for any specified level of significance α_0 ($0 < \alpha_0 < 1$), there is a UMP test of the hypotheses (12). We shall then determine the power function of the UMP test.

It is known from Example 2 that the joint p.d.f. of $X_1, \ldots, X_n$ has a monotone likelihood ratio in the statistic $\overline{X}_n$. Therefore, by Theorem 1, a test procedure δ_1 which rejects H_0 when $\overline{X}_n \geqslant c$ will be a UMP test of the hypotheses (12). The level of significance of this test will be $\alpha_0 = \Pr(\overline{X}_n \geqslant c \mid \mu = \mu_0)$.

Let Z denote a random variable having a standard normal distribution; and for any specified value of α_0, let ζ_{α_0} denote the number such that $\Pr(Z \geqslant \zeta_{\alpha_0}) = \alpha_0$. For example, if $\alpha_0 = 0.05$, then $\zeta_{\alpha_0} = 1.645$. When $\mu = \mu_0$, the random variable $Z = n^{1/2}(\overline{X}_n - \mu_0)/\sigma$ will have a standard normal distribution, and

$$\Pr(\overline{X}_n \geqslant c \mid \mu = \mu_0) = \Pr\left[Z \geqslant \frac{n^{1/2}(c - \mu_0)}{\sigma}\right].$$

If the preceding probability is to be equal to α_0, then it must be true that $n^{1/2}(c - \mu_0)/\sigma = \zeta_{\alpha_0}$ or, equivalently,

$$c = \mu_0 + \zeta_{\alpha_0}\sigma n^{-1/2}. \tag{13}$$

We shall now determine the power function $\pi(\mu \mid \delta_1)$ of this UMP test. By definition,

$$\pi(\mu \mid \delta_1) = \Pr(\text{Rejecting } H_0 \mid \mu) = \Pr\left(\overline{X}_n \geqslant \mu_0 + \zeta_{\alpha_0}\sigma n^{-1/2} \mid \mu\right). \tag{14}$$

For any given value of μ, the random variable $Z' = n^{1/2}(\overline{X}_n - \mu)/\sigma$ will have a standard normal distribution. Therefore, if Φ denotes the d.f. of the standard normal distribution, then

$$\pi(\mu \mid \delta_1) = \Pr\left[Z' \geqslant \zeta_{\alpha_0} + \frac{n^{1/2}(\mu_0 - \mu)}{\sigma}\right]$$

$$= 1 - \Phi\left[\zeta_{\alpha_0} + \frac{n^{1/2}(\mu_0 - \mu)}{\sigma}\right] = \Phi\left[\frac{n^{1/2}(\mu - \mu_0)}{\sigma} - \zeta_{\alpha_0}\right]. \tag{15}$$

The power function $\pi(\mu \mid \delta_1)$ is sketched in Fig. 8.2. $\square$

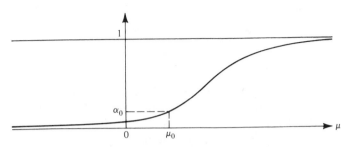

Figure 8.2 The power function $\pi(\mu \mid \delta_1)$ for the UMP test of the hypotheses (12).

In each of the pairs of hypotheses (9), (11), and (12), the alternative hypothesis H_1 is called a *one-sided alternative* because the set of possible values of the parameter under H_1 lies entirely on one side of the set of possible values under the null hypothesis H_0. In particular, for the hypotheses (9), (11), or (12), every possible value of the parameter under H_1 is larger than every possible value under H_0.

Suppose now that instead of testing the hypotheses (12) in Example 4, we are interested in testing the following hypotheses:

$$H_0: \quad \mu \geqslant \mu_0,$$
$$H_1: \quad \mu < \mu_0. \tag{16}$$

In this case the hypothesis H_1 is again a one-sided alternative, and it can be shown (see Exercise 11) that there exists a UMP test of the hypotheses (16) at any specified level of significance α_0 $(0 < \alpha_0 < 1)$. By analogy with Eq. (13), the UMP test δ_2 will reject H_0 when $\overline{X}_n \leqslant c$, where

$$c = \mu_0 - \zeta_{\alpha_0} \sigma n^{-1/2}. \tag{17}$$

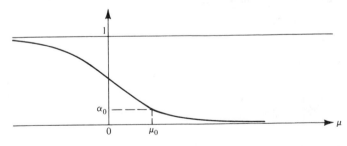

Figure 8.3 The power function $\pi(\mu \mid \delta_2)$ for the UMP test of the hypotheses (16).

The power function $\pi(\mu \mid \delta_2)$ of the test δ_2 will be

$$\pi(\mu \mid \delta_2) = \Pr(\overline{X}_n \leqslant c \mid \mu) = \Phi\left[\frac{n^{1/2}(\mu_0 - \mu)}{\sigma} - \zeta_{\alpha_0}\right]. \tag{18}$$

This function is sketched in Fig. 8.3.

Two-Sided Alternatives

Suppose, finally, that instead of testing either the hypotheses (12) in Example 4, or the hypotheses (16), we are interested in testing the following hypotheses:

$$\begin{align} H_0: &\quad \mu = \mu_0, \\ H_1: &\quad \mu \neq \mu_0. \end{align} \tag{19}$$

In this case, H_0 is a simple hypothesis and H_1 is a two-sided alternative. Since H_0 is a simple hypothesis, the level of significance α_0 of any test procedure δ will simply be equal to the value $\pi(\mu_0 \mid \delta)$ of the power function at the point $\mu = \mu_0$.

It follows from the discussion in this section that there will not be a UMP test of the hypotheses (19) for any given level of significance α_0 ($0 < \alpha_0 < 1$). In fact, for any particular value of μ such that $\mu > \mu_0$, the value of $\pi(\mu \mid \delta)$ will be maximized by the test procedure δ_1; whereas for any particular value of μ such that $\mu < \mu_0$, the value of $\pi(\mu \mid \delta)$ will be maximized by the test procedure δ_2. In the next section, we shall discuss the selection of an appropriate test procedure in this problem, in which there is no UMP test.

EXERCISES

1. Suppose that $X_1, \ldots, X_n$ form a random sample from a Poisson distribution for which the mean λ is unknown ($\lambda > 0$). Show that the joint p.f. of $X_1, \ldots, X_n$ has a monotone likelihood ratio in the statistic $\sum_{i=1}^{n} X_i$.

2. Suppose that $X_1, \ldots, X_n$ form a random sample from a normal distribution for which the value of the mean μ is known and the value of the variance σ^2 is unknown ($\sigma^2 > 0$). Show that the joint p.d.f. of $X_1, \ldots, X_n$ has a monotone likelihood ratio in the statistic $\sum_{i=1}^{n} (X_i - \mu)^2$.

3. Suppose that $X_1, \ldots, X_n$ form a random sample from a gamma distribution for which the value of the parameter α is unknown ($\alpha > 0$) and the value of the parameter β is known. Show that the joint p.d.f. of $X_1, \ldots, X_n$ has a monotone likelihood ratio in the statistic $\prod_{i=1}^{n} X_i$.

4. Suppose that $X_1, \ldots, X_n$ form a random sample from a gamma distribution for which the value of the parameter α is known and the value of the

parameter β is unknown ($\beta > 0$). Show that the joint p.d.f. of $X_1, \ldots, X_n$ has a monotone likelihood ratio in the statistic $-\bar{X}_n$.

5. Suppose that $X_1, \ldots, X_n$ form a random sample from a distribution which belongs to an exponential family, as defined in Exercise 11 of Sec. 6.7; and that the p.d.f. or the p.f. of this distribution is $f(x \mid \theta)$ as given in that exercise. Suppose also that $c(\theta)$ is a strictly increasing function of θ. Show that the joint p.d.f. or the joint p.f. of $X_1, \ldots, X_n$ has a monotone likelihood ratio in the statistic $\sum_{i=1}^n d(X_i)$.

6. Suppose that $X_1, \ldots, X_n$ form a random sample from a distribution involving a parameter θ whose value is unknown, and suppose that it is desired to test the following hypotheses:

$$H_0: \quad \theta \leqslant \theta_0,$$
$$H_1: \quad \theta > \theta_0.$$

Suppose also that the test procedure to be used ignores the observed values in the sample and, instead, depends only on an auxiliary randomization in which an unbalanced coin is tossed so that a head will be obtained with probability 0.05 and a tail will be obtained with probability 0.95. If a head is obtained, then H_0 is rejected; and if a tail is obtained, then H_0 is accepted. Describe the power function of this randomized test procedure.

7. Suppose that $X_1, \ldots, X_n$ form a random sample from a normal distribution for which the mean is 0 and the variance σ^2 is unknown, and suppose that it is desired to test the following hypotheses:

$$H_0: \quad \sigma^2 \leqslant 2,$$
$$H_1: \quad \sigma^2 > 2.$$

Show that there exists a UMP test of these hypotheses at any given level of significance α_0 ($0 < \alpha_0 < 1$).

8. Show that the UMP test in Exercise 7 rejects H_0 when $\sum_{i=1}^n X_i^2 \geqslant c$, and determine the value of c when $n = 10$ and $\alpha_0 = 0.05$.

9. Suppose that $X_1, \ldots, X_n$ form a random sample from a Bernoulli distribution for which the parameter p is unknown, and suppose that it is desired to test the following hypotheses:

$$H_0: \quad p \leqslant \frac{1}{2},$$
$$H_1: \quad p > \frac{1}{2}.$$

Show that if the sample size is $n = 20$, then there exists a nonrandomized UMP test of these hypotheses at the level of significance $\alpha_0 = 0.0577$ and at the level of significance $\alpha_0 = 0.0207$.

10. Suppose that $X_1, \ldots, X_n$ form a random sample from a Poisson distribution for which the mean λ is unknown, and suppose that it is desired to test the following hypotheses:

$H_0: \quad \lambda \leqslant 1,$
$H_1: \quad \lambda > 1.$

Show that if the sample size is $n = 10$, then there exists a nonrandomized UMP test of these hypotheses at the level of significance $\alpha_0 = 0.0143$.

11. Suppose that $X_1, \ldots, X_n$ form a random sample from a distribution that involves a parameter θ whose value is unknown; and that the joint p.d.f. or the joint p.f. $f_n(x \mid \theta)$ has a monotone likelihood ratio in the statistic $T = r(X)$. Let θ_0 be a specified value of θ, and suppose that the following hypotheses are to be tested:

$H_0: \quad \theta \geqslant \theta_0,$
$H_1: \quad \theta < \theta_0.$

Let c be a constant such that $\Pr(T \leqslant c \mid \theta = \theta_0) = \alpha_0$. Show that the test procedure which rejects H_0 if $T \leqslant c$ is a UMP test at the level of significance α_0.

12. Suppose that four observations are taken at random from a normal distribution for which the mean μ is unknown and the variance is 1. Suppose also that the following hypotheses are to be tested:

$H_0: \quad \mu \geqslant 10,$
$H_1: \quad \mu < 10.$

(a) Determine a UMP test at the level of significance $\alpha_0 = 0.1$.
(b) Determine the power of this test when $\mu = 9$.
(c) Determine the probability of accepting H_0 if $\mu = 11$.

13. Suppose that $X_1, \ldots, X_n$ form a random sample from a Poisson distribution for which the mean λ is unknown, and suppose that it is desired to test the following hypotheses:

$H_0: \quad \lambda \geqslant 1,$
$H_1: \quad \lambda < 1.$

Suppose also that the sample size is $n = 10$. At what levels of significance α_0 in the interval $0 < \alpha_0 < 0.03$ do there exist nonrandomized UMP tests?

14. Suppose that $X_1, \ldots, X_n$ form a random sample from an exponential distribution for which the value of the parameter β is unknown, and suppose that it is desired to test the following hypotheses:

$$H_0: \quad \beta \geqslant \frac{1}{2},$$

$$H_1: \quad \beta < \frac{1}{2}.$$

Show that at any given level of significance α_0 $(0 < \alpha < 1)$, there exists a UMP test which specifies rejecting H_0 when $\overline{X}_n > c$, for some constant c.

15. Consider again the conditions of Exercise 14, and suppose that the sample size is $n = 10$. Determine the value of the constant c which defines the UMP test at the level of significance $\alpha_0 = 0.05$. *Hint:* Use the table of the χ^2 distribution.

16. Consider a single observation X from a Cauchy distribution with an unknown location parameter θ; that is, from a distribution for which the p.d.f. $f(x \mid \theta)$ is as follows:

$$f(x \mid \theta) = \frac{1}{\pi \left[1 + (x - \theta)^2\right]} \qquad \text{for } -\infty < x < \infty.$$

Suppose that it is desired to test the following hypotheses:

$$H_0: \quad \theta = 0,$$
$$H_1: \quad \theta > 0.$$

Show that there does not exist a UMP test of these hypotheses at any specified level of significance α_0 $(0 < \alpha_0 < 1)$.

17. Suppose that $X_1, \ldots, X_n$ form a random sample from a normal distribution for which the mean μ is unknown and the variance is 1. Suppose also that the following hypotheses are to be tested:

$$H_0: \quad \mu \leqslant 0,$$
$$H_1: \quad \mu > 0.$$

Let δ^* denote the UMP test of these hypotheses at the level of significance $\alpha_0 = 0.025$, and let $\pi(\mu \mid \delta^*)$ denote the power function of δ^*.

(a) Determine the smallest value of the sample size n for which $\pi(\mu \mid \delta^*) \geqslant 0.9$ for $\mu \geqslant 0.5$.

(b) Determine the smallest value of n for which $\pi(\mu \mid \delta^*) \leqslant 0.001$ for $\mu \leqslant -0.1$.

8.5.   SELECTING A TEST PROCEDURE

General Form of the Procedure

We shall suppose here, as at the end of Sec. 8.4, that the variables $X_1, \ldots, X_n$ form a random sample from a normal distribution for which the mean μ is unknown and the variance σ^2 is known, and that it is desired to test the following hypotheses:

$$
\begin{aligned}
H_0: & \quad \mu = \mu_0, \\
H_1: & \quad \mu \neq \mu_0.
\end{aligned}
\tag{1}
$$

It was shown at the end of Sec. 8.4 that there is no UMP test of the hypotheses (1) at any specified level of significance α_0 ($0 < \alpha_0 < 1$). Neither the test procedure δ_1 nor the procedure δ_2 described in Sec. 8.4 is appropriate for testing the hypotheses (1), since those procedures are designed for one-sided alternatives and the alternative hypothesis H_1 to be considered here is two-sided. However, the properties of the procedures δ_1 and δ_2 given in Sec. 8.4 and the fact that the sample mean $\bar{X}_n$ is the M.L.E. of μ suggest that a reasonable test of the hypotheses (1) would be to accept H_0 if $\bar{X}_n$ is sufficiently close to μ_0 and to reject H_0 if $\bar{X}_n$ is far from μ_0. In other words, it seems reasonable to use a test procedure δ that rejects H_0 if either $\bar{X}_n \leqslant c_1$ or $\bar{X}_n \geqslant c_2$, where c_1 and c_2 are two suitably chosen constants.

If the size of the test is to be α_0, then the values of c_1 and c_2 must be chosen so as to satisfy the following relation:

$$
\Pr(\bar{X}_n \leqslant c_1 \mid \mu = \mu_0) + \Pr(\bar{X}_n \geqslant c_2 \mid \mu = \mu_0) = \alpha_0.
\tag{2}
$$

There are an infinite number of pairs of values of c_1 and c_2 that will satisfy Eq. (2). When $\mu = \mu_0$, the random variable $n^{1/2}(\bar{X}_n - \mu_0)/\sigma$ has a standard normal distribution. If, as usual, we let Φ denote the d.f. of the standard normal distribution, then it follows that Eq. (2) is equivalent to the following relation:

$$
\Phi\left[\frac{n^{1/2}(c_1 - \mu_0)}{\sigma}\right] + 1 - \Phi\left[\frac{n^{1/2}(c_2 - \mu_0)}{\sigma}\right] = \alpha_0.
\tag{3}
$$

Corresponding to any pair of positive numbers α_1 and α_2 such that $\alpha_1 + \alpha_2 = \alpha_0$, there exist a pair of numbers c_1 and c_2 such that $\Phi[n^{1/2}(c_1 - \mu_0)/\sigma] = \alpha_1$ and $1 - \Phi[n^{1/2}(c_2 - \mu_0)/\sigma] = \alpha_2$. Every such pair of values of c_1 and c_2 will satisfy Eq. (2).

For example, suppose that $\alpha_0 = 0.05$. Then choosing $\alpha_1 = 0.025$ and $\alpha_2 = 0.025$ yields a test procedure δ_3 which is defined by the values

$c_1 = \mu_0 - 1.96\sigma n^{-1/2}$ and $c_2 = \mu_0 + 1.96\sigma n^{-1/2}$. Also, choosing $\alpha_1 = 0.01$ and $\alpha_2 = 0.04$ yields a test procedure δ_4 which is defined by the values $c_1 = \mu_0 - 2.33\sigma n^{-1/2}$ and $c_2 = \mu_0 + 1.75\sigma n^{-1/2}$. The power functions $\pi(\mu \mid \delta_3)$ and $\pi(\mu \mid \delta_4)$ of these test procedures δ_3 and δ_4 are sketched in Fig. 8.4, along with the power functions $\pi(\mu \mid \delta_1)$ and $\pi(\mu \mid \delta_2)$ which had previously been sketched in Figs. 8.2 and 8.3.

As the values of c_1 and c_2 in Eq. (2) or Eq. (3) are decreased, the power function $\pi(\mu \mid \delta)$ will become smaller for $\mu < \mu_0$ and become larger for $\mu > \mu_0$. For $\alpha_0 = 0.05$, the limiting case is obtained by choosing $c_1 = -\infty$ and $c_2 = \mu_0 + 1.645\sigma n^{-1/2}$. The test procedure defined by these values is just the procedure δ_1. Similarly, as the values of c_1 and c_2 in Eq. (2) or Eq. (3) are increased, the power function $\pi(\mu \mid \delta)$ will become larger for $\mu < \mu_0$ and become smaller for $\mu > \mu_0$. For $\alpha_0 = 0.05$, the limiting case is obtained by choosing $c_2 = \infty$ and $c_1 = \mu_0 - 1.645\sigma n^{-1/2}$. The test procedure defined by these values is just the procedure δ_2.

Selection of the Test Procedure

For a given sample size n, the values of the constants c_1 and c_2 in Eq. (2) should be chosen so that the size and shape of the power function are appropriate for the particular problem to be solved. In some problems it is important not to reject the null hypothesis unless the data strongly indicate that μ differs greatly from μ_0. In such problems, a small value of α_0 should be used. In other problems, accepting the null hypothesis H_0 when μ is slightly larger than μ_0 is a more serious error than accepting H_0 when μ is slightly less than μ_0. Then it is better to select a test having a power function such as $\pi(\mu \mid \delta_4)$ in Fig. 8.4 than to select a test having a symmetric function such as $\pi(\mu \mid \delta_3)$.

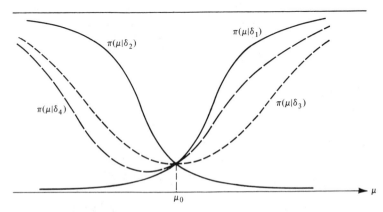

Figure 8.4   The power functions of four test procedures.

In general, the choice of a particular test procedure in a given problem should be based both on the cost of rejecting H_0 when $\mu = \mu_0$ and on the cost, for each possible value of μ, of accepting H_0 when $\mu \neq \mu_0$. Also, when a test is being selected, the relative likelihood of different values of μ should be considered. For example, if it is more likely that μ will be greater than μ_0 than that μ will be less than μ_0, then it is better to select a test for which the power function is large when $\mu > \mu_0$ and not so large when $\mu < \mu_0$ than to select one for which these relations are reversed.

Composite Null Hypothesis

From one point of view, it is silly to carry out a test of the hypotheses (1) in which the null hypothesis H_0 specifies a single exact value μ_0 for the parameter μ. Since it is inconceivable that μ will be *exactly* equal to μ_0 in any real problem, we know that the hypothesis H_0 cannot be true. Therefore, H_0 should be rejected as soon as it has been formulated.

This criticism is valid when it is interpreted literally. In many problems, however, the experimenter is interested in testing the null hypothesis H_0 that the value of μ is close to some specified value μ_0 against the alternative hypothesis that μ is not close to μ_0. In some of these problems, the simple hypothesis H_0 that $\mu = \mu_0$ can be used as an idealization or simplification for the purpose of choosing a decision. At other times, it is worthwhile to use a more realistic composite null hypothesis which specifies that μ lies in an explicit interval around the value μ_0. We shall now consider hypotheses of this type.

Suppose that $X_1, \ldots, X_n$ form a random sample from a normal distribution for which the mean μ is unknown and the variance σ^2 is 1, and suppose that the following hypotheses are to be tested:

$$H_0: \quad 9 \leqslant \mu \leqslant 10,$$
$$H_1: \quad \mu < 9 \text{ or } \mu > 10. \tag{4}$$

Since the alternative hypothesis H_1 is two-sided, it is again appropriate to use a test procedure δ that rejects H_0 if either $\overline{X}_n \leqslant c_1$ or $\overline{X}_n \geqslant c_2$. We shall determine the values of c_1 and c_2 for which the probability of rejecting H_0, when either $\mu = 9$ or $\mu = 10$, will be 0.05.

Let $\pi(\mu \mid \delta)$ denote the power function of δ. When $\mu = 9$, the random variable $n^{1/2}(\overline{X}_n - 9)$ has a standard normal distribution. Therefore,

$$\pi(9 \mid \delta) = \Pr(\text{Rejecting } H_0 \mid \mu = 9)$$

$$= \Pr(\overline{X}_n \leqslant c_1 \mid \mu = 9) + \Pr(\overline{X}_n \geqslant c_2 \mid \mu = 9) \tag{5}$$

$$= \Phi\left[n^{1/2}(c_1 - 9)\right] + 1 - \Phi\left[n^{1/2}(c_2 - 9)\right].$$

Similarly, when $\mu = 10$, the random variable $n^{1/2}(\overline{X}_n - 10)$ has a standard normal distribution and

$$\pi(10 \mid \delta) = \Phi\left[n^{1/2}(c_1 - 10)\right] + 1 - \Phi\left[n^{1/2}(c_2 - 10)\right]. \tag{6}$$

Both $\pi(9 \mid \delta)$ and $\pi(10 \mid \delta)$ must be made equal to 0.05. Because of the symmetry of the normal distribution, it follows that if the values of c_1 and c_2 are chosen symmetrically with respect to the value 9.5, then the power function $\pi(\mu \mid \delta)$ will be symmetric with respect to the point $\mu = 9.5$. In particular, it will then be true that $\pi(9 \mid \delta) = \pi(10 \mid \delta)$.

Accordingly, let $c_1 = 9.5 - c$ and $c_2 = 9.5 + c$. Then it follows from Eqs. (5) and (6) that

$$\pi(9 \mid \delta) = \pi(10 \mid \delta) = \Phi\left[n^{1/2}(0.5 - c)\right] + 1 - \Phi\left[n^{1/2}(0.5 + c)\right]. \tag{7}$$

The value of c must be chosen so that $\pi(9 \mid \delta) = \pi(10 \mid \delta) = 0.05$. Therefore, c must be chosen so that

$$\Phi\left[n^{1/2}(0.5 + c)\right] - \Phi\left[n^{1/2}(0.5 - c)\right] = 0.95. \tag{8}$$

For any given value of n, the value of c which satisfies Eq. (8) can be found by trial and error from a table of the standard normal distribution.

For example, if $n = 16$, then c must be chosen so that

$$\Phi(2 + 4c) - \Phi(2 - 4c) = 0.95. \tag{9}$$

After trying various values of c, we find that Eq. (9) will be satisfied when $c = 0.911$. Hence,

$$c_1 = 9.5 - 0.911 = 8.589 \text{ and } c_2 = 9.5 + 0.911 = 10.411.$$

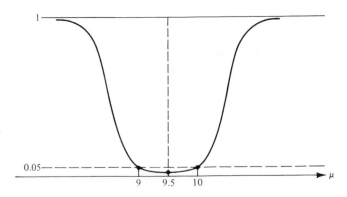

Figure 8.5   The power function $\pi(\mu \mid \delta)$ for a test of the hypotheses (4).

Thus, when $n = 16$, the procedure δ rejects H_0 when either $\overline{X}_n \leqslant 8.589$ or $\overline{X}_n \geqslant 10.411$. This procedure has a power function $\pi(\mu \mid \delta)$ which is symmetric with respect to the point $\mu = 9.5$ and for which $\pi(9 \mid \delta) = \pi(10 \mid \delta) = 0.05$. Furthermore, it is true that $\pi(\mu \mid \delta) < 0.05$ for $9 < \mu < 10$ and $\pi(\mu \mid \delta) > 0.05$ for $\mu < 9$ or $\mu > 10$. The function $\pi(\mu \mid \delta)$ is sketched in Fig. 8.5.

Unbiased Tests

Consider the general problem of testing the following hypotheses:

$$H_0: \quad \theta \in \Omega_0,$$
$$H_1: \quad \theta \in \Omega_1.$$

As usual, let $\pi(\theta \mid \delta)$ denote the power function of an arbitrary test procedure δ. The procedure δ is said to be *unbiased* if, for every pair of parameter values θ and θ' such that $\theta \in \Omega_0$ and $\theta' \in \Omega_1$, it is true that

$$\pi(\theta \mid \delta) \leqslant \pi(\theta' \mid \delta). \tag{10}$$

In other words, δ is unbiased if its power function throughout Ω_1 is at least as large as it is throughout Ω_0.

The notion of an unbiased test is appealing. Since the goal of a test procedure is to accept H_0 when $\theta \in \Omega_0$ and to reject H_0 when $\theta \in \Omega_1$, it seems desirable that the probability of rejecting H_0 should be at least as large when $\theta \in \Omega_1$ as it is whenever $\theta \in \Omega_0$. It can be seen that the test δ for which the power function is sketched in Fig. 8.5 is an unbiased test of the hypotheses (4). Also, among the four tests for which the power functions are sketched in Fig. 8.4, only δ_3 is an unbiased test of the hypotheses (1).

The requirement that a test is to be unbiased can sometimes narrow the selection of a test procedure. However, unbiased procedures should be sought only under relatively special circumstances. For example, when testing the hypotheses (4), the statistician should use the unbiased test δ represented in Fig. 8.5 only under the following conditions: He believes that, for any value $a > 0$, it is just as important to reject H_0 when $\theta = 10 + a$ as to reject H_0 when $\theta = 9 - a$, and he also believes that these two values of θ are equally likely. In practice, the statistician might very well forego the use of an unbiased test in order to use a biased test which has higher power in certain regions of Ω_1 that he regards as particularly important or most likely to contain the true value of θ.

Equivalence of Confidence Sets and Tests

Suppose again that a random sample $X_1, \ldots, X_n$ is to be taken from a distribution for which the p.d.f. is $f(x \mid \theta)$, where the value of the parameter θ is unknown but must lie in a specified parameter space Ω. The parameter θ may be either a real number or a vector. The concept of a confidence set for θ, which we shall now introduce, is a generalization of the concept of a confidence interval introduced in Sec. 7.5.

In an estimation problem, after we observe the values of $X_1, \ldots, X_n$, we choose a single point in the parameter space Ω to be our estimate of θ. In the present discussion, instead of choosing just one point, we shall choose an entire subset of Ω in which we think θ is likely to lie. We shall let $\omega(x_1, \ldots, x_n)$ denote the subset of Ω that is chosen after the values $x_1, \ldots, x_n$ have been observed.

Before the values of $X_1, \ldots, X_n$ have been observed, we can consider $\omega(X_1, \ldots, X_n)$ to be a random subset of Ω. For any given value $\theta_0 \in \Omega$, we can calculate the probability that the subset $\omega(X_1, \ldots, X_n)$ will contain the point θ_0 when θ_0 is the actual value of θ. Suppose that this probability has the same value for every point $\theta_0 \in \Omega$; that is, suppose that there is a number γ $(0 < \gamma < 1)$ such that, for every point $\theta_0 \in \Omega$,

$$\Pr[\theta_0 \in \omega(X_1, \ldots, X_n) \mid \theta = \theta_0] = \gamma. \tag{11}$$

In this case, after the values of $x_1, \ldots, x_n$ in the sample have been observed, it is said that the particular subset $\omega(x_1, \ldots, x_n)$ determined by these values is a *confidence set* for θ with confidence coefficient γ.

We shall now indicate why the theory of confidence sets and the theory of testing hypotheses are essentially equivalent theories. For any given point $\theta_0 \in \Omega$, consider testing the following hypotheses:

$$\begin{aligned} H_0: & \quad \theta = \theta_0, \\ H_1: & \quad \theta \neq \theta_0. \end{aligned} \tag{12}$$

Suppose that for every point $\theta_0 \in \Omega$ and every value of α $(0 < \alpha < 1)$, we can construct a test $\delta(\theta_0)$ of the hypotheses (12) for which the size is α; that is, we can construct a test $\delta(\theta_0)$ such that

$$\Pr[\text{Rejecting } H_0 \text{ when using the test } \delta(\theta_0) \mid \theta = \theta_0] = \alpha. \tag{13}$$

For each possible set of values $x_1, \ldots, x_n$ that might be observed in the random sample, let $\omega(x_1, \ldots, x_n)$ denote the set of all points $\theta_0 \in \Omega$ for which the test $\delta(\theta_0)$ specifies accepting the hypothesis H_0 when the observed values are $x_1, \ldots, x_n$. We shall now show that the set $\omega(x_1, \ldots, x_n)$ is a confidence set for θ with confidence coefficient $1 - \alpha$.

We must show that if we let $\gamma = 1 - \alpha$, then Eq. (11) is satisfied for every point $\theta_0 \in \Omega$. By the definition of $\omega(X_1, \ldots, X_n)$, a given point θ_0 will lie in the subset $\omega(x_1, \ldots, x_n)$ if and only if the observed values $x_1, \ldots, x_n$ lead to acceptance of the hypothesis that $\theta = \theta_0$ when the test $\delta(\theta_0)$ is used. Therefore, by Eq. (13),

$$\Pr\left[\theta_0 \in \omega(X_1, \ldots, X_n) \mid \theta = \theta_0\right]$$

$$= \Pr\left[\text{Accepting } H_0 \text{ when using the test } \delta(\theta_0) \mid \theta = \theta_0\right] = 1 - \alpha. \tag{14}$$

Conversely, suppose that $\omega(x_1, \ldots, x_n)$ is a confidence set for θ with confidence coefficient γ. Then for any given point $\theta_0 \in \Omega$, it can be shown that the test procedure which accepts the hypothesis H_0 if and only if $\theta_0 \in \omega(x_1, \ldots, x_n)$ will have size $1 - \gamma$.

We have now demonstrated that constructing a confidence set for θ with confidence coefficient γ is equivalent to constructing a family of tests of the hypotheses (12) such that there is one test for each value of $\theta_0 \in \Omega$ and each test has size $1 - \gamma$.

EXERCISES

1. Suppose that $X_1, \ldots, X_n$ form a random sample from a normal distribution for which the mean μ is unknown and the variance is 1; and that it is desired to test the following hypotheses for a given number μ_0:

   H_0: $\mu = \mu_0$,
   H_1: $\mu \neq \mu_0$.

   Consider a test procedure δ such that the hypothesis H_0 is rejected if either $\overline{X}_n \leq c_1$ or $\overline{X}_n \geq c_2$, and let $\pi(\mu \mid \delta)$ denote the power function of δ. Determine the values of the constants c_1 and c_2 such that $\pi(\mu_0 \mid \delta) = 0.10$ and the function $\pi(\mu \mid \delta)$ is symmetric with respect to the point $\mu = \mu_0$.

2. Consider again the conditions of Exercise 1. Determine the values of the constants c_1 and c_2 such that $\pi(\mu_0 \mid \delta) = 0.10$ and δ is unbiased.

3. Consider again the conditions of Exercise 1, and suppose that

   $$c_1 = \mu_0 - 1.96 n^{-1/2}.$$

   Determine the value of c_2 such that $\pi(\mu_0 \mid \delta) = 0.10$.

4. Consider again the conditions of Exercise 1 and also the test procedure described in that exercise. Determine the smallest value of n for which $\pi(\mu_0 \mid \delta) = 0.10$ and $\pi(\mu_0 + 1 \mid \delta) = \pi(\mu_0 - 1 \mid \delta) \geq 0.95$.

5. Suppose that $X_1, \ldots, X_n$ form a random sample from a normal distribution for which the mean μ is unknown and the variance is 1; and that it is desired to test the following hypotheses:

   H_0: $0.1 \leq \mu \leq 0.2$,
   H_1: $\mu < 0.1$ or $\mu > 0.2$.

Consider a test procedure δ such that the hypothesis H_0 is rejected if either $\overline{X}_n \leqslant c_1$ or $\overline{X}_n \geqslant c_2$, and let $\pi(\mu \mid \delta)$ denote the power function of δ. Suppose that the sample size is $n = 25$. Determine the values of the constants c_1 and c_2 such that $\pi(0.1 \mid \delta) = \pi(0.2 \mid \delta) = 0.07$.

6. Consider again the conditions of Exercise 5 and suppose also that $n = 25$. Determine the values of the constants c_1 and c_2 such that $\pi(0.1 \mid \delta) = 0.02$ and $\pi(0.2 \mid \delta) = 0.05$.

7. Suppose that $X_1, \ldots, X_n$ form a random sample from a uniform distribution on the interval $(0, \theta)$, where the value of θ is unknown; and that it is desired to test the following hypotheses:

$H_0: \quad \theta \leqslant 3,$
$H_1: \quad \theta > 3.$

(a) Show that for any given level of significance α_0 $(0 \leqslant \alpha_0 < 1)$, there exists a UMP test which specifies that H_0 should be rejected if $\max(X_1, \ldots, X_n) \geqslant c$.

(b) Determine the value of c for each possible value of α_0.

8. For a given sample size n and a given value of α_0, sketch the power function of the UMP test found in Exercise 7.

9. Suppose that $X_1, \ldots, X_n$ form a random sample from the uniform distribution described in Exercise 7, but suppose now that it is desired to test the following hypotheses:

$H_0: \quad \theta \geqslant 3,$
$H_1: \quad \theta < 3.$

(a) Show that at any given level of significance α_0 $(0 < \alpha_0 < 1)$, there exists a UMP test which specifies that H_0 should be rejected if $\max(X_1, \ldots, X_n) \leqslant c$.

(b) Determine the value of c for each possible value of α_0.

10. For a given sample size n and a given value of α_0, sketch the power function of the UMP test found in Exercise 9.

11. Suppose that $X_1, \ldots, X_n$ form a random sample from the uniform distribution described in Exercise 7, but suppose now that it is desired to test the following hypotheses:

$H_0: \quad \theta = 3,$
$H_1: \quad \theta \neq 3.$

Consider a test procedure δ such that the hypothesis H_0 is rejected if either $\max(X_1, \ldots, X_n) \leqslant c_1$ or $\max(X_1, \ldots, X_n) \geqslant c_2$, and let $\pi(\theta \mid \delta)$ denote the power function of δ. Determine the values of the constants c_1 and c_2 such that $\pi(3 \mid \delta) = 0.05$ and δ is unbiased.

8.6.   THE *t* TEST

Testing Hypotheses About the Mean of a Normal Distribution When the Variance Is Unknown

In this section we shall consider the problem of testing hypotheses about the mean of a normal distribution when both the mean and the variance are unknown. Specifically, we shall suppose that the variables $X_1, \ldots, X_n$ form a random sample from a normal distribution for which the mean μ and the variance σ^2 are unknown, and we shall consider testing the following hypotheses:

$$
\begin{aligned}
H_0&: \quad \mu \leqslant \mu_0, \\
H_1&: \quad \mu > \mu_0.
\end{aligned}
\tag{1}
$$

The parameter space Ω in this problem comprises every two-dimensional vector (μ, σ^2), where $-\infty < \mu < \infty$ and $\sigma^2 > 0$. The null hypothesis H_0 specifies that the vector (μ, σ^2) lies in the subset Ω_0 of Ω comprising all vectors for which $\mu \leqslant \mu_0$ and $\sigma^2 > 0$, as illustrated in Fig. 8.6. The alternative hypothesis H_1 specifies that (μ, σ^2) belongs to the subset Ω_1 of Ω comprising all the vectors that do not belong to Ω_0.

If it is desired to test the hypotheses (1) at a given level of significance α_0 $(0 < \alpha_0 < 1)$, then we must try to find a test procedure δ whose power function

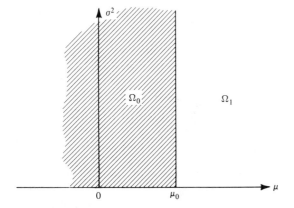

Figure 8.6   The subsets Ω_0 and Ω_1 of the parameter space Ω for the hypotheses (1).

$\pi(\mu, \sigma^2 \mid \delta)$ meets the following two requirements: First,

$$\pi(\mu, \sigma^2 \mid \delta) \leqslant \alpha_0 \qquad \text{for every point } (\mu, \sigma^2) \in \Omega_0. \tag{2}$$

Second, $\pi(\mu, \sigma^2 \mid \delta)$ should be as large as possible for every point $(\mu, \sigma^2) \in \Omega_1$. For $\alpha_0 < 1/2$, however, it can be shown that among all test procedures which satisfy the relation (2), there is no single procedure for which $\pi(\mu, \sigma^2 \mid \delta)$ will be maximized at every point $(\mu, \sigma^2) \in \Omega_1$. In other words, for $\alpha_0 < 1/2$, there does not exist a UMP test of the hypotheses (1) at the level of significance α_0.

Although there is no UMP test in this problem, it is common practice to use one particular procedure, called the t test. The t test, which we shall now derive, satisfies the relation (2) and also has the following five properties:

(i) $\pi(\mu, \sigma^2 \mid \delta) = \alpha_0$ when $\mu = \mu_0$,

(ii) $\pi(\mu, \sigma^2 \mid \delta) < \alpha_0$ when $\mu < \mu_0$,

(iii) $\pi(\mu, \sigma^2 \mid \delta) > \alpha_0$ when $\mu > \mu_0$,

(iv) $\pi(\mu, \sigma^2 \mid \delta) \to 0$ as $\mu \to -\infty$,

(v) $\pi(\mu, \sigma^2 \mid \delta) \to 1$ as $\mu \to \infty$.

It follows from properties (i), (ii), and (iii) that the t test is an unbiased test, as defined in Eq. (10) of Sec. 8.5. Furthermore, it can be shown that the t test is actually a UMP test within the class of all unbiased tests. Finally, it can be shown that for any level of significance $\alpha_0 \geqslant 1/2$, the t test is a UMP test among all tests, both biased and unbiased. The proofs of these last two statements are beyond the scope of this book and will not be presented [see Lehmann (1959)]. Of course, in any practical problem the level of significance α_0 will typically be much smaller than $1/2$.

Derivation of the t Test

After the values $x_1, \ldots, x_n$ in the random sample have been observed, the likelihood function is

$$f_n(x \mid \mu, \sigma^2) = \frac{1}{(2\pi\sigma^2)^{n/2}} \exp\left[-\frac{1}{2\sigma^2} \sum_{i=1}^{n} (x_i - \mu)^2 \right]. \tag{3}$$

A reasonable procedure for deciding whether to accept H_0 or to accept H_1 is to compare the following two values: the maximum value attained by the likelihood function (3) when the point (μ, σ^2) varies over all values in Ω_0 and the maximum value attained by the likelihood function when (μ, σ^2) varies over all values in Ω_1. Therefore, we shall consider the following ratio:

$$r(x) = \frac{\sup_{(\mu, \sigma^2) \in \Omega_1} f_n(x \mid \mu, \sigma^2)}{\sup_{(\mu, \sigma^2) \in \Omega_0} f_n(x \mid \mu, \sigma^2)}. \tag{4}$$

The procedure to be used then specifies that H_0 should be rejected if $r(x) \geq k$, where k is some chosen constant, and that H_0 should be accepted if $r(x) < k$. A test procedure derived in this way is called a *likelihood ratio test procedure*. We shall now determine more explicitly the form of the procedure in this problem.

As in Sec. 6.5, we shall let $\hat{\mu}$ and $\hat{\sigma}^2$ denote the M.L.E.'s of μ and σ^2 when it is known only that the point (μ, σ^2) belongs to the parameter space Ω. It was shown in Example 3 of Sec. 6.5 that

$$\hat{\mu} = \bar{x}_n \qquad \text{and} \qquad \hat{\sigma}^2 = \frac{1}{n} \sum_{i=1}^n (x_i - \bar{x}_n)^2.$$

Similarly, we shall let $\hat{\mu}_0$ and $\hat{\sigma}_0^2$ denote the M.L.E.'s of μ and σ^2 when the point (μ, σ^2) is constrained to lie in the subset Ω_0, and we shall let $\hat{\mu}_1$ and $\hat{\sigma}_1^2$ denote the M.L.E.'s when the point (μ, σ^2) is constrained to lie in the subset Ω_1. It follows that

$$\sup_{(\mu, \sigma^2) \in \Omega_0} f_n(x \mid \mu, \sigma^2) = f_n(x \mid \hat{\mu}_0, \hat{\sigma}_0^2) \tag{5}$$

and

$$\sup_{(\mu, \sigma^2) \in \Omega_1} f_n(x \mid \mu, \sigma^2) = f_n(x \mid \hat{\mu}_1, \hat{\sigma}_1^2). \tag{6}$$

Suppose first that the observed sample values are such that $\bar{x}_n > \mu_0$. Then the point $(\hat{\mu}, \hat{\sigma}^2)$ will lie in Ω_1. Since this point maximizes the value of $f_n(x \mid \mu, \sigma^2)$ among all points $(\mu, \sigma^2) \in \Omega$, it certainly maximizes the value of $f_n(x \mid \mu, \sigma^2)$ among all points (μ, σ^2) in the subset Ω_1. Hence, $\hat{\mu}_1 = \hat{\mu}$ and $\hat{\sigma}_1^2 = \hat{\sigma}^2$.

Furthermore, when $\bar{x}_n > \mu_0$, it can be shown that $f_n(x \mid \mu, \sigma^2)$ attains its maximum value among all points $(\mu, \sigma^2) \in \Omega_0$ if μ is chosen to be as close as possible to $\bar{x}_n$. The value of μ closest to $\bar{x}_n$ among all points in the subset Ω_0 is $\mu = \mu_0$. Hence, $\hat{\mu}_0 = \mu_0$. In turn, it can be shown, as in Example 3 of Sec. 6.5, that the M.L.E. of σ^2 will be

$$\hat{\sigma}_0^2 = \frac{1}{n} \sum_{i=1}^n (x_i - \hat{\mu}_0)^2 = \frac{1}{n} \sum_{i=1}^n (x_i - \mu_0)^2.$$

By substituting the values of $\hat{\mu}_0$, $\hat{\sigma}_0^2$, $\hat{\mu}_1$, and $\hat{\sigma}_1^2$ that have just been obtained into Eqs. (3) to (6), we find that

$$r(x) = \left(\frac{\hat{\sigma}_0^2}{\hat{\sigma}_1^2}\right)^{n/2}. \tag{7}$$

If we use the relation

$$\sum_{i=1}^{n} (x_i - \mu_0)^2 = \sum_{i=1}^{n} (x_i - \bar{x}_n)^2 + n(\bar{x}_n - \mu_0)^2,$$

then $r(x)$ can be written as follows:

$$r(x) = \left[1 + \frac{n(\bar{x}_n - \mu_0)^2}{\sum_{i=1}^{n}(x_i - \bar{x}_n)^2}\right]^{n/2}. \tag{8}$$

Now consider any specified constant $k > 1$. It follows from Eq. (8) that $r(x) \geq k$ if and only if

$$\frac{n(\bar{x}_n - \mu_0)^2}{\sum_{i=1}^{n}(x_i - \bar{x}_n)^2} \geq k', \tag{9}$$

where k' is another constant whose value can be derived from k. Finally, for $\bar{x}_n > \mu_0$, the relation (9) is equivalent to the relation

$$\frac{n^{1/2}(\bar{x}_n - \mu_0)}{\left[\dfrac{\sum_{i=1}^{n}(x_i - \bar{x}_n)^2}{n - 1}\right]^{1/2}} \geq c, \tag{10}$$

where $c = [(n - 1)k']^{1/2}$.

Similarly, if the observed sample values are such that $\bar{x}_n < \mu_0$ and if $k < 1$, then it can again be shown that $r(x) \geq k$ if and only if the relation (10) is satisfied, where c is a constant whose value can be derived from the value of k.

As in Sec. 7.4, we shall let $S_n^2 = \sum_{i=1}^{n}(X_i - \bar{X}_n)^2$, and we shall let U denote the following statistic:

$$U = \frac{n^{1/2}(\bar{X}_n - \mu_0)}{\left[\dfrac{S_n^2}{n - 1}\right]^{1/2}}. \tag{11}$$

The likelihood ratio test procedure which we have just derived specifies that the hypothesis H_0 should be rejected if $U \geq c$ and that H_0 should be accepted if $U < c$. In this problem, this procedure is called a *t* test, for reasons that will now be explained.

Properties of the *t* Test

When $\mu = \mu_0$, it follows from Sec. 7.4 that the distribution of the statistic U defined by Eq. (11) will be a *t* distribution with $n - 1$ degrees of freedom, regardless of the value of σ^2. Therefore, when $\mu = \mu_0$, it is possible to use a table of the *t* distribution to choose a constant c such that $\Pr(U \geq c) = \alpha_0$ regardless of the value of σ^2.

Now suppose that $\mu < \mu_0$, and let

$$U^* = \frac{n^{1/2}(\overline{X}_n - \mu)}{\left[\dfrac{S_n^2}{n-1}\right]^{1/2}} \quad \text{and} \quad W = \frac{n^{1/2}(\mu_0 - \mu)}{\left[\dfrac{S_n^2}{n-1}\right]^{1/2}}.$$

Then the random variable U^* will have a *t* distribution with $n - 1$ degrees of freedom, and the value of the random variable W will be positive. Furthermore, it can be seen from Eq. (11) that $U = U^* - W$. Therefore, for any given values of μ and σ^2 such that $\mu < \mu_0$,

$$\Pr(U \geq c) = \Pr(U^* - W \geq c) = \Pr(U^* \geq c + W) < \Pr(U^* \geq c) = \alpha_0.$$

Thus, when $\mu < \mu_0$, it is seen that $\Pr(U \geq c) < \alpha_0$, regardless of the value of σ^2. It follows that the size of this test, which rejects H_0 when $U \geq c$, is α_0 and the power function of the test satisfies the relation (2).

Also, an argument similar to the one just given shows that when $\mu > \mu_0$, then $\Pr(U \geq c) > \alpha_0$, regardless of the value of σ^2.

Finally, when μ is very large, the numerator of U will tend to be very large; and the probability of rejecting H_0 will be close to 1. Formally, it can be shown that for any value of $\sigma^2 > 0$,

$$\lim_{\mu \to \infty} \pi(\mu, \sigma^2 \mid \delta) = 1.$$

Similarly, it can be shown that for any value of $\sigma^2 > 0$,

$$\lim_{\mu \to -\infty} \pi(\mu, \sigma^2 \mid \delta) = 0.$$

Example 1: Lengths of Fibers. Suppose that the lengths, in millimeters, of metal fibers produced by a certain process have a normal distribution for which both the mean μ and the variance σ^2 are unknown; and that the following hypotheses are to be tested:

$$H_0: \quad \mu \leqslant 5.2,$$
$$H_1: \quad \mu > 5.2. \tag{12}$$

Suppose that the lengths of 15 fibers selected at random are measured, and it is found that the sample mean $\overline{X}_{15}$ is 5.4 and $S_{15}^2 = \sum_{i=1}^{15}(X_i - \overline{X}_{15})^2 = 2.5$. Based on these measurements, we shall carry out a t test at the level of significance $\alpha_0 = 0.05$.

Since $n = 15$ and $\mu_0 = 5.2$, the statistic U defined by Eq. (11) will have a t distribution with 14 degrees of freedom when $\mu = 5.2$. It is found in the table of the t distribution that $\Pr(U \geqslant 1.761) = 0.05$. Hence, the null hypothesis H_0 should be rejected if $U \geqslant 1.761$. Since the numerical value of U calculated from Eq. (11) is 1.833, H_0 should be rejected. $\square$

Testing with a Two-Sided Alternative

We shall continue to assume that the variables $X_1, \ldots, X_n$ form a random sample from a normal distribution for which both the mean μ and the variance σ^2 are unknown, but we shall suppose now that the following hypotheses are to be tested:

$$H_0: \quad \mu = \mu_0,$$
$$H_1: \quad \mu \neq \mu_0. \tag{13}$$

Here, the alternative hypothesis H_1 is two-sided.

By analogy with the t test which has been derived for a one-sided alternative, the standard procedure for testing the hypotheses (13) is to reject H_0 if either $U \leqslant c_1$ or $U \geqslant c_2$, where c_1 and c_2 are appropriately chosen constants.

For any given values of μ and σ^2 that satisfy the null hypothesis H_0, that is, for $\mu = \mu_0$ and $\sigma^2 > 0$, the statistic U has a t distribution with $n - 1$ degrees of freedom. Therefore, by using a table of this distribution, it is possible to choose $c_1 < 0$ and $c_2 > 0$ such that, when H_0 is true,

$$\Pr(U < c_1) + \Pr(U > c_2) = \alpha_0. \tag{14}$$

As discussed in Sec. 8.5, there will be many pairs of values of c_1 and c_2 that satisfy Eq. (14). In most experiments, it is convenient to choose c_1 and c_2 symmetrically with respect to 0. For this choice, $c_1 = -c$ and $c_2 = c$; and when H_0 is true,

$$\Pr(U < -c) = \Pr(U > c) = \frac{1}{2}\alpha_0. \tag{15}$$

For this symmetric choice, the *t* test will be an unbiased test of the hypotheses (13).

Example 2: Lengths of Fibers. We shall consider again the problem discussed in Example 1; but we shall suppose now that, instead of the hypotheses (12), the following hypotheses are to be tested:

$$H_0: \quad \mu = 5.2,$$
$$H_1: \quad \mu \neq 5.2. \tag{16}$$

We shall again assume that the lengths of 15 fibers are measured, and that the value of U calculated from the observed values is 1.833. We shall test the hypotheses (16) at the level of significance $\alpha_0 = 0.05$ by using a symmetric *t* test of the type specified in Eq. (15).

Since $\alpha_0 = 0.05$, each tail of the critical region is to have probability 0.025. Therefore, by using the column for $p = 0.975$ in the table of the *t* distribution with 14 degrees of freedom, we find that the *t* test specifies rejecting H_0 if either $U < -2.145$ or $U > 2.145$. Since $U = 1.833$, the hypothesis H_0 should be accepted. □

The numerical values in Examples 1 and 2 emphasize the importance of deciding whether the appropriate alternative hypothesis in a given problem is one-sided or two-sided. When the hypotheses (12) were tested at the level of significance 0.05, the hypothesis H_0 that $\mu \leqslant 5.2$ was rejected. When the hypotheses (16) were tested at the same level of significance and the same data were used, the hypothesis H_0 that $\mu = 5.2$ was accepted.

Confidence Intervals from the *t* Test

We can obtain a confidence interval for μ from the *t* test that we have developed for the hypotheses (13) by using the method described at the end of Sec. 8.5. The confidence interval for μ that would be obtained by this method is the same as the confidence interval for μ that was given in Sec. 7.5. Hence, we shall not consider this topic further at this time.

EXERCISES

1. Suppose that 9 observations are selected at random from a normal distribution for which both the mean μ and the variance σ^2 are unknown; and that for these 9 observations it is found that $\overline{X}_n = 22$ and $S_n^2 = 72$.

   (a) Carry out a test of the following hypotheses at the level of significance 0.05:
   $$H_0: \quad \mu \leqslant 20,$$
   $$H_1: \quad \mu > 20.$$

(b) Carry out a test of the following hypotheses at the level of significance 0.05 by using a symmetric test with probability 0.025 in each tail:

$$H_0: \quad \mu = 20,$$
$$H_1: \quad \mu \neq 20.$$

(c) From the data, construct a confidence interval for μ with confidence coefficient 0.95.

2. The manufacturer of a certain type of automobile claims that under typical urban driving conditions the automobile will travel, on the average, at least 20 miles per gallon of gasoline. The owner of an automobile of this type notes the mileages that he has obtained in his own urban driving when he fills his automobile's tank with gasoline on 9 different occasions. He finds that the results, in miles per gallon, for the different tankfuls were as follows: 15.6, 18.6, 18.3, 20.1, 21.5, 18.4, 19.1, 20.4, and 19.0. Test the manufacturer's claim by carrying out a test at the level of significance $\alpha_0 = 0.05$. List carefully the assumptions you must make.

3. Suppose that a random sample of 8 observations $X_1, \ldots, X_8$ is taken from a normal distribution for which both the mean μ and the variance σ^2 are unknown; and that it is desired to test the following hypotheses:

$$H_0: \quad \mu = 0,$$
$$H_1: \quad \mu \neq 0.$$

Suppose also that the sample data are such that $\sum_{i=1}^{8} X_i = -11.2$ and $\sum_{i=1}^{8} X_i^2 = 43.7$. If a symmetric t test is performed at the level of significance 0.10, so that each tail of the critical region has probability 0.05, should the hypothesis H_0 be accepted or rejected?

4. Consider again the conditions of Exercise 3, and suppose again that a t test is to be performed at the level of significance 0.10. Suppose now, however, that the test is not to be symmetric and that the hypothesis H_0 is to be rejected if the statistic U is such that either $U < c_1$ or $U > c_2$, where $\Pr(U < c_1) = 0.01$ and $\Pr(U > c_2) = 0.09$. For the sample data specified in Exercise 3, should H_0 be accepted or rejected?

5. Suppose that the variables $X_1, \ldots, X_n$ form a random sample from a normal distribution for which both the mean μ and the variance σ^2 are unknown; and that a t test at a given level of significance α_0 is to be carried out to test the following hypotheses:

$$H_0: \quad \mu \leqslant \mu_0,$$
$$H_1: \quad \mu > \mu_0.$$

Let $\pi(\mu, \sigma^2 | \delta)$ denote the power function of this t test, and assume that

(μ_1, σ_1^2) and (μ_2, σ_2^2) are values of the parameters such that

$$\frac{\mu_1 - \mu_0}{\sigma_1} = \frac{\mu_2 - \mu_0}{\sigma_2}.$$

Show that $\pi(\mu_1, \sigma_1^2 \mid \delta) = \pi(\mu_2, \sigma_2^2 \mid \delta)$.

6. Consider a normal distribution for which both the mean μ and the variance σ^2 are unknown, and suppose that it is desired to test the following hypotheses:

$$H_0: \quad \mu \leqslant \mu_0,$$
$$H_1: \quad \mu > \mu_0.$$

Suppose that it is possible to observe only a single value of X from this distribution, but that an independent random sample of n observations $Y_1, \ldots, Y_n$ is available from another normal distribution for which the variance is also σ^2 and it is known that the mean is 0. Show how to carry out a test of the hypotheses H_0 and H_1 based on the t distribution with n degrees of freedom.

7. Suppose that the variables $X_1, \ldots, X_n$ form a random sample from a normal distribution for which both the mean μ and the variance σ^2 are unknown. Let σ_0^2 be a given positive number, and suppose that it is desired to test the following hypotheses at a specified level of significance α_0 ($0 < \alpha_0 < 1$):

$$H_0: \quad \sigma^2 \leqslant \sigma_0^2,$$
$$H_1: \quad \sigma^2 > \sigma_0^2.$$

Let $S_n^2 = \sum_{i=1}^{n}(X_i - \overline{X}_n)^2$, and suppose that the test procedure to be used specifies that H_0 should be rejected if $S_n^2 / \sigma_0^2 \geqslant c$. Also, let $\pi(\mu, \sigma^2 \mid \delta)$ denote the power function of this procedure. Explain how to choose the constant c so that, regardless of the value of μ, the following requirements are satisfied: $\pi(\mu, \sigma^2 \mid \delta) < \alpha_0$ if $\sigma^2 < \sigma_0^2$; $\pi(\mu, \sigma^2 \mid \delta) = \alpha_0$ if $\sigma^2 = \sigma_0^2$; and $\pi(\mu, \sigma^2 \mid \delta) > \alpha_0$ if $\sigma^2 > \sigma_0^2$.

8. Suppose that a random sample of 10 observations $X_1, \ldots, X_{10}$ is taken from a normal distribution for which both the mean μ and the variance σ^2 are unknown; and that it is desired to test the following hypotheses:

$$H_0: \quad \sigma^2 \leqslant 4,$$
$$H_1: \quad \sigma^2 > 4.$$

Suppose that a test of the form described in Exercise 7 is to be carried out at the level of significance $\alpha_0 = 0.05$. If the observed value of S_n^2 is 60, should the hypothesis H_0 be accepted or rejected?

9. Suppose again, as in Exercise 8, that a random sample of 10 observations is taken from a normal distribution for which both the mean μ and the variance

σ^2 are unknown; but suppose now that the following hypotheses are to be tested at the level of significance 0.05:

$$H_0: \quad \sigma^2 = 4,$$
$$H_1: \quad \sigma^2 \neq 4.$$

Suppose that the null hypothesis H_0 is to be rejected if either $S_n^2 \leqslant c_1$ or $S_n^2 \geqslant c_2$, where the constants c_1 and c_2 are to be chosen so that, when the hypothesis H_0 is true,

$$\Pr(S_n^2 \leqslant c_1) = \Pr(S_n^2 \geqslant c_2) = 0.025.$$

Determine the values of c_1 and c_2.

8.7. DISCUSSION OF THE METHODOLOGY OF TESTING HYPOTHESES

In many respects, the theory of testing hypotheses, as it has developed in statistical methodology, is misleading. According to this theory, in a problem of testing hypotheses only two decisions are available to the experimenter. He must either accept the null hypothesis H_0 or reject H_0. It is true that there are problems of this type in statistical practice, and the theory of testing hypotheses can properly and usefully be applied to such problems. It is also true, however, that the methodology of testing hypotheses is applied to many situations in which the experimenter is mainly interested in determining the likelihood that the hypothesis H_0 is true and in which he does not necessarily have to choose one of two decisions. In this section we shall discuss the *methodology* of testing hypotheses as it is commonly practiced in various fields of application of statistics, in contrast to the *theory* of testing hypotheses which is presented elsewhere in this chapter.

Tail Areas

In order to facilitate the discussion in this section, we shall again consider Example 1 of Sec. 8.6, in which the hypotheses (12) are to be tested by using an appropriate one-sided t test based on the statistic U. We shall assume that the sample data are as given in the example.

In this example, the experimenter will typically realize that simply reporting whether the hypothesis H_0 was accepted or rejected by the t test, carried out at the level of significance 0.05, does not convey all the information contained in the sample data in regard to the likelihood that H_0 is true. If he reports only that H_0 was rejected, then he is simply reporting that the observed value of U exceeded the critical value 1.761. The result of the test would be more useful if he reports whether this observed value was only slightly larger than 1.761 or was very much larger than 1.761.

Furthermore, the decision to accept H_0 or to reject H_0 in a given problem obviously depends on the level of significance α_0 that is used in the problem. In most applications α_0 is chosen to be either 0.05 or 0.01, but there is no strong reason other than tradition for using one of these particular values. In our example, it was assumed that $\alpha_0 = 0.05$ and H_0 was rejected. If the value $\alpha_0 = 0.01$ had been used, instead of 0.05, then H_0 would have been accepted.

For these reasons, an experimenter does not typically choose a value of α_0 in advance of the experiment and then simply report whether H_0 was accepted or rejected on the basis of the observed value of U. In many fields of application it has become standard practice to report the observed value of U and *all* the values of α_0 for which this observed value of U would lead to the rejection of H_0. Thus, if the observed value of U is 1.833, as in our example, it is found from the table of the t distribution given at the end of this book that the hypothesis H_0 would be rejected for any level of significance $\alpha_0 \geqslant 0.05$ and that H_0 would not be rejected for any level of significance $\alpha_0 \leqslant 0.025$.

From a more complete table of the t distribution than the one given in this book, it can be found that the probability lying to the right of the value 1.833 in the tail of the t distribution with 14 degrees of freedom is 0.044. In other words, if Z denotes a random variable that has a t distribution with 14 degrees of freedom, then $\Pr(Z \geqslant 1.833) = 0.044$. The value 0.044 is called the *tail area* or the *p-value* corresponding to the observed value of the statistic U. Thus, if the observed value of U is 1.833, the hypothesis H_0 should be rejected for any value of $\alpha_0 > 0.044$ and should be accepted for any value of $\alpha_0 < 0.044$.

An experimenter, in his analysis of this experiment, would typically report that the observed value of U was 1.833 and the corresponding tail area is 0.044. It is then said that the observed value of U is *just significant* at the level of significance 0.044. One advantage to the experimenter, when he reports his experimental results in this manner, is that he does not have to select beforehand an arbitrary level of significance α_0 at which the t test is to be carried out. Also, when a reader of the experimenter's report learns that the observed value of U was just significant at the level of significance 0.044, he immediately knows that H_0 should be rejected for any larger value of α_0 and should not be rejected for any smaller value.

Tail Areas for a Two-Sided Alternative Hypothesis

Now consider Example 2 of Sec. 8.6, in which the hypotheses (16) are to be tested by using a symmetric two-sided t test based on the statistic U. In this example, the hypothesis H_0 is rejected if either $U \leqslant -c$ or $U \geqslant c$. If the test is to be carried out at the level of significance 0.05, the value of c is chosen to be 2.145, because $\Pr(U \leqslant -2.145) + \Pr(U \geqslant 2.145) = 0.05$.

Because the appropriate test procedure in this example is a two-sided t test, the appropriate tail area corresponding to the observed value $U = 1.833$ will be the sum of the following two probabilities: (i) the probability lying to the right of 1.833 in the right-hand tail of the t distribution and (ii) the probability lying to the left of -1.833 in the left-hand tail of this distribution. Because of the symmetry of the t distribution, these two probabilities are equal. Therefore, the tail area corresponding to the observed value $U = 1.833$ is $2(0.044) = 0.088$.

In other words, the observed value of U is just significant at the level of significance 0.088. The hypothesis H_0: $\mu = 5.2$ should be rejected for any level of significance $\alpha_0 > 0.088$ and should be accepted for any level of significance $\alpha_0 < 0.088$.

It should be emphasized that when the null hypothesis H_0 is accepted at a specified level of significance, it does not necessarily mean that the experimenter has become convinced that H_0 is true. Rather, it usually means merely that the data do not provide strong evidence that H_0 is not true.

Statistically Significant Results

We shall continue to consider Example 2 of Sec. 8.6, in which the hypotheses (16) are to be tested. As the value of the tail area corresponding to the observed value of U decreases, we regard the weight of evidence against the hypothesis H_0 as becoming greater. Thus, if we had found that the observed value of U was just significant at the level of significance 0.00088, rather than at the level 0.088, then we would have regarded the sample as providing much stronger evidence against H_0. It is often said that an observed value of U is *statistically significant* if the corresponding tail area is smaller than the traditional value 0.05 or 0.01. Although an experimenter does not know with certainty whether or not the hypothesis H_0 in a given problem is true, he would conclude that a statistically significant observed value of U provides, at the very least, strong evidence against H_0.

It is extremely important for the experimenter to distinguish between an observed value of U that is statistically significant and an actual value of the parameter μ that is significantly different from the value $\mu = 5.2$ specified by the null hypothesis H_0. Although a statistically significant observed value of U provides strong evidence that μ is not equal to 5.2, it does not necessarily provide strong evidence that the actual value of μ is *significantly* different from 5.2. In a given problem, the tail area corresponding to the observed value of U might be very small; and yet the actual value of μ might be so close to 5.2 that, for practical purposes, the experimenter would not regard μ as being significantly different from 5.2.

The situation just described can arise when the statistic U is based on a very large random sample. Suppose, for instance, that in Example 2 of Sec. 8.6 the lengths of 20,000 fibers in a random sample are measured, rather than the lengths of only 15 fibers. For a given level of significance, say $\alpha_0 = 0.05$, let $\pi(\mu, \sigma^2 \mid \delta)$ denote the power function of the t test based on these 20,000 observations. Then $\pi(5.2, \sigma^2 \mid \delta) = 0.05$ for every value of $\sigma^2 > 0$. However, because of the very large number of observations on which the test is based, the power $\pi(\mu, \sigma^2 \mid \delta)$ will be very close to 1 for any value of μ that differs only slightly from 5.2 and for a moderate value of σ^2. In other words, even if the value of μ differs only slightly from 5.2, the probability is close to 1 that the observed value of U will be statistically significant.

As explained in Sec. 8.5, it is inconceivable that the mean length μ of all the fibers in the entire population will be exactly 5.2. However, μ may be very close to 5.2; and when it is, the experimenter will want to accept the null hypothesis H_0. Nevertheless, it is very likely that the t test based on the sample of 20,000 fibers will lead to a statistically significant value of U. Therefore, when an experimenter analyzes a powerful test based on a very large sample, he must exercise caution in interpreting the actual significance of a "statistically significant" result. He knows in advance that there is a high probability of rejecting H_0 even when the true value of μ differs only slightly from the value 5.2 specified under H_0.

One way to handle this situation, as discussed in Sec. 8.2, is to recognize that a level of significance much smaller than the traditional value of 0.05 or 0.01 is appropriate for a problem with a large sample size. Another way is to regard the statistical problem as one of estimation rather than one of testing hypotheses.

When a large random sample is available, the sample mean and the sample variance will be excellent estimators of the parameters μ and σ^2. Before the experimenter chooses any decision involving the unknown values of μ and σ^2, he should calculate and consider the values of these estimators as well as the value of the statistic U.

The Bayesian Approach

When an experimenter or a statistician is testing hypotheses, he is most interested in the use of the sample data to determine the probability that H_0 is true. It must be emphasized that the methodology of testing hypotheses which has been discussed in this section does not, and cannot, deal directly with this probability. The tail area or p-value that is calculated from the observed sample provides no indication, by itself, of the probability that H_0 is true, although the p-value is sometimes misinterpreted in this way. In fact, it is sometimes incorrectly stated by experimenters that the rejection of H_0 at some specified level of significance α_0 indicates that the probability that H_0 is true is less than α_0.

No such interpretation can logically be made. In order to be able to determine the probability that H_0 is true, the experimenter must adopt a Bayesian approach. If a prior distribution is assigned to the parameter θ being tested, it is possible to calculate the posterior distribution of θ, given the sample data, and $\Pr(H_0$ is true$) = \Pr(\theta \in \Omega_0)$ can be determined from this posterior distribution.

Although the discussion on the methodology of testing hypotheses has been presented here in the context of the t test, it should be emphasized that this discussion pertains quite generally to all problems of testing hypotheses.

EXERCISES

1. Suppose that a random sample $X_1, \ldots, X_n$ is to be taken from a normal distribution for which both the mean μ and the variance σ^2 are unknown; and that the following hypotheses are to be tested:

   H_0: $\mu \leqslant 3$,
   H_1: $\mu > 3$.

   Suppose also that the sample size n is 17, and it is found from the observed values in the sample that $\bar{X}_n = 3.2$ and $(1/n)\sum_{i=1}^{n}(X_i - \bar{X}_n)^2 = 0.09$. Calculate the value of the statistic U, and find the value of the corresponding tail area.

2. Consider again the conditions of Exercise 1, but suppose now that the sample size n is 170 and that it is again found from the observed values in the sample that $\bar{X}_n = 3.2$ and $(1/n)\sum_{i=1}^{n}(X_i - \bar{X}_n)^2 = 0.09$. Calculate the value of the statistic U, and find the value of the corresponding tail area.

3. Consider again the conditions of Exercise 1, but suppose now that the following hypotheses are to be tested:

   H_0: $\mu = 3.1$,
   H_1: $\mu \neq 3.1$.

   Suppose, as in Exercise 1, that the sample size n is 17, and it is found from the observed values in the sample that $\bar{X}_n = 3.2$ and $(1/n)\sum_{i=1}^{n}(X_i - \bar{X}_n)^2 = 0.09$. Calculate the value of the statistic U, and find the value of the corresponding tail area.

4. Consider again the conditions of Exercise 3, but suppose now that the sample size n is 170 and that it is again found from the observed values in the sample that $\bar{X}_n = 3.2$ and $(1/n)\sum_{i=1}^{n}(X_i - \bar{X}_n)^2 = 0.09$. Calculate the value of the statistic U, and find the value of the corresponding tail area.

5. Consider again the conditions of Exercise 3. Suppose, as in Exercise 3, that the sample size n is 17, but suppose now that it is found from the observed values in the sample that $\bar{X}_n = 3.0$ and $(1/n)\sum_{i=1}^{n}(X_i - \bar{X}_n)^2 = 0.09$. Calculate the value of the statistic U, and find the value of the corresponding tail area.

6. Suppose that a single observation X is taken from a normal distribution for which the mean μ is unknown and the variance is 1. Suppose that it is known that the value of μ must be -5, 0, or 5; and that it is desired to test the following hypotheses at the level of significance 0.05:

$$H_0: \quad \mu = 0$$
$$H_1: \quad \mu = -5 \text{ or } \mu = 5.$$

Suppose also that the test procedure to be used specifies rejecting H_0 when $|X| > c$, where the constant c is chosen so that $\Pr(|X| > c \mid \mu = 0) = 0.05$.

(a) Find the value of c; and show that if $X = 2$, then H_0 will be rejected.

(b) Show that if $X = 2$, then the value of the likelihood function at $\mu = 0$ is 12.2 times as large as its value at $\mu = 5$ and is 5.9×10^9 times as large as its value at $\mu = -5$.

7. Suppose that a random sample of 10,000 observations is taken from a normal distribution for which the mean μ is unknown and the variance is 1; and that it is desired to test the following hypotheses at the level of significance 0.05:

$$H_0: \quad \mu = 0,$$
$$H_1: \quad \mu \neq 0.$$

Suppose also that the test procedure specifies rejecting H_0 when $|\overline{X}_n| > c$, where the constant c is chosen so that $\Pr(|\overline{X}_n| > c \mid \mu = 0) = 0.05$. Find the probability that the test will reject H_0 if (a) the actual value of μ is 0.01 and (b) the actual value of μ is 0.02.

8. Consider again the conditions of Exercise 7, but suppose now that it is desired to test the following hypotheses:

$$H_0: \quad \mu \leqslant 0,$$
$$H_1: \quad \mu > 0.$$

Suppose also that in the random sample of 10,000 observations, the sample mean $\overline{X}_n$ is 0.03. At what level of significance is this result just significant?

8.8.   THE F DISTRIBUTION

Definition of the F Distribution

In this section we shall introduce a probability distribution, called the F distribution, that arises in many important problems of testing hypotheses in which two or more normal distributions are to be compared on the basis of random samples from each of the distributions. We shall begin by defining the F distribution and deriving its p.d.f.

Consider two independent random variables Y and Z such that Y has a χ^2 distribution with m degrees of freedom and Z has a χ^2 distribution with n degrees of freedom, where m and n are given positive integers. Define a new random variable X as follows:

$$X = \frac{Y/m}{Z/n} = \frac{nY}{mZ}. \tag{1}$$

Then the distribution of X is called an *F distribution with m and n degrees of freedom.*

We shall next show that if a random variable X has an F distribution with m and n degrees of freedom, then its p.d.f. $f(x)$ is as follows, for $x > 0$:

$$f(x) = \frac{\Gamma\left[\frac{1}{2}(m+n)\right]m^{m/2}n^{n/2}}{\Gamma\left(\frac{1}{2}m\right)\Gamma\left(\frac{1}{2}n\right)} \cdot \frac{x^{(m/2)-1}}{(mx+n)^{(m+n)/2}}. \tag{2}$$

Of course, $f(x) = 0$ for $x \leqslant 0$.

Since the random variables Y and Z are independent, their joint p.d.f. $g(y, z)$ will be the product of their individual p.d.f.'s. Furthermore, since both Y and Z have χ^2 distributions, it follows from the p.d.f. of the χ^2 distribution, as given in Sec. 7.2, that $g(y, z)$ has the following form, for $y > 0$ and $z > 0$:

$$g(y, z) = cy^{(m/2)-1}z^{(n/2)-1}e^{-(y+z)/2}, \tag{3}$$

where

$$c = \frac{1}{2^{(m+n)/2}\Gamma\left(\frac{1}{2}m\right)\Gamma\left(\frac{1}{2}n\right)}. \tag{4}$$

We shall now change variables from Y and Z to X and Z, where X is defined by Eq. (1). The joint p.d.f. $h(x, z)$ of X and Z is obtained by first replacing y in Eq. (3) with its expression in terms of x and z and then multiplying the result by $|\partial y/\partial x|$. It follows from Eq. (1) that $y = (m/n)xz$ and $\partial y/\partial x = (m/n)z$. Hence, the joint p.d.f. $h(x, z)$ has the following form, for $x > 0$ and $z > 0$:

$$h(x, z) = c\left(\frac{m}{n}\right)^{m/2}x^{(m/2)-1}z^{[(m+n)/2]-1}\exp\left[-\frac{1}{2}\left(\frac{m}{n}x + 1\right)z\right]. \tag{5}$$

Here, the constant c is again given by Eq. (4).

The marginal p.d.f. $f(x)$ of X can be obtained for any value of $x > 0$ from the relation

$$f(x) = \int_0^\infty h(x, z) \, dz. \tag{6}$$

It follows from Eq. (8) of Sec. 5.9 that

$$\int_0^\infty z^{[(m+n)/2]-1} \exp\left[-\frac{1}{2}\left(\frac{m}{n}x + 1\right)z \right] dz = \frac{\Gamma\left[\frac{1}{2}(m+n)\right]}{\left[\frac{1}{2}\left(\frac{m}{n}x + 1\right)\right]^{(m+n)/2}}. \tag{7}$$

From Eqs. (4) to (7), we can conclude that the p.d.f. $f(x)$ has the form given in Eq. (2).

Properties of the *F* Distribution

When we speak of an *F* distribution with m and n degrees of freedom, the order in which the numbers m and n are given is important, as can be seen from the definition of X in Eq. (1). When $m \neq n$, the *F* distribution with m and n degrees of freedom and the *F* distribution with n and m degrees of freedom are two different distributions. In fact, if a random variable X has an *F* distribution with m and n degrees of freedom, then its reciprocal $1/X$ will have an *F* distribution with n and m degrees of freedom. This statement follows from the representation of X as a ratio of two random variables, as in Eq. (1).

The *F* distribution is related to the *t* distribution in the following way: If a random variable X has a *t* distribution with n degrees of freedom, then X^2 will have an *F* distribution with 1 and n degrees of freedom. This result follows from the representation of X in Eq. (1) of Sec. 7.4.

Two short tables of probabilities for the *F* distribution are given at the end of this book. In these tables, we give only the 0.95 quantile and the 0.975 quantile for different possible pairs of values of m and n. In other words, if G denotes the d.f. of the *F* distribution with m and n degrees of freedom, then the tables give the values of x_1 and x_2 such that $G(x_1) = 0.95$ and $G(x_2) = 0.975$. By applying the relation between the *F* distribution for X and the *F* distribution for $1/X$, it is possible to use the tables to obtain also the 0.05 and 0.025 quantiles of an *F* distribution.

Example 1: Determining the 0.05 Quantile of an F Distribution. Suppose that a random variable X has an *F* distribution with 6 and 12 degrees of freedom. We shall determine the value of x such that $\Pr(X < x) = 0.05$.

If we let $Y = 1/X$, then Y will have an F distribution with 12 and 6 degrees of freedom. It can be found from the table given at the end of this book that $\Pr(Y > 4.00) = 0.05$. Since the relation $Y > 4.00$ is equivalent to the relation $X < 0.25$, it follows that $\Pr(X < 0.25) = 0.05$. Hence, $x = 0.25$. □

Comparing the Variances of Two Normal Distributions

We shall now consider a problem of testing hypotheses which uses the F distribution. Suppose that the random variables $X_1, \ldots, X_m$ form a random sample of m observations from a normal distribution for which both the mean μ_1 and the variance σ_1^2 are unknown; and also that the random variables $Y_1, \ldots, Y_n$ form an independent random sample of n observations from another normal distribution for which both the mean μ_2 and the variance σ_2^2 are unknown. Suppose finally that the following hypotheses are to be tested at a specified level of significance α_0 $(0 < \alpha_0 < 1)$:

$$
\begin{aligned}
H_0: \quad & \sigma_1^2 \leqslant \sigma_2^2, \\
H_1: \quad & \sigma_1^2 > \sigma_2^2.
\end{aligned}
\tag{8}
$$

For any test procedure δ, we shall let $\pi(\mu_1, \mu_2, \sigma_1^2, \sigma_2^2 \mid \delta)$ denote the power function of δ. We must find a procedure δ such that $\pi(\mu_1, \mu_2, \sigma_1^2, \sigma_2^2 \mid \delta) \leqslant \alpha_0$ for $\sigma_1^2 \leqslant \sigma_2^2$ and such that $\pi(\mu_1, \mu_2, \sigma_1^2, \sigma_2^2 \mid \delta)$ is as large as possible for $\sigma_1^2 > \sigma_2^2$. There is no UMP test of the hypotheses (8), but it is common practice to use one particular procedure, called the F test. The F test, which we shall now derive, has the specified level of significance α_0 and also has the following five properties:

(i) $\pi(\mu_1, \mu_2, \sigma_1^2, \sigma_2^2 \mid \delta) = \alpha_0$ when $\sigma_1^2 = \sigma_2^2$,

(ii) $\pi(\mu_1, \mu_2, \sigma_1^2, \sigma_2^2 \mid \delta) < \alpha_0$ when $\sigma_1^2 < \sigma_2^2$,

(iii) $\pi(\mu_1, \mu_2, \sigma_1^2, \sigma_2^2 \mid \delta) > \alpha_0$ when $\sigma_1^2 > \sigma_2^2$,

(iv) $\pi(\mu_1, \mu_2, \sigma_1^2, \sigma_2^2 \mid \delta) \to 0$ as $\sigma_1^2 / \sigma_2^2 \to 0$,

(v) $\pi(\mu_1, \mu_2, \sigma_1^2, \sigma_2^2 \mid \delta) \to 1$ as $\sigma_1^2 / \sigma_2^2 \to \infty$.

It follows immediately from properties (i), (ii), and (iii) that the F test is unbiased. Furthermore, it can be shown that the F test is actually a UMP test within the class of all unbiased tests. The proof is beyond the scope of this book [see Lehmann (1959)].

Derivation of the F Test

After the values $x_1, \ldots, x_m$ and $y_1, \ldots, y_n$ in the two samples have been observed, the likelihood function $g(x, y \mid \mu_1, \mu_2, \sigma_1^2, \sigma_2^2)$ is

$$
g(x, y \mid \mu_1, \mu_2, \sigma_1^2, \sigma_2^2) = f_m(x \mid \mu_1, \sigma_1^2) f_n(y \mid \mu_2, \sigma_2^2).
\tag{9}
$$

Here, both $f_m(x \mid \mu_1, \sigma_1^2)$ and $f_n(y \mid \mu_2, \sigma_2^2)$ have the general form given in Eq. (3) of Sec. 8.6. The likelihood ratio test procedure is based on a comparison of the following two values: the maximum value attained by the likelihood function (9) when the point $(\mu_1, \mu_2, \sigma_1^2, \sigma_2^2)$ varies over the subset Ω_0 specified by H_0, and the maximum value attained by the likelihood function when $(\mu_1, \mu_2, \sigma_1^2, \sigma_2^2)$ varies over the subset Ω_1 specified by H_1. For the hypotheses (8), Ω_0 contains every point $(\mu_1, \mu_2, \sigma_1^2, \sigma_2^2)$ such that $\sigma_1^2 \leqslant \sigma_2^2$, and Ω_1 contains every point such that $\sigma_1^2 > \sigma_2^2$. Therefore, we shall consider the following ratio:

$$r(x, y) = \frac{\sup_{(\mu_1, \mu_2, \sigma_1^2, \sigma_2^2) \in \Omega_1} g(x, y \mid \mu_1, \mu_2, \sigma_1^2, \sigma_2^2)}{\sup_{(\mu_1, \mu_2, \sigma_1^2, \sigma_2^2) \in \Omega_0} g(x, y \mid \mu_1, \mu_2, \sigma_1^2, \sigma_2^2)}. \tag{10}$$

The procedure then specifies that H_0 should be rejected if $r(x, y) \geqslant k$, where k is some chosen constant, and that H_0 should be accepted if $r(x, y) < k$.

It can be shown by methods similar to those given in Sec. 8.6 that $r(x, y) \geqslant k$ if and only if

$$\frac{\sum_{i=1}^{m}(x_i - \bar{x}_m)^2/(m-1)}{\sum_{i=1}^{n}(y_i - \bar{y}_n)^2/(n-1)} \geqslant c, \tag{11}$$

where c is another constant whose value can be derived from k. The details of this derivation will not be given here.

Now we shall let

$$S_X^2 = \sum_{i=1}^{m}(X_i - \bar{X}_m)^2 \quad \text{and} \quad S_Y^2 = \sum_{i=1}^{n}(Y_i - \bar{Y}_n)^2. \tag{12}$$

Also, in accordance with the relation (11), we shall let the statistic V be defined by the following relation:

$$V = \frac{S_X^2/(m-1)}{S_Y^2/(n-1)}. \tag{13}$$

The likelihood ratio test procedure which we have just described specifies that the hypothesis H_0 should be rejected if $V \geqslant c$ and that H_0 should be accepted if $V < c$. In this problem, this procedure is called an F test, for reasons that will now be explained.

Properties of the F Test

We know from Sec. 7.3 that the random variable S_X^2/σ_1^2 has a χ^2 distribution with $m-1$ degrees of freedom and the random variable S_Y^2/σ_2^2 has a χ^2 distribution with $n-1$ degrees of freedom. Furthermore, these two random variables will be independent, since they are calculated from two different samples. Therefore, the following random variable $V*$ will have an F distribution with $m - 1$ and $n - 1$ degrees of freedom:

$$V* = \frac{S_X^2/[(m-1)\sigma_1^2]}{S_Y^2/[(n-1)\sigma_2^2]}. \tag{14}$$

It can be seen from Eqs. (13) and (14) that $V = (\sigma_1^2/\sigma_2^2)V*$. If $\sigma_1^2 = \sigma_2^2$, then $V = V*$. Therefore, when $\sigma_1^2 = \sigma_2^2$ the statistic V will have an F distribution with $m - 1$ and $n - 1$ degrees of freedom. In this case it is possible to use a table of the F distribution to choose a constant c such that $\Pr(V \geqslant c) = \alpha_0$, regardless of the common value of σ_1^2 and σ_2^2 and regardless of the values of μ_1 and μ_2.

Now suppose that $\sigma_1^2 < \sigma_2^2$. Then

$$\Pr(V \geqslant c) = \Pr\left(\frac{\sigma_1^2}{\sigma_2^2}V* \geqslant c\right) = \Pr\left(V* \geqslant \frac{\sigma_2^2}{\sigma_1^2}c\right) < \Pr(V* \geqslant c) = \alpha_0. \tag{15}$$

Thus, when $\sigma_1^2 < \sigma_2^2$, it is seen that $\Pr(V \geqslant c) < \alpha_0$, regardless of the values of μ_1 and μ_2. It follows that the size of this test, which rejects H_0 when $V \geqslant c$, is α_0.

Also, when $\sigma_1^2 > \sigma_2^2$, an argument similar to that in Eq. (15) shows that $\Pr(V \geqslant c) > \alpha_0$, regardless of the values of μ_1 and μ_2. This argument also shows that $\Pr(V \geqslant c) \to 0$ as $\sigma_1^2/\sigma_2^2 \to 0$, and that $\Pr(V \geqslant c) \to 1$ as $\sigma_1^2/\sigma_2^2 \to \infty$.

Example 2: Performing an F Test. Suppose that 6 observations $X_1, \ldots, X_6$ are selected at random from a normal distribution for which both the mean μ_1 and the variance σ_1^2 are unknown; and that it is found that $\sum_{i=1}^{6}(X_i - \bar{X}_6)^2 = 30$. Suppose also that 21 observations $Y_1, \ldots, Y_{21}$ are selected at random from another normal distribution for which both the mean μ_2 and the variance σ_2^2 are unknown; and that it is found that $\sum_{i=1}^{21}(Y_i - \bar{Y}_{21})^2 = 40$. We shall carry out an F test of the hypotheses (8).

In this example, $m = 6$ and $n = 21$. Therefore, when H_0 is true, the statistic V defined by Eq. (13) will have an F distribution with 5 and 20 degrees of freedom. It follows from Eqs. (12) and (13) that the value of V for the given samples is

$$V = \frac{30/5}{40/20} = 3.$$

It is found from the tables given at the end of this book that the 0.95 quantile of the F distribution with 5 and 20 degrees of freedom is 2.71, and the 0.975 quantile of that distribution is 3.29. Hence, the tail area corresponding to the value $V = 3$ is less than 0.05 and greater than 0.025. The hypothesis H_0 that $\sigma_1^2 \leqslant \sigma_2^2$ should therefore be rejected at the level of significance $\alpha_0 = 0.05$, and H_0 should be accepted at the level of significance $\alpha_0 = 0.025$. $\square$

EXERCISES

1. Suppose that a random variable X has an F distribution with 3 and 8 degrees of freedom. Determine the value of c such that $\Pr(X > c) = 0.975$.

2. Suppose that a random variable X has an F distribution with 1 and 8 degrees of freedom. Use the table of the t distribution to determine the value of c such that $\Pr(X > c) = 0.3$.

3. Suppose that a random variable X has an F distribution with m and n degrees of freedom $(n > 2)$. Show that $E(X) = n/(n-2)$. *Hint:* Find the value of $E(1/Z)$, where Z has a χ^2 distribution with n degrees of freedom.

4. What is the value of the median of an F distribution with m and n degrees of freedom when $m = n$?

5. Suppose that a random variable X has an F distribution with m and n degrees of freedom. Show that the random variable $mX/(mX + n)$ has a beta distribution with parameters $\alpha = m/2$ and $\beta = n/2$.

6. Consider two different normal distributions for which both the means μ_1 and μ_2 and the variances σ_1^2 and σ_2^2 are unknown; and suppose that it is desired to test the following hypotheses:

$$H_0: \quad \sigma_1^2 \leqslant \sigma_2^2,$$
$$H_1: \quad \sigma_1^2 > \sigma_2^2.$$

Suppose further that a random sample consisting of 16 observations from the first normal distribution yields the values $\sum_{i=1}^{16} X_i = 84$ and $\sum_{i=1}^{16} X_i^2 = 563$; and that an independent random sample consisting of 10 observations from the second normal distribution yields the values $\sum_{i=1}^{10} Y_i = 18$ and $\sum_{i=1}^{10} Y_i^2 = 72$.

(a) What are the M.L.E.'s of σ_1^2 and σ_2^2?

(b) If an F test is carried out at the level of significance 0.05, is the hypothesis H_0 accepted or rejected?

7. Consider again the conditions of Exercise 6, but suppose now that it is desired to test the following hypotheses:

$$H_0: \quad \sigma_1^2 \leqslant 3\sigma_2^2,$$
$$H_1: \quad \sigma_1^2 > 3\sigma_2^2.$$

Describe how to carry out an F test of these hypotheses.

8. Consider again the conditions of Exercise 6, but suppose now that it is desired to test the following hypotheses:

$$H_0: \quad \sigma_1^2 = \sigma_2^2,$$
$$H_1: \quad \sigma_1^2 \neq \sigma_2^2.$$

Suppose also that the statistic V is defined by Eq. (13); and it is desired to reject H_0 if either $V < c_1$ or $V > c_2$, where the constants c_1 and c_2 are chosen so that when H_0 is true, $\Pr(V < c_1) = \Pr(V > c_2) = 0.025$. Determine the values of c_1 and c_2 when $m = 16$ and $n = 10$, as in Exercise 6.

9. Suppose that a random sample consisting of 16 observations is available from a normal distribution for which both the mean μ_1 and the variance σ_1^2 are unknown; and that an independent random sample consisting of 10 observations is available from another normal distribution for which both the mean μ_2 and the variance σ_2^2 are also unknown. For any given constant $r > 0$, use the results of Exercise 8 to construct a test of the following hypotheses at the level of significance 0.05:

$$H_0: \quad \frac{\sigma_1^2}{\sigma_2^2} = r,$$

$$H_1: \quad \frac{\sigma_1^2}{\sigma_2^2} \neq r.$$

10. Consider again the conditions of Exercise 9. Use the results of that exercise to construct a confidence interval for σ_1^2/σ_2^2 with confidence coefficient 0.95.

11. Suppose that a random variable Y has a χ^2 distribution with m_0 degrees of freedom, and let c be a constant such that $\Pr(Y > c) = 0.05$. Explain why, in the table of the 0.95 quantile of the F distribution, the entry for $m = m_0$ and $n = \infty$ will be equal to c/m_0.

12. The final column in the table of the 0.95 quantile of the F distribution contains values for which $m = \infty$. Explain how to derive the entries in this column from a table of the χ^2 distribution.

8.9.   COMPARING THE MEANS OF TWO NORMAL DISTRIBUTIONS

Derivation of the Two-Sample t Test

We shall now consider a problem in which random samples are available from two normal distributions with a common unknown variance, and it is desired to determine which distribution has the larger mean. Specifically, we shall assume that the variables $X_1, \ldots, X_m$ form a random sample of m observations from a normal distribution for which both the mean μ_1 and the variance σ^2 are unknown; and that the variables $Y_1, \ldots, Y_n$ form an independent random sample of n observations from another normal distribution for which both the mean μ_2 and the variance σ^2 are unknown. We shall assume that the variance σ^2 is the same for both distributions, even though the value of σ^2 is unknown.

Suppose that it is desired to test the following hypotheses at a specified level of significance α_0 ($0 < \alpha_0 < 1$):

$$H_0: \quad \mu_1 \leq \mu_2,$$
$$H_1: \quad \mu_1 > \mu_2. \tag{1}$$

For any test procedure δ, we shall let $\pi(\mu_1, \mu_2, \sigma^2 \mid \delta)$ denote the power function of δ. We must find a procedure δ such that $\pi(\mu_1, \mu_2, \sigma^2 \mid \delta) \leq \alpha_0$ for $\mu_1 \leq \mu_2$ and such that $\pi(\mu_1, \mu_2, \sigma^2 \mid \delta)$ is as large as possible for $\mu_1 > \mu_2$.

There is no UMP test of the hypotheses (1), but it is common practice to use a certain t test. This t test, which we shall now derive, has the specified level of significance α_0 and also has the following five properties:

(i) $\pi(\mu_1, \mu_2, \sigma^2 \mid \delta) = \alpha_0$ when $\mu_1 = \mu_2$,

(ii) $\pi(\mu_1, \mu_2, \sigma^2 \mid \delta) < \alpha_0$ when $\mu_1 < \mu_2$,

(iii) $\pi(\mu_1, \mu_2, \sigma^2 \mid \delta) > \alpha_0$ when $\mu_1 > \mu_2$,

(iv) $\pi(\mu_1, \mu_2, \sigma^2 \mid \delta) \to 0$ as $\mu_1 - \mu_2 \to -\infty$,

(v) $\pi(\mu_1, \mu_2, \sigma^2 \mid \delta) \to 1$ as $\mu_1 - \mu_2 \to \infty$.

It follows immediately from properties (i), (ii), and (iii) that the test δ is unbiased. Furthermore, it can be shown that δ is actually UMP within the class of all unbiased tests. The proof is beyond the scope of this book [see Lehmann (1959)].

After the values $x_1, \ldots, x_m$ and $y_1, \ldots, y_n$ in the two samples have been observed, the likelihood function $g(x, y \mid \mu_1, \mu_2, \sigma^2)$ is

$$g(x, y \mid \mu_1, \mu_2, \sigma^2) = f_m(x \mid \mu_1, \sigma^2) f_n(y \mid \mu_2, \sigma^2). \tag{2}$$

Here, both $f_m(x \mid \mu_1, \sigma^2)$ and $f_n(y \mid \mu_2, \sigma^2)$ have the form given in Eq. (3) of Sec. 8.6, and the value of σ^2 is the same in both terms. The likelihood ratio test procedure is based on a comparison of the following two values: the maximum value attained by the likelihood function (2) when the point (μ_1, μ_2, σ^2) varies over the subset Ω_0 specified by H_0, and the maximum value attained by the likelihood function when (μ_1, μ_2, σ^2) varies over the subset Ω_1 specified by H_1. For the hypotheses (1), Ω_0 contains every point (μ_1, μ_2, σ^2) such that $\mu_1 \leq \mu_2$, and Ω_1 contains every point such that $\mu_1 > \mu_2$. Therefore, we shall consider the following ratio:

$$r(x, y) = \frac{\sup_{(\mu_1, \mu_2, \sigma^2) \in \Omega_1} g(x, y \mid \mu_1, \mu_2, \sigma^2)}{\sup_{(\mu_1, \mu_2, \sigma^2) \in \Omega_0} g(x, y \mid \mu_1, \mu_2, \sigma^2)}. \tag{3}$$

The procedure then specifies that H_0 should be rejected if $r(x, y) \geq k$, where k is some chosen constant, and that H_0 should be accepted if $r(x, y) < k$.

It can be shown by methods similar to those given in Sec. 8.6 that $r(x, y) \geq k$ if and only if

$$\frac{(m + n - 2)^{1/2}(\bar{x}_m - \bar{y}_n)}{\left(\frac{1}{m} + \frac{1}{n}\right)^{1/2}\left[\sum_{i=1}^{m}(x_i - \bar{x}_m)^2 + \sum_{i=1}^{n}(y_i - \bar{y}_n)^2\right]^{1/2}} \geq c, \tag{4}$$

where c is another constant whose value can be derived from k. The details of this derivation will not be given here.

We shall again let S_X^2 and S_Y^2 be the sums of squares defined in Eq. (12) of Sec. 8.8. Also, in accordance with the relation (4), we shall let the statistic U be defined by the following relation:

$$U = \frac{(m + n - 2)^{1/2}(\bar{X}_m - \bar{Y}_n)}{\left(\frac{1}{m} + \frac{1}{n}\right)^{1/2}(S_X^2 + S_Y^2)^{1/2}}. \tag{5}$$

The likelihood ratio test procedure which we have just described specifies that the hypothesis H_0 should be rejected if $U \geq c$, and that H_0 should be accepted if $U < c$. In this problem, this procedure is called a two-sample t test, for reasons that will now be explained.

Properties of the Two-Sample t Test

We shall derive the distribution of the statistic U when $\mu_1 = \mu_2$. For any given values of μ_1, μ_2, and σ^2, the sample mean $\bar{X}_m$ has a normal distribution with mean μ_1 and variance σ^2/m, and the sample mean $\bar{Y}_n$ has a normal distribution with mean μ_2 and variance σ^2/n. Since $\bar{X}_m$ and $\bar{Y}_n$ are independent, it follows that the difference $\bar{X}_m - \bar{Y}_n$ has a normal distribution with mean $\mu_1 - \mu_2$ and variance $[(1/m) + (1/n)]\sigma^2$. Therefore, when $\mu_1 = \mu_2$, the following random variable Z_1 will have a standard normal distribution:

$$Z_1 = \frac{\bar{X}_m - \bar{Y}_n}{\left(\frac{1}{m} + \frac{1}{n}\right)^{1/2}\sigma}. \tag{6}$$

Also, for any values of μ_1, μ_2, and σ^2, the random variable S_X^2/σ^2 has a χ^2 distribution with $m-1$ degrees of freedom; the random variable S_Y^2/σ^2 has a χ^2 distribution with $n-1$ degrees of freedom; and these two random variables will be independent. Therefore, the following random variable Z_2 has a χ^2 distribution with $m+n-2$ degrees of freedom:

$$Z_2 = \frac{S_X^2 + S_Y^2}{\sigma^2}. \tag{7}$$

Furthermore, the four random variables $\overline{X}_m$, $\overline{Y}_n$, S_X^2, and S_Y^2 are independent. This result is implied by the following two facts: (i) If one random variable is a function of $X_1, \ldots, X_m$ only and if another random variable is a function of $Y_1, \ldots, Y_n$ only, then these two variables must be independent. (ii) By Theorem 1 of Sec. 7.3, $\overline{X}_m$ and S_X^2 are independent, and $\overline{Y}_n$ and S_Y^2 are also independent. It follows that the two random variables Z_1 and Z_2 are independent. The random variable Z_1 has a standard normal distribution when $\mu_1 = \mu_2$, and the random variable Z_2 has a χ^2 distribution with $m+n-2$ degrees of freedom for any values of μ_1, μ_2, and σ^2. The statistic U can now be represented in the form

$$U = \frac{Z_1}{[Z_2/(m+n-2)]^{1/2}}. \tag{8}$$

Therefore, when $\mu_1 = \mu_2$, it follows from the definition of the t distribution given in Sec. 7.4 that U will have a t distribution with $m+n-2$ degrees of freedom. Thus, when $\mu_1 = \mu_2$, it is possible to use a table of this t distribution to choose a constant c such that $\Pr(U \geq c) = \alpha_0$, regardless of the common value of μ_1 and μ_2 and regardless of the value of σ^2.

If $\mu_1 < \mu_2$, then it can be shown by an argument similar to that given in Sec. 8.6 that $\Pr(U \geq c) < \alpha_0$, regardless of the value of σ^2. It follows that the size of this test, which rejects H_0 when $U \geq c$, is α_0. Also, it can be shown that if $\mu_1 > \mu_2$, then $\Pr(U \geq c) > \alpha_0$, regardless of the value of σ^2. Furthermore, $\Pr(U \geq c) \to 0$ as $\mu_1 - \mu_2 \to -\infty$ and $\Pr(U \geq c) \to 1$ as $\mu_1 - \mu_2 \to \infty$.

Example 1: Performing a Two-Sample t Test. Suppose that a random sample of 8 specimens of ore is collected from a certain location in a copper mine, and that the amount of copper in each of the specimens is measured in grams. We shall denote these 8 amounts by $X_1, \ldots, X_8$, and shall suppose that the observed values are such that $\overline{X}_8 = 2.6$ and $S_X^2 = \sum_{i=1}^8 (X_i - \overline{X}_8)^2 = 0.32$. Suppose also that a second random sample of 10 specimens of ore is collected from another part of the mine. We shall denote the amounts of copper in these specimens by $Y_1, \ldots, Y_{10}$, and shall suppose that the observed values in grams are such that $\overline{Y}_{10} = 2.3$, and $S_Y^2 = \sum_{i=1}^{10} (Y_i - \overline{Y}_{10})^2 = 0.22$. Let μ_1 denote the mean amount of copper in all the ore at the first location in the mine; let μ_2 denote the mean amount of copper in all the ore at the second location; and suppose that the hypotheses (1) are to be tested.

We shall assume that all the observations have a normal distribution and that the variance is the same at both locations in the mine, even though the means may be different. In this example, the sample sizes are $m = 8$ and $n = 10$, and the value of the statistic U defined by Eq. (5) is 3.442. Also, by use of a table of the t distribution with 16 degrees of freedom, it is found that the tail area corresponding to this observed value of U is less than 0.005. Hence, the null hypothesis will be rejected for any specified level of significance $\alpha_0 \geqslant 0.005$. $\square$

Two-Sided Alternatives and Unequal Variances

The procedure based on the statistic U which has just been described can easily be adapted to testing the following hypotheses at a specified level of significance α_0:

$$H_0: \quad \mu_1 = \mu_2, \qquad\qquad (9)$$
$$H_1: \quad \mu_1 \neq \mu_2.$$

Since the alternative hypothesis in this case is two-sided, the test procedure would be to reject H_0 if either $U < c_1$ or $U > c_2$, where the constants c_1 and c_2 are chosen so that when H_0 is true, $\Pr(U < c_1) + \Pr(U > c_2) = \alpha_0$.

More importantly, the basic procedure can be extended to a problem in which the variances of the two normal distributions are not equal, but the ratio of one variance to the other is known. Specifically, suppose that $X_1, \ldots, X_m$ form a random sample from a normal distribution with mean μ_1 and variance σ_1^2; and that $Y_1, \ldots, Y_n$ form an independent random sample from another normal distribution with mean μ_2 and variance σ_2^2. Suppose also that the values of μ_1, μ_2, σ_1^2, and σ_2^2 are unknown but that $\sigma_2^2 = k\sigma_1^2$, where k is a known positive constant. Then it can be shown (see Exercise 3 at the end of this section) that when $\mu_1 = \mu_2$, the following random variable U will have a t distribution with $m + n - 2$ degrees of freedom:

$$U = \frac{(m + n - 2)^{1/2}(\overline{X}_m - \overline{Y}_n)}{\left(\dfrac{1}{m} + \dfrac{k}{n}\right)^{1/2}\left(S_X^2 + \dfrac{S_Y^2}{k}\right)^{1/2}}. \qquad (10)$$

Hence, the statistic U defined by Eq. (10) can be used for testing either the hypotheses (1) or the hypotheses (9).

Finally, if the values of all four parameters μ_1, μ_2, σ_1^2, and σ_2^2 are unknown and if the value of the ratio σ_1^2/σ_2^2 is also unknown, then the problem of testing the hypotheses (1) or the hypotheses (9) becomes very difficult. This problem is known as the *Behrens-Fisher problem*. Various test procedures have been proposed, but most of them have been the subject of controversy in regard to their appropriateness or usefulness. No single procedure has found widespread acceptance among statisticians. This problem will not be discussed in this book.

EXERCISES

1. Suppose that a certain drug A was administered to eight patients selected at random; and that after a fixed time period, the concentration of the drug in certain body cells of each patient was measured in appropriate units. Suppose that these concentrations for the eight patients were found to be as follows:

   1.23, 1.42, 1.41, 1.62, 1.55, 1.51, 1.60, and 1.76.

   Suppose also that a second drug B was administered to six different patients selected at random; and that when the concentration of drug B was measured in a similar way for these six patients, the results were as follows:

   1.76, 1.41, 1.87, 1.49, 1.67, and 1.81.

   Assuming that all the observations have a normal distribution with a common unknown variance, test the following hypotheses at the level of significance 0.10: The null hypothesis is that the mean concentration of drug A among all patients is at least as large as the mean concentration of drug B. The alternative hypothesis is that the mean concentration of drug B is larger than that of drug A.

2. Consider again the conditions of Exercise 1, but suppose now that it is desired to test the following hypotheses: The null hypothesis is that the mean concentration of drug A among all patients is the same as the mean concentration of drug B. The alternative hypothesis, which is two-sided, is that the mean concentrations of the two drugs are not the same. Suppose that the test procedure specifies rejecting H_0 if either $U < c_1$ or $U > c_2$, where U is defined by Eq. (5) and c_1 and c_2 are chosen so that when H_0 is true, $\Pr(U < c_1) = \Pr(U > c_2) = 0.05$. Determine the values of c_1 and c_2, and determine whether the hypothesis H_0 will be accepted or rejected when the sample data are as given in Exercise 1.

3. Suppose that $X_1, \ldots, X_m$ form a random sample from a normal distribution with mean μ_1 and variance σ_1^2; and that $Y_1, \ldots, Y_n$ form an independent random sample from a normal distribution with mean μ_2 and variance σ_2^2. Show that if $\mu_1 = \mu_2$ and $\sigma_2^2 = k\sigma_1^2$, then the random variable U defined by Eq. (10) has a t distribution with $m + n - 2$ degrees of freedom.

4. Consider again the conditions and observed values of Exercise 1. However, suppose now that each observation for drug A has an unknown variance σ_1^2 and each observation for drug B has an unknown variance σ_2^2, but it is known that $\sigma_2^2 = (6/5)\sigma_1^2$. Test the hypotheses described in Exercise 1 at the level of significance 0.10.

5. Suppose that $X_1, \ldots, X_m$ form a random sample from a normal distribution with unknown mean μ_1 and unknown variance σ^2; and that $Y_1, \ldots, Y_n$ form an independent random sample from another normal distribution with unknown mean μ_2 and the same unknown variance σ^2. For any given constant λ $(-\infty < \lambda < \infty)$, construct a t test of the following hypotheses with $m + n - 2$ degrees of freedom:

$$H_0: \quad \mu_1 - \mu_2 = \lambda,$$
$$H_1: \quad \mu_1 - \mu_2 \neq \lambda.$$

6. Consider again the conditions of Exercise 1. Let μ_1 denote the mean of each observation for drug A, and let μ_2 denote the mean of each observation for drug B. It is assumed, as in Exercise 1, that all the observations have a common unknown variance. Use the results of Exercise 5 to construct a confidence interval for $\mu_1 - \mu_2$ with confidence coefficient 0.90.

8.10. SUPPLEMENTARY EXERCISES

1. Suppose that a sequence of Bernoulli trials is to be carried out with an unknown probability θ of success on each trial, and that the following hypotheses are to be tested:

$$H_0: \quad \theta = 0.1,$$
$$H_1: \quad \theta = 0.2.$$

Let X denote the number of trials required to obtain a success, and suppose that H_0 is to be rejected if $X \leq 5$. Determine the probabilities of errors of type 1 and type 2.

2. Consider again the conditions of Exercise 1. Suppose that the losses from errors of type 1 and type 2 are equal, and that the prior probabilities that H_0 and H_1 are true are equal. Determine the Bayes test procedure based on the observation X.

3. Suppose that a single observation X is to be drawn from the following p.d.f.:

$$f(x \mid \theta) = \begin{cases} 2(1 - \theta)x + \theta & \text{for } 0 \leq x \leq 1, \\ 0 & \text{otherwise,} \end{cases}$$

where the value of θ is unknown ($0 \leqslant \theta \leqslant 2$). Suppose also that the following hypotheses are to be tested:

H_0: $\theta = 2$,
H_1: $\theta = 0$.

Determine the test procedure δ for which $\alpha(\delta) + 2\beta(\delta)$ is a minimum, and calculate this minimum value.

4. Consider again the conditions of Exercise 3, and suppose that the value of α is given ($0 < \alpha < 1$). Determine the test procedure δ for which $\beta(\delta)$ will be a minimum. and calculate this minimum value.

5. Consider again the conditions of Exercise 3, but suppose now that the following hypotheses are to be tested:

H_0: $\theta \geqslant 1$,
H_1: $\theta < 1$.

(a) Determine the power function of the test δ that specifies rejecting H_0 if $X > 0.9$.

(b) What is the size of the test δ?

6. Consider again the conditions of Exercise 3. Show that the p.d.f. $f(x \mid \theta)$ has a monotone likelihood ratio in the statistic $r(X) = -X$, and determine a UMP test of the following hypotheses at the level of significance $\alpha_0 = 0.05$:

H_0: $\theta \leqslant \dfrac{1}{2}$,

H_1: $\theta > \dfrac{1}{2}$.

7. Suppose that a box contains a large number of chips of three different colors, red, brown, and blue; and that it is desired to test the null hypothesis H_0 that chips of the three colors are present in equal proportions against the alternative hypothesis H_1 that they are not present in equal proportions. Suppose that three chips are to be drawn at random from the box, and that H_0 is to be rejected if and only if at least two of the chips have the same color.

(a) Determine the size of the test.

(b) Determine the power of the test if $1/7$ of the chips are red, $2/7$ are brown, and $4/7$ are blue.

8. Suppose that a single observation X is to be drawn from an unknown distribution P, and that the following simple hypotheses are to be tested:

H_0: P is a uniform distribution on the interval $(0, 1)$,
H_1: P is a standard normal distribution.

Determine the most powerful test of size 0.01 and calculate the power of the test when H_1 is true.

9. Suppose that the 12 observations $X_1, \ldots, X_{12}$ form a random sample from a normal distribution for which both the mean μ and variance σ^2 are unknown. Describe how to carry out a t test of the following hypotheses at the level of significance $\alpha_0 = 0.005$:

$$H_0: \quad \mu \geqslant 3,$$
$$H_1: \quad \mu < 3.$$

10. Suppose that $X_1, \ldots, X_n$ form a random sample from a normal distribution with unknown mean θ and variance 1; and that it is desired to test the following hypotheses:

$$H_0: \quad \theta \leqslant 0,$$
$$H_1: \quad \theta > 0.$$

Suppose also that it is decided to use a UMP test for which the power is 0.95 when $\theta = 1$. Determine the size of this test if $n = 16$.

11. Suppose that eight observations $X_1, \ldots, X_8$ are drawn at random from a distribution with the following p.d.f.:

$$f(x \mid \theta) = \begin{cases} \theta x^{\theta-1} & \text{for } 0 < x < 1, \\ 0 & \text{otherwise.} \end{cases}$$

Suppose also that the value of θ is unknown ($\theta > 0$), and that it is desired to test the following hypotheses:

$$H_0: \quad \theta \leqslant 1,$$
$$H_1: \quad \theta > 1.$$

Show that the UMP test at the level of significance $\alpha_0 = 0.05$ specifies rejecting H_0 if $\sum_{i=1}^{8} \log X_i \geqslant -3.981$.

12. Suppose that $X_1, \ldots, X_n$ form a random sample from a χ^2 distribution for which the degrees of freedom θ is unknown ($\theta = 1, 2, \ldots$), and that it is desired to test the following hypotheses at a given level of significance α_0 ($0 < \alpha_0 < 1$):

$$H_0: \quad \theta \leqslant 8,$$
$$H_1: \quad \theta \geqslant 9.$$

Show that there exists a UMP test, and that the test specifies rejecting H_0 if $\sum_{i=1}^{n} \log X_i \geqslant k$ for some appropriate constant k.

13. Suppose that $X_1, \ldots, X_{10}$ form a random sample from a normal distribution for which both the mean and the variance are unknown. Construct a statistic that does not depend on any unknown parameters and that has an F distribution with 3 and 5 degrees of freedom.

14. Suppose that $X_1, \ldots, X_m$ form a random sample from a normal distribution for which both the mean μ_1 and the variance σ_1^2 are unknown; and that $Y_1, \ldots, Y_n$ form an independent random sample from another normal distribution for which both the mean μ_2 and the variance σ_2^2 are unknown. Suppose also that it is desired to test the following hypotheses with the usual F test at the level of significance $\alpha_0 = 0.05$:

$$H_0: \quad \sigma_1^2 \leqslant \sigma_2^2,$$
$$H_1: \quad \sigma_1^2 > \sigma_2^2.$$

Assuming that $m = 16$ and $n = 21$ show that the power of the test when $\sigma_1^2 = 2\sigma_2^2$ is given by $\Pr(V^* \geqslant 1.1)$, where V^* is a random variable having an F distribution with 15 and 20 degrees of freedom.

15. Suppose that the 9 observations $X_1, \ldots, X_9$ form a random sample from a normal distribution with unknown mean μ_1 and unknown variance σ^2; and that the 9 observations $Y_1, \ldots, Y_9$ form an independent random sample from another normal distribution with unknown mean μ_2 and the same unknown variance σ^2. Let S_X^2 and S_Y^2 be as defined in Eq. (12) of Sec. 8.8 (with $m = n = 9$), and let

$$T = \max\left\{ \frac{S_X^2}{S_Y^2}, \frac{S_Y^2}{S_X^2} \right\}.$$

Determine the value of the constant c such that $\Pr(T > c) = 0.05$.

16. An unethical experimenter desires to test the following hypotheses:

$$H_0: \quad \theta = \theta_0,$$
$$H_1: \quad \theta \neq \theta_0.$$

He draws a random sample $X_1, \ldots, X_n$ from a distribution with the p.d.f. $f(x \mid \theta)$, and carries out a test of size α. If this test does not reject H_0, he discards the sample, draws a new independent random sample of n observations, and repeats the test based on the new sample. He continues drawing new independent samples in this way until he obtains a sample for which H_0 is rejected.

(a) What is the overall size of this testing procedure?

(b) If H_0 is true, what is the expected number of samples that the experimenter will have to draw until he rejects H_0?

17. Suppose that $X_1, \ldots, X_n$ form a random sample from a normal distribution for which both the mean μ and the precision τ are unknown; and that the following hypotheses are to be tested:

$H_0: \quad \mu \leq 3,$
$H_1: \quad \mu > 3.$

Suppose that the prior joint distribution of μ and τ is a normal-gamma distribution, as described in Theorem 1 of Sec. 7.6, with $\mu_0 = 3$, $\lambda_0 = 1$, $\alpha_0 = 1$, and $\beta_0 = 1$. Suppose finally that $n = 17$, and it is found from the observed values in the sample that $\bar{X}_n = 3.2$ and $\sum_{i=1}^{n}(X_i - \bar{X}_n)^2 = 17$. Determine both the prior probability and the posterior probability that H_0 is true.

18. Consider a problem of testing hypotheses in which the following hypotheses about an arbitrary parameter θ are to be tested:

$H_0: \quad \theta \in \Omega_0,$
$H_1: \quad \theta \in \Omega_1.$

Suppose that δ is a test procedure of size α ($0 < \alpha < 1$) based on some vector of observations X, and let $\pi(\theta \mid \delta)$ denote the power function of δ. Show that if δ is unbiased, then $\pi(\theta \mid \delta) \geq \alpha$ at every point $\theta \in \Omega_1$.

19. Consider again the conditions of Exercise 18. Suppose now that θ is a two-dimensional vector $\theta = (\theta_1, \theta_2)$, where θ_1 and θ_2 are real-valued parameters. Suppose also that A is a particular circle in the $\theta_1\theta_2$-plane, and that the hypotheses to be tested are as follows:

$H_0: \quad \theta \in A,$
$H_1: \quad \theta \notin A.$

Show that if the test procedure δ is unbiased and of size α and if its power function $\pi(\theta \mid \delta)$ is a continuous function of θ, then it must be true that $\pi(\theta \mid \delta) = \alpha$ at each point θ on the boundary of the circle A.

20. Consider again the conditions of Exercise 18. Suppose now that θ is a real-valued parameter, and that the following hypotheses are to be tested:

$H_0: \quad \theta = \theta_0,$
$H_1: \quad \theta \neq \theta_0.$

Assume that θ_0 is an interior point of the parameter space Ω. Show that if the test procedure δ is unbiased and if its power function $\pi(\theta \mid \delta)$ is a differentiable function of θ, then $\pi'(\theta_0 \mid \delta) = 0$, where $\pi'(\theta_0 \mid \delta)$ denotes the derivative of $\pi(\theta \mid \delta)$ evaluated at the point $\theta = \theta_0$.

21. Suppose that the differential brightness θ of a certain star has an unknown value, and that it is desired to test the following simple hypotheses:

H_0: $\theta = 0$,
H_1: $\theta = 10$.

The statistician knows that when he goes to the observatory at midnight to measure θ there is probability $1/2$ that the meteorological conditions will be good and he will be able to obtain a measurement X having a normal distribution with mean θ and variance 1. He also knows that there is probability $1/2$ that the meteorological conditions will be poor and he will obtain a measurement Y having a normal distribution with mean θ and variance 100.

(a) Construct the most powerful test that has conditional size $\alpha = 0.05$, given good meteorological conditions; and that has conditional size $\alpha = 0.05$, given poor meteorological conditions.

(b) Construct the most powerful test that has conditional size $\alpha = 0.000001$, given good meteorological conditions; and that has conditional size $\alpha = 0.099999$, given poor meteorological conditions.

(c) Show that the overall size of both the test found in part (a) and the test found in part (b) is 0.05, and determine the power of each of these two tests.

22. Discuss the relative merits of the tests found in parts (a) and (b) of Exercise 21. In particular, show that although the test found in part (b) has greater power, it violates the likelihood principle. *Hint:* Show that the likelihood function for θ obtained when meteorological conditions are good is the same as the likelihood function that would be obtained if the statistician knew in advance with certainty that he would obtain an observation X having a normal distribution with mean θ and variance 1. However, for some values of X, the decisions specified by the most powerful tests of size 0.05 will be different.

Categorical Data and Nonparametric Methods

9

9.1. TESTS OF GOODNESS-OF-FIT

Description of Nonparametric Problems

In each of the problems of estimation and testing hypotheses which we have considered in Chapters 6, 7, and 8, we have assumed that the observations which are available to the statistician come from distributions for which the exact form is known, even though the values of some parameters are unknown. For example, it might be assumed that the observations form a random sample from a Poisson distribution for which the mean is unknown, or it might be assumed that the observations come from two normal distributions for which the means and variances are unknown. In other words, we have assumed that the observations come from a certain *parametric* family of distributions and a statistical inference must be made about the values of the parameters defining that family.

In many of the problems to be discussed in this chapter, we shall not assume that the available observations come from a particular parametric family of distributions. Rather, we shall study inferences that can be made about the distribution from which the observations come, without making special assumptions about the form of that distribution. As one example, we might assume simply that the observations form a random sample from a continuous distribution, without specifying the form of this distribution any further; and we might then investigate the possibility that this distribution is a normal distribution. As a second example, we might be interested in making an inference about the value of the median of the distribution from which the sample was drawn, and we might assume only that this distribution is continuous. As a third example, we might be interested in investigating the possibility that two independent random samples actually come from the same distribution, and we might assume only that both distributions from which the samples are taken are continuous.

Problems in which the possible distributions of the observations are not restricted to a specific parametric family are called *nonparametric problems*, and the statistical methods that are applicable in such problems are called *nonparametric methods*.

Categorical Data

In this section and the next four sections we shall consider statistical problems based on data such that each observation can be classified as belonging to one of a finite number of possible categories or types. Observations of this type are called *categorical data*. Since there are only a finite number of possible categories in these problems, and since we are interested in making inferences about the probabilities of these categories, these problems actually involve just a finite number of parameters. However, as we shall see, methods based on categorical data can be usefully applied in both parametric and nonparametric problems.

The χ^2 Test

Suppose that a large population consists of items of k different types, and let p_i denote the probability that an item selected at random will be of type i $(i = 1, \ldots, k)$. It is assumed that $p_i \geqslant 0$ for $i = 1, \ldots, k$ and that $\sum_{i=1}^{k} p_i = 1$. Let $p_1^0, \ldots, p_k^0$ be specific numbers such that $p_i^0 > 0$ for $i = 1, \ldots, k$ and $\sum_{i=1}^{k} p_i^0 = 1$, and suppose that the following hypotheses are to be tested:

$$
\begin{aligned}
H_0: \quad & p_i = p_i^0 \quad && \text{for } i = 1, \ldots, k, \\
H_1: \quad & p_i \neq p_i^0 \quad && \text{for at least one value of } i.
\end{aligned}
\tag{1}
$$

We shall assume that a random sample of size n is to be taken from the given population. That is, n independent observations are to be taken, and there is probability p_i that any particular observation will be of type i $(i = 1, \ldots, k)$. On the basis of these n observations, the hypotheses (1) are to be tested.

For $i = 1, \ldots, k$, we shall let N_i denote the number of observations in the random sample which are of type i. Thus, $N_1, \ldots, N_k$ are nonnegative integers such that $\sum_{i=1}^{k} N_i = n$. When the null hypothesis H_0 is true, the expected number of observations of type i is np_i^0 $(i = 1, \ldots, k)$. The difference between the actual number of observations N_i and the expected number np_i^0 will tend to be smaller when H_0 is true than when H_0 is not true. It seems reasonable, therefore, to base a test of the hypotheses (1) on values of the differences $N_i - np_i^0$ for $i = 1, \ldots, k$, and to reject H_0 when the magnitudes of these differences are relatively large.

In 1900, Karl Pearson proposed the use of the following statistic:

$$Q = \sum_{i=1}^{k} \frac{\left(N_i - np_i^0\right)^2}{np_i^0}. \tag{2}$$

Furthermore, Pearson showed that if the hypothesis H_0 is true, then as the sample size $n \to \infty$, the d.f. of Q converges to the d.f. of the χ^2 distribution with $k - 1$ degrees of freedom. Thus, if H_0 is true and the sample size n is large, the distribution of Q will be approximately a χ^2 distribution with $k - 1$ degrees of freedom. The discussion which we have presented indicates that H_0 should be rejected when $Q > c$, where c is an appropriate constant. If it is desired to carry out the test at the level of significance α_0, then c should be chosen so that $\Pr(Q > c) = \alpha_0$ when Q has a χ^2 distribution with $k - 1$ degrees of freedom. This test is called the χ^2 *test of goodness-of-fit*.

Whenever the value of each expectation np_i^0 $(i = 1, \ldots, k)$ is not too small, the χ^2 distribution will be a good approximation to the actual distribution of Q. Specifically, the approximation will be very good if $np_i^0 \geq 5$ for $i = 1, \ldots, k$, and the approximation should still be satisfactory if $np_i^0 \geq 1.5$ for $i = 1, \ldots, k$.

We shall now illustrate the use of the χ^2 test of goodness-of-fit by considering two examples.

Testing Hypotheses About a Proportion

Suppose that the proportion p of defective items in a large population of manufactured items is unknown and that the following hypotheses are to be tested:

$$\begin{aligned} H_0: &\quad p = 0.1, \\ H_1: &\quad p \neq 0.1. \end{aligned} \tag{3}$$

Suppose also that in a random sample of 100 items, it is found that 16 are defective. We shall test the hypotheses (3) by carrying out a χ^2 test of goodness-of-fit.

Since there are only two types of items in this example, namely, defective items and nondefective items, we know that $k = 2$. Furthermore, if we let p_1 denote the unknown proportion of defective items and let p_2 denote the unknown proportion of nondefective items, then the hypotheses (3) can be rewritten in the following form:

$$\begin{aligned} H_0: &\quad p_1 = 0.1 \quad \text{and} \quad p_2 = 0.9. \\ H_1: &\quad \text{The hypothesis } H_0 \text{ is not true.} \end{aligned} \tag{4}$$

For the sample size $n = 100$, the expected number of defective items if H_0 is true is $np_1^0 = 10$, and the expected number of nondefective items is $np_2^0 = 90$. Let N_1 denote the number of defective items in the sample, and let N_2 denote the number of nondefective items in the sample. Then, when H_0 is true, the distribution of the statistic Q defined by Eq. (2) will be approximately a χ^2 distribution with one degree of freedom.

In this example, $N_1 = 16$ and $N_2 = 84$; and it is found that the value of Q is 4. It can now be determined from a table of the χ^2 distribution with one degree of freedom that the tail area corresponding to the value $Q = 4$ lies between 0.025 and 0.05. Hence, the null hypothesis H_0 should be rejected at the level of significance $\alpha_0 = 0.05$, but it should be accepted at the level $\alpha_0 = 0.025$.

Testing Hypotheses About a Continuous Distribution

Consider a random variable X which takes values in the interval $0 < X < 1$ but which has an unknown p.d.f. over this interval. Suppose that a random sample of 100 observations is taken from this unknown distribution; and that it is desired to test the null hypothesis that the distribution is a uniform distribution on the interval $(0, 1)$ against the alternative hypothesis that the distribution is not uniform. This problem is a nonparametric problem, since the distribution of X might be any continuous distribution on the interval $(0, 1)$. However, as we shall now show, the χ^2 test of goodness-of-fit can be applied to this problem.

Suppose that we divide the interval from 0 to 1 into 20 subintervals of equal length, namely, the interval from 0 to 0.05, the interval from 0.05 to 0.10, and so on. If the actual distribution is a uniform distribution, then the probability that any particular observation will fall within the ith subinterval is $1/20$, for $i = 1, \ldots, 20$. Since the sample size in this example is $n = 100$, it follows that the expected number of observations in each subinterval is 5. If N_i denotes the number of observations in the sample which actually fall within the ith subinterval, then the statistic Q defined by Eq. (2) can be rewritten simply as follows:

$$Q = \frac{1}{5} \sum_{i=1}^{20} (N_i - 5)^2. \tag{5}$$

If the null hypothesis is true and the distribution from which the observations were taken is indeed a uniform distribution, then Q will have approximately a χ^2 distribution with 19 degrees of freedom.

The method that has been presented in this example obviously can be applied to any continuous distribution. To test whether a random sample of observations comes from a particular distribution, the following procedure can be adopted:

(i) Partition the entire real line, or any particular interval which has probability 1, into a finite number k of disjoint subintervals.

(ii) Determine the probability p_i^0 that the particular hypothesized distribution would assign to the ith subinterval, and calculate the expected number np_i^0 of observations in the ith subinterval ($i = 1, \ldots, k$).

(iii) Count the number N_i of observations in the sample which fall within the ith subinterval ($i = 1, \ldots, k$).

(iv) Calculate the value of Q as defined by Eq. (2). If the hypothesized distribution is correct, Q will have approximately a χ^2 distribution with $k - 1$ degrees of freedom.

One arbitrary feature of the procedure just described is the way in which the subintervals are chosen. Two statisticians working on the same problem might very well choose the subintervals in two different ways. Generally speaking, it is a good policy to choose the subintervals so that the expected numbers of observations in the individual subintervals are approximately equal; and also to choose as many subintervals as possible without allowing the expected number of observations in any subinterval to become small.

Discussion of the Test Procedure

The χ^2 test of goodness-of-fit is subject to the criticisms of tests of hypotheses that were presented in Sec. 8.7. In particular, the null hypothesis H_0 in the χ^2 test specifies the distribution of the observations exactly, but it is not likely that the actual distribution of the observations will be exactly the same as that of a random sample from this specific distribution. Therefore, if the χ^2 test is based on a very large number of observations, we can be almost certain that the tail area corresponding to the observed value of Q will be very small. For this reason, a very small tail area should not be regarded as strong evidence against the hypothesis H_0 without further analysis. Before a statistician concludes that the hypothesis H_0 is unsatisfactory, he should be certain that there exist *reasonable* alternative distributions for which the observed values provide a much better fit. For example, the statistician might calculate the values of the statistic Q for a few reasonable alternative distributions in order to be certain that, for at least one of these distributions, the tail area corresponding to the calculated value of Q is substantially larger than it is for the distribution specified by H_0.

A particular feature of the χ^2 test of goodness-of-fit is that the procedure is designed to test the null hypothesis H_0 that $p_i = p_i^0$ for $i = 1, \ldots, k$ against the general alternative that H_0 is not true. If it is desired to use a test procedure that is especially effective for detecting certain types of deviations of the actual values of $p_1, \ldots, p_k$ from the hypothesized values $p_1^0, \ldots, p_k^0$, then the statistician should design special tests which have higher power for these types of alternatives and lower power for alternatives of lesser interest. This topic will not be discussed in this book.

Since the random variables $N_1, \ldots, N_k$ in Eq. (2) are discrete, the χ^2 approximation to the distribution of Q can sometimes be improved by introducing a correction for continuity of the type described in Sec. 5.8. However, we shall not use the correction in this book.

EXERCISES

1. Show that if $p_i^0 = 1/k$ for $i = 1, \ldots, k$, then the statistic Q defined by Eq. (2) can be written in the form

$$Q = \left(\frac{k}{n} \sum_{i=1}^{k} N_i^2 \right) - n.$$

2. Investigate the "randomness" of the table of random digits given at the end of this book by regarding the 200 digits in the first five rows of the table as sample data and carrying out a χ^2 test of the hypothesis that each of the ten digits $0, 1, \ldots, 9$ has the same probability of occurring at each place in the table.

3. According to a simple genetic principle, if both the mother and the father of a child have genotype Aa, then there is probability $1/4$ that the child will have genotype AA; there is probability $1/2$ that it will have genotype Aa; and there is probability $1/4$ that it will have genotype aa. In a random sample of 24 children having both parents with genotype Aa, it is found that 10 have genotype AA, 10 have genotype Aa, and 4 have genotype aa. Investigate whether the simple genetic principle is correct by carrying out a χ^2 test of goodness-of-fit.

4. Suppose that in a sequence of n Bernoulli trials, the probability p of success on any given trial is unknown. Suppose also that p_0 is a given number in the interval $0 < p_0 < 1$, and that it is desired to test the following hypotheses:

$H_0: \quad p = p_0,$
$H_1: \quad p \neq p_0.$

Let $\overline{X}_n$ denote the proportion of successes in the n trials, and suppose that the given hypotheses are to be tested by using a χ^2 test of goodness-of-fit.

(a) Show that the statistic Q defined by Eq. (2) can be written in the form

$$Q = \frac{n(\overline{X}_n - p_0)^2}{p_0(1 - p_0)}.$$

(b) Assuming that H_0 is true, prove that as $n \to \infty$, the d.f. of Q converges to the d.f. of the χ^2 distribution with one degree of freedom. *Hint:* Show that $Q = Z^2$, where it is known from the central limit theorem that Z is a random variable whose d.f. converges to the d.f. of the standard normal distribution.

5. It is known that 30 percent of small steel rods produced by a standard process will break when subjected to a load of 3000 pounds. In a random sample of 50 similar rods produced by a new process, it was found that 21 of them broke when subjected to a load of 3000 pounds. Investigate the hypothesis that the breakage rate for the new process is the same as the rate for the old process by carrying out a χ^2 test of goodness-of-fit.

6. In a random sample of 1800 observed values from the interval $(0, 1)$, it was found that 391 values were between 0 and 0.2; that 490 values were between 0.2 and 0.5; that 580 values were between 0.5 and 0.8; and that 339 values were between 0.8 and 1. Test the hypothesis that the random sample was drawn from a uniform distribution on the interval $(0, 1)$ by carrying out a χ^2 test of goodness-of-fit at the level of significance 0.01.

7. Suppose that the distribution of the heights of men who reside in a certain large city is a normal distribution for which the mean is 68 inches and the standard deviation is 1 inch. Suppose also that when the heights of 500 men who reside in a certain neighborhood of the city were measured, the following distribution was obtained:

Height	Number of men
Less than 66 in.	18
Between 66 in. and 67.5 in.	177
Between 67.5 in. and 68.5 in.	198
Between 68.5 in. and 70 in.	102
Greater than 70 in.	5

Test the hypothesis that, with regard to height, these 500 men form a random sample from all the men who reside in the city.

Table 9.1

-1.28	-1.22	-0.45	-0.35	0.72
-0.32	-0.80	-1.66	1.39	0.38
-1.38	-1.26	0.49	-0.14	-0.85
2.33	-0.34	-1.96	-0.64	-1.32
-1.14	0.64	3.44	-1.67	0.85
0.41	-0.01	0.67	-1.13	-0.41
-0.49	0.36	-1.24	-0.04	-0.11
1.05	0.04	0.76	0.61	-2.04
0.35	2.82	-0.46	-0.63	-1.61
0.64	0.56	-0.11	0.13	-1.81

8. The 50 values in Table 9.1 are intended to be a random sample from a standard normal distribution.

   (a) Carry out a χ^2 test of goodness-of-fit by dividing the real line into five intervals each of which has probability 0.2 under the standard normal distribution.

   (b) Carry out a χ^2 test of goodness-of-fit by dividing the real line into ten intervals each of which has probability 0.1 under the standard normal distribution.

9.2.   GOODNESS-OF-FIT FOR COMPOSITE HYPOTHESES

Composite Null Hypotheses

We shall consider again a large population which consists of items of k different types, and shall again let p_i denote the probability that an item selected at random will be of type i $(i = 1, \ldots, k)$. We shall suppose now, however, that instead of testing the simple null hypothesis that the parameters $p_1, \ldots, p_k$ have specific values, we are interested in testing the composite null hypothesis that the values of $p_1, \ldots, p_k$ belong to some specified subset of possible values. In particular, we shall consider problems in which the null hypothesis specifies that the parameters $p_1, \ldots, p_k$ can actually be represented as functions of a smaller number of parameters.

For example, in certain genetics problems, each individual in a given population must have one of three possible genotypes, and it is assumed that the probabilities p_1, p_2, and p_3 of the three different genotypes can be represented in the following form:

$$p_1 = \theta^2, \qquad p_2 = 2\theta(1 - \theta), \qquad p_3 = (1 - \theta)^2. \tag{1}$$

Here, the value of the parameter θ is unknown and can lie anywhere in the interval $0 < \theta < 1$. For any value of θ in this interval, it can be seen that $p_i > 0$ for $i = 1$, 2, or 3 and that $p_1 + p_2 + p_3 = 1$. In this problem, a random sample is taken from the population, and the statistician must use the observed numbers of individuals who have each of the three genotypes to determine whether it is reasonable to believe that p_1, p_2, and p_3 can be represented in the hypothesized form (1) for *some* value of θ in the interval $0 < \theta < 1$.

In other genetics problems, each individual in the population must have one of six possible genotypes, and it is assumed that the probabilities $p_1, \ldots, p_6$ of the different genotypes can be represented in the following form, for *some* values of

θ_1 and θ_2 such that $\theta_1 > 0$, $\theta_2 > 0$, and $\theta_1 + \theta_2 < 1$:

$$p_1 = \theta_1^2, \qquad p_2 = \theta_2^2, \qquad p_3 = (1 - \theta_1 - \theta_2)^2, \qquad p_4 = 2\theta_1\theta_2,$$

$$p_5 = 2\theta_1(1 - \theta_1 - \theta_2), \qquad p_6 = 2\theta_2(1 - \theta_1 - \theta_2). \tag{2}$$

Again, for any values of θ_1 and θ_2 satisfying the stated conditions, it can be verified that $p_i > 0$ for $i = 1, \ldots, 6$ and $\sum_{i=1}^{6} p_i = 1$. On the basis of the observed numbers $N_1, \ldots, N_6$ of individuals having each genotype in a random sample, the statistician must decide either to accept or to reject the null hypothesis that the probabilities $p_1, \ldots, p_6$ can be represented in the form (2) for some values of θ_1 and θ_2.

In formal terms, in a problem of the type being considered, we are interested in testing the hypothesis that for $i = 1, \ldots, k$, each probability p_i can be represented as a particular function $\pi_i(\boldsymbol{\theta})$ of a vector of parameters $\boldsymbol{\theta} = (\theta_1, \ldots, \theta_s)$. It is assumed that $s < k - 1$ and that no component of the vector $\boldsymbol{\theta}$ can be expressed as a function of the other $s - 1$ components. We shall let Ω denote the s-dimensional parameter space of all possible values of $\boldsymbol{\theta}$. Furthermore, we shall assume that the functions $\pi_1(\boldsymbol{\theta}), \ldots, \pi_k(\boldsymbol{\theta})$ always form a feasible set of values of $p_1, \ldots, p_k$ in the sense that for every value of $\boldsymbol{\theta} \in \Omega$, $\pi_i(\boldsymbol{\theta}) > 0$ for $i = 1, \ldots, k$ and $\sum_{i=1}^{k} \pi_i(\boldsymbol{\theta}) = 1$.

The hypotheses to be tested can be written in the following form:

H_0:   There exists a value of $\boldsymbol{\theta} \in \Omega$ such that

$$p_i = \pi_i(\boldsymbol{\theta}) \text{ for } i = 1, \ldots, k, \tag{3}$$

H_1:   The hypothesis H_0 is not true.

The assumption that $s < k - 1$ guarantees that the hypothesis H_0 actually restricts the values of $p_1, \ldots, p_k$ to a proper subset of the set of all possible values of these probabilities. In other words, as the vector $\boldsymbol{\theta}$ runs through all the values in the set Ω, the vector $[\pi_1(\boldsymbol{\theta}), \ldots, \pi_k(\boldsymbol{\theta})]$ runs through only a proper subset of the possible values of $(p_1, \ldots, p_k)$.

The χ^2 Test for Composite Null Hypotheses

In order to carry out a χ^2 test of goodness-of-fit of the hypotheses (3), the statistic Q defined by Eq. (2) of Sec. 9.1 must be modified because the expected number np_i^0 of observations of type i in a random sample of n observations is no longer completely specified by the null hypothesis H_0. The modification that is used is simply to replace np_i^0 by the M.L.E. of this expected number under the assumption that H_0 is true. In other words, if $\hat{\boldsymbol{\theta}}$ denotes the M.L.E. of the parameter vector θ based on the observed numbers $N_1, \ldots, N_k$, then the statistic Q is defined as follows:

$$Q = \sum_{i=1}^{k} \frac{\left[N_i - n\pi_i(\hat{\theta}) \right]^2}{n\pi_i(\hat{\theta})}. \tag{4}$$

Again, it is reasonable to base a test of the hypotheses (3) on this statistic Q by rejecting H_0 if $Q > c$, where c is an appropriate constant. In 1924, R. A. Fisher showed that if the null hypothesis H_0 is true and certain regularity conditions are satisfied, then as the sample size $n \to \infty$, the d.f. of Q converges to the d.f. of the χ^2 distribution with $k - 1 - s$ degrees of freedom.

Thus, when the sample size n is large and the null hypothesis H_0 is true, the distribution of Q will be approximately a χ^2 distribution. To determine the number of degrees of freedom, we must subtract s from the number $k - 1$ used in Sec. 9.1 because we are now estimating the s parameters $\theta_1, \ldots, \theta_s$ when we compare the observed number N_i with the expected number $n\pi_i(\hat{\theta})$ for $i = 1, \ldots, k$. In order that this result will hold, it is necessary to satisfy the following regularity conditions: First, the M.L.E. $\hat{\theta}$ of the vector θ must be found in the usual way by taking the partial derivatives of the likelihood function with respect to each of the parameters $\theta_1, \ldots, \theta_s$; setting each of these s partial derivatives equal to 0; and then solving the resulting set of s equations for $\hat{\theta}_1, \ldots, \hat{\theta}_s$. Furthermore, these partial derivatives must satisfy certain conditions of the type alluded to in Sec. 7.8 when we discussed the asymptotic properties of M.L.E.'s.

To illustrate the use of the statistic Q defined by Eq. (4), we shall consider again the two types of genetics problems described earlier in this section. In a problem of the first type, $k = 3$ and it is desired to test the null hypothesis H_0 that the probabilities p_1, p_2, and p_3 can be represented in the form (1) against the alternative H_1 that H_0 is not true. In this problem, $s = 1$. Therefore, when H_0 is true, the distribution of the statistic Q defined by Eq. (4) will be approximately a χ^2 distribution with one degree of freedom.

In a problem of the second type, $k = 6$ and it is desired to test the null hypothesis H_0 that the probabilities $p_1, \ldots, p_6$ can be represented in the form (2) against the alternative H_1 that H_0 is not true. In this problem, $s = 2$. Therefore, when H_0 is true, the distribution of Q will be approximately a χ^2 distribution with 3 degrees of freedom.

Determining the Maximum Likelihood Estimates

When the null hypothesis H_0 in (3) is true, the likelihood function $L(\theta)$ for the observed numbers $N_1, \ldots, N_k$ will be

$$L(\theta) = \left[\pi_1(\theta) \right]^{N_1} \cdots \left[\pi_k(\theta) \right]^{N_k}. \tag{5}$$

Thus,

$$\log L(\theta) = \sum_{i=1}^{k} N_i \log \pi_i(\theta). \tag{6}$$

The M.L.E. $\hat{\theta}$ will be the value of θ for which $\log L(\theta)$ is a maximum.

For example, when $k = 3$ and H_0 specifies that the probabilities p_1, p_2, and p_3 can be represented in the form (1), then

$$\log L(\theta) = N_1 \log(\theta^2) + N_2 \log[2\theta(1 - \theta)] + N_3 \log[(1 - \theta)^2]$$
$$= (2N_1 + N_2)\log \theta + (2N_3 + N_2)\log(1 - \theta) + N_2 \log 2. \tag{7}$$

It can be found by differentiation that the value of θ for which $\log L(\theta)$ is a maximum is

$$\hat{\theta} = \frac{2N_1 + N_2}{2(N_1 + N_2 + N_3)} = \frac{2N_1 + N_2}{2n}. \tag{8}$$

The value of the statistic Q defined by Eq. (4) can now be calculated from the observed numbers N_1, N_2, and N_3. As previously mentioned, when H_0 is true and n is large, the distribution of Q will be approximately a χ^2 distribution with one degree of freedom. Hence, the tail area corresponding to the observed value of Q can be found from that χ^2 distribution.

Testing Whether a Distribution is Normal

Consider now a problem in which a random sample $X_1, \ldots, X_n$ is taken from some continuous distribution for which the p.d.f. is unknown; and it is desired to test the null hypothesis H_0 that this distribution is a normal distribution against the alternative hypothesis H_1 that the distribution is not normal. A χ^2 test of goodness-of-fit can be applied to this problem if we divide the real line into k subintervals and count the number N_i of observations in the random sample which fall into the ith subinterval ($i = 1, \ldots, k$).

If H_0 is true and if μ and σ^2 denote the unknown mean and variance of the normal distribution, then the parameter vector θ is the two-dimensional vector $\theta = (\mu, \sigma^2)$. The probability $\pi_i(\theta)$, or $\pi_i(\mu, \sigma^2)$, that an observation will fall within the ith subinterval is the probability assigned to that subinterval by the normal distribution with mean μ and variance σ^2. In other words, if the ith subinterval is the interval from a_i to b_i, then

$$\pi_i(\mu, \sigma^2) = \int_{a_i}^{b_i} \frac{1}{(2\pi)^{1/2}\sigma} \exp\left[-\frac{(x - \mu)^2}{2\sigma^2}\right] dx. \tag{9}$$

It is important to note that in order to calculate the value of the statistic Q defined by Eq. (4), the M.L.E.'s $\hat{\mu}$ and $\hat{\sigma}^2$ must be found by using the numbers $N_1, \ldots, N_k$ of observations in the different subintervals; they should not be found by using the observed values of $X_1, \ldots, X_n$ themselves. In other words, $\hat{\mu}$ and $\hat{\sigma}^2$ will be the values of μ and σ^2 which maximize the likelihood function

$$L(\mu, \sigma^2) = \left[\pi_1(\mu, \sigma^2) \right]^{N_1} \cdots \left[\pi_k(\mu, \sigma^2) \right]^{N_k}. \tag{10}$$

Because of the complicated nature of the function $\pi_i(\mu, \sigma^2)$, as given by Eq. (9), a lengthy numerical computation would usually be required to determine the values of μ and σ^2 which maximize $L(\mu, \sigma^2)$. On the other hand, we know that the M.L.E.'s of μ and σ^2 based on the n observed values $X_1, \ldots, X_n$ in the original sample are simply the sample mean $\overline{X}_n$ and the sample variance S_n^2/n. Furthermore, if the estimators which maximize the likelihood function $L(\mu, \sigma^2)$ are used to calculate the statistic Q, then we know that when H_0 is true, the distribution of Q will be approximately a χ^2 distribution with $k - 3$ degrees of freedom. On the other hand, if the M.L.E.'s $\overline{X}_n$ and S_n^2/n based on the observed values in the original sample are used to calculate Q, then this χ^2 approximation to the distribution of Q will not be appropriate. Because of the simple nature of the estimators $\overline{X}_n$ and S_n^2/n, we shall use these estimators to calculate Q; but we shall describe how their use modifies the distribution of Q.

In 1954, H. Chernoff and E. L. Lehmann established the following result: If the M.L.E.'s $\overline{X}_n$ and S_n^2/n are used to calculate the statistic Q, and if the null hypothesis H_0 is true, then as $n \to \infty$, the d.f. of Q converges to a d.f. which lies between the d.f. of the χ^2 distribution with $k - 3$ degrees of freedom and the d.f. of the χ^2 distribution with $k - 1$ degrees of freedom. It follows that if the value of Q is calculated in this simplified way, then the tail area corresponding to this value of Q is actually larger than the tail area found from a table of the χ^2 distribution with $k - 3$ degrees of freedom. In fact, the appropriate tail area lies somewhere between the tail area found from a table of the χ^2 distribution with $k - 3$ degrees of freedom and the larger tail area found from a table of the χ^2 distribution with $k - 1$ degrees of freedom. Thus, when the value of Q is calculated in this simplified way, the corresponding tail area will be bounded by two values that can be obtained from a table of the χ^2 distribution.

Testing Composite Hypotheses About an
Arbitrary Distribution

The procedure just described can be applied quite generally. Consider again a problem in which a random sample of n observations is taken from some continuous distribution for which the p.d.f. is unknown. Suppose now that it is desired to test the null hypothesis H_0 that this distribution belongs to a certain family of distributions indexed by the s-dimensional parameter vector $\theta = (\theta_1, \ldots, \theta_s)$ against the alternative hypothesis H_1 that the distribution does not belong to that particular family. Furthermore, suppose as usual that the real line is divided into k subintervals.

If the null hypothesis H_0 is true and the vector θ is estimated by maximizing the likelihood function $L(\theta)$ given by Eq. (5), then the statistic Q will have approximately a χ^2 distribution with $k - 1 - s$ degrees of freedom. However, if H_0 is true and the M.L.E. of θ that is found from the n observed values in the original sample is used to calculate the statistic Q, then the appropriate approximation to the distribution of Q is a distribution which lies between the χ^2 distribution with $k - 1 - s$ degrees of freedom and the χ^2 distribution with $k - 1$ degrees of freedom. Therefore, corresponding to this calculated value of Q, the tail area found from a table of the χ^2 distribution with $k - 1 - s$ degrees of freedom will be a lower bound for the appropriate tail area, and the tail area found from a table of the χ^2 distribution with $k - 1$ degrees of freedom will be an upper bound for the appropriate tail area.

The results just described also apply to discrete distributions. Suppose, for example, that a random sample of n observations is taken from a discrete distribution for which the possible values are the nonnegative integers $0, 1, 2, \ldots$. Suppose also that it is desired to test the null hypothesis H_0 that this distribution is a Poisson distribution against the alternative hypothesis H_1 that the distribution is not Poisson. Finally, suppose that the nonnegative integers $0, 1, 2, \ldots$ are divided into k classes such that each observation will lie in one of these classes.

It is known from Exercise 4 of Sec. 6.5 that if H_0 is true, then the sample mean $\overline{X}_n$ is the M.L.E. of the unknown mean θ of the Poisson distribution based on the n observed values in the original sample. Therefore, if the estimator $\hat{\theta} = \overline{X}_n$ is used to calculate the statistic Q defined by Eq. (4), then the approximate distribution of Q when H_0 is true lies between a χ^2 distribution with $k - 2$ degrees of freedom and a χ^2 distribution with $k - 1$ degrees of freedom.

EXERCISES

1. At the fifth hockey game of the season at a certain arena, 200 people were selected at random and asked how many of the previous four games they had attended. The results are given in Table 9.2. Test the hypothesis that these 200 observed values can be regarded as a random sample from a binomial distribution; that is, that there exists a number θ $(0 < \theta < 1)$ such that the probabilities are as follows:

$$p_0 = (1 - \theta)^4, \qquad p_1 = 4\theta(1 - \theta)^3, \qquad p_2 = 6\theta^2(1 - \theta)^2,$$
$$p_3 = 4\theta^3(1 - \theta), \qquad p_4 = \theta^4.$$

Table 9.2

Number of games previously attended	Number of people
0	33
1	67
2	66
3	15
4	19

2. Consider a genetics problem in which each individual in a certain population must have one of six genotypes, and it is desired to test the null hypothesis H_0 that the probabilities of the six genotypes can be represented in the form specified in Eq. (2).

   (a) Suppose that in a random sample of n individuals, the observed numbers of individuals having the six genotypes are $N_1, \ldots, N_6$. Find the M.L.E.'s of θ_1 and θ_2 when the null hypothesis H_0 is true.

   (b) Suppose that in a random sample of 150 individuals, the observed numbers are as follows:

   $$N_1 = 2, \quad N_2 = 36, \quad N_3 = 14, \quad N_4 = 36, \quad N_5 = 20, \quad N_6 = 42.$$

   Determine the value of Q and the corresponding tail area.

3. Consider again the sample consisting of the heights of 500 men given in Exercise 7 of Sec. 9.1. Suppose that before these heights were grouped into the intervals given in that exercise, it was found that for the 500 observed heights in the original sample, the sample mean was $\bar{X}_n = 67.6$ and the sample variance was $S_n^2/n = 1.00$. Test the hypothesis that these observed heights form a random sample from a normal distribution.

Table 9.3

Number of tickets purchased	Number of persons
0	52
1	60
2	55
3	18
4	8
5 or more	7

4. In a large city 200 persons were selected at random, and each person was asked how many tickets he purchased that week in the state lottery. The results are given in Table 9.3. Suppose that among the seven persons who had purchased 5 or more tickets, three persons had purchased exactly 5 tickets; two persons had purchased 6 tickets; one person had purchased 7 tickets; and one person had purchased 10 tickets. Test the hypothesis that these 200 observations form a random sample from a Poisson distribution.

Table 9.4

Number of particles emitted	Number of time periods
0	54
1	143
2	218
3	231
4	174
5	110
6	39
7	20
8	4
9	1
10	3
11	1
12	1
13	0
14	1
15 or more	0
Total	$\overline{1000}$

Table 9.5

9.69	8.93	7.61	8.12	−2.74
2.78	7.47	8.46	7.89	5.93
5.21	2.62	0.22	−0.59	8.77
4.07	5.15	8.32	6.01	0.68
9.81	5.61	13.98	10.22	7.89
0.52	6.80	2.90	2.06	11.15
10.22	5.05	6.06	14.51	13.05
9.09	9.20	7.82	8.67	7.52
3.03	5.29	8.68	11.81	7.80
16.80	8.07	0.66	4.01	8.64

5. The number of alpha particles emitted by a certain mass of radium was observed for 1000 disjoint time periods, each of which lasted 6 seconds. The results are given in Table 9.4. Test the hypothesis that these 1000 observations form a random sample from a Poisson distribution.

6. Test the hypothesis that the 50 observations in Table 9.5 form a random sample from a normal distribution.

7. Test the hypothesis that the 50 observations in Table 9.6 form a random sample from an exponential distribution.

Table 9.6

0.91	1.22	1.28	0.02	2.33
0.90	0.86	1.45	1.22	0.55
0.16	2.02	1.59	1.73	0.49
1.62	0.56	0.53	0.50	0.24
1.28	0.06	0.19	0.29	0.74
1.16	0.22	0.91	0.04	1.41
3.65	3.41	0.07	0.51	1.27
0.61	0.31	0.22	0.37	0.06
1.75	0.89	0.79	1.28	0.57
0.76	0.05	1.53	1.86	1.28

9.3. CONTINGENCY TABLES

Independence in Contingency Tables

Suppose that 200 students are selected at random from the entire enrollment at a large university, and that each student in the sample is classified both according to the curriculum in which he is enrolled and according to his preference for either of two candidates A and B in a forthcoming election. Suppose that the results are as presented in Table 9.7.

A table in which each observation is classified in two or more ways is called a *contingency table*. In Table 9.7 only two classifications are considered for each student, namely, the curriculum in which he is enrolled and the candidate he prefers. Such a table is called a *two-way* contingency table.

When a statistician analyzes a contingency table, he is often interested in testing the hypothesis that the different classifications are independent. Thus, for Table 9.7, he might be interested in testing the hypothesis that the curriculum in which a student is enrolled and the candidate he prefers are independent

Table 9.7

Curriculum	Candidate preferred			
	A	B	Undecided	Totals
Engineering and science	24	23	12	59
Humanities and social sciences	24	14	10	48
Fine arts	17	8	13	38
Industrial and public administration	27	19	9	55
Totals	92	64	44	

variables. In precise terms, in this problem, the hypothesis of independence states that if a student is selected at random from the entire enrollment at the university, then the probability that he is enrolled in a particular curriculum i *and* prefers a certain candidate j is equal to the probability that he is enrolled in curriculum i *times* the probability that he prefers candidate j.

In general, we shall consider a two-way contingency table containing R rows and C columns. For $i = 1, \ldots, R$ and $j = 1, \ldots, C$, we shall let p_{ij} denote the probability that an individual selected at random from a given population will be classified in the ith row and the jth column of the table. Furthermore, we shall let p_{i+} denote the marginal probability that the individual will be classified in the ith row of the table, and we shall let p_{+j} denote the marginal probability that the individual will be classified in the jth column of the table. Thus,

$$p_{i+} = \sum_{j=1}^{C} p_{ij} \quad \text{and} \quad p_{+j} = \sum_{i=1}^{R} p_{ij}. \tag{1}$$

Furthermore, since the sum of the probabilities for all the cells of the table must be 1, we have

$$\sum_{i=1}^{R} \sum_{j=1}^{C} p_{ij} = \sum_{i=1}^{R} p_{i+} = \sum_{j=1}^{C} p_{+j} = 1. \tag{2}$$

Suppose now that a random sample of n individuals is taken from the given population. For $i = 1, \ldots, R$ and $j = 1, \ldots, C$, we shall let N_{ij} denote the number of individuals who are classified in the ith row and the jth column of the table. Furthermore, we shall let N_{i+} denote the total number of individuals classified in the ith row, and we shall let N_{+j} denote the total number of individuals classified in the jth column. Thus,

$$N_{i+} = \sum_{j=1}^{C} N_{ij} \quad \text{and} \quad N_{+j} = \sum_{i=1}^{R} N_{ij}. \tag{3}$$

Also,

$$\sum_{i=1}^{R} \sum_{j=1}^{C} N_{ij} = \sum_{i=1}^{R} N_{i+} = \sum_{j=1}^{C} N_{+j} = n. \tag{4}$$

On the basis of these observations, the following hypotheses are to be tested:

$$H_0: \quad p_{ij} = p_{i+}p_{+j} \quad \text{for } i = 1, \ldots, R \text{ and } j = 1, \ldots, C, \tag{5}$$
$$H_1: \quad \text{The hypothesis } H_0 \text{ is not true.}$$

The χ^2 Test of Independence

The χ^2 tests described in Sec. 9.2 can be applied to the problem of testing the hypotheses (5). Each individual in the population from which the sample is taken must belong in one of the RC cells of the contingency table. Under the null hypothesis H_0, the unknown probabilities p_{ij} of these cells have been expressed as functions of the unknown parameters p_{i+} and p_{+j}. Since $\sum_{i=1}^{R} p_{i+} = 1$ and $\sum_{j=1}^{C} p_{+j} = 1$, the actual number of unknown parameters to be estimated when H_0 is true is $(R - 1) + (C - 1)$, or $R + C - 2$.

For $i = 1, \ldots, R$ and $j = 1, \ldots, C$, let $\hat{E}_{ij}$ denote the M.L.E., when H_0 is true, of the expected number of observations that will be classified in the ith row and the jth column of the table. In this problem, the statistic Q defined by Eq. (4) of Sec. 9.2 will have the following form:

$$Q = \sum_{i=1}^{R} \sum_{j=1}^{C} \frac{\left(N_{ij} - \hat{E}_{ij}\right)^2}{\hat{E}_{ij}}. \tag{6}$$

Furthermore, since the contingency table contains RC cells, and since $R + C - 2$ parameters are to be estimated when H_0 is true, it follows that when H_0 is true and $n \to \infty$, the d.f. of Q converges to the d.f. of the χ^2 distribution for which the number of degrees of freedom is $RC - 1 - (R + C - 2) = (R - 1)(C - 1)$.

Next, we shall consider the form of the estimator $\hat{E}_{ij}$. The expected number of observations in the ith row and the jth column is simply np_{ij}. When H_0 is true, $p_{ij} = p_{i+}p_{+j}$. Therefore, if $\hat{p}_{i+}$ and $\hat{p}_{+j}$ denote the M.L.E.'s of p_{i+} and p_{+j}, then it follows that $\hat{E}_{ij} = n\hat{p}_{i+}\hat{p}_{+j}$. Next, since p_{i+} is the probability that an

Table 9.8

27.14	18.88	12.98
22.08	15.36	10.56
17.48	12.16	8.36
25.30	17.60	12.10

observation will be classified in the ith row, $\hat{p}_{i+}$ is simply the proportion of observations in the sample that are classified in the ith row; that is, $\hat{p}_{i+} = N_{i+}/n$. Similarly, $\hat{p}_{+j} = N_{+j}/n$, and it follows that

$$\hat{E}_{ij} = n\left(\frac{N_{i+}}{n}\right)\left(\frac{N_{+j}}{n}\right) = \frac{N_{i+}N_{+j}}{n}. \tag{7}$$

If we substitute this value of $\hat{E}_{ij}$ into Eq. (6), we can calculate the value of Q from the observed values of N_{ij}. The null hypothesis H_0 should be rejected if $Q > c$, where c is an appropriately chosen constant. When H_0 is true and the sample size n is large, the distribution of Q will be approximately a χ^2 distribution with $(R - 1)(C - 1)$ degrees of freedom.

For example, suppose that we wish to test the hypotheses (5) on the basis of the data in Table 9.7. By using the totals given in the table, we find that $N_{1+} = 59$, $N_{2+} = 48$, $N_{3+} = 38$, and $N_{4+} = 55$; and also that $N_{+1} = 92$, $N_{+2} = 64$, and $N_{+3} = 44$. Since $n = 200$, it follows from Eq. (7) that the 4×3 table of values of $\hat{E}_{ij}$ is as given in Table 9.8.

The values of N_{ij} given in Table 9.7 can now be compared with the values of $\hat{E}_{ij}$ in Table 9.8. The value of Q defined by Eq. (6) turns out to be 6.68. Since $R = 4$ and $C = 3$, the corresponding tail area is to be found from a table of the χ^2 distribution with $(R - 1)(C - 1) = 6$ degrees of freedom. Its value is larger than 0.3. Therefore, in the absence of other information, the observed values do not provide any evidence that H_0 is not true.

EXERCISES

1. Show that the statistic Q defined by Eq. (6) can be rewritten in the form

$$Q = \left(\sum_{i=1}^{R}\sum_{j=1}^{C}\frac{N_{ij}^2}{\hat{E}_{ij}}\right) - n.$$

Table 9.9

	Wears a moustache	Does not wear a moustache
Between 18 and 30	12	28
Over 30	8	52

2. Show that if $C = 2$, the statistic Q defined by Eq. (6) can be rewritten in the form

$$Q = \frac{n}{N_{+2}} \left(\sum_{i=1}^{R} \frac{N_{i1}^2}{\hat{E}_{i1}} - N_{+1} \right).$$

3. Suppose that an experiment is carried out to see if there is any relation between a man's age and whether he wears a moustache. Suppose that 100 men, 18 years of age or older, are selected at random; and that each man is classified according to whether or not he is between 18 and 30 years of age, and also according to whether or not he wears a moustache. The observed numbers are given in Table 9.9. Test the hypothesis that there is no relationship between a man's age and whether he wears a moustache.

4. Suppose that 300 persons are selected at random from a large population; and that each person in the sample is classified according to whether his blood type is O, A, B, or AB, and also according to whether his blood type is Rh positive or Rh negative. The observed numbers are given in Table 9.10. Test the hypothesis that the two classifications of blood types are independent.

5. Suppose that a store carries two different brands, A and B, of a certain type of breakfast cereal. Suppose that during a one-week period the store noted whether each package of this type of cereal that was purchased was of brand A or of brand B and also noted whether the purchaser was a man or a woman. (A purchase made by a child or by a man and a woman together was

Table 9.10

	O	A	B	AB
Rh positive	82	89	54	19
Rh negative	13	27	7	9

Table 9.11

	Brand A	Brand B
Men	9	6
Women	13	16

not counted.) Suppose that 44 packages were purchased, and that the results were as given in Table 9.11. Test the hypothesis that the brand purchased and the sex of the purchaser are independent.

6. Consider a two-way contingency table with three rows and three columns. Suppose that, for $i = 1, 2, 3$, and $j = 1, 2, 3$, the probability p_{ij} that an individual selected at random from a given population will be classified in the ith row and the jth column of the table is as given in Table 9.12

   (a) Show that the rows and columns of this table are independent by verifying that the values p_{ij} satisfy the null hypothesis H_0 in Eq. (5).

   (b) Generate a random sample of 300 observations from the given population by choosing 300 *pairs* of digits from a table of random digits and classifying each pair in some cell of the contingency table in the following manner: Since $p_{11} = 0.15$, classify a pair of digits in the first cell if it is one of the first fifteen pairs $01, 02, \ldots, 15$. Since $p_{12} = 0.09$, classify a pair of digits in the second cell if it is one of the next nine pairs $16, 17, \ldots, 24$. Continue in this way for all nine cells. Thus, since the last cell of the table has probability $p_{33} = 0.08$, a pair of digits will be classified in that cell if it is one of the last eight pairs $93, 94, \ldots, 99, 00$.

   (c) Consider the 3×3 table of observed values N_{ij} generated in part (b). Pretend that the probabilities p_{ij} were unknown, and test the hypotheses (5).

7. If all the students in a class carry out Exercise 6 independently of each other and use different sets of random digits, then the different values of the statistic Q obtained by the different students should form a random sample from a χ^2 distribution with 4 degrees of freedom. If the values of Q for all

Table 9.12

0.15	0.09	0.06
0.15	0.09	0.06
0.20	0.12	0.08

the students in the class are available to you, test the hypothesis that these values form such a random sample.

8. Consider a three-way contingency table of size $R \times C \times T$. For $i = 1, \ldots, R$, $j = 1, \ldots, C$, and $k = 1, \ldots, T$, let p_{ijk} denote the probability that an individual selected at random from a given population will fall into the (i, j, k) cell of the table. Let

$$p_{i++} = \sum_{j=1}^{C} \sum_{k=1}^{T} p_{ijk}, \qquad p_{+j+} = \sum_{i=1}^{R} \sum_{k=1}^{T} p_{ijk}, \qquad p_{++k} = \sum_{i=1}^{R} \sum_{j=1}^{C} p_{ijk}.$$

On the basis of a random sample of n observations from the given population, construct a test of the following hypotheses:

H_0: $p_{ijk} = p_{i++}p_{+j+}p_{++k}$ for all values of i, j, and k,

H_1:   The hypothesis H_0 is not true.

9. Consider again the conditions of Exercise 8. For $i = 1, \ldots, R$ and $j = 1, \ldots, C$, let

$$p_{ij+} = \sum_{k=1}^{T} p_{ijk}.$$

On the basis of a random sample of n observations from the given population, construct a test of the following hypotheses:

H_0: $p_{ijk} = p_{ij+}p_{++k}$ for all values of i, j, and k,

H_1:   The hypothesis H_0 is not true.

9.4. TESTS OF HOMOGENEITY

Samples from Several Populations

Consider again the problem described at the beginning of Sec. 9.3 in which each student in a random sample from the entire enrollment at a large university is classified in a contingency table according to the curriculum in which he is enrolled and according to his preference for either of two political candidates A and B. The results for a random sample of 200 students were presented in Table 9.7.

Suppose that we are still interested in investigating whether there is a relationship between the curriculum in which a student is enrolled and the candidate he prefers. Now suppose, however, that instead of selecting 200 students at random from the entire enrollment at the university and classifying them in a contingency table, the experiment is carried out in the following manner:

First, 59 students are selected at random from all of those who are enrolled in engineering and science, and each of these 59 students in the random sample is classified according to whether he prefers candidate A, prefers candidate B, or is undecided. We shall suppose, for convenience, that the results are as given in the first row of Table 9.7. Also, 48 students are selected at random from all of those who are enrolled in humanities and social sciences. Again, for convenience, we shall suppose that the results are as given in the second row of Table 9.7. Similarly, 38 students are selected at random from all of those enrolled in fine arts, and 55 students are selected at random from all of those enrolled in industrial and public administration; and we shall suppose that the results are as given in the bottom two rows of Table 9.7.

Thus, we are assuming that we have again obtained a table of values identical to Table 9.7; but we are assuming now that this table was obtained by taking four different random samples from the different populations of students defined by the four rows of the table. In this context, we are interested in testing the hypothesis that, in all four populations, the same proportion of students prefers candidate A, the same proportion prefers candidate B, and the same proportion is undecided.

In general, we shall consider a problem in which random samples are taken from R different populations and each observation in each sample can be classified as one of C different types. Thus, the data obtained from the R samples can be represented in an $R \times C$ table. For $i = 1, \ldots, R$ and $j = 1, \ldots, C$, we shall let p_{ij} denote the probability that an observation chosen at random from the ith population will be of type j. Thus,

$$\sum_{j=1}^{C} p_{ij} = 1 \qquad \text{for } i = 1, \ldots, R. \tag{1}$$

The hypotheses to be tested are as follows:

$$\begin{aligned} H_0: \quad & p_{1j} = p_{2j} = \cdots = p_{Rj} \qquad \text{for } j = 1, \ldots, C, \\ H_1: \quad & \text{The hypothesis } H_0 \text{ is not true.} \end{aligned} \tag{2}$$

The null hypothesis H_0 in (2) states that all the distributions from which the R different samples are drawn are actually alike; that is, that the R distributions are homogeneous. For this reason, a test of the hypotheses (2) is called a test of homogeneity of the R distributions.

For $i = 1, \ldots, R$, we shall let N_{i+} denote the number of observations in the random sample from the ith population; and for $j = 1, \ldots, C$, we shall let N_{ij}

denote the number of observations in this random sample that are of type j. Thus,

$$\sum_{j=1}^{C} N_{ij} = N_{i+} \qquad \text{for } i = 1, \ldots, R. \tag{3}$$

Furthermore, if we let n denote the total number of observations in all R samples and we let N_{+j} denote the total number of observations of type j in the R samples, then all the relations in Eqs. (3) and (4) of Sec. 9.3 will again be satisfied.

The χ^2 Test of Homogeneity

We shall now develop a test procedure for the hypotheses (2). Suppose for the moment that the probabilities p_{ij} are known; and consider the following statistic calculated from the observations in the ith random sample:

$$\sum_{j=1}^{C} \frac{\left(N_{ij} - N_{i+}p_{ij}\right)^2}{N_{i+}p_{ij}}. \tag{4}$$

This statistic is just the standard χ^2 statistic, introduced in Eq. (2) of Sec. 9.1, for the random sample of N_{i+} observations from the ith population. Therefore, when the sample size N_{i+} is large, the distribution of this statistic will be approximately a χ^2 distribution with $C - 1$ degrees of freedom.

If we now sum this statistic over the R different samples, we obtain the following statistic:

$$\sum_{i=1}^{R} \sum_{j=1}^{C} \frac{\left(N_{ij} - N_{i+}p_{ij}\right)^2}{N_{i+}p_{ij}}. \tag{5}$$

Since the observations in the R samples are drawn independently, the distribution of the statistic (5) will be the distribution of the sum of R independent random variables, each of which has approximately a χ^2 distribution with $C - 1$ degrees of freedom. Hence, the distribution of the statistic (5) will be approximately a χ^2 distribution with $R(C - 1)$ degrees of freedom.

Since the probabilities p_{ij} are not actually known, their values must be estimated from the observed numbers in the R random samples. When the null hypothesis H_0 is true, the R random samples are actually drawn from the same distribution. Therefore, the M.L.E. of the probability that an observation in any one of these samples will be of type j is simply the proportion of all the observations in the R samples that are of type j. In other words, the M.L.E. of p_{ij}

is the same for all values of i ($i = 1, \ldots, R$), and this estimator is $\hat{p}_{ij} = N_{+j}/n$. When this M.L.E. is substituted into (5), we obtain the statistic

$$Q = \sum_{i=1}^{R} \sum_{j=1}^{C} \frac{\left(N_{ij} - \hat{E}_{ij}\right)^2}{\hat{E}_{ij}}, \tag{6}$$

where

$$\hat{E}_{ij} = \frac{N_{i+}N_{+j}}{n}. \tag{7}$$

It can be seen that Eqs. (6) and (7) are precisely the same as Eqs. (6) and (7) of Sec. 9.3. Thus, the statistic Q to be used for the test of homogeneity in this section is precisely the same as the statistic Q to be used for the test of independence in Sec. 9.3. We shall now show that the number of degrees of freedom is also precisely the same for the test of homogeneity as for the test of independence.

Since the distributions of the R populations are alike when H_0 is true, and since $\sum_{j=1}^{C} p_{ij} = 1$ for this common distribution, we have estimated $C - 1$ parameters in this problem. Therefore, the statistic Q will have approximately a χ^2 distribution with $R(C - 1) - (C - 1) = (R - 1)(C - 1)$ degrees of freedom. This number is the same as that found in Sec. 9.3.

In summary, consider Table 9.7 again. The statistical analysis of this table will be the same for either of the following two procedures: The 200 observations are drawn as a single random sample from the entire enrollment of the university, and a test of independence is carried out; or the 200 observations are drawn as separate random samples from four different groups of students, and a test of homogeneity is carried out. In either case, in a problem of this type with R rows and C columns, we should calculate the statistic Q defined by Eqs. (6) and (7); and we should assume that its distribution when H_0 is true will be approximately a χ^2 distribution with $(R - 1)(C - 1)$ degrees of freedom.

Comparing Two or More Proportions

Consider a problem in which it is desired to find out whether the proportion of adults who watched a certain television program was the same in R different cities ($R \geqslant 2$). Suppose that for $i = 1, \ldots, R$, a random sample of N_{i+} adults is selected from city i; that the number in the sample who watched the program is N_{i1}; and that the number who did not watch the program is $N_{i2} = N_{i+} - N_{i1}$. These data can be presented in an $R \times 2$ table such as Table 9.13. The hypotheses to be tested will have the same form as the hypotheses (2). Hence,

Table 9.13

City	Watched program	Did not watch	Sample size
1	N_{11}	N_{12}	N_{1+}
2	N_{21}	N_{22}	N_{2+}
⋮	⋮	⋮	⋮
R	N_{R1}	N_{R2}	N_{R+}

when the null hypothesis H_0 is true, that is, when the proportion of adults who watched the program is the same in all R cities, the statistic Q defined by Eqs. (6) and (7) will have approximately a χ^2 distribution with $R - 1$ degrees of freedom.

Correlated 2 × 2 Tables

We shall now describe a type of problem in which the use of the χ^2 test of homogeneity would not be appropriate. Suppose that 100 persons were selected at random in a certain city, and that each person was asked whether he thought the service provided by the Fire Department in the city was satisfactory. Shortly after this survey was carried out, a large fire occurred in the city. Suppose that after this fire, the same 100 persons were again asked whether they thought that the service provided by the Fire Department was satisfactory. The results are presented in Table 9.14.

Table 9.14 has the same general appearance as other tables we have been considering in this section. However, it would not be appropriate to carry out a χ^2 test of homogeneity for this table because the observations taken before the fire and the observations taken after the fire are not independent. Although the total number of observations in Table 9.14 is 200, only 100 independently chosen persons were questioned in the surveys. It is reasonable to believe that a particular person's opinion before the fire and his opinion after the fire are dependent. For this reason, Table 9.14 is called a correlated 2 × 2 table.

Table 9.14

	Satisfactory	Unsatisfactory
Before the fire	80	20
After the fire	72	28

Table 9.15

		After the fire	
		Satisfactory	Unsatisfactory
Before the fire {	Satisfactory	70	10
	Unsatisfactory	2	18

The proper way to display the opinions of the 100 persons in the random sample is shown in Table 9.15. It is not possible to construct Table 9.15 from the data in Table 9.14 alone. The entries in Table 9.14 are simply the marginal totals of Table 9.15. However, in order to construct Table 9.15, it is necessary to go back to the original data and, for each person in the sample, to consider both his opinion before the fire and his opinion after the fire.

Furthermore, it usually is not appropriate to carry out either a χ^2 test of independence or a χ^2 test of homogeneity for Table 9.15, because the hypotheses that are tested by either of these procedures usually are not those in which a statistician would be interested in this type of problem. In fact, in this problem a statistician would basically be interested in the answers to one or both of the following two questions: First, what proportion of the persons in the city changed their opinions about the Fire Department after the fire occurred? Second, among those persons in the city who did change their opinions after the fire, were the changes predominantly in one direction rather than the other?

Table 9.15 provides information pertaining to both these questions. According to Table 9.15, the number of persons in the sample who changed their opinions after the fire was $10 + 2 = 12$. Furthermore, among the 12 persons who did change their opinions, the opinions of 10 of them were changed from satisfactory to unsatisfactory and the opinions of 2 of them were changed from unsatisfactory to satisfactory. On the basis of these statistics, it is possible to make inferences about the corresponding proportions for the entire population of the city.

In this example, the M.L.E. $\hat{\theta}$ of the proportion of the population who changed their opinions after the fire is 0.12. Also, among those who did change their opinions, the M.L.E. $\hat{p}_{12}$ of the proportion who changed from satisfactory to unsatisfactory is $5/6$. Of course, if $\hat{\theta}$ is very small in a particular problem, then there is little interest in the value of $\hat{p}_{12}$.

EXERCISES

1. An examination was given to 500 high-school seniors in each of two large cities, and their grades were recorded as low, medium, or high. The results are

Table 9.16

	Low	Medium	High
City A	103	145	252
City B	140	136	224

Table 9.17

	Number of lectures attended				
	0	1	2	3	4
Freshmen	10	16	27	6	11
Sophomores	14	19	20	4	13
Juniors	15	15	17	4	9
Seniors	19	8	6	5	12

as given in Table 9.16. Test the hypothesis that the distributions of scores among seniors in the two cities are the same.

2. Every Tuesday afternoon during the school year, a certain university brought in a visiting speaker to present a lecture on some topic of current interest. On the day after the fourth lecture of the year, random samples of 70 freshmen, 70 sophomores, 60 juniors, and 50 seniors were selected from the student body at the university; and each of these students was asked how many of the four lectures he had attended. The results are given in Table 9.17. Test the hypothesis that freshmen, sophomores, juniors, and seniors at the university attended the lectures with equal frequency.

3. Suppose that five persons shoot at a target. Suppose also that for $i = 1, \ldots, 5$, person i shoots n_i times and hits the target y_i times; and that the values of n_i and y_i are as given in Table 9.18. Test the hypothesis that the five persons are equally good marksmen.

4. A manufacturing plant has preliminary contracts with three different suppliers of machines. Each supplier delivered 15 machines, which were used in the plant for 4 months in preliminary production. It turned out that one of the machines from supplier 1 was defective; that seven of the machines from supplier 2 were defective; and that seven of the machines from supplier 3 were defective. The plant statistician decided to test the null hypothesis H_0

Table 9.18

i	n_i	y_i
1	17	8
2	16	4
3	10	7
4	24	13
5	16	10

Table 9.19

	Supplier		
	1	2	3
Number of defectives N_i	1	7	7
Expected number of defectives E_i under H_0	5	5	5
$\dfrac{(N_i - E_i)^2}{E_i}$	$\dfrac{16}{5}$	$\dfrac{4}{5}$	$\dfrac{4}{5}$

that the three suppliers provided the same quality. Therefore, he set up Table 9.19 and carried out a χ^2 test. By summing the values in the bottom row of Table 9.19, he found that the value of the χ^2 statistic was $24/5$ with 2 degrees of freedom. He then found from a table of the χ^2 distribution that H_0 should be accepted when the level of significance is 0.05. Criticize this procedure, and provide a meaningful analysis of the observed data.

5. Suppose that 100 students in a physical-education class shoot at a target with a bow and arrow, and that 27 students hit the target. These 100 students are then given a demonstration on the proper technique for shooting with the bow and arrow. After the demonstration, they again shoot at the target. This time 35 students hit the target. Investigate the hypothesis that the demonstration was helpful. What additional information, if any, is needed?

6. As people entered a certain meeting, n persons were selected at random; and each was asked either to name one of two political candidates he favored in a forthcoming election or to say "undecided" if he had no real preference. During the meeting, the people heard a speech on behalf of one of the

candidates. After the meeting, each of the same n persons was again asked to express his opinion. Describe a method for evaluating the effectiveness of the speaker.

9.5. SIMPSON'S PARADOX

Comparing Treatments

Suppose that an experiment is carried out in order to compare a new treatment for a particular disease with the standard treatment for the disease. In the experiment 80 subjects suffering from the disease are treated, 40 subjects receiving the new treatment and 40 receiving the standard treatment. After a certain period of time, it is observed how many of the subjects in each group have improved and how many have not. Suppose that the overall results for all 80 patients are as shown in Table 9.20.

According to this table, 20 of the 40 subjects who received the new treatment improved, and 24 of the 40 subjects who received the standard treatment improved. Thus, 50 percent of the subjects improved under the new treatment, whereas 60 percent improved under the standard treatment. On the basis of these results, the new treatment appears inferior to the standard treatment.

Aggregation and Disaggregation

In order to investigate the efficacy of the new treatment more carefully, we might compare it with the standard treatment just for the men in the sample and, separately, just for the women in the sample. The results in Table 9.20 can thus be partitioned into two tables, one pertaining just to men and the other just to women. This process of splitting the overall data into disjoint components pertaining to different subgroups of the population is called *disaggregation*.

Suppose that, when the values in Table 9.20 are disaggregated by considering the men and the women separately, the results are as given in Table 9.21. It can be verified that when the data in these separate tables are combined, or *aggre-*

Table 9.20

All patients	Improved	Not improved	Percent improved
New treatment	20	20	50
Standard treatment	24	16	60

Table 9.21

Men only	Improved	Not improved	Percent improved
New treatment	12	18	40
Standard treatment	3	7	30

Women only	Improved	Not improved	Percent improved
New treatment	8	2	80
Standard treatment	21	9	70

gated, we again obtain Table 9.20. However, Table 9.21 contains a big surprise because the new treatment appears to be superior to the standard treatment both for men and for women. Specifically, 40 percent of the men (12 out of 30) who received the new treatment improved, but only 30 percent of the men (3 out of 10) who received the standard treatment improved. Furthermore, 80 percent of the women (8 out of 10) who received the new treatment improved, but only 70 percent of the women (21 out of 30) who received the standard treatment improved.

Thus, Tables 9.20 and 9.21 together yield somewhat anomalous results. According to Table 9.21, the new treatment is superior to the standard treatment both for men and for women; but according to Table 9.20, the new treatment is inferior to the standard treatment when all the subjects are aggregated. This type of result is known as *Simpson's paradox*.

It should be emphasized that Simpson's paradox is *not* a phenomenon that occurs because we are working with small samples. The small numbers in Tables 9.20 and 9.21 were used merely for convenience in this explanation. Each of the entries in these tables could be multiplied by 1000 or by 1,000,000 without changing the results.

The Paradox Explained

Of course, Simpson's paradox is not actually a paradox; it is merely a result that is surprising and puzzling to someone who has not seen or thought about it before. It can be seen from Table 9.21 that in the example we are considering, women have a higher rate of improvement from the disease than men have, regardless of which treatment they receive. Furthermore, among the 40 men in the sample, 30 received the new treatment and only 10 received the standard treatment; whereas among the 40 women in the sample, these numbers are reversed. Thus, although the numbers of men and women in the experiment were equal, a high proportion of the women and a low proportion of the men received the standard treatment. Since women have a much higher rate of improvement than men, it is found in the aggregated Table 9.20 that the standard treatment manifests a higher overall rate of improvement than does the new treatment.

Simpson's paradox demonstrates dramatically the dangers in making inferences from an aggregated table like Table 9.20. To make sure that Simpson's paradox cannot occur in an experiment like that just described, the relative proportions of men and women among the subjects who receive the new treatment must be the same, or approximately the same, as the relative proportions of men and women among the subjects who receive the standard treatment. It is *not* necessary that there be equal numbers of men and women in the sample.

We can express Simpson's paradox in probability terms. Let A denote the event that a subject chosen for the experiment will be a man, and let A^c denote the event that the subject will be a woman. Also, let B denote the event that a subject will receive the new treatment, and let B^c denote the event that the subject will receive the standard treatment. Finally, let I denote the event that a subject will improve. Simpson's paradox then reflects the fact that it is possible for all three of the following inequalities to hold simultaneously:

$$\Pr(I \mid A, B) > \Pr(I \mid A, B^c),$$
$$\Pr(I \mid A^c, B) > \Pr(I \mid A^c, B^c), \tag{1}$$
$$\Pr(I \mid B) < \Pr(I \mid B^c).$$

The discussion that we have just given in regard to the prevention of Simpson's paradox can be expressed as follows: If $\Pr(A \mid B) = \Pr(A \mid B^c)$, then it is not possible for all three inequalities in (1) to hold (see Exercise 4).

The possibility of Simpson's paradox lurks within any contingency table. Even though we might take care to design a particular experiment so that Simpson's paradox cannot occur when we disaggregate with respect to men and women, it is always possible that there is some other variable, such as the age of the subject or the intensity and the stage of the disease, with respect to which disaggregation would lead us to a conclusion directly opposite to that indicated by the aggregated table.

EXERCISES

1. Consider two populations I and II. Suppose that 80 percent of the men and 30 percent of the women in population I have a certain characteristic; and that only 60 percent of the men and 10 percent of the women in population II have the characteristic. Explain how, under these conditions, it might be

true that the proportion of population II having the characteristic is larger than the proportion of population I having the characteristic.

2. Suppose that A and B are events such that $0 < \Pr(A) < 1$ and $0 < \Pr(B) < 1$. Show that $\Pr(A \mid B) = \Pr(A \mid B^c)$ if and only if $\Pr(B \mid A) = \Pr(B \mid A^c)$.

3. Suppose that each adult subject in an experiment is given either treatment I or treatment II. Prove that the proportion of men among the subjects who receive treatment I is equal to the proportion of men among the subjects who receive treatment II if and only if the proportion of all men in the experiment who receive treatment I is equal to the proportion of all women who receive treatment I.

4. Show that all three inequalities in (1) cannot hold if $\Pr(A \mid B) = \Pr(A \mid B^c)$.

5. It was believed that a certain university was discriminating against women in its admissions policy because 30 percent of all the male applicants to the university were admitted, whereas only 20 percent of all the female applicants were admitted. In order to determine which of the five colleges in the university were most responsible for this discrimination, the admissions rates for each college were analyzed separately. Surprisingly, it was found that in each college the proportion of female applicants who were admitted to the college was actually larger than the proportion of male applicants who were admitted. Discuss and explain this result.

6. In an experiment involving 800 subjects, each subject received either treatment I or treatment II, and each subject was classified into one of the following four categories: older males, younger males, older females, and younger females. At the end of the experiment, it was determined for each subject whether the treatment that the subject had received was helpful or not. The results for each of the four categories of subjects are given in Table 9.22.

Table 9.22

Older males	Helpful	Not
Treatment I	120	120
Treatment II	20	10

Younger males	Helpful	Not
Treatment I	60	20
Treatment II	40	10

Older females	Helpful	Not
Treatment I	10	50
Treatment II	20	50

Younger females	Helpful	Not
Treatment I	10	10
Treatment II	160	90

(a) Show that treatment II is more helpful than treatment I within each of the four categories of subjects.

(b) Show that if these four categories are aggregated into just the two categories older subjects and younger subjects, then treatment I is more helpful than treatment II within each of these categories.

(c) Show that if the two categories in part (b) are aggregated into a single category containing all 800 subjects, then treatment II again appears to be more helpful than treatment I.

*9.6. KOLMOGOROV-SMIRNOV TESTS

The Sample Distribution Function

Suppose that the random variables $X_1, \ldots, X_n$ form a random sample from some continuous distribution, and let $x_1, \ldots, x_n$ denote the observed values of $X_1, \ldots, X_n$. Since the observations come from a continuous distribution, there is probability 0 that any two of the observed values $x_1, \ldots, x_n$ will be equal. Therefore, we shall assume for simplicity that all n values are different. We shall consider now a function $F_n(x)$ which is constructed from the values $x_1, \ldots, x_n$ in the following way.

For each number x $(-\infty < x < \infty)$, the value of $F_n(x)$ is defined as the proportion of observed values in the sample which are less than or equal to x. In other words, if exactly k of the observed values in the sample are less than or equal to x, then $F_n(x) = k/n$. The function $F_n(x)$ defined in this way is called the *sample distribution function* or simply the *sample d.f.* Sometimes $F_n(x)$ is called the *empirical d.f.*

The sample d.f. $F_n(x)$ can be regarded as the d.f. of a discrete distribution which assigns probability $1/n$ to each of the n values $x_1, \ldots, x_n$. Thus, $F_n(x)$ will be a step function with a jump of magnitude $1/n$ at each point x_i $(i = 1, \ldots, n)$. If we let $y_1 < y_2 < \cdots < y_n$ denote the values of the order statistics of the sample, as defined in Sec. 6.8, then $F_n(x) = 0$ for $x < y_1$; $F_n(x)$ jumps to the value $1/n$ at $x = y_1$ and remains at $1/n$ for $y_1 \leqslant x < y_2$; $F_n(x)$ jumps to the value $2/n$ at $x = y_2$ and remains at $2/n$ for $y_2 \leqslant x < y_3$; and so on. A typical sample d.f. $F_n(x)$ is sketched in Fig. 9.1 for a few different values of n.

Now let $F(x)$ denote the d.f. of the distribution from which the random sample $X_1, \ldots, X_n$ was drawn. For any given number x $(-\infty < x < \infty)$, the probability that any particular observation X_i will be less than or equal to x is $F(x)$. Therefore it follows from the law of large numbers that as $n \to \infty$, the proportion $F_n(x)$ of observations in the sample which are less than or equal to x will converge in probability to $F(x)$. In symbols,

$$\underset{n \to \infty}{\text{plim}} \, F_n(x) = F(x) \qquad \text{for } -\infty < x < \infty. \tag{1}$$

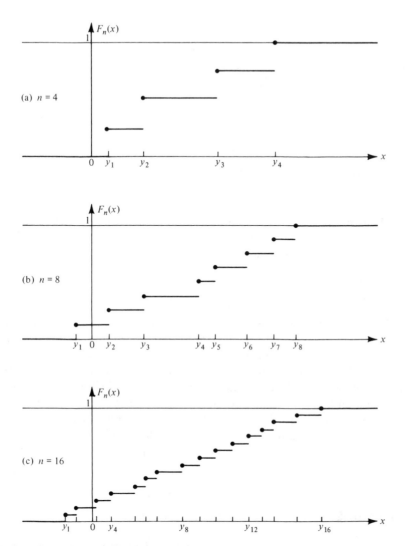

Figure 9.1   The sample d.f. $F_n(x)$ for $n = 4, 8, 16$.

The relation (1) expresses the fact that at each point x, the sample d.f. $F_n(x)$ will converge to the actual d.f. $F(x)$ of the distribution from which the random sample was taken. An even stronger result, known as the Glivenko-Cantelli lemma, states that $F_n(x)$ will converge to $F(x)$ uniformly over all values of x. More precisely, let

$$D_n = \sup_{-\infty < x < \infty} |F_n(x) - F(x)|. \tag{2}$$

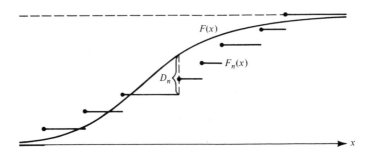

Figure 9.2   The value of D_n.

A value of D_n is illustrated in Fig. 9.2 for a typical example. Before the values of $X_1, \ldots, X_n$ have been observed, the value of D_n is a random variable. The Glivenko-Cantelli lemma states that

$$\operatorname*{plim}_{n \to \infty} D_n = 0. \tag{3}$$

The relation (3) implies that when the sample size n is large, the sample d.f. $F_n(x)$ is quite likely to be close to the d.f. $F(x)$ over the entire real line. In this sense, when the d.f. $F(x)$ is unknown, the sample d.f. $F_n(x)$ can be considered to be an estimator of $F(x)$. In another sense, however, $F_n(x)$ is not a very reasonable estimator of $F(x)$. As we explained earlier, $F_n(x)$ will be the d.f. of a discrete distribution that is concentrated on n points, whereas we are assuming in this section that the unknown d.f. $F(x)$ is the d.f. of a continuous distribution. Some type of smoothed version of $F_n(x)$, from which the jumps have been removed, might yield a reasonable estimator of $F(x)$, but we shall not pursue this topic further here.

The Kolmogorov-Smirnov Test of a Simple Hypothesis

Suppose now that we wish to test the simple null hypothesis that the unknown d.f. $F(x)$ is actually a particular continuous d.f. $F^*(x)$ against the general alternative that the actual d.f. is not $F^*(x)$. In other words, suppose that we wish to test the following hypotheses:

$$\begin{aligned} H_0&: \quad F(x) = F^*(x) \qquad \text{for } -\infty < x < \infty, \\ H_1&: \quad \text{The hypothesis } H_0 \text{ is not true.} \end{aligned} \tag{4}$$

This problem is a nonparametric problem because the unknown distribution from which the random sample is taken might be any continuous distribution.

In Sec. 9.1 we described how the χ^2 test of goodness-of-fit can be used to test hypotheses having the form (4). That test, however, requires grouping the observations into a finite number of intervals in an arbitrary manner. We shall now describe a test of the hypotheses (4) that does not require such grouping.

As before, we shall let $F_n(x)$ denote the sample d.f. Also, we shall now let D_n^* denote the following statistic:

$$D_n^* = \sup_{-\infty < x < \infty} |F_n(x) - F^*(x)|. \tag{5}$$

In other words, D_n^* is the maximum difference between the sample d.f. $F_n(x)$ and the hypothesized d.f. $F^*(x)$. When the null hypothesis H_0 in (4) is true, the probability distribution of D_n^* will be a certain distribution which is the same for every possible continuous d.f. $F^*(x)$ and that does not depend on the particular d.f. $F^*(x)$ being studied in a specific problem. Tables of this distribution, for various values of the sample size n, have been developed and are presented in many published collections of statistical tables.

It follows from the Glivenko-Cantelli lemma that the value of D_n^* will tend to be small if the null hypothesis H_0 is true, and D_n^* will tend to be larger if the actual d.f. $F(x)$ is different from $F^*(x)$. Therefore, a reasonable test procedure for the hypotheses (4) is to reject H_0 if $n^{1/2}D_n^* > c$, where c is an appropriate constant.

Table 9.23

t	$H(t)$	t	$H(t)$
0.30	0.0000	1.20	0.8878
0.35	0.0003	1.25	0.9121
0.40	0.0028	1.30	0.9319
0.45	0.0126	1.35	0.9478
0.50	0.0361	1.40	0.9603
0.55	0.0772	1.45	0.9702
0.60	0.1357	1.50	0.9778
0.65	0.2080	1.60	0.9880
0.70	0.2888	1.70	0.9938
0.75	0.3728	1.80	0.9969
0.80	0.4559	1.90	0.9985
0.85	0.5347	2.00	0.9993
0.90	0.6073	2.10	0.9997
0.95	0.6725	2.20	0.9999
1.00	0.7300	2.30	0.9999
1.05	0.7798	2.40	1.0000
1.10	0.8223	2.50	1.0000
1.15	0.8580		

It is convenient to express the test procedure in terms of $n^{1/2}D_n^*$, rather than simply D_n^*, because of the following result, which was established in the 1930's by A. N. Kolmogorov and N. V. Smirnov:

If the null hypothesis H_0 is true, then for any given value $t > 0$,

$$\lim_{n \to \infty} \Pr\left(n^{1/2}D_n^* \leqslant t\right) = 1 - 2 \sum_{i=1}^{\infty} (-1)^{i-1} e^{-2i^2 t^2}. \tag{6}$$

Thus, if the null hypothesis H_0 is true, then as $n \to \infty$, the d.f. of $n^{1/2}D_n^*$ will converge to the d.f. given by the infinite series on the right side of Eq. (6). For any value of $t > 0$, we shall let $H(t)$ denote the value on the right side of Eq. (6). The values of $H(t)$ are given in Table 9.23.

A test procedure which rejects H_0 when $n^{1/2}D_n^* > c$ is called a *Kolmogorov-Smirnov test*. It follows from Eq. (6) that when the sample size n is large, the constant c can be chosen from Table 9.23 to achieve, at least approximately, any specified level of significance α_0 $(0 < \alpha_0 < 1)$. For example, it is found from Table 9.23 that $H(1.36) = 0.95$. Therefore, if the null hypothesis H_0 is true, then $\Pr(n^{1/2}D_n^* > 1.36) = 0.05$. It follows that the level of significance of a Kolmogorov-Smirnov test with $c = 1.36$ will be 0.05.

Example 1: Testing Whether a Sample Comes from a Standard Normal Distribution.

Suppose that it is desired to test the null hypothesis that a certain random sample of 25 observations was drawn from a standard normal distribution against the alternative that the random sample was drawn from some other continuous distribution. The 25 observed values in the sample, in order from the smallest to the largest, are designated as $y_1, \ldots, y_{25}$ and are listed in Table 9.24. The table also includes the value $F_n(y_i)$ of the sample d.f. and the value $\Phi(y_i)$ of the d.f. of the standard normal distribution.

By examining the values in Table 9.24, we find that D_n^*, which is the largest difference between $F_n(x)$ and $\Phi(x)$, occurs when we pass from $i = 4$ to $i = 5$; that is, as x increases from the point $x = -0.99$ toward the point $x = -0.42$. The comparison of $F_n(x)$ and $\Phi(x)$ over this interval is illustrated in Fig. 9.3, from which we see that $D_n^* = 0.3372 - 0.16 = 0.1772$. Since $n = 25$ in this example, it follows that $n^{1/2}D_n^* = 0.886$. From Table 9.23, we find that $H(0.886) = 0.6$. Hence, the tail area corresponding to the observed value of $n^{1/2}D_n^*$ is 0.4, and we may conclude that the sample d.f. $F_n(x)$ agrees very closely with the hypothesized d.f. $\Phi(x)$. □

It is important to emphasize again that when the sample size n is large, even a small value of the tail area corresponding to the observed value of $n^{1/2}D_n^*$ would not necessarily indicate that the true d.f. $F(x)$ was much different from the hypothesized d.f. $\Phi(x)$. When n itself is large, even a small difference between the d.f. $F(x)$ and the d.f. $\Phi(x)$ would be sufficient to generate a large value of

Table 9.24

i	y_i	$F_n(y_i)$	$\Phi(y_i)$
1	− 2.46	0.04	0.0069
2	− 2.11	0.08	0.0174
3	− 1.23	0.12	0.1093
4	− 0.99	0.16	0.1611
5	− 0.42	0.20	0.3372
6	− 0.39	0.24	0.3483
7	− 0.21	0.28	0.4168
8	− 0.15	0.32	0.4404
9	− 0.10	0.36	0.4602
10	− 0.07	0.40	0.4721
11	− 0.02	0.44	0.4920
12	0.27	0.48	0.6064
13	0.40	0.52	0.6554
14	0.42	0.56	0.6628
15	0.44	0.60	0.6700
16	0.70	0.64	0.7580
17	0.81	0.68	0.7910
18	0.88	0.72	0.8106
19	1.07	0.76	0.8577
20	1.39	0.80	0.9177
21	1.40	0.84	0.9192
22	1.47	0.88	0.9292
23	1.62	0.92	0.9474
24	1.64	0.96	0.9495
25	1.76	1.00	0.9608

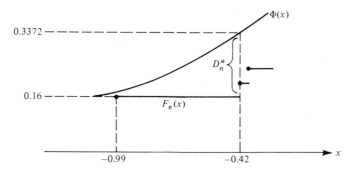

Figure 9.3 The value of D_n^* in Example 1.

$n^{1/2}D_n^*$. Therefore, before a statistician rejects the null hypothesis, he should make certain that there is a plausible alternative d.f. with which the sample $F_n(x)$ provides closer agreement.

The Kolmogorov-Smirnov Test for Two Samples

Consider a problem in which a random sample of m observations $X_1, \ldots, X_m$ is taken from a distribution for which the d.f. $F(x)$ is unknown, and an independent random sample of n observations $Y_1, \ldots, Y_n$ is taken from another distribution for which the d.f. $G(x)$ is also unknown. We shall assume that both $F(x)$ and $G(x)$ are continuous functions; and that it is desired to test the hypothesis that these functions are identical, without specifying their common form. Thus, the following hypotheses are to be tested:

$$H_0: \quad F(x) = G(x) \quad \text{for } -\infty < x < \infty,$$
$$H_1: \quad \text{The hypothesis } H_0 \text{ is not true.} \tag{7}$$

We shall let $F_m(x)$ denote the sample d.f. calculated from the observed values of $X_1, \ldots, X_m$, and shall let $G_n(x)$ denote the sample d.f. calculated from the observed values of $Y_1, \ldots, Y_n$. Furthermore, we shall consider the statistic D_{mn}, which is defined as follows:

$$D_{mn} = \sup_{-\infty < x < \infty} |F_m(x) - G_n(x)|. \tag{8}$$

The value of D_{mn} is illustrated in Fig. 9.4 for a typical example in which $m = 5$ and $n = 3$.

When the null hypothesis H_0 is true and $F(x)$ and $G(x)$ are identical functions, the sample d.f.'s $F_m(x)$ and $G_n(x)$ will tend to be close to each other.

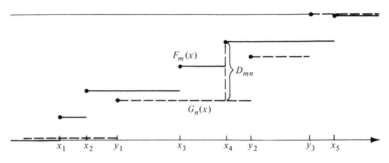

Figure 9.4   A representation of $F_m(x)$, $G_n(x)$, and D_{mn} for $m = 5$ and $n = 3$.

In fact, when H_0 is true, it follows from the Glivenko-Cantelli lemma that

$$\operatorname*{plim}_{\substack{m \to \infty \\ n \to \infty}} D_{mn} = 0. \tag{9}$$

It seems reasonable, therefore, to use a test procedure which specifies rejecting H_0 when

$$\left(\frac{mn}{m+n} \right)^{1/2} D_{mn} > c, \tag{10}$$

where c is an appropriate constant.

It is convenient to work with the statistic on the left side of the relation (10), rather than simply with D_{mn}, because of the following result:

For any value of $t > 0$, let $H(t)$ denote the right side of Eq. (6). If the null hypothesis H_0 is true, then

$$\lim_{\substack{m \to \infty \\ n \to \infty}} \Pr\left[\left(\frac{mn}{m+n} \right)^{1/2} D_{mn} \leqslant t \right] = H(t). \tag{11}$$

A test procedure which rejects H_0 when the relation (10) is satisfied is called a *Kolmogorov-Smirnov two-sample test*. Values of the function $H(t)$ are given in Table 9.23. Hence, when the sample sizes m and n are large, the constant c in the relation (10) can be chosen from Table 9.23 to achieve, at least approximately, any specified level of significance. For example, if m and n are large and the test is to be carried out at the level of significance 0.05, then it is found from Table 9.23 that we should choose $c = 1.36$.

EXERCISES

1. Suppose that the ordered values in a random sample of five observations are $y_1 < y_2 < y_3 < y_4 < y_5$. Let $F_n(x)$ denote the sample d.f. constructed from these values; let $F(x)$ be a continuous d.f.; and let D_n be defined by Eq. (2). Prove that the minimum possible value of D_n is 0.1; and prove that $D_n = 0.1$ if and only if $F(y_1) = 0.1$, $F(y_2) = 0.3$, $F(y_3) = 0.5$, $F(y_4) = 0.7$, and $F(y_5) = 0.9$.

2. Consider again the conditions of Exercise 1. Prove that $D_n \leqslant 0.2$ if and only if $F(y_1) \leqslant 0.2 \leqslant F(y_2) \leqslant 0.4 \leqslant F(y_3) \leqslant 0.6 \leqslant F(y_4) \leqslant 0.8 \leqslant F(y_5)$.

3. Test the hypothesis that the 25 values in Table 9.25 form a random sample from a uniform distribution on the interval $(0, 1)$.

Table 9.25

0.42	0.06	0.88	0.40	0.90
0.38	0.78	0.71	0.57	0.66
0.48	0.35	0.16	0.22	0.08
0.11	0.29	0.79	0.75	0.82
0.30	0.23	0.01	0.41	0.09

4. Test the hypothesis that the 25 values given in Exercise 3 form a random sample from a continuous distribution for which the p.d.f. $f(x)$ is as follows:

$$f(x) = \begin{cases} \dfrac{3}{2} & \text{for } 0 < x \leqslant \dfrac{1}{2}, \\ \dfrac{1}{2} & \text{for } \dfrac{1}{2} < x < 1, \\ 0 & \text{otherwise.} \end{cases}$$

5. Consider again the conditions of Exercises 3 and 4. Suppose that the prior probability is $1/2$ that the 25 values given in Table 9.25 were obtained from a uniform distribution on the interval $(0, 1)$, and the prior probability is $1/2$ that they were obtained from a distribution for which the p.d.f. is as given in Exercise 4. Find the posterior probability that they were obtained from a uniform distribution.

6. Test the hypothesis that the 50 values in Table 9.26 form a random sample from a normal distribution for which the mean is 26 and the variance is 4.

7. Test the hypothesis that the 50 values given in Table 9.26 form a random sample from a normal distribution for which the mean is 24 and the variance is 4.

8. Suppose that 25 observations are selected at random from a distribution for which the d.f. $F(x)$ is unknown, and that the values given in Table 9.27 are

Table 9.26

25.088	26.615	25.468	27.453	23.845
25.996	26.516	28.240	25.980	30.432
26.560	25.844	26.964	23.382	25.282
24.432	23.593	24.644	26.849	26.801
26.303	23.016	27.378	25.351	23.601
24.317	29.778	29.585	22.147	28.352
29.263	27.924	21.579	25.320	28.129
28.478	23.896	26.020	23.750	24.904
24.078	27.228	27.433	23.341	28.923
24.466	25.153	25.893	26.796	24.743

Table 9.27

0.61	0.29	0.06	0.59	−1.73
−0.74	0.51	−0.56	−0.39	1.64
0.05	−0.06	0.64	−0.82	0.31
1.77	1.09	−1.28	2.36	1.31
1.05	−0.32	−0.40	1.06	−2.47

Table 9.28

2.20	1.66	1.38	0.20
0.36	0.00	0.96	1.56
0.44	1.50	−0.30	0.66
2.31	3.29	−0.27	−0.37
0.38	0.70	0.52	−0.71

obtained. Suppose also that 20 observations are selected at random from another distribution for which the d.f. $G(x)$ is unknown, and that the values given in Table 9.28 are obtained. Test the hypothesis that $F(x)$ and $G(x)$ are identical functions.

9. Consider again the conditions of Exercise 8. Let X denote a random variable for which the d.f. is $F(x)$, and let Y denote a random variable for which the d.f. is $G(x)$. Test the hypothesis that the random variables $X + 2$ and Y have the same distribution.

10. Consider again the conditions of Exercises 8 and 9. Test the hypothesis that the random variables X and $3Y$ have the same distribution.

*9.7. INFERENCES ABOUT THE MEDIAN AND OTHER QUANTILES

Confidence Intervals and Tests for the Median

Suppose that the random variables $X_1, \ldots, X_n$ form a random sample from a continuous distribution for which the d.f. $F(x)$ is unknown, and let θ denote a point such that $F(\theta) = 1/2$. In other words, let θ denote a median of this distribution. For many continuous distributions, there will be a unique value of θ such that $F(\theta) = 1/2$. For some distributions, however, this relation will be satisfied by an entire interval of values of θ. In this case, the discussion to be presented here pertains to any value of θ such that $F(\theta) = 1/2$. It will be convenient in this section to refer to θ as *the* median of the distribution, regardless of whether or not θ is unique.

Next, let $Y_1 < Y_2 < \cdots < Y_n$ denote the order statistics of the sample. Thus, the random variable Y_1 is the smallest of the observations $X_1, \ldots, X_n$; the random variable Y_2 is the second smallest; and so on. It should be noted that for $r = 1, \ldots, n$, the relation $Y_r < \theta$ is satisfied if and only if at least r of the observations in the random sample are less than θ. Similarly, for $s = 1, \ldots, n$, the relation $Y_s > \theta$ is satisfied if and only if fewer than s observations in the sample are less than θ. Together, these statements imply that for $1 \leqslant r < s \leqslant n$, the relation $Y_r < \theta < Y_s$ will be satisfied if and only if the number of observations that are less than θ is at least r but smaller than s.

Since $F(\theta) = 1/2$, the probability that any single observation will be less than θ is $1/2$. Therefore, the probability that exactly k of the n observations will be less than θ is $\binom{n}{k} \left(\dfrac{1}{2} \right)^n$. It now follows from the development given here that

$$\Pr(Y_r < \theta < Y_s) = \sum_{k=r}^{s-1} \binom{n}{k} \left(\frac{1}{2} \right)^n. \tag{1}$$

This relation reveals that regardless of the distribution from which the random sample was drawn, and regardless of the value of the median θ, there is a fixed probability that θ will lie between any two specified order statistics Y_r and Y_s; and this probability is given by the sum on the right side of Eq. (1).

It follows from the relation (1) that after the values $y_1 < y_2 < \cdots < y_n$ of the order statistics have been observed, the interval $y_r < \theta < y_s$ will be a confidence interval for θ for which the confidence coefficient is given by the sum on the right side of Eq. (1). Different choices of r and s yield different confidence intervals for θ, and these intervals will have different confidence coefficients.

For any given value of n, the value of the sum on the right side of Eq. (1) can be found from a table of the binomial distribution for $p = 1/2$. For example, if $n = 10$, it is found from the table at the end of this book that $\Pr(Y_3 < \theta < Y_8) = 0.891$. Thus, in a random sample of 10 observations, the interval from the third smallest observed value to the third largest observed value will be a confidence interval for θ with confidence coefficient 0.891.

Next, let θ_0 be a specified number, and suppose that we wish to test the following hypotheses about the median θ:

$$\begin{aligned} H_0: & \quad \theta = \theta_0, \\ H_1: & \quad \theta \neq \theta_0. \end{aligned} \tag{2}$$

Because of the equivalence between confidence sets and tests described in Sec. 8.5, the discussion presented here for confidence intervals can be adapted to tests of the hypotheses (2). For given values of n, r, and s, let γ denote the value on the right side of Eq. (1); and consider the test procedure which accepts H_0 if $Y_r < \theta_0 < Y_s$ and rejects H_0 otherwise. Then it follows from the results in Sec. 8.5 that the size of this test will be $1 - \gamma$; that is, the probability of rejecting H_0 when in fact $\theta = \theta_0$ will be $1 - \gamma$.

For example, suppose again that $n = 10$, and that we use a test procedure which accepts H_0 if θ_0 lies in the interval $Y_3 < \theta_0 < Y_8$ and rejects H_0 otherwise. Then the size of this test is 0.109.

Confidence Intervals and Tests for Quantiles

We shall continue to assume that $X_1, \ldots, X_n$ form a random sample from a continuous distribution for which the d.f. is $F(x)$. We shall let p be any given number in the interval $0 < p < 1$, and shall let ζ_p denote a point such that $F(\zeta_p) = p$. The point ζ_p is called a *quantile*, or a *percentile*, of the distribution. For example, if $p = 0.95$, we say that the point ζ_p is a 0.95 quantile or a 95th percentile of the distribution. In this terminology, a median of the distribution is a 0.5 quantile. For a continuous distribution and a given value of p, either there is a unique value of ζ_p such that $F(\zeta_p) = p$, or this relationship is satisfied for an entire interval of values of ζ_p.

The discussion that we have given here for the median can easily be extended to the development of confidence intervals and tests for any quantile ζ_p. If we again let $Y_1 < Y_2 < \cdots < Y_n$ denote the order statistics of the random sample, then the relation $Y_r < \zeta_p < Y_s$ will again be satisfied if and only if the number of observations that are less than ζ_p is at least r and smaller than s. Since the probability that any single observation will be less than ζ_p is p, the probability that exactly k of the observations will be less than ζ_p is $\binom{n}{k} p^k (1 - p)^{n-k}$. Hence,

$$\Pr(Y_r < \zeta_p < Y_s) = \sum_{k=r}^{s-1} \binom{n}{k} p^k (1 - p)^{n-k}. \tag{3}$$

For given values of n, p, r, and s, the value of the sum on the right side of Eq. (3) can be found from a table of the binomial distribution with parameters n and p. If the value of this sum is γ, then after the values $y_1 < y_2 < \cdots < y_n$ of the order statistics have been observed, the interval from y_r to y_s will be a confidence interval for ζ_p with confidence coefficient γ.

For example, if $n = 15$ and $p = 0.2$, it is found from the table of the binomial distribution in the back of this book that $\Pr(Y_1 < \zeta_p < Y_6) = 0.90$. Hence, the interval between the smallest observation Y_1 and the sixth smallest observation Y_6 will form a confidence interval for the 0.2 quantile of the distribution with confidence coefficient 0.90.

This discussion of confidence intervals for ζ_p can easily be adapted to tests of hypotheses about the value of ζ_p.

EXERCISES

Note. In each of the following exercises, it is assumed that the random variables $X_1, \ldots, X_n$ form a random sample from a continuous distribution for which the d.f. $F(x)$ is unknown; that θ denotes a median of this distribution; and that $Y_1 < Y_2 < \cdots < Y_n$ denote the order statistics of the sample.

1. Find the probability that all n observations in the random sample will be smaller than θ.

2. Evaluate (a) $\Pr(Y_1 > \theta)$ and (b) $\Pr(Y_2 > \theta)$.

3. Find the smallest value of n for which $\Pr(Y_1 < \theta < Y_n) \geq 0.99$.

4. Suppose that $n = 7$ and that the seven observed values in the sample are 7.11, 5.12, 8.44, 7.13, 7.12, 12.96, and 4.07. Determine the endpoints and the confidence coefficient for each of the following two confidence intervals for θ:

   I_1: The interval between Y_2 and Y_4;
   I_2: The interval between Y_3 and Y_5.

5. Consider again the conditions of Exercise 4. Discuss the practical interpretation of the term "confidence" in this problem, in which the confidence coefficient for the interval I_2 is greater than the confidence coefficient for I_1, but the length of I_1 is 100 times the length of I_2.

6. Find the probability that all n observations in the random sample will be smaller than the 0.3 quantile of the distribution.

7. Find the probability that all n observations will lie between the 0.3 quantile and the 0.8 quantile of the distribution.

8. Let A denote the 0.3 quantile of the distribution. Find the smallest value of n for which $\Pr(Y_1 < A < Y_n) \geq 0.95$.

9. Let B denote the 0.7 quantile of the distribution. Find the smallest value of n for which $\Pr(Y_1 < B < Y_n) \geq 0.95$.

10. Suppose that $n = 20$, and let B denote the 0.7 quantile of the distribution. Evaluate $\Pr(Y_{12} < B < Y_{17})$.

*9.8. ROBUST ESTIMATION

Estimating the Median

Suppose that the random variables $X_1, \ldots, X_n$ form a random sample from a continuous distribution for which the p.d.f. $f(x)$ is unknown, but may be assumed to be a symmetric function with respect to some unknown point θ $(-\infty < \theta < \infty)$. Because of this symmetry, the point θ will be a median of the unknown distribution. We shall estimate the value of θ from the observations $X_1, \ldots, X_n$.

If we know that the observations actually come from a normal distribution, then the sample mean $\bar{X}_n$ will be the M.L.E. of θ. Without any strong prior information indicating that the value of θ might be quite different from the observed value of $\bar{X}_n$, we may assume that the value of $\bar{X}_n$ will be a reasonable estimate of θ. Suppose, however, that the observations might come from a distribution for which the p.d.f. $f(x)$ has much thicker tails than the p.d.f. of a normal distribution; that is, suppose that as $x \rightarrow \infty$ or $x \rightarrow -\infty$, the p.d.f. $f(x)$ might come down to 0 much more slowly than does the p.d.f. of a normal distribution. In this case, the sample mean $\bar{X}_n$ may be a poor estimator of θ because its M.S.E. may be much larger than that of some other possible estimator.

For example, if the underlying distribution is actually a Cauchy distribution centered at an unknown point θ, as defined in Example 3 of Sec. 6.6, then the M.S.E. of $\bar{X}_n$ will actually be infinite. In this case the M.L.E. of θ will have a finite M.S.E. and will be a much better estimator than $\bar{X}_n$. In fact, for a large value of n, the M.S.E. of the M.L.E. is approximately $2/n$, no matter what the true value of θ is. However, as pointed out in Example 3 of Sec. 6.6, this estimator is very complicated and must be determined by a numerical calculation for each given set of observations. A relatively simple and reasonable estimator for this problem is the *sample median*, which was defined in Example 1 of Sec. 6.9. It can be shown that the M.S.E. of the sample median for a large value of n is approximately $2.47/n$.

It follows from this discussion that if we could assume that the underlying distribution is normal or nearly normal, then we might use the sample mean as an estimator of θ. On the other hand, if we believe that the underlying distribution is Cauchy or nearly Cauchy, then we might use the sample median. However, we typically do not know whether the underlying distribution is nearly normal, is nearly Cauchy, or does not correspond closely to either of these types of distributions. For this reason, we should try to find an estimator of θ that will have a small M.S.E. for several different possible types of distributions. An estimator which performs well for several different types of distributions, even though it may not be the best available estimator for any particular type of distribution, is called a *robust estimator*. In this section, we shall define and study a special type of robust estimator known as a *trimmed mean*. The term *robust* was introduced by G. E. P. Box in 1953, and the term *trimmed mean* was introduced by J. W. Tukey in 1962. However, the first mathematical treatment of trimmed means was given by P. Daniell in 1920.

Trimmed Means

We shall continue to suppose that $X_1, \ldots, X_n$ form a random sample from an unknown continuous distribution for which the p.d.f. $f(x)$ is assumed to be symmetric with respect to an unknown point θ. As usual, we shall let $Y_1 < Y_2 < \cdots < Y_n$ denote the order statistics of the sample. The sample mean $\overline{X}_n$ is simply the average of these n order statistics. However, if we suspect that the p.d.f. $f(x)$ might have thicker tails than a normal distribution has, then we may wish to estimate θ by using a weighted average of the order statistics which assigns less weight to the extreme observations such as Y_1, Y_2, Y_{n-1}, and Y_n, and assigns more weight to the middle observations. The sample median is a special example of a weighted average. When n is odd, it assigns zero weight to every observation except the middle one. When n is even, it assigns the weight $1/2$ to each of the two middle observations and it assigns zero weight to every other observation.

Another estimator which is a weighted average of the order statistics is called a *trimmed mean*. It is constructed as follows: For a suitable positive integer k such that $k < n/2$, the k smallest observations $Y_1, \dots, Y_k$ and the k largest observations $Y_n, Y_{n-1}, \dots, Y_{n-k+1}$ are deleted from the sample. The average of the remaining $n - 2k$ intermediate observations is called the kth *level trimmed mean*. Clearly, the kth level trimmed mean can be represented as a weighted average of the order statistics having the form

$$\frac{1}{n-2k} \sum_{i=k+1}^{n-k} Y_i. \tag{1}$$

The sample median is an example of a trimmed mean. When n is odd, the sample median is the $[(n-1)/2]$th level trimmed mean. When n is even, it is the $[(n-2)/2]$th level trimmed mean.

Comparison of the Estimators

We have mentioned the desirability of using an estimator such as a trimmed mean in a situation in which it is suspected that the observations $X_1, \dots, X_n$ may form a random sample from a distribution for which the tails of the p.d.f. are thicker than the tails of the p.d.f. of a normal distribution. The use of a trimmed mean is also desirable when a few of the observations in the sample appear to be unusually large or unusually small. In this situation a statistician might suspect that most of the observations in the sample came from one normal distribution, whereas the few extreme observations may have come from a different normal distribution with a much larger variance than the first one. The extreme observations, which are called *outliers*, will substantially affect the value of $\overline{X}_n$ and make it an unreliable estimator of θ. Since the values of these outliers would be omitted in a trimmed mean, the trimmed mean will usually be a more reliable estimator than $\overline{X}_n$.

It is acknowledged that a trimmed mean will perform better than $\overline{X}_n$ in a situation of the type just described. However, if $X_1, \dots, X_n$ actually do form a random sample from a normal distribution, then $\overline{X}_n$ will perform better than a trimmed mean. Since we are typically not certain which situation prevails in a particular problem, it is important to know how much larger the M.S.E. of a trimmed mean will be than the M.S.E. of $\overline{X}_n$ when the actual distribution is normal. In other words, it is important to know how much is lost if we use a trimmed mean when the actual distribution is normal. We shall now consider this question.

When $X_1, \ldots, X_n$ form a random sample from a normal distribution with mean θ and variance σ^2, both the probability distribution of $\overline{X}_n$ and the probability distribution of a trimmed mean will be symmetric with respect to the value θ. Therefore, the mean of each of these estimators will be θ; the M.S.E. of each estimator will be equal to its variance; and this M.S.E. will have a certain constant value for each estimator regardless of the true value of θ. The values of several of these M.S.E.'s for a normal distribution when the sample size n is 10 or 20 are presented in Table 9.29. It should be noted that when $n = 10$, the trimmed mean for $k = 4$ and the sample median are the same estimator.

It can be seen from Table 9.29 that when the underlying distribution is actually a normal distribution, the M.S.E.'s of trimmed means are not much larger than the M.S.E. of $\overline{X}_n$. In fact, when $n = 20$, the M.S.E. of the second-level trimmed mean ($k = 2$), in which four of the 20 observed values in the sample are omitted, is only 1.06 times as large as the M.S.E. of $\overline{X}_n$. Even the M.S.E. of the sample median is only 1.5 times that of $\overline{X}_n$. These values indicate that trimmed means can be regarded as robust estimators of θ.

We shall now consider the improvement in the M.S.E. that can be achieved by using a trimmed mean when the underlying distribution is not normal. If $X_1, \ldots, X_n$ form a random sample of size n from a Cauchy distribution, then the M.S.E. of $\overline{X}_n$ is infinite. The M.S.E.'s of trimmed means for a Cauchy distribution when the sample size n is 10 or 20 are given in Table 9.30.

Table 9.29

Estimator	$n = 10$	$n = 20$
Sample mean $\overline{X}_n$	1.00	1.00
Trimmed mean for $k = 1$	1.05	1.02
Trimmed mean for $k = 2$	1.12	1.06
Trimmed mean for $k = 3$	1.21	1.10
Trimmed mean for $k = 4$	1.37	1.14
Sample median	1.37	1.50

For a normal distribution the M.S.E. is equal to the tabulated value multiplied by σ^2/n.

Table 9.30

Estimator	$n = 10$	$n = 20$
Sample mean $\overline{X}_n$	∞	∞
Trimmed mean for $k = 1$	27.22	23.98
Trimmed mean for $k = 2$	8.57	7.32
Trimmed mean for $k = 3$	3.86	4.57
Trimmed mean for $k = 4$	3.66	3.58
Sample median	3.66	2.88

For a Cauchy distribution the M.S.E. is equal to the tabulated value divided by n.

Finally, the M.S.E.'s for two other situations are presented in Table 9.31. In the first situation we shall assume that in a sample of 20 independent observations, 19 are drawn from a normal distribution with mean θ and variance σ^2, and the other one is drawn from a normal distribution with mean θ and variance $100\sigma^2$. The M.S.E.'s of $\overline{X}_n$ and the trimmed means in this situation are given in column (1) of Table 9.31. In the second situation, we shall assume that in a sample of 20 independent observations, 18 are drawn from a normal distribution with mean θ and variance σ^2, and the other two are drawn from a normal distribution with mean θ and variance $100\sigma^2$. The M.S.E.'s in this situation are given in column (2) of Table 9.31.

It can be seen from Tables 9.30 and 9.31 that the M.S.E. of a trimmed mean can be substantially smaller than that of $\overline{X}_n$. When a trimmed mean is to be used as an estimator of θ, it is evident that a specific value of k must be chosen. No

Table 9.31

Estimator	(1)	(2)
Sample mean $\overline{X}_n$	5.95	10.90
Trimmed mean for $k = 1$	1.23	2.90
Trimmed mean for $k = 2$	1.20	1.46
Trimmed mean for $k = 3$	1.21	1.43
Trimmed mean for $k = 4$	1.24	1.44
Sample median	1.55	1.80

For a mixed sample of 20 observations from certain normal distributions, the M.S.E. is equal to the tabulated value multiplied by $\sigma^2/20$.

general rule for choosing k will be best under all conditions. If there is reason to believe that the p.d.f. $f(x)$ is approximately normal, then θ might be estimated by using a trimmed mean which is obtained by omitting about 10 or 15 percent of the observed values at each end of the ordered sample. If the p.d.f. $f(x)$ might be far from normal or if several of the observations might be outliers, then the sample median might be used to estimate θ.

Large-Sample Properties of the Sample Median

It can be shown that if $X_1, \ldots, X_n$ form a large random sample from a continuous distribution for which the p.d.f. is $f(x)$ and for which there is a unique median θ, then the distribution of the sample median will be approximately a normal distribution. Specifically, it must be assumed that the p.d.f. $f(x)$ has a derivative $f'(x)$ which is continuous at the point θ and that $f(\theta) > 0$.

Let $\tilde{X}_n$ denote the sample median. Then, as $n \to \infty$, the d.f. of $n^{1/2}(\tilde{X}_n - \theta)$ will converge to the d.f. of a normal distribution with mean 0 and variance $1/[2f(\theta)]^2$. In other words, when n is large, the distribution of the sample median $\tilde{X}_n$ will be approximately a normal distribution with mean θ and variance $1/[4nf^2(\theta)]$.

EXERCISES

1. Suppose that a sample comprises the 14 observed values in Table 9.32. Calculate the values of (a) the sample mean; (b) the trimmed means for $k = 1, 2, 3,$ and 4; and (c) the sample median.

2. Suppose that a sample comprises the 15 observed values in Table 9.33. Calculate the values of (a) the sample mean; (b) the trimmed means for $k = 1, 2, 3,$ and 4; and (c) the sample median.

3. Describe how to use the table of random digits given at the end of this book to obtain a random sample of n observations from a uniform distribution on the interval $(-b, b)$, where b is a given positive number.

Table 9.32

1.24	0.36	0.23
0.24	1.78	−2.00
−0.11	0.69	0.24
0.10	0.03	0.00
−2.40	0.12	

Table 9.33

23.0	21.5	63.0
22.5	2.1	22.1
22.4	2.2	21.7
21.7	22.2	22.9
21.3	21.8	22.1

4. Use the table of random digits given at the end of this book to obtain a sample of 15 independent observations in which 13 of the 15 are drawn from the uniform distribution on the interval $(-1, 1)$ and the other two are drawn from the uniform distribution on the interval $(-10, 10)$. For the 15 values that are obtained, calculate the values of (a) the sample mean; (b) the trimmed means for $k = 1, 2, 3,$ and 4; and (c) the sample median. Which of these estimators is closest to 0?

5. Repeat Exercise 4 ten times, using different random digits each time. In other words, construct ten independent samples each of which contains 15 observations and each of which satisfies the conditions of Exercise 4.

   (a) For each sample, which of the estimators listed in Exercise 4 is closest to 0?

   (b) For each of the estimators listed in Exercise 4, determine the square of the distance between the estimator and 0 in each of the ten samples, and determine the average of these ten squared distances. For which of the estimators is this average squared distance from 0 smallest?

6. Suppose that a random sample of 100 observations is taken from a normal distribution for which the mean θ is unknown and the variance is 1, and let $\tilde{X}_n$ denote the sample median. Determine the value of $\Pr(|\tilde{X}_n - \theta| \leq 0.1)$.

7. Suppose that a random sample of 100 observations is taken from a Cauchy distribution centered at an unknown point θ, and let $\tilde{X}_n$ denote the sample median. Determine the value of $\Pr(|\tilde{X}_n - \theta| \leq 0.1)$.

8. Let $f(x)$ denote the p.d.f. of a normal distribution for which the mean θ is unknown and the variance is 1. Also, let $g(x)$ denote the p.d.f. of a normal distribution for which the mean has the same unknown value θ and the variance is 4. Finally, let the p.d.f. $h(x)$ be defined as follows:

$$h(x) = \frac{1}{2}[f(x) + g(x)].$$

Suppose that 100 observations are selected at random from a distribution for which the p.d.f. is $h(x)$. Determine the M.S.E. of the sample mean and the M.S.E. of the sample median.

*9.9.   PAIRED OBSERVATIONS

Comparative Experiments and Matched Pairs

In this section and the next one we shall discuss experiments in which two different methods or treatments are to be compared in order to learn which one is better or more effective. Suppose, for example, that two different drugs A and B are to be administered to patients having a certain illness in order to learn which drug stimulates greater enzyme activity of a particular type. We shall assume that the experiment is to be carried out by administering drug A to n patients and drug B to n other patients; and after a specified period of time, measuring the amount of enzyme activity exhibited by each patient. This amount for a patient will be called his *response* to the drug.

Because of the wide variation among patients in regard to their personal characteristics and their medical histories, it is typically better in an experiment of this kind to try to select patients for the two samples in such a way that the two groups will resemble each other as closely as possible, rather than simply to select two random samples having the same size from the available population of patients. If the patients who are to receive each drug are simply selected at random, it is possible that the n patients who receive drug A may systematically differ from the patients who receive drug B in regard to some characteristic that is related to the enzyme activity being studied. It might happen, for instance, that most of the patients in one sample are women while most of those in the other sample are men; or that one sample contains a large number of elderly patients while the other contains a large proportion of young patients. Such differences between the two samples can lead to Simpson's paradox, as described in Sec. 9.5, and would obviously vitiate any comparisons between the two drugs that do not take these differences into account.

Furthermore, regardless of whether or not there are systematic differences between the two samples, the wide variation among patients implies that the probability distribution of the response will be different for each patient in the sample. Hence, when we take into account the different characteristics of the patients who appear in each of the samples, the responses for either sample cannot be considered to form a random sample of observations from some common distribution. Standard statistical methods would therefore be inappropriate, and the analysis would become difficult.

The difficulties that have just been described can be avoided or greatly reduced by designing the experiment carefully and eliminating as many chance factors as possible. One way to design the experiment is to select pairs of patients that are matched as closely as possible in regard to such physical characteristics as age, sex, medical history, and present physical condition. If n pairs of patients are selected in this way, then the characteristics of the patients in each pair will be quite similar, even though there may be wide variations in the characteristics among the n patients in each sample.

In each pair, one patient is treated with drug A and the other is treated with drug B. Since these two patients will not have identical characteristics, even though they are carefully matched, the choice of which patient receives drug A and which one receives drug B should be made by some method, such as an auxiliary randomization, which ensures that the experimenter will not subconsciously exploit any preferences or biases that he might have. If the drugs are equally effective, then in each pair the patient who receives drug A and the patient who receives drug B are equally likely to exhibit the greater response. Moreover, if drug A is more effective than drug B, then in each pair the patient who receives drug A is more likely to exhibit a greater response than is the patient who receives drug B. The statistical analysis should be carried out by comparing the responses of the two patients in each of the n pairs.

In some experiments of this type, the same patient can be used to obtain a pair of observations. For example, if two types of sleeping pills are to be compared, it is often possible to give a patient one type of pill on one night and then give the same patient the other type of pill under similar conditions on another night. In this experiment, the n pairs of observations would be obtained from only n different patients. The analysis would proceed, however, just as in an experiment in which there were n matched pairs of patients. When a *crossover design* of this type is used, it is important to verify that the responses of a particular patient to the two different types of pills do not depend on which type the patient receives first.

The Sign Test

For $i = 1, \ldots, n$, let p_i denote the probability that in the ith pair of patients, the patient who receives drug A will exhibit a greater response than the patient who receives drug B. Since the patients in each pair have been matched as carefully as possible, we shall assume that this probability p_i has the same value p for each of the n pairs. Suppose that we wish to test the null hypothesis that drug A is not more effective than drug B against the alternative hypothesis that drug A is more effective than drug B. These hypotheses can now be expressed in the following form:

$$H_0: \quad p \leqslant \frac{1}{2},$$
$$H_1: \quad p > \frac{1}{2}. \tag{1}$$

It is assumed that there is zero probability that the two patients in any pair will have exactly the same response; that is, it is assumed that in each pair either one drug or the other will appear to be better. Under these conditions, the n pairs represent n Bernoulli trials, for each of which there is probability p that drug A will yield the larger response. Therefore, the number of pairs X in which drug A yields the larger response will have a binomial distribution with parameters n and p. If the null hypothesis H_0 is true, then $p \leqslant 1/2$; and if the alternative hypothesis H_1 is true, then $p > 1/2$. Hence, a reasonable procedure is to reject H_0 if $X > c$, where c is an appropriate constant. This procedure is called the *sign test*.

In summary, a sign test is carried out as follows: In each pair, the difference between the response to drug A and the response to drug B is measured, and the number of pairs for which this difference is positive is counted. The decision to accept or reject H_0 is then based solely on this number of positive differences.

For example, suppose that the number of pairs is 15, and it is found that drug A yields a larger response than drug B in exactly 11 of the 15 pairs. Then $n = 15$ and $X = 11$. When $p = 1/2$, it is found from the table of the binomial distribution that the corresponding tail area is 0.0593. Thus, the null hypothesis H_0 should be rejected at any level of significance greater than this number.

The only information which the sign test utilizes from each pair of observations is the sign of the difference between the two responses. To apply the sign test, the experimenter only has to be able to observe whether the response to drug A or drug B is larger. He does not have to be able to obtain a numerical measurement of the magnitude of the difference between the two responses. However, if the magnitude of the difference for each pair can be measured, it is useful to apply a test procedure which not only considers the sign of the difference but also recognizes the fact that a large difference between the responses is more important than a small difference. We shall now describe a procedure based on the relative magnitudes of the differences.

The Wilcoxon Signed-Ranks Test

We shall continue to assume that drugs A and B are administered to n matched pairs of patients, and we shall assume that the response of each patient can be measured in appropriate units on some numerical scale. For $i = 1, \ldots, n$, we shall let A_i denote the response of the patient in pair i who receives drug A; we shall let B_i denote the response of the patient in pair i who receives drug B; and we shall let $D_i = A_i - B_i$.

Since the n differences $D_1, \ldots, D_n$ pertain to different pairs of patients, these differences will be independent random variables. Furthermore, because the patients in each pair have been carefully matched, we shall assume that all the differences $D_1, \ldots, D_n$ have the same continuous distribution. Finally, we shall assume that this distribution is symmetric with respect to some unknown point θ. In summary, we shall assume that the differences $D_1, \ldots, D_n$ are i.i.d. and form a random sample from a continuous distribution which is symmetric with respect to the point θ.

The null hypothesis H_0 that drug A is not more effective than drug B is equivalent to the statement that $\Pr(D_i \leqslant 0) \geqslant 1/2$. In turn, this statement is equivalent to the statement that $\theta \leqslant 0$. Similarly, the alternative hypothesis H_1 that drug A is more effective than drug B is equivalent to the statement that $\theta > 0$. Thus, we must test the following hypotheses:

$$H_0: \quad \theta \leqslant 0, \qquad\qquad (2)$$
$$H_1: \quad \theta > 0.$$

Since θ is the median of the distribution of each difference D_i, confidence intervals for θ and tests of the hypotheses (2) can be developed by using the methods described in Sec. 9.7. We shall now describe a different procedure for testing the hypotheses (2) which was proposed by F. Wilcoxon in 1945 and is known as the *Wilcoxon signed-ranks test*.

First, the absolute values $|D_1|, \ldots, |D_n|$ are arranged in order from the smallest absolute value to the largest. It is assumed that no two of these absolute values are equal, and that each is different from 0.

Second, each absolute value $|D_i|$ is assigned a rank corresponding to its position in this ordering. Thus, the smallest absolute value is assigned the rank 1; the second smallest absolute value is assigned the rank 2; and so on. The largest absolute value is assigned the rank n.

Third, each of the ranks $1, \ldots, n$ is assigned either a plus sign or a minus sign, the assignment depending on whether the original difference D_i which yielded that rank was positive or negative.

Finally, the statistic S_n is defined to be the sum of those ranks to which a plus sign was assigned. The Wilcoxon signed-ranks test is based on the value of the statistic S_n.

Before we proceed further in our description of the procedure for carrying out this test, we shall consider an example. Suppose that the number of pairs is 15, and that the observed responses A_i and B_i, for $i = 1, \ldots, 15$, are as given in Table 9.34. The absolute values $|D_i|$ have been ranked from 1 to 15 in column (5) of the table. Then in column (6) each rank has been given the same sign as the corresponding value of D_i. The value of the statistic S_n is the sum of the positive ranks in column (6). In this example, it is found that $S_n = 93$.

We shall now consider the distribution of the statistic S_n. Suppose that the null hypothesis H_0 is true. If $\theta = 0$, then the drugs A and B are equally effective. In this case, each of the ranks $1, \ldots, n$ is equally likely to be given a plus sign or a minus sign, and the n assignments of plus signs and minus signs are independent of each other. Furthermore, if $\theta < 0$, then drug A is less effective than drug B. In

Table 9.34

i	(1) A_i	(2) B_i	(3) D_i	(4) $|D_i|$	(5) Rank	(6) Signed rank
1	3.84	3.03	0.81	0.81	4	+4
2	6.27	4.91	1.36	1.36	7	+7
3	8.75	7.65	1.10	1.10	6	+6
4	4.39	5.00	−0.61	0.61	3	−3
5	9.24	7.42	1.82	1.82	10	+10
6	6.59	4.20	2.39	2.39	13	+13
7	9.73	7.21	2.52	2.52	14	+14
8	5.61	7.59	−1.98	1.98	11	−11
9	2.75	3.64	−0.89	0.89	5	−5
10	8.83	6.23	2.60	2.60	15	+15
11	4.41	4.34	0.07	0.07	1	+1
12	3.82	5.27	−1.45	1.45	8	−8
13	7.66	5.33	2.33	2.33	12	+12
14	2.96	2.82	0.14	0.14	2	+2
15	2.86	1.14	1.72	1.72	9	+9

this case, each rank is more likely to receive a minus sign than a plus sign, and the statistic S_n will tend to be smaller than it would be if $\theta = 0$.

On the other hand, if the alternative hypothesis H_1 is true and $\theta > 0$, then drug A is actually more effective than drug B and each rank is more likely to receive a plus sign than a minus sign. In this case, the statistic S_n will tend to be larger than it would be under the hypothesis H_0. For this reason, the Wilcoxon signed-ranks test specifies rejecting H_0 when $S_n \geq c$, where the constant c is chosen appropriately.

When $\theta = 0$, the mean and the variance of S_n can be derived as follows: For $i = 1, \ldots, n$, let $W_i = 1$ if the rank i receives a plus sign and let $W_i = 0$ if the rank i receives a minus sign. Then the statistic S_n can be represented in the form

$$S_n = \sum_{i=1}^{n} iW_i. \tag{3}$$

If $\theta = 0$, we have $\Pr(W_i = 0) = \Pr(W_i = 1) = 1/2$. Hence, $E(W_i) = 1/2$ and $\mathrm{Var}(W_i) = 1/4$. Furthermore, the random variables $W_1, \ldots, W_n$ are independent. It now follows from Eq. (3) that when $\theta = 0$,

$$E(S_n) = \sum_{i=1}^{n} iE(W_i) = \frac{1}{2} \sum_{i=1}^{n} i \tag{4}$$

and

$$\text{Var}(S_n) = \sum_{i=1}^{n} i^2 \text{Var}(W_i) = \frac{1}{4} \sum_{i=1}^{n} i^2. \tag{5}$$

The final sum in Eq. (4), which is the sum of the integers from 1 to n, is equal to $(1/2)n(n + 1)$. The final sum in Eq. (5), which is the sum of the squares of the integers from 1 to n, is equal to $(1/6)n(n + 1)(2n + 1)$. Therefore, when $\theta = 0$,

$$E(S_n) = \frac{n(n + 1)}{4} \quad \text{and} \quad \text{Var}(S_n) = \frac{n(n + 1)(2n + 1)}{24}. \tag{6}$$

Furthermore, it can be shown that as the number of pairs $n \to \infty$, the distribution of S_n converges to a normal distribution. More precisely, if μ_n and σ_n^2 represent the mean and the variance of S_n as given by the relations (6), then as $n \to \infty$, the d.f. of $(S_n - \mu_n)/\sigma_n$ converges to the d.f. of the standard normal distribution. The practical interpretation of this result is as follows: When the number of pairs n is large and $\theta = 0$, the distribution of S_n will be approximately a normal distribution for which the mean and variance are given by the relations (6).

Suppose that the hypotheses (2) are to be tested at a specified level of significance α_0 ($0 < \alpha_0 < 1$); and that the constant c is determined so that when $\theta = 0$, $\Pr(S_n \geq c) = \alpha_0$. Then the procedure which rejects H_0 when $S_n \geq c$ will satisfy the specified level of significance.

For example, consider again the data in Table 9.34. In this example, $n = 15$ and it is found from the relations (6) that $E(S_n) = 60$ and $\text{Var}(S_n) = 310$. Hence, the standard deviation of S_n is $\sigma_n = \sqrt{310} = 17.6$. It follows that when $\theta = 0$, the random variable $Z_n = (S_n - 60)/17.6$ will have approximately a standard normal distribution. If we suppose that it is desired to test the hypotheses (2) at the level of significance 0.05, then H_0 should be rejected if $Z_n \geq 1.645$.

For the data in Table 9.34 it was found that $S_n = 93$. Therefore, $Z_n = 1.875$, and it follows that the null hypothesis H_0 should be rejected. In fact, the tail area corresponding to this observed value of S_n is found from the table of the standard normal distribution to be 0.03.

For a small value of n, the normal approximation is not applicable. In this case, the exact distribution of S_n when $\theta = 0$ is given in many published collections of statistical tables.

Ties

The theory under discussion is based on the assumption that the values of $D_1, \ldots, D_n$ will be distinct nonzero numbers. Since the measurements in an actual experiment may be made with only limited precision, however, there may actually be ties or zeros among the observed values of $D_1, \ldots, D_n$. Suppose that a sign test is to be performed and it is found that $D_i = 0$ for one or more values of i. In this case, the sign test should be carried out twice. In the first test, it should be assumed that each 0 is actually a positive difference. In the second test, each 0 should be treated as a negative difference. If the tail areas found from the two tests are roughly equal, then the zeros are a relatively unimportant part of the data. If, on the other hand, the tail areas are quite different, then the zeros can seriously affect the inferences that are to be made. In this case the experimenter should try to obtain additional measurements or more refined measurements.

Similar comments pertain to the Wilcoxon signed-rank test. If $D_i = 0$ for one or more values of i, these zeros should be assigned the lowest ranks and the test should be carried out twice. In the first test, plus signs should be assigned to these ranks. In the second test, minus signs should be assigned to them. A small difference in the tail areas would indicate that the zeros are relatively unimportant. A large difference would indicate that the data may be too unreliable to be used.

The same type of reasoning can be applied to two differences D_i and D_j which have different signs but the same absolute value. These pairs will occupy successive positions, say k and $k + 1$, in the ranking of absolute values. However, since there is a tie, it is not clear which of the two ranks should be assigned the plus sign. Therefore, the test should be carried out twice. First, the rank $k + 1$ should be assigned a plus sign and the rank k should be assigned a minus sign. Then these signs should be interchanged.

Other reasonable methods for handling ties have been proposed. When two or more absolute values are the same, one simple method is to consider the successive ranks that are to be assigned to these absolute values and then to assign the average of these ranks to each of the tied values. The plus and minus signs are then assigned in the usual way. When this method is used, the value of $\text{Var}(S_n)$ must be corrected because of the ties.

EXERCISES

1. In an experiment to compare two different types of long-lasting razor blades, A and B, 20 men were asked to shave with a blade of type A for one week and then with a blade of type B for one week. At the end of this period, 15 men thought that blade A gave a smoother shave than blade B, and the other 5 men thought that blade B gave a smoother shave. Test the null hypothesis that blade A does not tend to give a smoother shave than blade B against the alternative that blade A tends to give a smoother shave than blade B.

2. Consider again the conditions of Exercise 1. Discuss how the design of this experiment might be improved by considering how the 20 men who participate might be selected and also by considering the possible effect of

having each man shave first with a blade of type B and then with a blade of type A.

3. Consider the data presented in Table 9.34, and assume that the 15 differences $D_1, \ldots, D_{15}$ form a random sample from a normal distribution for which both the mean μ and the variance σ^2 are unknown. Carry out a t test of the following hypotheses:

H_0: $\mu \leqslant 0$,
H_1: $\mu > 0$.

4. Consider again the data presented in Table 9.34.

   (a) Compare the tail areas that are obtained from applying the sign test, the Wilcoxon signed-ranks test, and the t test to the differences $D_1, \ldots, D_{15}$.

   (b) Discuss the assumptions that are needed to apply each of these three tests.

   (c) Discuss the inferences that can be drawn in regard to the relative effectiveness of drug A and drug B because the three tail areas obtained in part (a) have approximately the same magnitude; and discuss the inferences that could be drawn if these tail areas were widely different.

5. In an experiment to compare two different diets A and B for pigs, a pair of pigs were selected from each of 20 different litters. One pig in each pair was selected at random and was fed diet A for a fixed period of time, while the other pig in the pair was fed diet B. At the end of the fixed period, the gain in weight of each pig was noted. The results are presented in Table 9.35. Test the null hypothesis that pigs do not tend to gain more weight from diet A than from diet B against the alternative that pigs do tend to gain more weight from diet A, by using (a) the sign test and (b) the Wilcoxon signed-ranks test.

Table 9.35

Pair	Gain from diet A	Gain from diet B	Pair	Gain from diet A	Gain from diet B
1	21.5	14.7	11	19.0	19.4
2	18.0	18.1	12	18.8	13.6
3	14.7	15.2	13	19.0	19.2
4	19.3	14.6	14	15.8	9.1
5	21.7	17.5	15	19.6	13.2
6	22.9	15.6	16	22.0	16.6
7	22.3	24.8	17	13.4	10.8
8	19.1	20.3	18	16.8	13.3
9	13.3	12.0	19	18.4	15.4
10	19.8	20.9	20	24.9	21.7

6. Consider again the experiment described in Exercise 5 and the data presented in Table 9.35.

   (a) Test the hypotheses described in Exercise 5 by assuming that in each of the 20 pairs of pigs, the difference between the gain from diet A and the gain from diet B has a normal distribution with an unknown mean μ and an unknown variance σ^2.

   (b) Test the hypotheses described in Exercise 5 by assuming that the gain in weight for each pig that is fed diet A has a normal distribution with an unknown mean μ_A and an unknown variance σ^2, and the gain in weight for each pig that is fed diet B has a normal distribution with an unknown mean μ_B and the same unknown variance σ^2.

   (c) Compare the results obtained in parts (a) and (b) of this exercise and parts (a) and (b) of Exercise 5; and discuss the interpretation of these results.

   (d) Discuss methods for investigating whether the assumptions made in part (b) of this exercise are reasonable.

7. In an experiment to compare two different materials A and B that might be used for manufacturing the heels of men's dress shoes, 15 men were selected and fitted with a new pair of shoes on which one heel was made of material A and one heel was made of material B. At the beginning of the experiment, each heel was 10 millimeters thick. After the shoes had been worn for one

Table 9.36

Pair	Material A	Material B
1	6.6	7.4
2	7.0	5.4
3	8.3	8.8
4	8.2	8.0
5	5.2	6.8
6	9.3	9.1
7	7.9	6.3
8	8.5	7.5
9	7.8	7.0
10	7.5	6.6
11	6.1	4.4
12	8.9	7.7
13	6.1	4.2
14	9.4	9.4
15	9.1	9.1

month, the remaining thickness of each heel was measured. The results are given in Table 9.36. Test the null hypothesis that material A is not more durable than material B against the alternative that material A is more durable than material B, by using (a) the sign test and (b) the Wilcoxon signed-ranks test.

8. Consider again the conditions of Exercise 7, and suppose that for each pair of shoes it was decided by an auxiliary randomization whether the left shoe or the right shoe would receive the heel made of material A.

   (a) Discuss this method of designing the experiment, and consider in particular the possibility that in every pair the left shoe receives the heel made of material A.

   (b) Discuss methods for improving the design of this experiment. In addition to the data presented in Table 9.36, would it be helpful to know which shoe had the heel made of material A, and would it be helpful to have each man in the experiment also wear a pair of shoes in which both heels were made of material A or both were made of material B?

*9.10.   RANKS FOR TWO SAMPLES

Comparing Two Distributions

In this section we shall consider a problem in which a random sample of m observations $X_1, \ldots, X_m$ is taken from a continuous distribution for which the p.d.f. $f(x)$ is unknown, and an independent random sample of n observations $Y_1, \ldots, Y_n$ is taken from another continuous distribution for which the p.d.f. $g(x)$ is also unknown. We shall assume that either the distribution of each observation Y_i in the second sample is the same as the distribution of each observation X_i in the first sample, or else there exists a constant θ such that the distribution of each random variable $Y_i + \theta$ is the same as the distribution of each X_i. In other words, we shall assume that either $f(x) = g(x)$ for all values of x, or else there exists a constant θ such that $f(x) = g(x - \theta)$ for all values of x. Finally, we shall assume that the following hypotheses are to be tested:

$$
\begin{aligned}
&H_0: \quad f(x) = g(x) \qquad \text{for } -\infty < x < \infty, \\
&H_1: \quad \text{There exists a constant } \theta \ (\theta \neq 0) \text{ such that} \\
&\qquad f(x) = g(x - \theta) \qquad \text{for } -\infty < x < \infty.
\end{aligned}
\tag{1}
$$

It should be noted that the common form of the p.d.f.'s $f(x)$ and $g(x)$ is not specified by the hypothesis H_0, and that the value of θ is not specified by the hypothesis H_1.

Two methods for testing the hypotheses (1) have already been proposed in this chapter. One method is to use the χ^2 test of homogeneity described in Sec. 9.4, which can be applied by grouping the observations in each sample into C intervals. The other method is to use the Kolmogorov-Smirnov test for two samples described in Sec. 9.6. Furthermore, if we are willing to assume that the two samples are actually drawn from normal distributions, then testing the hypotheses (1) is the same as testing whether two normal distributions have the same mean when it is assumed that they have the same unknown variance. Therefore, under this assumption, we could use a t test based on $m + n - 2$ degrees of freedom as described in Sec. 8.9.

In this section we shall present another procedure for testing the hypotheses (1) that does not require any assumptions about the form of the distributions from which the samples are drawn. This procedure, which was introduced separately by F. Wilcoxon and by H. B. Mann and D. R. Whitney in the 1940's, is known as the *Wilcoxon-Mann-Whitney ranks test*.

The Wilcoxon-Mann-Whitney Ranks Test

In this procedure we begin by arranging the $m + n$ observations in the two samples in a single sequence from the smallest value that appears in the two samples to the largest value that appears. Since all the observations come from continuous distributions, it may be assumed that no two of the $m + n$ observations have the same value. Thus, a total ordering of these $m + n$ values can be obtained. Each observation in this total ordering is then assigned a rank corresponding to its position in the ordering. Thus, the smallest observation among the $m + n$ observations is assigned the rank 1 and the largest observation is assigned the rank $m + n$.

The Wilcoxon-Mann-Whitney ranks test is based on the property that if the null hypothesis H_0 is true and the two samples are actually drawn from the same distribution, then the observations $X_1, \ldots, X_m$ will tend to be dispersed throughout the ordering of all $m + n$ observations, rather than be concentrated among the smaller values or among the larger values. In fact, when H_0 is true, the ranks that are assigned to the m observations $X_1, \ldots, X_m$ will be the same as if they were a random sample of m ranks drawn at random without replacement from a box containing the $m + n$ ranks $1, 2, \ldots, m + n$.

Let S denote the sum of the ranks that are assigned to the m observations $X_1, \ldots, X_m$. Since the average of the ranks $1, 2, \ldots, m + n$ is $(1/2)(m + n + 1)$, it follows from the discussion just given that when H_0 is true,

$$E(S) = \frac{m(m + n + 1)}{2}. \tag{2}$$

Also, it can be shown that when H_0 is true,

$$\text{Var}(S) = \frac{mn(m + n + 1)}{12}. \tag{3}$$

Furthermore, when the sample sizes m and n are large and H_0 is true, the distribution of S will be approximately a normal distribution for which the mean and the variance are given by Eqs. (2) and (3).

Suppose now that the alternative hypothesis H_1 is true. If $\theta < 0$, then the observations $X_1, \ldots, X_m$ will tend to be smaller than the observations $Y_1, \ldots, Y_n$. Therefore, the ranks that are assigned to the observations $X_1, \ldots, X_m$ will tend to be among the smaller ranks, and the random variable S will tend to be smaller than it would be if H_0 were true. Similarly, if $\theta > 0$, then the ranks that are assigned to the observations $X_1, \ldots, X_m$ will tend to be among the larger ranks, and the random variable S will tend to be larger than it would be if H_0 were true. Because of these properties, the Wilcoxon-Mann-Whitney ranks test specifies rejecting H_0 if the value of S deviates very far from its mean value given by Eq. (2). In other words, the test specifies rejecting H_0 if $|S - (1/2)m(m + n + 1)| \geq c$, where the constant c is chosen appropriately. In particular, when the approximate normal distribution of S is used, the constant c can be chosen so that the test is carried out at any specified level of significance α_0.

Example 1: Carrying out a Wilcoxon-Mann-Whitney Ranks Test. Suppose that the size m of the first sample is 20 and that the observed values are given in Table 9.37. Suppose also that the size n of the second sample is 36 and that these observed values are given in Table 9.38. We shall test the hypotheses (1) by carrying out a Wilcoxon-Mann-Whitney ranks test.

The 56 values in the two samples are ordered from smallest to largest in Table 9.39. Each observed value in the first sample is identified by the symbol x, and each observed value in the second sample is identified by the symbol y. The sum S of the ranks of the 20 observed values in the first sample is found to be 494.

Since $m = 20$ and $n = 36$ in this example, it follows from Eqs. (2) and (3) that if H_0 is true, then S has approximately a normal distribution with mean 570 and variance 3420. The standard deviation of S is therefore $(3420)^{1/2} = 58.48$.

Table 9.37

0.730	1.033	0.362	0.859	0.911
1.411	1.420	1.073	1.427	1.166
0.039	1.352	1.171	−0.174	1.214
0.247	−0.779	0.477	1.016	0.273

Table 9.38

1.520	0.876	1.148	1.633.	0.566
0.931	0.664	1.952	0.482	0.279
1.268	1.039	0.912	2.632	1.267
0.756	2.589	1.281	0.274	−0.078
0.542	1.532	−1.079	1.676	0.789
1.705	1.277	0.065	1.733	0.709
−0.127	1.160	1.010	1.428	1.372
0.939				

Table 9.39

Rank	Observed value	Sample	Rank	Observed value	Sample
1	−1.079	y	29	1.016	x
2	−0.779	x	30	1.033	x
3	−0.174	x	31	1.039	y
4	−0.127	y	32	1.073	x
5	−0.078	y	33	1.148	y
6	0.039	x	34	1.160	y
7	0.065	y	35	1.166	x
8	0.247	x	36	1.171	x
9	0.273	x	37	1.214	x
10	0.274	y	38	1.267	y
11	0.279	y	39	1.268	y
12	0.362	x	40	1.277	y
13	0.477	x	41	1.281	y
14	0.482	y	42	1.352	x
15	0.542	y	43	1.372	y
16	0.566	y	44	1.411	x
17	0.664	y	45	1.420	x
18	0.709	y	46	1.427	x
19	0.730	x	47	1.428	y
20	0.756	y	48	1.520	y
21	0.789	y	49	1.532	y
22	0.859	x	50	1.633	y
23	0.876	y	51	1.676	y
24	0.911	x	52	1.705	y
25	0.912	y	53	1.733	y
26	0.931	y	54	1.952	y
27	0.939	y	55	2.589	y
28	1.010	y	56	2.632	y

Hence, if H_0 is true, the random variable $Z = (S - 570)/(58.48)$ will have approximately a standard normal distribution. Since $S = 494$ in this example, it follows that $Z = -1.300$. In other words, the observed value of S lies 1.3 standard deviations to the left of its mean.

The Wilcoxon-Mann-Whitney ranks test specifies rejecting H_0 if $|Z| > c$, where c is an appropriate constant. Therefore, the tail area corresponding to any observed value of Z is the sum of the area of the standard normal distribution to the right of $|Z|$ and the area of that distribution to the left of $-|Z|$. The tail area corresponding to the observed value $Z = -1.3$ is found in this way from a table of the standard normal distribution to be 0.1936. Hence, the null hypothesis H_0 should be accepted at any level of significance $\alpha_0 < 0.1936$. $\square$

For small values of m and n, the normal approximation to the distribution of S will not be appropriate. Tables of the exact distributions of S for small sample sizes are given in many published collections of statistical tables.

Ties

Since we have again assumed that both samples come from continuous distributions, there is probability 0 that any two observations will be equal. Nevertheless, because the measurements in an actual problem are made with only limited precision, it may be found that some of the recorded observed values are equal. Suppose that a group of two or more tied values includes at least one x and one y. One procedure for handling these tied values is to assign to each observation in the group the average of the ranks that would be assigned to these observations. When this procedure is used, the value of Var(S) as given in Eq. (3) must be changed to take into account the ties that are present in the data.

We shall not consider this procedure further here. Rather, we shall repeat the recommendation made at the end of Sec. 9.9 that the test be carried out twice. In the first test, the smaller ranks in each group of tied observations should be assigned to the x's and the larger ranks should be assigned to the y's. In the second test, these assignments should be reversed. If the decision to accept or reject H_0 is different for the two assignments, or if the calculated tail areas are very different, the data must be regarded as inconclusive.

EXERCISES

1. Consider again the data in Example 1. Test the hypotheses (1) by applying the Kolmogorov-Smirnov test for two samples.

2. Consider again the data in Example 1. Test the hypotheses (1) by assuming that the observations are taken from two normal distributions with the same variance, and applying a t test of the type described in Sec. 8.9.

3. In an experiment to compare the effectiveness of two drugs A and B in reducing blood glucose concentrations, drug A was administered to 25 patients and drug B was administered to 15 patients. The reductions in blood glucose concentrations for the 25 patients who received drug A are given in Table 9.40. The reductions in concentrations for the 15 patients who received drug B are given in Table 9.41. Test the hypothesis that the two drugs are equally effective in reducing blood glucose concentrations by using the Wilcoxon-Mann-Whitney ranks test.

4. Consider again the data in Exercise 3. Test the hypothesis that the two drugs are equally effective by applying the Kolmogorov-Smirnov test for two samples.

5. Consider again the data in Exercise 3. Test the hypothesis that the two drugs are equally effective by assuming that the observations are taken from two normal distributions with the same variance and applying a t test of the type described in Sec. 8.9.

6. Suppose that $X_1, \ldots, X_m$ form a random sample of m observations from a continuous distribution for which the p.d.f. $f(x)$ is unknown; and that $Y_1, \ldots, Y_n$ form an independent random sample of n observations from another continuous distribution for which the p.d.f. $g(x)$ is also unknown. Suppose also that $f(x) = g(x - \theta)$ for $-\infty < x < \infty$, where the value of the parameter θ is unknown $(-\infty < \theta < \infty)$. Describe how to carry out a one-sided Wilcoxon-Mann-Whitney ranks test of the following hypotheses:

$H_0: \quad \theta \leq 0,$
$H_1: \quad \theta > 0.$

Table 9.40

0.35	1.12	1.54	0.13	0.77
0.16	1.20	0.40	1.38	0.39
0.58	0.04	0.44	0.75	0.71
1.64	0.49	0.90	0.83	0.28
1.50	1.73	1.15	0.72	0.91

Table 9.41

1.78	1.25	1.01
1.82	1.95	1.81
0.68	1.48	1.59
0.89	0.86	1.63
1.26	1.07	1.31

7. Consider again the conditions of Exercise 6. Describe how to use the Wilcoxon-Mann-Whitney ranks test to test the following hypotheses for a specified value of θ_0:

$$H_0: \quad \theta = \theta_0,$$
$$H_1: \quad \theta \neq \theta_0.$$

8. Consider again the conditions of Exercises 6 and 7. Describe how to use the Wilcoxon-Mann-Whitney ranks test to determine a confidence interval for θ with confidence coefficient γ ($0 < \gamma < 1$).

9. Consider again the conditions of Exercises 6 and 7. Determine a confidence interval for θ with confidence coefficient 0.90 based on the values given in Example 1.

10. Let $X_1, \ldots, X_m$ and $Y_1, \ldots, Y_n$ be the observations in two samples, and suppose that no two of these observations are equal. Consider the mn pairs

$$(X_1, Y_1), \ldots, (X_1, Y_n),$$
$$(X_2, Y_1), \ldots, (X_2, Y_n),$$
$$\vdots \qquad \qquad \vdots$$
$$(X_m, Y_1), \ldots, (X_m, Y_n).$$

Let U denote the number of these pairs for which the value of the X component is greater than the value of the Y component. Show that

$$U = S - \frac{1}{2} m(m + 1),$$

where S is the sum of the ranks assigned to $X_1, \ldots, X_m$, as defined in this section.

9.11. SUPPLEMENTARY EXERCISES

1. Suppose that 400 persons are chosen at random from a large population, and that each person in the sample specifies which one of five breakfast cereals he most prefers. For $i = 1, \ldots, 5$, let p_i denote the proportion of the population that prefers cereal i, and let N_i denote the number of persons in the sample who prefer cereal i. It is desired to test the following hypotheses at the level of significance 0.01:

$$H_0: \quad p_1 = p_2 = \cdots = p_5,$$
$$H_1: \quad \text{The hypothesis } H_0 \text{ is not true.}$$

For what values of $\sum_{i=1}^{5} N_i^2$ should H_0 be rejected?

2. Consider a large population of families that have exactly three children; and suppose that it is desired to test the null hypothesis H_0 that the distribution of the number of boys in each family is a binomial distribution with parameters $n = 3$ and $p = 1/2$ against the general alternative H_1 that H_0 is not true. Suppose also that in a random sample of 128 families it is found that 26 families have no boys, 32 families have one boy, 40 families have two boys, and 30 families have three boys. At what levels of significance should H_0 be rejected?

3. Consider again the conditions of Exercise 2, including the observations in the random sample of 128 families, but suppose now that it is desired to test the composite null hypothesis H_0 that the distribution of the number of boys in each family is a binomial distribution for which $n = 3$ and the value of p is not specified against the general alternative H_1 that H_0 is not true. At what levels of significance should H_0 be rejected?

4. In order to study the genetic history of three different large groups of native Americans, a random sample of 50 persons is drawn from group 1; a random sample of 100 persons is drawn from group 2; and a random sample of 200 persons is drawn from group 3. The blood type of each person in the samples is classified as A, B, AB, or O, and the results are as given in Table 9.42. Test the hypothesis that the distribution of blood types is the same in all three groups at the level of significance 0.1.

5. Consider again the conditions of Exercise 4. Explain how to change the numbers in Table 9.42 in such a way that each row total and each column total remains unchanged, but the value of the χ^2 test statistic is increased.

6. Consider a χ^2 test of independence that is to be applied to the elements of a 2×2 contingency table. Show that the quantity $(N_{ij} - \hat{E}_{ij})^2$ has the same value for each of the four cells of the table.

7. Consider again the conditions of Exercise 6. Show that the χ^2 statistic Q can be written in the form

$$Q = \frac{n(N_{11}N_{22} - N_{12}N_{21})^2}{N_{1+}N_{2+}N_{+1}N_{+2}}.$$

Table 9.42

	A	B	AB	O	
Group 1	24	6	5	15	50
Group 2	43	24	7	26	100
Group 3	69	47	22	62	200

Table 9.43

$n + a$	$n - a$
$n - a$	$n + a$

Table 9.44

αn	$(1 - \alpha)n$
$(1 - \alpha)n$	αn

8. Suppose that a χ^2 test of independence at the level of significance 0.01 is to be applied to the elements of a 2×2 contingency table containing $4n$ observations; and that the data have the form given in Table 9.43. For what values of a should the null hypothesis be rejected?

9. Suppose that a χ^2 test of independence at the level of significance 0.05 is to be applied to the elements of a 2×2 contingency table containing $2n$ observations; and that the data have the form given in Table 9.44. For what values of α should the null hypothesis be rejected?

10. In a study of the health effects of air pollution, it was found that the proportion of the total population of city A that suffered from respiratory diseases was larger than the proportion for city B. Since city A was generally regarded as being less polluted and more healthful than city B, this result was considered surprising. Therefore, separate investigations were made for the younger population (under age 40) and for the older population (age 40 or older). It was found that the proportion of the younger population suffering from respiratory diseases was smaller for city A than for city B; and also that the proportion of the older population suffering from respiratory diseases was smaller for city A than for city B. Discuss and explain these results.

11. Suppose that an achievement test in mathematics was given to students from two different high schools A and B. When the results of the test were tabulated, it was found that the average score for the freshmen at school A was higher than the average for the freshmen at school B; and that the same relationship existed for the sophomores at the two schools, for the juniors, and for the seniors. On the other hand, it was found also that the average score of all the students at school A was lower than that of all the students at school B. Discuss and explain these results.

12. A random sample of 100 hospital patients suffering from depression received a particular treatment over a period of three months. Prior to the beginning

of the treatment, each patient was classified as being at one of five levels of depression, where level 1 represented the most severe level of depression and level 5 represented the mildest level. At the end of the treatment each patient was again classified according to the same five levels of depression. The results are given in Table 9.45. Discuss the use of this table for determining whether the treatment has been helpful in alleviating depression.

13. Suppose that a random sample of three observations is drawn from a distribution with the following p.d.f.:

$$f(x) = \begin{cases} \theta x^{\theta-1} & \text{for } 0 < x < 1, \\ 0 & \text{otherwise,} \end{cases}$$

where $\theta > 0$. Determine the p.d.f. of the sample median.

14. Suppose that a random sample of n observations is drawn from a distribution for which the p.d.f. is as given in Exercise 13. Determine the asymptotic distribution of the sample median.

15. Suppose that a random sample of n observations is drawn from a t distribution with α degrees of freedom. Show that the asymptotic distribution of both the sample mean $\overline{X}_n$ and the sample median $\tilde{X}_n$ is normal; and determine the positive integers α for which the variance of this asymptotic distribution is smaller for $\overline{X}_n$ than for $\tilde{X}_n$.

16. Suppose that $X_1, \ldots, X_n$ form a large random sample from a distribution for which the p.d.f. is $h(x \mid \theta) = \alpha f(x \mid \theta) + (1 - \alpha)g(x \mid \theta)$. Here $f(x \mid \theta)$ is the p.d.f. of the normal distribution with unknown mean θ and variance 1; $g(x \mid \theta)$ is the p.d.f. of the normal distribution with the same unknown mean θ and variance σ^2; and $0 \leqslant \alpha \leqslant 1$. Let $\overline{X}_n$ and $\tilde{X}_n$ denote the sample mean and the sample median, respectively.

(a) For $\sigma^2 = 100$, determine the values of α for which the M.S.E. of $\tilde{X}_n$ will be smaller than the M.S.E. of $\overline{X}_n$.

Table 9.45

		Level of depression after treatment				
		1	2	3	4	5
Level of depression before treatment	1	7	3	0	0	0
	2	1	27	14	2	0
	3	0	0	19	8	2
	4	0	1	2	12	0
	5	0	0	1	1	0

(b) For $\alpha = 1/2$, determine the values of σ^2 for which the M.S.E. of $\tilde{X}_n$ will be smaller than the M.S.E. of $\overline{X}_n$.

17. Suppose that $X_1, \ldots, X_n$ form a random sample from a distribution with p.d.f. $f(x)$, and let $Y_1 < Y_2 < \cdots < Y_n$ denote the order statistics of the sample. Prove that the joint p.d.f. of $Y_1, \ldots, Y_n$ is as follows:

$$g(y_1, \ldots, y_n) = \begin{cases} n! f(y_1) \cdots f(y_n) & \text{for } y_1 < y_2 < \cdots < y_n, \\ 0 & \text{otherwise.} \end{cases}$$

18. Let $Y_1 < Y_2 < Y_3$ denote the order statistics of a random sample of three observations from a uniform distribution on the interval $(0, 1)$. Determine the conditional distribution of Y_2 given that $Y_1 = y_1$ and $Y_3 = y_3$ $(0 < y_1 < y_3 < 1)$.

19. Suppose that a random sample of 20 observations is drawn from an unknown continuous distribution, and let $Y_1 < \cdots < Y_{20}$ denote the order statistics of the sample. Also, let θ denote the 0.3 quantile of the distribution, and suppose that it is desired to present a confidence interval for θ which has the form (Y_r, Y_{r+3}). Determine the value of r $(r = 1, 2, \ldots, 17)$ for which this interval will have the largest confidence coefficient γ, and determine the value of γ.

20. Suppose that $X_1, \ldots, X_m$ form a random sample from a continuous distribution for which the p.d.f. $f(x)$ is unknown; that $Y_1, \ldots, Y_n$ form an independent random sample from another continuous distribution for which the p.d.f. $g(x)$ also is unknown; and that $f(x) = g(x - \theta)$ for $-\infty < x < \infty$, where the value of the parameter θ is unknown $(-\infty < \theta < \infty)$. Suppose that it is desired to carry out a Wilcoxon-Mann-Whitney ranks test of the following hypotheses at a specified level of significance α $(0 < \alpha < 1)$:

$H_0: \quad \theta = \theta_0,$
$H_1: \quad \theta \neq \theta_0.$

Assume that no two of the observations are equal; and let U_{θ_0} denote the number of pairs (X_i, Y_j) such that $X_i - Y_j > \theta_0$, where $i = 1, \ldots, m$ and $j = 1, \ldots, n$. Show that for large values of m and n, the hypothesis H_0 should be accepted if and only if

$$\frac{mn}{2} - c\left(1 - \frac{\alpha}{2}\right)\left[\frac{mn(m + n + 1)}{12}\right]^{1/2}$$

$$< U_{\theta_0} < \frac{mn}{2} + c\left(1 - \frac{\alpha}{2}\right)\left[\frac{mn(m + n + 1)}{12}\right]^{1/2},$$

where $c(\gamma)$ denotes the γ quantile of the standard normal distribution. *Hint:* See Exercise 10 of Sec. 9.10.

21. Consider again the conditions of Exercise 20. Show that a confidence interval for θ with confidence coefficient $1 - \alpha$ can be obtained by the following procedure: Let k be the largest integer less than or equal to

$$\frac{mn}{2} - c\left(1 - \frac{\alpha}{2}\right)\left[\frac{mn(m + n + 1)}{12}\right]^{1/2}.$$

Also, let A be the kth smallest of the mn differences $X_i - Y_j$, where $i = 1, \ldots, m$ and $j = 1, \ldots, n$; and let B be the kth largest of these mn differences. Then the interval $A < \theta < B$ is a confidence interval of the required type.

Linear Statistical Models

10

10.1. THE METHOD OF LEAST SQUARES

Fitting a Straight Line

Suppose that each of ten patients is treated first with a certain amount of a standard drug A and then with an equal amount of a new drug B, and that the change in blood pressure induced by each drug is observed for each patient. This change in blood pressure will be called the *response* of the patient. For $i = 1, \ldots, 10$, we shall let x_i denote the response, measured in appropriate units, of the ith patient when he receives drug A, and we shall let y_i denote his response when he receives drug B. We shall suppose also that the observed values of the responses are as given in Table 10.1. The ten points (x_i, y_i) for $i = 1, \ldots, 10$ are plotted in Fig. 10.1.

Suppose now that we are interested in describing the relationship between the response y of a patient to drug B and his response x to drug A. In order to obtain a simple expression for this relationship, we might wish to fit a straight line to the ten points plotted in Fig. 10.1. Although these ten points obviously do not lie exactly on a straight line, we might believe that the deviations from such a line are caused by the fact that the observed change in the blood pressure of each patient is affected not only by the two drugs but also by various other factors. In other words, we might believe that if it were possible to control all these other factors, the observed points would actually lie on a straight line. We might believe further that if we measured the responses to the two drugs for a very large number of patients, instead of for just ten patients, we would then find that the observed points tend to cluster along a straight line. Perhaps we might also wish

593

Table 10.1

i	x_i	y_i
1	1.9	0.7
2	0.8	−1.0
3	1.1	−0.2
4	0.1	−1.2
5	−0.1	−0.1
6	4.4	3.4
7	4.6	0.0
8	1.6	0.8
9	5.5	3.7
10	3.4	2.0

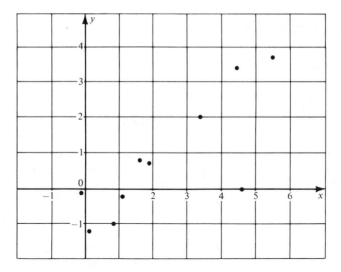

Figure 10.1   A plot of the observed values in Table 10.1.

to be able to predict the response y of a future patient to the new drug B on the basis of his response x to the standard drug A. One procedure for making such a prediction would be to fit a straight line to the points in Fig. 10.1, and to use this line for predicting the value of y corresponding to any given value of x.

It can be seen from Fig. 10.1 that if we did not have to consider the point (4.6, 0.0), which is obtained from the patient for whom $i = 7$ in Table 10.1, then the other nine points lie roughly along a straight line. One arbitrary line which fits reasonably well to these nine points is sketched in Fig. 10.2. However, if we wish

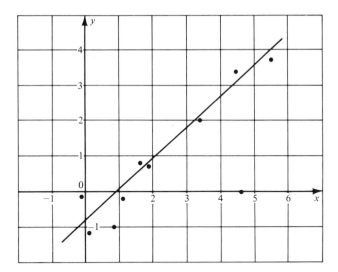

Figure 10.2 A straight line fitted to nine of the points in Table 10.1.

to fit a straight line to all ten points, it is not clear just how much the line in Fig. 10.2 should be adjusted in order to accommodate the anomalous point. We shall now describe a method for fitting such a line.

The Least-Squares Line

We shall assume here that we are interested in fitting a straight line to the points plotted in Fig. 10.1 in order to obtain a simple mathematical relationship for expressing the response y of a patient to the new drug B as a function of his response x to the standard drug A. In other words, our main objective is to be able to predict closely a patient's response y to drug B from his response x to drug A. We are interested, therefore, in constructing a straight line such that, for each observed response x_i, the corresponding value of y on the straight line will be as close as possible to the actual observed response y_i. The vertical deviations of the ten plotted points from the line drawn in Fig. 10.2 are sketched in Fig. 10.3.

One method of constructing a straight line to fit the observed values is called *the method of least squares*. According to this method, the line should be drawn so that the sum of the squares of the vertical deviations of all the points from the line is a minimum. We shall now study this method in more detail.

Consider an arbitrary straight line $y = \beta_1 + \beta_2 x$, in which the values of the constants β_1 and β_2 are to be determined. When $x = x_i$, the height of this line is

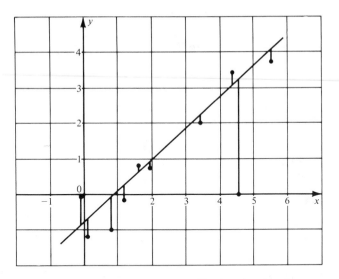

Figure 10.3   Vertical deviations of the plotted points from a straight line.

$\beta_1 + \beta_2 x_i$. Therefore, the vertical distance between the point (x_i, y_i) and the line is $|y_i - (\beta_1 + \beta_2 x_i)|$. Suppose that the line is to be fitted to n points, and let Q denote the sum of the squares of the vertical distances at the n points. Then

$$Q = \sum_{i=1}^{n} [y_i - (\beta_1 + \beta_2 x_i)]^2. \tag{1}$$

The method of least squares specifies that the values of β_1 and β_2 must be chosen so that the value of Q is minimized.

It is not difficult to minimize the value of Q with respect to β_1 and β_2. We have

$$\frac{\partial Q}{\partial \beta_1} = -2 \sum_{i=1}^{n} (y_i - \beta_1 - \beta_2 x_i) \tag{2}$$

and

$$\frac{\partial Q}{\partial \beta_2} = -2 \sum_{i=1}^{n} (y_i - \beta_1 - \beta_2 x_i) x_i. \tag{3}$$

By setting each of these two partial derivatives equal to 0, we obtain the following pair of equations:

$$\beta_1 n + \beta_2 \sum_{i=1}^{n} x_i = \sum_{i=1}^{n} y_i,$$

$$\beta_1 \sum_{i=1}^{n} x_i + \beta_2 \sum_{i=1}^{n} x_i^2 = \sum_{i=1}^{n} x_i y_i. \tag{4}$$

The equations (4) are called the *normal equations* for β_1 and β_2. By considering the second-order derivatives of Q, we can show that the values of β_1 and β_2 which satisfy the normal equations will be the values for which the sum of squares Q in Eq. (1) is minimized. If we denote these values by $\hat{\beta}_1$ and $\hat{\beta}_2$, then the equation of the straight line obtained by the method of least squares will be $y = \hat{\beta}_1 + \hat{\beta}_2 x$. This line is called the *least-squares line*.

As usual, we shall let $\bar{x}_n = (1/n)\sum_{i=1}^n x_i$ and $\bar{y}_n = (1/n)\sum_{i=1}^n y_i$. By solving the normal equations (4) for β_1 and β_2, we obtain the following results:

$$\hat{\beta}_2 = \frac{\sum_{i=1}^n x_i y_i - n\bar{x}_n \bar{y}_n}{\sum_{i=1}^n x_i^2 - n\bar{x}_n^2},$$

$$\hat{\beta}_1 = \bar{y}_n - \hat{\beta}_2 \bar{x}_n. \tag{5}$$

For the values given in Table 10.1, $n = 10$ and it is found from Eq. (5) that $\hat{\beta}_1 = -0.786$ and $\hat{\beta}_2 = 0.685$. Hence, the equation of the least-squares line is $y = -0.786 + 0.685x$. This line is sketched in Fig. 10.4.

Fitting a Polynomial by the Method of Least Squares

Suppose now that instead of simply fitting a straight line to n plotted points, we wish to fit a polynomial of degree k $(k \geqslant 2)$. Such a polynomial will have the

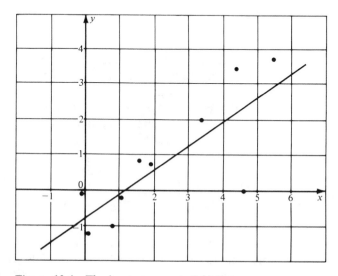

Figure 10.4   The least-squares straight line.

following form:

$$y = \beta_1 + \beta_2 x + \beta_3 x^2 + \cdots + \beta_{k+1} x^k. \tag{6}$$

The method of least squares specifies that the constants $\beta_1, \ldots, \beta_{k+1}$ should be chosen so that the sum Q of the squares of the vertical deviations of the points from the curve is a minimum. In other words, these constants should be chosen so as to minimize the following expression for Q:

$$Q = \sum_{i=1}^{n} \left[y_i - \left(\beta_1 + \beta_2 x_i + \cdots + \beta_{k+1} x_i^k \right) \right]^2. \tag{7}$$

If we calculate the $k + 1$ partial derivatives $\partial Q / \partial \beta_1, \ldots, \partial Q / \partial \beta_{k+1}$ and we set each of these derivatives equal to 0, we obtain the following $k + 1$ linear equations involving the $k + 1$ unknown values $\beta_1, \ldots, \beta_{k+1}$:

$$\beta_1 n + \beta_2 \sum_{i=1}^{n} x_i + \cdots + \beta_{k+1} \sum_{i=1}^{n} x_i^k = \sum_{i=1}^{n} y_i,$$

$$\beta_1 \sum_{i=1}^{n} x_i + \beta_2 \sum_{i=1}^{n} x_i^2 + \cdots + \beta_{k+1} \sum_{i=1}^{n} x_i^{k+1} = \sum_{i=1}^{n} x_i y_i,$$

$$\vdots \qquad \qquad \vdots \tag{8}$$

$$\beta_1 \sum_{i=1}^{n} x_i^k + \beta_2 \sum_{i=1}^{n} x_i^{k+1} + \cdots + \beta_{k+1} \sum_{i=1}^{n} x_i^{2k} = \sum_{i=1}^{n} x_i^k y_i.$$

As before, these equations are called the *normal equations*. There will be a unique set of values of $\beta_1, \ldots, \beta_{k+1}$ which satisfy the normal equations if and only if the determinant of the $(k + 1) \times (k + 1)$ matrix formed from the coefficients of $\beta_1, \ldots, \beta_{k+1}$ is not zero. If there are at least $k + 1$ different values among the n observed values $x_1, \ldots, x_n$, then this determinant will not be zero and there will be a unique solution to the normal equations. We shall assume that this condition is satisfied. It can be shown by the methods of advanced calculus that the unique values of $\beta_1, \ldots, \beta_{k+1}$ which satisfy the normal equations will then be the values which minimize the value of Q in Eq. (7). If we denote these values by $\hat{\beta}_1, \ldots, \hat{\beta}_{k+1}$, then the least-squares polynomial will have the form $y = \hat{\beta}_1 + \hat{\beta}_2 x + \cdots + \hat{\beta}_{k+1} x^k$.

Example 1: Fitting a Parabola. Suppose that we wish to fit a polynomial of the form $y = \beta_1 + \beta_2 x + \beta_3 x^2$ (which represents a parabola) to the ten points given in Table 10.1. In this example, it is found that the normal equations (8) are as follows:

$$10\beta_1 + 23.3\beta_2 + 90.37\beta_3 = 8.1,$$

$$23.3\beta_1 + 90.37\beta_2 + 401.0\beta_3 = 43.59, \tag{9}$$

$$90.37\beta_1 + 401.0\beta_2 + 1892.7\beta_3 = 204.55.$$

The unique values of β_1, β_2, and β_3 that satisfy these three equations are $\hat{\beta}_1 = -0.744$, $\hat{\beta}_2 = 0.616$, and $\hat{\beta}_3 = 0.013$. Hence, the least-squares parabola is

$$y = -0.744 + 0.616x + 0.013x^2. \tag{10}$$

This curve is sketched in Fig. 10.5 together with the least-squares straight line. Because the coefficient of x^2 in Eq. (10) is so small, the least-squares parabola and the least-squares straight line are very close together over the range of values included in Fig. 10.5. □

Fitting a Linear Function of Several Variables

We shall now consider an extension of the example discussed at the beginning of this section, in which we were interested in representing a patient's response to a new drug B as a linear function of his response to drug A. Suppose that we wish

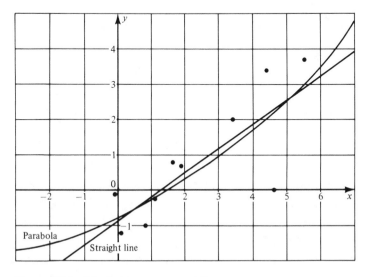

Figure 10.5 The least-squares parabola.

to represent a patient's response to drug B as a linear function involving not only his response to drug A but also some other relevant variables. For example, we may wish to represent the patient's response y to drug B as a linear function involving his response x_1 to drug A, his heart rate x_2, his blood pressure x_3 before he receives any drugs, and other relevant variables $x_4, \ldots, x_k$.

Suppose that for each patient i ($i = 1, \ldots, n$) we measure his response y_i to drug B, his response x_{i1} to drug A, and also his values $x_{i2}, \ldots, x_{ik}$ for the other variables. Suppose also that in order to fit these observed values for the n patients, we wish to consider a linear function having the form

$$y = \beta_1 + \beta_2 x_1 + \cdots + \beta_{k+1} x_k. \tag{11}$$

In this case, also, the values of $\beta_1, \ldots, \beta_{k+1}$ can be determined by the method of least squares. For each given set of observed values $x_{i1}, \ldots, x_{ik}$, we again consider the difference between the observed response y_i and the value $\beta_1 + \beta_2 x_{i1} + \cdots + \beta_{k+1} x_{ik}$ of the linear function given in Eq. (11). As before, it is required to minimize the sum Q of the squares of these differences. Here,

$$Q = \sum_{i=1}^{n} \left[y_i - (\beta_1 + \beta_2 x_{i1} + \cdots + \beta_{k+1} x_{ik}) \right]^2. \tag{12}$$

A set of $k + 1$ normal equations can again be obtained by setting each of the partial derivatives $\partial Q / \partial \beta_j$ equal to 0 for $j = 1, \ldots, k + 1$. These equations will have the following form:

$$\beta_1 n + \beta_2 \sum_{i=1}^{n} x_{i1} + \cdots + \beta_{k+1} \sum_{i=1}^{n} x_{ik} = \sum_{i=1}^{n} y_i,$$

$$\beta_1 \sum_{i=1}^{n} x_{i1} + \beta_2 \sum_{i=1}^{n} x_{i1}^2 + \cdots + \beta_{k+1} \sum_{i=1}^{n} x_{i1} x_{ik} = \sum_{i=1}^{n} x_{i1} y_i,$$

$$\vdots \qquad\qquad\qquad\qquad\qquad\qquad \vdots \tag{13}$$

$$\beta_1 \sum_{i=1}^{n} x_{ik} + \beta_2 \sum_{i=1}^{n} x_{ik} x_{i1} + \cdots + \beta_{k+1} \sum_{i=1}^{n} x_{ik}^2 = \sum_{i=1}^{n} x_{ik} y_i.$$

If the determinant of the $(k + 1) \times (k + 1)$ matrix formed from the coefficients of $\beta_1, \ldots, \beta_{k+1}$ in these equations is not zero, then there will be a unique solution $\hat{\beta}_1, \ldots, \hat{\beta}_{k+1}$ of the equations. The least-squares linear function will then be $y = \hat{\beta}_1 + \hat{\beta}_2 x_1 + \cdots + \hat{\beta}_{k+1} x_k$.

Example 2: Fitting a Linear Function of Two Variables. Suppose that we expand Table 10.1 to include the values given in the third column in Table 10.2. Here, for

Table 10.2

i	x_{i1}	x_{i2}	y_i
1	1.9	66	0.7
2	0.8	62	−1.0
3	1.1	64	−0.2
4	0.1	61	−1.2
5	−0.1	63	−0.1
6	4.4	70	3.4
7	4.6	68	0.0
8	1.6	62	0.8
9	5.5	68	3.7
10	3.4	66	2.0

each patient i ($i = 1, \ldots, 10$), x_{i1} denotes his response to the standard drug A; x_{i2} denotes his heart rate; and y_i denotes his response to the new drug B. Suppose also that to these values we wish to fit a linear function having the form $y = \beta_1 + \beta_2 x_1 + \beta_3 x_2$.

In this example, it is found that the normal equations (13) are as follows:

$$10\beta_1 + 23.3\beta_2 + 650\beta_3 = 8.1,$$

$$23.3\beta_1 + 90.37\beta_2 + 1563.6\beta_3 = 43.59, \tag{14}$$

$$650\beta_1 + 1563.6\beta_2 + 42{,}334\beta_3 = 563.1.$$

The unique values of β_1, β_2, and β_3 which satisfy these three equations are $\hat{\beta}_1 = -11.4527$, $\hat{\beta}_2 = 0.4503$, and $\hat{\beta}_3 = 0.1725$. Hence, the least-squares linear function is

$$y = -11.4527 + 0.4503x_1 + 0.1725x_2. \quad \square \tag{15}$$

It should be noted that the problem of fitting a polynomial of degree k involving only one variable, as specified by Eq. (6), can be regarded as a special case of the problem of fitting a linear function involving several variables, as specified by Eq. (11). To make Eq. (11) applicable to the problem of fitting a polynomial having the form given in Eq. (6), we define the k variables $x_1, \ldots, x_k$ simply as $x_1 = x$, $x_2 = x^2, \ldots, x_k = x^k$.

A polynomial involving more than one variable can also be represented in the form of Eq. (11). For example, suppose that the values of four variables r, s, t, and y are observed for several different patients, and that we wish to fit to these observed values a function having the following form:

$$y = \beta_1 + \beta_2 r + \beta_3 r^2 + \beta_4 rs + \beta_5 s^2 + \beta_6 t^3 + \beta_7 rst. \tag{16}$$

We can regard the function in Eq. (16) as a linear function having the form given in Eq. (11) with $k = 6$ if we define the six variables $x_1, \ldots, x_6$ as follows: $x_1 = r$, $x_2 = r^2$, $x_3 = rs$, $x_4 = s^2$, $x_5 = t^3$, and $x_6 = rst$.

EXERCISES

1. Show that the value of $\hat{\beta}_2$ in Eq. (5) can be rewritten in each of the following three forms:

   (a) $\hat{\beta}_2 = \dfrac{\sum_{i=1}^n (x_i - \bar{x}_n)(y_i - \bar{y}_n)}{\sum_{i=1}^n (x_i - \bar{x}_n)^2}$,

   (b) $\hat{\beta}_2 = \dfrac{\sum_{i=1}^n (x_i - \bar{x}_n)y_i}{\sum_{i=1}^n (x_i - \bar{x}_n)^2}$,

   (c) $\hat{\beta}_2 = \dfrac{\sum_{i=1}^n x_i(y_i - \bar{y}_n)}{\sum_{i=1}^n (x_i - \bar{x}_n)^2}$.

2. Show that the least-squares line $y = \hat{\beta}_1 + \hat{\beta}_2 x$ passes through the point $(\bar{x}_n, \bar{y}_n)$.

3. For $i = 1, \ldots, n$, let $\hat{y}_i = \hat{\beta}_1 + \hat{\beta}_2 x_i$. Show that $\hat{\beta}_1$ and $\hat{\beta}_2$, as given by Eq. (5), are the unique values of β_1 and β_2 such that

$$\sum_{i=1}^n (y_i - \hat{y}_i) = 0 \quad \text{and} \quad \sum_{i=1}^n x_i(y_i - \hat{y}_i) = 0.$$

4. Fit a straight line to the observed values given in Table 10.1 so that the sum of the squares of the *horizontal* deviations of the points from the line is a minimum. Sketch on the same graph both this line and the least-squares line given in Fig. 10.4.

5. Suppose that both the least-squares line and the least-squares parabola were fitted to the same set of points. Explain why the sum of the squares of the deviations of the points from the parabola cannot be larger than the sum of the squares of the deviations of the points from the straight line.

6. Suppose that eight specimens of a certain type of alloy were produced at different temperatures, and that the durability of each specimen was then observed. The observed values are as given in Table 10.3, where x_i denotes the temperature (in coded units) at which specimen i was produced and y_i denotes the durability (in coded units) of that specimen.

   (a) Fit a straight line of the form $y = \beta_1 + \beta_2 x$ to these values by the method of least squares.

   (b) Fit a parabola of the form $y = \beta_1 + \beta_2 x + \beta_3 x^2$ to these values by the method of least squares.

Table 10.3

i	x_i	y_i
1	0.5	40
2	1.0	41
3	1.5	43
4	2.0	42
5	2.5	44
6	3.0	42
7	3.5	43
8	4.0	42

(c) Sketch on the same graph the eight data points, the line found in part (a), and the parabola found in part (b).

7. Let (x_i, y_i) for $i = 1, \ldots, k + 1$, denote $k + 1$ given points in the xy-plane such that no two of these points have the same x-coordinate. Show that there is a unique polynomial having the form $y = \beta_1 + \beta_2 x + \cdots + \beta_{k+1}x^k$ which passes through these $k + 1$ points.

8. The resilience y of a certain type of plastic is to be represented as a linear function of both the temperature x_1 at which the plastic is baked and the number of minutes x_2 for which it is baked. Suppose that ten pieces of plastic are prepared by using different values of x_1 and x_2, and that the observed values in appropriate units are as given in Table 10.4. Fit a function having the form $y = \beta_1 + \beta_2 x_1 + \beta_3 x_2$ to these observed values by the method of least squares.

Table 10.4

i	x_{i1}	x_{i2}	y_i
1	100	1	113
2	100	2	118
3	110	1	127
4	110	2	132
5	120	1	136
6	120	2	144
7	120	3	138
8	130	1	146
9	130	2	156
10	130	3	149

9. Consider again the observed values presented in Table 10.4. Fit a function having the form $y = \beta_1 x_1 + \beta_2 x_2 + \beta_3 x_2^2$ to these values by the method of least squares.

10. Consider again the observed values presented in Table 10.4, and consider also the two functions which were fitted to these values in Exercises 8 and 9. Which of these two functions fits the observed values better?

10.2. REGRESSION

Regression Functions

In this section, we shall describe some statistical problems in which the method of least squares can be used to derive estimators of various parameters. In particular, we shall study problems in which we are interested in learning about the conditional distribution of some random variable Y for given values of some other variables $X_1, \ldots, X_k$. The variables $X_1, \ldots, X_k$ may be random variables whose values are to be observed in an experiment along with the values of Y, or they may be *control variables* whose values are to be chosen by the experimenter. In general, some of these variables might be random variables and some might be control variables. In any case, we can study the conditional distribution of Y given $X_1, \ldots, X_k$.

The conditional expectation of Y for any given values of $X_1, \ldots, X_k$ is called the *regression function of Y on $X_1, \ldots, X_k$*, or simply the *regression of Y on $X_1, \ldots, X_k$*. The regression of Y on $X_1, \ldots, X_k$ is a function of the values $x_1, \ldots, x_k$ of $X_1, \ldots, X_k$. In symbols, this function is $E(Y \mid x_1, \ldots, x_k)$.

We shall assume that the regression function $E(Y \mid x_1, \ldots, x_k)$ is a linear function having the following form:

$$E(Y \mid x_1, \ldots, x_k) = \beta_1 + \beta_2 x_1 + \cdots + \beta_{k+1} x_k. \tag{1}$$

The coefficients $\beta_1, \ldots, \beta_{k+1}$ in Eq. (1) are called *regression coefficients*. We shall suppose that these regression coefficients are unknown. Therefore, they are to be regarded as parameters whose values are to be estimated. We shall suppose also that n vectors of observations are obtained. For $i = 1, \ldots, n$, we shall assume that the ith vector $(x_{i1}, \ldots, x_{ik}, y_i)$ consists of a set of controlled or observed values of $X_1, \ldots, X_k$ and the corresponding observed value of Y.

One set of estimators of the regression coefficients $\beta_1, \ldots, \beta_{k+1}$ that can be calculated from these observations is the set of values $\hat{\beta}_1, \ldots, \hat{\beta}_{k+1}$ that are obtained by the method of least squares, as described in Sec. 10.1. These estimators are called the *least-squares estimators* of $\beta_1, \ldots, \beta_{k+1}$. We shall now specify some further assumptions about the conditional distribution of Y given

$X_1, \ldots, X_k$ in order to be able to determine in greater detail the properties of these least-squares estimators.

Simple Linear Regression

We shall consider first a problem in which we wish to study the regression of Y on just a single variable X. We shall assume that for any given value $X = x$, the random variable Y can be represented in the form $Y = \beta_1 + \beta_2 x + \varepsilon$, where ε is a random variable that has a normal distribution with mean 0 and variance σ^2. It follows from this assumption that the conditional distribution of Y given $X = x$ will be a normal distribution with mean $\beta_1 + \beta_2 x$ and variance σ^2.

A problem of this type is called a problem of *simple linear regression*. Here the term *simple* refers to the fact that we are considering the regression of Y on just a single variable X, rather than on more than one variable; and the term *linear* refers to the fact that the regression function $E(Y \mid x) = \beta_1 + \beta_2 x$ is a linear function of x.

Suppose now that n pairs of observations $(x_1, Y_1), \ldots, (x_n, Y_n)$ are obtained and that the following assumptions are made: For any given values $x_1, \ldots, x_n$, the random variables $Y_1, \ldots, Y_n$ are independent. Also, for $i = 1, \ldots, n$, the distribution of Y_i is a normal distribution with mean $\beta_1 + \beta_2 x_i$ and variance σ^2. Thus, for given values of the vector $x = (x_1, \ldots, x_n)$ and the parameters $\beta_1, \beta_2,$ and σ^2, the joint p.d.f. of $Y_1, \ldots, Y_n$ will be

$$f_n(y \mid x, \beta_1, \beta_2, \sigma^2) = \frac{1}{(2\pi\sigma^2)^{n/2}} \exp\left[-\frac{1}{2\sigma^2} \sum_{i=1}^{n} (y_i - \beta_1 - \beta_2 x_i)^2 \right]. \qquad (2)$$

For any observed vector $y = (y_1, \ldots, y_n)$, this function will be the likelihood function of the parameters $\beta_1, \beta_2,$ and σ^2. We shall now use this likelihood function to determine the M.L.E.'s of $\beta_1, \beta_2,$ and σ^2. It can be seen from Eq. (2) that regardless of the value of σ^2, the values of β_1 and β_2 for which the likelihood function is a maximum will be the values for which the following sum of squares is a minimum:

$$\sum_{i=1}^{n} (y_i - \beta_1 - \beta_2 x_i)^2. \qquad (3)$$

But this sum of squares is precisely the same as the sum of squares Q given in Eq. (1) of Sec. 10.1, which is to be minimized by the method of least squares. Thus, the M.L.E.'s of the regression coefficients β_1 and β_2 are precisely the same as the least-squares estimators of β_1 and β_2. The exact form of these estimators $\hat{\beta}_1$ and $\hat{\beta}_2$ was given in Eq. (5) of Sec. 10.1.

Finally, the M.L.E. of σ^2 can be found by first replacing β_1 and β_2 in Eq. (2) by their M.L.E.'s $\hat{\beta}_1$ and $\hat{\beta}_2$, and then maximizing the resulting expression with respect to σ^2. The result (see Exercise 1 at the end of this section) is

$$\hat{\sigma}^2 = \frac{1}{n} \sum_{i=1}^{n} \left(y_i - \hat{\beta}_1 - \hat{\beta}_2 x_i \right)^2. \tag{4}$$

Besides assuming that the regression of Y on X is a linear function having the form $E(Y \mid x) = \beta_1 + \beta_2 x$, we have made three further assumptions about the joint distribution of $Y_1, \ldots, Y_n$ for any given values of $x_1, \ldots, x_n$. These assumptions may be summarized as follows:

(i) *Normality.* We have assumed that each variable Y_i has a normal distribution.

(ii) *Independence.* We have assumed that the variables $Y_1, \ldots, Y_n$ are independent.

(iii) *Homoscedasticity.* We have assumed that the variables $Y_1, \ldots, Y_n$ have the same variance σ^2. This assumption is called the assumption of *homoscedasticity.* In general, it is said that random variables having the same variance are *homoscedastic,* and random variables having different variances are *heteroscedastic.*

Under these assumptions we have shown that the M.L.E.'s of β_1 and β_2 are given by Eq. (5) of Sec. 10.1, and that the M.L.E. of σ^2 is given by Eq. (4) in this section.

The Distribution of the Least-Squares Estimators

We shall now discuss the joint distribution of the estimators $\hat{\beta}_1$ and $\hat{\beta}_2$ when they are regarded as functions of the random variables $Y_1, \ldots, Y_n$ for given values of $x_1, \ldots, x_n$. It should be kept in mind that as long as the three assumptions we have just described are satisfied, the analysis to be presented here is correct either in a problem where the values of $x_1, \ldots, x_n$ are actually chosen by the experimenter or in a problem where the values of $x_1, \ldots, x_n$ are observed along with the values of $Y_1, \ldots, Y_n$.

To determine the distribution of $\hat{\beta}_2$, it is convenient to write $\hat{\beta}_2$ as follows (see Exercise 1 at the end of Sec. 10.1):

$$\hat{\beta}_2 = \frac{\sum_{i=1}^{n}(x_i - \bar{x}_n) Y_i}{\sum_{i=1}^{n}(x_i - \bar{x}_n)^2}. \tag{5}$$

It can be seen from Eq. (5) that $\hat{\beta}_2$ is a linear function of $Y_1, \ldots, Y_n$. Since the random variables $Y_1, \ldots, Y_n$ are independent and each has a normal distribution, it follows that $\hat{\beta}_2$ will also have a normal distribution. Furthermore, the mean of this distribution will be

$$E(\hat{\beta}_2) = \frac{\sum_{i=1}^n (x_i - \bar{x}_n) E(Y_i)}{\sum_{i=1}^n (x_i - \bar{x}_n)^2}. \tag{6}$$

Since $E(Y_i) = \beta_1 + \beta_2 x_i$ for $i = 1, \ldots, n$, it can now be found (see Exercise 2 at the end of this section) that

$$E(\hat{\beta}_2) = \beta_2. \tag{7}$$

Thus, $\hat{\beta}_2$ is an unbiased estimator of β_2.

Furthermore, since the random variables $Y_1, \ldots, Y_n$ are independent and since each has variance σ^2, it follows from Eq. (5) that

$$\mathrm{Var}(\hat{\beta}_2) = \frac{\sum_{i=1}^n (x_i - \bar{x}_n)^2 \mathrm{Var}(Y_i)}{\left[\sum_{i=1}^n (x_i - \bar{x}_n)^2\right]^2}$$

$$= \frac{\sigma^2}{\sum_{i=1}^n (x_i - \bar{x}_n)^2}. \tag{8}$$

We shall now consider the distribution of $\hat{\beta}_1$. We know from Eq. (5) of Sec. 10.1 that $\hat{\beta}_1 = \bar{Y}_n - \hat{\beta}_2 \bar{x}_n$. Since both $\bar{Y}_n$ and $\hat{\beta}_2$ are linear functions of $Y_1, \ldots, Y_n$, it follows that $\hat{\beta}_1$ is also a linear function of $Y_1, \ldots, Y_n$. Hence $\hat{\beta}_1$ will have a normal distribution. The mean of $\hat{\beta}_1$ can be determined from the relation $E(\hat{\beta}_1) = E(\bar{Y}_n) - \bar{x}_n E(\hat{\beta}_2)$. It can be shown (see Exercise 3) that

$$E(\hat{\beta}_1) = \beta_1. \tag{9}$$

Thus, $\hat{\beta}_1$ is an unbiased estimator of β_1.

Furthermore, it can be shown (see Exercise 4) that

$$\mathrm{Var}(\hat{\beta}_1) = \frac{\left(\sum_{i=1}^n x_i^2\right) \sigma^2}{n \sum_{i=1}^n (x_i - \bar{x}_n)^2}. \tag{10}$$

Next, consider the covariance of $\hat{\beta}_1$ and $\hat{\beta}_2$. It can be shown (see Exercise 5) that the value of this covariance is as follows:

$$\mathrm{Cov}(\hat{\beta}_1, \hat{\beta}_2) = -\frac{\bar{x}_n \sigma^2}{\sum_{i=1}^n (x_i - \bar{x}_n)^2}. \tag{11}$$

To complete the description of the joint distribution of $\hat{\beta}_1$ and $\hat{\beta}_2$, it can be shown that this joint distribution is a bivariate normal distribution for which the means $E(\hat{\beta}_1)$ and $E(\hat{\beta}_2)$, the variances $\text{Var}(\hat{\beta}_1)$ and $\text{Var}(\hat{\beta}_2)$, and the covariance $\text{Cov}(\hat{\beta}_1, \hat{\beta}_2)$ are as given in Eqs. (7) to (11).

The Gauss-Markov Theorem for Simple Linear Regression

It can be shown, although we shall not present a proof in this book, that among all unbiased estimators of the regression coefficient β_1, the least-squares estimator $\hat{\beta}_1$ has the smallest variance for all possible values of the parameters β_1, β_2, and σ^2. Also, among all unbiased estimators of the regression coefficient β_2, the least-squares estimator $\hat{\beta}_2$ has the smallest variance for all possible values of β_1, β_2, and σ^2.

In fact, a somewhat stronger result can be established. Suppose that it is desired to estimate the value of some particular linear combination of the parameters β_1 and β_2 having the form $\theta = c_1\beta_1 + c_2\beta_2 + c_3$, where c_1, c_2, and c_3 are given constants. Since $\hat{\beta}_1$ and $\hat{\beta}_2$ are unbiased estimators of β_1 and β_2, then $\hat{\theta} = c_1\hat{\beta}_1 + c_2\hat{\beta}_2 + c_3$ will be an unbiased estimator of θ. It can be shown, in fact, that among all unbiased estimators of θ, the estimator $\hat{\theta}$ has the smallest variance for all possible values of β_1, β_2, and σ^2.

A similar result, known as the *Gauss-Markov theorem*, can be established even without assuming that the observations $Y_1, \ldots, Y_n$ have normal distributions. Specifically, suppose that for any given values $x_1, \ldots, x_n$, the observations $Y_1, \ldots, Y_n$ are uncorrelated; and for $i = 1, \ldots, n$, $E(Y_i) = \beta_1 + \beta_2 x_i$ and $\text{Var}(Y_i) = \sigma^2$. No further assumptions about the distribution of $Y_1, \ldots, Y_n$ need be made. Under these conditions, the Gauss-Markov theorem can be stated as follows:

> Let $\theta = c_1\beta_1 + c_2\beta_2 + c_3$, where c_1, c_2, and c_3 are any given constants. Then among all unbiased estimators of θ that are linear combinations of the observations $Y_1, \ldots, Y_n$, the estimator $\hat{\theta} = c_1\hat{\beta}_1 + c_2\hat{\beta}_2 + c_3$ has the smallest variance for all possible values of the parameters β_1, β_2, and σ^2.

The distinction between the Gauss-Markov theorem and the result that we previously stated should be carefully noted. When it is assumed only that the observations $Y_1, \ldots, Y_n$ are uncorrelated, as in the Gauss-Markov theorem, it can be concluded only that $\hat{\theta}$ has the smallest variance among all *linear* unbiased estimators of θ. When it is also assumed that the observations $Y_1, \ldots, Y_n$ are independent and have normal distributions, it can be concluded that $\hat{\theta}$ has the smallest variance among *all* unbiased estimators of θ.

The variance of $\hat{\theta}$ can be found by substituting the values of $\text{Var}(\hat{\beta}_1)$, $\text{Var}(\hat{\beta}_2)$, and $\text{Cov}(\hat{\beta}_1, \hat{\beta}_2)$ given in Eqs. (10), (8), and (11) in the following relation:

$$\mathrm{Var}(\hat{\theta}) = c_1^2\mathrm{Var}(\hat{\beta}_1) + c_2^2\mathrm{Var}(\hat{\beta}_2) + 2c_1c_2\mathrm{Cov}(\hat{\beta}_1, \hat{\beta}_2). \tag{12}$$

When these substitutions have been made, the result can be written in the following form:

$$\mathrm{Var}(\hat{\theta}) = \frac{\sum_{i=1}^n (c_1x_i - c_2)^2}{n\sum_{i=1}^n (x_i - \bar{x}_n)^2}\sigma^2. \tag{13}$$

When $c_1 = 1$ and $c_2 = c_3 = 0$, we have $\hat{\theta} = \hat{\beta}_1$. In this case, it can be seen that the value of $\mathrm{Var}(\hat{\theta})$ given in Eq. (13) reduces to the value of $\mathrm{Var}(\hat{\beta}_1)$ given in Eq. (10). Similarly, when $c_2 = 1$ and $c_1 = c_3 = 0$, we have $\hat{\theta} = \hat{\beta}_2$, and it can be seen that the value of $\mathrm{Var}(\hat{\theta})$ reduces to the value of $\mathrm{Var}(\hat{\beta}_2)$ given in Eq. (8).

Design of the Experiment

Consider a problem of simple linear regression in which the variable X is a control variable whose values $x_1, \ldots, x_n$ can be chosen by the experimenter. We shall discuss methods for choosing these values so as to obtain good estimators of the regression coefficients β_1 and β_2.

Suppose first that the values $x_1, \ldots, x_n$ are to be chosen so as to minimize the M.S.E. of the least-squares estimator $\hat{\beta}_1$. Since $\hat{\beta}_1$ is an unbiased estimator of β_1, the M.S.E. of $\hat{\beta}_1$ is equal to $\mathrm{Var}(\hat{\beta}_1)$, as given in Eq. (10). Also, since $\sum_{i=1}^n x_i^2 \geq \sum_{i=1}^n (x_i - \bar{x}_n)^2$ for any values $x_1, \ldots, x_n$, and there is equality in this relation if and only if $\bar{x}_n = 0$, it follows from Eq. (10) that $\mathrm{Var}(\hat{\beta}_1) \geq \sigma^2/n$ for any values $x_1, \ldots, x_n$ and there will be equality in this relation if and only if $\bar{x}_n = 0$. Hence, $\mathrm{Var}(\hat{\beta}_1)$ will attain its minimum value σ^2/n for any values $x_1, \ldots, x_n$ such that $\bar{x}_n = 0$.

Suppose next that the values $x_1, \ldots, x_n$ are to be chosen so as to minimize the M.S.E. of the estimator $\hat{\beta}_2$. Again, the M.S.E. of $\hat{\beta}_2$ will be equal to $\mathrm{Var}(\hat{\beta}_2)$, as given in Eq. (8). It can be seen from Eq. (8) that $\mathrm{Var}(\hat{\beta}_2)$ will be minimized by choosing the values $x_1, \ldots, x_n$ so that the value of $\sum_{i=1}^n (x_i - \bar{x}_n)^2$ is maximized. If the values $x_1, \ldots, x_n$ must be chosen from some bounded interval (a, b) of the real line and if n is an even integer, then the value of $\sum_{i=1}^n (x_i - \bar{x}_n)^2$ will be maximized by choosing $x_i = a$ for exactly $n/2$ values and choosing $x_i = b$ for the other $n/2$ values. If n is an odd integer, all the values should again be chosen at the endpoints a and b, but one endpoint must now receive one more observation than the other endpoint.

It follows from this discussion that if the experiment is to be designed so as to minimize both the M.S.E. of $\hat{\beta}_1$ and the M.S.E. of $\hat{\beta}_2$, then the values $x_1, \ldots, x_n$

should be chosen so that exactly, or approximately, $n/2$ values are equal to some number c that is as large as is feasible in the given experiment and the remaining values are equal to $-c$. In this way, the value of $\bar{x}_n$ will be exactly, or approximately, equal to 0, and the value of $\sum_{i=1}^{n}(x_i - \bar{x}_n)^2$ will be as large as possible.

Finally, suppose that the linear combination $\theta = c_1\beta_1 + c_2\beta_2 + c_3$ is to be estimated, where $c_1 \neq 0$; and that the experiment is to be designed so as to minimize the M.S.E. of $\hat{\theta}$, that is, to minimize $\text{Var}(\hat{\theta})$. It follows from Eq. (13) that we can write $\text{Var}(\hat{\theta})$ in the following form:

$$\text{Var}(\hat{\theta}) = \frac{\sum_{i=1}^{n}\left(x_i - \dfrac{c_2}{c_1}\right)^2}{\sum_{i=1}^{n}(x_i - \bar{x}_n)^2} \cdot \frac{c_1^2\sigma^2}{n}. \tag{14}$$

Since $\sum_{i=1}^{n}[x_i - (c_2/c_1)]^2 \geqslant \sum_{i=1}^{n}(x_i - \bar{x}_n)^2$ for any values $x_1, \ldots, x_n$, and since there is equality in this relation if and only if $\bar{x}_n = c_2/c_1$, it follows that $\text{Var}(\hat{\theta})$ will attain its minimum value $c_1^2\sigma^2/n$ if and only if the values $x_1, \ldots, x_n$ are chosen so that $\bar{x}_n = c_2/c_1$.

In practice, an experienced statistician would not usually choose all the values $x_1, \ldots, x_n$ at a single point or at just the two endpoints of the interval (a, b), as the optimal designs that we have just derived would dictate. The reason is that when all n observations are taken at just one or two values of X, the experiment provides no possibility of checking the assumption that the regression of Y on X is a linear function. In order to check this assumption without unduly increasing the M.S.E. of the least-squares estimators, many of the values $x_1, \ldots, x_n$ should be chosen at the endpoints a and b, but at least some of the values should be chosen at a few interior points of the interval. Linearity can then be checked by visual inspection of the plotted points and the fitting of a polynomial of degree two or higher.

Prediction

Suppose that n pairs of observations $(x_1, Y_1), \ldots, (x_n, Y_n)$ are to be obtained in a problem of simple linear regression; and that on the basis of these n pairs, it is necessary to predict the value of an independent observation Y that will be obtained when a certain specified value x is assigned to the control variable. Since the observation Y will have a normal distribution with mean $\beta_1 + \beta_2 x$ and variance σ^2, it is natural to use the value $\hat{Y} = \hat{\beta}_1 + \hat{\beta}_2 x$ as the predicted value of Y. We shall now determine the M.S.E. $E[(\hat{Y} - Y)^2]$ of this prediction, where both $\hat{Y}$ and Y are random variables.

In this problem, $E(\hat{Y}) = E(Y) = \beta_1 + \beta_2 x$. Thus, if we let $\mu = \beta_1 + \beta_2 x$, then

$$E\left[(\hat{Y} - Y)^2\right] = E\left\{[(\hat{Y} - \mu) - (Y - \mu)]^2\right\}$$

$$= \text{Var}(\hat{Y}) + \text{Var}(Y) - 2\,\text{Cov}(\hat{Y}, Y). \tag{15}$$

However, the random variables $\hat{Y}$ and Y are independent, since $\hat{Y}$ is a function of the first n pairs of observations and Y is an independent observation. Therefore, $\text{Cov}(\hat{Y}, Y) = 0$ and it follows that

$$E\left[(\hat{Y} - Y)^2\right] = \text{Var}(\hat{Y}) + \text{Var}(Y). \tag{16}$$

Finally, since $\hat{Y} = \hat{\beta}_1 + \hat{\beta}_2 x$, the value of $\text{Var}(\hat{Y})$ is given by Eq. (13) with $c_1 = 1$ and $c_2 = x$. Since $\text{Var}(Y) = \sigma^2$, we have

$$E\left[(\hat{Y} - Y)^2\right] = \left[\frac{\sum_{i=1}^{n}(x_i - x)^2}{n\sum_{i=1}^{n}(x_i - \bar{x}_n)^2} + 1\right]\sigma^2. \tag{17}$$

EXERCISES

1. Show that the M.L.E. of σ^2 is given by Eq. (4).
2. Show that $E(\hat{\beta}_2) = \beta_2$.
3. Show that $E(\hat{\beta}_1) = \beta_1$.
4. Show that $\text{Var}(\hat{\beta}_1)$ is as given in Eq. (10).
5. Show that $\text{Cov}(\hat{\beta}_1, \hat{\beta}_2)$ is as given in Eq. (11).
6. Show that in a problem of simple linear regression, the estimators $\hat{\beta}_1$ and $\hat{\beta}_2$ will be independent if $\bar{x}_n = 0$.
7. Consider a problem of simple linear regression in which a patient's response Y to a new drug B is to be related to his response X to a standard drug A. Suppose that the 10 pairs of observed values given in Table 10.1 are obtained. Determine the values of the M.L.E.'s $\hat{\beta}_1$, $\hat{\beta}_2$, and $\hat{\sigma}^2$, and also the values of $\text{Var}(\hat{\beta}_1)$ and $\text{Var}(\hat{\beta}_2)$.
8. For the conditions of Exercise 7, determine the value of the correlation of $\hat{\beta}_1$ and $\hat{\beta}_2$.
9. Consider again the conditions of Exercise 7, and suppose that it is desired to estimate the value of $\theta = 3\beta_1 - 2\beta_2 + 5$. Let $\hat{\theta}$ denote the unbiased estimator of θ that has the smallest variance among all unbiased estimators. Determine the value of $\hat{\theta}$ and also the M.S.E. of $\hat{\theta}$.

10. Consider again the conditions of Exercise 7, and let $\theta = 3\beta_1 + c_2\beta_2$, where c_2 is a constant. Let $\hat{\theta}$ denote the unbiased estimator of θ that has the smallest variance among all unbiased estimators. For what value of c_2 will the M.S.E. of $\hat{\theta}$ be smallest?

11. Consider again the conditions of Exercise 7. If a particular patient's response to drug A has the value $x = 2$, what is the predicted value of his response to drug B and what is the M.S.E. of this prediction?

12. Consider again the conditions of Exercise 7. For what value x of a patient's response to drug A can his response to drug B be predicted with the smallest M.S.E.?

13. Consider a problem of simple linear regression in which the durability Y of a certain type of alloy is to be related to the temperature X at which it was produced. Suppose that the eight pairs of observed values given in Table 10.3 are obtained. Determine the values of the M.L.E.'s $\hat{\beta}_1$, $\hat{\beta}_2$, and $\hat{\sigma}^2$, and also the values of $\text{Var}(\hat{\beta}_1)$ and $\text{Var}(\hat{\beta}_2)$.

14. For the conditions of Exercise 13, determine the value of the correlation of $\hat{\beta}_1$ and $\hat{\beta}_2$.

15. Consider again the conditions of Exercise 13, and suppose that it is desired to estimate the value of $\theta = 5 - 4\beta_1 + \beta_2$. Let $\hat{\theta}$ denote the unbiased estimator of θ that has the smallest variance among all unbiased estimators. Determine the value of $\hat{\theta}$ and the M.S.E. of $\hat{\theta}$.

16. Consider again the conditions of Exercise 13, and let $\theta = c_2\beta_2 - \beta_1$, where c_2 is a constant. Let $\hat{\theta}$ denote the unbiased estimator of θ that has the smallest variance among all unbiased estimators. For what value of c_2 will the M.S.E. of $\hat{\theta}$ be smallest?

17. Consider again the conditions of Exercise 13. If a specimen of the alloy is to be produced at the temperature $x = 3.25$, what is the predicted value of the durability of the specimen and what is the M.S.E. of this prediction?

18. Consider again the conditions of Exercise 13. For what value of the temperature x can the durability of a specimen of the alloy be predicted with the smallest M.S.E.?

10.3. TESTS OF HYPOTHESES AND CONFIDENCE INTERVALS IN SIMPLE LINEAR REGRESSION

Joint Distribution of the Estimators

It was stated in Sec. 10.2 that in a problem of simple linear regression, the joint distribution of the M.L.E.'s $\hat{\beta}_1$ and $\hat{\beta}_2$ will be a bivariate normal distribution for which the means, the variances, and the covariance are specified by Eqs. (8) to (11). In this section we shall also consider the M.L.E. $\hat{\sigma}^2$, which was presented in Eq. (4) of Sec. 10.2, and we shall derive the joint distribution of $\hat{\beta}_1$, $\hat{\beta}_2$, and $\hat{\sigma}^2$. In particular, we shall show that the estimator $\hat{\sigma}^2$ is independent of the estimators $\hat{\beta}_1$ and $\hat{\beta}_2$.

For given values $x_1, \ldots, x_n (n \geq 3)$, we shall assume again that the observations $Y_1, \ldots, Y_n$ are independent, and that Y_i has a normal distribution with mean $\beta_1 + \beta_2 x_i$ and variance σ^2. The derivation of the joint distribution of $\hat{\beta}_1$, $\hat{\beta}_2$, and $\hat{\sigma}^2$ which we shall present is based on the properties of orthogonal matrices, as described in Sec. 7.3.

For convenience of notation, let

$$s_x = \left[\sum_{i=1}^{n} (x_i - \bar{x}_n)^2 \right]^{1/2}. \tag{1}$$

Also, let $a_1 = (a_{11}, \ldots, a_{1n})$ and $a_2 = (a_{21}, \ldots, a_{2n})$ be n-dimensional vectors, which are defined as follows:

$$a_{1j} = \frac{1}{n^{1/2}} \qquad \text{for } j = 1, \ldots, n, \tag{2}$$

and

$$a_{2j} = \frac{1}{s_x}(x_j - \bar{x}_n) \qquad \text{for } j = 1, \ldots, n. \tag{3}$$

It is easily verified that $\sum_{j=1}^{n} a_{1j}^2 = 1$, $\sum_{j=1}^{n} a_{2j}^2 = 1$, and $\sum_{j=1}^{n} a_{1j}a_{2j} = 0$.

Because the vectors a_1 and a_2 have these properties, it is possible to construct an $n \times n$ orthogonal matrix A such that the components of a_1 form the first row of A and the components of a_2 form the second row of A. We shall assume that such a matrix A has been constructed:

$$A = \begin{bmatrix} a_{11} & \cdots & a_{1n} \\ a_{21} & \cdots & a_{2n} \\ \vdots & & \vdots \\ a_{n1} & \cdots & a_{nn} \end{bmatrix}. \tag{4}$$

We shall now define a new random vector Z by the relation $Z = AY$, where

$$Y = \begin{bmatrix} Y_1 \\ \vdots \\ Y_n \end{bmatrix} \qquad \text{and} \qquad Z = \begin{bmatrix} Z_1 \\ \vdots \\ Z_n \end{bmatrix}. \tag{5}$$

The joint distribution of $Z_1, \ldots, Z_n$ can be found from the following theorem, which is an extension of Theorem 2 of Sec. 7.3.

Theorem 1. *Suppose that the random variables $Y_1, \ldots, Y_n$ are independent and that each has a normal distribution with the same variance σ^2. If A is an orthogonal $n \times n$ matrix and $Z = AY$, then the random variables $Z_1, \ldots, Z_n$ also are independent and each has a normal distribution with variance σ^2.*

Proof. Let $E(Y_i) = \mu_i$ for $i = 1, \ldots, n$ (it is not assumed in the theorem that $Y_1, \ldots, Y_n$ have the same mean), and let

$$\mu = \begin{bmatrix} \mu_1 \\ \vdots \\ \mu_n \end{bmatrix}.$$

Also, let $X = (1/\sigma)(Y - \mu)$. Since it is assumed that the components of the random vector Y are independent, then the components of the random vector X will also be independent. Furthermore, each component of X will have a standard normal distribution. Therefore, it follows from Theorem 2 of Sec. 7.3 that the components of the n-dimensional random vector AX will also be independent, and each will have a standard normal distribution.

But

$$AX = \frac{1}{\sigma}A(Y - \mu) = \frac{1}{\sigma}Z - \frac{1}{\sigma}A\mu.$$

Hence,

$$Z = \sigma AX + A\mu. \tag{6}$$

Since the components of the random vector AX are independent and each has a standard normal distribution, then the components of the random vector σAX will also be independent and each will have a normal distribution with mean 0 and variance σ^2. When the vector $A\mu$ is added to the random vector σAX, the mean of each component will be shifted; but the components will remain independent, and the variance of each component will be unchanged. It now follows from Eq. (6) that the components of the random vector Z will be independent and each will have a normal distribution with variance σ^2. $\square$

In a problem of simple linear regression, the observations $Y_1, \ldots, Y_n$ satisfy the conditions of Theorem 1. Therefore, the components of the random vector $Z = AY$ will be independent, and each will have a normal distribution with variance σ^2.

The first two components Z_1 and Z_2 of the random vector $\mathbf{Z}$ can easily be derived. The first component is

$$Z_1 = \sum_{j=1}^{n} a_{1j} Y_j = \frac{1}{n^{1/2}} \sum_{j=1}^{n} Y_j = n^{1/2} \overline{Y}_n. \tag{7}$$

Since $\hat{\beta}_1 = \overline{Y}_n - \overline{x}_n \hat{\beta}_2$, we may also write

$$Z_1 = n^{1/2} (\hat{\beta}_1 + \overline{x}_n \hat{\beta}_2). \tag{8}$$

The second component is

$$Z_2 = \sum_{j=1}^{n} a_{2j} Y_j = \frac{1}{s_x} \sum_{j=1}^{n} (x_j - \overline{x}_n) Y_j. \tag{9}$$

By Eq. (5) of Sec. 10.2, we may also write

$$Z_2 = s_x \hat{\beta}_2. \tag{10}$$

Now let the random variable S^2 be defined as follows:

$$S^2 = \sum_{i=1}^{n} (Y_i - \hat{\beta}_1 - \hat{\beta}_2 x_i)^2. \tag{11}$$

We shall show that S^2 and the random vector $(\hat{\beta}_1, \hat{\beta}_2)$ are independent. Since $\hat{\beta}_1 = \overline{Y}_n - \overline{x}_n \hat{\beta}_2$, we may rewrite S^2 as follows:

$$\begin{aligned} S^2 &= \sum_{i=1}^{n} \left[Y_i - \overline{Y}_n - \hat{\beta}_2 (x_i - \overline{x}_n) \right]^2 \\ &= \sum_{i=1}^{n} (Y_i - \overline{Y}_n)^2 - 2\hat{\beta}_2 \sum_{i=1}^{n} (x_i - \overline{x}_n)(Y_i - \overline{Y}_n) + \hat{\beta}_2^2 s_x^2. \end{aligned} \tag{12}$$

It now follows from Exercise 1(a) of Sec. 10.1, that

$$S^2 = \sum_{i=1}^{n} Y_i^2 - n\overline{Y}_n^2 - s_x^2 \hat{\beta}_2^2. \tag{13}$$

Since $\mathbf{Z} = \mathbf{A}\mathbf{Y}$, where $\mathbf{A}$ is an orthogonal matrix, we know from Sec. 7.3 that $\sum_{i=1}^{n} Y_i^2 = \sum_{i=1}^{n} Z_i^2$. By using this fact, we can now obtain the following relation from Eqs. (7), (10), and (13):

$$S^2 = \sum_{i=1}^{n} Z_i^2 - Z_1^2 - Z_2^2 = \sum_{i=3}^{n} Z_i^2. \tag{14}$$

The variables $Z_1, \ldots, Z_n$ are independent, and we have now shown that S^2 is equal to the sum of the squares of only the variables $Z_3, \ldots, Z_n$. It follows, therefore, that S^2 and the random vector (Z_1, Z_2) are independent. But $\hat{\beta}_1$ and $\hat{\beta}_2$ are functions of Z_1 and Z_2 only. In fact, by Eqs. (8) and (10),

$$\hat{\beta}_1 = \frac{Z_1}{n^{1/2}} - \frac{\bar{x}_n Z_2}{s_x} \quad \text{and} \quad \hat{\beta}_2 = \frac{Z_2}{s_x}.$$

Hence, S^2 and the random vector $(\hat{\beta}_1, \hat{\beta}_2)$ are independent.

We shall now derive the distribution of S^2. For $i = 3, \ldots, n$, we have $Z_i = \sum_{j=1}^n a_{ij} Y_j$. Hence

$$
\begin{aligned}
E(Z_i) &= \sum_{j=1}^n a_{ij} E(Y_j) = \sum_{j=1}^n a_{ij} (\beta_1 + \beta_2 x_j) \\
&= \sum_{j=1}^n a_{ij} \left[\beta_1 + \beta_2 \bar{x}_n + \beta_2 (x_j - \bar{x}_n) \right] \qquad (15) \\
&= (\beta_1 + \beta_2 \bar{x}_n) \sum_{j=1}^n a_{ij} + \beta_2 \sum_{j=1}^n a_{ij} (x_j - \bar{x}_n).
\end{aligned}
$$

Since the matrix A is orthogonal, the sum of the products of the corresponding terms in any two different rows must be 0. In particular, for $i = 3, \ldots, n$,

$$\sum_{j=1}^n a_{ij} a_{1j} = 0 \quad \text{and} \quad \sum_{j=1}^n a_{ij} a_{2j} = 0.$$

It now follows from the expressions for a_{1j} and a_{2j} given in Eqs. (2) and (3) that for $i = 3, \ldots, n$,

$$\sum_{j=1}^n a_{ij} = 0 \quad \text{and} \quad \sum_{j=1}^n a_{ij} (x_j - \bar{x}_n) = 0. \qquad (16)$$

When these values are substituted into Eq. (15), it is found that $E(Z_i) = 0$ for $i = 3, \ldots, n$.

We now know that the $n - 2$ random variables $Z_3, \ldots, Z_n$ are independent, and that each has a normal distribution with mean 0 and variance σ^2. Since $S^2 = \sum_{i=3}^n Z_i^2$, it follows that the random variable S^2/σ^2 will have a χ^2 distribution with $n - 2$ degrees of freedom.

Now consider the M.L.E. $\hat{\sigma}^2$. Since $\hat{\sigma}^2 = S^2/n$, it follows from the results we have just obtained that the estimator $\hat{\sigma}^2$ is independent of the estimators $\hat{\beta}_1$ and

$\hat{\beta}_2$, and that the distribution of $n\hat{\sigma}^2/\sigma^2$ is a χ^2 distribution with $n - 2$ degrees of freedom.

Tests of Hypotheses about the Regression Coefficients

Tests of Hypotheses about β_1.   Let β_1^* be a specified number $(-\infty < \beta_1^* < \infty)$, and suppose that it is desired to test the following hypotheses about the regression coefficient β_1:

$$H_0: \quad \beta_1 = \beta_1^*, \qquad\qquad (17)$$
$$H_1: \quad \beta_1 \neq \beta_1^*.$$

We shall construct a t test of these hypotheses which will reject H_0 when the least-squares estimator $\hat{\beta}_1$ is far from the hypothesized value β_1^* and will accept H_0 otherwise.

It follows from Eqs. (9) and (10) of Sec. 10.2 that when the null hypothesis H_0 is true, the following random variable W_1 will have a standard normal distribution:

$$W_1 = \left[\frac{n\sum_{i=1}^{n}(x_i - \bar{x}_n)^2}{\sum_{i=1}^{n}x_i^2} \right]^{1/2} \left(\frac{\hat{\beta}_1 - \beta_1^*}{\sigma} \right). \qquad (18)$$

Since the value of σ is unknown, a test of the hypotheses (17) cannot be based simply on the random variable W_1. However, the random variable S^2/σ^2 has a χ^2 distribution with $n - 2$ degrees of freedom for all possible values of the parameters β_1, β_2, and σ^2. Moreover, since $\hat{\beta}_1$ and S^2 are independent random variables, it follows that W_1 and S^2 are also independent. Hence, when the hypothesis H_0 is true, the following random variable U_1 will have a t distribution with $n - 2$ degrees of freedom:

$$U_1 = \frac{W_1}{\left[\left(\frac{1}{n-2} \right) \left(\frac{S^2}{\sigma^2} \right) \right]^{1/2}}$$

or

$$U_1 = \left[\frac{n(n-2)\sum_{i=1}^{n}(x_i - \bar{x}_n)^2}{\sum_{i=1}^{n}x_i^2} \right]^{1/2} \frac{(\hat{\beta}_1 - \beta_1^*)}{\left[\sum_{i=1}^{n}(Y_i - \hat{\beta}_1 - \hat{\beta}_2 x_i)^2 \right]^{1/2}}. \qquad (19)$$

It can be seen from Eq. (19) that the random variable U_1 is a statistic, since it is a function of only the observations $(x_1, Y_1), \ldots, (x_n, Y_n)$ and it is not a function of the parameters β_1, β_2, and σ^2. Thus, a reasonable test of the hypotheses (17) specifies rejecting H_0 if $|U_1| > c_1$, where c_1 is a suitable constant whose value can be chosen to obtain any specified level of significance α_0 $(0 < \alpha_0 < 1)$.

This same test procedure will also be the likelihood ratio test procedure for the hypotheses (17). After the values $x_1, y_1, \ldots, x_n, y_n$ have been observed, the likelihood function $f_n(\, y \,|\, x, \beta_1, \beta_2, \sigma^2)$ is as given in Eq. (2) of Sec. 10.2. The likelihood ratio test procedure is to compare the following two values: the maximum value attained by this likelihood function when β_2 and σ^2 vary over all their possible values but β_1 can only have the value β_1^*; and the maximum value attained by the likelihood function when all three parameters β_1, β_2, and σ^2 vary over all their possible values. Therefore, we consider the following ratio:

$$r(\, y \,|\, x) = \frac{\sup_{\beta_1, \beta_2, \sigma^2} f_n(\, y \,|\, x, \beta_1, \beta_2, \sigma^2)}{\sup_{\beta_2, \sigma^2} f_n(\, y \,|\, x, \beta_1^*, \beta_2, \sigma^2)}. \tag{20}$$

The procedure then specifies that H_0 should be rejected if $r(\, y \,|\, x) > k$, where k is some chosen constant, and that H_0 should be accepted if $r(\, y \,|\, x) \leqslant k$. It can be shown that this procedure is equivalent to the procedure which specifies rejecting H_0 if $|U_1| > c_1$. The derivation of this result will not be given.

As an illustration of the use of this test procedure, suppose that in a problem of simple linear regression we are interested in testing the null hypothesis that the regression line $y = \beta_1 + \beta_2 x$ passes through the origin against the alternative hypothesis that the line does not pass through the origin. These hypotheses can be stated in the following form:

$$\begin{aligned} H_0 &: \quad \beta_1 = 0, \\ H_1 &: \quad \beta_1 \neq 0. \end{aligned} \tag{21}$$

Here the hypothesized value β_1^* is 0.

Let u_1 denote the value of U_1 calculated from a given set of observed values (x_i, y_i) for $i = 1, \ldots, n$. Then the tail area corresponding to this value is the two-sided tail area

$$\Pr(U_1 > |u_1|) + \Pr(U_1 < -|u_1|). \tag{22}$$

For example, suppose that $n = 20$ and the calculated value of U_1 is 2.1. It is found from a table of the t distribution with 18 degrees of freedom that the corresponding tail area is 0.05. Hence, the null hypothesis H_0 should be accepted at any level of significance $\alpha_0 < 0.05$, and it should be rejected at any level of significance $\alpha_0 > 0.05$.

Tests of Hypotheses about β_2.   Let β_2^* be a specified number $(-\infty < \beta_2^* < \infty)$, and suppose that it is desired to test the following hypotheses about the regression coefficient β_2:

$$H_0: \quad \beta_2 = \beta_2^*,$$
$$H_1: \quad \beta_2 \neq \beta_2^*. \tag{23}$$

It follows from Eqs. (7) and (8) of Sec. 10.2 that when H_0 is true, the following random variable W_2 will have a standard normal distribution:

$$W_2 = \left[\sum_{i=1}^{n} (x_i - \bar{x}_n)^2 \right]^{1/2} \left(\frac{\hat{\beta}_2 - \beta_2^*}{\sigma} \right). \tag{24}$$

Since W_2 and S^2 are independent, it follows that when H_0 is true, the following random variable U_2 will have a t distribution with $n - 2$ degrees of freedom:

$$U_2 = \frac{W_2}{\left[\left(\dfrac{1}{n-2} \right) \left(\dfrac{S^2}{\sigma^2} \right) \right]^{1/2}}$$

or

$$U_2 = \left[\frac{(n-2)\sum_{i=1}^{n}(x_i - \bar{x}_n)^2}{\sum_{i=1}^{n}(Y_i - \hat{\beta}_1 - \hat{\beta}_2 x_i)^2} \right]^{1/2} (\hat{\beta}_2 - \beta_2^*). \tag{25}$$

The test of the hypotheses (23) specifies that the null hypothesis H_0 should be rejected if $|U_2| > c_2$, where c_2 is a suitable constant whose value can be chosen to obtain any specified level of significance α_0 $(0 < \alpha_0 < 1)$.

Again it can be shown that this same test will also be the likelihood ratio test procedure for the hypotheses (23). The derivation will not be given here.

As an illustration of the use of this test procedure, suppose that in a problem of simple linear regression we are interested in testing the hypothesis that the variable Y is actually unrelated to the variable X. Under the assumptions of normality and homoscedasticity described in Sec. 10.2, this hypothesis is equivalent to the hypothesis that the regression function $E(Y \mid x)$ is constant and is not actually a function of x. Since it is assumed that the regression function has the linear form $E(Y \mid x) = \beta_1 + \beta_2 x$, this hypothesis is in turn equivalent to the hypothesis that $\beta_2 = 0$. Thus, the problem is one of testing the following hypotheses:

$$H_0: \quad \beta_2 = 0,$$
$$H_1: \quad \beta_2 \neq 0. \tag{26}$$

Here, the hypothesized value β_2^* is 0.

Let u_2 denote the value of U_2 calculated from a given set of observed values (x_i, y_i) for $i = 1, \ldots, n$. Then the tail area corresponding to this value is the two-sided tail area

$$\Pr(U_2 > |u_2|) + \Pr(U_2 < -|u_2|). \tag{27}$$

Tests of Hypotheses about a Linear Combination of β_1 and β_2. Next, let a_1, a_2, and b be specified numbers, where $a_1 \neq 0$ and $a_2 \neq 0$, and suppose that we are interested in testing the following hypotheses:

$$
\begin{aligned}
H_0: & \quad a_1\beta_1 + a_2\beta_2 = b, \\
H_1: & \quad a_1\beta_1 + a_2\beta_2 \neq b.
\end{aligned} \tag{28}
$$

In general,

$$E(a_1\hat{\beta}_1 + a_2\hat{\beta}_2) = a_1\beta_1 + a_2\beta_2$$

and

$$\text{Var}(a_1\hat{\beta}_1 + a_2\hat{\beta}_2) = a_1^2\text{Var}(\hat{\beta}_1) + a_2^2\text{Var}(\hat{\beta}_2) + 2a_1a_2\text{Cov}(\hat{\beta}_1, \hat{\beta}_2).$$

The value of $\text{Var}(a_1\hat{\beta}_1 + a_2\hat{\beta}_2)$ can be obtained from Eq. (13) of Sec. 10.2 simply by replacing c_1 and c_2 in that expression with a_1 and a_2. Therefore, when H_0 is true, the following random variable W_{12} will have a standard normal distribution:

$$W_{12} = \left[\frac{n\sum_{i=1}^{n}(x_i - \bar{x}_n)^2}{\sum_{i=1}^{n}(a_1x_i - a_2)^2} \right]^{1/2} \left(\frac{a_1\hat{\beta}_1 + a_2\hat{\beta}_2 - b}{\sigma} \right). \tag{29}$$

It follows that when H_0 is true, the following random variable U_{12} will have a t distribution with $n - 2$ degrees of freedom:

$$U_{12} = \frac{W_{12}}{\left[\left(\dfrac{1}{n-2} \right) \left(\dfrac{S^2}{\sigma^2} \right) \right]^{1/2}}$$

or

$$U_{12} = \left[\frac{n(n-2)\sum_{i=1}^{n}(x_i - \bar{x}_n)^2}{\sum_{i=1}^{n}(a_1x_i - a_2)^2} \right]^{1/2} \frac{(a_1\hat{\beta}_1 + a_2\hat{\beta}_2 - b)}{\left[\sum_{i=1}^{n}(Y_i - \hat{\beta}_1 - \hat{\beta}_2x_i)^2 \right]^{1/2}}. \tag{30}$$

The test of the hypotheses (28) specifies that the null hypothesis H_0 should be rejected if $|U_{12}| > c_{12}$, where c_{12} is a suitable constant whose value can be chosen to obtain any specified level of significance α_0 $(0 < \alpha_0 < 1)$.

As before, it can be shown that this same test procedure will also be the likelihood ratio test procedure for the hypotheses (28). Again the derivation will be omitted.

As an illustration of the use of this test procedure, suppose that we are interested in testing the hypothesis that the regression line $y = \beta_1 + \beta_2 x$ passes through a particular point (x^*, y^*), where $x^* \neq 0$. In other words, suppose that we are interested in testing the following hypotheses:

$$\begin{aligned} H_0: & \quad \beta_1 + \beta_2 x^* = y^*, \\ H_1: & \quad \beta_1 + \beta_2 x^* \neq y^*. \end{aligned} \tag{31}$$

These hypotheses have the same form as the hypotheses (28) with $a_1 = 1$, $a_2 = x^*$, and $b = y^*$. Hence, they can be tested by carrying out a t test with $n - 2$ degrees of freedom that is based on the statistic W_{12}.

Tests of Hypotheses about Both β_1 and β_2. Suppose next that β_1^* and β_2^* are given numbers, and that we are interested in testing the following hypotheses about the values of β_1 and β_2:

$$\begin{aligned} H_0: & \quad \beta_1 = \beta_1^* \quad \text{and} \quad \beta_2 = \beta_2^*, \\ H_1: & \quad \text{The hypothesis } H_0 \text{ is not true.} \end{aligned} \tag{32}$$

We shall derive the likelihood ratio test procedure for the hypotheses (32).

The likelihood function $f_n(y \mid x, \beta_1, \beta_2, \sigma^2)$ is given by Eq. (2) of Sec. 10.2. When the null hypothesis H_0 is true, the values of β_1 and β_2 must be β_1^* and β_2^*, respectively. For these values of β_1 and β_2, the maximum value of $f_n(y \mid x, \beta_1^*, \beta_2^*, \sigma^2)$ over all the possible values of σ^2 will be attained when σ^2 has the following value $\hat{\sigma}_0^2$:

$$\hat{\sigma}_0^2 = \frac{1}{n} \sum_{i=1}^{n} \left(y_i - \beta_1^* - \beta_2^* x_i \right)^2. \tag{33}$$

When the alternative hypothesis H_1 is true, we know from Sec. 10.2 that the likelihood function $f_n(y \mid x, \beta_1, \beta_2, \sigma^2)$ will attain its maximum value when β_1, β_2, and σ^2 are equal to the M.L.E.'s $\hat{\beta}_1$, $\hat{\beta}_2$, and $\hat{\sigma}^2$, as given by Eq. (5) of Sec. 10.1 and Eq. (4) of Sec. 10.2.

Now consider the statistic

$$r(y \mid x) = \frac{\sup_{\beta_1, \beta_2, \sigma^2} f_n(y \mid x, \beta_1, \beta_2, \sigma^2)}{\sup_{\sigma^2} f_n(y \mid x, \beta_1^*, \beta_2^*, \sigma^2)}. \tag{34}$$

By using the results that have just been described, it can be shown that

$$r(y \mid x) = \left(\frac{\hat{\sigma}_0^2}{\hat{\sigma}^2} \right)^{n/2} = \left[\frac{\sum_{i=1}^{n}(y_i - \beta_1^* - \beta_2^* x_i)^2}{\sum_{i=1}^{n}(y_i - \hat{\beta}_1 - \hat{\beta}_2 x_i)^2} \right]^{n/2}. \tag{35}$$

The numerator of the final expression in Eq. (35) can be rewritten as follows:

$$\sum_{i=1}^{n} \left(y_i - \beta_1^* - \beta_2^* x_i \right)^2 \tag{36}$$

$$= \sum_{i=1}^{n} \left[(y_i - \hat{\beta}_1 - \hat{\beta}_2 x_i) + (\hat{\beta}_1 - \beta_1^*) + (\hat{\beta}_2 - \beta_2^*) x_i \right]^2.$$

To simplify this expression further, let the statistic S^2 be defined by Eq. (11), and let the statistic Q^2 be defined as follows:

$$Q^2 = n(\hat{\beta}_1 - \beta_1^*)^2 + \left(\sum_{i=1}^{n} x_i^2 \right)(\hat{\beta}_2 - \beta_2^*)^2 \tag{37}$$

$$+ 2n\bar{x}_n(\hat{\beta}_1 - \beta_1^*)(\hat{\beta}_2 - \beta_2^*).$$

We shall now expand the right side of Eq. (36) and shall use the following relations, which were established in Exercise 3 of Sec. 10.1:

$$\sum_{i=1}^{n} (y_i - \hat{\beta}_1 - \hat{\beta}_2 x_i) = 0 \quad \text{and} \quad \sum_{i=1}^{n} x_i(y_i - \hat{\beta}_1 - \hat{\beta}_2 x_i) = 0.$$

We then obtain the relation

$$\sum_{i=1}^{n} \left(y_i - \beta_1^* - \beta_2^* x_i \right)^2 = S^2 + Q^2. \tag{38}$$

It now follows from Eq. (35) that

$$r(y \mid x) = \left(\frac{S^2 + Q^2}{S^2} \right)^{n/2} = \left(1 + \frac{Q^2}{S^2} \right)^{n/2}. \tag{39}$$

The likelihood ratio test procedure specifies rejecting H_0 when $r(y \mid x) \geq k$. It can be seen from Eq. (39) that this procedure is equivalent to rejecting H_0 when $Q^2/S^2 \geq k'$, where k' is a suitable constant. To put this procedure in a more standard form, we shall let the statistic U^2 be defined as follows:

$$U^2 = \frac{\frac{1}{2}Q^2}{\left(\frac{1}{n-2}\right)S^2}. \tag{40}$$

Then the likelihood ratio test procedure specifies rejecting H_0 when $U^2 > \gamma$, where γ is a suitable constant.

We shall now determine the distribution of the statistic U^2 when the hypothesis H_0 is true. It can be shown (see Exercises 8 and 9) that when H_0 is true, the random variable Q^2/σ^2 will have a χ^2 distribution with two degrees of freedom. Also, since the random variable S^2 and the random vector $(\hat{\beta}_1, \hat{\beta}_2)$ are independent, and since Q^2 is a function of $\hat{\beta}_1$ and $\hat{\beta}_2$, it follows that the random variables Q^2 and S^2 are independent. Finally, we know that S^2/σ^2 has a χ^2 distribution with $n-2$ degrees of freedom. Therefore, when H_0 is true, the statistic U^2 defined by Eq. (40) will have an F distribution with 2 and $n-2$ degrees of freedom. Since the null hypothesis H_0 is to be rejected if $U^2 > \gamma$, the value of γ corresponding to any specified level of significance α_0 $(0 < \alpha_0 < 1)$ can be determined from a table of this F distribution.

Confidence Intervals and Confidence Sets

A confidence interval for β_1 can be obtained from the test of the hypotheses (17) based on the statistic U_1 defined by Eq. (19). It was explained in Sec. 8.5 that for any given observed values (x_i, y_i) for $i = 1, \ldots, n$, the set of all values of β_1^* for which the null hypothesis H_0 in (17) would be accepted at the level of significance α_0 will form a confidence interval for β_1 with confidence coefficient $1 - \alpha_0$. Specifically, let $g_{n-2}(x)$ denote the p.d.f. of the t distribution with $n-2$ degrees of freedom, and let c denote a constant such that

$$\int_{-c}^{c} g_{n-2}(x)\, dx = 1 - \alpha_0. \tag{41}$$

Then the set of all values of β_1^* such that $|U_1| < c$ will form a confidence interval for β_1 with confidence coefficient $1 - \alpha_0$.

Similarly, if U_2 is defined by Eq. (25) and if c satisfies Eq. (41), then the set of all values of β_2^* such that $|U_2| < c$ will form a confidence interval for β_2 with confidence coefficient $1 - \alpha_0$.

A confidence interval for a particular linear combination having the form $a_1\beta_1 + a_2\beta_2$ can be constructed in the same way from the statistic U_{12} defined by Eq. (30). Specifically, suppose that we wish to construct a confidence interval for the height $b = \beta_1 + \beta_2 x$ of the regression line at a given point x. If we let $a_1 = 1$ and $a_2 = x$ in Eq. (30), then the set of all values of b such that $|U_{12}| < c$ will form a confidence interval for b with confidence coefficient $1 - \alpha_0$. It follows from some algebra that for each value of x, the upper and lower limits of this confidence interval will lie on the curves defined by the following relations (note the sign $\pm$ before c):

$$y = \hat{\beta}_1 + \hat{\beta}_2 x \pm c \left[\frac{S^2}{n(n-2)} \cdot \frac{\sum_{i=1}^n (x_i - x)^2}{\sum_{i=1}^n (x_i - \bar{x}_n)^2} \right]^{1/2}, \tag{42}$$

where c and S^2 are defined by Eqs. (41) and (11). In other words, with confidence coefficient $1 - \alpha_0$ for any given value of x, the actual value $\beta_1 + \beta_2 x$ of the regression line will lie between the value obtained by using the plus sign in (42) and the value obtained by using the minus sign.

Next, consider the problem of constructing a confidence set for the pair of unknown regression coefficients β_1 and β_2. Such a confidence set can be obtained from the statistic U^2 defined by Eq. (40), which was used to test the hypotheses (32). Specifically, let $h_{2,n-2}(x)$ denote the p.d.f. of the F distribution with 2 and $n - 2$ degrees of freedom, and let γ be a constant such that

$$\int_0^\gamma h_{2,n-2}(x)\,dx = 1 - \alpha_0. \tag{43}$$

Then the set of all pairs of values of β_1^* and β_2^* such that $U^2 < \gamma$ will form a confidence set for the pair β_1 and β_2 with confidence coefficient $1 - \alpha_0$. It can be shown (see Exercise 17) that this confidence set will contain all the points (β_1, β_2) inside a certain ellipse in the $\beta_1\beta_2$-plane. In other words, this confidence set will actually be a confidence ellipse.

The confidence ellipse that has just been derived for β_1 and β_2 can be used to construct a confidence set for the entire regression line $y = \beta_1 + \beta_2 x$. Corresponding to each point (β_1, β_2) inside the ellipse, we can draw a straight line $y = \beta_1 + \beta_2 x$ in the xy-plane. The collection of all these straight lines corresponding to all the points (β_1, β_2) inside the ellipse will be a confidence set with confidence coefficient $1 - \alpha_0$ for the actual regression line. A rather lengthy and detailed analysis, which will not be presented here [see Kendall and Stuart (1973)], shows that the upper and lower limits of this confidence set are the curves defined by the following relations:

$$y = \hat{\beta}_1 + \hat{\beta}_2 x \pm \left[\frac{2\gamma S^2}{n(n-2)} \cdot \frac{\sum_{i=1}^n (x_i - x)^2}{\sum_{i=1}^n (x_i - \bar{x}_n)^2} \right]^{1/2}, \tag{44}$$

where γ and S^2 are defined by Eqs. (43) and (11). In other words, with confidence coefficient $1 - \alpha_0$, the actual regression line $y = \beta_1 + \beta_2 x$ will lie

Table 10.5

α_0	$n-2$	c	$(2\gamma)^{1/2}$	$(2\gamma)^{1/2}/c$
0.05	2	4.30	6.16	1.43
	5	2.57	3.40	1.32
	10	2.23	2.86	1.29
	20	2.09	2.64	1.26
	60	2.00	2.51	1.25
	120	1.98	2.48	1.25
	∞	1.96	2.45	1.25
0.025	2	6.21	8.83	1.42
	5	3.16	4.11	1.30
	10	2.63	3.30	1.25
	20	2.42	2.99	1.24
	60	2.30	2.80	1.22
	120	2.27	2.76	1.22
	∞	2.24	2.72	1.21

between the curve obtained by using the plus sign in (44) and the curve obtained by using the minus sign in (44). The region between these curves is often called a *confidence band* or *confidence belt* for the regression line.

It is interesting to compare the confidence limits defined in (44) for the entire regression line with the limits defined in (42) that are appropriate when we desire a confidence interval at just a particular point x. It can be seen that the curves defined by (44) have the same form as those defined by (42). The only difference is that the constant c in (42) is replaced by the constant $(2\gamma)^{1/2}$ in (44). It follows from Eq. (41) that c is the $1-(1/2)\alpha_0$ quantile of the t distribution with $n-2$ degrees of freedom; and it follows from Eq. (43) that γ is the $1-\alpha_0$ quantile of the F distribution with 2 and $n-2$ degrees of freedom. Some values of c and $(2\gamma)^{1/2}$ are presented in Table 10.5. It can be seen from the last column of this table that, for $\alpha_0 = 0.05$, the width of the confidence belt for the entire regression line at any point x is typically only about 25 or 30 percent greater than the width of the confidence interval for the value of the regression line at that particular point. For $\alpha_0 = 0.025$, the ratio of these widths is even smaller.

The Analysis of Residuals

Whenever a statistical analysis is carried out, it is important to verify that the observed data appear to satisfy the assumptions on which the analysis is based. For example, in the statistical analysis of a problem of simple linear regression, we have assumed that the regression of Y on X is a linear function and that the observations $Y_1, \ldots, Y_n$ are independent. The estimators of β_1 and β_2 and the tests of hypotheses about β_1 and β_2 were developed on the basis of these assumptions, but the data were not examined to find out whether or not these assumptions were reasonable.

One way to make a quick and informal check of these assumptions is to examine the observed values of the *residuals* $z_1, \ldots, z_n$, which are defined as follows:

$$z_i = y_i - \hat{\beta}_1 - \hat{\beta}_2 x_i \qquad \text{for } i = 1, \ldots, n. \tag{45}$$

Specifically, suppose that the n points (x_i, z_i), for $i = 1, \ldots, n$ are plotted in the xz-plane. It must be true (see Exercise 3 at the end of Sec. 10.1) that $\sum_{i=1}^{n} z_i = 0$ and $\sum_{i=1}^{n} x_i z_i = 0$. However, subject to these restrictions, the positive and negative residuals should be scattered randomly among the points (x_i, z_i). If the positive residuals z_i tend to be concentrated at either the smaller values of x_i or the larger values of x_i, then either the assumption that the regression of Y on X is a linear function or the assumption that the observations $Y_1, \ldots, Y_n$ are independent may be violated. In fact, if the plot of the points (x_i, z_i) exhibits any type of regular pattern, the assumptions may be violated.

EXERCISES

1. Determine a value of c such that in a problem of simple linear regression, the statistic $c\sum_{i=1}^{n}(Y_i - \hat{\beta}_1 - \hat{\beta}_2 x_i)^2$ will be an unbiased estimator of σ^2.

2. Suppose that in a problem of simple linear regression, the ten pairs of observed values of x_i and y_i given in Table 10.6 are obtained. Test the following hypotheses at the level of significance 0.05:

$H_0: \quad \beta_1 = 0.7,$
$H_1: \quad \beta_1 \neq 0.7.$

Table 10.6

i	x_i	y_i	i	x_i	y_i
1	0.3	0.4	6	1.0	0.8
2	1.4	0.9	7	2.0	0.7
3	1.0	0.4	8	-1.0	-0.4
4	-0.3	-0.3	9	-0.7	-0.2
5	-0.2	0.3	10	0.7	0.7

3. For the data presented in Table 10.6, test at the level of significance 0.05 the hypothesis that the regression line passes through the origin in the xy-plane.

4. For the data presented in Table 10.6, test at the level of significance 0.05 the hypothesis that the slope of the regression line is 1.

5. For the data presented in Table 10.6, test at the level of significance 0.05 the hypothesis that the regression line is horizontal.

6. For the data presented in Table 10.6, test the following hypotheses at the level of significance 0.10:

$$H_0: \quad \beta_2 = 5\beta_1,$$
$$H_1: \quad \beta_2 \neq 5\beta_1.$$

7. For the data presented in Table 10.6, test at the level of significance 0.01 the hypothesis that when $x = 1$, the height of the regression line is $y = 1$.

8. In a problem of simple linear regression, let $D = \hat{\beta}_1 + \hat{\beta}_2 \bar{x}_n$. Show that the random variables $\hat{\beta}_2$ and D are uncorrelated, and explain why $\hat{\beta}_2$ and D must therefore be independent.

9. Let the random variable D be defined as in Exercise 8, and let the random variable Q^2 be defined by Eq. (37) in this section.

   (a) Show that

   $$\frac{Q^2}{\sigma^2} = \frac{\left(\hat{\beta}_2 - \beta_2^*\right)^2}{\text{Var}(\hat{\beta}_2)} + \frac{\left(D - \beta_i^* - \beta_2^* \bar{x}_n\right)^2}{\text{Var}(D)}.$$

   (b) Explain why the random variable Q^2/σ^2 will have a χ^2 distribution with two degrees of freedom when the hypothesis H_0 in (32) is true.

10. For the data presented in Table 10.6, test the following hypotheses at the level of significance 0.05:

$$H_0: \quad \beta_1 = 0 \text{ and } \beta_2 = 1,$$
$$H_1: \quad \text{At least one of the values } \beta_1 = 0 \text{ and } \beta_2 = 1 \text{ is incorrect.}$$

11. For the data presented in Table 10.6, construct a confidence interval for β_1 with confidence coefficient 0.95.

12. For the data presented in Table 10.6, construct a confidence interval for β_2 with confidence coefficient 0.95.

13. For the data presented in Table 10.6, construct a confidence interval for $5\beta_1 - \beta_2 + 4$ with confidence coefficient 0.90.

14. For the data presented in Table 10.6, construct a confidence interval with confidence coefficient 0.99 for the height of the regression line at the point $x = 1$.

15. For the data presented in Table 10.6, construct a confidence interval with confidence coefficient 0.99 for the height of the regression line at the point $x = 0.42$.

16. Suppose that in a problem of simple linear regression, a confidence interval with confidence coefficient $1 - \alpha_0$ ($0 < \alpha_0 < 1$) is constructed for the height of the regression line at a given value of x. Show that the length of this confidence interval is shortest when $x = \bar{x}_n$.

17. Let the statistic U^2 be as defined by Eq. (40), and let γ be any fixed positive constant. Show that for any given observed values (x_i, y_i), for $i = 1, \ldots, n$, the set of points (β_1^*, β_2^*) such that $U^2 < \gamma$ is the interior of an ellipse in the $\beta_1^* \beta_2^*$-plane.

18. For the data presented in Table 10.6, construct a confidence ellipse for β_1 and β_2 with confidence coefficient 0.95.

19. (a) For the data presented in Table 10.6, sketch a confidence band in the xy-plane for the regression line with confidence coefficient 0.95.

    (b) On the same graph, sketch the curves which specify the limits at each point x of a confidence interval with confidence coefficient 0.95 for the value of the regression line at the point x.

10.4.   THE REGRESSION FALLACY

Use of the Term "Regression"

The use of the term "regression" in Secs. 10.2 and 10.3 to describe the methodology of fitting straight lines to statistical data is a holdover from some of the early applications of this methodology by Francis Galton, who studied the inheritance of physical characteristics in the late 1880's. In particular, Galton found that the sons of tall men tended to be taller than average, but also tended to be shorter than their fathers. Thus, the heights of the sons were closer to the mean height of the population. Similarly, he found that the sons of short men tended to be shorter than average, but also tended to be taller than their fathers. Thus, the heights of these sons also were closer to the mean height of the population. From these observations, one might conclude that the variability of height decreases over successive generations; both tall persons and short persons tend to be eliminated; and the population "regresses" toward some average height. This conclusion is an example of the *regression fallacy*, which we shall discuss in some detail in this section.

Another example of the regression fallacy comes from professional sports. It has been found that athletes who perform very well during their first year in a major league tend not to perform so well during their second year. This phenomenon is sometimes called the "sophomore jinx." It has also been found that athletes who perform rather poorly during their first year, but are allowed to continue for a second year, tend to perform better during the second year. Again there would appear to be a regression in the population toward some mean level of performance.

One way to understand why these conclusions may be fallacious is to consider unusually tall sons and unusually short sons. For such men it is found that the heights of their fathers tend to be closer to average; and one might therefore reach the opposite conclusion that the variability of height is increasing over successive generations. Similarly, if we consider athletes who performed very well or very poorly during their second year, we find that their performances during their first year tended to be closer to average. The following numerical example illustrates these concepts.

Example 1: Examination Scores. Suppose that the students in a certain class were given two examinations during a semester. Suppose that the results on the first examination were as follows: one-third of the students scored 90; another one-third scored 60; and the remaining one-third scored 30. Thus, the mean score was 60. Also, the results on the second examination were as follows: Among the students who scored 90 on the first examination, two-thirds again scored 90 and one-third scored 60; among the students who scored 60 on the first examination, one-third scored 90, one-third scored 60, and the remaining one-third scored 30; and among the students who scored 30 on the first examination, one-third scored 60 and the other two-thirds again scored 30.

If we let X_1 denote the score on the first examination and X_2 the score on the second, then the joint distribution of the values of X_1 and X_2 just described can be represented as follows:

$$\Pr(X_1 = 90) = \Pr(X_1 = 60) = \Pr(X_1 = 30) = \frac{1}{3}, \tag{1}$$

and

$$\Pr(X_2 = 90 \mid X_1 = 90) = \frac{2}{3}, \qquad \Pr(X_2 = 60 \mid X_1 = 90) = \frac{1}{3},$$

$$\Pr(X_2 = 90 \mid X_1 = 60) = \Pr(X_2 = 60 \mid X_1 = 60)$$

$$= \Pr(X_2 = 30 \mid X_1 = 60) = \frac{1}{3}, \tag{2}$$

$$\Pr(X_2 = 60 \mid X_1 = 30) = \frac{1}{3}, \qquad \Pr(X_2 = 30 \mid X_1 = 30) = \frac{2}{3}.$$

It follows from the conditional probabilities (2) that

$$E(X_2 | X_1 = 90) = 80,$$

$$E(X_2 | X_1 = 60) = 60, \tag{3}$$

$$E(X_2 | X_1 = 30) = 40.$$

Thus, among those students with either a high score ($X_1 = 90$) or a low score ($X_1 = 30$) on the first examination, the scores on the second examination tended to be closer to average. For those with the mean score 60 on the first examination, the average score on the second examination was again 60.

As these comments indicate, the conditional expectations (3) reflect the conditions under which the regression fallacy can occur. Specifically, we might conclude from (3) that the scores X_2 on the second examination are more concentrated around the value 60 than are the scores X_1 on the first examination. However, this conclusion is wrong. It is found from Eqs. (1) and (2) that the marginal distribution of X_2 is as follows:

$$\Pr(X_2 = 90) = \Pr(X_2 = 60) = \Pr(X_2 = 30) = \frac{1}{3}. \tag{4}$$

In other words, the distribution of X_2 is precisely the same as the distribution of X_1. □

The Normal Distribution

We shall now show that if the joint distribution of the heights X_1 of the fathers and the heights X_2 of their sons is a bivariate normal distribution, then the conditional expectation $E(X_2 | X_1)$ will be such that the regression fallacy can occur. Accordingly, suppose that X_1 and X_2 are random variables having a bivariate normal distribution with means μ_1 and μ_2, variances σ_1^2 and σ_2^2, and correlation ρ ($0 < \rho < 1$), as described in Sec. 5.12. Then, by Eq. (6) of Sec. 5.12,

$$\frac{E(X_2 | X_1) - \mu_2}{\sigma_2} = \rho \left(\frac{X_1 - \mu_1}{\sigma_1} \right). \tag{5}$$

Since $0 < \rho < 1$, it follows from Eq. (5) that $E(X_2 | X_1)$ is larger than its mean μ_2 if and only if X_1 is greater than its mean μ_1. Thus, if a father is taller than average, then he can expect his son also to be taller than average, even though the mean height μ_2 for the population of sons may be different from the mean μ_1 for the population of fathers.

Consider the special case where $\sigma_1 = \sigma_2$. Since $0 < \rho < 1$, it can be seen from Eq. (5) that in this case $E(X_2 \mid X_1)$ will always be closer to its mean μ_2 than X_1 is to its mean μ_1. Thus, regardless of the height X_1 of the father, he can expect that the height of his son will be closer to average than was his own height.

The regression fallacy might lead us to conclude from the foregoing considerations that the variance in the population of sons must be smaller than the variance in the population of fathers. However, the assumption that $\sigma_1 = \sigma_2$ is precisely the assumption that the variance is the same in these two populations.

10.5. MULTIPLE REGRESSION

The General Linear Model

In this section, we shall study regression problems in which the observations $Y_1, \ldots, Y_n$ again satisfy the assumptions of normality, independence, and homoscedasticity that were made in Secs. 10.2 and 10.3. In other words, we shall again assume that each observation Y_i has a normal distribution; that the observations $Y_1, \ldots, Y_n$ are independent; and that the observations $Y_1, \ldots, Y_n$ have the same variance σ^2. In addition, we shall assume that the mean of each observation Y_i is a linear combination of p unknown parameters $\beta_1, \ldots, \beta_p$. Specifically, for $i = 1, \ldots, n$, we shall assume that $E(Y_i)$ has the following form:

$$E(Y_i) = z_{i1}\beta_1 + z_{i2}\beta_2 + \cdots + z_{ip}\beta_p. \tag{1}$$

Here $z_{i1}, \ldots, z_{ip}$ are known numbers. Any value z_{ij} either may be fixed by the experimenter before the experiment is started or may be observed in the experiment along with the value of Y_i.

The form for $E(Y_i)$ given in Eq. (1) is general enough to include many different types of regression problems. For example, in a problem of simple linear regression, $E(Y_i) = \beta_1 + \beta_2 x_i$ for $i = 1, \ldots, n$. This expectation can be represented in the form given in Eq. (1), with $p = 2$, by letting $z_{i1} = 1$ and $z_{i2} = x_i$ for $i = 1, \ldots, n$. Similarly, if the regression of Y on X is a polynomial of degree k, then, for $i = 1, \ldots, n$,

$$E(Y_i) = \beta_1 + \beta_2 x_i + \cdots + \beta_{k+1} x_i^k. \tag{2}$$

In this case, $p = k + 1$ and $E(Y_i)$ can be represented in the form given in Eq. (1) by letting $z_{ij} = x_i^{j-1}$ for $j = 1, \ldots, k + 1$.

As a final example, consider a problem in which the regression of Y on k variables $X_1, \ldots, X_k$ is a linear function like that given in Eq. (1) of Sec. 10.2. A problem of this type is called a problem of *multiple linear regression* because we are considering the regression of Y on k variables $X_1, \ldots, X_k$, rather than on just a single variable X, and are assuming also that this regression is a linear function of the values of $X_1, \ldots, X_k$. In a problem of multiple linear regression, we obtain n vectors of observations $(x_{i1}, \ldots, x_{ik}, Y_i)$, for $i = 1, \ldots, n$, and $E(Y_i)$ is given by the relation

$$E(Y_i) = \beta_1 + \beta_2 x_{i1} + \cdots + \beta_{k+1} x_{ik}. \tag{3}$$

This expectation can also be represented in the form given in Eq. (1), with $p = k + 1$, by letting $z_{i1} = 1$ and $z_{ij} = x_{i, j-1}$ for $j = 2, \ldots, k + 1$.

The statistical model we are considering in this section, in which the observations $Y_1, \ldots, Y_n$ satisfy the assumptions of normality, independence, and homoscedasticity, and in which their expectations have the form given in Eq. (1), is often called the *general linear model*. Here, the term *linear* refers to the fact that the expectation of each observation Y_i is a linear function of the unknown parameters $\beta_1, \ldots, \beta_p$. Our discussion has indicated that the general linear model is general enough to include problems of simple and multiple linear regression, problems in which the regression function is a polynomial, problems in which the regression function has the form given in Eq. (16) of Sec. 10.1, and many other problems.

Two books devoted to regression and other linear models are Draper and Smith (1980) and Guttman (1982).

Maximum Likelihood Estimators

We shall now describe a procedure for determining the M.L.E.'s of $\beta_1, \ldots, \beta_p$ in the general linear model. Since $E(Y_i)$ is given by Eq. (1) for $i = 1, \ldots, n$, the likelihood function for any observed values $y_1, \ldots, y_n$ will have the following form:

$$\frac{1}{(2\pi\sigma^2)^{n/2}} \exp\left[-\frac{1}{2\sigma^2} \sum_{i=1}^{n} (y_i - z_{i1}\beta_1 - \cdots - z_{ip}\beta_p)^2 \right]. \tag{4}$$

Since the M.L.E.'s are the values which maximize the likelihood function (4), it can be seen that the estimates $\hat{\beta}_1, \ldots, \hat{\beta}_p$ will be the values of $\beta_1, \ldots, \beta_p$ for which the following sum of squares Q is minimized:

$$Q = \sum_{i=1}^{n} (y_i - z_{i1}\beta_1 - \cdots - z_{ip}\beta_p)^2. \tag{5}$$

Since Q is the sum of the squares of the deviations of the observed values from the linear function given in Eq. (1), it follows that the M.L.E.'s $\hat{\beta}_1, \ldots, \hat{\beta}_p$ will be the same as the least-squares estimates.

To determine the values of $\hat{\beta}_1, \ldots, \hat{\beta}_p$, we can calculate the p partial derivatives $\partial Q / \partial \beta_j$ for $j = 1, \ldots, p$, and can set each of these derivatives equal to 0. The resulting p equations, which are called the *normal equations*, will form a set of p linear equations in $\beta_1, \ldots, \beta_p$. We shall assume that the $p \times p$ matrix formed by the coefficients of $\beta_1, \ldots, \beta_p$ in the normal equations is nonsingular. Then these equations will have a unique solution $\hat{\beta}_1, \ldots, \hat{\beta}_p$; and $\hat{\beta}_1, \ldots, \hat{\beta}_p$ will be both the M.L.E.'s and the least-squares estimates of $\beta_1, \ldots, \beta_p$.

For a problem of polynomial regression in which $E(Y_i)$ is given by Eq. (2), the normal equations were presented as the relations (8) of Sec. 10.1. For a problem of multiple linear regression in which $E(Y_i)$ is given by Eq. (3), the normal equations were presented as the relations (13) of Sec. 10.1.

Finally, it can be shown (see Exercise 1 at the end of this section) that the M.L.E. of σ^2 in the general linear model will be

$$\hat{\sigma}^2 = \frac{1}{n} \sum_{i=1}^{n} \left(Y_i - z_{i1}\hat{\beta}_1 - \cdots - z_{ip}\hat{\beta}_p \right)^2. \tag{6}$$

Explicit Form of the Estimators

In order to derive the explicit form and the properties of the estimators $\hat{\beta}_1, \ldots, \hat{\beta}_p$, it is convenient to use the notation and techniques of vectors and matrices. We shall let the $n \times p$ matrix Z be defined as follows:

$$Z = \begin{bmatrix} z_{11} & \cdots & z_{1p} \\ z_{21} & \cdots & z_{2p} \\ \vdots & & \vdots \\ z_{n1} & \cdots & z_{np} \end{bmatrix}. \tag{7}$$

This matrix Z distinguishes one regression problem from another, because the entries in Z determine the particular linear combinations of the unknown parameters $\beta_1, \ldots, \beta_p$ that are relevant in a given problem. The matrix Z for a particular problem is sometimes called the *design matrix* of the problem, because the entries in Z are often chosen by the experimenter to achieve a well-designed experiment. It should be kept in mind, however, that some or all of the entries in Z may be simply the observed values of certain variables, and may not actually be controlled by the experimenter.

We shall also let y be the $n \times 1$ vector of observed values of $Y_1, \ldots, Y_n$; let β be the $p \times 1$ vector of parameters; and let $\hat{\beta}$ be the $p \times 1$ vector of estimates.

These vectors may be represented as follows:

$$
y = \begin{bmatrix} y_1 \\ \vdots \\ y_n \end{bmatrix}, \quad \beta = \begin{bmatrix} \beta_1 \\ \vdots \\ \beta_p \end{bmatrix}, \quad \text{and} \quad \hat{\beta} = \begin{bmatrix} \hat{\beta}_1 \\ \vdots \\ \hat{\beta}_p \end{bmatrix}. \tag{8}
$$

The transpose of any vector or matrix v will be denoted by v'.

The sum of squares Q given in Eq. (5) can now be written in the following form:

$$
Q = (y - Z\beta)'(y - Z\beta). \tag{9}
$$

Also, it can be shown that the set of p normal equations can be written in matrix form as follows:

$$
Z'Z\beta = Z'y. \tag{10}
$$

Since it is assumed that the $p \times p$ matrix $Z'Z$ is nonsingular, the vector of estimates $\hat{\beta}$ will be the unique solution of Eq. (10). In order for $Z'Z$ to be nonsingular, the number of observations n must be at least p, and there must be at least p linearly independent rows in the matrix Z. When this assumption is satisfied, it follows from Eq. (10) that $\hat{\beta} = (Z'Z)^{-1}Z'y$. Thus, if we replace the vector y of observed values by the vector Y of random variables, the form for the vector of estimators $\hat{\beta}$ will be

$$
\hat{\beta} = (Z'Z)^{-1}Z'Y. \tag{11}
$$

It follows from Eq. (11) that each of the estimators $\hat{\beta}_1, \ldots, \hat{\beta}_p$ will be a linear combination of the components $Y_1, \ldots, Y_n$ of the vector Y. Since each of these components has a normal distribution, it follows that each estimator $\hat{\beta}_j$ will also have a normal distribution. We shall now derive the means, variances, and covariances of these estimators.

Mean Vector and Covariance Matrix

Suppose that Y is an n-dimensional random vector with components $Y_1, \ldots, Y_n$. Thus,

$$
Y = \begin{bmatrix} Y_1 \\ \vdots \\ Y_n \end{bmatrix}. \tag{12}
$$

The expectation $E(Y)$ of this random vector is defined to be the n-dimensional vector whose components are the expectations of the individual components of Y. Hence,

$$E(Y) = \begin{bmatrix} E(Y_1) \\ \vdots \\ E(Y_n) \end{bmatrix}. \tag{13}$$

The vector $E(Y)$ is called the *mean vector* of Y.

The *covariance matrix* of the random vector Y is defined to be the $n \times n$ matrix such that, for $i = 1, \ldots, n$ and $j = 1, \ldots, n$, the element in the ith row and jth column is $\mathrm{Cov}(Y_i, Y_j)$. We shall let $\mathrm{Cov}(Y)$ denote this covariance matrix. Thus, if $\mathrm{Cov}(Y_i, Y_j) = \sigma_{ij}$, then

$$\mathrm{Cov}(Y) = \begin{bmatrix} \sigma_{11} & \cdots & \sigma_{1n} \\ \vdots & & \vdots \\ \sigma_{n1} & \cdots & \sigma_{nn} \end{bmatrix}. \tag{14}$$

For $i = 1, \ldots, n$, $\mathrm{Var}(Y_i) = \mathrm{Cov}(Y_i, Y_i) = \sigma_{ii}$. Therefore, the n diagonal elements of the matrix $\mathrm{Cov}(Y)$ are the variances of $Y_1, \ldots, Y_n$. Furthermore, since $\mathrm{Cov}(Y_i, Y_j) = \mathrm{Cov}(Y_j, Y_i)$, then $\sigma_{ij} = \sigma_{ji}$. Therefore, the matrix $\mathrm{Cov}(Y)$ must be symmetric.

The mean vector and the covariance matrix of the random vector Y in the general linear model can easily be determined. It follows from Eq. (1) that

$$E(Y) = Z\beta. \tag{15}$$

Also, the components $Y_1, \ldots, Y_n$ of Y are independent, and the variance of each of these components is σ^2. Therefore,

$$\mathrm{Cov}(Y) = \sigma^2 I, \tag{16}$$

where I is the $n \times n$ identity matrix.

Theorem 1. *Suppose that Y is an n-dimensional random vector as specified by Eq. (12), for which the mean vector $E(Y)$ and the covariance matrix $\mathrm{Cov}(Y)$ exist. Suppose also that A is a $p \times n$ matrix whose elements are constants, and that W is a p-dimensional random vector defined by the relation $W = AY$. Then $E(W) = AE(Y)$ and $\mathrm{Cov}(W) = A\,\mathrm{Cov}(Y)A'$.*

Proof. Let the elements of the matrix A be denoted as follows:

$$A = \begin{bmatrix} a_{11} & \cdots & a_{1n} \\ \vdots & & \vdots \\ a_{p1} & \cdots & a_{pn} \end{bmatrix}. \tag{17}$$

Then the ith component of the vector $E(W)$ is

$$E(W_i) = E\left(\sum_{j=1}^{n} a_{ij}Y_j\right) = \sum_{j=1}^{n} a_{ij}E(Y_j). \tag{18}$$

It can be seen that the final summation in Eq. (18) is the ith component of the vector $AE(Y)$. Hence, $E(W) = AE(Y)$.

Next, for $i = 1, \ldots, p$ and $j = 1, \ldots, p$, the element in the ith row and jth column of the $p \times p$ matrix $\text{Cov}(W)$ is

$$\text{Cov}(W_i, W_j) = \text{Cov}\left(\sum_{r=1}^{n} a_{ir}Y_r, \sum_{s=1}^{n} a_{js}Y_s\right). \tag{19}$$

Therefore, by Exercise 7 of Sec. 4.6,

$$\text{Cov}(W_i, W_j) = \sum_{r=1}^{n} \sum_{s=1}^{n} a_{ir}a_{js}\text{Cov}(Y_r, Y_s). \tag{20}$$

It can be verified that the right side of Eq. (20) is the element in the ith row and jth column of the $p \times p$ matrix $A\,\text{Cov}(Y)A'$. Hence, $\text{Cov}(W) = A\,\text{Cov}(Y)A'$. $\square$

The means, the variances, and the covariances of the estimators $\hat{\beta}_1, \ldots, \hat{\beta}_p$ can be obtained by applying Theorem 1. It is known from Eq. (11) that $\hat{\beta}$ can be represented in the form $\hat{\beta} = AY$, where $A = (Z'Z)^{-1}Z'$. Therefore, it follows from Theorem 1 and Eq. (15) that

$$E(\hat{\beta}) = (Z'Z)^{-1}Z'E(Y) = (Z'Z)^{-1}Z'Z\beta = \beta. \tag{21}$$

In other words, $E(\hat{\beta}_j) = \beta_j$ for $j = 1, \ldots, p$.

Also, it follows from Theorem 1 and Eq. (16) that

$$\text{Cov}(\hat{\beta}) = (Z'Z)^{-1}Z'\text{Cov}(Y)Z(Z'Z)^{-1}$$

$$= (Z'Z)^{-1}Z'(\sigma^2 I)Z(Z'Z)^{-1} \tag{22}$$

$$= \sigma^2(Z'Z)^{-1}.$$

Thus, for $j = 1, \ldots, n$, $\text{Var}(\hat{\beta}_j)$ will be equal to σ^2 times the jth diagonal entry of the matrix $(Z'Z)^{-1}$. Also, for $i \neq j$, $\text{Cov}(\hat{\beta}_i, \hat{\beta}_j)$ will be equal to σ^2 times the entry in the ith row and jth column of the matrix $(Z'Z)^{-1}$.

The Gauss-Markov Theorem for the General Linear Model

Suppose that the observations $Y_1, \ldots, Y_n$ are uncorrelated; that $E(Y_i)$ is given by Eq. (1); that $\mathrm{Var}(Y_i) = \sigma^2$ for $i = 1, \ldots, n$; and that no further assumptions are made about the distribution of $Y_1, \ldots, Y_n$. In particular, it is not assumed that $Y_1, \ldots, Y_n$ necessarily have normal distributions. Suppose also that it is desired to estimate the value of $\theta = c_1\beta_1 + \cdots + c_p\beta_p + c_{p+1}$, where $c_1, \ldots, c_{p+1}$ are given constants, and consider the estimator $\hat{\theta} = c_1\hat{\beta}_1 + \cdots + c_p\hat{\beta}_p + c_{p+1}$. In this case, just as in the problem of simple linear regression discussed in Sec. 10.2, the following result, which is known as the Gauss-Markov theorem, can be established:

> *Among all unbiased estimators of θ which are linear combinations of the observations $Y_1, \ldots, Y_n$, the estimator $\hat{\theta}$ has the smallest variance for all possible values of $\beta_1, \ldots, \beta_p$ and σ^2.*

In particular, for $j = 1, \ldots, p$, the least-squares estimator $\hat{\beta}_j$ will have the smallest variance among all linear unbiased estimators of β_j.

Furthermore, if it is also assumed that the observations $Y_1, \ldots, Y_n$ are independent and have normal distributions, as in the general linear model, then $\hat{\theta}$ will have the smallest variance among *all* unbiased estimators of θ, including unbiased estimators that are not linear functions of $Y_1, \ldots, Y_n$.

The Joint Distribution of the Estimators

We shall let the entries in the symmetric $p \times p$ matrix $(Z'Z)^{-1}$ be denoted as follows:

$$(Z'Z)^{-1} = \begin{bmatrix} \zeta_{11} & \cdots & \zeta_{1p} \\ \vdots & & \vdots \\ \zeta_{p1} & \cdots & \zeta_{pp} \end{bmatrix}. \tag{23}$$

It has been shown earlier in this section that the joint distribution of the estimators $\hat{\beta}_1, \ldots, \hat{\beta}_p$ has the following properties: For $j = 1, \ldots, n$, the estimator $\hat{\beta}_j$ has a normal distribution with mean β_j and variance $\zeta_{jj}\sigma^2$. Furthermore, for $i \neq j$, we have $\mathrm{Cov}(\hat{\beta}_i, \hat{\beta}_j) = \zeta_{ij}\sigma^2$.

For $i = 1, \ldots, n$, we shall let $\hat{Y}_i$ denote the M.L.E. of $E(Y_i)$. It follows from Eq. (1) that

$$\hat{Y}_i = z_{i1}\hat{\beta}_1 + \cdots + z_{ip}\hat{\beta}_p. \tag{24}$$

Also, we shall let the random variable S^2 be defined as follows:

$$S^2 = \sum_{i=1}^{n} (Y_i - \hat{Y}_i)^2. \tag{25}$$

This sum of squares S^2 can also be represented in the following form:

$$S^2 = (Y - Z\hat{\beta})'(Y - Z\hat{\beta}). \tag{26}$$

It can be shown by methods beyond the scope of this book that the random variable S^2/σ^2 has a χ^2 distribution with $n - p$ degrees of freedom. Furthermore, it can be shown that the random variable S^2 and the random vector $\hat{\beta}$ are independent.

From Eq. (6), we see that $\hat{\sigma}^2 = S^2/n$. Hence, the random variable $n\hat{\sigma}^2/\sigma^2$ has a χ^2 distribution with $n - p$ degrees of freedom; and the estimators $\hat{\sigma}^2$ and $\hat{\beta}$ are independent. The description of the joint distribution of the estimators $\hat{\beta}_1, \ldots, \hat{\beta}_p$ and $\hat{\sigma}^2$ is now complete.

Testing Hypotheses

Suppose that it is desired to test the hypothesis that one of the regression coefficients β_j has a particular value β_j^*. In other words, suppose that the following hypotheses are to be tested:

$$\begin{aligned} H_0: & \quad \beta_j = \beta_j^*, \\ H_1: & \quad \beta_j \neq \beta_j^*. \end{aligned} \tag{27}$$

Since $\text{Var}(\hat{\beta}_j) = \zeta_{jj}\sigma^2$, it follows that when H_0 is true, the following random variable W_j will have a standard normal distribution:

$$W_j = \frac{(\hat{\beta}_j - \beta_j^*)}{\zeta_{jj}^{1/2}\sigma}. \tag{28}$$

Furthermore, since the random variable S^2/σ^2 has a χ^2 distribution with $n - p$ degrees of freedom, and since S^2 and $\hat{\beta}_j$ are independent, it follows that when H_0 is true, the following random variable U_j will have a t distribution with $n - p$ degrees of freedom:

$$U_j = \frac{W_j}{\left[\dfrac{1}{n - p} \left(\dfrac{S^2}{\sigma^2} \right) \right]^{1/2}}$$

or

$$U_j = \left(\frac{n-p}{\hat{\zeta}_{jj}} \right)^{1/2} \frac{\left(\hat{\beta}_j - \beta_j^* \right)}{\left[(Y - Z\hat{\beta})'(Y - Z\hat{\beta}) \right]^{1/2}}. \tag{29}$$

The test of the hypotheses (27) specifies that the null hypothesis H_0 should be rejected if $|U_j| > c$, where c is a suitable constant whose value can be chosen to obtain any specified level of significance α_0 ($0 < \alpha_0 < 1$). Thus, if u is the value of U_j observed in a given problem, the corresponding tail area is the two-sided area

$$\Pr(U_j > |u|) + \Pr(U_j < -|u|). \tag{30}$$

Problems of testing hypotheses which specify the values of two coefficients β_i and β_j are discussed in Exercises 16 to 20 at the end of this section.

Multiple Linear Regression

In a problem of multiple linear regression, in which the regression of Y on the k variables $X_1, \ldots, X_k$ is given by Eq. (1) of Sec. 10.2, the mean $E(Y_i)$, for $i = 1, \ldots, n$, is given by Eq. (3) of this section. We are often interested in testing the hypothesis that a particular one of the variables $X_1, \ldots, X_k$ does not actually appear in the regression function. In other words, we are often interested in testing the following hypotheses for some particular value of j ($j = 2, \ldots, k + 1$):

$$\begin{aligned} H_0: & \quad \beta_j = 0, \\ H_1: & \quad \beta_j \neq 0. \end{aligned} \tag{31}$$

Because of this interest in knowing whether or not $\beta_j = 0$, it has become standard practice in the analysis of a problem of multiple linear regression to present not only the M.L.E.'s $\hat{\beta}_1, \ldots, \hat{\beta}_{k+1}$ but also the values of the statistics $U_2, \ldots, U_{k+1}$ and the corresponding tail areas that are found from the expression (30).

Also, in a problem of multiple linear regression, we are typically interested in determining how well the variables $X_1, \ldots, X_k$ explain the observed variation in the random variable Y. The variation among the n observed values $y_1, \ldots, y_n$ of Y can be measured by the value of $\sum_{i=1}^{n}(y_i - \bar{y}_n)^2$, which is the sum of the squares of the deviations of $y_1, \ldots, y_n$ from the average $\bar{y}_n$. Similarly, after the regression of Y on $X_1, \ldots, X_k$ has been fitted from the data, the variation among the n observed values of Y that is still present can be measured by the sum of the squares of the deviations of $y_1, \ldots, y_n$ from the fitted regression. This sum of squares will be equal to the value of S^2 calculated from the observed values. It can be written in the form $(y - Z\hat{\beta})'(y - Z\hat{\beta})$.

It now follows that the proportion of the variation among the observed values $y_1, \ldots, y_n$ that remains unexplained by the fitted regression is

$$\frac{(y - Z\hat{\beta})'(y - Z\hat{\beta})}{\sum_{i=1}^{n}(y_i - \bar{y}_n)^2}. \tag{32}$$

In turn, the proportion of the variation among the observed values $y_1, \ldots, y_n$ that *is explained* by the fitted regression is given by the following value R^2:

$$R^2 = 1 - \frac{(y - Z\hat{\beta})'(y - Z\hat{\beta})}{\sum_{i=1}^{n}(y_i - \bar{y}_n)^2}. \tag{33}$$

The value of R^2 must lie in the interval $0 \leqslant R^2 \leqslant 1$. When $R^2 = 0$, the least-squares estimates have the values $\hat{\beta}_1 = \bar{y}_n$ and $\hat{\beta}_2 = \cdots = \hat{\beta}_{k+1} = 0$. In this case, the fitted regression function is just the constant function $y = \bar{y}_n$. When R^2 is close to 1, the variation of the observed values of Y around the fitted regression function is much smaller than their variation around $\bar{y}_n$.

Screening Regression Equations

A common practice in many areas of application is as follows:

    (i)   Begin with a large number of variables $X_1, \ldots, X_k$.

   (ii)   Calculate the regression of Y on $X_1, \ldots, X_k$.

  (iii)   Drop out each variable X_i for which the regression estimate $\hat{\beta}_i$ is relatively small.

  (iv)   Recalculate the regression of Y on just the variables that remain.

Under this two-stage process, the vector $\hat{\beta}$ of estimators at the second stage can no longer appropriately be regarded as having the mean vector and the covariance matrix developed in this section. If the hypotheses (31) are tested by using the usual t test of size α, then the probability of rejecting H_0 will in fact be larger than α, even if Y and $X_1, \ldots, X_k$ are independent.

A similar comment pertains to the common practice of trying several different forms for the multiple regression. A typical first step might be to include X_i^2 and X_i^3, as well as X_i, or to replace X_i with $\log X_i$. The final step is to report the regression equation that provides the "best fit" in some sense. If the usual t test of the hypotheses (31) is applied to a regression coefficient in the equation that is ultimately selected, the probability of rejecting H_0 will again be larger than the nominal size α of the test, even when Y and $X_1, \ldots, X_k$ are independent. In summary, if we screen many different regression equations involving Y and

$X_1, \ldots, X_k$, and if we select just one equation that seems to provide a good fit, then this equation will usually suggest that there is a much stronger relation between Y and $X_1, \ldots, X_k$ than actually exists.

EXERCISES

1. Show that the M.L.E. of σ^2 in the general linear model is given by Eq. (6).

2. Consider a regression problem in which, for any given value x of a certain variable X, the random variable Y has a normal distribution with mean βx and variance σ^2, where the values of β and σ^2 are unknown. Suppose that n independent pairs of observations (x_i, Y_i) are obtained. Show that the M.L.E. of β is

$$\hat{\beta} = \frac{\sum_{i=1}^{n} x_i Y_i}{\sum_{i=1}^{n} x_i^2}.$$

3. For the conditions of Exercise 2, show that $E(\hat{\beta}) = \beta$ and $\mathrm{Var}(\hat{\beta}) = \sigma^2/(\sum_{i=1}^{n} x_i^2)$.

4. Suppose that when a small amount x of an insulin preparation is injected into a rabbit, the percentage decrease Y in blood sugar has a normal distribution with mean βx and variance σ^2, where the values of β and σ^2 are unknown. Suppose that when independent observations are made on ten different rabbits, the observed values of x_i and Y_i for $i = 1, \ldots, 10$ are as given in Table 10.7. Determine the values of the M.L.E.'s $\hat{\beta}$ and $\hat{\sigma}^2$, and the value of $\mathrm{Var}(\hat{\beta})$.

5. For the conditions of Exercise 4 and the data in Table 10.7, carry out a test of the following hypotheses:

$$H_0: \quad \beta = 10,$$
$$H_1: \quad \beta \neq 10.$$

Table 10.7

i	x_i	y_i	i	x_i	y_i
1	0.6	8	6	2.2	19
2	1.0	3	7	2.8	9
3	1.7	5	8	3.5	14
4	1.7	11	9	3.5	22
5	2.2	10	10	4.2	22

6. Consider a regression problem in which a patient's response Y to a new drug B is to be related to his response X to a standard drug A. Suppose that for any given value x of X, the regression function is a polynomial of the form $E(Y) = \beta_1 + \beta_2 x + \beta_3 x^2$. Suppose also that ten pairs of observed values are as given in Table 10.1. Under the standard assumptions of normality, independence, and homoscedasticity of the observations, determine the values of the M.L.E.'s $\hat{\beta}_1$, $\hat{\beta}_2$, $\hat{\beta}_3$, and $\hat{\sigma}^2$.

7. For the conditions of Exercise 6 and the data in Table 10.1, determine the values of $\text{Var}(\hat{\beta}_1)$, $\text{Var}(\hat{\beta}_2)$, $\text{Var}(\hat{\beta}_3)$, $\text{Cov}(\hat{\beta}_1, \hat{\beta}_2)$, $\text{Cov}(\hat{\beta}_1, \hat{\beta}_3)$, and $\text{Cov}(\hat{\beta}_2, \hat{\beta}_3)$.

8. For the conditions of Exercise 6 and the data in Table 10.1, carry out a test of the following hypotheses:

H_0: $\beta_3 = 0$,
H_1: $\beta_3 \neq 0$.

9. For the conditions of Exercise 6 and the data in Table 10.1, carry out a test of the following hypotheses:

H_0: $\beta_2 = 4$,
H_1: $\beta_2 \neq 4$.

10. For the conditions of Exercise 6 and the data given in Table 10.1, determine the value of R^2, as defined by Eq. (33).

11. Consider a problem of multiple linear regression in which a patient's response Y to a new drug B is to be related to his response X_1 to a standard drug A and to his heart rate X_2. Suppose that for any given values $X_1 = x_1$ and $X_2 = x_2$, the regression function has the form $E(Y) = \beta_1 + \beta_2 x_1 + \beta_3 x_2$; and that the values of ten sets of observations (x_{i1}, x_{i2}, Y_i) are as given in Table 10.2. Under the standard assumptions of multiple linear regression, determine the values of the M.L.E.'s $\hat{\beta}_1$, $\hat{\beta}_2$, $\hat{\beta}_3$, and $\hat{\sigma}^2$.

12. For the conditions of Exercise 11 and the data in Table 10.2, determine the values of $\text{Var}(\hat{\beta}_1)$, $\text{Var}(\hat{\beta}_2)$, $\text{Var}(\hat{\beta}_3)$, $\text{Cov}(\hat{\beta}_1, \hat{\beta}_2)$, $\text{Cov}(\hat{\beta}_1, \hat{\beta}_3)$, and $\text{Cov}(\hat{\beta}_2, \hat{\beta}_3)$.

13. For the conditions of Exercise 11 and the data in Table 10.2, carry out a test of the following hypotheses:

H_0: $\beta_2 = 0$,
H_1: $\beta_2 \neq 0$.

14. For the conditions of Exercise 11 and the data in Table 10.2, carry out a test of the following hypotheses:

H_0: $\beta_3 = -1$,
H_1: $\beta_3 \neq -1$.

15. For the conditions of Exercise 11 and the data in Table 10.2, determine the value of R^2, as defined by Eq. (33).

16. Consider the general linear model in which the observations $Y_1, \ldots, Y_n$ are independent and have normal distributions with the same variance σ^2, and in which $E(Y_i)$ is given by Eq. (1). Let the matrix $(Z'Z)^{-1}$ be defined by Eq. (23). For any given values of i and j such that $i \neq j$, let the random variable A_{ij} be defined as follows:

$$A_{ij} = \hat{\beta}_i - \frac{\zeta_{ij}}{\zeta_{jj}} \hat{\beta}_j.$$

Show that $\text{Cov}(\hat{\beta}_j, A_{ij}) = 0$, and explain why $\hat{\beta}_j$ and A_{ij} are therefore independent.

17. For the conditions of Exercise 16, show that $\text{Var}(A_{ij}) = [\zeta_{ii} - (\zeta_{ij}^2/\zeta_{jj})]\sigma^2$. Also show that the following random variable W^2 has a χ^2 distribution with two degrees of freedom:

$$W^2 = \frac{\zeta_{jj}(\hat{\beta}_i - \beta_i)^2 + \zeta_{ii}(\hat{\beta}_j - \beta_j)^2 - 2\zeta_{ij}(\hat{\beta}_i - \beta_i)(\hat{\beta}_j - \beta_j)}{(\zeta_{ii}\zeta_{jj} - \zeta_{ij}^2)\sigma^2}.$$

Hint: Show that

$$W^2 = \frac{(\hat{\beta}_j - \beta_j)^2}{\zeta_{jj}\sigma^2} + \frac{[A_{ij} - E(A_{ij})]^2}{\text{Var}(A_{ij})}.$$

18. Consider again the conditions of Exercises 16 and 17, and let the random variable S^2 be as defined by Eq. (26).

    (a) Show that the random variable $(n - p)\sigma^2 W^2/(2S^2)$ has an F distribution with 2 and $n - p$ degrees of freedom.

    (b) For any two given numbers β_i^* and β_j^*, describe how to carry out a test of the following hypotheses:

    H_0: $\beta_i = \beta_i^*$ and $\beta_j = \beta_j^*$,
    H_1: The hypothesis H_0 is not true.

19. For the conditions of Exercise 6 and the data in Table 10.1, carry out a test of the following hypotheses:

    H_0: $\beta_2 = \beta_3 = 0$,
    H_1: The hypothesis H_0 is not true.

20. For the conditions of Exercise 11 and the data in Table 10.2, carry out a test of the following hypotheses:

H_0: $\beta_2 = 1$ and $\beta_3 = 0$,
H_1: The hypothesis H_0 is not true.

21. Consider a problem of simple linear regression as described in Sec. 10.2, and let R^2 be defined by Eq. (33) of this section. Show that

$$R^2 = \frac{\left[\sum_{i=1}^{n}(x_i - \bar{x}_n)(y_i - \bar{y}_n)\right]^2}{\left[\sum_{i=1}^{n}(x_i - \bar{x}_n)^2\right]\left[\sum_{i=1}^{n}(y_i - \bar{y}_n)^2\right]}.$$

22. Suppose that X and Y are n-dimensional random vectors for which the mean vectors $E(X)$ and $E(Y)$ exist. Show that $E(X + Y) = E(X) + E(Y)$.

23. Suppose that X and Y are independent n-dimensional random vectors for which the covariance matrices $\text{Cov}(X)$ and $\text{Cov}(Y)$ exist. Show that $\text{Cov}(X + Y) = \text{Cov}(X) + \text{Cov}(Y)$.

24. Suppose that Y is a three-dimensional random vector with components Y_1, Y_2, and Y_3, and suppose that the covariance matrix of Y is as follows:

$$\text{Cov}(Y) = \begin{bmatrix} 9 & -3 & 0 \\ -3 & 4 & 0 \\ 0 & 0 & 5 \end{bmatrix}.$$

Determine the value of $\text{Var}(3Y_1 + Y_2 - 2Y_3 + 8)$.

10.6. ANALYSIS OF VARIANCE

The One-Way Layout

In this section and in the remainder of the book, we shall study a topic known as the *analysis of variance*, abbreviated ANOVA. Problems of ANOVA are actually problems of multiple regression in which the design matrix Z has a very special form. In other words, the study of ANOVA can be placed within the framework of the general linear model, and the basic assumptions for such a model still hold: The observations that are obtained are independent and normally distributed; all these observations have the same variance σ^2; and the mean of each observation can be represented as a linear combination of certain unknown parameters. The theory and methodology of ANOVA were mainly developed by R. A. Fisher during the 1920's.

We shall begin our study of ANOVA by considering a problem known as the one-way layout. In this problem, it is assumed that random samples from p different normal distributions are available; that each of these distributions has the same variance σ^2; and that the means of the p distributions are to be compared on the basis of the observed values in the samples. This problem was considered for two populations ($p = 2$) in Sec. 8.9, and the results to be presented here for an arbitrary value of p will generalize those presented in Sec. 8.9. Specifically, we shall now make the following assumption: For $i = 1, \ldots, p$, the random variables $Y_{i1}, \ldots, Y_{in_i}$ form a random sample of n_i observations from a normal distribution with mean β_i and variance σ^2, and the values of $\beta_1, \ldots, \beta_p$ and σ^2 are unknown.

In this problem, the sample sizes $n_1, \ldots, n_p$ are not necessarily the same. We shall let $n = \sum_{i=1}^{p} n_i$ denote the total number of observations in the p samples, and we shall assume that all n observations are independent.

It follows from the assumptions we have just made that for $j = 1, \ldots, n_i$ and $i = 1, \ldots, p$, we have $E(Y_{ij}) = \beta_i$ and $\text{Var}(Y_{ij}) = \sigma^2$. Since the expectation $E(Y_{ij})$ of each observation is equal to one of the p parameters $\beta_1, \ldots, \beta_p$, it is obvious that each of these expectations can be regarded as a linear combination of $\beta_1, \ldots, \beta_p$. Furthermore, we can regard the n observations Y_{ij} as the elements of a single long n-dimensional vector Y, which can be written as follows:

$$Y = \begin{bmatrix} Y_{11} \\ \vdots \\ Y_{1n_1} \\ \vdots \\ Y_{p1} \\ \vdots \\ Y_{pn_p} \end{bmatrix}. \tag{1}$$

This one-way layout therefore satisfies the conditions of the general linear model. The $n \times p$ design matrix Z, as defined by Eq. (7) of Sec. 10.5, will have the following form:

$$Z = \begin{bmatrix} 1 & 0 & \cdots & 0 \\ \vdots & \vdots & & \vdots \\ 1 & 0 & \cdots & 0 \\ \vdots & \vdots & & \vdots \\ 0 & 0 & \cdots & 1 \\ \vdots & \vdots & & \vdots \\ 0 & 0 & \cdots & 1 \end{bmatrix} \begin{array}{l} \left.\vphantom{\begin{matrix}1\\ \vdots \\ 1\end{matrix}}\right\} n_1 \text{ rows} \\ \\ \left.\vphantom{\begin{matrix}0\\ \vdots \\ 0\end{matrix}}\right\} n_p \text{ rows} \end{array} \tag{2}$$

For $i = 1, \ldots, p$, we shall let $\overline{Y}_{i+}$ denote the mean of the n_i observations in the ith sample. Thus,

$$\overline{Y}_{i+} = \frac{1}{n_i} \sum_{j=1}^{n_i} Y_{ij}. \tag{3}$$

It can then be shown (see Exercise 1) that $\overline{Y}_{i+}$ is the M.L.E., or least-squares estimator, of β_i for $i = 1, \ldots, p$. Also, the M.L.E. of σ^2 is

$$\hat{\sigma}^2 = \frac{1}{n} \sum_{i=1}^{p} \sum_{j=1}^{n_i} \left(Y_{ij} - \overline{Y}_{i+} \right)^2. \tag{4}$$

Partitioning a Sum of Squares

In a one-way layout, we are often interested in testing the hypothesis that the p distributions from which the samples were drawn are actually the same; that is, we desire to test the following hypotheses:

$$\begin{aligned} H_0&: \quad \beta_1 = \cdots = \beta_p, \\ H_1&: \quad \text{The hypothesis } H_0 \text{ is not true.} \end{aligned} \tag{5}$$

Before we develop an appropriate test procedure, we shall carry out some preparatory algebraic manipulations in which we shall partition the sum of squares $\sum_{i=1}^{p} \sum_{j=1}^{n_i} (Y_{ij} - \beta_i)^2$ into various smaller sums of squares. Each of these smaller sums of squares can be associated with a certain type of variation among the n observations. The test of the hypotheses (5) that we shall develop will be based on an analysis of these different types of variations, that is, on a comparison of these sums of squares. For this reason, the name *analysis of variance* has been applied to this problem and to other related problems.

If we consider only the n_i observations in sample i, then the sum of squares for those values can be written as follows:

$$\sum_{j=1}^{n_i} \frac{\left(Y_{ij} - \beta_i \right)^2}{\sigma^2} = \sum_{j=1}^{n_i} \frac{\left(Y_{ij} - \overline{Y}_{i+} \right)^2}{\sigma^2} + \frac{n_i \left(\overline{Y}_{i+} - \beta_i \right)^2}{\sigma^2}. \tag{6}$$

The sum on the left side of Eq. (6) is the sum of squares of n_i independent random variables, each of which has a standard normal distribution. This sum, therefore, has a χ^2 distribution with n_i degrees of freedom. Furthermore, it follows from Theorem 1 of Sec. 7.3 that the sum forming the first term on the right side of Eq. (6) has a χ^2 distribution with $n_i - 1$ degrees of freedom; that the other term on the right side of Eq. (6) has a χ^2 distribution with one degree of freedom; and that these two terms on the right side are independent.

If we now sum each of the terms in Eq. (6) over the values of i, we obtain the relation

$$\sum_{i=1}^{p} \sum_{j=1}^{n_i} \frac{(Y_{ij} - \beta_i)^2}{\sigma^2} = \sum_{i=1}^{p} \sum_{j=1}^{n_i} \frac{(Y_{ij} - \overline{Y}_{i+})^2}{\sigma^2} + \sum_{i=1}^{p} \frac{n_i(\overline{Y}_{i+} - \beta_i)^2}{\sigma^2}. \tag{7}$$

Since all the observations in the p samples are independent, each of the three summations from $i = 1$ to $i = p$ in Eq. (7) is the sum of p independent random variables having χ^2 distributions. Hence, each sum in Eq. (7) will itself have a χ^2 distribution. In particular, the sum on the left side will have a χ^2 distribution with $\sum_{i=1}^{p} n_i = n$ degrees of freedom. Also, the first sum on the right side of Eq. (7) will have a χ^2 distribution with $\sum_{i=1}^{p}(n_i - 1) = n - p$ degrees of freedom; and the second sum on the right side will have a χ^2 distribution with p degrees of freedom. Furthermore, the two sums on the right side will still be independent.

If we let Q_1, Q_2, and Q_3 denote the three summations appearing in Eq. (7), then $Q_1 = Q_2 + Q_3$. The variable Q_1 is the sum over n observations of the square of the deviation of each observation Y_{ij} from its mean β_i. Hence, Q_1 can be regarded as the total variation of the observations around their means. Similarly, Q_2 can be regarded as the total variation of the observations around the sample means, or the total residual variation within the samples. Also, Q_3 can be regarded as the total variation of the sample means around the actual means. Thus, the total variation Q_1 has been partitioned into two independent components, Q_2 and Q_3, which represent different types of variations. We shall now partition Q_3 further.

We shall let $\overline{Y}_{++}$ denote the average of all n observations. Thus,

$$\overline{Y}_{++} = \frac{1}{n} \sum_{i=1}^{p} \sum_{j=1}^{n_i} Y_{ij} = \frac{1}{n} \sum_{i=1}^{p} n_i \overline{Y}_{i+}. \tag{8}$$

Also, we shall define β as follows:

$$\beta = E(\overline{Y}_{++}) = \frac{1}{n} \sum_{i=1}^{p} \sum_{j=1}^{n_i} E(Y_{ij}) = \frac{1}{n} \sum_{i=1}^{p} n_i \beta_i. \tag{9}$$

Finally, we shall let

$$\alpha_i = \beta_i - \beta \qquad \text{for } i = 1, \ldots, p. \tag{10}$$

The parameter α_i is called the *effect* of the ith distribution. It follows from Eqs. (9) and (10) that $\sum_{i=1}^{p} n_i \alpha_i = 0$. Furthermore, the hypotheses (5) can now be rewritten as follows:

$$H_0: \quad \alpha_i = 0 \qquad \text{for } i = 1, \ldots, p,$$
$$H_1: \quad \text{The hypothesis } H_0 \text{ is not true.} \tag{11}$$

For $i = 1, \ldots, p$, the least-squares estimator of the effect α_i will be $\hat{\alpha}_i = \overline{Y}_{i+} - \overline{Y}_{++}$.

If we replace $\overline{Y}_{i+} - \beta_i$ with $(\overline{Y}_{i+} - \overline{Y}_{++} - \alpha_i) + (\overline{Y}_{++} - \beta)$ in the expression for the sum of squares Q_3, then Q_3 can be written as follows (see Exercise 2):

$$\sum_{i=1}^{p} \frac{n_i(\overline{Y}_{i+} - \beta_i)^2}{\sigma^2} = \sum_{i=1}^{p} \frac{n_i(\overline{Y}_{i+} - \overline{Y}_{++} - \alpha_i)^2}{\sigma^2} + \frac{n(\overline{Y}_{++} - \beta)^2}{\sigma^2}. \tag{12}$$

If we denote the two terms on the right side of Eq. (12) by Q_4 and Q_5, then $Q_3 = Q_4 + Q_5$. It can be shown that the random variables Q_4 and Q_5 are independent; that Q_4 has a χ^2 distribution with $p - 1$ degrees of freedom; and that Q_5 has a χ^2 distribution with one degree of freedom. Thus, we have partitioned the sum of squares Q_3, which has a χ^2 distribution with p degrees of freedom, into the sum of two independent random variables, each of which has a χ^2 distribution. Since the sum of squares Q_4 is equal to $\sum_{i=1}^{p} n_i(\hat{\alpha}_i - \alpha_i)^2/\sigma^2$, this sum can be regarded as the total variation among the estimators of the effects. Moreover, Q_5 can be regarded as the variation of the overall sample mean $\overline{Y}_{++}$ from the actual overall mean β.

By combining Eqs. (7) and (12), we obtain a relation having the form $Q_1 = Q_2 + Q_4 + Q_5$. Here, the total variation Q_1 of the observations around their means, which has a χ^2 distribution with n degrees of freedom, has been partitioned into the sum of three other independent random variables, each of which itself has a χ^2 distribution. This partitioning is often summarized in a table which is called the ANOVA table for the one-way layout and is presented here as Table 10.8.

This table differs from the ANOVA tables that are presented in most textbooks because the parameters β and $\alpha_1, \ldots, \alpha_p$ are included in the sums of squares that are given in this table. In most ANOVA tables, the values of all the parameters are taken to be 0 in the sums of squares. Because of this difference, Table 10.8 is more general than most ANOVA tables and is more flexible for testing hypotheses about the parameters.

Table 10.8

Source of variation	Degrees of freedom	Sum of squares
Overall mean	1	$n(\bar{Y}_{++} - \beta)^2$
Effects	$p - 1$	$\sum_{i=1}^{p} n_i(\bar{Y}_{i+} - \bar{Y}_{++} - \alpha_i)^2$
Residuals	$n - p$	$\sum_{i=1}^{p} \sum_{j=1}^{n_i} (Y_{ij} - \bar{Y}_{i+})^2$
Total	n	$\sum_{i=1}^{p} \sum_{j=1}^{n_i} (Y_{ij} - \beta_i)^2$

Testing Hypotheses

We are now ready to describe a test of the hypotheses (5) or, equivalently, of the hypotheses (11). If the null hypothesis H_0 is true and $\alpha_i = 0$ for $i = 1, \ldots, p$, then it follows from the results given here that Q_4 has the form

$$Q_4^0 = \sum_{i=1}^{p} n_i \frac{(\bar{Y}_{i+} - \bar{Y}_{++})^2}{\sigma^2},$$

and that Q_4^0 has a χ^2 distribution with $p - 1$ degrees of freedom. Moreover, regardless of whether or not H_0 is true, Q_2 has a χ^2 distribution with $n - p$ degrees of freedom, and Q_4^0 and Q_2 are independent. It therefore follows that when H_0 is true, the following random variable U^2 will have an F distribution with $p - 1$ and $n - p$ degrees of freedom:

$$U^2 = \frac{Q_4^0/(p - 1)}{Q_2/(n - p)} \tag{13}$$

or

$$U^2 = \frac{(n - p)\sum_{i=1}^{p} n_i(\bar{Y}_{i+} - \bar{Y}_{++})^2}{(p - 1)\sum_{i=1}^{p}\sum_{j=1}^{n_i}(Y_{ij} - \bar{Y}_{i+})^2}. \tag{14}$$

When the null hypothesis H_0 is not true and the value of $\alpha_i = E(\bar{Y}_{i+} - \bar{Y}_{++})$ is different from 0 for at least one value of i, then the expectation of the numerator of U^2 will be larger than it would be if H_0 were true. The distribution of the denominator of U^2 remains the same regardless of whether or not H_0 is true. It can be shown that the likelihood ratio test procedure for the hypotheses (11) specifies rejecting H_0 if $U^2 > c$, where c is a suitable constant whose value for any specified level of significance can be determined from a table of the F distribution with $p - 1$ and $n - p$ degrees of freedom.

In some problems, a useful form of the statistic U^2 for computational purposes is as follows (see Exercises 3 and 4):

$$U^2 = \frac{(n - p)\left(\sum_{i=1}^{p} n_i \overline{Y}_{i+}^2 - n\overline{Y}_{++}^2\right)}{(p - 1)\left(\sum_{i=1}^{p}\sum_{j=1}^{n_i} Y_{ij}^2 - \sum_{i=1}^{p} n_i \overline{Y}_{i+}^2\right)}. \tag{15}$$

EXERCISES

1. In a one-way layout, show that $\overline{Y}_{i+}$ is the least-squares estimator of β_i by showing that the ith component of the vector $(Z'Z)^{-1}Z'Y$ is $\overline{Y}_{i+}$ for $i = 1, \ldots, p$.

2. Verify Eq. (12).

3. Show that

$$\sum_{i=1}^{p} n_i(\overline{Y}_{i+} - \overline{Y}_{++})^2 = \sum_{i=1}^{p} n_i \overline{Y}_{i+}^2 - n\overline{Y}_{++}^2.$$

4. Show that

$$\sum_{i=1}^{p}\sum_{j=1}^{n_i}(Y_{ij} - \overline{Y}_{i+})^2 = \sum_{i=1}^{p}\sum_{j=1}^{n_i} Y_{ij}^2 - \sum_{i=1}^{p} n_i \overline{Y}_{i+}^2.$$

5. Specimens of milk from a number of dairies in three different districts were analyzed, and the concentration of the radioactive isotope Strontium-90 was measured in each specimen. Suppose that specimens were obtained from four dairies in the first district, from six dairies in the second district, and from three dairies in the third district; and that the results measured in picocuries per liter were as follows:

District 1: 6.4, 5.8, 6.5, 7.7,
District 2: 7.1, 9.9, 11.2, 10.5, 6.5, 8.8,
District 3: 9.5, 9.0, 12.1.

(a) Assuming that the variance of the concentration of Strontium-90 is the same for the dairies in all three districts, determine the M.L.E. of the mean concentration in each of the districts and the M.L.E. of the common variance.

(b) Test the hypothesis that the three districts have identical concentrations of Strontium-90.

6. A random sample of 10 students was selected from the senior class at each of four large high schools, and the score of each of these 40 students on a certain mathematics examination was observed. Suppose that for the 10 students from each school, the sample mean and the sample variance of the scores were as given in Table 10.9. Test the hypothesis that the senior classes at all four high schools would perform equally well on this examination. Discuss carefully the assumptions that you are making in carrying out this test.

7. Suppose that a random sample of size n is taken from a normal distribution with mean μ and variance σ^2; and that the sample is divided into p groups of observations of sizes $n_1, \ldots, n_p$, where $n_i \geq 2$ for $i = 1, \ldots, p$ and $\sum_{i=1}^{p} n_i = n$. For $i = 1, \ldots, p$, let Q_i denote the sum of the squares of the deviations of the n_i observations in the ith group from the sample mean of those n_i observations. Find the distribution of the sum $Q_1 + \cdots + Q_p$ and the distribution of the ratio Q_1/Q_p.

8. Verify that the t test presented in Sec. 8.9 for comparing the means of two normal distributions is the same as the test presented in this section for the one-way layout with $p = 2$ by verifying that if U is defined by Eq. (5) of Sec. 8.9, then U^2 is equal to the expression given in Eq. (14) of this section.

9. Show that in a one-way layout the following statistic is an unbiased estimator of σ^2:

$$\frac{1}{n - p} \sum_{i=1}^{p} \sum_{j=1}^{n_i} \left(Y_{ij} - \overline{Y}_{i+} \right)^2.$$

Table 10.9

School	Sample mean	Sample variance
1	105.7	30.3
2	102.0	54.4
3	93.5	25.0
4	110.8	36.4

10. In a one-way layout, show that for any values of i, i', and j, where $j = 1, \ldots, n_i$, $i = 1, \ldots, p$, and $i' = 1, \ldots, p$, the following three random variables W_1, W_2, and W_3 are uncorrelated with each other:

$$W_1 = Y_{ij} - \overline{Y}_{i+}, \qquad W_2 = \overline{Y}_{i'+} - \overline{Y}_{++}, \qquad W_3 = \overline{Y}_{++}.$$

10.7.   THE TWO-WAY LAYOUT

The Two-Way Layout with One Observation in Each Cell

We shall now consider problems of ANOVA in which the value of the random variable being observed is affected by two factors. For example, suppose that in an experiment to measure the concentration of a certain radioactive isotope in milk, specimens of milk are obtained from four different dairies and the concentration of the isotope in each specimen is measured by three different methods. If we let Y_{ij} denote the measurement that is made for the specimen from the ith dairy by using the jth method, for $i = 1, 2, 3, 4$ and $j = 1, 2, 3$, then in this example there will be a total of 12 measurements. A problem of this type is called a two-way layout.

In the general two-way layout, there are two factors, which we shall call A and B. We shall assume that there are I possible different values, or different *levels*, of factor A; and there are J possible different values, or different *levels*, of factor B. For $i = 1, \ldots, I$ and $j = 1, \ldots, J$, an observation Y_{ij} of the variable being studied is obtained when factor A has the value i and factor B has the value j. If the IJ observations are arranged in a matrix as in Table 10.10, then Y_{ij} is the observation in the (i, j) cell of the matrix.

Table 10.10

		Factor B		
	1	2	$\cdots$	J
1	Y_{11}	Y_{12}	$\cdots$	Y_{1J}
2	Y_{21}	Y_{22}	$\cdots$	Y_{2J}
$\vdots$	$\vdots$	$\vdots$		$\vdots$
I	Y_{I1}	Y_{I2}	$\cdots$	Y_{IJ}

Factor A (labels the rows: 1, 2, $\ldots$, I)

We shall continue to make the three standard assumptions of independence, normality, and homoscedasticity of the observations for the two-way layout. Thus, we shall assume that all the observations Y_{ij} are independent; that each observation has a normal distribution; and that all the observations have the same variance σ^2. In addition, we shall make the following special assumption about the mean $E(Y_{ij})$: We shall assume not only that $E(Y_{ij})$ depends on the values i and j of the two factors, but also that there exist numbers $\theta_1, \ldots, \theta_I$ and $\psi_1, \ldots, \psi_J$ such that

$$E(Y_{ij}) = \theta_i + \psi_j \quad \text{for } i = 1, \ldots, I \quad \text{and } j = 1, \ldots, J. \tag{1}$$

Thus, Eq. (1) states that the value of $E(Y_{ij})$ is the sum of the following two effects: an effect θ_i due to factor A having the value i, and an effect ψ_j due to factor B having the value j. For this reason, the assumption that $E(Y_{ij})$ has the form given in Eq. (1) is called an assumption of additivity of the effects of the factors.

The meaning of the assumption of additivity can be clarified by the following example. Consider the sales of I different magazines at J different newsstands. Suppose that a particular newsstand sells on the average 30 more copies per week of magazine 1 than of magazine 2. Then by the assumption of additivity, it must also be true that each of the other $J - 1$ newsstands sells on the average 30 more copies per week of magazine 1 than of magazine 2. Similarly, suppose that the sales of a particular magazine are on the average 50 more copies per week at newsstand 1 than at newsstand 2. Then by the assumption of additivity, it must also be true that the sales of each of the other $I - 1$ magazines are on the average 50 more copies per week at newsstand 1 than at newsstand 2. The assumption of additivity is a very restrictive assumption because it does not allow for the possibility that a particular magazine may sell unusually well at some particular newsstand.

Even though we assume in the general two-way layout that the effects of the factors A and B are additive, the numbers θ_i and ψ_j which satisfy Eq. (1) are not uniquely defined. We can add an arbitrary constant c to each of the numbers $\theta_1, \ldots, \theta_I$ and subtract the same constant c from each of the numbers $\psi_1, \ldots, \psi_J$ without changing the value of $E(Y_{ij})$ for any of the IJ observations. Hence, it does not make sense to try to estimate the value of θ_i or ψ_j from the given observations, since neither θ_i nor ψ_j is uniquely defined. In order to avoid this difficulty, we shall express $E(Y_{ij})$ in terms of different parameters. The following assumption is equivalent to the assumption of additivity.

We shall assume that there exist numbers $\mu, \alpha_1, \ldots, \alpha_I$, and $\beta_1, \ldots, \beta_J$ such that

$$\sum_{i=1}^{I} \alpha_i = 0 \quad \text{and} \quad \sum_{j=1}^{J} \beta_j = 0, \tag{2}$$

and

$$E(Y_{ij}) = \mu + \alpha_i + \beta_j \qquad \text{for } i = 1, \ldots, I \text{ and } j = 1, \ldots, J. \tag{3}$$

There is an advantage in expressing $E(Y_{ij})$ in this way. If the values of $E(Y_{ij})$ for $i = 1, \ldots, I$ and $j = 1, \ldots, J$ are a set of numbers that satisfy Eq. (1) for *some* set of values of $\theta_1, \ldots, \theta_I$ and $\psi_1, \ldots, \psi_J$, then there exists a *unique* set of values of μ, $\alpha_1, \ldots, \alpha_I$, and $\beta_1, \ldots, \beta_J$ that satisfy Eqs. (2) and (3) (see Exercise 2).

The parameter μ is called the *overall mean*, or the *grand mean*, since it follows from Eqs. (2) and (3) that

$$\mu = \frac{1}{IJ} \sum_{i=1}^{I} \sum_{j=1}^{J} E(Y_{ij}). \tag{4}$$

The parameters $\alpha_1, \ldots, \alpha_I$ are called the *effects of factor A*, and the parameters $\beta_1, \ldots, \beta_J$ are called the *effects of factor B*.

It follows from Eq. (2) that $\alpha_I = -\sum_{i=1}^{I-1} \alpha_i$ and $\beta_J = -\sum_{j=1}^{J-1} \beta_j$. Hence, each expectation $E(Y_{ij})$ in Eq. (3) can be expressed as a particular linear combination of the $I + J - 1$ parameters $\mu, \alpha_1, \ldots, \alpha_{I-1}$, and $\beta_1, \ldots, \beta_{J-1}$. Therefore, if we regard the IJ observations as elements of a single long IJ-dimensional vector, then the two-way layout satisfies the conditions of the general linear model. In a practical problem, however, it is not convenient to actually replace α_I and β_J with their expressions in terms of the other α_i's and β_j's, because this replacement would destroy the symmetry that is present in the experiment among the different levels of each factor.

Estimating the Parameters

We shall let

$$\overline{Y}_{i+} = \frac{1}{J} \sum_{j=1}^{J} Y_{ij} \qquad \text{for } i = 1, \ldots, I,$$

$$\overline{Y}_{+j} = \frac{1}{I} \sum_{i=1}^{I} Y_{ij} \qquad \text{for } j = 1, \ldots, J, \tag{5}$$

$$\overline{Y}_{++} = \frac{1}{IJ} \sum_{i=1}^{I} \sum_{j=1}^{J} Y_{ij} = \frac{1}{I} \sum_{i=1}^{I} \overline{Y}_{i+} = \frac{1}{J} \sum_{j=1}^{J} \overline{Y}_{+j}.$$

It can then be shown that the M.L.E.'s, or least-squares estimators, of $\mu, \alpha_1, \ldots, \alpha_I$, and $\beta_1, \ldots, \beta_J$ are as follows:

$$\hat{\mu} = \overline{Y}_{++},$$

$$\hat{\alpha}_i = \overline{Y}_{i+} - \overline{Y}_{++} \qquad \text{for } i = 1, \ldots, I, \tag{6}$$

$$\hat{\beta}_j = \overline{Y}_{+j} - \overline{Y}_{++} \qquad \text{for } j = 1, \ldots, J.$$

It is easily verified (see Exercise 5) that $\sum_{i=1}^{I}\hat{\alpha}_i = \sum_{j=1}^{J}\hat{\beta}_j = 0$; that $E(\hat{\mu}) = \mu$; that $E(\hat{\alpha}_i) = \alpha_i$ for $i = 1, \ldots, I$; and that $E(\hat{\beta}_j) = \beta_j$ for $j = 1, \ldots, J$. Finally, the M.L.E. of σ^2 will be

$$\hat{\sigma}^2 = \frac{1}{IJ} \sum_{i=1}^{I} \sum_{j=1}^{J} \left(Y_{ij} - \hat{\mu} - \hat{\alpha}_i - \hat{\beta}_j\right)^2$$

or, equivalently,

$$\hat{\sigma}^2 = \frac{1}{IJ} \sum_{i=1}^{I} \sum_{j=1}^{J} \left(Y_{ij} - \overline{Y}_{i+} - \overline{Y}_{+j} + \overline{Y}_{++}\right)^2. \tag{7}$$

Partitioning the Sum of Squares

Consider now the following sum Q_1:

$$Q_1 = \sum_{i=1}^{I} \sum_{j=1}^{J} \frac{\left(Y_{ij} - \mu - \alpha_i - \beta_j\right)^2}{\sigma^2}. \tag{8}$$

Since Q_1 is the sum of the squares of IJ independent random variables, each of which has a standard normal distribution, then Q_1 has a χ^2 distribution with IJ degrees of freedom. We shall now partition the sum of squares Q_1 into various smaller sums of squares. Each of these smaller sums of squares will be associated with a certain type of variation among the observations Y_{ij}. Each of them will have a χ^2 distribution, and they will be mutually independent. Therefore, just as in the one-way layout, we can construct tests of various hypotheses based on an analysis of variance, that is, on an analysis of these different types of variations. We shall begin by rewriting Q_1 as follows:

$$Q_1 = \frac{1}{\sigma^2} \sum_{i=1}^{I} \sum_{j=1}^{J} \left[\left(Y_{ij} - \hat{\mu} - \hat{\alpha}_i - \hat{\beta}_j\right) + \left(\hat{\mu} - \mu\right) + \left(\hat{\alpha}_i - \alpha_i\right) + \left(\hat{\beta}_j - \beta_j\right)\right]^2. \tag{9}$$

By expanding the right side of Eq. (9), we obtain (see Exercise 7) the relation

$$Q_1 = \sum_{i=1}^{I} \sum_{j=1}^{J} \frac{\left(Y_{ij} - \hat{\mu} - \hat{\alpha}_i - \hat{\beta}_j\right)^2}{\sigma^2} + \frac{IJ(\hat{\mu} - \mu)^2}{\sigma^2} + J \sum_{i=1}^{I} \frac{(\hat{\alpha}_i - \alpha_i)^2}{\sigma^2}$$

$$+ I \sum_{j=1}^{J} \frac{(\hat{\beta}_j - \beta_j)^2}{\sigma^2}.$$

(10)

If we denote the four terms on the right side of Eq. (10) by Q_2, Q_3, Q_4, and Q_5, then $Q_1 = Q_2 + Q_3 + Q_4 + Q_5$. When the expressions for $\hat{\mu}$, $\hat{\alpha}_i$, and $\hat{\beta}_j$ given in the relations (6) are used, we obtain the following forms:

$$Q_2 = \sum_{i=1}^{I} \sum_{j=1}^{J} \frac{\left(Y_{ij} - \overline{Y}_{i+} - \overline{Y}_{+j} + \overline{Y}_{++}\right)^2}{\sigma^2},$$

$$Q_3 = \frac{IJ(\overline{Y}_{++} - \mu)^2}{\sigma^2},$$

(11)

$$Q_4 = J \sum_{i=1}^{I} \frac{\left(\overline{Y}_{i+} - \overline{Y}_{++} - \alpha_i\right)^2}{\sigma^2},$$

$$Q_5 = I \sum_{j=1}^{J} \frac{\left(\overline{Y}_{+j} - \overline{Y}_{++} - \beta_j\right)^2}{\sigma^2}.$$

It can be shown that the random variables Q_2, Q_3, Q_4, and Q_5 are independent (see Exercise 8 for a related result). Furthermore, it can be shown that Q_2 has a χ^2 distribution with $IJ - (I + J - 1) = (I - 1)(J - 1)$ degrees of freedom; that Q_3 has a χ^2 distribution with one degree of freedom; that Q_4 has a χ^2 distribution with $I - 1$ degrees of freedom; and that Q_5 has a χ^2 distribution with $J - 1$ degrees of freedom. Thus, we have partitioned the sum of squares Q_1, which has a χ^2 distribution with IJ degrees of freedom, into the sum of four independent random variables, each of which itself has a χ^2 distribution and each of which can be associated with a particular type of variation, as seen from Eq. (9) or Eq. (10). These properties are summarized in Table 10.11, which is the ANOVA table for the two-way layout. As discussed in Sec. 10.6, this table is different from, and more flexible than, the ANOVA tables presented in most textbooks because the parameters are included in the sums of squares given in this table.

Table 10.11

Source of variation	Degrees of freedom	Sum of squares
Overall mean	1	$IJ(\overline{Y}_{++} - \mu)^2$
Effects of factor A	$I - 1$	$J \sum_{i=1}^{I} (\overline{Y}_{i+} - \overline{Y}_{++} - \alpha_i)^2$
Effects of factor B	$J - 1$	$I \sum_{j=1}^{J} (\overline{Y}_{+j} - \overline{Y}_{++} - \beta_j)^2$
Residuals	$(I - 1)(J - 1)$	$\sum_{i=1}^{I} \sum_{j=1}^{J} (Y_{ij} - \overline{Y}_{i+} - \overline{Y}_{+j} + \overline{Y}_{++})^2$
Total	IJ	$\sum_{i=1}^{I} \sum_{j=1}^{J} (Y_{ij} - \mu - \alpha_i - \beta_j)^2$

Testing Hypotheses

In a problem involving a two-way layout, we are often interested in testing the hypothesis that one of the factors has no effect on the distribution of the observations. In other words, we are often interested either in testing the hypothesis that each of the effects $\alpha_1, \ldots, \alpha_I$ of factor A is equal to 0 or in testing the hypothesis that each of the effects $\beta_1, \ldots, \beta_J$ of factor B is equal to 0. For example, consider again the problem described at the beginning of this section, in which specimens of milk are collected from four different dairies and the concentration of a radioactive isotope in each specimen is measured by three different methods. If we regard the dairy as factor A and the method as factor B, then the hypothesis that $\alpha_i = 0$ for $i = 1, \ldots, I$ means that for any given method of measurement, the concentration of the isotope has the same distribution for all four dairies. In other words, there are no differences among the dairies. Similarly, the hypothesis that $\beta_j = 0$ for $j = 1, \ldots, J$ means that for each dairy, the three methods of measurement yield the same distribution for the concentration of the isotope. However, this hypothesis does not state that regardless of which of the three different methods is applied to a particular specimen of milk, the same value would be obtained. Because of the inherent variability of the measurements, the hypothesis states only that the values yielded by the three methods have the same normal distribution.

Consider now the following hypotheses:

$$H_0: \quad \alpha_i = 0 \quad \text{for } i = 1, \ldots, I,$$
$$H_1: \quad \text{The hypothesis } H_0 \text{ is not true.} \tag{12}$$

It can be seen from Eq. (11) that when the null hypothesis H_0 is true, the random variable Q_4 has the form $Q_4^0 = J\sum_{i=1}^{I}(\bar{Y}_{i+} - \bar{Y}_{++})^2/\sigma^2$, and Q_4^0 has a χ^2 distribution with $I - 1$ degrees of freedom. It follows therefore from the discussion leading up to Table 10.11 that when H_0 is true, the following random variable U_A^2 will have an F distribution with $I - 1$ and $(I - 1)(J - 1)$ degrees of freedom:

$$U_A^2 = \frac{Q_4^0/(I - 1)}{Q_2/[(I - 1)(J - 1)]} = \frac{(J - 1)Q_4^0}{Q_2} \tag{13}$$

or

$$U_A^2 = \frac{J(J - 1)\sum_{i=1}^{I}(\bar{Y}_{i+} - \bar{Y}_{++})^2}{\sum_{i=1}^{I}\sum_{j=1}^{J}(Y_{ij} - \bar{Y}_{i+} - \bar{Y}_{+j} + \bar{Y}_{++})^2}. \tag{14}$$

When the null hypothesis H_0 is not true, the value of $\alpha_i = E(\bar{Y}_{i+} - \bar{Y}_{++})$ is not 0 for at least one value of i. Hence, the expectation of the numerator of U_A^2 will be larger than it would be when H_0 is true. The distribution of the denominator of U_A^2 remains the same regardless of whether or not H_0 is true. It can be shown that the likelihood ratio test procedure for the hypotheses (12) specifies rejecting H_0 if $U_A^2 > c$, where c is a suitable constant whose value for any specified level of significance can be determined from a table of the F distribution with $I - 1$ and $(I - 1)(J - 1)$ degrees of freedom.

In some problems, a form for the statistic U_A^2 that is useful for computational purposes is as follows (see Exercises 9 and 10):

$$U_A^2 = \frac{J(J - 1)\left(\sum_{i=1}^{I}\bar{Y}_{i+}^2 - I\bar{Y}_{++}^2\right)}{\sum_{i=1}^{I}\sum_{j=1}^{J}Y_{ij}^2 - J\sum_{i=1}^{I}\bar{Y}_{i+}^2 - I\sum_{j=1}^{J}\bar{Y}_{+j}^2 + IJ\bar{Y}_{++}^2}. \tag{15}$$

Similarly, suppose next that the following hypotheses are to be tested:

$$H_0: \quad \beta_j = 0 \quad \text{for } j = 1, \ldots, J,$$
$$H_1: \quad \text{The hypothesis } H_0 \text{ is not true.} \tag{16}$$

When the null hypothesis H_0 is true, the following statistic U_B^2 will have an F distribution with $J - 1$ and $(I - 1)(J - 1)$ degrees of freedom:

$$U_B^2 = \frac{I(I - 1)\sum_{j=1}^{J}(\bar{Y}_{+j} - \bar{Y}_{++})^2}{\sum_{i=1}^{I}\sum_{j=1}^{J}(Y_{ij} - \bar{Y}_{i+} - \bar{Y}_{+j} + \bar{Y}_{++})^2}. \tag{17}$$

The hypothesis H_0 should be rejected if $U_B^2 > c$, where c is a suitable constant. An expression analogous to that in Eq. (15) can also be given for U_B^2.

Table 10.12

		Method		
		1	2	3
	1	6.4	3.2	6.9
Dairy	2	8.5	7.8	10.1
	3	9.3	6.0	9.6
	4	8.8	5.6	8.4

Example 1: Estimating the Parameters. Suppose that in the problem we have been discussing in this section, the concentrations of a radioactive isotope measured in picocuries per liter by three different methods in specimens of milk from four dairies are as given in Table 10.12. It is found from Table 10.12 that the row averages are $\overline{Y}_{1+} = 5.5$, $\overline{Y}_{2+} = 8.8$, $\overline{Y}_{3+} = 8.3$, and $\overline{Y}_{4+} = 7.6$; that the column averages are $\overline{Y}_{+1} = 8.25$, $\overline{Y}_{+2} = 5.65$, and $\overline{Y}_{+3} = 8.75$; and that the average of all the observations is $\overline{Y}_{++} = 7.55$. Hence, by Eq. (6), the values of the M.L.E.'s are $\hat{\mu} = 7.55$, $\hat{\alpha}_1 = -2.05$, $\hat{\alpha}_2 = 1.25$, $\hat{\alpha}_3 = 0.75$, $\hat{\alpha}_4 = 0.05$, $\hat{\beta}_1 = 0.70$, $\hat{\beta}_2 = -1.90$, and $\hat{\beta}_3 = 1.20$.

Because $E(Y_{ij}) = \mu + \alpha_i + \beta_j$, the M.L.E. of $E(Y_{ij})$ is $\overline{Y}_{i+} + \overline{Y}_{+j} - \overline{Y}_{++} = \hat{\mu} + \hat{\alpha}_i + \hat{\beta}_j$. The values of these estimates are given in Table 10.13. By comparing the observed values in Table 10.12 with the estimated expectations in Table 10.13, we see that the differences between corresponding terms are generally small. These small differences indicate that the model used in the two-way layout, which assumes the additivity of the effects of the two factors, provides a good fit for the observed values. Finally, it is found from Tables 10.12 and 10.13 that

$$\sum_{i=1}^{I} \sum_{j=1}^{J} \left(Y_{ij} - \overline{Y}_{i+} - \overline{Y}_{+j} + \overline{Y}_{++} \right)^2 = 2.74.$$

Hence, by Eq. (7), $\hat{\sigma}^2 = 2.74/12 = 0.228$. □

Table 10.13

		Method		
		1	2	3
	1	6.2	3.6	6.7
Dairy	2	9.5	6.9	10.0
	3	9.0	6.4	9.5
	4	8.3	5.7	8.8

Example 2: Testing for Differences Among the Dairies. Suppose now that it is desired to use the observed values in Table 10.12 to test the hypothesis that there are no differences among the dairies, that is, to test the hypotheses (12). In this example, the statistic U_A^2 defined by Eq. (14) has an F distribution with 3 and 6 degrees of freedom. We find that $U_A^2 = 13.86$. Since the corresponding tail area is much smaller than 0.025, the hypothesis that there are no differences among the dairies would be rejected at the level of significance 0.025 or larger.   □

Example 3: Testing for Differences Among the Methods of Measurement. Suppose next that it is desired to use the observed values in Table 10.12 to test the hypothesis that each of the effects of the different methods of measurement is equal to 0, that is, to test the hypotheses (16). In this example, the statistic U_B^2 defined by Eq. (17) has an F distribution with 2 and 6 degrees of freedom. We find that $U_B^2 = 24.26$. Since the corresponding tail area is again much less than 0.025, the hypothesis that there are no differences among the methods would be rejected at the level of significance 0.025 or larger.   □

EXERCISES

1. Consider a two-way layout in which the values of $E(Y_{ij})$ for $i = 1, \ldots, I$ and $j = 1, \ldots, J$ are as given in each of the following four matrices. For each matrix, state whether or not the effects of the factors are additive.

(a)

Factor B

	1	2
Factor A 1	5	7
2	10	14

(b)

Factor B

	1	2
Factor A 1	3	6
2	4	7

(c)

Factor B

	1	2	3	4
Factor A 1	3	−1	0	3
2	8	4	5	8
3	4	0	1	4

(d)

Factor B

	1	2	3	4
Factor A 1	1	2	3	4
2	2	4	6	8
3	3	6	9	12

2. Show that if the effects of the factors in a two-way layout are additive, then there exist unique numbers μ, $\alpha_1, \ldots, \alpha_I$, and $\beta_1, \ldots, \beta_J$ that satisfy Eqs. (2) and (3).

3. Suppose that in a two-way layout, with $I = 2$ and $J = 2$, the values of $E(Y_{ij})$ are as given in part (b) of Exercise 1. Determine the values of μ, α_1, α_2, β_1, and β_2 that satisfy Eqs. (2) and (3).

4. Suppose that in a two-way layout, with $I = 3$ and $J = 4$, the values of $E(Y_{ij})$ are as given in part (c) of Exercise 1. Determine the values of μ, α_1, α_2, α_3, and $\beta_1, \ldots, \beta_4$ that satisfy Eqs. (2) and (3).

5. Verify that if $\hat{\mu}$, $\hat{\alpha}_i$, and $\hat{\beta}_j$ are defined by Eq. (6), then $\sum_{i=1}^{I} \hat{\alpha}_i = \sum_{j=1}^{J} \hat{\beta}_j = 0$; $E(\hat{\mu}) = \mu$; $E(\hat{\alpha}_i) = \alpha_i$ for $i = 1, \ldots, I$; and $E(\hat{\beta}_j) = \beta_j$ for $j = 1, \ldots, J$.

6. Show that if $\hat{\mu}$, $\hat{\alpha}_i$, and $\hat{\beta}_j$ are defined by Eq. (6), then

$$\mathrm{Var}(\hat{\mu}) = \frac{1}{IJ}\sigma^2,$$

$$\mathrm{Var}(\hat{\alpha}_i) = \frac{I-1}{IJ}\sigma^2 \qquad \text{for } i = 1, \ldots, I,$$

$$\mathrm{Var}(\hat{\beta}_j) = \frac{J-1}{IJ}\sigma^2 \qquad \text{for } j = 1, \ldots, J.$$

7. Show that the right sides of Eqs. (9) and (10) are equal.

8. Show that in a two-way layout, for any values of i, j, i', and j' (i and $i' = 1, \ldots, I$; j and $j' = 1, \ldots, J$), the following four random variables W_1, W_2, W_3, and W_4 are uncorrelated with one another:

$$W_1 = Y_{ij} - \bar{Y}_{i+} - \bar{Y}_{+j} + \bar{Y}_{++},$$

$$W_2 = \bar{Y}_{i'+} - \bar{Y}_{++}, \qquad W_3 = \bar{Y}_{+j'} - \bar{Y}_{++},$$

$$W_4 = \bar{Y}_{++}.$$

9. Show that

$$\sum_{i=1}^{I} (\bar{Y}_{i+} - \bar{Y}_{++})^2 = \sum_{i=1}^{I} \bar{Y}_{i+}^2 - I\bar{Y}_{++}^2$$

and

$$\sum_{j=1}^{J} (\bar{Y}_{+j} - \bar{Y}_{++})^2 = \sum_{j=1}^{J} \bar{Y}_{+j}^2 - J\bar{Y}_{++}^2.$$

Table 10.14

		Type of surface				
		1	2	3	4	5
Type of paint	1	14.5	13.6	16.3	23.2	19.4
	2	14.6	16.2	14.8	16.8	17.3
	3	16.2	14.0	15.5	18.7	21.0

10. Show that

$$\sum_{i=1}^{I}\sum_{j=1}^{J}\left(Y_{ij}-\overline{Y}_{i+}-\overline{Y}_{+j}+\overline{Y}_{++}\right)^2$$

$$=\sum_{i=1}^{I}\sum_{j=1}^{J}Y_{ij}^2-J\sum_{i=1}^{I}\overline{Y}_{i+}^2-I\sum_{j=1}^{J}\overline{Y}_{+j}^2+IJ\overline{Y}_{++}^2.$$

11. In a study to compare the reflective properties of various paints and various plastic surfaces, three different types of paint were applied to specimens of five different types of plastic surfaces. Suppose that the observed results in appropriate coded units were as given in Table 10.14. Determine the values of $\hat{\mu}$, $\hat{\alpha}_1$, $\hat{\alpha}_2$, $\hat{\alpha}_3$, and $\hat{\beta}_1, \ldots, \hat{\beta}_5$.

12. For the conditions of Exercise 11 and the data in Table 10.14, determine the value of the least-squares estimate of $E(Y_{ij})$ for $i = 1, 2, 3$, and $j = 1, \ldots, 5$; and determine the value of $\hat{\sigma}^2$.

13. For the conditions of Exercise 11 and the data in Table 10.14, test the hypothesis that the reflective properties of the three different types of paint are the same.

14. For the conditions of Exercise 11 and the data in Table 10.14, test the hypothesis that the reflective properties of the five different types of plastic surfaces are the same.

10.8. THE TWO-WAY LAYOUT WITH REPLICATIONS

The Two-Way Layout with K Observations in Each Cell

We shall continue to consider problems of ANOVA involving a two-way layout. Now, however, instead of having just a single observation Y_{ij} for each combination of i and j, we shall have K independent observations Y_{ijk} for $k = 1, \ldots, K$.

In other words, instead of having just one observation in each cell of Table 10.10, we have K i.i.d. observations. The K observations in each cell are obtained under similar experimental conditions and are called *replications*. The total number of observations in this two-way layout with replications is IJK. We continue to assume that all the observations are independent; that each observation has a normal distribution; and that all the observations have the same variance σ^2.

We shall let θ_{ij} denote the mean of each of the K observations in the (i, j) cell. Thus, for $i = 1, \ldots, I$, $j = 1, \ldots, J$, and $k = 1, \ldots, K$, we have

$$E(Y_{ijk}) = \theta_{ij}. \tag{1}$$

In a two-way layout with replications, it is not necessary to assume, as we did in Sec. 10.7, that the effects of the two factors are additive. Here we can assume that the expectations θ_{ij} are arbitrary numbers. As we shall see later in this section, we can then test the hypothesis that the effects are additive.

It is easy to verify that the M.L.E., or least-squares estimator, of θ_{ij} is simply the sample mean of the K observations in the (i, j) cell. Thus,

$$\hat{\theta}_{ij} = \frac{1}{K} \sum_{k=1}^{K} Y_{ijk} = \overline{Y}_{ij+}. \tag{2}$$

The M.L.E. of σ^2 is therefore

$$\hat{\sigma}^2 = \frac{1}{IJK} \sum_{i=1}^{I} \sum_{j=1}^{J} \sum_{k=1}^{K} \left(Y_{ijk} - \overline{Y}_{ij+} \right)^2. \tag{3}$$

In order to identify and discuss the effects of the two factors, and to examine the possibility that these effects are additive, it is helpful to replace the parameters θ_{ij}, for $i = 1, \ldots, I$ and $j = 1, \ldots, J$, with a new set of parameters μ, α_i, β_j, and γ_{ij}. These new parameters are defined by the following relations:

$$\theta_{ij} = \mu + \alpha_i + \beta_j + \gamma_{ij} \qquad \text{for } i = 1, \ldots, I \text{ and } j = 1, \ldots, J, \tag{4}$$

and

$$\sum_{i=1}^{I} \alpha_i = 0, \qquad \sum_{j=1}^{J} \beta_j = 0,$$

$$\sum_{i=1}^{I} \gamma_{ij} = 0 \qquad \text{for } j = 1, \ldots, J, \tag{5}$$

$$\sum_{j=1}^{J} \gamma_{ij} = 0 \qquad \text{for } i = 1, \ldots, I.$$

It can be shown (see Exercise 1) that corresponding to any given numbers θ_{ij} for $i = 1, \ldots, I$ and $j = 1, \ldots, J$, there exist unique numbers μ, α_i, β_j, and γ_{ij} that satisfy Eqs. (4) and (5).

The parameter μ is called the *overall mean* or the *grand mean*. The parameters $\alpha_1, \ldots, \alpha_I$ are called the *main effects of factor A*, and the parameters $\beta_1, \ldots, \beta_J$ are called the *main effects of factor B*. The parameters γ_{ij}, for $i = 1, \ldots, I$ and $j = 1, \ldots, J$, are called the *interactions*. It can be seen from Eqs. (1) and (4) that the effects of the factors A and B are additive if and only if all the interactions vanish, that is, if and only if $\gamma_{ij} = 0$ for every combination of values of i and j.

The notation which has been developed in Secs. 10.6 and 10.7 will again be used here. We shall replace a subscript of Y_{ijk} with a plus sign to indicate that we have summed the values of Y_{ijk} over all possible values of that subscript. If we have made two or three summations, we shall use two or three plus signs. We shall then place a bar over Y to indicate that we have divided this sum by the number of terms in the summation and have thereby obtained an average of the values of Y_{ijk} for the subscript or subscripts involved in the summation. For example,

$$\overline{Y}_{i+k} = \frac{1}{J} \sum_{j=1}^{J} Y_{ijk},$$

$$\overline{Y}_{+j+} = \frac{1}{IK} \sum_{i=1}^{I} \sum_{k=1}^{K} Y_{ijk},$$

and $\overline{Y}_{+++}$ denotes the average of all IJK observations.

It can be shown that the M.L.E.'s, or least-squares estimators, of μ, α_i, and β_j are as follows:

$$\hat{\mu} = \overline{Y}_{+++},$$

$$\hat{\alpha}_i = \overline{Y}_{i++} - \overline{Y}_{+++} \qquad \text{for } i = 1, \ldots, I, \tag{6}$$

$$\hat{\beta}_j = \overline{Y}_{+j+} - \overline{Y}_{+++} \qquad \text{for } j = 1, \ldots, J.$$

Also, for $i = 1, \ldots, I$ and $j = 1, \ldots, J$,

$$\hat{\gamma}_{ij} = \overline{Y}_{ij+} - \left(\hat{\mu} + \hat{\alpha}_i + \hat{\beta}_j \right)$$

$$= \overline{Y}_{ij+} - \overline{Y}_{i++} - \overline{Y}_{+j+} + \overline{Y}_{+++}. \tag{7}$$

For all values of i and j, it can then be verified (see Exercise 3) that $E(\hat{\mu}) = \mu$, $E(\hat{\alpha}_i) = \alpha_i$, $E(\hat{\beta}_j) = \beta_j$, and $E(\hat{\gamma}_{ij}) = \gamma_{ij}$.

Partitioning the Sum of Squares

Consider now the following sum Q_1:

$$Q_1 = \sum_{i=1}^{I} \sum_{j=1}^{J} \sum_{k=1}^{K} \frac{\left(Y_{ijk} - \mu - \alpha_i - \beta_j - \gamma_{ij}\right)^2}{\sigma^2}. \tag{8}$$

Since Q_1 is the sum of squares of IJK independent random variables, each of which has a standard normal distribution, then Q_1 has a χ^2 distribution with IJK degrees of freedom. We shall now indicate how Q_1 can be partitioned into five smaller independent sums of squares, each of which itself has a χ^2 distribution and each of which is associated with a particular type of variation among the observations.

These sums of squares are defined as follows:

$$Q_2 = \frac{1}{\sigma^2} \sum_{i=1}^{I} \sum_{j=1}^{J} \sum_{k=1}^{K} \left(Y_{ijk} - \hat{\mu} - \hat{\alpha}_i - \hat{\beta}_j - \hat{\gamma}_{ij}\right)^2,$$

$$Q_3 = \frac{IJK}{\sigma^2} (\hat{\mu} - \mu)^2,$$

$$Q_4 = \frac{JK}{\sigma^2} \sum_{i=1}^{I} (\hat{\alpha}_i - \alpha_i)^2, \tag{9}$$

$$Q_5 = \frac{IK}{\sigma^2} \sum_{j=1}^{J} (\hat{\beta}_j - \beta_j)^2,$$

$$Q_6 = \frac{K}{\sigma^2} \sum_{i=1}^{I} \sum_{j=1}^{J} (\hat{\gamma}_{ij} - \gamma_{ij})^2.$$

It can be shown (see Exercise 6) that

$$Q_1 = Q_2 + Q_3 + Q_4 + Q_5 + Q_6. \tag{10}$$

Each random variable Q_i has a χ^2 distribution. Since $\hat{\mu}$ has a normal distribution with mean μ and variance $\sigma^2/(IJK)$, it follows that Q_3 has one degree of freedom. Also, Q_4 has $I - 1$ degrees of freedom, and Q_5 has $J - 1$ degrees of freedom.

The number of degrees of freedom for Q_6 can be determined as follows: Although there are IJ estimators $\hat{\gamma}_{ij}$ of the interactions, these estimators satisfy the following $I + J$ equations (see Exercise 3):

$$\sum_{j=1}^{J} \hat{\gamma}_{ij} = 0 \qquad \text{for } i = 1, \ldots, I,$$

$$\sum_{i=1}^{I} \hat{\gamma}_{ij} = 0 \qquad \text{for } j = 1, \ldots, J.$$

However, if any $I + J - 1$ of these $I + J$ equations are satisfied, then the remaining equation must also be satisfied. Therefore, since the IJ estimators $\hat{\gamma}_{ij}$ actually must satisfy $I + J - 1$ constraints, we are, in effect, estimating only $IJ - (I + J - 1) = (I - 1)(J - 1)$ interactions. It can be shown that Q_6 has a χ^2 distribution with $(I - 1)(J - 1)$ degrees of freedom.

It remains to determine the degrees of freedom for Q_2. Since Q_1 has IJK degrees of freedom and the sum of the degrees of freedom for Q_3, Q_4, Q_5, and Q_6 is IJ, it is anticipated that the degrees of freedom remaining for Q_2 must be $IJK - IJ = IJ(K - 1)$. Since Q_2 is a sum over IJ independent samples, each of size K, it can be shown that Q_2 has a χ^2 distribution with $IJ(K - 1)$ degrees of freedom.

Finally, it can be shown that the random variables Q_2, Q_3, Q_4, Q_5, and Q_6 are independent (see Exercise 7 for a related result).

These properties are summarized in Table 10.15, which is the ANOVA table for the two-way layout with K observations per cell. As discussed in Sec. 10.6,

Table 10.15

Source of variation	Degrees of freedom	Sum of squares
Overall mean	1	$IJK(\overline{Y}_{+++} - \mu)^2$
Main effects of A	$I - 1$	$JK \sum\limits_{i=1}^{I} (\overline{Y}_{i++} - \overline{Y}_{+++} - \alpha_i)^2$
Main effects of B	$J - 1$	$IK \sum\limits_{j=1}^{J} (\overline{Y}_{+j+} - \overline{Y}_{+++} - \beta_j)^2$
Interactions	$(I - 1)(J - 1)$	$K \sum\limits_{i=1}^{I} \sum\limits_{j=1}^{J} (\overline{Y}_{ij+} - \overline{Y}_{i++} - \overline{Y}_{+j+} + \overline{Y}_{+++} - \gamma_{ij})^2$
Residuals	$IJ(K - 1)$	$\sum\limits_{i=1}^{I} \sum\limits_{j=1}^{J} \sum\limits_{k=1}^{K} (Y_{ijk} - \overline{Y}_{ij+})^2$
Total	IJK	$\sum\limits_{i=1}^{I} \sum\limits_{j=1}^{J} \sum\limits_{k=1}^{K} (Y_{ijk} - \mu - \alpha_i - \beta_j - \gamma_{ij})^2$

this table is different from, and more flexible than, the ANOVA tables presented in most textbooks, because the parameters are included in the sums of squares given in this table.

Testing Hypotheses

As mentioned before, the effects of the factors A and B are additive if and only if all the interactions γ_{ij} vanish. Hence, to test whether the effects of the factors are additive, we must test the following hypotheses:

$$H_0: \quad \gamma_{ij} = 0 \quad \text{for } i = 1, \ldots, I \text{ and } j = 1, \ldots, J,$$
$$H_1: \quad \text{The hypothesis } H_0 \text{ is not true.} \tag{11}$$

It follows from Table 10.15 and the discussion leading up to it that when the null hypothesis H_0 is true, the random variable $K \sum_{i=1}^{I} \sum_{j=1}^{J} \hat{\gamma}_{ij}^2 / \sigma^2$ has a χ^2 distribution with $(I-1)(J-1)$ degrees of freedom. Furthermore, regardless of whether or not H_0 is true, the independent random variable

$$\frac{\sum_{i=1}^{I} \sum_{j=1}^{J} \sum_{k=1}^{K} \left(Y_{ijk} - \overline{Y}_{ij+} \right)^2}{\sigma^2}$$

has a χ^2 distribution with $IJ(K-1)$ degrees of freedom. Thus, when H_0 is true, the following random variable U_{AB}^2 has an F distribution with $(I-1)(J-1)$ and $IJ(K-1)$ degrees of freedom:

$$U_{AB}^2 = \frac{IJK(K-1) \sum_{i=1}^{I} \sum_{j=1}^{J} \left(\overline{Y}_{ij+} - \overline{Y}_{i++} - \overline{Y}_{+j+} + \overline{Y}_{+++} \right)^2}{(I-1)(J-1) \sum_{i=1}^{I} \sum_{j=1}^{J} \sum_{k=1}^{K} \left(Y_{ijk} - \overline{Y}_{ij+} \right)^2}. \tag{12}$$

The null hypothesis H_0 should be rejected if $U_{AB}^2 > c$, where c is a suitable constant which is chosen to obtain any specified level of significance. An alternate form for U_{AB}^2 that is sometimes useful for computation is (see Exercise 8):

$$U_{AB}^2 = \frac{IJK(K-1) \left(\sum_{i=1}^{I} \sum_{j=1}^{J} \overline{Y}_{ij+}^2 - J \sum_{i=1}^{I} \overline{Y}_{i++}^2 - I \sum_{j=1}^{J} \overline{Y}_{+j+}^2 + IJ \overline{Y}_{+++}^2 \right)}{(I-1)(J-1) \left(\sum_{i=1}^{I} \sum_{j=1}^{J} \sum_{k=1}^{K} Y_{ijk}^2 - K \sum_{i=1}^{I} \sum_{j=1}^{J} \overline{Y}_{ij+}^2 \right)}.$$

$$\tag{13}$$

If the null hypothesis H_0 in (11) is rejected, then we conclude that at least some of the interactions γ_{ij} are not 0. Therefore, the means of the observations for certain combinations of i and j will be larger than the means of the observations for other combinations, and both factor A and factor B affect these means. In this case, since both factor A and factor B affect the means of the observations, there is not usually any further interest in testing whether either the main effects $\alpha_1, \ldots, \alpha_I$ or the main effects $\beta_1, \ldots, \beta_J$ are zero.

On the other hand, if the null hypothesis H_0 in (11) is not rejected, then it is possible that all the interactions are 0. If, in addition, all the main effects $\alpha_1, \ldots, \alpha_I$ were 0, then the mean value of each observation would not depend in any way on the value of i. In this case, factor A would have no effect on the observations. Therefore, if the null hypothesis H_0 in (11) is not rejected, we might be interested in testing the following hypotheses:

$$H_0: \quad \alpha_i = 0 \text{ and } \gamma_{ij} = 0 \qquad \text{for } i = 1, \ldots, I \text{ and } j = 1, \ldots, J,$$
$$H_1: \quad \text{The hypothesis } H_0 \text{ is not true.} \tag{14}$$

It follows from Table 10.15 and the discussion leading up to it that when H_0 is true, the following random variable will have a χ^2 distribution with $(I - 1) + (I - 1)(J - 1) = (I - 1)J$ degrees of freedom:

$$\frac{JK \sum_{i=1}^{I} \hat{\alpha}_i^2 + K \sum_{i=1}^{I} \sum_{j=1}^{J} \hat{\gamma}_{ij}^2}{\sigma^2}. \tag{15}$$

Also, regardless of whether or not H_0 is true, the independent random variable $\sum_{i=1}^{I} \sum_{j=1}^{J} \sum_{k=1}^{K} (Y_{ijk} - \overline{Y}_{ij+})^2 / \sigma^2$ has a χ^2 distribution with $IJ(K - 1)$ degrees of freedom. Hence, when H_0 is true, the following random variable U_A^2 has an F distribution with $(I - 1)J$ and $IJ(K - 1)$ degrees of freedom:

$$U_A^2 =$$

$$\frac{IK(K - 1)\left[J\sum_{i=1}^{I}(\overline{Y}_{i++} - \overline{Y}_{+++})^2 + \sum_{i=1}^{I}\sum_{j=1}^{J}(\overline{Y}_{ij+} - \overline{Y}_{i++} - \overline{Y}_{+j+} + \overline{Y}_{+++})^2 \right]}{(I - 1)\sum_{i=1}^{I}\sum_{j=1}^{J}\sum_{k=1}^{K}(Y_{ijk} - \overline{Y}_{ij+})^2}. \tag{16}$$

The null hypothesis H_0 should be rejected if $U_A^2 > c$, where c is a suitable constant.

Similarly, we may want to find out whether all the main effects of factor B, as well as the interactions, are 0. In this case, we would test the following hypotheses:

$$H_0: \quad \beta_j = 0 \text{ and } \gamma_{ij} = 0 \qquad \text{for } i = 1, \ldots, I \text{ and } j = 1, \ldots, J,$$
$$H_1: \quad \text{The hypothesis } H_0 \text{ is not true.} \tag{17}$$

By analogy with Eq. (16), it follows that when H_0 is true, the following random variable U_B^2 has an F distribution with $I(J-1)$ and $IJ(K-1)$ degrees of freedom:

$$U_B^2 =$$

$$\frac{JK(K-1)\left[I\sum_{j=1}^{J}\left(\overline{Y}_{+j+}-\overline{Y}_{+++}\right)^2 + \sum_{i=1}^{I}\sum_{j=1}^{J}\left(\overline{Y}_{ij+}-\overline{Y}_{i++}-\overline{Y}_{+j+}+\overline{Y}_{+++}\right)^2\right]}{(J-1)\sum_{i=1}^{I}\sum_{j=1}^{J}\sum_{k=1}^{K}\left(Y_{ijk}-\overline{Y}_{ij+}\right)^2}.$$

$$(18)$$

Again, the hypothesis H_0 should be rejected if $U_B^2 > c$.

In a given problem, if the null hypothesis in (11) is not rejected and the null hypotheses in both (14) and (17) are rejected, then we may be willing to proceed with further studies and experimentation by using a model in which it is assumed that the effects of factor A and factor B are approximately additive and the effects of both factors are important.

One further consideration should be emphasized. Suppose that the hypotheses (14) or the hypotheses (17) are tested after the null hypothesis in (11) has been accepted at some given level of significance α_0. Then the size of this second test should no longer be regarded simply as the usual value α chosen by the experimenter. More appropriately, the size should now be regarded as the *conditional* probability that H_0 in (14) or (17) will be rejected by a test procedure of nominal size α, given that H_0 is true *and* the sample data are such that the null hypothesis in (11) was accepted by the first test.

Example 1: Estimating the Parameters in a Two-Way Layout with Replications. Suppose that an experiment is carried out by an automobile manufacturer to investigate whether a certain device, installed on the carburetor of an automobile, affects the amount of gasoline consumed by the automobile. The manufacturer produces three different models of automobiles, namely, a compact model, an intermediate model, and a standard model. Five cars of each model, which were equipped with this device, were driven over a fixed route through city traffic; and the gasoline consumption of each car was measured. Also, five cars of each model, which were not equipped with this device, were driven over the same route; and the gasoline consumption of each of these cars was measured. The results, in liters of gasoline consumed, are given in Table 10.16.

In this example, $I = 2$, $J = 3$, and $K = 5$. The average value $\overline{Y}_{ij+}$ for each of the six cells in Table 10.16 is presented in Table 10.17, which also gives the average value $\overline{Y}_{i++}$ for each of the two rows, the average value $\overline{Y}_{+j+}$ for each of the three columns, and the average value $\overline{Y}_{+++}$ of all 30 observations.

Table 10.16

	Compact model	Intermediate model	Standard model
Equipped with device	8.3 8.9 7.8 8.5 9.4	9.2 10.2 9.5 11.3 10.4	11.6 10.2 10.7 11.9 11.0
Not equipped with device	8.7 10.0 9.7 7.9 8.4	8.2 10.6 10.1 11.3 10.8	12.4 11.7 10.0 11.1 11.8

Table 10.17

	Compact model	Intermediate model	Standard model	Average for row
Equipped with device	$\overline{Y}_{11+} = 8.58$	$\overline{Y}_{12+} = 10.12$	$\overline{Y}_{13+} = 11.08$	$\overline{Y}_{1++} = 9.9267$
Not equipped with device	$\overline{Y}_{21+} = 8.94$	$\overline{Y}_{22+} = 10.20$	$\overline{Y}_{23+} = 11.40$	$\overline{Y}_{2++} = 10.1800$
Average for column	$\overline{Y}_{+1+} = 8.76$	$\overline{Y}_{+2+} = 10.16$	$\overline{Y}_{+3+} = 11.24$	$\overline{Y}_{+++} = 10.0533$

It follows from Table 10.17 and Eqs. (6) and (7) that the values of the M.L.E.'s, or least-squares estimators, in this example are:

$$\hat{\mu} = 10.0533, \qquad \hat{\alpha}_1 = -0.1267, \qquad \hat{\alpha}_2 = 0.1267,$$

$$\hat{\beta}_1 = -1.2933, \qquad \hat{\beta}_2 = 0.1067, \qquad \hat{\beta}_3 = 1.1867,$$

$$\hat{\gamma}_{11} = -0.0533, \qquad \hat{\gamma}_{12} = 0.0867, \qquad \hat{\gamma}_{13} = -0.0333,$$

$$\hat{\gamma}_{21} = 0.0533, \qquad \hat{\gamma}_{22} = -0.0867, \qquad \hat{\gamma}_{23} = 0.0333.$$

In this example, the estimates of the interactions $\hat{\gamma}_{ij}$ are small for all values of i and j. □

Example 2: Testing for Additivity. Suppose now that it is desired to use the observed values in Table 10.16 to test the null hypothesis that the effects of equipping a car with the device and of using a particular model are additive, against the alternative that these effects are not additive. In other words, suppose that it is desired to test the hypotheses (11). It is found from Eq. (12) that $U_{AB}^2 = 0.076$. The corresponding tail area, as found from a table of the F distribution with 2 and 24 degrees of freedom, is much larger than 0.05. Hence, the null hypothesis that the effects are additive would not be rejected at the usual levels of significance. □

Example 3: Testing for an Effect on Gasoline Consumption. Suppose next that it is desired to test the null hypothesis that the device has no effect on gasoline consumption for any of the models tested, against the alternative that the device does affect gasoline consumption. In other words, suppose that it is desired to test the hypotheses (14). It is found from Eq. (16) that $U_A^2 = 0.262$. The corresponding tail area, as found from a table of the F distribution with 3 and 24 degrees of freedom, is much larger than 0.05. Hence, the null hypothesis would not be rejected at the usual levels of significance. Of course, this analysis does not take into account the conditioning effect, described just before Example 1, of testing the hypotheses (14) after first testing the hypotheses (11) with the same data. □

The results obtained in Example 3 do not provide any indication that the device is effective. Nevertheless, it can be seen from Table 10.17 that for each of the three models, the average consumption of gasoline for the cars that were equipped with the device is smaller than the average consumption for the cars that were not so equipped. If we assume that the effects of the device and the model of automobile are additive, then regardless of the model of the automobile that is used, the M.L.E. of the reduction in gasoline consumption over the given route that is achieved by equipping an automobile with the device is $\hat{\alpha}_2 - \hat{\alpha}_1 = 0.2534$ liter.

The Two-Way Layout with Unequal Numbers of Observations in the Cells

Consider again a two-way layout with I rows and J columns; but suppose now that instead of there being K observations in each cell, some cells have more observations than others. For $i = 1, \ldots, I$ and $j = 1, \ldots, J$, we shall let K_{ij} denote the number of observations in the (i, j) cell. Thus, the total number of observations is $\sum_{i=1}^{I} \sum_{j=1}^{J} K_{ij}$. We shall assume that every cell contains at least one observation, and we shall again let Y_{ijk} denote the kth observation in the (i, j)

cell. For any given values of i and j, the values of the subscript k are $1, \ldots, K_{ij}$. We shall also assume, as before, that all the observations Y_{ijk} are independent; that each has a normal distribution; that $\mathrm{Var}(Y_{ijk}) = \sigma^2$ for all values of i, j, and k; and that $E(Y_{ijk}) = \mu + \alpha_i + \beta_j + \gamma_{ij}$, where these parameters satisfy the conditions given in Eq. (5).

As usual, we shall let $\overline{Y}_{ij+}$ denote the average of the observations in the (i, j) cell. It can then be shown that for $i = 1, \ldots, I$ and $j = 1, \ldots, J$, the M.L.E.'s, or least-squares estimators, are as follows:

$$\hat{\mu} = \frac{1}{IJ} \sum_{i=1}^{I} \sum_{j=1}^{J} \overline{Y}_{ij+}, \qquad \hat{\alpha}_i = \frac{1}{J} \sum_{j=1}^{J} \overline{Y}_{ij+} - \hat{\mu},$$

$$\hat{\beta}_j = \frac{1}{I} \sum_{i=1}^{I} \overline{Y}_{ij+} - \hat{\mu}, \qquad \hat{\gamma}_{ij} = \overline{Y}_{ij+} - \hat{\mu} - \hat{\alpha}_i - \hat{\beta}_j.$$

(19)

These estimators are intuitively reasonable and analogous to those given in Eqs. (6) and (7).

Suppose now, however, that it is desired to test hypotheses such as (11), (14), or (17). The construction of appropriate tests becomes somewhat more difficult because, in general, the sums of squares analogous to those given in Eq. (9) will not be independent when there are unequal numbers of observations in the different cells. Hence, the test procedures presented earlier in this section cannot directly be copied here. It is necessary to develop other sums of squares that will be independent and that will reflect the different types of variations in the data in which we are interested. We shall not consider this problem further in this book. This problem and other problems of ANOVA are described in the advanced book by Scheffé (1959).

EXERCISES

1. Show that for any given set of numbers θ_{ij} ($i = 1, \ldots, I$ and $j = 1, \ldots, J$), there exists a unique set of numbers μ, α_i, β_j, and γ_{ij} ($i = 1, \ldots, I$ and $j = 1, \ldots, J$) that satisfy Eqs. (4) and (5).

2. Suppose that in a two-way layout, the values of θ_{ij} are as given in each of the four matrices presented in parts (a), (b), (c), and (d) of Exercise 1 of Sec. 10.7. For each matrix, determine the values of μ, α_i, β_j, and γ_{ij} that satisfy Eqs. (4) and (5).

3. Verify that if $\hat{\alpha}_i$, $\hat{\beta}_j$, and $\hat{\gamma}_{ij}$ are as given by Eqs. (6) and (7), then $\sum_{i=1}^{I} \hat{\alpha}_i = 0$; $\sum_{j=1}^{J} \hat{\beta}_j = 0$; $\sum_{i=1}^{I} \hat{\gamma}_{ij} = 0$ for $j = 1, \ldots, J$; and $\sum_{j=1}^{J} \hat{\gamma}_{ij} = 0$ for $i = 1, \ldots, I$.

4. Verify that if $\hat{\mu}$, $\hat{\alpha}_i$, $\hat{\beta}_j$, and $\hat{\gamma}_{ij}$ are as given by Eqs. (6) and (7), then $E(\hat{\mu}) = \mu$; $E(\hat{\alpha}_i) = \alpha_i$; $E(\hat{\beta}_j) = \beta_j$; and $E(\hat{\gamma}_{ij}) = \gamma_{ij}$ for all values of i and j.

5. Show that if $\hat{\mu}$, $\hat{\alpha}_i$, $\hat{\beta}_j$, and $\hat{\gamma}_{ij}$ are as given by Eqs. (6) and (7), then the following results are true for all values of i and j:

$$\mathrm{Var}(\hat{\mu}) = \frac{I}{IJK}\sigma^2, \qquad \mathrm{Var}(\hat{\alpha}_i) = \frac{(I-1)}{IJK}\sigma^2,$$

$$\mathrm{Var}(\hat{\beta}_j) = \frac{(J-1)}{IJK}\sigma^2, \qquad \mathrm{Var}(\hat{\gamma}_{ij}) = \frac{(I-1)(J-1)}{IJK}\sigma^2.$$

6. Verify Eq. (10).

7. In a two-way layout with K observations in each cell, show that for any values of i, i_1, i_2, j, j_1, j_2, and k, the following five random variables are uncorrelated with one another:

$$Y_{ijk} - \overline{Y}_{ij+}, \hat{\alpha}_{j_1}, \hat{\beta}_{j_1}, \hat{\gamma}_{i_2 j_2}, \text{ and } \hat{\mu}.$$

8. Verify the fact that the right sides of Eqs. (12) and (13) are equal.

9. Suppose that in an experimental study to determine the combined effects of receiving both a stimulant and a tranquilizer, three different types of stimulants and four different types of tranquilizers are administered to a group of rabbits. Each rabbit in the experiment receives one of the stimulants and then, 20 minutes later, receives one of the tranquilizers. After 1 hour, the response of the rabbit is measured in appropriate units. In order that each possible pair of drugs may be administered to two different rabbits, 24 rabbits are used in the experiment. The responses of these 24 rabbits are given in Table 10.18. Determine the values of $\hat{\mu}$, $\hat{\alpha}_i$, $\hat{\beta}_j$, and $\hat{\gamma}_{ij}$ for $i = 1, 2, 3$ and $j = 1, 2, 3, 4$, and determine also the value of $\hat{\sigma}^2$.

10. For the conditions of Exercise 9 and the data in Table 10.18, test the hypothesis that every interaction between a stimulant and a tranquilizer is 0.

11. For the conditions of Exercise 9 and the data in Table 10.18, test the hypothesis that all three stimulants yield the same responses.

12. For the conditions of Exercise 9 and the data in Table 10.18, test the hypothesis that all four tranquilizers yield the same responses.

13. For the conditions of Exercise 9 and the data in Table 10.18, test the following hypotheses:

$$H_0: \quad \mu = 8,$$
$$H_1: \quad \mu \neq 8.$$

Table 10.18

		Tranquilizer			
		1	2	3	4
Stimulant	1	11.2	7.4	7.1	9.6
		11.6	8.1	7.0	7.6
	2	12.7	10.3	8.8	11.3
		14.0	7.9	8.5	10.8
	3	10.1	5.5	5.0	6.5
		9.6	6.9	7.3	5.7

14. For the conditions of Exercise 9 and the data in Table 10.18, test the following hypotheses:

$H_0: \quad \alpha_2 \leqslant 1,$
$H_1: \quad \alpha_2 > 1.$

15. In a two-way layout with unequal numbers of observations in the cells, show that if $\hat{\mu}$, $\hat{\alpha}_i$, $\hat{\beta}_j$, and $\hat{\gamma}_{ij}$ are as given by Eq. (19), then $E(\hat{\mu}) = \mu$, $E(\hat{\alpha}_i) = \alpha_i$, $E(\hat{\beta}_j) = \beta_j$, and $E(\hat{\gamma}_{ij}) = \gamma_{ij}$ for all values of i and j.

16. Verify that if $\hat{\mu}$, $\hat{\alpha}_i$, $\hat{\beta}_j$, and $\hat{\gamma}_{ij}$ are as given by Eq. (19), then $\sum_{i=1}^{I} \hat{\alpha}_i = 0$; $\sum_{j=1}^{J} \hat{\beta}_j = 0$; $\sum_{i=1}^{I} \hat{\gamma}_{ij} = 0$ for $j = 1, \dots, J$; and $\sum_{j=1}^{J} \hat{\gamma}_{ij} = 0$ for $i = 1, \dots, I$.

17. Show that if $\hat{\mu}$ and $\hat{\alpha}_i$ are as given by Eq. (19), then for $i = 1, \dots, I$,

$$\text{Cov}(\hat{\mu}, \hat{\alpha}_i) = \frac{\sigma^2}{IJ^2} \left[\sum_{j=1}^{J} \frac{1}{K_{ij}} - \frac{1}{I} \sum_{r=1}^{I} \sum_{j=1}^{J} \frac{1}{K_{rj}} \right].$$

10.9. SUPPLEMENTARY EXERCISES

1. Suppose that (X_i, Y_i), $i = 1, \dots, n$, form a random sample of size n from a bivariate normal distribution with means μ_1 and μ_2, variances σ_1^2 and σ_2^2, and correlation ρ; and let $\hat{\mu}_i$, $\hat{\sigma}_i^2$, and $\hat{\rho}$ denote their M.L.E.'s Also, let $\hat{\beta}_2$ denote the M.L.E. of β_2 in the regression of Y on X. Show that

$$\hat{\beta}_2 = \hat{\rho}\hat{\sigma}_2/\hat{\sigma}_1.$$

Hint: See Exercise 13 of Sec. 6.5.

2. Suppose that (X_i, Y_i), $i = 1, \dots, n$, form a random sample of size n from a bivariate normal distribution with means μ_1 and μ_2, variances σ_1^2 and σ_2^2,

and correlation ρ. Determine the mean and the variance of the following statistic T, given the observed values $X_1 = x_1, \ldots, X_n = x_n$:

$$T = \frac{\sum_{i=1}^{n}(x_i - \bar{x}_n)Y_i}{\sum_{i=1}^{n}(x_i - \bar{x}_n)^2}.$$

3. Let θ_1, θ_2, and θ_3 denote the unknown angles of a triangle, measured in degrees ($\theta_i > 0$ for $i = 1, 2, 3$, and $\theta_1 + \theta_2 + \theta_3 = 180$). Suppose that each angle is measured by an instrument that is subject to error, and that the measured values of θ_1, θ_2, and θ_3 are found to be $y_1 = 83$, $y_2 = 47$, and $y_3 = 56$, respectively. Determine the least-squares estimates of θ_1, θ_2, and θ_3.

4. Suppose that a straight line is to be fitted to n points $(x_1, y_1), \ldots, (x_n, y_n)$ such that $x_2 = x_3 = \cdots = x_n$ but $x_1 \neq x_2$. Show that the least-squares line will pass through the point (x_1, y_1).

5. Suppose that a least-squares line is fitted to the n points $(x_1, y_1), \ldots, (x_n, y_n)$ in the usual way by minimizing the sum of squares of the vertical deviations of the points from the line; and that another least-squares line is fitted by minimizing the sum of squares of the horizontal deviations of the points from the line. Under what conditions will these two lines coincide?

6. Suppose that a straight line $y = \beta_1 + \beta_2 x$ is to be fitted to the n points $(x_1, y_1), \ldots, (x_n, y_n)$ in such a way that the sum of the squared perpendicular (or orthogonal) distances from the points to the line is a minimum. Determine the optimal values of β_1 and β_2.

7. Suppose that twin sisters are each to take a certain mathematics examination. They know that the scores they will obtain on the examination have the same mean μ, the same variance σ^2, and positive correlation ρ. Assuming that their scores have a bivariate normal distribution, show that after each twin learns her own score, she expects her sister's score to be closer to μ.

8. Suppose that a sample of n observations is formed from k subsamples containing $n_1, \ldots, n_k$ observations $(n_1 + \cdots + n_k = n)$. Let x_{ij} ($j = 1, \ldots, n_i$) denote the observations in the ith subsample; and let $\bar{x}_{i+}$ and v_i^2 denote the sample mean and the sample variance of that subsample:

$$\bar{x}_{i+} = \frac{1}{n_i} \sum_{j=1}^{n_i} x_{ij}, \qquad v_i^2 = \frac{1}{n_i} \sum_{j=1}^{n_i} (x_{ij} - \bar{x}_{i+})^2.$$

Finally, let $\bar{x}_{++}$ and v^2 denote the sample mean and the sample variance of the entire sample of n observations:

$$\bar{x}_{++} = \frac{1}{n} \sum_{i=1}^{k} \sum_{j=1}^{n_i} x_{ij}, \qquad v^2 = \frac{1}{n} \sum_{i=1}^{n} \sum_{j=1}^{n_i} (x_{ij} - \bar{x}_{++})^2.$$

Determine an expression for v^2 in terms of $\bar{x}_{++}$, $\bar{x}_{i+}$, and v_i^2 ($i = 1, \ldots, k$).

9. Consider the linear regression model

$$Y_i = \beta_1 w_i + \beta_2 x_i + \varepsilon_i \quad \text{for } i = 1, \ldots, n,$$

where $(w_1, x_1), \ldots, (w_n, x_n)$ are given pairs of constants and $\varepsilon_1, \ldots, \varepsilon_n$ are i.i.d. random variables, each of which has a normal distribution with mean 0 and variance σ^2. Determine explicitly the M.L.E.'s of β_1 and β_2.

10. Determine an unbiased estimator of σ^2 in a two-way layout with K observations in each cell ($K \geq 2$).

11. In a two-way layout with one observation in each cell, construct a test of the null hypothesis that all the effects of both factor A and factor B are 0.

12. In a two-way layout with K observations in each cell ($K \geq 2$), construct a test of the null hypothesis that all the main effects for both factor A and factor B, and also all the interactions, are 0.

13. Suppose that each of two different varieties of corn is treated with two different types of fertilizer in order to compare the yields, and that K independent replications are obtained for each of the four combinations. Let X_{ijk} denote the yield on the kth replication of the combination of variety i with fertilizer j ($i = 1, 2$; $j = 1, 2$; $k = 1, \ldots, K$). Assume that all the observations are independent and normally distributed; that each distribution has the same unknown variance; and that $E(X_{ijk}) = \mu_{ij}$ for $k = 1, \ldots, K$. Describe how to carry out a test of the following hypotheses:

$$H_0: \quad \mu_{11} - \mu_{12} = \mu_{21} - \mu_{22},$$
$$H_1: \quad \text{The hypothesis } H_0 \text{ is not true.}$$

14. Suppose that W_1, W_2, and W_3 are independent random variables, each of which has a normal distribution with the following mean and variance:

$$E(W_1) = \theta_1 + \theta_2, \qquad \text{Var}(W_1) = \sigma^2,$$
$$E(W_2) = \theta_1 + \theta_2 - 5, \qquad \text{Var}(W_2) = \sigma^2,$$
$$E(W_3) = 2\theta_1 - 2\theta_2, \qquad \text{Var}(W_3) = 4\sigma^2.$$

Determine the M.L.E.'s of θ_1, θ_2, and σ^2, and determine also the joint distribution of these estimators.

15. Suppose that it is desired to fit a curve of the form $y = \alpha x^\beta$ to a given set of n points (x_i, y_i) with $x_i > 0$ and $y_i > 0$ for $i = 1, \ldots, n$. Explain how this curve can be fitted either by direct application of the method of least squares or by first transforming the problem into one of fitting a straight line to the n points $(\log x_i, \log y_i)$ and then applying the method of least squares. Discuss the conditions under which each of these methods is appropriate.

16. Consider a problem of simple linear regression, and let $Z_i = Y_i - \hat{\beta}_1 - \hat{\beta}_2 x_i$ denote the residual of the observation Y_i ($i = 1, \ldots, n$), as defined by Eq. (45) of Sec. 10.3. Evaluate $\text{Var}(Z_i)$ for given values of $x_1, \ldots, x_n$, and show that it is a decreasing function of the distance between x_i and $\bar{x}_n$.

17. Consider a general linear model with $n \times p$ design matrix $\mathbf{Z}$, and let $\mathbf{W} = \mathbf{Y} - \mathbf{Z}\hat{\boldsymbol{\beta}}$ denote the vector of residuals. (In other words, the ith component of $\mathbf{W}$ is $Y_i - \hat{Y}_i$, where $\hat{Y}_i$ is given by Eq. (24) of Sec. 10.5.)

    (a) Show that $\mathbf{W} = \mathbf{D}\mathbf{Y}$, where

    $$\mathbf{D} = \mathbf{I} - \mathbf{Z}(\mathbf{Z}'\mathbf{Z})^{-1}\mathbf{Z}'.$$

    (b) Show that the matrix $\mathbf{D}$ is *idempotent*; that is, $\mathbf{D}\mathbf{D} = \mathbf{D}$.

    (c) Show that $\text{Cov}(\mathbf{W}) = \sigma^2 \mathbf{D}$.

18. Consider a two-way layout in which the effects of the factors are additive, so that Eq. (1) of Sec. 10.7 is satisfied; and let $v_1, \ldots, v_I$ and $w_1, \ldots, w_J$ be arbitrary given positive numbers. Show that there exist unique numbers μ, $\alpha_1, \ldots, \alpha_I$, and $\beta_1, \ldots, \beta_J$ such that

    $$\sum_{i=1}^{I} v_i \alpha_i = \sum_{j=1}^{J} w_j \beta_j = 0$$

    and

    $$E(Y_{ij}) = \mu + \alpha_i + \beta_j \quad \text{for } i = 1, \ldots, I \text{ and } j = 1, \ldots, J.$$

19. Consider a two-way layout in which the effects of the factors are additive, as in Exercise 18; and suppose that there are K_{ij} observations per cell, where $K_{ij} > 0$ for $i = 1, \ldots, I$ and $j = 1, \ldots, J$. Let $v_i = K_{i+}$ for $i = 1, \ldots, I$, and $w_j = K_{+j}$ for $j = 1, \ldots, J$. Assume that $E(Y_{ijk}) = \mu + \alpha_i + \beta_j$ for $k = 1, \ldots, K_{ij}$, $i = 1, \ldots, j$, and $j = 1, \ldots, J$, where $\sum_{i=1}^{I} v_i \alpha_i = \sum_{j=1}^{J} w_j \beta_j = 0$, as in Exercise 18. Verify that the least-squares estimators of μ, α_i, and β_j are as follows:

    $$\hat{\mu} = \bar{Y}_{+++},$$

    $$\hat{\alpha}_i = \frac{1}{K_{i+}} Y_{i++} - \bar{Y}_{+++} \quad \text{for } i = 1, \ldots, I,$$

    $$\hat{\beta}_j = \frac{1}{K_{+j}} Y_{+j+} - \bar{Y}_{+++} \quad \text{for } j = 1, \ldots, J.$$

20. Consider again the conditions of Exercises 18 and 19, and let the estimators $\hat{\mu}$, $\hat{\alpha}_i$, and $\hat{\beta}_j$ be as given in Exercise 19. Show that $\text{Cov}(\hat{\mu}, \hat{\alpha}_i) = \text{Cov}(\hat{\mu}, \hat{\beta}_j) = 0$.

21. Consider again the conditions of Exercises 18 and 19, and suppose that the numbers K_{ij} have the following proportionality property:

$$K_{ij} = \frac{K_{i+}K_{+j}}{n} \qquad \text{for } i = 1, \ldots, I \quad \text{and } j = 1, \ldots, J.$$

Show that $\text{Cov}(\hat{\alpha}_i, \hat{\beta}_j) = 0$, where the estimators $\hat{\alpha}_i$ and $\hat{\beta}_j$ are as given in Exercise 19.

22. In a three-way layout with one observation in each cell, the observations Y_{ijk} $(i = 1, \ldots, I; \ j = 1, \ldots, J; \ k = 1, \ldots, K)$ are assumed to be independent and normally distributed, with a common variance σ^2. Suppose that $E(Y_{ijk}) = \theta_{ijk}$. Show that for any given set of numbers θ_{ijk}, there exists a unique set of numbers μ, α_i^A, α_j^B, α_k^C, β_{ij}^{AB}, β_{ik}^{AC}, β_{jk}^{BC}, and γ_{ijk} $(i = 1, \ldots, I; \ j = 1, \ldots, J; \ k = 1, \ldots, K)$ such that

$$\alpha_+^A = \alpha_+^B = \alpha_+^C = 0,$$

$$\beta_{i+}^{AB} = \beta_{+j}^{AB} = \beta_{i+}^{AC} = \beta_{+k}^{AC} = \beta_{j+}^{BC} = \beta_{+k}^{BC} = 0,$$

$$\gamma_{ij+} = \gamma_{i+k} = \gamma_{+jk} = 0,$$

and

$$\theta_{ijk} = \mu + \alpha_i^A + \alpha_j^B + \alpha_k^C + \beta_{ij}^{AB} + \beta_{ik}^{AC} + \beta_{jk}^{BC} + \gamma_{ijk}$$

for all values of i, j, and k.

References

Bickel, P. J., and Doksum, K. A. (1977). *Mathematical Statistics: Basic Ideas and Selected Topics*. Holden-Day, San Francisco.

Brunk, H. D. (1975). *An Introduction to Mathematical Statistics*, 3rd ed. Xerox College Publishing, Lexington, MA.

Cramér, H. (1946). *Mathematical Methods of Statistics*. Princeton University Press, Princeton, NJ.

David, F. N. (1962). *Games, Gods, and Gambling*. Hafner Publishing Co., New York.

DeGroot, M. H. (1970). *Optimal Statistical Decisions*. McGraw-Hill Book Co., Inc., New York.

Devore, J. L. (1982). *Probability and Statistics for Engineering and the Sciences*. Brooks/Cole Publishing Co., Monterey, CA.

Draper, N. R., and Smith, H. (1980). *Applied Regression Analysis*, 2nd ed. John Wiley and Sons, Inc., New York.

Feller, W. (1968). *An Introduction to Probability Theory and Its Applications*, Vol. 1, 3rd ed. John Wiley and Sons, Inc., New York.

Ferguson, T. S. (1967). *Mathematical Statistics: A Decision Theoretic Approach*. Academic Press, Inc., New York.

Fraser, D. A. S. (1976). *Probability and Statistics*. Duxbury Press, Boston.

Freund, J. E., and Walpole, R. E. (1980). *Mathematical Statistics*, 3rd ed. Prentice-Hall, Inc., Englewood Cliffs, NJ.

Guttman, I. (1982). *Linear Models: An Introduction*. John Wiley and Sons, Inc., New York.

Hoel, P. G., Port, S., and Stone, C. L. (1971). *Introduction to Probability Theory*. Houghton-Mifflin, Inc., Boston.

Hogg, R. V., and Craig, A. T. (1978). *Introduction to Mathematical Statistics*, 4th ed. The Macmillan Co., New York.

Kempthorne, O., and Folks, L. (1971). *Probability, Statistics, and Data Analysis*. Iowa State University Press, Ames, IA.

Kendall, M. G., and Stuart, A. (1973). *The Advanced Theory of Statistics*, Vol 2, 3rd ed. Hafner Publishing Co., New York.

Kennedy, W. J., Jr., and Gentle, J. E. (1980). *Statistical Computing*. Marcel Dekker, Inc., New York.

Larson, H. J. (1974). *Introduction to Probability Theory and Statistical Inference*, 2nd ed. John Wiley and Sons, Inc., New York.

Lehmann, E. L. (1959). *Testing Statistical Hypotheses*. John Wiley and Sons, Inc., New York.

Lehmann, E. L. (1983). *Theory of Point Estimation*. John Wiley and Sons, Inc., New York.

Lindgren, B. W. (1976). *Statistical Theory*, 3rd ed. The Macmillan Co., New York.

Mendenhall, W., Scheaffer, R. L., and Wackerly, D. D. (1981). *Mathematical Statistics with Applications*, 2nd ed. Duxbury Press, Boston.

Meyer, P. L. (1970). *Introductory Probability and Statistical Applications*, 2nd ed. Addison-Wesley Publishing Co., Reading, MA.

Mood, A. M., Graybill, F. A., and Boes, D. C. (1974). *Introduction to the Theory of Statistics*, 3rd ed. McGraw-Hill Book Co., New York.

Olkin, I., Gleser, L. J., and Derman, C. (1980). *Probability Models and Applications*. The Macmillan Co., New York.

Ore, O. (1960). Pascal and the invention of probability theory. *American Mathematical Monthly*, Vol. 67, pp. 409–419.

Rao, C. R. (1973). *Linear Statistical Inference and Its Applications*, 2nd ed. John Wiley and Sons, Inc., New York.

Rohatgi, V. K. (1976). *An Introduction to Probability Theory and Mathematical Statistics*. John Wiley and Sons, Inc., New York.

Rubinstein, R. Y. (1981). *Simulation and the Monte Carlo Method*. John Wiley and Sons, Inc., New York.

Scheffé, H. (1959). *The Analysis of Variance*. John Wiley and Sons, Inc., New York.

Todhunter, I. (1865). *A History of the Mathematical Theory of Probability from the Time of Pascal to That of Laplace*. Reprinted by G. E. Stechert and Co., New York, 1931.

Zacks, S. (1971). *The Theory of Statistical Inference*. John Wiley and Sons, Inc., New York.

Zacks, S. (1981). *Parametric Statistical Inference*. Pergamon Press, New York.

Tables

Table of Binomial Probabilities

$$\Pr(X = k) = \binom{n}{k} p^k (1 - p)^{n-k}$$

n	k	p = 0.1	p = 0.2	p = 0.3	p = 0.4	p = 0.5
2	0	.8100	.6400	.4900	.3600	.2500
	1	.1800	.3200	.4200	.4800	.5000
	2	.0100	.0400	.0900	.1600	.2500
3	0	.7290	.5120	.3430	.2160	.1250
	1	.2430	.3840	.4410	.4320	.3750
	2	.0270	.0960	.1890	.2880	.3750
	3	.0010	.0080	.0270	.0640	.1250
4	0	.6561	.4096	.2401	.1296	.0625
	1	.2916	.4096	.4116	.3456	.2500
	2	.0486	.1536	.2646	.3456	.3750
	3	.0036	.0256	.0756	.1536	.2500
	4	.0001	.0016	.0081	.0256	.0625
5	0	.5905	.3277	.1681	.0778	.0312
	1	.3280	.4096	.3602	.2592	.1562
	2	.0729	.2048	.3087	.3456	.3125
	3	.0081	.0512	.1323	.2304	.3125
	4	.0005	.0064	.0284	.0768	.1562
	5	.0000	.0003	.0024	.0102	.0312
6	0	.5314	.2621	.1176	.0467	.0156
	1	.3543	.3932	.3025	.1866	.0938
	2	.0984	.2458	.3241	.3110	.2344
	3	.0146	.0819	.1852	.2765	.3125
	4	.0012	.0154	.0595	.1382	.2344
	5	.0001	.0015	.0102	.0369	.0938
	6	.0000	.0001	.0007	.0041	.0156
7	0	.4783	.2097	.0824	.0280	.0078
	1	.3720	.3670	.2471	.1306	.0547
	2	.1240	.2753	.3176	.2613	.1641
	3	.0230	.1147	.2269	.2903	.2734
	4	.0026	.0287	.0972	.1935	.2734
	5	.0002	.0043	.0250	.0774	.1641
	6	.0000	.0004	.0036	.0172	.0547
	7	.0000	.0000	.0002	.0016	.0078

continued

Table of Binomial Probabilities (*Continued*)

n	k	$p = 0.1$	$p = 0.2$	$p = 0.3$	$p = 0.4$	$p = 0.5$
8	0	.4305	.1678	.0576	.0168	.0039
	1	.3826	.3355	.1977	.0896	.0312
	2	.1488	.2936	.2965	.2090	.1094
	3	.0331	.1468	.2541	.2787	.2188
	4	.0046	.0459	.1361	.2322	.2734
	5	.0004	.0092	.0467	.1239	.2188
	6	.0000	.0011	.0100	.0413	.1094
	7	.0000	.0001	.0012	.0079	.0312
	8	.0000	.0000	.0001	.0007	.0039
9	0	.3874	.1342	.0404	.0101	.0020
	1	.3874	.3020	.1556	.0605	.0176
	2	.1722	.3020	.2668	.1612	.0703
	3	.0446	.1762	.2668	.2508	.1641
	4	.0074	.0661	.1715	.2508	.2461
	5	.0008	.0165	.0735	.1672	.2461
	6	.0001	.0028	.0210	.0743	.1641
	7	.0000	.0003	.0039	.0212	.0703
	8	.0000	.0000	.0004	.0035	.0176
	9	.0000	.0000	.0000	.0003	.0020
10	0	.3487	.1074	.0282	.0060	.0010
	1	.3874	.2684	.1211	.0403	.0098
	2	.1937	.3020	.2335	.1209	.0439
	3	.0574	.2013	.2668	.2150	.1172
	4	.0112	.0881	.2001	.2508	.2051
	5	.0015	.0264	.1029	.2007	.2461
	6	.0001	.0055	.0368	.1115	.2051
	7	.0000	.0008	.0090	.0425	.1172
	8	.0000	.0001	.0014	.0106	.0439
	9	.0000	.0000	.0001	.0016	.0098
	10	.0000	.0000	.0000	.0001	.0010

continued

Table of Binomial Probabilities (*Continued*)

n	k	p = 0.1	p = 0.2	p = 0.3	p = 0.4	p = 0.5
15	0	.2059	.0352	.0047	.0005	.0000
	1	.3432	.1319	.0305	.0047	.0005
	2	.2669	.2309	.0916	.0219	.0032
	3	.1285	.2501	.1700	.0634	.0139
	4	.0428	.1876	.2186	.1268	.0417
	5	.0105	.1032	.2061	.1859	.0916
	6	.0019	.0430	.1472	.2066	.1527
	7	.0003	.0138	.0811	.1771	.1964
	8	.0000	.0035	.0348	.1181	.1964
	9	.0000	.0007	.0116	.0612	.1527
	10	.0000	.0001	.0030	.0245	.0916
	11	.0000	.0000	.0006	.0074	.0417
	12	.0000	.0000	.0001	.0016	.0139
	13	.0000	.0000	.0000	.0003	.0032
	14	.0000	.0000	.0000	.0000	.0005
	15	.0000	.0000	.0000	.0000	.0000
20	0	.1216	.0115	.0008	.0000	.0000
	1	.2701	.0576	.0068	.0005	.0000
	2	.2852	.1369	.0278	.0031	.0002
	3	.1901	.2054	.0716	.0123	.0011
	4	.0898	.2182	.1304	.0350	.0046
	5	.0319	.1746	.1789	.0746	.0148
	6	.0089	.1091	.1916	.1244	.0370
	7	.0020	.0545	.1643	.1659	.0739
	8	.0003	.0222	.1144	.1797	.1201
	9	.0001	.0074	.0654	.1597	.1602
	10	.0000	.0020	.0308	.1171	.1762
	11	.0000	.0005	.0120	.0710	.1602
	12	.0000	.0001	.0039	.0355	.1201
	13	.0000	.0000	.0010	.0146	.0739
	14	.0000	.0000	.0002	.0049	.0370
	15	.0000	.0000	.0000	.0013	.0148
	16	.0000	.0000	.0000	.0003	.0046
	17	.0000	.0000	.0000	.0000	.0011
	18	.0000	.0000	.0000	.0000	.0002
	19	.0000	.0000	.0000	.0000	.0000
	20	.0000	.0000	.0000	.0000	.0000

Table of Random Digits

2671	4690	1550	2262	2597	8034	0785	2978	4409	0237
9111	0250	3275	7519	9740	4577	2064	0286	3398	1348
0391	6035	9230	4999	3332	0608	6113	0391	5789	9926
2475	2144	1886	2079	3004	9686	5669	4367	9306	2595
5336	5845	2095	6446	5694	3641	1085	8705	5416	9066
6808	0423	0155	1652	7897	4335	3567	7109	9690	3739
8525	0577	8940	9451	6726	0876	3818	7607	8854	3566
0398	0741	8787	3043	5063	0617	1770	5048	7721	7032
3623	9636	3638	1406	5731	3978	8068	7238	9715	3363
0739	2644	4917	8866	3632	5399	5175	7422	2476	2607
6713	3041	8133	8749	8835	6745	3597	3476	3816	3455
7775	9315	0432	8327	0861	1515	2297	3375	3713	9174
8599	2122	6842	9202	0810	2936	1514	2090	3067	3574
7955	3759	5254	1126	5553	4713	9605	7909	1658	5490
4766	0070	7260	6033	7997	0109	5993	7592	5436	1727
5165	1670	2534	8811	8231	3721	7947	5719	2640	1394
9111	0513	2751	8256	2931	7783	1281	6531	7259	6993
1667	1084	7889	8963	7018	8617	6381	0723	4926	4551
2145	4587	8585	2412	5431	4667	1942	7238	9613	2212
2739	5528	1481	7528	9368	1823	6979	2547	7268	2467
8769	5480	9160	5354	9700	1362	2774	7980	9157	8788
6531	9435	3422	2474	1475	0159	3414	5224	8399	5820
2937	4134	7120	2206	5084	9473	3958	7320	9878	8609
1581	3285	3727	8924	6204	0797	0882	5945	9375	9153
6268	1045	7076	1436	4165	0143	0293	4190	7171	7932
4293	0523	8625	1961	1039	2856	4889	4358	1492	3804
6936	4213	3212	7229	1230	0019	5998	9206	6753	3762
5334	7641	3258	3769	1362	2771	6124	9813	7915	8960
9373	1158	4418	8826	5665	5896	0358	4717	8232	4859
6968	9428	8950	5346	1741	2348	8143	5377	7695	0685
4229	0587	8794	4009	9691	4579	3302	7673	9629	5246
3807	7785	7097	5701	6639	0723	4819	0900	2713	7650
4891	8829	1642	2155	0796	0466	2946	2970	9143	6590
1055	2968	7911	7479	8199	9735	8271	5339	7058	2964
2983	2345	0568	4125	0894	8302	0506	6761	7706	4310
4026	3129	2968	8053	2797	4022	9838	9611	0975	2437
4075	0260	4256	0337	2355	9371	2954	6021	5783	2827
8488	5450	1327	7358	2034	8060	1788	6913	6123	9405
1976	1749	5742	4098	5887	4567	6064	2777	7830	5668
2793	4701	9466	9554	8294	2160	7486	1557	4769	2781

Table of Random Digits (*Continued*)

0916	6272	6825	7188	9611	1181	2301	5516	5451	6832
5961	1149	7946	1950	2010	0600	5655	0796	0569	4365
3222	4189	1891	8172	8731	4769	2782	1325	4238	9279
1176	7834	4600	9992	9449	5824	5344	1008	6678	1921
2369	8971	2314	4806	5071	8908	8274	4936	3357	4441
0041	4329	9265	0352	4764	9070	7527	7791	1094	2008
0803	8302	6814	2422	6351	0637	0514	0246	1845	8594
9965	7804	3930	8803	0268	1426	3130	3613	3947	8086
0011	2387	3148	7559	4216	2946	2865	6333	1916	2259
1767	9871	3914	5790	5287	7915	8959	1346	5482	9251
2604	3074	0504	3828	7881	0797	1094	4098	4940	7067
6930	4180	3074	0060	0909	3187	8991	0682	2385	2307
6160	9899	9084	5704	5666	3051	0325	4733	5905	9226
4884	1857	2847	2581	4870	1782	2980	0587	8797	5545
7294	2009	9020	0006	4309	3941	5645	6238	5052	4150
3478	4973	1056	3687	3145	5988	4214	5543	9185	9375
1764	7860	4150	2881	9895	2531	7363	8756	3724	9359
3025	0890	6436	3461	1411	0303	7422	2684	6256	3495
1771	3056	6630	4982	2386	2517	4747	5505	8785	8708
0254	1892	9066	4890	8716	2258	2452	3913	6790	6331
8537	9966	8224	9151	1855	8911	4422	1913	2000	1482
1475	0261	4465	4803	8231	6469	9935	4256	0648	7768
5209	5569	8410	3041	4325	7290	3381	5209	5571	9458
5456	5944	6038	3210	7165	0723	4820	1846	0005	3865
5043	6694	4853	8425	5871	1322	1052	1452	2486	1669
1719	0148	6977	1244	6443	5955	7945	1218	9391	6485
7432	2955	3933	8110	8585	1893	9218	7153	7566	6040
4926	4761	7812	7439	6436	3145	5934	7852	9095	9497
0769	0683	3768	1048	8519	2987	0124	3064	1881	3177
0805	3139	8514	5014	3274	6395	0549	3858	0820	6406
0204	7273	4964	5475	2648	6977	1371	6971	4850	6873
0092	1733	2349	2648	6609	5676	6445	3271	8867	3469
3139	4867	3666	9783	5088	4852	4143	7923	3858	0504
2033	7430	4389	7121	9982	0651	9110	9731	6421	4731
3921	0530	3605	8455	4205	7363	3081	3931	9331	1313
4111	9244	8135	9877	9529	9160	4407	9077	5306	0054
6573	1570	6654	3616	2049	7001	5185	7108	9270	6550
8515	8029	6880	4329	9367	1087	9549	1684	4838	5686
3590	2106	3245	1989	3529	3828	8091	6054	5656	3035
7212	9909	5005	7660	2620	6406	0690	4240	4070	6549

Table of Random Digits (*Continued*)

6701	0154	8806	1716	7029	6776	9465	8818	2886	3547
3777	9532	1333	8131	2929	6987	2408	0487	9172	6177
2495	3054	1692	0089	4090	2983	2136	8947	4625	7177
2073	8878	9742	3012	0042	3996	9930	1651	4982	9645
2252	8004	7840	2105	3033	8749	9153	2872	5100	8674
2104	2224	4052	2273	4753	4505	7156	5417	9725	7599
2371	0005	3844	6654	3246	4853	4301	8886	5217	1153
3270	1214	9649	1872	6930	9791	0248	2687	8126	1501
6209	7237	1966	5541	4224	7080	7630	6422	1160	5675
1309	9126	2920	4359	1726	0562	9654	4182	4097	7493
2406	8013	3634	6428	8091	5925	3923	1686	6097	9670
7365	9859	9378	7084	9402	9201	1815	7064	4324	7081
2889	4738	9929	1476	0785	3832	1281	5821	3690	9185
7951	3781	4755	6986	1659	5727	8108	9816	5759	4188
4548	6778	7672	9101	3911	8127	1918	8512	4197	6402
5701	8342	2852	4278	3343	9830	1756	0546	6717	3114
2187	7266	1210	3797	1636	7917	9933	3518	6923	6349
9360	6640	1315	6284	8265	7232	0291	3467	1088	7834
7850	7626	0745	1992	4998	7349	6451	6186	8916	4292
6186	9233	6571	0925	1748	5490	5264	3820	9829	1335

Table of Poisson Probabilities

$$\Pr(X = k) = \frac{e^{-\lambda}\lambda^k}{k!}$$

k	λ = .1	.2	.3	.4	.5	.6	.7	.8	.9	1.0
0	.9048	.8187	.7408	.6703	.6065	.5488	.4966	.4493	.4066	.3679
1	.0905	.1637	.2222	.2681	.3033	.3293	.3476	.3595	.3659	.3679
2	.0045	.0164	.0333	.0536	.0758	.0988	.1217	.1438	.1647	.1839
3	.0002	.0011	.0033	.0072	.0126	.0198	.0284	.0383	.0494	.0613
4	.0000	.0001	.0003	.0007	.0016	.0030	.0050	.0077	.0111	.0153
5	.0000	.0000	.0000	.0001	.0002	.0004	.0007	.0012	.0020	.0031
6	.0000	.0000	.0000	.0000	.0000	.0000	.0001	.0002	.0003	.0005
7	.0000	.0000	.0000	.0000	.0000	.0000	.0000	.0000	.0000	.0001
8	.0000	.0000	.0000	.0000	.0000	.0000	.0000	.0000	.0000	.0000

k	λ = 1.5	2	3	4	5	6	7	8	9	10
0	.2231	.1353	.0498	.0183	.0067	.0025	.0009	.0003	.0001	.0000
1	.3347	.2707	.1494	.0733	.0337	.0149	.0064	.0027	.0011	.0005
2	.2510	.2707	.2240	.1465	.0842	.0446	.0223	.0107	.0050	.0023
3	.1255	.1804	.2240	.1954	.1404	.0892	.0521	.0286	.0150	.0076
4	.0471	.0902	.1680	.1954	.1755	.1339	.0912	.0573	.0337	.0189
5	.0141	.0361	.1008	.1563	.1755	.1606	.1277	.0916	.0607	.0378
6	.0035	.0120	.0504	.1042	.1462	.1606	.1490	.1221	.0911	.0631
7	.0008	.0034	.0216	.0595	.1044	.1377	.1490	.1396	.1171	.0901
8	.0001	.0009	.0081	.0298	.0653	.1033	.1304	.1396	.1318	.1126
9	.0000	.0002	.0027	.0132	.0363	.0688	.1014	.1241	.1318	.1251
10	.0000	.0000	.0008	.0053	.0181	.0413	.0710	.0993	.1186	.1251
11	.0000	.0000	.0002	.0019	.0082	.0225	.0452	.0722	.0970	.1137
12	.0000	.0000	.0001	.0006	.0034	.0113	.0264	.0481	.0728	.0948
13	.0000	.0000	.0000	.0002	.0013	.0052	.0142	.0296	.0504	.0729
14	.0000	.0000	.0000	.0001	.0005	.0022	.0071	.0169	.0324	.0521
15	.0000	.0000	.0000	.0000	.0002	.0009	.0033	.0090	.0194	.0347
16	.0000	.0000	.0000	.0000	.0000	.0003	.0014	.0045	.0109	.0217
17	.0000	.0000	.0000	.0000	.0000	.0001	.0006	.0021	.0058	.0128
18	.0000	.0000	.0000	.0000	.0000	.0000	.0002	.0009	.0029	.0071
19	.0000	.0000	.0000	.0000	.0000	.0000	.0001	.0004	.0014	.0037
20	.0000	.0000	.0000	.0000	.0000	.0000	.0000	.0002	.0006	.0019
21	.0000	.0000	.0000	.0000	.0000	.0000	.0000	.0001	.0003	.0009
22	.0000	.0000	.0000	.0000	.0000	.0000	.0000	.0000	.0001	.0004
23	.0000	.0000	.0000	.0000	.0000	.0000	.0000	.0000	.0000	.0002
24	.0000	.0000	.0000	.0000	.0000	.0000	.0000	.0000	.0000	.0001
25	.0000	.0000	.0000	.0000	.0000	.0000	.0000	.0000	.0000	.0000

Table of the Standard Normal Distribution Function

$$\Phi(x) = \int_{-\infty}^{x} \frac{1}{(2\pi)^{1/2}} \exp\left(-\frac{1}{2}u^2\right) du$$

x	$\Phi(x)$	x	$\Phi(x)$	x	$\Phi(x)$	x	$\Phi(x)$	x	$\Phi(x)$
0.00	0.5000	0.60	0.7257	1.20	0.8849	1.80	0.9641	2.40	0.9918
0.01	0.5040	0.61	0.7291	1.21	0.8869	1.81	0.9649	2.41	0.9920
0.02	0.5080	0.62	0.7324	1.22	0.8888	1.82	0.9656	2.42	0.9922
0.03	0.5120	0.63	0.7357	1.23	0.8907	1.83	0.9664	2.43	0.9925
0.04	0.5160	0.64	0.7389	1.24	0.8925	1.84	0.9671	2.44	0.9927
0.05	0.5199	0.65	0.7422	1.25	0.8944	1.85	0.9678	2.45	0.9929
0.06	0.5239	0.66	0.7454	1.26	0.8962	1.86	0.9686	2.46	0.9931
0.07	0.5279	0.67	0.7486	1.27	0.8980	1.87	0.9693	2.47	0.9932
0.08	0.5319	0.68	0.7517	1.28	0.8997	1.88	0.9699	2.48	0.9934
0.09	0.5359	0.69	0.7549	1.29	0.9015	1.89	0.9706	2.49	0.9936
0.10	0.5398	0.70	0.7580	1.30	0.9032	1.90	0.9713	2.50	0.9938
0.11	0.5438	0.71	0.7611	1.31	0.9049	1.91	0.9719	2.52	0.9941
0.12	0.5478	0.72	0.7642	1.32	0.9066	1.92	0.9726	2.54	0.9945
0.13	0.5517	0.73	0.7673	1.33	0.9082	1.93	0.9732	2.56	0.9948
0.14	0.5557	0.74	0.7704	1.34	0.9099	1.94	0.9738	2.58	0.9951
0.15	0.5596	0.75	0.7734	1.35	0.9115	1.95	0.9744	2.60	0.9953
0.16	0.5636	0.76	0.7764	1.36	0.9131	1.96	0.9750	2.62	0.9956
0.17	0.5675	0.77	0.7794	1.37	0.9147	1.97	0.9756	2.64	0.9959
0.18	0.5714	0.78	0.7823	1.38	0.9162	1.98	0.9761	2.66	0.9961
0.19	0.5753	0.79	0.7852	1.39	0.9177	1.99	0.9767	2.68	0.9963
0.20	0.5793	0.80	0.7881	1.40	0.9192	2.00	0.9773	2.70	0.9965
0.21	0.5832	0.81	0.7910	1.41	0.9207	2.01	0.9778	2.72	0.9967
0.22	0.5871	0.82	0.7939	1.42	0.9222	2.02	0.9783	2.74	0.9969
0.23	0.5910	0.83	0.7967	1.43	0.9236	2.03	0.9788	2.76	0.9971
0.24	0.5948	0.84	0.7995	1.44	0.9251	2.04	0.9793	2.78	0.9973
0.25	0.5987	0.85	0.8023	1.45	0.9265	2.05	0.9798	2.80	0.9974
0.26	0.6026	0.86	0.8051	1.46	0.9279	2.06	0.9803	2.82	0.9976
0.27	0.6064	0.87	0.8079	1.47	0.9292	2.07	0.9808	2.84	0.9977
0.28	0.6103	0.88	0.8106	1.48	0.9306	2.08	0.9812	2.86	0.9979
0.29	0.6141	0.89	0.8133	1.49	0.9319	2.09	0.9817	2.88	0.9980
0.30	0.6179	0.90	0.8159	1.50	0.9332	2.10	0.9821	2.90	0.9981
0.31	0.6217	0.91	0.8186	1.51	0.9345	2.11	0.9826	2.92	0.9983
0.32	0.6255	0.92	0.8212	1.52	0.9357	2.12	0.9830	2.94	0.9984
0.33	0.6293	0.93	0.8238	1.53	0.9370	2.13	0.9834	2.96	0.9985
0.34	0.6331	0.94	0.8264	1.54	0.9382	2.14	0.9838	2.98	0.9986
0.35	0.6368	0.95	0.8289	1.55	0.9394	2.15	0.9842	3.00	0.9987
0.36	0.6406	0.96	0.8315	1.56	0.9406	2.16	0.9846	3.05	0.9989
0.37	0.6443	0.97	0.8340	1.57	0.9418	2.17	0.9850	3.10	0.9990
0.38	0.6480	0.98	0.8365	1.58	0.9429	2.18	0.9854	3.15	0.9992
0.39	0.6517	0.99	0.8389	1.59	0.9441	2.19	0.9857	3.20	0.9993
0.40	0.6554	1.00	0.8413	1.60	0.9452	2.20	0.9861	3.25	0.9994
0.41	0.6591	1.01	0.8437	1.61	0.9463	2.21	0.9864	3.30	0.9995
0.42	0.6628	1.02	0.8461	1.62	0.9474	2.22	0.9868	3.35	0.9996
0.43	0.6664	1.03	0.8485	1.63	0.9485	2.23	0.9871	3.40	0.9997
0.44	0.6700	1.04	0.8508	1.64	0.9495	2.24	0.9875	3.45	0.9997
0.45	0.6736	1.05	0.8531	1.65	0.9505	2.25	0.9878	3.50	0.9998
0.46	0.6772	1.06	0.8554	1.66	0.9515	2.26	0.9881	3.55	0.9998
0.47	0.6808	1.07	0.8577	1.67	0.9525	2.27	0.9884	3.60	0.9998
0.48	0.6844	1.08	0.8599	1.68	0.9535	2.28	0.9887	3.65	0.9999
0.49	0.6879	1.09	0.8621	1.69	0.9545	2.29	0.9890	3.70	0.9999
0.50	0.6915	1.10	0.8643	1.70	0.9554	2.30	0.9893	3.75	0.9999
0.51	0.6950	1.11	0.8665	1.71	0.9564	2.31	0.9896	3.80	0.9999
0.52	0.6985	1.12	0.8686	1.72	0.9573	2.32	0.9898	3.85	0.9999
0.53	0.7019	1.13	0.8708	1.73	0.9582	2.33	0.9901	3.90	1.0000
0.54	0.7054	1.14	0.8729	1.74	0.9591	2.34	0.9904	3.95	1.0000
0.55	0.7088	1.15	0.8749	1.75	0.9599	2.35	0.9906	4.00	1.0000
0.56	0.7123	1.16	0.8770	1.76	0.9608	2.36	0.9909		
0.57	0.7157	1.17	0.8790	1.77	0.9616	2.37	0.9911		
0.58	0.7190	1.18	0.8810	1.78	0.9625	2.38	0.9913		
0.59	0.7224	1.19	0.8830	1.79	0.9633	2.39	0.9916		

Donald B. Owen, HANDBOOK OF STATISTICAL TABLES, © 1962. Addison-Wesley Publishing Company, Reading, Massachusetts. Reprinted with permission.

Table of the χ^2 Distribution

If X has a χ^2 distribution with n degrees of freedom, this table gives the value of x such that $\Pr(X \leqslant x) = p$.

n \ p	.005	.01	.025	.05	.10	.20	.25	.30	.40
1	.0000	.0002	.0010	.0039	.0158	.0642	.1015	.1484	.2750
2	.0100	.0201	.0506	.1026	.2107	.4463	.5754	.7133	1.022
3	.0717	.1148	.2158	.3518	.5844	1.005	1.213	1.424	1.869
4	.2070	.2971	.4844	.7107	1.064	1.649	1.923	2.195	2.753
5	.4117	.5543	.8312	1.145	1.610	2.343	2.675	3.000	3.655
6	.6757	.8721	1.237	1.635	2.204	3.070	3.455	3.828	4.570
7	.9893	1.239	1.690	2.167	2.833	3.822	4.255	4.671	5.493
8	1.344	1.647	2.180	2.732	3.490	4.594	5.071	5.527	6.423
9	1.735	2.088	2.700	3.325	4.168	5.380	5.899	6.393	7.357
10	2.156	2.558	3.247	3.940	4.865	6.179	6.737	7.267	8.295
11	2.603	3.053	3.816	4.575	5.578	6.989	7.584	8.148	9.237
12	3.074	3.571	4.404	5.226	6.304	7.807	8.438	9.034	10.18
13	3.565	4.107	5.009	5.892	7.042	8.634	9.299	9.926	11.13
14	4.075	4.660	5.629	6.571	7.790	9.467	10.17	10.82	12.08
15	4.601	5.229	6.262	7.261	8.547	10.31	11.04	11.72	13.03
16	5.142	5.812	6.908	7.962	9.312	11.15	11.91	12.62	13.98
17	5.697	6.408	7.564	8.672	10.09	12.00	12.79	13.53	14.94
18	6.265	7.015	8.231	9.390	10.86	12.86	13.68	14.43	15.89
19	6.844	7.633	8.907	10.12	11.65	13.72	14.56	15.35	16.85
20	7.434	8.260	9.591	10.85	12.44	14.58	15.45	16.27	17.81
21	8.034	8.897	10.28	11.59	13.24	15.44	16.34	17.18	18.77
22	8.643	9.542	10.98	12.34	14.04	16.31	17.24	18.10	19.73
23	9.260	10.20	11.69	13.09	14.85	17.19	18.14	19.02	20.69
24	9.886	10.86	12.40	13.85	15.66	18.06	19.04	19.94	21.65
25	10.52	11.52	13.12	14.61	16.47	18.94	19.94	20.87	22.62
30	13.79	14.95	16.79	18.49	20.60	23.36	24.48	25.51	27.44
40	20.71	22.16	24.43	26.51	29.05	32.34	33.66	34.87	36.16
50	27.99	29.71	32.36	34.76	37.69	41.45	42.94	44.31	46.86
60	35.53	37.48	40.48	43.19	46.46	50.64	52.29	53.81	56.62
70	43.27	45.44	48.76	51.74	55.33	59.90	61.70	63.35	66.40
80	51.17	53.54	57.15	60.39	64.28	69.21	71.14	72.92	76.19
90	59.20	61.75	65.65	69.13	73.29	78.56	80.62	82.51	85.99
100	67.33	70.06	74.22	77.93	82.86	87.95	90.13	92.13	95.81

Adapted with permission from *Biometrika Tables for Statisticians*, Vol. 1, 3rd ed., Cambridge University Press, 1966, edited by E. S. Pearson and H. O. Hartley; and from "A new table of percentage points of the chi-square distribution," *Biometrika*, Vol. 51(1964), pp. 231–239, by H. L. Harter, Aerospace Research Laboratories.

Table of the χ^2 Distribution (*Continued*)

.50	.60	.70	.75	.80	.90	.95	.975	.99	.995
.4549	.7083	1.074	1.323	1.642	2.706	3.841	5.024	6.635	7.879
1.386	1.833	2.408	2.773	3.219	4.605	5.991	7.378	9.210	10.60
2.366	2.946	3.665	4.108	4.642	6.251	7.815	9.348	11.34	12.84
3.357	4.045	4.878	5.385	5.989	7.779	9.488	11.14	13.28	14.86
4.351	5.132	6.064	6.626	7.289	9.236	11.07	12.83	15.09	16.75
5.348	6.211	7.231	7.841	8.558	10.64	12.59	14.45	16.81	18.55
6.346	7.283	8.383	9.037	9.803	12.02	14.07	16.01	18.48	20.28
7.344	8.351	9.524	10.22	11.03	13.36	15.51	17.53	20.09	21.95
8.343	9.414	10.66	11.39	12.24	14.68	16.92	19.02	21.67	23.59
9.342	10.47	11.78	12.55	13.44	15.99	18.31	20.48	23.21	25.19
10.34	11.53	12.90	13.70	14.63	17.27	19.68	21.92	24.72	26.76
11.34	12.58	14.01	14.85	15.81	18.55	21.03	23.34	26.22	28.30
12.34	13.64	15.12	15.98	16.98	19.81	22.36	24.74	27.69	29.82
13.34	14.69	16.22	17.12	18.15	21.06	23.68	26.12	29.14	31.32
14.34	15.73	17.32	18.25	19.31	22.31	25.00	27.49	30.58	32.80
15.34	16.78	18.42	19.37	20.47	23.54	26.30	28.85	32.00	34.27
16.34	17.82	19.51	20.49	21.61	24.77	27.59	30.19	33.41	35.72
17.34	18.87	20.60	21.60	22.76	25.99	28.87	31.53	34.81	37.16
18.34	19.91	21.69	22.72	23.90	27.20	30.14	32.85	36.19	38.58
19.34	20.95	22.77	23.83	25.04	28.41	31.41	34.17	37.57	40.00
20.34	21.99	23.86	24.93	26.17	29.62	32.67	35.48	38.93	41.40
21.34	23.03	24.94	26.04	27.30	30.81	33.92	36.78	40.29	42.80
22.34	24.07	26.02	27.14	28.43	32.01	35.17	38.08	41.64	44.18
23.34	25.11	27.10	28.24	29.55	33.20	36.42	39.36	42.98	45.56
24.34	26.14	28.17	29.34	30.68	34.38	37.65	40.65	44.31	46.93
29.34	31.32	33.53	34.80	36.25	40.26	43.77	46.98	50.89	53.67
39.34	41.62	44.16	45.62	47.27	51.81	55.76	59.34	63.69	66.77
49.33	51.89	54.72	56.33	58.16	63.17	67.51	71.42	76.15	79.49
59.33	62.13	65.23	66.98	68.97	74.40	79.08	83.30	88.38	91.95
69.33	72.36	75.69	77.58	79.71	85.53	90.53	95.02	100.4	104.2
79.33	82.57	86.12	88.13	90.41	96.58	101.9	106.6	112.3	116.3
89.33	92.76	96.52	98.65	101.1	107.6	113.1	118.1	124.1	128.3
99.33	102.9	106.9	109.1	111.7	118.5	124.3	129.6	135.8	140.2

Table of the *t* Distribution

If X has a t distribution with n degrees of freedom, the table gives the value of x such that $\Pr(X \leq x) = p$.

n	p = .55	.60	.65	.70	.75	.80	.85	.90	.95	.975	.99	.995
1	.158	.325	.510	.727	1.000	1.376	1.963	3.078	6.314	12.706	31.821	63.657
2	.142	.289	.445	.617	.816	1.061	1.386	1.886	2.920	4.303	6.965	9.925
3	.137	.277	.424	.584	.765	.978	1.250	1.638	2.353	3.182	4.541	5.841
4	.134	.271	.414	.569	.741	.941	1.190	1.533	2.132	2.776	3.747	4.604
5	.132	.267	.408	.559	.727	.920	1.156	1.476	2.015	2.571	3.365	4.032
6	.131	.265	.404	.553	.718	.906	1.134	1.440	1.943	2.447	3.143	3.707
7	.130	.263	.402	.549	.711	.896	1.119	1.415	1.895	2.365	2.998	3.499
8	.130	.262	.399	.546	.706	.889	1.108	1.397	1.860	2.306	2.896	3.355
9	.129	.261	.398	.543	.703	.883	1.100	1.383	1.833	2.262	2.821	3.250
10	.129	.260	.397	.542	.700	.879	1.093	1.372	1.812	2.228	2.764	3.169
11	.129	.260	.396	.540	.697	.876	1.088	1.363	1.796	2.201	2.718	3.106
12	.128	.259	.395	.539	.695	.873	1.083	1.356	1.782	2.179	2.681	3.055
13	.128	.259	.394	.538	.694	.870	1.079	1.350	1.771	2.160	2.650	3.012
14	.128	.258	.393	.537	.692	.868	1.076	1.345	1.761	2.145	2.624	2.977
15	.128	.258	.393	.536	.691	.866	1.074	1.341	1.753	2.131	2.602	2.947
16	.128	.258	.392	.535	.690	.865	1.071	1.337	1.746	2.120	2.583	2.921
17	.128	.257	.392	.534	.689	.863	1.069	1.333	1.740	2.110	2.567	2.898
18	.127	.257	.392	.534	.688	.862	1.067	1.330	1.734	2.101	2.552	2.878
19	.127	.257	.391	.533	.688	.861	1.066	1.328	1.729	2.093	2.539	2.861
20	.127	.257	.391	.533	.687	.860	1.064	1.325	1.725	2.086	2.528	2.845

Table of the t Distribution (*Continued*)

n	$p = .55$	.60	.65	.70	.75	.80	.85	.90	.95	.975	.99	.995
21	.127	.257	.391	.532	.686	.859	1.063	1.323	1.721	2.080	2.518	2.831
22	.127	.256	.390	.532	.686	.858	1.061	1.321	1.717	2.074	2.508	2.819
23	.127	.256	.390	.532	.685	.858	1.060	1.319	1.714	2.069	2.500	2.807
24	.127	.256	.390	.531	.685	.857	1.059	1.318	1.711	2.064	2.492	2.797
25	.127	.256	.390	.531	.684	.856	1.058	1.316	1.708	2.060	2.485	2.787
26	.127	.256	.390	.531	.684	.856	1.058	1.315	1.706	2.056	2.479	2.779
27	.127	.256	.389	.531	.684	.855	1.057	1.314	1.703	2.052	2.473	2.771
28	.127	.256	.389	.530	.683	.855	1.056	1.313	1.701	2.048	2.467	2.763
29	.127	.256	.389	.530	.683	.854	1.055	1.311	1.699	2.045	2.462	2.756
30	.127	.256	.389	.530	.683	.854	1.055	1.310	1.697	2.042	2.457	2.750
40	.126	.255	.388	.529	.681	.851	1.050	1.303	1.684	2.021	2.423	2.704
60	.126	.254	.387	.527	.679	.848	1.046	1.296	1.671	2.000	2.390	2.660
120	.126	.254	.386	.526	.677	.845	1.041	1.289	1.658	1.980	2.358	2.617
∞	.126	.253	.385	.524	.674	.842	1.036	1.282	1.645	1.960	2.326	2.576

This table is taken from Table III of Fisher & Yates: *Statistical Tables for Biological, Agricultural and Medical Research*, published by Longman Group Ltd. London (previously published by Oliver and Boyd Ltd., Edinburgh) and by permission of the authors and publishers.

Table of the 0.95 Quantile of the F Distribution

If X has an F distribution with m and n degrees of freedom, the table gives the value of x such that $\Pr(X \le x) = 0.95$.

n \ m	1	2	3	4	5	6	7	8	9	10	15	20	30	40	60	120	∞
1	161.4	199.5	215.7	224.6	230.2	234.0	236.8	238.9	240.5	241.9	245.9	248.0	250.1	251.1	252.2	253.3	254.3
2	18.51	19.00	19.16	19.25	19.30	19.33	19.35	19.37	19.38	19.40	19.43	19.45	19.46	19.47	19.48	19.49	19.50
3	10.13	9.55	9.28	9.12	9.01	8.94	8.89	8.85	8.81	8.79	8.70	8.66	8.62	8.59	8.57	8.55	8.53
4	7.71	6.94	6.59	6.39	6.26	6.16	6.09	6.04	6.00	5.96	5.86	5.80	5.75	5.72	5.69	5.66	5.63
5	6.61	5.79	5.41	5.19	5.05	4.95	4.88	4.82	4.77	4.74	4.62	4.56	4.50	4.46	4.43	4.40	4.36
6	5.99	5.14	4.76	4.53	4.39	4.28	4.21	4.15	4.10	4.06	3.94	3.87	3.81	3.77	3.74	3.70	3.67
7	5.59	4.74	4.35	4.12	3.97	3.87	3.79	3.73	3.68	3.64	3.51	3.44	3.38	3.34	3.30	3.27	3.23
8	5.32	4.46	4.07	3.84	3.69	3.58	3.50	3.44	3.39	3.35	3.22	3.15	3.08	3.04	3.01	2.97	2.93
9	5.12	4.26	3.86	3.63	3.48	3.37	3.29	3.23	3.18	3.14	3.01	2.94	2.86	2.83	2.79	2.75	2.71
10	4.96	4.10	3.71	3.48	3.33	3.22	3.14	3.07	3.02	2.98	2.85	2.77	2.70	2.66	2.62	2.58	2.54
15	4.54	3.68	3.29	3.06	2.90	2.79	2.71	2.64	2.59	2.54	2.40	2.33	2.25	2.20	2.16	2.11	2.07
20	4.35	3.49	3.10	2.87	2.71	2.60	2.51	2.45	2.39	2.35	2.20	2.12	2.04	1.99	1.95	1.90	1.84
30	4.17	3.32	2.92	2.69	2.53	2.42	2.33	2.27	2.21	2.16	2.01	1.93	1.84	1.79	1.74	1.68	1.62
40	4.08	3.23	2.84	2.61	2.45	2.34	2.25	2.18	2.12	2.08	1.92	1.84	1.74	1.69	1.64	1.58	1.51
60	4.00	3.15	2.76	2.53	2.37	2.25	2.17	2.10	2.04	1.99	1.84	1.75	1.65	1.59	1.53	1.47	1.39
120	3.92	3.07	2.68	2.45	2.29	2.17	2.09	2.02	1.96	1.91	1.75	1.66	1.55	1.50	1.43	1.35	1.25
∞	3.84	3.00	2.60	2.37	2.21	2.10	2.01	1.94	1.88	1.83	1.67	1.57	1.46	1.39	1.32	1.22	1.00

Adapted with permission from *Biometrika Tables for Statisticians*, *Vol.* 1, 3rd ed., Cambridge University Press, 1966, edited by E.S. Pearson and H. O. Hartley.

Table of the 0.975 Quantile of the F Distribution

If X has an F distribution with m and n degrees of freedom, the table gives the value of x such that $\Pr(X \leqslant x) = 0.975$.

n \ m	1	2	3	4	5	6	7	8	9	10	15	20	30	40	60	120	∞
1	647.8	799.5	864.2	899.6	921.8	937.1	948.2	956.7	963.3	968.6	984.9	993.1	1001	1006	1010	1014	1018
2	38.51	39.00	39.17	39.25	39.30	39.33	39.36	39.37	39.39	39.40	39.43	39.45	39.46	39.47	39.48	39.49	39.50
3	17.44	16.04	15.44	15.10	14.88	14.73	14.62	14.54	14.47	14.42	14.25	14.17	14.08	14.04	13.99	13.95	13.90
4	12.22	10.65	9.98	9.60	9.36	9.20	9.07	8.98	8.90	8.84	8.66	8.56	8.46	8.41	8.36	8.31	8.26
5	10.01	8.43	7.76	7.39	7.15	6.98	6.85	6.76	6.68	6.62	6.43	6.33	6.23	6.18	6.12	6.07	6.02
6	8.81	7.26	6.60	6.23	5.99	5.82	5.70	5.60	5.52	5.46	5.27	5.17	5.07	5.01	4.96	4.90	4.85
7	8.07	6.54	5.89	5.52	5.29	5.12	4.99	4.90	4.82	4.76	4.57	4.47	4.36	4.31	4.25	4.20	4.14
8	7.57	6.06	5.42	5.05	4.82	4.65	4.53	4.43	4.36	4.30	4.10	4.00	3.89	3.84	3.78	3.73	3.67
9	7.21	5.71	5.08	4.72	4.48	4.32	4.20	4.10	4.03	3.96	3.77	3.67	3.56	3.51	3.45	3.39	3.33
10	6.94	5.46	4.83	4.47	4.24	4.07	3.95	3.85	3.78	3.72	3.52	3.42	3.31	3.26	3.20	3.14	3.08
15	6.20	4.77	4.15	3.80	3.58	3.41	3.29	3.20	3.12	3.06	2.86	2.76	2.64	2.59	2.52	2.46	2.40
20	5.87	4.46	3.86	3.51	3.29	3.13	3.01	2.91	2.84	2.77	2.57	2.46	2.35	2.29	2.22	2.16	2.09
30	5.57	4.18	3.59	3.25	3.03	2.87	2.75	2.65	2.57	2.51	2.31	2.20	2.07	2.01	1.94	1.87	1.79
40	5.42	4.05	3.46	3.13	2.90	2.74	2.62	2.53	2.45	2.39	2.18	2.07	1.94	1.88	1.80	1.72	1.64
60	5.29	3.93	3.34	3.01	2.79	2.63	2.51	2.41	2.33	2.27	2.06	1.94	1.82	1.74	1.67	1.58	1.48
120	5.15	3.80	3.23	2.89	2.67	2.52	2.39	2.30	2.22	2.16	1.94	1.82	1.69	1.61	1.53	1.43	1.31
∞	5.02	3.69	3.12	2.79	2.57	2.41	2.29	2.19	2.11	2.05	1.83	1.71	1.57	1.48	1.39	1.27	1.00

Adapted with permission from *Biometrika Tables for Statisticians*, *Vol.* 1, 3rd ed., Cambridge University Press, 1966, edited by E. S. Pearson and H. O. Hartley.

Answers to Even-Numbered Exercises

Chapter 1

Sec. 1.4

6. (a) $\{x: x < 1 \text{ or } x > 5\}$; (b) $\{x: 1 \leqslant x \leqslant 7\}$; (c) B; (d) $\{x: 0 < x < 1 \text{ or } x > 7\}$; (e) $\emptyset$.

Sec. 1.5

2. $\frac{2}{5}$. 4. 0.4. 6. (a) $\frac{1}{2}$; (b) $\frac{1}{6}$; (c) $\frac{3}{8}$.

8. 0.4 if $A \subset B$ and 0.1 if $\Pr(A \cup B) = 1$.

10. (a) $1 - \frac{\pi}{4}$; (b) $\frac{3}{4}$; (c) $\frac{2}{3}$; (d) 0.

Sec. 1.6

2. $\frac{4}{7}$.   4. $\frac{1}{2}$.   6. $\frac{2}{3}$.

Sec. 1.7

2. $5!$.   4. $\frac{5}{18}$.   6. $\frac{20!}{8!20^{12}}$.   8. $\frac{(3!)^2}{6!}$.

Sec. 1.8

2. They are equal.

4. This number is $\binom{4251}{97}$ and therefore it must be an integer.

6. $\dfrac{n+1-k}{\dbinom{n}{k}}$.   8. $\dfrac{n+1}{\dbinom{2n}{n}}$.   10. $\dfrac{\dbinom{98}{10}}{\dbinom{100}{12}}$.   12. $\dfrac{\dbinom{20}{6}+\dbinom{20}{10}}{\dbinom{24}{10}}$.

16. $\dfrac{4\dbinom{13}{4}}{\dbinom{52}{4}}$.

Sec. 1.9

2. $\dfrac{300!}{5!8!287!}$.

4. $\dfrac{n!}{6^n n_1! n_2! \cdots n_6!}$.

6. $\dfrac{\dfrac{12!}{6!2!4!} \cdot \dfrac{13!}{4!6!3!}}{\dfrac{25!}{10!8!7!}}$.

8. $\dfrac{4!(13!)^4}{52!}$.

Sec. 1.10

2. 45 percent.   4. $\dfrac{9}{24}$.

6. $1 - \dfrac{1}{\dbinom{100}{15}} \left\{ \left[\dbinom{90}{15} + \dbinom{80}{15} + \dbinom{70}{15} + \dbinom{60}{15} \right] \right.$

$\qquad - \left[\dbinom{70}{15} + \dbinom{60}{15} + \dbinom{50}{15} + \dbinom{50}{15} + \dbinom{40}{15} + \dbinom{30}{15} \right]$

$\qquad \left. + \left[\dbinom{40}{15} + \dbinom{30}{15} + \dbinom{20}{15} \right] \right\}$.

8. $n = 10$.

10. $\dfrac{\dbinom{5}{r}\dbinom{5}{5-r}}{\dbinom{10}{5}}$, where $r = \dfrac{x}{2}$ and $x = 0, 2, \ldots, 10$.

Sec. 1.11

4. $1 - \dfrac{1}{10^6}$.   6. 0.92.   8. $\dfrac{1}{7}$.   10. $10(0.01)(0.99)^9$.

12. $n > \dfrac{\log(0.2)}{\log(0.99)}$.   14. $\dfrac{1}{12}$.

16. $[(0.8)^{10} + (0.7)^{10}] - [(0.2)^{10} + (0.3)^{10}]$.

Sec. 1.13

2. No.

4. (a) If and only if $A \cup B = S$. (b) Always.

6. $\frac{1}{6}$.

8. $1 - \left(\frac{49}{50}\right)^{50}$.

10. (a) 0.93. (b) 0.38.

12. $\binom{6}{3} p^3 (1-p)^3$.

14. $\frac{4}{81}$.

16. 0.067.

18. $\frac{1}{\binom{r+w}{r}}$.

20. $\dfrac{\binom{7}{j}\binom{3}{5-j}}{\binom{10}{5}}$, where $k = 2j - 2$ and $j = 2, 3, 4, 5$.

22. $p_1 + p_2 + p_3 - p_1 p_2 - p_2 p_3 - p_1 p_3 + p_1 p_2 p_3$, where

$$p_1 = \frac{\binom{6}{1}}{\binom{8}{3}}, \ p_2 = \frac{\binom{6}{2}}{\binom{8}{4}}, \ p_3 = \frac{\binom{6}{3}}{\binom{8}{5}}.$$

24. $\Pr(A \text{ wins}) = \frac{4}{7}$; $\Pr(B \text{ wins}) = \frac{2}{7}$; $\Pr(C \text{ wins}) = \frac{1}{7}$.

Chapter 2

Sec. 2.1

2. $\Pr(A)$.

4. $\dfrac{r(r+k)(r+2k)b}{(r+b)(r+b+k)(r+b+2k)(r+b+3k)}$.

6. $\frac{2}{3}$.

8. $\frac{1}{3}$

10. (a) $\frac{3}{4}$; (b) $\frac{3}{5}$.

12. (a) $p_2 + p_3 + p_4$, where $p_j = \binom{4}{j}\left(\frac{1}{4}\right)^j\left(\frac{3}{4}\right)^{4-j}$

Sec. 2.2

2. 0.47.

6. 0.301.

8. $\frac{18}{59}$.

10. (a) $0, \frac{1}{10}, \frac{2}{10}, \frac{3}{10}, \frac{4}{10}$; (b) $\frac{3}{4}$; (c) $\frac{1}{4}$.

Sec. 2.3

2. (a) 0.667; (b) 0.666.

4. (a) 0.38; (b) 0.338; (c) 0.3338.

6. (a) 0.632; (b) 0.605.

8. (a) $\frac{1}{8}$; (b) $\frac{1}{8}$.

10. (a) $\frac{40}{81}$; (b) $\frac{41}{81}$.

12.

	HHH	HHT	HTH	THH	TTH	THT	HTT	TTT
HHH	0	1	0	0	0	0	0	0
HHT	0	0	$\frac{1}{2}$	0	0	0	$\frac{1}{2}$	0
HTH	0	0	0	$\frac{1}{2}$	0	$\frac{1}{2}$	0	0
THH	$\frac{1}{2}$	$\frac{1}{2}$	0	0	0	0	0	0
TTH	0	0	0	$\frac{1}{2}$	0	$\frac{1}{2}$	0	0
THT	0	0	$\frac{1}{2}$	0	0	0	$\frac{1}{2}$	0
HTT	0	0	0	0	$\frac{1}{2}$	0	0	$\frac{1}{2}$
TTT	0	0	0	0	1	0	0	0

Sec. 2.4

2. Condition (a).   4. $i \geqslant 198$.   8. $\frac{2}{3}$.

Sec. 2.6

4. $\frac{11}{12}$.   6. $\dfrac{1}{\binom{10}{3}}$.   8. 0.372.

10. (a) 0.659. (b) 0.051.   12. $\dfrac{1 - \left(\frac{1}{2}\right)^{n-1}}{1 - \left(\frac{1}{2}\right)^{n}}$.

14. (a) $\dfrac{1 - p_0 - p_1}{1 - p_0}$, where $p_0 = \dfrac{\binom{48}{13}}{\binom{52}{13}}$ and $p_1 = \dfrac{4\binom{48}{12}}{\binom{52}{13}}$.

(b) $1 - p_1$.

18. $\frac{7}{9}$.

22. (a) The second condition. (b) The first condition.
(c) Equal probability under both conditions.

Chapter 3

Sec. 3.1

2. $f(0) = \frac{1}{6}$, $f(1) = \frac{5}{18}$, $f(2) = \frac{2}{9}$, $f(3) = \frac{1}{6}$, $f(4) = \frac{1}{9}$, $f(5) = \frac{1}{18}$.

4. $f(x) = \begin{cases} \dfrac{\binom{7}{x}\binom{3}{5-x}}{\binom{10}{5}} & \text{for } x = 2,3,4,5, \\ 0 & \text{otherwise.} \end{cases}$

6. 0.806.   8. $\dfrac{6}{\pi^2}$.

Sec. 3.2

2. (a) $\dfrac{1}{2}$; (b) $\dfrac{13}{27}$; (c) $\dfrac{2}{27}$.

4. (a) $t = 2$; (b) $t = \sqrt{8}$.

6. $f(x) = \begin{cases} \dfrac{1}{10} & \text{for } -2 \leqslant x \leqslant 8, \\ 0 & \text{otherwise,} \end{cases}$ and probability is $\dfrac{7}{10}$.

Sec. 3.3

4. $f(x) = (2/9)x$ for $0 \leqslant x \leqslant 3$;
   $f(x) = 0$ otherwise.

6. $F(x) = \begin{cases} 0 & \text{for } x < -2, \\ \dfrac{1}{10}(x + 2) & \text{for } -2 \leqslant x \leqslant 8, \\ 1 & \text{for } x > 8. \end{cases}$

Sec. 3.4

2. (a) $\dfrac{1}{40}$; (b) $\dfrac{1}{20}$; (c) $\dfrac{7}{40}$; (d) $\dfrac{7}{10}$.

4. (a) $\dfrac{5}{4}$; (b) $\dfrac{79}{256}$; (c) $\dfrac{13}{16}$; (d) 0.

6. (a) 0.55; (b) 0.8.

Sec. 3.5

2. (a) $f_1(x) = \begin{cases} \dfrac{1}{2} & \text{for } 0 \leqslant x \leqslant 2, \\ 0 & \text{otherwise.} \end{cases}$ $f_2(y) = \begin{cases} 3y^2 & \text{for } 0 \leqslant y \leqslant 1, \\ 0 & \text{otherwise.} \end{cases}$
   (b) Yes; (c) Yes.

4. (a) $f(x, y) = \begin{cases} p_x p_y & \text{for } x = 0,1,2,3 \text{ and } y = 0,1,2,3, \\ 0 & \text{otherwise.} \end{cases}$
   (b) 0.3; (c) 0.35.

6. Yes.

8. (a) $f(x, y) = \begin{cases} \dfrac{1}{6} & \text{for } (x, y) \in S, \\ 0 & \text{otherwise.} \end{cases}$

   $f_1(x) = \begin{cases} \dfrac{1}{2} & \text{for } 0 \leqslant x \leqslant 2, \\ 0 & \text{otherwise.} \end{cases}$ $f_2(y) = \begin{cases} \dfrac{1}{3} & \text{for } 1 \leqslant y \leqslant 4, \\ 0 & \text{otherwise.} \end{cases}$
   (b) Yes.

10. $\dfrac{11}{36}$.

Sec. 3.6

2. (a) For $-2 < x < 4$,

$$g_2(y|x) = \begin{cases} \dfrac{1}{2\left[9 - (x-1)^2\right]^{1/2}} & \text{for } (y+2)^2 < 9 - (x-1)^2, \\ 0 & \text{otherwise.} \end{cases}$$

(b) $\dfrac{2 - \sqrt{2}}{4}$.

4. (a) For $0 < y < 1$, $g_1(x|y) = \begin{cases} \dfrac{-1}{(1-x)\log(1-y)} & \text{for } 0 < x < y, \\ 0 & \text{otherwise.} \end{cases}$

(b) $\dfrac{1}{2}$.

6. (a) For $0 < x < 2$, $g_2(y|x) = \begin{cases} \dfrac{4 - 2x - y}{2(2-x)^2} & \text{for } 0 < y < 4 - 2x, \\ 0 & \text{otherwise.} \end{cases}$

(b) $\dfrac{1}{9}$.

8. (a) $f_1(x) = \begin{cases} \dfrac{1}{2}x(2 + 3x) & \text{for } 0 < x < 1, \\ 0 & \text{otherwise.} \end{cases}$

(b) $\dfrac{8}{11}$.

Sec. 3.7

2. (a) 6; (b) $f_{13}(x_1, x_3) = \begin{cases} 3e^{-(x_1 + 3x_3)} & \text{for } x_i > 0 \,(i = 1, 3), \\ 0 & \text{otherwise.} \end{cases}$ (c) $1 - \dfrac{1}{e}$.

6. $\displaystyle\sum_{i=k}^{n} \binom{n}{i} p^i (1-p)^{n-i}$, where $p = \displaystyle\int_a^b f(x)\,dx$.

Sec. 3.8

2. $G(y) = 1 - (1-y)^{1/2}$ for $0 < y < 1$;

$$g(y) = \begin{cases} \dfrac{1}{2(1-y)^{1/2}} & \text{for } 0 < y < 1, \\ 0 & \text{otherwise.} \end{cases}$$

6. (a) $g(y) = \begin{cases} \dfrac{1}{2}y^{-1/2} & \text{for } 0 < y < 1, \\ 0 & \text{otherwise.} \end{cases}$

(b) $g(y) = \begin{cases} \dfrac{1}{3}|y|^{-2/3} & \text{for } -1 < y < 0, \\ 0 & \text{otherwise.} \end{cases}$

(c) $g(y) = \begin{cases} 2y & \text{for } 0 < y < 1, \\ 0 & \text{otherwise.} \end{cases}$

8. $Y = 2X^{1/3}$.

Sec. 3.9

2. $g(y) = \begin{cases} y & \text{for } 0 < y \leqslant 1, \\ 2 - y & \text{for } 1 < y < 2, \\ 0 & \text{otherwise.} \end{cases}$

4. $g(z) = \begin{cases} \dfrac{1}{3}(z + 1) & \text{for } 0 < z \leqslant 1, \\ \dfrac{1}{3z^3}(z + 1) & \text{for } z > 1, \\ 0 & \text{for } z \leqslant 0. \end{cases}$

6. $g(y) = \dfrac{1}{2}e^{-|y|}$ for $-\infty < y < \infty$.   8. $(0.8)^n - (0.7)^n$.   10. $\left(\dfrac{1}{3}\right)^n + \left(\dfrac{2}{3}\right)^n$.

12. $f(z) = \begin{cases} \dfrac{n(n-1)}{8}\left(\dfrac{z}{8}\right)^{n-2}\left(1 - \dfrac{z}{8}\right) & \text{for } 0 < z < 8, \\ 0 & \text{otherwise.} \end{cases}$

Sec. 3.11

2. $f(x) = \begin{cases} \dfrac{2}{5} & \text{for } 0 < x < 1, \\ \dfrac{3}{5} & \text{for } 1 < x < 2, \\ 0 & \text{otherwise.} \end{cases}$

4. $\dfrac{\pi}{4}$.   6. $1 - \dfrac{1}{2^{p-1}} + \dfrac{1}{2^{2p-1}}$.   8. $\dfrac{1}{10}$.

10. $Y = 5(1 - e^{-2X})$ or $Y = 5e^{-2X}$.

12. The sets (c) and (d).   14. 0.3715.

16. $f_2(y) = -9y^2 \log y$ for $0 < y < 1$.

$g_1(x|y) = -\dfrac{1}{x \log y}$ for $0 < y < x < 1$.

18. $f_1(x) = 3(1 - x)^2$ for $0 < x < 1$,
$f_2(y) = 6y(1 - y)$ for $0 < y < 1$,
$f_3(z) = 3z^2$ for $0 < z < 1$.

20. (a) $g(u, v) = \begin{cases} ve^{-v} & \text{for } 0 < u < 1, v > 0, \\ 0 & \text{otherwise.} \end{cases}$
(b) Yes.

22. $h(y_1|y_n) = \dfrac{(n-1)(e^{-y_1} - e^{-y_n})^{n-2}e^{-y_1}}{(1 - e^{-y_n})^{n-1}}$ for $0 < y_1 < y_n$.

Chapter 4

Sec. 4.1

2. 18.92.   4. 4.867.   8. $\dfrac{3}{4}$.   10. $\dfrac{1}{n+1}$ and $\dfrac{n}{n+1}$.

Sec. 4.2

2. $\frac{1}{2}$.

4. $n \int_a^b f(x)\, dx.$

6. $c \left(\frac{5}{4} \right)^n.$

8. $n(2p - 1).$

10. $2k.$

Sec. 4.3

2. $\frac{1}{12}(b - a)^2.$

6. (a) 6; (b) 39.

Sec. 4.4

2. 0.

6. $\mu = \frac{1}{2},\ \sigma^2 = \frac{3}{4}.$

8. $E(Y) = c\mu;\ \text{Var}(Y) = c(\sigma^2 + \mu^2).$

10. $f(1) = \frac{1}{5};\ f(4) = \frac{2}{5};\ f(8) = \frac{2}{5}.$

Sec. 4.5

2. $m = \log 2.$

4. (a) $\frac{1}{2}(\mu_f + \mu_g)$; (b) Any number m such that $1 \leqslant m \leqslant 2.$

6. (a) $\frac{7}{12}$; (b) $\frac{1}{2}(\sqrt{5} - 1).$

8. (a) 0.1; (b) 1.   10. $Y.$

Sec. 4.6

10. The value of $\rho(X, Y)$ would be less than $-1.$

12. (a) 11; (b) 51.

14. $n + \dfrac{n(n - 1)}{4}.$

Sec. 4.7

4. $1 - \dfrac{1}{2^n}.$

6. $E(Y|X) = \dfrac{3X + 2}{3(2X + 1)};\ \text{Var}(Y|X) = \dfrac{1}{36}\left[3 - \dfrac{1}{(2X + 1)^2} \right].$

8. $\dfrac{1}{12} - \dfrac{\log 3}{144}.$

12. (a) $\frac{3}{5}$; (b) $\dfrac{\sqrt{29} - 3}{4}.$

14. (a) $\dfrac{18}{31}$; (b) $\dfrac{\sqrt{5} - 1}{2}.$

Sec. 4.8

4. 25.

12. (a) Yes; (b) No.

Sec. 4.9

2. $Z.$

4. $\frac{2}{3}$

6. $p.$

8. $a = 1$ if $p > \frac{1}{2}$; $a = 0$ if $p < \frac{1}{2}$; a can be chosen arbitrarily if $p = \frac{1}{2}.$

12. $b = A$ if $p > \frac{1}{2}$; $b = 0$ if $p < \frac{1}{2}$; b can be chosen arbitrarily if $p = \frac{1}{2}.$

14. $x_0 > \dfrac{4}{(\alpha + 1)^{1/\alpha}}.$

Sec. 4.10

4. $a = \pm \dfrac{1}{\sigma}, \; b = -a\mu.$

6. $\dfrac{3}{2}.$

10. Order an amount s such that

$$\int_0^s f(x)\, dx = \frac{g}{g+c}.$$

12. (a) and (b) $E(Z) = 29$; $\mathrm{Var}(Z) = 109.$
    (c) $E(Z) = 29$; $\mathrm{Var}(Z) = 94.$

16. 1.

20. $-\dfrac{1}{2}.$

24. (a) 0.1333. (b) 0.1414.

26. $a = pm.$

Chapter 5

Sec. 5.2

4. 0.5000.

6. $\dfrac{113}{64}.$

8. $\dfrac{k}{n}.$

10. $n(n-1)p^2.$

Sec. 5.3

2. $E(\overline{X}) = \dfrac{1}{3}$; $\mathrm{Var}(\overline{X}) = \dfrac{8}{441}.$

6. $\dfrac{T-1}{2}$ or $\dfrac{T+1}{2}$ if T is odd, and $\dfrac{T}{2}$ if T is even.

8. (a) $\dfrac{\dbinom{0.7T}{10} + 0.3T\dbinom{0.7T}{9}}{\dbinom{T}{10}}$; (b) $(0.7)^{10} + 10(0.3)(0.7)^9.$

Sec. 5.4

2. 0.0165.

4. $\displaystyle\sum_{x=m}^{n} \binom{n}{x} \left(\sum_{i=k+1}^{\infty} \frac{e^{-\lambda}\lambda^i}{i!} \right)^x \left(\sum_{i=0}^{k} \frac{e^{-\lambda}\lambda^i}{i!} \right)^{n-x}.$

6. $\displaystyle\sum_{x=21}^{\infty} \frac{e^{-30} 30^x}{x!}.$

8. Poisson distribution with mean $p\lambda.$

10. If λ is not an integer, mode is the greatest integer less than λ. If λ is an integer, modes are λ and $\lambda - 1.$

12. 0.3476.

Sec. 5.5

2. (a) 150; (b) 4350.

8. Geometric distribution with parameter $p = 1 - \displaystyle\prod_{i=1}^{n} q_i.$

Sec. 5.6

2. Normal with $\mu = 20$ and $\sigma = \dfrac{20}{9}$.

4. $(0.1360)^3$.   6. 0.6826.   8. $n = 1083$.   10. 0.3811.

12. (a) $\dfrac{\exp\left\{-\dfrac{1}{2}(x-25)^2\right\}}{\exp\left\{-\dfrac{1}{2}(x-25)^2\right\} + 9\exp\left\{-\dfrac{1}{2}(x-20)^2\right\}}$; (b) $x > 22.5 + \dfrac{1}{5}\log 9$.

14. $f(x) = \dfrac{1}{(2\pi)^{1/2}\sigma x}\exp\left\{-\dfrac{1}{2\sigma^2}(\log x - \mu)^2\right\}$ for $x > 0$, and $f(x) = 0$ for $x \leqslant 0$.

Sec. 5.7

2. 0.9938 4. $n \geqslant 542$.   6. 0.7385.

8. (a) 0.36; (b) 0.7888.   10. 0.9938.

Sec. 5.8

2. 0.0012.   4. 0.9938.   6. 0.7539.

Sec. 5.9

6. 0.1587.   8. $\dfrac{1}{e}$.   10. $\left(\dfrac{1}{n} + \dfrac{1}{n-1} + \dfrac{1}{n-2}\right)\dfrac{1}{\beta}$.

12. $1 - e^{-5/2}$.   14. $e^{-5/4}$ 16. $1 \cdot 3 \cdot 5 \cdots (2n-1)\sigma^{2n}$.

Sec. 5.10

4. $\dfrac{\alpha(\alpha+1)\cdots(\alpha+r-1)\beta(\beta+1)\cdots(\beta+s-1)}{(\alpha+\beta)(\alpha+\beta+1)\cdots(\alpha+\beta+r+s-1)}$.

Sec. 5.11

2. $\dfrac{2424}{6^5}$.   4. 0.0501.

Sec. 5.12

2. 0.1562.   4. 90 and 36.

6. $\mu_1 = 4$, $\mu_2 = -2$, $\sigma_1 = 1$, $\sigma_2 = 2$, $\rho = -0.3$.

Sec. 5.13

2. 0.0404.   6. $3\mu\sigma^2 + \mu^3$.   8. 0.8152.

10. $\dfrac{15}{7}$.   12. 8.00.

14. (a) Exponential, parameter $\beta = 5$.
    (b) Gamma, parameters $\alpha = k$ and $\beta = 5$.
    (c) $e^{-5(k-1)/3}$.

22. Without continuity correction, 0.473; with continuity correction, 0.571; exact probability, 0.571.

24. (a) $\rho(X_i, X_j) = -\left(\dfrac{p_i}{1 - p_i} \cdot \dfrac{p_j}{1 - p_j} \right)^{1/2}$,

    where p_i is the proportion of students in class i.
    (b) $i = 1$, $j = 2$.   (c) $i = 3$, $j = 4$.

26. Normal with $\mu = -3$ and $\sigma^2 = 16$; $\rho(X, Y) = \dfrac{1}{2}$.

Chapter 6

Sec. 6.2

2. $\xi(1.0 | X = 3) = 0.2456$; $\xi(1.5 | X = 3) = 0.7544$.

4. The p.d.f. of a beta distribution with parameters $\alpha = 3$ and $\beta = 6$.

6. Beta distribution with parameters $\alpha = 4$ and $\beta = 7$.

8. Beta distribution with parameters $\alpha = 4$ and $\beta = 6$.

10. Uniform distribution on the interval (11.2, 11.4).

Sec. 6.3

2. Beta distribution with parameters $\alpha = 5$ and $\beta = 297$.

4. Gamma distribution with parameters $\alpha = 16$ and $\beta = 6$.

6. Normal distribution with mean 69.07 and variance 0.286.

8. Normal distribution with mean 0 and variance $\dfrac{1}{5}$.

12. $n \geqslant 100$.

16. $\xi(\theta | x) = \begin{cases} \dfrac{6(8^6)}{\theta^7} & \text{for } \theta > 8, \\ 0 & \text{for } \theta \leqslant 8. \end{cases}$

18. $\dfrac{\alpha + n}{\beta - \sum_{i=1}^{n} \log x_i}$ and $\dfrac{\alpha + n}{\left(\beta - \sum_{i=1}^{n} \log x_i \right)^2}$.

Sec. 6.4

2. (a) 12 or 13; (b) 0.   4. $\dfrac{8}{3}$.   8. $n \geqslant 396$.

12. $\dfrac{\alpha + n}{\alpha + n - 1}\max(x_0, X_1, \ldots, X_n)$.

Sec. 6.5

2. $\dfrac{2}{3}$.   4. (a) $\hat{\theta} = \bar{x}_n$.   6. $\hat{\beta} = \dfrac{1}{\bar{X}_n}$.

8. $\hat{\theta} = -\dfrac{n}{\sum_{i=1}^{n}\log X_i}$.

10. $\hat{\theta}_1 = \min(X_1, \ldots, X_n)$; $\hat{\theta}_2 = \max(X_1, \ldots, X_n)$.   12. $\hat{\mu}_1 = \bar{X}_n$; $\hat{\mu}_2 = \bar{Y}_n$.

Sec. 6.6

2. $\hat{m} = \bar{X}_n\log 2$.   4. $\hat{\mu} = \dfrac{1}{2}[\min\{X_1, \ldots, X_n\} + \max\{X_1, \ldots, X_n\}]$.

6. $\hat{\nu} = \Phi\left(\dfrac{\hat{\mu} - 2}{\hat{\sigma}}\right)$.   8. $\bar{X}_n$.   14. $\hat{\mu} = 6.75$.   16. $\hat{p} = \dfrac{2}{5}$.

Sec. 6.8

8. Yes.   10. No.   12. Yes.   14. Yes   16. Yes.

Sec. 6.9

2. $R(\theta, \delta_1) = \dfrac{\theta^2}{3n}$.   4. $c^* = \dfrac{n + 2}{n + 1}$.   6. $R(\beta, \delta) = (\beta - 3)^2$.

10. $\hat{\theta} = \delta_0$.   12. $\left(\dfrac{n - 1}{n}\right)^T$.

Sec. 6.10

2. $\dfrac{6}{17}$.   4. $\dfrac{\sigma_2^2 b_1 x_1 + \sigma_1^2 b_2 x_2}{\sigma_2^2 b_1^2 + \sigma_1^2 b_2^2}$.

6. (a) $\dfrac{1}{3}\left(X_1 + \dfrac{1}{2}X_2 + \dfrac{1}{3}X_3\right)$.

   (b) Gamma distribution, parameters $\alpha + 3$ and $\beta + x_1 + \dfrac{1}{2}x_2 + \dfrac{1}{3}x_3$.

8. (a) $x + 1$. (b) $x + \log 2$.

10. $\hat{p} = 2\left(\hat{\theta} - \dfrac{1}{4}\right)$, where

$$\hat{\theta} = \begin{cases} \dfrac{X}{n} & \text{if } \dfrac{1}{4} \leqslant \dfrac{X}{n} \leqslant \dfrac{3}{4}, \\[2mm] \dfrac{1}{4} & \text{if } \dfrac{X}{n} < \dfrac{1}{4}, \\[2mm] \dfrac{3}{4} & \text{if } \dfrac{X}{n} > \dfrac{3}{4}. \end{cases}$$

12. $2^{1/5}$.   14. $\min(X_1, \ldots, X_n)$.

16. $\hat{x}_0 = \min(X_1, \ldots, X_n)$, and $\hat{\alpha} = \left(\frac{1}{n}\sum_{i=1}^{n}\log x_i - \log \hat{x}_0\right)^{-1}$.

18. The smallest integer greater than $\frac{x}{p} - 1$. If $\frac{x}{p} - 1$ is itself an integer, both $\frac{x}{p} - 1$ and $\frac{x}{p}$ are M.L.E.'s.

20. 16.

Chapter 7

Sec. 7.1

2. $n \geqslant 255$.   4. $n = 10$.   6. $n \geqslant 16$.

Sec. 7.2

4. 0.20.   8. χ^2 distribution with one degree of freedom.

10. $\dfrac{2^{1/2}\Gamma[(n+1)/2]}{\Gamma(n/2)}$.

Sec. 7.3

6. (a) $n = 21$; (b) $n = 13$.   8. The same for both samples.

Sec. 7.4

4. $c = \sqrt{3/2}$.   6. 0.70.

Sec. 7.5

2. (a) $6.16\sigma^2$; (b) $2.05\sigma^2$; (c) $0.56\sigma^2$; (d) $1.80\sigma^2$; (e) $2.80\sigma^2$; (f) $6.12\sigma^2$.

Sec. 7.6

4. $\mu_0 = -5$; $\lambda_0 = 4$; $\alpha_0 = 2$; $\beta_0 = 4$.

6. The conditions imply that $\alpha_0 = \frac{1}{4}$, and $E(\mu)$ exists only for $\alpha_0 > \frac{1}{2}$.

8. (a) (7.084, 7.948); (b) (7.031, 7.969).   10. (0.446, 1.530).   12. (0.724, 3.336).

Sec. 7.7

2. $\dfrac{1}{n}\sum_{i=1}^{n}X_i^2 - \dfrac{1}{n-1}\sum_{i=1}^{n}(X_i - \bar{X}_n)^2$.   4. $\delta(X) = 2^X$.

10. (a) All values; (b) $\alpha = \dfrac{m}{m + 4n}$.   14. (c) $c_0 = \dfrac{1}{3}(1 + \theta_0)$.

Sec. 7.8

2. $I(\theta) = \dfrac{1}{\theta}$.   4. $I(\sigma^2) = \dfrac{1}{2\sigma^4}$.

16. Normal with mean θ^3 and variance $\dfrac{9\theta^4\sigma^2}{n}$.

Sec. 7.9

6. (a) For $\alpha(m-1) + 2\beta(n-1) = 1$. (b) $\alpha = \dfrac{1}{m+n-2}$, $\beta = \dfrac{1}{2(m+n-2)}$.

8. $\dfrac{Y}{2\left[\dfrac{S_n^2}{n-1}\right]^{1/2}}$.

10. $\overline{X}_n - c\left[\dfrac{S_n^2}{n(n-1)}\right]^{1/2}$, where c is the 0.99 quantile of the t distribution with $n-1$ degrees of freedom.

12. (a) $(\mu_1 - 1.96\nu_1, \mu_1 + 1.96\nu_1)$, where μ_1 and ν_1 are given by Eqs. (1) and (2) of Sec. 6.3.

14. Normal, with mean θ and variance $\dfrac{\theta^2}{n}$.

Chapter 8

Sec. 8.1

2. (a) $\pi(0) = 1$, $\pi(0.1) = 0.3941$, $\pi(0.2) = 0.1558$,
   $\pi(0.3) = 0.3996$, $\pi(0.4) = 0.7505$, $\pi(0.5) = 0.9423$,
   $\pi(0.6) = 0.9935$, $\pi(0.7) = 0.9998$, $\pi(0.8) = 1.0000$,
   $\pi(0.9) = 1.0000$, $\pi(1) = 1.0000$; (b) 0.1558.

4. (a) Simple; (b) Composite; (c) Composite; (d) Composite.

Sec. 8.2

2. (b) 1.
4. (a) Reject H_0 when $\overline{X}_n > 5 - 1.645n^{-1/2}$;
   (b) $\alpha(\delta) = 0.0877$.

6. (b) $c = 31.02$.
8. $\beta(\delta) = \left(\dfrac{1}{2}\right)^n$.

10. (a) 0.6170; (b) 0.3174; (c) 0.0454; (d) 0.0026.
12. $X > 50.653$.

14. Decide that failure was caused by a major defect if $\displaystyle\sum_{i=1}^n X_i > \dfrac{4n + \log(0.64)}{\log(7/3)}$.

Sec. 8.3

2. (b) $\xi_1 = \dfrac{8}{17}$ and $\xi_2 = \dfrac{9}{17}$ or $\xi_1 = \dfrac{2}{3}$ and $\xi_2 = \dfrac{1}{3}$.
4. $X \leqslant 51.40$.

Sec. 8.4

6. The power function is 0.05 for every value of θ.
8. $c = 36.62$.

12. (a) Reject H_0 if $\overline{X}_n \leqslant 9.359$; (b) 0.7636; (c) 0.9995.

Sec. 8.5

2. $c_1 = \mu_0 - 1.645n^{-1/2}$ and $c_2 = \mu_0 + 1.645n^{-1/2}$.
4. $n = 11$.
6. $c_1 = -0.424$ and $c_2 = 0.531$.

Sec. 8.6

2. Since $U = -1.809$, do not reject the claim.   4. Accept H_0.

8. Since $\dfrac{S_n^2}{4} < 16.92$, accept H_0.

Sec. 8.7

2. $U = \dfrac{26}{3}$; the corresponding tail area is very small.

4. $U = \dfrac{13}{3}$; the corresponding tail area is very small.

6. (a) $c = 1.96$.   8. 0.0013.

Sec. 8.8

2. $c = 1.228$.   4. 1.   6. (a) $\hat{\sigma}_1^2 = 7.625$ and $\hat{\sigma}_2^2 = 3.96$;
                                              (b) Accept H_0.

8. $c_1 = 0.321$ and $c_2 = 3.77$.   10. $0.265V < r < 3.12V$.

Sec. 8.9

2. $c_1 = -1.782$ and $c_2 = 1.782$; H_0 will be accepted.

4. Since $U = -1.672$, reject H_0.   6. $-0.320 < \mu_1 - \mu_2 < 0.008$.

Sec. 8.10

2. Reject H_0 for $X \leqslant 6$.

4. Reject H_0 for $X > 1 - \alpha^{1/2}$; $\beta(\delta) = (1 - \alpha^{1/2})^2$.

6. Reject H_0 for $X \leqslant \dfrac{1}{2}[(1.4)^{1/2} - 1]$.

8. Reject H_0 for $X \leqslant 0.01$ or $X \geqslant 1$; power is 0.6627.

10. 0.0093.   16. (a) 1. (b) $\dfrac{1}{\alpha}$.

Chapter 9

Sec. 9.1

2. $Q = 7.4$; corresponding tail area is 0.6.

6. $Q = 11.5$; reject the hypothesis.

8. (a) $Q = 5.4$ and corresponding tail area is 0.25; (b) $Q = 8.8$ and corresponding tail area is between 0.4 and 0.5.

Sec. 9.2

2. (a) $\hat{\theta}_1 = \dfrac{2N_1 + N_4 + N_5}{2n}$ and $\hat{\theta}_2 = \dfrac{2N_2 + N_4 + N_6}{2n}$.

(b) $Q = 4.37$ and corresponding tail area is 0.226.

4. $\hat{\theta} = 1.5$ and $Q = 7.56$; corresponding tail area lies between 0.1 and 0.2.

Sec. 9.3

4. $Q = 8.6$; corresponding tail area lies between 0.025 and 0.05.

Sec. 9.4

2. $Q = 18.9$; corresponding tail area is between 0.1 and 0.05.

4. Correct value of Q is 7.2, for which the corresponding tail area is less than 0.05.

Sec. 9.5

6. (b)

	Proportion helped	
	Older subjects	Younger subjects
Treatment I	0.433	0.700
Treatment II	0.400	0.667

(c)

	Proportion helped, all subjects
Treatment I	0.500
Treatment II	0.600

Sec. 9.6

4. $D_n^* = 0.15$; corresponding tail area is 0.63.

6. $D_n^* = 0.065$; corresponding tail area is approximately 0.98.

8. $D_{mn} = 0.27$; corresponding tail area is 0.39.

10. $D_{mn} = 0.50$; corresponding tail area is 0.008.

Sec. 9.7

2. (a) $\left(\frac{1}{2}\right)^n$; (b) $(n + 1)\left(\frac{1}{2}\right)^n$.

4. $I_1 = (5.12, 7.12)$ and confidence coefficient is 0.4375;
    $I_2 = (7.11, 7.13)$ and confidence coefficient is 0.5469.

6. $(0.3)^n$.   8. $n = 9$.   10. 0.78.

Sec. 9.8

2. (a) 22.17; (b) 20.57, 22.02, 22.00, 22.00; (c) 22.10.

6. 0.575.   8. M.S.E. $(\bar{X}_n) = 0.025$ and M.S.E. $(\tilde{X}_n) = 0.028$.

Sec. 9.9

6. (a) Tail area is smaller than 0.005; (b) Tail area is slightly larger than 0.01.

Sec. 9.10

2. $U = -1.341$; corresponding (two-sided) tail area is between 0.10 and 0.20.

4. $D_{mn} = 0.5333$; corresponding tail area is 0.010.

Sec. 9.11

2. Any level greater than 0.005, the smallest probability given in the table in this book.

4. Do not reject the hypothesis.

8. $|a| > \dfrac{1}{2}(6.635n)^{1/2}$.

14. Normal, with mean $\left(\dfrac{1}{2}\right)^{1/\theta}$ and variance $\dfrac{1}{n\theta^2 4^{1/\theta}}$.

16. (a) $0.031 < \alpha < 0.994$. (b) $\sigma < 0.447$ or $\sigma > 2.237$.

18. Uniform on the interval (y_1, y_3).

Chapter 10

Sec. 10.1

4. $y = -1.670 + 1.064x$.

6. (a) $y = 40.893 + 0.548x$; (b) $y = 38.483 + 3.440x - 0.643x^2$.

8. $y = 3.7148 + 1.1013x_1 + 1.8517x_2$.

10. The sum of the squares of the deviations of the observed values from the fitted curve is smaller in Exercise 9.

Sec. 10.2

8. -0.775

14. -0.891.

10. $c_2 = 3\bar{x}_n = 6.99$.

16. $c_2 = -\bar{x}_n = -2.25$.

12. $x = \bar{x}_n = 2.33$.

18. $x = \bar{x}_n = 2.25$.

Sec. 10.3

2. Since $U_1 = -6.695$, reject H_0.

4. Since $U_2 = -6.894$, reject H_0.

6. Since $|U_{12}| = 0.664$, accept H_0.

10. Since $U^2 = 24.48$, reject H_0.

12. $0.246 < \beta_2 < 0.624$.

14. $0.284 < y < 0.880$.

18. $10(\beta_1 - 0.147)^2 + 10.16(\beta_2 - 0.435)^2 + 8.4(\beta_1 - 0.147)(\beta_2 - 0.435) < 0.503$.

Sec. 10.5

4. $\hat{\beta} = 5.126$, $\hat{\sigma}^2 = 16.994$, and $\text{Var}(\hat{\beta}) = 0.0150\sigma^2$.

6. $\hat{\beta} = -0.744$, $\hat{\beta}_2 = 0.616$, $\hat{\beta}_3 = 0.013$, $\hat{\sigma}^2 = 0.937$.

8. $U_3 = 0.095$; corresponding tail area is greater than 0.90.

10. $R^2 = 0.644$.

12. $\text{Var}(\hat{\beta}_1) = 222.7\sigma^2$, $\text{Var}(\hat{\beta}_2) = 0.1355\sigma^2$, $\text{Var}(\hat{\beta}_3) = 0.0582\sigma^2$, $\text{Cov}(\hat{\beta}_1, \hat{\beta}_2) = 4.832\sigma^2$, $\text{Cov}(\hat{\beta}_1, \hat{\beta}_3) = -3.598\sigma^2$, $\text{Cov}(\hat{\beta}_2, \hat{\beta}_3) = -0.0792\sigma^2$.

14. $U_3 = 4.319$; corresponding tail area is less than 0.01.

20. The value of the F statistic with 2 and 7 degrees of freedom is 1.615; corresponding tail area is greater than 0.05.

24. 87.

Sec. 10.6

6. $U^2 = 13.09$; corresponding tail area is less than 0.025.

Sec. 10.7

4. $\mu = 3.25$, $\alpha_1 = -2$, $\alpha_2 = 3$, $\alpha_3 = -1$, $\beta_1 = 1.75$, $\beta_2 = -2.25$, $\beta_3 = -1.25$, $\beta_4 = 1.75$.

12. $\hat{\sigma}^2 = 1.9647$.

14. $U_B^2 = 4.664$; corresponding tail area is between 0.05 and 0.025.

Sec. 10.8

2. (a) $\mu = 9$, $\alpha_1 = -3$, $\alpha_2 = 3$, $\beta_1 = -1.5$, $\beta_2 = 1.5$, $\gamma_{11} = \gamma_{22} = \dfrac{1}{2}$, $\gamma_{12} = \gamma_{21} = -\dfrac{1}{2}$.

(b) $\mu = 5$, $\alpha_1 = -\dfrac{1}{2}$, $\alpha_2 = \dfrac{1}{2}$, $\beta_1 = -\dfrac{3}{2}$, $\beta_2 = \dfrac{3}{2}$, $\gamma_{11} = \gamma_{12} = \gamma_{21} = \gamma_{22} = 0$.

(c) $\mu = 3\dfrac{1}{4}$, $\alpha_1 = -2$, $\alpha_2 = 3$, $\alpha_3 = -1$, $\beta_1 = 1\dfrac{3}{4}$, $\beta_2 = -2\dfrac{1}{4}$, $\beta_3 = -1\dfrac{1}{4}$, $\beta_4 = 1\dfrac{3}{4}$, $\gamma_{ij} = 0$ for all values of i and j.

(d) $\mu = 5$, $\alpha_1 = -2\dfrac{1}{2}$, $\alpha_2 = 0$, $\alpha_3 = 2\dfrac{1}{2}$, $\beta_1 = -3$, $\beta_2 = -1$, $\beta_3 = 1$, $\beta_4 = 3$, $\gamma_{11} = 1\dfrac{1}{2}$, $\gamma_{12} = \dfrac{1}{2}$, $\gamma_{13} = -\dfrac{1}{2}$, $\gamma_{14} = -1\dfrac{1}{2}$, $\gamma_{21} = \gamma_{22} = \gamma_{23} = \gamma_{24} = 0$, $\gamma_{31} = -1\dfrac{1}{2}$, $\gamma_{32} = -\dfrac{1}{2}$, $\gamma_{33} = \dfrac{1}{2}$, $\gamma_{34} = 1\dfrac{1}{2}$.

10. $U_{AB}^2 = 0.7047$; corresponding tail area is much larger than 0.05.

12. $U_B^2 = 9.0657$; corresponding tail area is less than 0.025.

14. The value of the appropriate statistic having a t distribution with 12 degrees of freedom is 2.8673; the corresponding tail area is between 0.01 and 0.005.

Sec. 10.9

2. $E(T) = \dfrac{\rho \sigma_2}{\sigma_1}$; $\mathrm{Var}(T) = \dfrac{(1 - \rho^2)\sigma_2^2}{\sum_{i=1}^{n}(x_i - \bar{x}_n)^2}$.

6. $\beta_2 = \dfrac{\sum_{i=1}^{n}(y_i'^2 - x_i'^2) \pm \left\{ \left[\sum_{i=1}^{n}(y_i'^2 - x_i'^2)\right]^2 + 4\left(\sum_{i=1}^{n}x_i' y_i'\right)^2 \right\}^{1/2}}{2\sum_{i=1}^{n}x_i' y_i'}$,

$\beta_1 = \bar{y}_n - \beta_2 \bar{x}_n$, where $x_i' = x_i - \bar{x}_n$ and $y_i' = y_i - \bar{y}_n$. Either the plus sign or the minus sign in β_2 should be used, depending on whether the optimal line has a positive or a negative slope.

8. $\dfrac{1}{n} \sum\limits_{i=1}^{k} n_i \left[v_i^2 + (\bar{x}_{i+} - \bar{x}_{++})^2 \right].$

10. $\dfrac{1}{IJ(K-1)} \sum\limits_{i,j,k} (Y_{ijk} - \bar{Y}_{ij+})^2.$

12. Let $A = JK\sum\limits_{i} (\bar{Y}_{i++} - \bar{Y}_{+++})^2,$

$B = IK\sum\limits_{j} (\bar{Y}_{+j+} - \bar{Y}_{+++})^2,$

$C = K\sum\limits_{i}\sum\limits_{j} (\bar{Y}_{ij+} - \bar{Y}_{i++} - \bar{Y}_{+j+} + \bar{Y}_{+++})^2,$

$R = \sum\limits_{i}\sum\limits_{j}\sum\limits_{k} (Y_{ijk} - \bar{Y}_{ij+})^2,$

$U = \dfrac{IJ(K-1)(A+B+C)}{(IJ-1)R}.$

Reject H_0 if $U \geqslant c$. Under H_0, U has an F distribution with $IJ - 1$ and $IJ(K-1)$ degrees of freedom.

14. $\hat{\theta}_1 = \dfrac{1}{4}(Y_1 + Y_2) + \dfrac{1}{2}Y_3,$

$\hat{\theta}_2 = \dfrac{1}{4}(Y_1 + Y_2) - \dfrac{1}{2}Y_3,$

$\hat{\sigma}^2 = \dfrac{1}{3}[(Y_1 - \hat{\theta}_1 - \hat{\theta}_2)^2 + (Y_2 - \hat{\theta}_1 - \hat{\theta}_2)^2 + (Y_3 - \hat{\theta}_1 + \hat{\theta}_2)^2],$

where $Y_1 = W_1$, $Y_2 = W_2 - 5$, $Y_3 = \dfrac{1}{2} W_3$; $(\hat{\theta}_1, \hat{\theta}_2)$ and $\hat{\sigma}^2$ are independent; $(\hat{\theta}_1, \hat{\theta}_2)$ has a bivariate normal distribution with mean vector (θ_1, θ_2) and covariance matrix

$$\begin{bmatrix} \dfrac{3}{8} & -\dfrac{1}{8} \\[2mm] -\dfrac{1}{8} & \dfrac{3}{8} \end{bmatrix} \sigma^2;$$

$\dfrac{3\hat{\sigma}^2}{\sigma^2}$ has a χ^2 distribution with one degree of freedom.

16. $\operatorname{Var}(Z_i) = \left[1 - \dfrac{1}{n} - \dfrac{(x_i - \bar{x}_n)^2}{\sum_{j=1}^{n}(x_j - \bar{x}_n)^2} \right] \sigma^2.$

18. $\mu = \bar{\theta} + \bar{\psi}$; $\alpha_i = \theta_i - \bar{\theta}$; and $\beta_j = \psi_j - \bar{\psi}$, where

$$\bar{\theta} = \dfrac{\sum_{i=1}^{I} v_i \theta_i}{v_+} \quad \text{and} \quad \bar{\psi} = \dfrac{\sum_{j=1}^{J} w_j \psi_j}{w_+}.$$

22. $\mu = \bar{\theta}_{+++}$,

$\alpha_i^A = \bar{\theta}_{i++} - \bar{\theta}_{+++}$,

$\alpha_j^B = \bar{\theta}_{+j+} - \bar{\theta}_{+++}$,

$\alpha_k^C = \bar{\theta}_{++k} - \bar{\theta}_{+++}$,

$\beta_{ij}^{AB} = \bar{\theta}_{ij+} - \bar{\theta}_{i++} - \bar{\theta}_{+j+} + \bar{\theta}_{+++}$,

$\beta_{ik}^{AC} = \bar{\theta}_{i+k} - \bar{\theta}_{i++} - \bar{\theta}_{++k} + \bar{\theta}_{+++}$,

$\beta_{jk}^{BC} = \bar{\theta}_{+jk} - \bar{\theta}_{+j+} - \bar{\theta}_{++k} + \bar{\theta}_{+++}$,

$\gamma_{ijk} = \theta_{ijk} - \bar{\theta}_{ij+} - \bar{\theta}_{i+k} - \bar{\theta}_{+jk} + \bar{\theta}_{i++} + \bar{\theta}_{+j+} + \bar{\theta}_{++k} - \bar{\theta}_{+++}$.

Index